THIRD EDITION

INTERMEDIATE ALGEBRA

IGNACIO BELLO
Hillsborough Community College

with
JACK R. BRITTON
University of South Florida
Emeritus

DELLEN PUBLISHING COMPANY
San Francisco

COLLIER MACMILLAN CANADA
Toronto

divisions of Macmillan, Inc.

IN MEMORY OF MY UNCLE

On the cover: The painting on the cover, executed by Ronald Davis in 1988, is vinyl-acrylic copolymer and dry pigments on canvas. Ronald Davis is involved in expressing the illusion of a three-dimensional object rendered on a flat surface. His work can be seen in leading museums throughout the world, including the San Francisco Museum of Modern Art, the Los Angeles County Museum of Art, and the Whitney Museum of American Art. In Los Angeles, Davis is represented by the Asher–Faure gallery.

Permissions: Dellen Publishing Company
                 400 Pacific Avenue
                 San Francisco, California 94133

Orders: Dellen Publishing Company
                 c/o Macmillan Publishing Company
                 Front and Brown Streets
                 Riverside, New Jersey 08075

LIBRARY OF CONGRESS CATALOGING IN PUBLICATION DATA

Bello, Ignacio.
    Intermediate algebra/Ignacio Bello with Jack R. Britton. 3rd ed.
       p.   cm.
    Includes index.
    ISBN 0-02-307931-2
    1. Algebra. I. Britton, Jack Rolf, 1908–   . II. Title.
QA152.2.B4517   1990   90-3348
512.9—dc20   CIP

Printing: 1 2 3 4 5 6 7 8 9   Year: 0 1 2 3 4

ISBN 0-02-307931-2

CONTENTS

CHAPTER 8   SYSTEMS OF LINEAR EQUATIONS   578

HOW TO USE THIS BOOK

1. Take the Pretest. If you score 80% or more, go to the next chapter.
2. If you score less than 80% on the Pretest, do the following:
   a. Read the Reviews at the beginning of each section.
   b. Read the Objectives in each section.
   c. Read and study the explanations and Examples.
   d. Do the Problems in the margin.
   e. Do the odd-numbered problems at the end of each section.
   f. Read the Summary at the end of the chapter.
3. Do part (a) of each problem in the Review Exercises. If you have the correct answer for 80% or more of the Review Exercises, take the Practice Test. Study the answers you missed and go on to the next chapter.

   If fewer than 80% of your answers are correct, consult your instructor or a qualified tutor, then do part (b) of the Review Exercises. If 80% or more of your answers are correct, take the Practice Test, study the answers you missed, and go on to the next chapter. If not, consult your instructor or a qualified tutor. You can also look at the complete solutions for the Practice Test in the *Student's Solutions Manual* or watch the videotape corresponding to the chapter you are working on. Keep trying the Review Exercises until 80% or more of your answers are correct, then take the Practice Test, study the answers you missed, and go on to the next chapter. The flowchart on the next page may simplify things for you.

We always welcome students' comments and suggestions. You may send them to us at the following address:

Ignacio Bello
Hillsborough Community College
P.O. Box 5096
Tampa, Florida 33675-5096

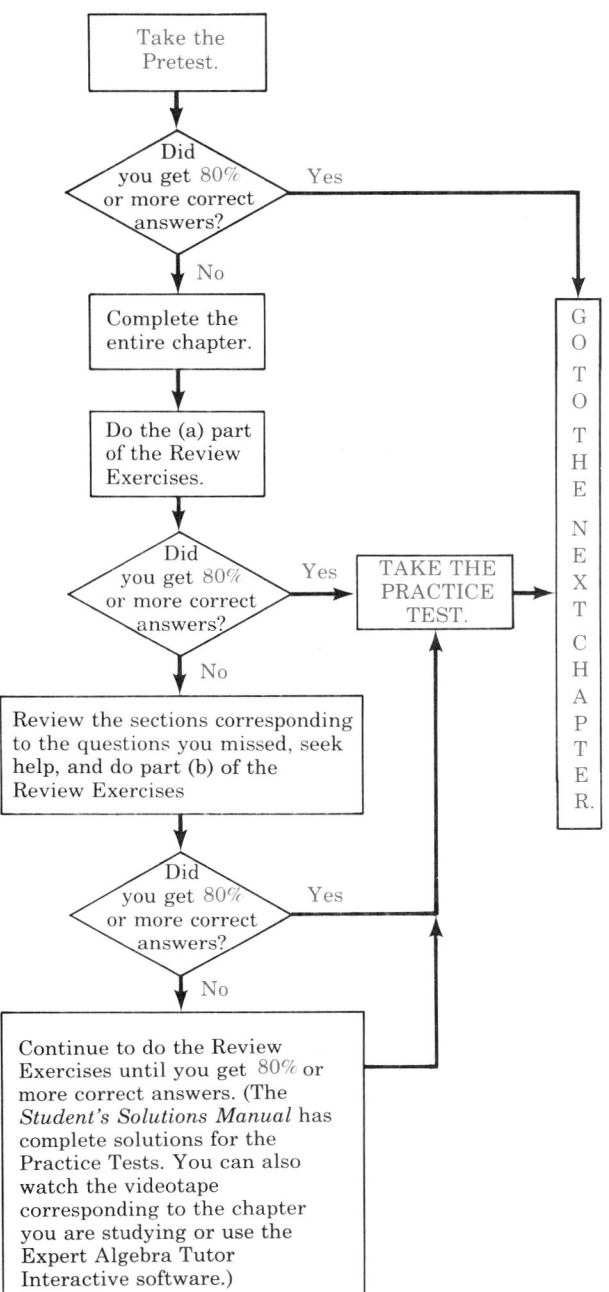

Take the Pretest.

Did you get 80% or more correct answers?

Yes

No

Complete the entire chapter.

Do the (a) part of the Review Exercises.

Did you get 80% or more correct answers?

Yes

No

TAKE THE PRACTICE TEST.

Review the sections corresponding to the questions you missed, seek help, and do part (b) of the Review Exercises

Did you get 80% or more correct answers?

Yes

No

Continue to do the Review Exercises until you get 80% or more correct answers. (The *Student's Solutions Manual* has complete solutions for the Practice Tests. You can also watch the videotape corresponding to the chapter you are studying or use the Expert Algebra Tutor Interactive software.)

GO TO THE NEXT CHAPTER.

PREFACE

This book is intended for students who need help with their intermediate algebra. For these students, the following features have been incorporated in the book.

Pretests   At the beginning of each chapter. Pinpoint student's strengths and weaknesses. With diagnostic answers.

Reviews   At the beginning of each section. Tell students what they need to know in order to complete the section.

Objectives   At the beginning of each section. Tell students what they must learn. Coordinated with the Examples and Exercises.

Examples   Illustrate the topics under discussion. Coordinated with the Objectives and Exercises (Objective A goes with Example 1).

Margin Problems   Correlated with the corresponding Examples and designed to check the student's facility in applying the concepts developed. The answers are given in the margin of following pages.

Exercises   To practice the principles being learned. Coordinated with the Objectives and Examples (Group A goes with Objective A).

Applications   Given at the end of most Exercise sets. Show students how the material is applied to other areas.

Skill Checkers   Test previous material needed in the succeeding section. For example: Review factorization before reducing fractional expressions.

Using Your Knowledge   Follows the Exercises and applies the ideas discussed in each section to different situations.

Calculators   The **Calculator Corner** section shows the students how the material under discussion can be handled using calculators.

Summary   At the end of each chapter. Highlights important concepts and their meaning. Examples are included when possible.

Review Exercises   At the end of each chapter. Review all objectives covered. Tell students in which section the question under consideration appears.

Practice Tests   With diagnostic answers. Tell students in which section, Example, and page the question under consideration appears.

Ancillary Materials	*Student's Solutions Manual* (solutions to odd-numbered problems and Pretests and Practice Tests)
	Instructor's Manual (tests and all answers)
	IPS Testing System (IBM and Apple)
	Videos (lectures; solutions to Practice Tests)
	Expert Algebra Tutor Interactive software

SUGGESTED COURSES USING THIS BOOK

We have class-tested the entire book with both standard and special classes at Hillsborough Community College. From this testing and the many suggestions received by the reviewers, it is evident that the book can be used in various ways:

1. As a textbook in a traditional lecture course. Simply skip the margin problems and lecture as usual.
2. As a textbook in an individualized study course. Assign the reading portion and the exercises one day; answer questions and collect homework the next day.
3. As a lab text. Use as any other textbook in the mathematics lab.
4. As a combination of these methods.

WHAT IS NEW IN THE THIRD EDITION?

We have followed the suggestions of many users and reviewers of the first two editions to clarify the exposition, expand coverage, and, in general, improve the book. Specifically, we have:

- Added the **Reviews** at the beginning of each section.
- Introduced subsections A, B, C, and so on, for ease in identifying the topic under discussion.
- Added the **Skill Checkers** at the end of the Exercises.
- Added new chapters and sections covering Quadratic Inequalities, Conic Sections, and Nonlinear Systems of Equations.
- Provided many new worked examples, Using Your Knowledge problems, and hundreds of new Applications and problems.

ACKNOWLEDGMENTS

The author would like to express his appreciation to the following persons:

REVIEWERS (PREVIOUS EDITIONS)
Carole Bauer, Triton College
Calvin Lathan, Monroe Community College
Ara B. Sullenberger, Tarrant County Jr. College District
George Kosan, Hillsborough Community College
Donald Rose II, Hillsborough Community College

In addition, Professors Debbie Ritchie and Patricia Stanley did a magnificent job reviewing the present edition.

Special thanks go to Professor Barbara Burrows who offered many ideas and suggestions for improving the material and to Professor Liana Fox who offered her several years of expertise in teaching Intermediate Algebra and contributed many teaching tips and techniques incorporated in the book. Josephine Rinaldo read every word in the book and worked the exercises through several stages of development and finally decided to do it right by writing the solutions manual as well. After this arduous task, polynomials (or even monomials), are not as cherished as they used to be. Finally, thanks to Professor Jack Britton, gentleman, scholar, and bon vivant, who has been trying to retire and make our books more readable for the last several years, and to the wonderful production department and staff provided by our editor Don Dellen. Without their help, this book would not have been possible.

The photographs for the Chapter Openings are the work of Betty Berenson.

THIRD EDITION

INTERMEDIATE ALGEBRA

CHAPTER 1

THE REAL NUMBERS

Why do we need algebra? To solve problems! If you have studied introductory algebra, this chapter is a review. If you have not, you will acquire the necessary background to learn how to solve the equations and problems of Chapter 2. We study sets of numbers, the terminology used, and the properties of the real numbers. Addition, subtraction, multiplication, and division with these numbers are discussed. The law of exponents, scientific notation, and the order of operations are also included.

(Answers on pages 6–7)

ANSWERS

1. Use set notation to list the natural numbers between 3 and 9.

    1. _____

2. Classify the given number by making a check mark (✔) in the appropriate place.

    2. _____

Set	0.3	0	$\dfrac{-3}{4}$	-5	$\sqrt{3}$
Natural number					
Whole number					
Integer					
Rational numbers					
Irrational number					
Real number					

3. Write $\frac{2}{3}$ as a decimal.

    3. _____

4. Write 0.31 as a fraction.

    4. _____

5. Graph the additive inverse of $\frac{2}{3}$ on the number line.

    5.

6. Find.
    a. $|-8|$
    b. $|0.4|$

    6. a. _____
       b. _____

7. Find.
    a. $-9 + 6$
    b. $-0.7 + (-0.9)$

    7. a. _____
       b. _____

8. Find.
    a. $-15 - 7$
    b. $-0.5 - (-0.4)$

    8. a. _____
       b. _____

9. Which law is being illustrated?
   a. $(7 + 2) + 5 = (2 + 7) + 5$
   b. $(2 + 9) + 4 = 2 + (9 + 4)$

9. a. _____
   b. _____

10. Find.
   a. $7 \cdot (-9)$
   b. $-3 \cdot (-1.2)$

10. a. _____
    b. _____

11. Find.

   a. $-\dfrac{1}{2} \cdot \dfrac{2}{7}$

   b. $-\dfrac{3}{4} \div \dfrac{9}{8}$

11. a. _____
    b. _____

12. Simplify.
   a. $-3(x + 7)$
   b. $5x - (2x + 1) + (3x + 2)$

12. a. _____
    b. _____

13. Simplify $[(4x^2 - 3) + (3x + 5)] - [(x - 3) + (2x^2 - 2)]$.

13. _____

14. Evaluate.
   a. $(-2)^4$
   b. -2^4

14. a. _____
    b. _____

15. Write as a fraction.
   a. 8^{-2}
   b. x^{-7}

15. a. _____
    b. _____

16. Perform the indicated operation and simplify.
   a. $(3x^3 y)(-4x^{-8} y^7)$

   b. $\dfrac{48x^4}{16x^{-7}}$

16. a. _____
    b. _____

17. Simplify.
   a. $(-2x^6 y^{-5})^3$

   b. $\left(\dfrac{x^5}{y^{-2}}\right)^{-3}$

17. a. _____
    b. _____

18. Do the calculation and write the answer in scientific notation.

$(7.1 \times 10^5) \times (3 \times 10^{-7})$

18. _____

19. Evaluate.
  a. $[-6(4 + 3)] + 9$

  b. $\dfrac{5 \cdot (140 - 32)}{9}$

19. **a.** _____
    b. _____

20. Evaluate $-3^3 + \dfrac{(6 - 10)}{2} + 15 \div 5$.

20. _____

IF YOU MISSED QUESTION	SECTION	EXAMPLES	PAGE	ANSWERS
1	1.1	1	10	**1.** $\{4, 5, 6, 7, 8\}$
2	1.1	2	11	**2.** (see table below)
3	1.1	3	11	**3.** $0.\overline{6}$
4	1.1	4	12	**4.** $\dfrac{31}{100}$
5	1.2	1, 2, 3	17–19	**5.** (number line below)
6a	1.2	4	19	**6. a.** 8
6b	1.2	4	19	**b.** 0.4
7a	1.3	1, 2	26–27	**7. a.** -3
7b	1.3	1, 2	26–27	**b.** -1.6
8a	1.3	3	28	**8. a.** -22
8b	1.3	3	28	**b.** -0.1
9a	1.3	4	29	**9. a.** Commutative law of addition
9b	1.3	4	29	**b.** Associative law of addition
10a	1.4	1, 2	36–37	**10. a.** -63
10b	1.4	1, 2	36–37	**b.** 3.6
11a	1.4	3	38	**11. a.** $-\dfrac{1}{7}$
11b	1.4	4, 5, 6	39–40	**b.** $-\dfrac{2}{3}$
12a	1.5	1	48	**12. a.** $-3x - 21$
12b	1.5	2, 3, 4	49–51	**b.** $6x + 1$
13	1.5	5	51	**13.** $2x^2 + 2x + 7$
14a	1.6	1	58	**14. a.** 16
14b	1.6	1	58	**b.** -16
15a	1.6	2	59	**15. a.** $\dfrac{1}{64}$
15b	1.6	2	59	**b.** $\dfrac{1}{x^7}$

2.

Set	0.3	0	$\dfrac{-3}{4}$	-5	$\sqrt{3}$
N					
W		✓			
I		✓		✓	
Rat.	✓	✓	✓	✓	
Irr.					✓
R	✓	✓	✓	✓	✓

5. $-\dfrac{2}{3}$ plotted on a number line between -1 and 0, with marks at $-2, -1, 0, 1, 2$.

REVIEW

IF YOU MISSED QUESTION	SECTION	EXAMPLES	PAGE	ANSWERS
16a	1.6	3	61	**16. a.** $-\dfrac{12y^8}{x^5}$
16b	1.6	4	62	**b.** $3x^{11}$
17a	1.6	5	63	**17. a.** $-\dfrac{8x^{18}}{y^{15}}$
17b	1.6	6	64	**b.** $\dfrac{1}{x^{15}y^6}$
18	1.6	7, 8, 9	65–66	**18.** 2.13×10^{-1}
19a	1.7	1, 2, 3, 4	71–72	**19. a.** -33
19b	1.7	5	73	**b.** 60
20	1.7	6	74	**20.** -26

1.1 THE NUMBERS OF ALGEBRA

OBJECTIVES

REVIEW

Before starting this section you should know:

1. What the counting numbers are.
2. How to read a decimal such as 0.3 or 0.15.

OBJECTIVES

You should be able to:

A. Use set notation to list the elements of a specified set.
B. Write a fraction as a decimal and a decimal as a fraction.

Have you been to a ballgame lately? How did you find your seat? You probably used the fact that the seats were numbered with the counting numbers 1, 2, 3, and so on. In this chapter we will study the set of real numbers and introduce you to intermediate algebra. We start by discussing sets of numbers used in arithmetic.

A. SETS OF NUMBERS

We use capital letters to denote sets and lowercase letters to denote elements (or members) of these sets. If possible, we "list" the elements of a set in braces {   } and separate them by commas. Thus $A = \{a, b, c\}$ is the set A that has $a, b,$ and c for its elements. If a set has no elements, it is called the **empty,** or **null,** set and is denoted by the symbol $\varnothing$ (read "the empty set").

In algebra, we use the set of *natural numbers*.

> The **natural numbers** are 1, 2, 3, and so on. In set notation, $N = \{1, 2, 3, \ldots\}$.

The three dots inside the braces indicate that the enumeration continues without end (in words we say "and so on"). Note that the number 0 is *not* a natural number; it is an element of the set of *whole numbers*.

> The **whole numbers** are 0, 1, 2, 3, and so on. Thus, $W = \{0, 1, 2, 3, \ldots\}$.

Another set you probably learned about is the set of *integers*.

> The **integers** are $\ldots, -2, -1, 0, 1, 2,$ and so on. Thus, $I = \{\ldots, -2, -1, 0, 1, 2, \ldots\}$.

The set of integers includes three important subsets:

1. The **positive integers,** $I^+ = \{+1, +2, +3, \ldots\}$
2. The **negative integers,** $I^- = \{\ldots, -3, -2, -1\}$
3. The set consisting of the single element 0, $\{0\}$.

Notice that the set of positive integers is the same as the set of natural numbers. We write $N = I^+$.

EXAMPLE 1 Use set notation to list:

a. The natural numbers between 2 and 7
b. The first three whole numbers
c. The first two negative integers
d. The only number that is neither positive nor negative

Solution

a. The set of natural numbers between 2 and 7 is $\{3, 4, 5, 6\}$. (2 and 7 are not included.)

b. The set of the first three whole numbers is $\{0, 1, 2\}$

c. The set of the first two negative integers is $\{-1, -2\}$

d. The set containing the only number that is neither positive (it is not in the set I^+) nor negative (it is not in the set I^-) is $\{0\}$. ▲

Integers can be written as **common fractions** of the form $\dfrac{a}{b}$ (or a/b), in which the numerator a and the denominator b are both integers and the denominator is not 0 ($b \neq 0$). For example, $5 = \frac{10}{2}$ and $1 = \frac{7}{7}$. If a number can be written in the form $\dfrac{a}{b}$, where a and b are integers and $b \neq 0$, the number is called a *rational number* (because it is a *ratio* of two integers). Thus $\dfrac{1}{5}, \dfrac{-4}{3}, \dfrac{4}{1}$, and $\dfrac{-7}{1}$ are rational numbers. We cannot list all the rational numbers, so we use a new notation, called **set-builder notation,** to define this set.

> The set Q of **rational numbers** consists of all the numbers that can be written as the ratio of two integers. Thus
>
> $$Q = \left\{ r \,\middle|\, r = \frac{a}{b}, \ a \text{ and } b \text{ integers}, \ b \neq 0 \right\}$$
>
> Read "Q equals the set of all r such that r equals a divided by b, a and b integers and b not equal to 0." Note that "|" is read as "such that."

There are some numbers such as $\sqrt{2}$ (the square root of 2) that are *not* rational. They are called *irrational* numbers. Here is the definition.

> The **irrational numbers** are numbers that *cannot* be written as ratios of two integers. We write $H = \{x \mid x \text{ is not rational}\}$.

Problem 1 Use set notation to list:
a. The natural numbers between 3 and 8
b. The first four whole numbers
c. The first four negative integers
d. The first whole number

The set consisting of all the rational and all the irrational numbers is called the set of **real numbers** and is denoted by the letter *R*.

Here are some real numbers:

$$5,\ 17,\ -4,\ -9,\ 0,\ \frac{3}{5},\ 0.6,\ \frac{1}{-10},\ -0.1,\ \frac{4}{-3},\ \sqrt{3},\ \pi$$

EXAMPLE 2 Classify the given number by making a check mark ($\checkmark$) in the appropriate row.

Set	0	$-\frac{4}{5}$	-4	$\sqrt{2}$	7	$-\pi$
Natural number					$\checkmark$	
Whole number	$\checkmark$				$\checkmark$	
Integer	$\checkmark$		$\checkmark$		$\checkmark$	
Rational number	$\checkmark$	$\checkmark$	$\checkmark$		$\checkmark$	
Irrational number				$\checkmark$		$\checkmark$
Real number	$\checkmark$	$\checkmark$	$\checkmark$	$\checkmark$	$\checkmark$	$\checkmark$

B. DECIMALS AND FRACTIONS

Any rational number $\dfrac{a}{b}$ can be written as a decimal by dividing a by b. The result is either a **terminating decimal** (as in $\frac{1}{2} = 0.5$) or a **nonterminating, repeating decimal** (as in $\frac{1}{3} = 0.333\ldots$). We often place a bar over the repeating digits in a nonterminating repeating decimal. Thus, $\frac{1}{3} = 0.\overline{3}$ and $\frac{2}{11} = 0.181818\ldots = 0.\overline{18}$.

EXAMPLE 3 Write as a decimal.

a. $\frac{4}{5}$ **b.** $\frac{3}{11}$

Solution

a. Dividing 4 by 5, we obtain

$$\begin{array}{r} 0.8 \\ 5\overline{)4.0} \\ \underline{4\,0} \\ 0 \end{array}$$

$$\frac{4}{5} = 0.8$$

b. Dividing 3 by 11, we have

$$\begin{array}{r} 0.27\ldots \\ 11\overline{)3.0} \\ \underline{2\,2} \\ 80 \\ \underline{77} \\ 3 \end{array}$$

After the second step the remainder is 3, the original dividend. Thus, the digits 27 will repeat without end to give

$$\frac{3}{11} = 0.272727\ldots = 0.\overline{27}$$ ▲

Problem 2 Classify the given number by making a check mark ($\checkmark$) in the appropriate row.

Set	$-\frac{8}{5}$	π	4	$\sqrt{3}$	-7
Natural number					
Whole number					
Integer					
Rational number					
Irrational number					
Real number					

Problem 3 Write as a decimal.
a. $\frac{3}{8}$ **b.** $\frac{5}{11}$

From the preceding discussion you can see that rational numbers can be written as terminating or repeating decimals. Is the converse true? That is, can we write terminating or repeating decimals in the form $\frac{a}{b}$? Let us start with terminating decimals. Terminating decimals can be converted to fractions of the form $\frac{a}{b}$ if you know how to read them. Thus, if you recall that 0.3 is read as "three tenths," you can then write $0.3 = \frac{3}{10}$. Similarly, 0.15 (fifteen hundredths) is written as $0.15 = \frac{15}{100}$, or $\frac{3}{20}$.

EXAMPLE 4 Write as a fraction.

a. 0.32   **b.** 0.414

Solution

a. 0.32 is thirty-two hundredths. Thus $0.32 = \frac{32}{100}$, or $\frac{8}{25}$.

b. 0.414 is four hundred fourteen thousandths. Thus $0.414 = \frac{414}{1000}$, or $\frac{207}{500}$.   ▲

Now we know how to write terminating decimals in the form $\frac{a}{b}$. What about repeating decimals such as $0.\overline{1}$, $0.\overline{2}$, or $0.\overline{13}$? Can we write them in the form $\frac{a}{b}$? The answer is yes! (See Using Your Knowledge 1.1).

> The set Q of rational numbers is the same as the set of terminating or repeating decimals.

Do you have a good idea of the relationship between the sets of numbers we have discussed? We can clarify the situation by using the idea of a *subset*. We say that A is a **subset** of B, denoted by $A \subseteq B$, when all the elements in A are also in B. Thus, since all natural numbers N are whole numbers, $N \subseteq W$ (read "N is a subset of W"). Also, since all whole numbers are integers, $W \subseteq I$. Here is the complete picture:

$$N \subseteq W \subseteq I \subseteq Q \subseteq R$$

The diagram shows the sets involved.

Real numbers

Rational numbers	Irrational numbers
$Q = \{r \mid r = a/b, \, a, b \text{ integers}, b \neq 0\}$	$H = \{x \mid x \text{ is not rational}\}$
(terminating or repeating decimals)	(nonterminating, nonrepeating decimals)

Integers	Rational nonintegers
$I = \{\dots, -2, -1, 0, 1, 2, \dots\}$	(common fractions)

Natural numbers	Zero	Negatives of natural numbers
$N = \{1, 2, 3, \dots\}$	(0)	$\{-1, -2, -3, \dots\}$

Problem 4 Write as a fraction.
a. 0.64   **b.** 0.212

ANSWERS

A. In Problems 1–10 use set notation to write the indicated set.

1. The first two natural numbers

2. The first six natural numbers

3. The natural numbers between 4 and 8

4. The natural numbers between 7 and 10

5. The first three negative integers

6. The negative integers between -4 and 7

7. The whole numbers between -3 and 4

8. The integers less than 0

9. The integers greater than 0

10. The nonnegative integers

1. _____
2. _____
3. _____
4. _____
5. _____
6. _____
7. _____
8. _____
9. _____
10. _____

In Problems 11–16 classify the given numbers by placing a check mark in the appropriate row.

Set	11. $\dfrac{-3}{8}$	12. 0	13. $\sqrt{8}$	14. $\dfrac{3}{7}$	15. $0.\bar{3}$	16. -9
Natural numbers						
Whole numbers						
Integers						
Rational numbers						
Irrational numbers						
Real numbers						

11. _____
12. _____
13. _____
14. _____
15. _____
16. _____

B. In Problems 17–24 write the given number as a decimal.

17. $\dfrac{2}{3}$

18. $\dfrac{1}{6}$

19. $\dfrac{7}{8}$

20. $\dfrac{5}{6}$

21. $\dfrac{5}{2}$

22. $\dfrac{4}{3}$

23. $\dfrac{7}{6}$

24. $\dfrac{9}{8}$

17. _____
18. _____
19. _____
20. _____
21. _____
22. _____
23. _____
24. _____

In Problems 25–30 write the given number as a fraction.

25. 0.31

26. 0.41

25. _____
26. _____

27. 0.34

28. 0.26

29. 0.418

30. 0.626

27. _____

28. _____

29. _____

30. _____

C. Applications

31. The batting average of a baseball player is the number of hits divided by the number of times at bat. Find the batting average of a player with 40 hits in 120 times at bat and write the answer as a decimal.

31. _____

32. On the average, Babe Ruth hit a home run $\frac{1}{12}$ of the times he came to bat. Write $\frac{1}{12}$ as a decimal.

32. _____

33. What is the fastest-growing of all plants? No, not grass, but bamboo. The growth of a bamboo plant was recorded at 36 in. in 24 h, that is, $\frac{36}{24}$ in./hr. Write $\frac{36}{24}$ as a decimal.

33. _____

34. Who can keep a secret? In a recent survey it was discovered that, on the average, 0.36 of the bankers, 0.64 of the lawyers, 0.66 of the doctors, and 0.74 of the clergy could do so. Write 0.36, 0.64, 0.66, and 0.74 as fractions.

34. _____

35. What are the occupations with the highest percentages of females? Secretaries, bookkeepers, and sewing-machine operators: 98 out of 100 secretaries, 91 out of 100 bookkeepers, and 90 out of 100 sewing-machine operators are female. Write these three numbers as decimals.

35. _____

36. What are the occupations with the highest percentage of males? Construction workers, truck drivers, and machinists: 99 out of 100 construction workers, 98 out of 100 truck drivers, and 97 out of 100 machinists are male. Write these three numbers as decimals.

36. _____

37. An optometrist must grind an optical lens to $\frac{3}{64}$ in. Write $\frac{3}{64}$ as a decimal. (The machine that grinds lenses is graduated in decimals.)

37. _____

38. Parts suppliers usually list the dimensions of parts in decimals. A mechanic needs to replace a piston ring that is $3\frac{7}{8}$ in. in diameter. How will this size be listed by the supplier?

38. _____

39. The total permissible variation in a dimension of a machine part is called its *tolerance*. For example, if the nominal size of a certain screw is 0.437 in. in diameter and it can be larger or smaller by 0.002 in., then the tolerance is ± 0.002 in. This means that an acceptable diameter is between 0.435 and 0.439 in., inclusive. If the nominal size of the diameter of a certain pin is $\frac{1}{4}$ in. with a tolerance of ± 0.002 in., what is the largest acceptable diameter for this pin? (Express your answer as a single decimal.)

39. _____

ANSWER (to problem on page 12)
4. **a.** $\frac{64}{100}$, or $\frac{16}{25}$ **b.** $\frac{212}{1000}$, or $\frac{53}{250}$

40. If the nominal diameter of the pin in Problem 39 is $\frac{3}{4}$ in., what is the smallest acceptable diameter for this pin? (Express your answer as a single decimal.)

40. _____

1.1 USING YOUR KNOWLEDGE

We have already seen that any rational number can be written as a decimal by dividing its numerator by its denominator. Can you tell if the result will be terminating or repeating? Here are some fractions and their decimal equivalents.

TERMINATING	**REPEATING**
$\dfrac{7}{8} = \dfrac{7}{2 \times 2 \times 2} = 0.875$	$\dfrac{1}{3} = \dfrac{1}{3} = 0.\bar{3}$
$\dfrac{3}{4} = \dfrac{3}{2 \times 2} = 0.75$	$\dfrac{2}{3} = \dfrac{2}{3} = 0.\bar{6}$
$\dfrac{5}{8} = \dfrac{5}{2 \times 2 \times 2} = 0.625$	$\dfrac{1}{6} = \dfrac{1}{2 \times 3} = 0.1\bar{6}$
$\dfrac{1}{2} = \dfrac{1}{2} = 0.5$	$\dfrac{1}{9} = \dfrac{1}{3 \times 3} = 0.\bar{1}$
$\dfrac{3}{8} = \dfrac{3}{2 \times 2 \times 2} = 0.375$	$\dfrac{1}{12} = \dfrac{1}{2 \times 2 \times 3} = 0.08\bar{3}$
$\dfrac{1}{5} = \dfrac{1}{5} = 0.2$	$\dfrac{1}{7} = \dfrac{1}{7} = 0.\overline{142857}$
$\dfrac{1}{10} = \dfrac{1}{2 \times 5} = 0.1$	$\dfrac{1}{11} = \dfrac{1}{11} = 0.\overline{09}$

1. Look at the denominators of the terminating fractions. Now look at the denominators of the repeating fractions. Can you make a conjecture (guess) about the denominators of the fractions that terminate?

1. _____

Look at the patterns:

$$0.\bar{1} = \frac{1}{9} \qquad 0.\overline{11} = \frac{11}{99}$$

$$0.\bar{2} = \frac{2}{9} \qquad 0.\overline{22} = \frac{22}{99}$$

$$0.\bar{3} = \frac{3}{9} \qquad 0.\overline{33} = \frac{33}{99}$$

2. If you follow this pattern, $0.\bar{4} =$ _____.

2. _____

3. If you follow this pattern, $0.\bar{5} =$ _____.

3. _____

4. If you follow this pattern, $0.\overline{44} =$ _____.

4. _____

5. If you follow this pattern, $0.\overline{55} =$ _____.

5. _____

You should now be convinced that any repeating decimal can be written as a fraction!

To convert a fraction to a decimal using a calculator is very simple. Thus to write $\frac{4}{5}$ as a decimal (Example 3), we simply recall that $\frac{4}{5}$ means $4 \div 5$. Pressing $\boxed{4}$ $\boxed{\div}$ $\boxed{5}$ $\boxed{=}$ will give us the correct answer, 0.8. (The calculator is nice enough to write the zero to the left of the decimal in the final answer.) You can also do Example 3 using division. Moreover, if your instructor permits it, you can do Problems 17–20 in Exercise 1.1.

1.2 SOME PROPERTIES OF REAL NUMBERS

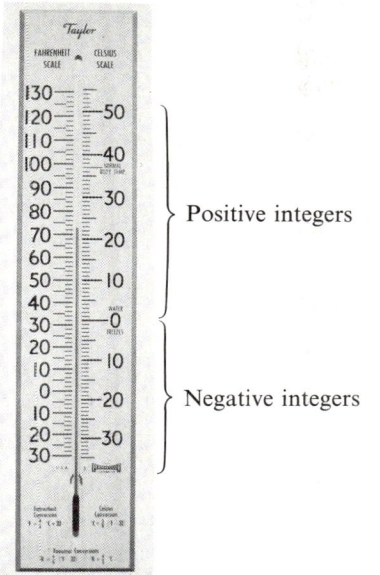

Positive integers

Negative integers

Look at the thermometer in the photo. A thermometer associates temperatures with the numbers on its scale. If we take this scale and turn it sideways so that the positive numbers are on the right, the resulting scale is called a *number line*. (See Figure 1.) On a number line the *positive* real numbers are to the *right* of 0, the *negative* real numbers are to the *left* of 0, and 0 itself is called the **origin.**

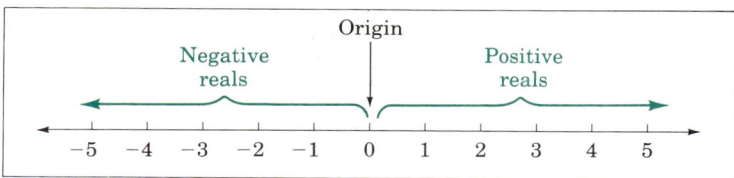

FIG. 1

The line is 5 units long on each side, but the arrows indicate that the line extends indefinitely in both directions.

A. GRAPHING POINTS ON THE NUMBER LINE

The number corresponding to a point on the number line is called the **coordinate** of the point, and the point is called the **graph** of the number. Note that there is exactly one real number for each point on the number line and one point for each real number. Thus the coordinate of point A is 2 and that of B is -1. The graphs of the two points are shown on the following number line.

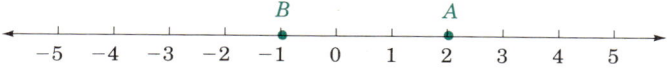

EXAMPLE 1 Graph the following points on the number line.

a. -4 **b.** 0   **c.** 3

Problem 1 Graph on the number line.

a. -3 **b.** 4

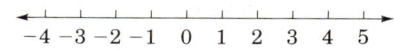

Solution The graphs of the three numbers are shown.

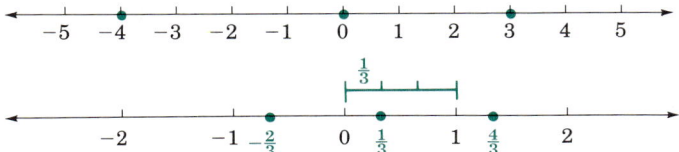

FIG. 2

Rational numbers can be graphed on the number line. For example, to graph $\frac{1}{3}$, divide the segment from 0 to 1 into *three* (3) equal parts. Each part is $\frac{1}{3}$ of a unit. The point of division closest to 0 corresponds to $\frac{1}{3}$. If we count 4 units to the right of the origin, we can graph $\frac{4}{3}$ and moving 2 units to the left of 0, we can graph $-\frac{2}{3}$, as shown in Figure 2.

EXAMPLE 2 Graph the points $\frac{1}{2}$ and $-\frac{3}{4}$ on the number line.

Solution To graph the point $\frac{1}{2}$, we divide the line segment from 0 to 1 into 2 parts. The point at the end of the first part corresponds to the point $\frac{1}{2}$; its graph is shown next.

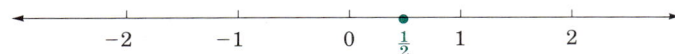

To graph $\dfrac{-3}{4}$, we divide the line segment from -1 to 0 into 4 equal parts. The graph of the point $\dfrac{-3}{4}$ is shown next.

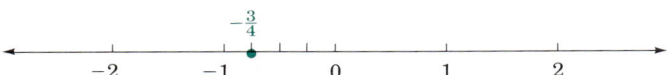

B. ADDITIVE INVERSES (OPPOSITES)

Each point on the number line has another point *opposite* to it with respect to 0. The numbers corresponding to these two points are called the *additive inverses (opposites)* of each other. Thus, 3 and -3 (read "the inverse of 3," or "negative 3") are additive inverses and -4 and 4 are additive inverses. (See Figure 3.)

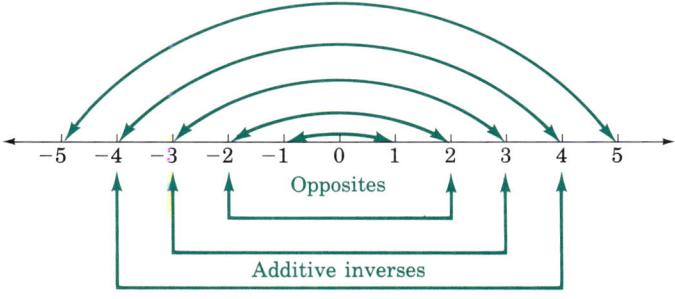

Opposites

Additive inverses

FIG. 3

Here is the definition.

The **additive inverse (opposite)** of a is $-a$.

Problem 2 Graph the points $-\frac{1}{2}$ and $\frac{3}{4}$ on the number line.

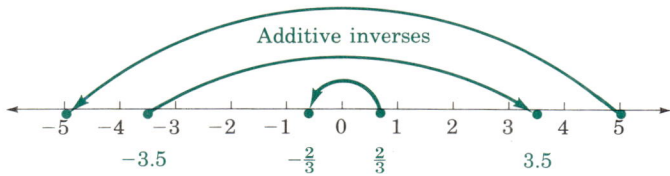

FIG. 4

EXAMPLE 3 Find the additive inverse.

a. 5   **b.** −3.5   **c.** $\frac{2}{3}$

Solution

a. The additive inverse of 5 is −5. (See Figure 4.)

b. The additive inverse of −3.5 is 3.5. (See Figure 4.) In symbols, −(−3.5) = 3.5. (Read "the inverse of negative 3.5 is 3.5," or "the inverse of the inverse of 3.5 is 3.5.")

c. The additive inverse of $\frac{2}{3}$ is −$\frac{2}{3}$. (See Figure 4.) ▲

Here are some facts about additive inverses.

> Every real number has an additive inverse.
>
> **1.** The additive inverse of a is $-a$.
>
> **2.** If the number is positive, its inverse is negative.
>
> **3.** If the number is negative, its inverse is positive.
>
> **4.** The inverse of 0 is 0.

C.   ABSOLUTE VALUES

Now, let us go back to the number line. What is the distance between 3 and 0? The answer is 3 units. What about between −3 and 0? The answer is *still* 3 units. The distance between any number a and 0 is called the *absolute value* of a and is denoted by $|a|$. Thus, $|-3| = 3$ and $|3| = 3$.

> The **absolute value** of a number a is its distance from 0 and is denoted by $|a|$.

EXAMPLE 4 Find.

a. $|-8|$ **b.** $|\frac{1}{7}|$ **c.** $|0|$ **d.** $|4.2|$ **e.** $|0.\overline{3}|$

Solution

a. $|-8| = 8$ −8 is 8 units from 0
b. $|\frac{1}{7}| = \frac{1}{7}$ $\frac{1}{7}$ is $\frac{1}{7}$ units from 0
c. $|0| = 0$ 0 is 0 units from 0
d. $|4.2| = 4.2$ 4.2 is 4.2 units from 0
e. $|0.\overline{3}| = 0.\overline{3}$ $0.\overline{3}$ is $0.\overline{3}$ units from 0

Problem 3 Find the additive inverse.
a. 3   **b.** −2.5   **c.** $\frac{3}{5}$

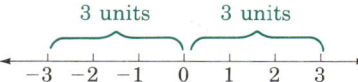

You can think of the absolute value of a number as the number of units with the sign disregarded.

Problem 4 Find.
a. $|-19|$ **b.** $|\frac{1}{6}|$
c. $|-0|$ **d.** $|3.1|$

D. ORDER

On the number line, numbers are shown in order; they *increase* as you move right and *decrease* as you move left. Thus, a number a is greater than another number b, denoted by $a > b$ (read "a is greater than b"), if a is to the right of b on the number line. A number a is less than another number b, denoted by $a < b$ (read "a is less than b"), if a is to the left of b on the number line. The symbols "$<$" and "$>$" are called **inequality signs,** and statements such as $a > b$ or $b < a$ are called **inequalities.** Thus, $3 < 4$ (and $4 > 3$) because 3 is to the left of 4 on the number line. $-3 < -2$ (and $-2 > -3$) because -3 is to the left of -2. Similarly, $3.14 > 3.13$ (and $3.13 < 3.14$) because 3.14 is to the right of 3.13 on the number line.

EXAMPLE 5 Fill in the blank with $<$ or $>$ to make the resulting statement true.

a. -3 _____ -2 **b.** -5 _____ -6 **c.** $\frac{1}{2}$ _____ 0

d. $\frac{1}{5}$ _____ $\frac{1}{3}$ **e.** -2.3 _____ -2.2

Solution

a. $-3 < -2$

b. $-5 > -6$

c. $\frac{1}{2} > 0$

d. $\frac{1}{5} < \frac{1}{3}$

e. $-2.3 < -2.2$

Problem 5 Fill in the blank with $<$ or $>$ to make the resulting statement true.

a. 3 _____ -2 **b.** -7 _____ -6

c. 0 _____ $\frac{1}{3}$ **d.** $\frac{1}{3}$ _____ $\frac{1}{2}$

e. -2.4 _____ -2.3

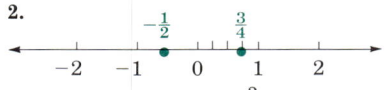

EXERCISE 1.2

NAME

CLASS

SECTION

A. In Problems 1–10 graph the given number on the number line.

1. 2

2. -2

3. -4

4. 0

5. $\dfrac{1}{4}$

6. $-\dfrac{1}{4}$

7. $-\dfrac{3}{4}$

8. $\dfrac{3}{4}$

9. $\dfrac{3}{2}$

10. $-\dfrac{3}{2}$

B. In Problems 11–26 find the additive inverse of the given number.

11. 8

12. -9

13. -7

14. 6

15. $\dfrac{3}{4}$

16. $-\dfrac{1}{4}$

17. $-\dfrac{1}{5}$

18. $\dfrac{2}{5}$

19. 0.5

20. -0.6

21. $0.\overline{2}$

22. $-0.\overline{3}$

23. $-1.\overline{36}$

24. $2.\overline{38}$

25. π

26. $-\pi$

C. In Problems 27–40 find each value.

27. $|10|$

28. $|-11|$

ANSWERS

1. _____

2. _____

3. _____

4. _____

5. _____

6. _____

7. _____

8. _____

9. _____

10. _____

11. _____

12. _____

13. _____

14. _____

15. _____

16. _____

17. _____

18. _____

19. _____

20. _____

21. _____

22. _____

23. _____

24. _____

25. _____

26. _____

27. _____

28. _____

EXERCISE 1.2

29. $|-17|$

30. $|18|$

31. $\left|\dfrac{3}{5}\right|$

32. $\left|-\dfrac{5}{7}\right|$

33. $|0.\bar{5}|$

34. $|-0.\bar{7}|$

35. $|-3.\overline{61}|$

36. $|2.\overline{48}|$

37. $|\sqrt{2}|$

38. $|-\sqrt{3}|$

39. $|-\pi|$

40. $|\pi|$

D. In Problems 41–50 fill in the blanks with $<$ or $>$ to make the resulting statement true.

41. -5 _____ 2

42. 3 _____ -4

43. -6 _____ -8

44. -7 _____ -5

45. $\dfrac{1}{2}$ _____ $\dfrac{1}{4}$

46. $\dfrac{1}{3}$ _____ $\dfrac{1}{2}$

47. $-\dfrac{3}{5}$ _____ $-\dfrac{1}{4}$

48. $-\dfrac{1}{3}$ _____ $-\dfrac{1}{4}$

49. -3.5 _____ -3.4

50. -3.2 _____ -3.1

29. _____

30. _____

31. _____

32. _____

33. _____

34. _____

35. _____

36. _____

37. _____

38. _____

39. _____

40. _____

41. _____

42. _____

43. _____

44. _____

45. _____

46. _____

47. _____

48. _____

49. _____

50. _____

✓ **SKILL CHECKER**

The **SKILL CHECKER** exercises that appear periodically in the exercise sets review skills previously studied. They will help you maintain skills you have already mastered.

In Problems 51–55 use set notation to write the indicated set.

51. The integers greater than 5

52. The integers less than 2

53. The whole numbers between -2 and 5

54. The whole numbers greater than -2

55. The natural numbers between -3.5 and 3.5

51. _____

52. _____

53. _____

54. _____

55. _____

EXERCISE 1.2

1.2 USING YOUR KNOWLEDGE

In this section we learned how to compare integers and decimals. Thus, we learned that $0.33 > 0.32$ and that $\frac{1}{3} < \frac{1}{2}$. To compare 0.33 and $\frac{1}{3}$ we write $\frac{1}{3}$ as a decimal by dividing the numerator by the denominator, obtaining $0.333\ldots$. We then write 0.33 as 0.33**0** (note the extra **0**) and write both numbers in a column with the decimal points aligned.

0.333 . . .

0.330
$\quad\quad\hookrightarrow 3 > 0$, so $0.333\ldots > 0.330$

Thus, $0.333 > 0.330$.

We can use this knowledge in doing some problems.

1. A McDonald's hamburger weighs 100 g and contains 11 g of fat; that is, $\frac{11}{100}$ is fat. A Burger King hamburger is 0.11009 fat. Write $\frac{11}{100}$ as a decimal and determine which hamburger has the most fat.

1. _____

2. Lawyers have to know about fractions and decimals too. In a court case called *The U.S. v. Forty Barrels and Twenty Kegs of Coca-Cola,* a chemical analysis indicated that $\frac{3}{7}$ of the Coca-Cola was water. A second analysis showed that 0.41 was water. Which of the two analyses indicated more water in the Coke?

2. _____

3. In a recent year the New York Yankees won the Eastern Division Championship with 100 wins in 160 games. The Los Angeles Dodgers had a 0.584 win-loss average. Write $\frac{100}{160}$ as a decimal and determine which of the two teams had the better average.

3. _____

Here are the heights of three of the tallest people in the world:

NAME	FEET	INCHES
(a) Sulaiman Ali Nashnush	8	$\frac{1}{25}$
(b) Gabriel Estavao Monjane	8	$\frac{3}{4}$
(c) Constantine	8	0.8

4. Which one is the tallest of the three?

4. _____

5. Which one is the second tallest?

5. _____

6. Which one is the shortest of the three?

6. _____

1.3 ADDITION AND SUBTRACTION OF REAL NUMBERS

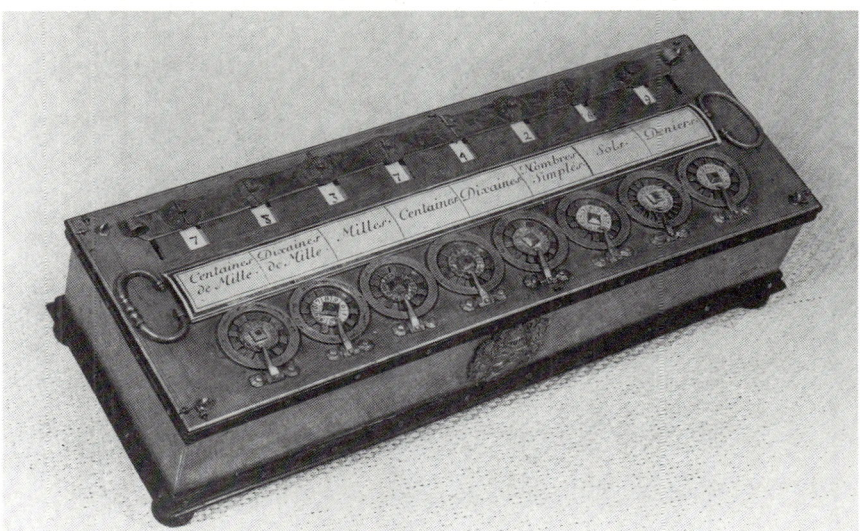

Courtesy of International Business Machines Corporation.

OBJECTIVES

REVIEW

Before starting this section, you should know:

1. How to find the absolute value of a number.
2. How to add and subtract natural numbers.
3. How to find the additive inverse of a number.

OBJECTIVES

You should be able to:

A. Add two real numbers.
B. Subtract two real numbers.
C. Identify uses of the commutative and associative laws of addition.
D. Solve applications using the concepts studied.

The illustration shows a photo of the world's first adding machine, which was invented by the French mathematician Blaise Pascal. We will now show you how to add real numbers without a machine.

A. ADDITION

The addition of two positive numbers in algebra is done just as in arithmetic. The addition $2 + 3 = 5$, for example, can be represented on the number line by starting at 0, moving 2 units in the positive direction (to the right), followed by moving 3 more units to the right, ending at 5 (see Figure 5). Thus, $2 + 3 = 5$.

To add -2 and -3, we follow a similar pattern. We start at 0, but this time we move 2 units in the *negative* (left) direction and follow by moving 3 more units to the left (see Figure 6). The result is -5. Thus $(-2) + (-3) = -5$. Note that we wrote $(-2) + (-3)$ instead of $-2 + -3$, which is confusing. Try not to write two signs together without parentheses.

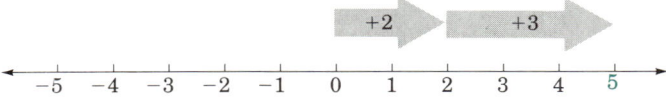

FIG. 5

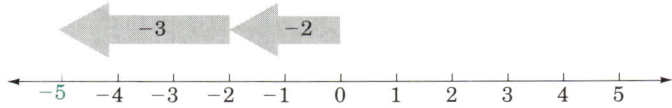

FIG. 6

From these and similar examples, we can deduce a procedure that uses *absolute values* for adding real numbers with the same sign. Here is the rule.

> To add real numbers with the *same* sign, *add* their absolute values and give the sum the common sign.

Thus, $3 + 9 = +(|3| + |9|) = 12$ and $(-5) + (-4) = -(|-5| + |-4|) = -9$.

EXAMPLE 1 Find.

a. $7 + 11$ **b.** $(-3) + (-7)$ **c.** $0.3 + 0.8$
d. $(-0.7) + (-0.8)$ **e.** $\frac{2}{3} + \frac{1}{6}$ **f.** $-\frac{1}{2} + (-\frac{1}{4})$

Solution

a. $7 + 11 = +(|7| + |11|) = 18$

b. $(-3) + (-7) = -(|-3| + |-7|) = -10$

c. $0.3 + 0.8 = +(|0.3| + |0.8|) = 1.1$

d. $(-0.7) + (-0.8) = -(|-0.7| + |-0.8|) = -1.5$

e. $\frac{2}{3} + \frac{1}{6} = +(|\frac{2}{3}|) + (|\frac{1}{6}|) = +(\frac{2}{3} + \frac{1}{6}) = +(\frac{4}{6} + \frac{1}{6}) = +\frac{5}{6}$

Note that we wrote $\frac{2}{3}$ as $\frac{4}{6}$ to add it to $\frac{1}{6}$.

f. $-\frac{1}{2} + (-\frac{1}{4}) = -(|-\frac{1}{2}| + |-\frac{1}{4}|) = -(\frac{1}{2} + \frac{1}{4}) = -(\frac{2}{4} + \frac{1}{4}) = -\frac{3}{4}$

Note that we wrote $\frac{1}{2}$ as $\frac{2}{4}$ before adding.   ▲

How do we add a positive and a negative number? To do this, we again consider the number line. For example, to add $5 + (-3)$, we begin at 0, move 5 units to the right, and then move 3 units to the left, obtaining 2. Thus $5 + (-3) = 2$, as shown in Figure 7.

Now consider the sum $(-5) + 3$. We begin at 0, move 5 units to the left, and then move 3 units to the right. The result is -2, as shown in Figure 8.

Note that $5 + (-3) = 5 - 3 = 2$, and that $(-5) + 3 = -(|-5| - |3|) = -2$. These two and similar examples lead to the following rule.

Problem 1 Find.
a. $3 + 8$ **b.** $(-4) + (-6)$
c. $0.2 + 0.6$ **d.** $(-0.6) + (-0.7)$
e. $\frac{1}{3} + \frac{1}{4}$ **f.** $-\frac{1}{2} + (-\frac{3}{4})$

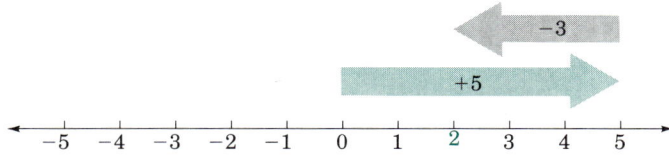

FIG. 7

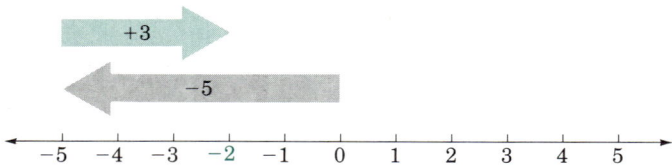

FIG. 8

TO ADD A POSITIVE AND A NEGATIVE REAL NUMBER

1. Find the absolute value of the numbers.
2. Subtract the number with the smaller absolute value from the one with the greater absolute value.
3. Use the sign of the number with the greater absolute value for the result obtained in Step 2.

Of course, if 0 is added to any number a, the result is the number a. Thus $3 + 0 = 3$ and $0 + \frac{1}{3} = \frac{1}{3}$. The number 0 is called the identity for addition.

For any real number a, $a + 0 = 0 = 0 + a$.

EXAMPLE 2   Find.

a. $6 + (-3)$ **b.** $(-8) + 4$ **c.** $(-3) + 9$
d. $5 + (-11)$ **e.** $0.8 + (-0.5)$ **f.** $-0.7 + (0.4)$
g. $\frac{4}{5} + (-\frac{2}{5})$ **h.** $-\frac{3}{8} + (\frac{1}{4})$

Problem 2   Find.
a. $7 + (-4)$ **b.** $(-9) + 6$
c. $(-2) + 9$ **d.** $6 + (-13)$
e. $0.9 + (-0.6)$ **f.** $-0.9 + (0.3)$
g. $\frac{4}{7} + (-\frac{3}{7})$ **h.** $-\frac{5}{8} + \frac{1}{4}$

Solution.

a. $6 + (-3) = |6| - |-3| = 6 - 3 = 3$

b. $(-8) + 4 = -(|-8| - |4|) = -4$

c. $(-3) + 9 = (|9| - |-3|) = 9 - 3 = 6$

d. $5 + (-11) = -(|-11| - |5|) = -(11 - 5) = -6$

e. $0.8 + (-0.5) = (|0.8| - |-0.5|) = 0.8 - 0.5 = 0.3$

f. $-0.7 + (0.4) = -(|0.7| - |0.4|) = -(0.7 - 0.4) = -0.3$

g. $\frac{4}{5} + (-\frac{2}{5}) = (|\frac{4}{5}| + |-\frac{2}{5}|) = (\frac{4}{5} - \frac{2}{5}) = \frac{2}{5}$

h. $-\frac{3}{8} + (\frac{1}{4}) = -\frac{3}{8} + \frac{2}{8} = -(|-\frac{3}{8}| - |\frac{2}{8}|) = -(\frac{3}{8} - \frac{2}{8}) = -\frac{1}{8}$ (We wrote $\frac{1}{4}$ as $\frac{2}{8}$ first, so we could determine the **sign** of the answer.) ▲

In actual practice, most of these operations are carried out mentally. Thus we write $6 + (-3) = 3$ and $-8 + 4 = -4$. Note that $-3 + 3 = 0$, $\frac{1}{2} + (-\frac{1}{2}) = 0$, and $-2.1 + 2.1 = 0$. We have the following.

For any real number a, $a + (-a) = -a + a = 0$.

B. SUBTRACTION

Subtraction of real numbers is defined as follows.

If a and b are real numbers, the *difference* of a and b, denoted by $a - b$, is the number c with the property that $(a = b + c)$, that is, $(a - b = c)$ means that $a = b + c$. (a is called the minuend and b the subtrahend.)

Thus $8 - 3 = 5$ because $8 = 3 + 5$, and $11 - 4 = 7$ because $11 = 4 + 7$. This definition is useful for checking subtraction but is not very practical otherwise.

Since we have developed a procedure to add real numbers, it will be convenient to do subtraction in terms of addition. Using the definition, it can be shown that the following is true.

If a and b are real numbers, $a - b = a + (-b)$.

This means that we can *subtract* by adding the additive inverse. Thus,

$$6 - (-3) = 6 + 3 = 9$$
$$-0.7 - 0.2 = -0.7 + (-0.2) = -0.9$$
$$-\frac{1}{5} - \left(-\frac{4}{5}\right) = -\frac{1}{5} + \frac{4}{5} = \frac{3}{5}$$

EXAMPLE 3 Find.

a. $-20 - 5$ **b.** $-0.6 - (-0.2)$ **c.** $-\frac{1}{7} - \left(-\frac{2}{7}\right)$

Solution

a. $-20 - 5 = -20 + (-5) = -25$
b. $-0.6 - (-0.2) = -0.6 + 0.2 = -0.4$
c. $-\frac{1}{7} - \left(-\frac{2}{7}\right) = -\frac{1}{7} + \frac{2}{7} = \frac{1}{7}$

Problem 3 Find.
a. $-15 - 3$
b. $-0.7 - (-0.3)$
c. $-\frac{1}{5} - \left(-\frac{3}{5}\right)$

C. THE COMMUTATIVE AND ASSOCIATIVE LAWS

How do you check this addition?

$$\begin{array}{r} 246 \\ + \ 329 \\ \hline 575 \end{array}$$

One way is to add from the bottom up $(329 + 246)$. The result should be unchanged because the sum of two numbers is the same no matter in which *order* you add them. Thus, $8 + 9 = 9 + 8$, $1.92 + 0.34 = 0.34 + 1.92$, and $\frac{1}{4} + \frac{7}{8} = \frac{7}{8} + \frac{1}{4}$. In general, we have the following.

THE COMMUTATIVE LAW OF ADDITION

For any real numbers a and b, $a + b = b + a$.

Now, suppose you are adding $8 + 4 + 6$. Would you add 8 and 4 first or 4 and 6 first? It does not matter, the answer is the same. You can indicate the manner in which you wish to do the addition by using **parentheses ( )**. By convention, a quantity enclosed in parentheses must be computed *first*. Thus if you wish to add 8 and 4 first in $8 + 4 + 6$, write $(8 + 4) + 6$, but if you want to add 4 and 6 first, then write $8 + (4 + 6)$. In either case the answer is 18. Here is the idea.

EXAMPLE 4   Which law is illustrated in each of the statements?

a. $(3 + 0.4) + \frac{1}{2} = 3 + (0.4 + \frac{1}{2})$ **b.** $(\frac{1}{5} + 3) + 7 = (3 + \frac{1}{5}) + 7$

Solution

a. We have changed the *grouping* of the numbers. We used the associative law.
b. We have changed the *order* inside the parentheses. Thus we have used the commutative law.

D.   APPLICATIONS

The ideas studied can be used in many areas. For example, in 1980 U.S. Treasury receipts amounted to $520 million. Unfortunately, outlays were about $580 million. Since $520 − $580 = −$60 million, the deficit is $60 million.

EXAMPLE 5   In a recent year receipts for the U.S. Treasury amounted to $854 million. Outlays were $1002 million. What was the amount of the deficit?

Solution

Since $854 − $1002 = −$148, the deficit was $148 million.

Problem 4   Which law is illustrated in each of the statements?
a. $5 + (0.7 + \frac{1}{9}) = 5 + (\frac{1}{9} + 0.7)$
b. $2 + (\frac{1}{3} + \frac{1}{4}) = (2 + \frac{1}{3}) + \frac{1}{4}$

Problem 5   In one year receipts were $769 million and outlays $990 million. What was the amount of the deficit?

NAME

CLASS

SECTION

A. In Problems 1–10 find the sum. Verify the answer using a number line.

ANSWERS

1. $\dfrac{3}{5} + \left(-\dfrac{1}{5}\right)$ **2.** $-0.4 + 0.9$

3. $(-0.3) + 0.2$ **4.** $(-8) + 5$

5. $(-4) + 6$ **6.** $(-0.2) + 0.3$

7. $(-0.5) + (-0.3)$ **8.** $(-7) + (-11)$

9. $\left(-\dfrac{1}{5}\right) + \dfrac{2}{5}$ **10.** $\left(-\dfrac{4}{7}\right) + \dfrac{2}{7}$

B. In Problems 11–32 find each difference.

11. $6 - 13$ **12.** $8 - 13$

13. $0.6 - 0.9$ **14.** $0.3 - 0.8$

15. $\dfrac{1}{7} - \dfrac{3}{7}$ **16.** $\dfrac{3}{8} - \dfrac{4}{8}$

17. $-8 - 6$ **18.** $-4 - 9$

19. $-0.4 - 0.2$ **20.** $-0.3 - 0.5$

21. $-\dfrac{3}{7} - \dfrac{2}{7}$ **22.** $-\dfrac{4}{9} - \dfrac{1}{9}$

23. $-6 - (-5)$ **24.** $-7 - (-9)$

25. $-8 - (-4)$ **26.** $-9 - (-2)$

27. $-0.7 - (-0.6)$ **28.** $-0.9 - (-0.3)$

1. _____
2. _____
3. _____
4. _____
5. _____
6. _____
7. _____
8. _____
9. _____
10. _____
11. _____
12. _____
13. _____
14. _____
15. _____
16. _____
17. _____
18. _____
19. _____
20. _____
21. _____
22. _____
23. _____
24. _____
25. _____
26. _____
27. _____
28. _____

29. $-0.3 - (-0.2)$ **30.** $-0.5 - (-0.3)$

29. _____

30. _____

31. $-\dfrac{2}{7} - \left(-\dfrac{4}{7}\right)$ **32.** $-\dfrac{3}{11} - \left(-\dfrac{5}{11}\right)$

31. _____

32. _____

C. In Problems 33–40 indicate which law is illustrated in the statement.

33. $5.6 + 9.2 = 9.2 + 5.6$

33. _____

34. $\dfrac{1}{3} + 2.5 = 2.5 + \dfrac{1}{3}$

34. _____

35. $1.2 + (3.3 + 1) = (3.3 + 1) + 1.2$

35. _____

36. $3.1 + (4.2 + 5) = (3.1 + 4.2) + 5$

36. _____

37. $\left(\dfrac{1}{5} + \dfrac{2}{7}\right) + \dfrac{1}{8} = \dfrac{1}{5} + \left(\dfrac{2}{7} + \dfrac{1}{8}\right)$

37. _____

38. $7.1 + (-2.1) = -2.1 + 7.1$

38. _____

39. $(-3.5 + 2) + 7.4 = -3.5 + (2 + 7.4)$

39. _____

40. $-\dfrac{3}{2} + \dfrac{1}{8} = \dfrac{1}{8} + \left(-\dfrac{3}{2}\right)$

40. _____

D. Applications

41. The temperature in the center core of the earth reaches $+5000°C$. In the thermosphere (a region in the upper atmosphere), the temperature is $+1500°C$. Find the difference in temperature between the center of the earth and the thermosphere.

41. _____

42. The record high temperature in Calgary, Alberta, is $+99°F$. The record low temperature is $-46°F$. Find the difference between these extremes.

42. _____

43. The price of a certain stock at the beginning of the week was $47. Here are the changes in price during the week: $+1$, $+2$, -1, -2, -1. What is the price of the stock at the end of the week?

43. _____

44. The price of a stock was $37. On Monday, the price went up $2; on Tuesday, it went down $3; and on Wednesday, it went down another $1. What was the price of the stock then?

44. _____

45. Here are the temperature changes (in degrees Celsius, C) by the hour in a certain city:

45. _____

1 P.M.	$+2$
2 P.M.	$+1$
3 P.M.	-1
4 P.M.	-3

If the temperature was initially 15°C, what was it at 4 P.M.?

EXERCISE 1.3

✓ SKILL CHECKER

Fill in the blank with the additive inverse.

46. $-3 +$ _____ $= 0$

47. $7 +$ _____ $= 0$

48. $-\dfrac{1}{5} +$ _____ $= 0$

49. _____ $+ (-2.3) = 0$

50. _____ $+ 7.5 = 0$

46. _____

47. _____

48. _____

49. _____

50. _____

1.3 USING YOUR KNOWLEDGE

A Little History   The accompanying chart contains some important historical dates.

IMPORTANT HISTORICAL DATES	
323 B.C.	Alexander the Great dies
216 B.C.	Hannibal defeats the Romans
A.D. 476	Fall of the Roman Empire
A.D. 1492	Columbus discovers America
A.D. 1776	The Declaration of Independence signed
A.D. 1939	World War II starts
A.D. 1969	First human on the moon
A.D. 1988	Reagan and Gorbachev hold summit

We can use negative integers to represent years B.C. For example, the year Alexander the Great died can be written as -323, whereas the Fall of the Roman Empire occurred in $+476$ (or simply 476). To find the number of years elapsed between the fall of the Roman Empire and their defeat by Hannibal, we write

$$476 - (-216) = 476 + 216 = 692$$

Fall of the Roman Empire (476 A.D.)   Hannibal defeats the Romans (216 B.C.)   Years elapsed

Use these ideas to find the number of years elapsed between the following:

1. The fall of the Roman Empire and the death of Alexander the Great

2. Columbus's discovery of America and Hannibal's defeat of the Romans

3. The discovery of America and the signing of the Declaration of Independence

4. The year a human landed on the moon and the discovery of America

5. The Reagan–Gorbachev summit and the start of World War II

1. _____

2. _____

3. _____

4. _____

5. _____

EXERCISE 1.3

Some calculators have a key that finds the additive inverse (opposite) of a given number. For example, to find the opposite of 5, press $\boxed{5}$ $\boxed{+/-}$ or $\boxed{5}$ $\boxed{\text{CHS}}$ and the correct answer, -5, will be displayed. Here are some of the examples in this section done with a calculator.

EXAMPLE 1b $(-3) + (-7)$ Press $\boxed{-}$ $\boxed{3}$ $\boxed{+}$ $\boxed{-}$ $\boxed{7}$ $\boxed{=}$ or

$\boxed{-}$ $\boxed{3}$ $\boxed{+}$ $\boxed{7}$ $\boxed{+/-}$ $\boxed{=}$

EXAMPLE 2a $6 + (-3)$ Press $\boxed{6}$ $\boxed{+}$ $\boxed{-}$ $\boxed{3}$ $\boxed{=}$ * or

$\boxed{6}$ $\boxed{+}$ $\boxed{3}$ $\boxed{+/-}$ $\boxed{=}$

EXAMPLE 2b $(-8) + 4$ Press $\boxed{-}$ $\boxed{8}$ $\boxed{+}$ $\boxed{4}$ $\boxed{=}$

EXAMPLE 3

a. $-20 - 5$ Press $\boxed{-}$ $\boxed{2}$ $\boxed{0}$ $\boxed{-}$ $\boxed{5}$ $\boxed{=}$

b. $-0.6 - (-0.2)$ presents a different problem. Here you must know that $-(-0.2)$ must be entered as $+0.2$ or as -0.2 $\boxed{+/-}$. You then press

$\boxed{-}$ $\boxed{.}$ $\boxed{6}$ $\boxed{+}$ $\boxed{.}$ $\boxed{2}$ $\boxed{=}$

or

$\boxed{-}$ $\boxed{.}$ $\boxed{6}$ $\boxed{-}$ $\boxed{.}$ $\boxed{2}$ $\boxed{+/-}$ $\boxed{=}$

Note: The calculator is no substitute for knowledge. Even the best calculator will not get the correct answer for Example 3b unless you know the arithmetic involved.

---

* Some calculators will indicate an *error* if two signs of operation are pressed consecutively (for example, $\boxed{+}$ $\boxed{-}$).

MOST ACTIVE STOCKS						
	OPEN	HIGH	LOW	CLOSE	CHG.	VOLUME
Citicorp	$26\frac{3}{4}$	$26\frac{3}{4}$	$24\frac{1}{2}$	$24\frac{1}{2}$	$-1\frac{1}{2}$	285,000
Am Medical	$3\frac{1}{8}$	$3\frac{1}{8}$	$2\frac{3}{4}$	3	$-\frac{1}{8}$	245,900
Southern Co	$10\frac{3}{8}$	$10\frac{3}{8}$	$10\frac{1}{8}$	$10\frac{1}{8}$	$-\frac{1}{4}$	211,700
FedNat Mtg	$12\frac{3}{4}$	$12\frac{7}{8}$	12	12	$-\frac{5}{8}$	191,900
FMC	$13\frac{1}{4}$	$13\frac{1}{4}$	$12\frac{1}{2}$	$12\frac{1}{2}$	$-\frac{5}{8}$	161,200
Kauf Broad	$2\frac{1}{2}$	$2\frac{5}{8}$	$2\frac{1}{4}$	$2\frac{1}{2}$	. . .	123,900
Dow Chem	$56\frac{3}{8}$	$56\frac{3}{8}$	$52\frac{3}{4}$	$52\frac{3}{4}$	$-3\frac{1}{8}$	122,900
Evans Pd	$3\frac{1}{4}$	$3\frac{1}{4}$	$2\frac{7}{8}$	3	. . .	119,400
MorganJP	$46\frac{3}{8}$	$46\frac{1}{2}$	44	44	$-2\frac{1}{8}$	113,200
East Kodak	$69\frac{7}{8}$	$69\frac{7}{8}$	$65\frac{7}{8}$	66	-3	110,500

Average closing price of most active stocks: 23.C3.

REVIEW

Before starting this section, you should know:

1. How to multiply two whole numbers.

2. How to divide two whole numbers.

OBJECTIVES

You should be able to:

A. Multiply positive and negative numbers.

B. Divide positive and negative numbers.

C. Identify uses of the commutative, associative, and identity laws.

In arithmetic, the product of a and b is written as $a \times b$. In algebra, however, the multiplication sign $\times$ can be mistaken for the letter x, so we use a different notation.

A. MULTIPLICATION

In algebra, the product of a and b can be written as follows.

Using a raised dot, $\cdot$	$a \cdot b$
Writing a and b next to each other	ab
Using parentheses	$(a)(b)$, $a(b)$, $(a)b$

The numbers represented by a and b are called **factors,** and the result of the multiplication is called the **product** of a and b.

Now, look at the preceding stock listing and suppose you own 4 shares of Kodak. The closing price was *down* \$3, written as -3. Your loss that day would be the product $4 \cdot (-3)$, with factors 4 and -3. What is this product? This multiplication is just the repeated addition of -3.

$$4 \cdot (-3) = \underbrace{(-3) + (-3) + (-3) + (-3)}_{\text{4 negative threes}} = -12$$

Next, look at $(-3) \cdot 4$. As with addition, multiplication of real numbers has the commutative property. (See Section C.) Thus

$$(-3) \cdot 4 = 4 \cdot (-3) = -12$$

We can generalize this idea to show that the product of any two numbers, one *positive* and the other *negative,* is a *negative number.* As in addition, we can state this result in terms of absolute values.

> To multiply a *positive* number by a *negative* number, multiply their absolute values. The product is *negative*.

Thus, $3 \cdot (-2) = -6$, $-4 \cdot 8 = -32$, and $2 \cdot (-3.5) = -7$. Note that the factors in each case, 3 and -2, -4 and 8, and 2 and -3.5, have *different* (*unlike*) signs.

EXAMPLE 1 Find.

a. $7 \cdot (-8)$ **b.** $-2.5 \cdot 4$

Solution

a. $7 \cdot (-8) = -56$ **b.** $-2.5 \cdot 4 = -10$ ▲

Problem 1 Find.
a. $9 \cdot (-7)$ b. $-4.5 \cdot 2$

What about the product of two negative integers, such as $-4 \cdot (-3)$? Look for the pattern.

This number decreases by 1.   This number increases by 3.

$$4 \cdot (-3) = -12$$
$$3 \cdot (-3) = -9$$
$$2 \cdot (-3) = -6$$
$$1 \cdot (-3) = -3$$
$$0 \cdot (-3) = 0$$
$$-1 \cdot (-3) = 3$$
$$-2 \cdot (-3) = 6$$
$$-3 \cdot (-3) = 9$$
$$-4 \cdot (-3) = 12$$

Thus, $-4 \cdot (-3) = 12$. Note that in this case -4 and -3 have the *same* sign $(-)$. When we multiply $4 \cdot 3$, 4 and 3 are both positive, so they also have the *same* sign $(+)$. We can summarize this discussion by the following rule:

TO MULTIPLY TWO NUMBERS WITH	THE PRODUCT IS
Like (same) signs	Positive $(+)$
Unlike (different) signs	Negative $(-)$

Here are some examples.

$$2.5 \cdot 3 = 7.5 \qquad\qquad 9 \cdot 4 = 36$$
$$(-2.5) \cdot (-3) = 7.5 \qquad (-9) \cdot (-4) = 36$$
Same signs, thus the answer is positive

$$(-4.5) \cdot 2 = -9 \qquad\qquad (-9) \cdot 6 = -54$$
$$3 \cdot (-6.1) = -18.3 \qquad 6 \cdot (-9) = -54$$
Opposite signs, thus the answer is negative

What will happen if we multiply any number a by 1? The answer will be a. Thus, $1 \cdot 4 = 4$, $1 \cdot (-3) = -3$, $2.5 \cdot 1 = 2.5$, $1 \cdot \frac{3}{5} = \frac{3}{5}$, and $(-\frac{2}{9}) \cdot 1 = -\frac{2}{9}$. Note that multiplying by 1 leaves the number unchanged. For this reason, 1 is called the **identity for multiplication.**

> **IDENTITY FOR MULTIPLICATION**
>
> For any real number **a**, $a \cdot 1 = 1 \cdot a = a$.

EXAMPLE 2 Find.

a. $7 \cdot 8$　　　　**b.** $(-8) \cdot 6$　　　　**c.** $4 \cdot (-3)$

d. $(-7) \cdot (-9)$　　**e.** $2 \cdot 4.5$　　　　**f.** $(-3) \cdot 6.1$

g. $4 \cdot (-2.1)$　　　**h.** $(-3) \cdot (-4.2)$

Solution

a. $7 \cdot 8 = 56$

b. $(-8) \cdot 6 = -48$

　　opposite signs　negative answer

c. $4 \cdot (-3) = -12$

　　opposite signs　negative answer

d. $(-7) \cdot (-9) = 63$

　　same signs　　positive answer
　　　　　　　　　$(63 = +63)$

e. $2 \cdot 4.5 = 9$

f. $(-3) \cdot 6.1 = -18.3$

　　opposite signs　negative answer

g. $4 \cdot (-2.1) = -8.4$

　　opposite signs　negative answer

h. $(-3) \cdot (-4.2) = 12.6$

　　same signs　　positive answer
　　　　　　　　　$(12.6 = +12.6)$　　　　　　　▲

What about fractions? To multiply fractions, we need the following definition:

$$\frac{a}{b} \cdot \frac{c}{d} = \frac{a \cdot c}{b \cdot d}$$

The same laws of signs apply. Thus

$$\left(-\frac{9}{5}\right) \cdot \frac{3}{4} = -\frac{9 \cdot 3}{5 \cdot 4} = -\frac{27}{20}$$

When multiplying fractions, it saves time if common factors are divided out before you multiply. Thus to multiply $\frac{2}{5} \cdot \frac{5}{7}$, we write

$$\frac{\overset{1}{\cancel{5}}}{7} \cdot \left(-\frac{2}{\underset{1}{\cancel{5}}}\right) = -\frac{1 \cdot 2}{7 \cdot 1} = -\frac{2}{7} \quad \text{Note that } \frac{5}{5} = 1.$$

Problem 2 Find.

a. $6 \cdot 9$　　　　**b.** $(-3) \cdot 11$

c. $9 \cdot (-10)$　　**d.** $(-5) \cdot (-11)$

e. $2 \cdot 3.5$　　　**f.** $(-3) \cdot 8.1$

g. $5 \cdot (-4.1)$　　**h.** $(-4) \cdot (-3.2)$

EXAMPLE 3 Find.

a. $\left(-\dfrac{3}{7}\right) \cdot \dfrac{7}{8}$ **b.** $\left(-\dfrac{5}{8}\right) \cdot \left(-\dfrac{4}{15}\right)$

Solution

a. $\left(-\dfrac{\overset{1}{3}}{7}\right) \cdot \dfrac{\overset{}{7}}{8}_{1} = -\dfrac{3 \cdot 1}{1 \cdot 8} = -\dfrac{3}{8}$ Unlike (different) signs; the answer is negative.

b. $\left(-\dfrac{\overset{1}{5}}{\underset{2}{8}}\right) \cdot \left(-\dfrac{\overset{1}{4}}{\underset{3}{15}}\right) = \dfrac{1 \cdot 1}{2 \cdot 3} = \dfrac{1}{6}$ Like (same) signs; the answer is positive.

B. DIVISION

What about the rules for division? As you recall, a division problem can always be checked by multiplication. Thus the division

$$
\begin{array}{r}
6 \\
3\overline{)18} \\
-18 \\
\hline
0
\end{array}
$$

that is,

$$\frac{18}{3} = 6$$

is correct because $18 = 3 \cdot 6$.

> If a and b are real numbers
>
> $\dfrac{a}{b} = c$ means $a = b \cdot c$
>
> a is called the **dividend,** b is the **divisor,** and c is the **quotient.**

Because of this, the same rules of sign that apply to the multiplication of real numbers also apply to the division of real numbers; that is, the quotient of two numbers with the *same* sign is *positive,* and the quotient of two numbers with *opposite* signs is *negative.* Here are some examples:

$$\left.\begin{array}{ll} \dfrac{24}{6} = 4 & \dfrac{3.2}{1.6} = 2 \\[2mm] \dfrac{-18}{-9} = 2 & \dfrac{-3.3}{-1.1} = 3 \end{array}\right\} \text{Same signs, positive answers}$$

$$\left.\begin{array}{ll} \dfrac{-32}{4} = -8 & \dfrac{-6.3}{0.9} = -7 \\[2mm] \dfrac{35}{-7} = -5 & \dfrac{4.5}{-0.5} = -9 \end{array}\right\} \text{Different signs, negative answers}$$

Problem 3 Find.

a. $\left(-\dfrac{2}{5}\right) \cdot \dfrac{5}{7}$ **b.** $\left(-\dfrac{3}{14}\right) \cdot \left(-\dfrac{7}{6}\right)$

EXAMPLE 4   Find.

a. $48 \div 6$ **b.** $\dfrac{54}{-9}$ **c.** $\dfrac{-63}{-7}$ **d.** $-28 \div 4$ **e.** $5 \div 0$

f. $3.4 \div 1.7$ **g.** $\dfrac{4.8}{-1.2}$ **h.** $\dfrac{-5.6}{-0.8}$ **i.** $0 \div 3.5$

Solution

a. $48 \div 6 = 8$ (48 and 6 have the same sign; the answer is positive.)

b. $\dfrac{54}{-9} = -6$ (54 and -9 have opposite signs; the answer is negative.)

c. $\dfrac{-63}{-7} = 9$ (-63 and -7 have the same sign; the answer is positive.)

d. $-28 \div 4 = -7$ (-28 and 4 have opposite signs; the answer is negative.)

e. $5 \div 0$ is not defined. Note that if you make $5 \div 0$ equal any number, say a, we have:

$$\frac{5}{0} = a \quad \text{which means } 5 = a \cdot 0 = 0$$

This is impossible. Thus $\frac{5}{0}$ is not defined.

f. $3.4 \div 1.7 = 2$ (3.4 and 1.7 have the same sign; the answer is positive.)

g. $\dfrac{4.8}{-1.2} = -4$ (4.8 and -1.2 have different signs; the answer is negative.)

h. $\dfrac{-5.6}{-0.8} = 7$ (-5.6 and -0.8 have the same signs, the answer is positive.)

i. $0 \div 3.5 = 0$.

We can check this using the definition of division:

$$0 \div 3.5 = \frac{0}{3.5} = 0 \text{ means } 0 = 3.5 \cdot 0, \text{ which is true.} \qquad \blacktriangle$$

Here are two rules to help you out.

For $a \neq 0$, $\dfrac{0}{a} = 0$ and $\dfrac{a}{0}$ is *not* defined.

Let us look at the division problem $2 \div 5$. We can write: $2 \div 5 = \frac{2}{5} = 2 \cdot \frac{1}{5}$. Thus to *divide* 2 by 5, we *multiply* 2 by $\frac{1}{5}$. The numbers 5 and $\frac{1}{5}$ are *reciprocals*. Here is the definition.

Every nonzero real number a has a **reciprocal** such that $a \cdot \dfrac{1}{a} = 1$.

Problem 4   Find.

a. $60 \div 10$ **b.** $\dfrac{48}{-3}$

c. $\dfrac{-18}{-2}$ **d.** $-14 \div 2$

e. $-4 \div 0$ **f.** $4.8 \div 1.6$

g. $\dfrac{4.2}{-2.1}$ **h.** $\dfrac{-3.8}{-1.9}$

i. $0 \div 9.2$

The reciprocal of 3 is $\frac{1}{3}$, the reciprocal of -6 is $\frac{1}{-6} = -\frac{1}{6}$, and the reciprocal of $\frac{2}{3}$ is $\frac{3}{2}$. Note that the reciprocal of a *positive* number is *positive* and the reciprocal of a *negative* number is *negative*.

EXAMPLE 5   Find the reciprocal:

a. $\frac{2}{3}$ **b.** $-\frac{4}{5}$ **c.** 0.2

Solution

a. The reciprocal of $\frac{2}{3}$ is $\frac{3}{2}$.

b. The reciprocal of $-\frac{4}{5}$ is $-\frac{5}{4}$. (Remember that the reciprocal of a negative number is negative.)

c. The reciprocal of 0.2 is $\dfrac{1}{0.2}$.   ▲

Division of fractions is done in terms of reciprocals. Here is the definition.

$$\frac{a}{b} \div \frac{c}{d} = \frac{a}{b} \cdot \frac{d}{c}$$

Thus, to divide by $\dfrac{c}{d}$, we multiply by the reciprocal, $\dfrac{d}{c}$.

EXAMPLE 6   Find.

a. $\dfrac{2}{5} \div \left(-\dfrac{3}{4} \right)$ **b.** $\left(-\dfrac{5}{6} \right) \div \left(-\dfrac{7}{2} \right)$ **c.** $\left(-\dfrac{3}{7} \right) \div \dfrac{6}{7}$

Solution

a. $\dfrac{2}{5} \div \left(-\dfrac{3}{4} \right) = \dfrac{2}{5} \cdot \left(-\dfrac{4}{3} \right) = -\dfrac{8}{15}$

b. $\left(-\dfrac{5}{6} \right) \div \left(-\dfrac{7}{2} \right) = \left(-\dfrac{5}{6} \right) \cdot \left(-\dfrac{2}{7} \right) = \dfrac{10}{42} = \dfrac{5}{21}$

c. $\left(-\dfrac{3}{7} \right) \div \dfrac{6}{7} = \left(-\dfrac{3}{7} \right) \cdot \dfrac{7}{6} = -\dfrac{21}{42} = -\dfrac{1}{2}$

C.   THE COMMUTATIVE AND ASSOCIATIVE LAWS OF MULTIPLICATION

As with addition, multiplication of real numbers is *commutative* and *associative*. Thus, we have

THE COMMUTATIVE LAW OF MULTIPLICATION

If a and b are real numbers, $a \cdot b = b \cdot a$.

Problem 5   Find each reciprocal.

a. $\frac{5}{7}$ **b.** $-\frac{7}{8}$ **c.** 0.5

Problem 6   Find.

a. $\dfrac{3}{5} \div \left(-\dfrac{4}{7} \right)$ **b.** $\left(-\dfrac{6}{7} \right) \div \left(-\dfrac{3}{5} \right)$

c. $\left(-\dfrac{4}{5} \right) \div \dfrac{8}{5}$

This law says that the *order* in which we multiply real numbers does not affect the result. Thus

$$3 \cdot 5 = 5 \cdot 3, \qquad 2.1 \cdot 3.7 = 3.7 \cdot 3.1 \quad \text{and} \quad \frac{1}{2} \cdot \frac{1}{3} = \frac{1}{3} \cdot \frac{1}{2}$$

We also have

THE ASSOCIATIVE LAW OF MULTIPLICATION

If a, b, and c are real numbers, $a \cdot (b \cdot c) = (a \cdot b) \cdot c$.

Here, the law says that the grouping of the numbers in a multiplication does not affect the result. Thus

$$3 \cdot (7 \cdot 5) = (3 \cdot 7) \cdot 5 \quad \text{and} \quad \frac{1}{3} \cdot \left(\frac{2}{5} \cdot \frac{3}{4} \right) = \left(\frac{1}{3} \cdot \frac{2}{5} \right) \cdot \frac{3}{4}. \text{ Also}$$

$$-4 \cdot \left(\frac{1}{2} \cdot \frac{1}{4} \right) = \left(-4 \cdot \frac{1}{2} \right) \cdot \frac{1}{4}$$

EXAMPLE 7 Which law is illustrated in each statement?

a. $(-3) \cdot \frac{7}{5} = \frac{7}{5} \cdot (-3)$ **b.** $3 \cdot (4 \cdot 7) = (4 \cdot 7) \cdot 3$
c. $(-4) \cdot \left(\frac{1}{2} \cdot \frac{1}{8} \right) = \left(-4 \cdot \frac{1}{2} \right) \cdot \frac{1}{8}$

Solution

a. Since we changed the order, the commutative law applies.

b. Here again, we changed the order (the 3 changed position). The commutative law was used.

c. This time, we grouped the numbers differently. The associative law was used. ▲

Note that the associative law helps you find the answer to a problem such as $(3)(-5)(6)$. To find the answer, you can write either of the following:

1. $(3)(-5)(6) = (-15)(6)$ Multiply $(3)(-5)$ first.
                  $= -90$ Multiply $(-15)(6)$ next.
2. $(3)(-5)(6) = (3)(-30)$ Multiply $(-5)(6)$ first.
                  $= -90$ Multiply $(3)(-30)$ next.

Of course, the answer is the same in both cases. We will practice with this type of problem in the exercises.

Problem 7 Which law is illustrated in each statement?
a. $2 \cdot (3 \cdot 5) = (3 \cdot 5) \cdot 2$
b. $(-4) \cdot \left(\frac{2}{5} \right) = \frac{2}{5} \cdot (-4)$
c. $7 \cdot \left(\frac{1}{4} \cdot \frac{1}{2} \right) = \left(7 \cdot \frac{1}{4} \right) \cdot \frac{1}{2}$

NAME

CLASS

SECTION

ANSWERS

A. Find.

1. $(-5)(8)$

2. $(-9)(6)$

3. $(4)(-3)$

4. $(6)(-8)$

5. $(-10)(-5)$

6. $(-6)(-9)$

7. $(-3)(4)(-5)$

8. $(-5)(2)(3)$

9. $(-4)(-2)(5)$

10. $(-2)(-5)(9)$

11. $(-3)(5)(-2)$

12. $(-3)(10)(-2)$

13. $(4)(-5)(2)$

14. $(10)(-3)(6)$

15. $-2.2(3.3)$

16. $-1.4(3.1)$

17. $-1.3(-2.2)$

18. $-1.5(-1.1)$

19. $\dfrac{5}{6}\left(-\dfrac{5}{7}\right)$

20. $\dfrac{3}{8}\left(-\dfrac{5}{7}\right)$

21. $-\dfrac{3}{5}\left(-\dfrac{5}{12}\right)$

22. $-\dfrac{4}{7}\left(-\dfrac{21}{8}\right)$

23. $-\dfrac{6}{7}\left(\dfrac{35}{8}\right)$

24. $-\dfrac{7}{5}\left(\dfrac{15}{28}\right)$

B. Find.

25. $\dfrac{-18}{9}$

26. $\dfrac{-32}{16}$

27. $\dfrac{20}{-5}$

28. $\dfrac{36}{-3}$

29. $\dfrac{-14}{-7}$

30. $\dfrac{-24}{-8}$

1. _____
2. _____
3. _____
4. _____
5. _____
6. _____
7. _____
8. _____
9. _____
10. _____
11. _____
12. _____
13. _____
14. _____
15. _____
16. _____
17. _____
18. _____
19. _____
20. _____
21. _____
22. _____
23. _____
24. _____
25. _____
26. _____
27. _____
28. _____
29. _____
30. _____

31. $\dfrac{0}{-3}$

32. $\dfrac{0}{-9}$

33. $\dfrac{4}{0}$

34. $\dfrac{-7}{0}$

35. $-\left(\dfrac{-4}{-2}\right)$

36. $-\left(\dfrac{-10}{-5}\right)$

37. $-\left(\dfrac{-27}{3}\right)$

38. $-\left(\dfrac{-9}{3}\right)$

39. $-\left(\dfrac{15}{-5}\right)$

40. $-\left(\dfrac{18}{-6}\right)$

41. $\dfrac{-3}{-3}$

42. $\dfrac{-18}{-9}$

43. $\dfrac{-16}{4}$

44. $\dfrac{-48}{6}$

45. $\dfrac{-56}{8}$

46. $\dfrac{-54}{6}$

47. $\dfrac{3}{5} \div \left(-\dfrac{4}{7}\right)$

48. $\dfrac{4}{9} \div \left(-\dfrac{1}{7}\right)$

49. $-\dfrac{2}{3} \div \left(-\dfrac{7}{6}\right)$

50. $-\dfrac{5}{6} \div \left(-\dfrac{25}{18}\right)$

51. $-\dfrac{5}{8} \div \dfrac{7}{8}$

52. $-\dfrac{4}{5} \div \dfrac{8}{15}$

53. $\dfrac{-3.1}{6.2}$

54. $\dfrac{1.2}{-4.8}$

55. $\dfrac{-1.6}{-9.6}$

56. $\dfrac{-9.8}{-1.4}$

C. In Problems 57–66, indicate which law is illustrated in each statement. a, b, and c represent real numbers.

57. $-3 \cdot 4 = 4 \cdot (-3)$

58. $5 \cdot (-2) = -2 \cdot 5$

59. $(-4 \cdot 2) \cdot 5 = -4 \cdot (2 \cdot 5)$

60. $-3 \cdot (a \cdot 2) = (-3 \cdot a) \cdot 2$

61. $(-3 \cdot a) \cdot 2 = 2 \cdot (-3 \cdot a)$

62. $5(3a) = (3a)5$

63. $1 \cdot (3 + b) = 3 + b$

64. $(a + b) \cdot 1 = a + b$

31. _____
32. _____
33. _____
34. _____
35. _____
36. _____
37. _____
38. _____
39. _____
40. _____
41. _____
42. _____
43. _____
44. _____
45. _____
46. _____
47. _____
48. _____
49. _____
50. _____
51. _____
52. _____
53. _____
54. _____
55. _____
56. _____

57. _____
58. _____
59. _____
60. _____
61. _____
62. _____
63. _____
64. _____

65. $a \cdot (bc) = (ab)c$

66. $(ab) \cdot c = a \cdot (b \cdot c)$

67. Sir John Robertson, who was five times prime minister of New South Wales, Australia, bought 3 pt of rum every morning for 35 years. Assuming that a year has 365 days, how many pints of rum did he buy? By the way, he did not drink it all himself! He drank one pint on the spot, gave one pint to his horse, and poured the third pint into his riding boot as a preventive against rheumatism.

67. _____

68. Can you pull 90 times your own weight? The Passalid beetle can. If the beetle's weight were to reach 2 oz and it could maintain its phenomenal strength, how much (in ounces) could it pull?

68. _____

69. When driving on a highway, you should stay about 18 ft from the car ahead of you for each 10 mi/hr of your speed. Find the distance (in feet) that you should keep from the car ahead when traveling
a. 30 mi/hr
b. 40 mi/hr
c. 55 mi/hr

69. a. _____
    b. _____
    c. _____

70. The price/earnings (P/E) ratio of a stock is defined as the market price of the stock divided by its actual or indicated annual earnings per share. If the market value of a share of American Construction is $54 and the stock earned $9 per share, find the P/E ratio for the stock.

70. _____

71. Do you know what your I.Q. (intelligence quotient) is? The highest I.Q. ever recorded is that of Kim Ung-Yong of South Korea. At the age of 4 yr, 8 mo, he had an I.Q. of 210. The I.Q. of a child is computed by dividing the mental age (M.A.) of the child by the chronological age (C.A.) and then multiplying by 100. If the mental age of a child is 8 years and his chronological age is 5 years, what is his I.Q.?

71. _____

72. The highest recorded shorthand writing speed under championship conditions is 1500 words in 5 min. How many words per minute is that?

72. _____

✓ **SKILL CHECKER**

In Problems 73–76 fill in the blanks.

	NUMBER	ADDITIVE INVERSE	RECIPROCAL
73.	7		
74.	-2.3		
75.	0		
76.	$-\dfrac{2}{3}$		

1.4 USING YOUR KNOWLEDGE

Have you met anybody *nice* today or did you have an *unpleasant* experience? Perhaps the person you met was *very nice* or your experience *very unpleasant*. Psychologists and linguists have a numerical way to indicate the difference between *nice* and *very nice* or between *unpleasant* and *very unpleasant*. Suppose you assign a positive number ($+2$, for example) to the adjective *nice*, and a negative number (say, -2) to *unpleasant*, and a positive number greater than 1 (say $+1.75$) to *very*. Then, very nice means

very nice

$$(1.75) \cdot (2) = 3.50$$

and very unpleasant means

very unpleasant

$$(1.75) \cdot (-2) = -3.50$$

Here are some adverbs and adjectives and their average numerical values as rated by a panel of college students.

ADVERBS		ADJECTIVES	
Slightly	0.54	Wicked	-2.5
Rather	0.84	Disgusting	-2.1
Decidedly	0.16	Average	-0.8
Very	1.25	Good	3.1
Extremely	1.45	Lovable	2.4

Find the value of:

1. Slightly wicked

2. Decidedly average

3. Extremely disgusting

4. Rather lovable

5. Very good

By the way, if you got all the answers correct, you are 4.495!

1. _____

2. _____

3. _____

4. _____

5. _____

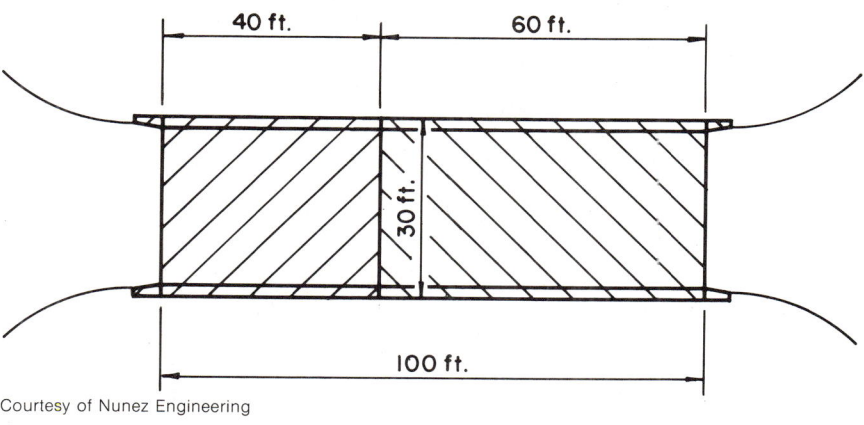

Courtesy of Nunez Engineering

The total area of the pavement in the bridge sketch is the sum of the areas of the two spans. The area in square feet is:

area of left span		area of right span	
$(30 \cdot 40)$	$+$	$(30 \cdot 60)$	Multiply first.
1200	$+$	1800	Add next.
	3000		

The area is also:

$30 \cdot (40 + 60)$	
$30 \cdot (100)$	Add first.
3000	Multiply next.

Thus, $30 \cdot (40 + 60) = (30 \cdot 40) + (30 \cdot 60)$. This is an example of the **distributive law of multiplication over addition.** Note that the parentheses in $(30 \cdot 40) + (30 \cdot 60)$ can be omitted as long as we agree that multiplications must be done *first*. With this convention, we have the following.

THE DISTRIBUTIVE LAW OF MULTIPLICATION OVER ADDITION

For any real numbers **a**, **b**, and **c**,

$$a(b + c) = ab + ac$$

(Recall that $a(b + c)$ means $a \cdot (b + c)$, ab means $a \cdot b$, and ac means $a \cdot c$.) We also have the following:

THE DISTRIBUTIVE LAW OF MULTIPLICATION OVER SUBTRACTION

For any real numbers a, b, and c,

$$a(b - c) = ab - ac$$

Thus, $5(7 - 3) = 5 \cdot 7 - 5 \cdot 3$, or 20, and $9(5 - 3) = 9 \cdot 5 - 9 \cdot 3$, or 18.

REVIEW

Before starting this section, you should know:

1. How to use the laws of signs to add, subtract, multiply, and divide real numbers.
2. The identity for multiplication.

OBJECTIVES

You should be able to:

A. Use the distributive law to simplify expressions.
B. Collect like terms.
C. Simplify expressions by removing grouping symbols and collecting like terms.

A. REMOVING PARENTHESES

In algebra, the distributive law is used to remove parentheses in expressions such as $3(x + 5)$ or $4(x - 7)$, where x is a real number. Thus,

$$3(x + 5) = 3x + 3 \cdot 5 = 3x + 15$$

and

$$4(x - 7) = 4x - 4 \cdot 7 = 4x - 28$$

EXAMPLE 1   Remove parentheses (simplify).

a. $-2(x + 8)$ **b.** $0.5(y - 7)$

Solution

a. $-2(x + 8) = -2x + (-2 \cdot 8)$
$\qquad\qquad\quad = -2x + (-16)$
$\qquad\qquad\quad = -2x - 16$ Recall that $a - b = a + (-b)$

b. $0.5(7 - y) = 0.5 \cdot 7 - 0.5y$
$\qquad\qquad\quad = 3.5 - 0.5y$ ▲

Expressions of the form $-(a + b)$ or $-(a - b)$, where a and b are called **terms,** require special consideration. We first recall the following:

> For any real number a, $a = 1 \cdot a$.

Since any real number has an additive inverse and the additive inverse of $a(= 1 \cdot a)$ is $-a$, the additive inverse of $1 \cdot a$ is $-1 \cdot a$.

> For any real number a, $-a = -1 \cdot a$.

Hence

$$-(a + b) = -1 \cdot (a + b)$$
$$\qquad\qquad = -1 \cdot a + (-1 \cdot b)$$
$$\qquad\qquad = -a - b$$

> $-(a + b) = -a - b$

Similarly,

> $-(a - b) = -a + b$

These rules tell us that to remove the parentheses in an expression preceded by a minus sign, we simply *change the sign of every term inside the parentheses* or, equivalently, *multiply each term inside the parentheses by* −1.

Problem 1   Simplify.
a. $-3(a + 5)$ **b.** $0.2(b - 6)$

EXAMPLE 2   Remove parentheses (simplify).

a. $-(x - 2)$ **b.** $-(ab + 3)$

Solution

a. $-(x - 2) = -1 \cdot (x - 2)$

$$= -1 \cdot x + (-1)(-2)$$

$$= -x + 2$$

Note that changing the signs inside the parentheses in $-(x - 2)$ will immediately yield

change sign

$$-(x - 2) = -x + 2$$

change sign

b. $-(ab + 3) = -1 \cdot (ab + 3) = -1 \cdot ab + -1 \cdot (3)$

$$= -ab + (-3)$$

$$= -ab - 3$$

Note that changing signs inside the parentheses will immediately yield the answer $-ab - 3$.   ▲

We can summarize this discussion by the following two facts:

1. If the factor in front of the parentheses has no written sign, multiply each term inside the parentheses by this factor; that is,

   $$a(b - c + d - e) = ab - ac + ad - ae$$

2. If the factor in front of the parentheses is preceded by a minus sign, multiply this factor by each of the terms inside the parentheses and change the sign of each of these terms; that is,

   $$-a(b - c + d - e) = -ab + ac - ad + ae$$

EXAMPLE 3   Remove parentheses (simplify).

a. $4(x - 2y + 3)$ **b.** $-5(2x + y - z)$
c. $0.4(-3x + 2y - 7z - 8)$ **d.** $0.5x(y + 3z - 5)$

Solution

a. $4(x - 2y + 3) = 4x - 8y + 12$
b. $-5(2x + y - z) = -10x - 5y + 5z$
c. $0.4(-3x + 2y - 7z - 8) = -1.2x + 0.8y - 2.8z - 3.2$
d. $0.5x(y + 3z - 5) = 0.5xy + 1.5xz - 2.5x$

B.   COMBINING LIKE TERMS

Suppose we wish to simplify $3x + 2(x + 5)$. We start by simplifying $2(x + 5)$, obtaining:

$$3x + 2(x + 5) = 3x + 2x + 10$$

Problem 2   Simplify.
a. $-(y - 6)$ **b.** $-(xy + 7)$

Problem 3   Simplify.
a. $5(a - 3b + 4)$
b. $-3(4x + y - z)$
c. $0.5(-2x + 3y - 6z - 4)$
d. $0.3x(y + 2z - 4)$

The terms $3x$ and $2x$ are called *like terms*. They differ only in their numerical parts (coefficients). Similarly, $-3y$ and $5y$ are like terms and $9z^2$ and $-3z^2$ are like terms (z^2 means $z \cdot z$). In general, we have the following.

> Two or more terms that differ only in their numerical coefficients are called **similar,** or **like,** terms.

We can *combine like terms* by using a variation of the distributive law. As you recall, for any real numbers a, b, and c,

$$a(b + c) = ab + ac$$

and

$$a(b - c) = ab - ac$$

Using the commutative law of multiplication, we can rewrite the two distributive laws as follows.

> $(b + c)a = ba + ca$ Distributive laws
> $(b - c)a = ba - ca$

Now,

$$3x + 2x = (3 + 2)x = 5x$$

Similarly,

$$7z^2 - 5z^2 = (7 - 5)z^2 = 2z^2$$

and

$$8xy + 3xy - 2xy = (8 + 3 - 2)xy$$
$$= 9xy$$

Note that $x + x = 1 \cdot x + 1 \cdot x = (1 + 1)x = 2x$. Thus, the coefficient of x is understood to be 1. Also, if an expression within parentheses is preceded by a plus sign, we can simply remove parentheses and combine any like terms. Thus

$$3x + (2 + 5x) = 3x + 2 + 5x$$
$$= 8x + 2$$

As you can see, you can combine like terms by simply adding or subtracting their coefficients. Use this idea in the next example.

EXAMPLE 4   Simplify.

a. $5x + 2(x - 4)$ **b.** $-3(x + 5) - 2x$ **c.** $5x - 2(x + 1) + (x + 3)$

Solution

a. $5x + 2(x - 4) = 5x + 2x - 8$
$$= 7x - 8$$

Problem 4   Simplify.
a. $8y + 3(y - 4)$
b. $-4(y + 2) - 3y$
c. $6y - 3(y + 2) + (y + 8)$

b. $-3(x + 5) - 2x = -3x - 15 - 2x$

$$= -5x - 15 \qquad \begin{array}{l} -3x - 2x = (-3 - 2)x \\ = -5x \end{array}$$

c. $5x - 2(x + 1) + (x + 3) = 5x - 2x - 2 + x + 3$

$$= 4x - 2 + 3$$
$$= 4x + 1$$

C. REMOVING OTHER GROUPING SYMBOLS

Sometimes parentheses occur within other parentheses. To avoid confusion, we do not write $((x + 5) + 3)$. Instead, we use a different grouping symbol, the brackets [ ], and write $[(x + 5) + 3]$. To simplify (combine like terms) in such expressions, the innermost grouping symbols are removed first. The procedure is illustrated in the next example.

EXAMPLE 5   Remove grouping symbols and simplify.

$$[(4x^2 - 1) + (2x + 5)] - [(x - 2) + (3x^2 - 3)]$$

Solution

We first remove the innermost parentheses and then add like terms. Thus

$$[(4x^2 - 1) + (2x + 5)] - [(x - 2) + (3x^2 - 3)]$$
$$= [4x^2 - 1 + 2x + 5] - [x - 2 + 3x^2 - 3] \qquad \text{Remove parentheses.}$$
$$= [4x^2 + 2x + 4] - [3x^2 + x - 5] \qquad \text{Add like terms.}$$
$$= 4x^2 + 2x + 4 - 3x^2 - x + 5 \qquad \text{Remove brackets.}$$
$$= x^2 + x + 9 \qquad \text{Add like terms.}$$

Problem 5   Remove grouping symbols and simplify.
$$[(2x^2 - 3) + (3x + 1)] - [(x - 1) + (x^2 - 2)]$$

NAME

CLASS

SECTION

A. Remove parentheses (simplify).

1. $4(x - y)$

2. $3(a - b)$

3. $-9(a - b)$

4. $-6(x - y)$

5. $0.3(4x - 2)$

6. $0.2(3a - 9)$

7. $-\left(\dfrac{3a}{2} - \dfrac{6}{7}\right)$

8. $-\left(\dfrac{2x}{3} - \dfrac{1}{5}\right)$

9. $-(2x - 6y)$

10. $-(3a - 6b)$

11. $-(2.1 + 3y)$

12. $-(5.4 + 4b)$

13. $-4(a + 5)$

14. $-6(x + 8)$

15. $-x(6 + y)$

16. $-y(2x + 3)$

17. $-8(x - y)$

18. $-9(a - b)$

19. $-3(2a - 7b)$

20. $-4(3x - 9y)$

21. $0.5(x + y - 2)$

22. $0.8(a + b - 6)$

23. $-\dfrac{6}{5}(a - b + 5)$

24. $-\dfrac{2}{3}(x - y - 4)$

25. $-2(x - y + 3z + 5)$

26. $-4(a - b + 2c + 8)$

27. $-0.3(x + y - 2z - 6)$

28. $-0.2(a + b - 3c - 4)$

29. $-\dfrac{5}{2}(a - 2b + c + 2d - 2)$

30. $-\dfrac{4}{7}(2a - b + 3c + 7d - 7)$

ANSWERS

1. _____
2. _____
3. _____
4. _____
5. _____
6. _____
7. _____
8. _____
9. _____
10. _____
11. _____
12. _____
13. _____
14. _____
15. _____
16. _____
17. _____
18. _____
19. _____
20. _____
21. _____
22. _____
23. _____
24. _____
25. _____
26. _____
27. _____
28. _____
29. _____
30. _____

B. Remove parentheses and combine like terms.

31. $6x + 3(x - 2)$ **32.** $8y + 6(y - 3)$

33. $-4(x + 2) - 5x$ **34.** $-5(x + 3) - 6x$

35. $(5L - 3W) - (W - 6L)$ **36.** $(2ab - 2ac) - (ab - 4ac)$

37. $5x - (8x + 1) + (x + 1)$ **38.** $3x - (7x + 2) + (x + 2)$

39. $\dfrac{2x}{9} - \left(\dfrac{x}{9} - 2\right)$ **40.** $\dfrac{5x}{7} - \left(\dfrac{2x}{7} - 3\right)$

41. $4a - (a + b) + 3(b + a)$ **42.** $8x - 3(x + y) - (x - y)$

43. $7x - 3(x + y) - (x + y)$

44. $4(b - a) + 3(b + a) - 2(a + b)$

45. $-(x + y - 2) + 3(x - y + 6) - (x + y - 16)$

C. Remove grouping symbols and simplify.

46. $[(a^2 - 4) + (2a^3 - 5)] + [(4a^3 + a) + (a^2 + 9)]$

47. $(x^2 + 7 - x) + [-2x^3 + (8x^2 - 2x) + 5]$

48. $[(0.4x - 7) + 0.6x^2] - [(0.3x^2 - 2) - 0.8x]$

49. $\left[\left(\dfrac{5}{7}x^2 + \dfrac{1}{5}x\right) - \dfrac{1}{8}\right] - \left[\left(\dfrac{3}{7}x^2 - \dfrac{3}{5}x\right) + \dfrac{5}{8}\right]$

50. $[3(x + 2) - 10] + [5 + 2(5 + x)]$

51. $[3(2a - 4) + 5] - [2(a - 1) + 6]$

52. $[6(a - b) + 2a] - [3b - 4(a - b)]$

53. $[4a - (3 + 2b)] - [6(a - 2b) + 5a]$

54. $-[-(x + y) + 3(x - y)] - [4(x + y) - (3x - 5y)]$

55. $-[-(0.2x + y) + 3(x - y)] - [2(x + 0.3y) - 5]$

✓ **SKILL CHECKER**

Find:

56. $[(-3)(-3)](-3)$ **57.** $[(-2)(-2)][(-2)(-2)]$

58. $(-2)[(-2)(-2)]$ **59.** $\dfrac{(-2)(-2)(-2)}{(-2)(-2)}$

60. $\dfrac{(-3)(-3)(-3)(-3)}{(-3)(-3)}$

31. _____
32. _____
33. _____
34. _____
35. _____
36. _____
37. _____
38. _____
39. _____
40. _____
41. _____
42. _____
43. _____
44. _____
45. _____

46. _____
47. _____
48. _____

49. _____

50. _____
51. _____
52. _____
53. _____
54. _____
55. _____

56. _____
57. _____
58. _____
59. _____
60. _____

1.5 USING YOUR KNOWLEDGE

The distributive law is used to solve problems in many areas. Use your knowledge of the distributive law to remove parentheses in Problems 1–5.

1. If your car is accelerating at a constant rate, and v_1 is the initial velocity and v_2 is the final velocity, the *average* velocity is

$$v_a = \frac{1}{2}(v_1 + v_2)$$

1. _____

2. The momentum M of a billiard ball is the product of its mass m and its velocity v. If two billiard balls of equal mass m and moving in the same straight line, with velocities v_1 and v_2, respectively, collide, the total momentum M is given by

$$M = m(v_1 + v_2)$$

2. _____

3. The total kinetic energy (K.E.) of the billiard balls in Problem 2 is given by

$$\text{K.E.} = \frac{1}{2}m(v_1^2 + v_2^2)$$

3. _____

4. The length of a belt L needed to connect two pulleys of radius r_1 and r_2, respectively, with centers d units apart is

$$L = \pi(r_1 + r_2) + 2d$$

4. _____

CALCULATOR CORNER

Did you know that the order of operations used in the distributive law is programmed into most calculators? As you recall $5(4 + 6) = 5 \cdot 4 + 5 \cdot 6$, with the provision that the multiplications on the right side ($5 \cdot 4$ and $5 \cdot 6$) must be performed first. This is because without a specific set of rules $5 \cdot 4 + 5 \cdot 6$ could have several meanings. If you have a set of parentheses keys, $\boxed{(}$ and $\boxed{)}$, find the answer for the several meanings of $5 \cdot 4 + 5 \cdot 6$ given below:

1. $5 \times (4 + 5) \times 6 = $ _____

1. _____

2. $(5 \times 4) + (5 \times 6) = $ _____

2. _____

3. $(5 \times 4 + 5) \times 6 = $ _____

3. _____

4. $5 \times (4 + (5 \times 6)) = $ _____ (Here, we need *two* sets of parentheses. If you key in $5 \times (4 + 5 \times 6)$, some calculators will not get the correct answer.)

4. _____

Now, try $5 \times 4 + 5 \times 6$. Which of the interpretations did your calculator choose? You can tell by looking at the answer you obtain. If it was answer 2, the distributive law is programmed into your calculator. If the answer was different from that obtained in 2, when using your calculator to find expressions such as $p \times q + r \times s$ you must enter the sequence

$$\boxed{(}\ \boxed{p}\ \boxed{\times}\ \boxed{q}\ \boxed{)}\ \boxed{+}\ \boxed{(}\ \boxed{r}\ \boxed{\times}\ \boxed{s}\ \boxed{)}\ \boxed{=}$$

1.6 PROPERTIES OF EXPONENTS

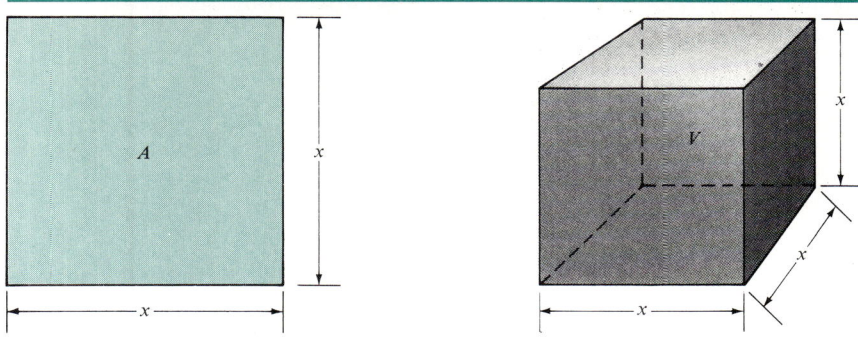

OBJECTIVES

REVIEW

Before starting this section, you should know how to multiply and divide signed numbers.

OBJECTIVES

You should be able to:

A. Evaluate an expression containing natural-number exponents.

B. Write an expression containing negative exponents as a fraction.

C. Multiply and divide expressions containing exponents.

D. Raise a power to a power.

E. Raise a quotient to a power.

F. Convert between ordinary decimal notation and scientific notation and use scientific notation in computations.

What is the area A of the square? The area is:

$A = x \cdot x = x^2$ Read "x squared, or x to the second power."

In the expression x^2, the *exponent* 2 indicates that the *base* x is to be used as a factor twice. What about the volume V of the cube? It is:

$V = x \cdot x \cdot x = x^3$ Read "x cubed, or x to the third power."

This time the *exponent* 3 indicates that the *base* x is used as a factor three times. In general, we have the following definition:

If a is a real number and n is a natural number,

$a^n = \underbrace{a \cdot a \cdot a \cdots a}_{n \text{ factors}}$

n is called the **exponent** and a is the **base.**

When $n = 1$, the exponent is usually omitted. Thus, $a^1 = a$, $b^1 = b$, and $c^1 = c$.

A. NATURAL-NUMBER EXPONENTS

We can **evaluate** (find the value of) expressions containing natural-number exponents. Thus

$4^2 = 4 \cdot 4 = 16$

$(-3)^2 = (-3)(-3) = 9$

$(-2)^3 = (-2)(-2)(-2)$

$\quad\quad = 4(-2)$ $(-2)(-2) = 4$

$\quad\quad = -8$

What about -2^3? As you recall $-a = -1 \cdot a$, so we interpret -2^3 to mean $-1 \cdot 2^3$. Hence,

$-2^3 = -1 \cdot 2^3$

$\quad\quad = -1 \cdot (2 \cdot 2 \cdot 2)$

$\quad\quad = -1 \cdot 8$

$\quad\quad = -8$

EXAMPLE 1 Evaluate.

a. $(-4)^2$ **b.** -4^2 **c.** $(-4)^3$ **d.** -4^3

Solution

a. $(-4)^2 = (-4)(-4) = 16$

b. $-4^2 = -1 \cdot 4^2$

       $= -1 \cdot 16$ Since $4^2 = 16$

       $= -16$

c. $(-4)^3 = (-4)(-4)(-4) = -64$

d. $-4^3 = -1 \cdot 4^3$

       $= -1 \cdot 64$ Since $4^3 = 4 \cdot 4 \cdot 4 = 64$

       $= -64$ ▲

Note that $(-4)^2 \neq -4^2$ because $(-4)^2$ is positive, while $-4^2 = -16$, which is negative. The placement of parentheses when using exponents is extremely important. We *must* interpret -4^3 as $-(4^3)$.

B. NEGATIVE AND ZERO EXPONENTS

In science and technology, negative numbers are used as exponents. For example, the diameter of a DNA molecule is 10^{-8} m, and the time it takes for an electron to go from source to screen in a TV tube is 10^{-6} sec. What do 10^{-8} and 10^{-6} mean? Look at the pattern obtained by dividing by 10 in each step.

$10^3 = 1000$

$10^2 = 100$

$10^1 = 10$

$10^0 = 1$

$$10^{-1} = \frac{1}{10} = \frac{1}{10}$$

$$10^{-2} = \frac{1}{100} = \frac{1}{10^2}$$

$$10^{-3} = \frac{1}{1000} = \frac{1}{10^3}$$

As you can see, this procedure yields $10^0 = 1$. In general, we make the following definition:

If a is a nonzero real number,

$a^0 = 1$

Thus $5^0 = 1$, $8^0 = 1$, and $9^0 = 1$. Note that 0^0 is **not** defined. Now look at the numbers in color. Note that we obtained

$$10^{-1} = \frac{1}{10}, \qquad 10^{-2} = \frac{1}{10^2}, \quad \text{and} \quad 10^{-3} = \frac{1}{10^3}$$

Thus, we make the following definition.

> If n is a positive integer,
>
> $$a^{-n} = \frac{1}{a^n}, \qquad a \neq 0$$

This definition says that a^{-n} and a^n are reciprocals since

$$a^{-n} \cdot a^n = \frac{1}{a^n} \cdot a^n = 1$$

By definition,

$$5^{-2} = \frac{1}{5^2} = \frac{1}{5 \cdot 5} = \frac{1}{25}$$

and

$$(-2)^{-3} = \frac{1}{(-2)^3} = \frac{1}{(-2) \cdot (-2) \cdot (-2)} = \frac{1}{-8} = -\frac{1}{8}$$

Similarly,

$$\frac{1}{4^2} = 4^{-2}$$

and

$$\frac{1}{3^4} = 3^{-4}$$

EXAMPLE 2 Write as a fraction.

a. 6^{-2} **b.** $(-4)^{-3}$ **c.** x^{-4}

Solution

a. $6^{-2} = \dfrac{1}{6^2} = \dfrac{1}{6 \cdot 6} = \dfrac{1}{36}$

b. $(-4)^{-3} = \dfrac{1}{(-4)^3} = \dfrac{1}{(-4) \cdot (-4) \cdot (-4)} = \dfrac{1}{-64} = -\dfrac{1}{64}$

c. $x^{-4} = \dfrac{1}{x^4}$

C. MULTIPLICATION AND DIVISION WITH EXPONENTS

Suppose we wish to multiply $x^2 \cdot x^3$. We write

$$\underbrace{x^2}_{} \cdot \underbrace{x^3}_{}$$
$$\underbrace{x \cdot x \cdot x \cdot x \cdot x}_{x^5}$$

Clearly, we have simply added the exponents of x^2 and x^3 to find the exponent of the result. Similarly,

$$a^3 \cdot a^4 = a^{3+4} = a^7$$
$$b^2 \cdot b^4 = b^{2+4} = b^6$$

Problem 2 Write as a fraction.
a. 5^{-2} **b.** $(-3)^{-3}$ **c.** x^{-5}

Now, let us consider $x^5 \cdot x^{-3}$. By the definition of exponents,

$$x^5 \cdot x^{-3} = x \cdot x \cdot x \cdot x \cdot x \cdot \frac{1}{x \cdot x \cdot x}$$

$$= x \cdot x \cdot \frac{x \cdot x \cdot x}{x \cdot x \cdot x}$$

$$= x^2$$

Adding exponents,

$$x^5 \cdot x^{-3} = x^{5+(-3)} = x^2 \quad \text{Same answer}$$

Let us try $x^{-2} \cdot x^{-3}$. This time we have

$$x^{-2} \cdot x^{-3} = \frac{1}{x \cdot x} \cdot \frac{1}{x \cdot x \cdot x}$$

$$= \frac{1}{x \cdot x \cdot x \cdot x \cdot x} \quad \text{Multiplying}$$

$$= x^{-5} \qquad \text{By the definition of negative exponent.}$$

Adding exponents,

$$x^{-2} \cdot x^{-3} = x^{-2+(-3)} = x^{-5} \quad \text{Same answer}$$

Thus, we have the following law.

FIRST LAW OF EXPONENTS

If a is a real number and m and n are integers,

$$a^m \cdot a^n = a^{m+n}$$

This law tells us that to *multiply* expressions with the *same* base, we *add* the exponents. Note that this law *does not* apply to expressions such as $x^m \cdot y^n$ because the bases are *different*. If the expressions involved have numerical coefficients, we multiply numbers by numbers and letters (variables) by letters using the commutative and associative laws that we have studied. Thus, to multiply $(-3x^2)(5x^3)$, we write:

$$(-3x^2)(5x^3) = (-3 \cdot 5)(x^2)(x^3)$$
$$= -15x^{2+3} = -15x^5$$

Similarly,

$$(-4x^{-2})(3x^5) = (-4 \cdot 3)(x^{-2+5})$$
$$= -12x^3 \qquad -2 + 5 = 3$$

and

$$(-2x^{-5})(-3x^2) = (-2)(-3)(x^{-5+2})$$
$$= 6x^{-3} \qquad -5 + 2 = -3$$
$$= 6 \cdot \frac{1}{x^3}$$
$$= \frac{6}{x^3}$$

Note that we wrote the answer without using negative exponents.

EXAMPLE 3 Multiply and simplify.

a. $(-4x^2)(5x^5)$ **b.** $(3x^7)(-2x^4)$ **c.** $(4x^3y)(-2x^{-8}y^6)$

Solution

a. $(-4x^2)(5x^5) = (-4 \cdot 5)(x^2 \cdot x^5)$
$$= -20x^{2+5}$$
$$= -20x^7$$

b. $(3x^7)(-2x^4) = (3)(-2)(x^7 \cdot x^4)$
$$= -6x^{7+4}$$
$$= -6x^{11}$$

c. $(4x^3y)(-2x^{-8}y^6) = (4)(-2)(x^3 \cdot x^{-8})(y^1 \cdot y^6)$
$$= -8(x^{3+(-8)})(y^{1+6})$$
$$= -8(x^{-5})(y^7)$$
$$= -8\left(\frac{1}{x^5}\right)(y^7)$$
$$= -\frac{8y^7}{x^5}$$

▲

Now consider the following division:

$$\frac{x^5}{x^3} = \frac{x \cdot x \cdot x \cdot x \cdot x}{x \cdot x \cdot x}$$
$$= x \cdot x \cdot \frac{x \cdot x \cdot x}{x \cdot x \cdot x}$$
$$= x^2 \cdot 1$$
$$= x^2$$
$$= x^{5-3}$$

As you can see, the exponent 2 in the answer can be obtained by *subtracting* $5 - 2 = 3$. Thus we have the second law of exponents.

SECOND LAW OF EXPONENTS

If a is a real number and m and n are integers,

$$\frac{a^m}{a^n} = a^{m-n}, \qquad a \neq 0$$

This law tells us that to *divide* expressions with the same base, we *subtract* the exponent of the denominator from that of the numerator. Thus

$$\frac{x^8}{x^5} = x^{8-5} = x^3$$

$$\frac{x^3}{x^8} = x^{3-8} = x^{-5} = \frac{1}{x^5}$$

$$\frac{x^{-3}}{x^{-8}} = x^{-3-(-8)} = x^{-3+8} = x^5$$

$$\frac{x^8}{x^8} = x^{8-8} = x^0 = 1$$

Problem 3 Multiply and simplify.
a. $(-3x^3)(5x^5)$ **b.** $(3x^8)(-2x^3)$
c. $(5x^3y^4)(-2x^{-7}y)$

EXAMPLE 4 Divide and simplify.

a. $\dfrac{4x^8}{2x^5}$ **b.** $\dfrac{-12x^4}{-3x^6}$ **c.** $\dfrac{5x^{-4}}{-15x^{-6}}$ **d.** $\dfrac{30x^3}{-15x^{-6}}$

Solution

a. $\dfrac{4x^8}{2x^5} = \dfrac{4}{2} \cdot \dfrac{x^8}{x^5}$

$= 2x^{8-5}$

$= 2x^3$

b. $\dfrac{-12x^4}{-3x^6} = \dfrac{-12}{-3} \cdot \dfrac{x^4}{x^6}$

$= 4x^{4-6}$

$= 4x^{-2} \qquad 4 - 6 = -2$

$= \dfrac{4}{x^2}$

c. $\dfrac{5x^{-4}}{-15x^{-6}} = \dfrac{5}{-15} \cdot \dfrac{x^{-4}}{x^{-6}}$

$= -\dfrac{1}{3} \cdot x^{-4-(-6)}$

$= -\dfrac{1}{3} \cdot x^{-4+6} \qquad -4 - (-6) = -4 + 6$

$= -\dfrac{1}{3} \cdot x^2$

$= -\dfrac{x^2}{3}$

d. $\dfrac{30x^3}{-15x^{-6}} = \dfrac{30}{-15} \cdot \dfrac{x^3}{x^{-6}}$

$= -2x^{3-(-6)}$

$= -2x^9 \qquad 3 - (-6) = 3 + 6 = 9$

D. RAISING A POWER TO A POWER

Now, let us consider $(5^3)^2$. By definition,

$$(5^3)^2 = (5^3)(5^3) = 5^{3+3} = 5^6$$

Notice that the exponent is $6 = 3 \cdot 2$, so we could have obtained the answer by multiplying exponents in the expression $(5^3)^2$. Similarly,

$$(3^{-2})^3 = \left(\dfrac{1}{3^2}\right)\left(\dfrac{1}{3^2}\right)\left(\dfrac{1}{3^2}\right) = \dfrac{1}{3^6} = 3^{-6}$$

Again, the exponent -6 could be obtained by multiplying the original exponents -2 and 3. Generalizing this result, we have the following.

THIRD LAW OF EXPONENTS

If a is a real number and m and n are integers,

$$(a^m)^n = a^{m \cdot n}$$

Problem 4 Divide and simplify.

a. $\dfrac{6x^9}{12x^3}$ **b.** $\dfrac{-18x^3}{-9x^8}$

c. $\dfrac{10x^{-6}}{-20x^{-8}}$ **d.** $\dfrac{45x^4}{-15x^{-6}}$

Thus, to raise a power to another power, we multiply the exponents that is,

$$(x^3)^4 = x^{3 \cdot 4} = x^{12}$$

$$(y^6)^{-3} = y^{6 \cdot (-3)} = y^{-18} = \frac{1}{y^{18}}$$

$$(z^{-5})^4 = z^{-5 \cdot 4} = z^{-20} = \frac{1}{z^{20}}$$

Now, let us consider $(5x^4)^3$. By definition,

$$
\begin{aligned}
(5x^4)^3 &= (5x^4)(5x^4)(5x^4) \\
&= (5 \cdot 5 \cdot 5)(x^4 \cdot x^4 \cdot x^4) \\
&= 5^3 \cdot x^{3 \cdot 4} \\
&= 5^3 \cdot x^{12}
\end{aligned}
$$

Note that the exponent 3 is applied to the 5 and the x^4; that is, if we raise several factors inside parentheses to a power, we raise each factor to the given power. In general, we have the following law.

If a and b are real numbers and m, n, and k are integers,

$$(a^m b^n)^k = a^{m \cdot k} b^{n \cdot k}$$

This result can be generalized further to apply to any number of factors inside the parentheses.

EXAMPLE 5   Simplify.

a. $(3x^5 y^3)^{-2}$ **b.** $(-2x^5 y^{-4})^3$

Solution

a. $(3x^5 y^3)^{-2} = (3)^{-2}(x^5)^{-2}(y^3)^{-2}$

$$= \frac{1}{3^2} \cdot x^{5 \cdot (-2)} \cdot y^{3 \cdot (-2)} \quad (3)^{-2} = \frac{1}{3^2}$$

$$= \frac{1}{3^2} \cdot x^{-10} \cdot y^{-6}$$

$$= \frac{1}{3^2} \cdot \frac{1}{x^{10}} \cdot \frac{1}{y^6} \qquad x^{-10} = \frac{1}{x^{10}}, \ y^{-6} = \frac{1}{y^6}$$

$$= \frac{1}{9x^{10}y^6}$$

b. $(-2x^5 y^{-4})^3 = (-2)^3(x^5)^3(y^{-4})^3$

$$= -8x^{5 \cdot 3} y^{-4 \cdot 3}$$

$$= -8x^{15} y^{-12} \qquad y^{-12} = \frac{1}{y^{12}}$$

$$= -\frac{8x^{15}}{y^{12}}$$

Problem 5   Simplify.
a. $(4x^3 y^4)^{-2}$ **b.** $(-3x^2 y^{-5})^3$

E. RAISING A QUOTIENT TO A POWER

We have already raised a product to a power. Can we raise a quotient to a power? Let us try $\left(\dfrac{2^3}{3^4}\right)^2$. By definition of exponents,

$$\left(\frac{2^3}{3^4}\right)^2 = \frac{2^3}{3^4} \cdot \frac{2^3}{3^4} = \frac{2^{3+3}}{3^{4+4}} = \frac{2^6}{3^8} = \frac{2^{3\cdot 2}}{3^{4\cdot 2}}$$

Note that the same answer is obtained by multiplying each of the exponents in the numerator and denominator by 2. Here is the general rule.

If a and b are real numbers and m, n, and k are integers ($b \neq 0$),

$$\left(\frac{a^m}{b^n}\right)^k = \frac{a^{m\cdot k}}{b^{n\cdot k}}$$

EXAMPLE 6 Simplify.

a. $\left(\dfrac{x^4}{y^{-3}}\right)^{-2}$ **b.** $\left(\dfrac{3x^{-3}y^2}{2y^3}\right)^3$

Solution

a. $\left(\dfrac{x^4}{y^{-3}}\right)^{-2} = \dfrac{x^{4\cdot(-2)}}{y^{-3\cdot(-2)}} = \dfrac{x^{-8}}{y^6}$ $x^{-8} = \dfrac{1}{x^8}$

$\qquad = \dfrac{1}{x^8 y^6}$

b. In this case, it is easier to do the operations inside the parentheses first. Thus

$\left(\dfrac{3x^{-3}y^2}{2y^3}\right)^3 = \left(\dfrac{3}{2} \cdot x^{-3} \cdot y^{2-3}\right)^3$ $\dfrac{y^2}{y^3} = y^{2-3}$

$\qquad = \left(\dfrac{3y^{-1}}{2x^3}\right)^3$ $x^{-3} = \dfrac{1}{x^3}$

$\qquad = \left(\dfrac{3}{2x^3 y}\right)^3$ $y^{-1} = \dfrac{1}{y}$

$\qquad = \dfrac{3^3}{(2x^3 y)^3}$

$\qquad = \dfrac{27}{8x^9 y^3}$ ▲

Since the reciprocal of $\dfrac{a}{b}$ is $\dfrac{b}{a}$, $\left(\dfrac{a}{b}\right)^{-1} = \left(\dfrac{b}{a}\right)$. Thus

$$\left[\left(\frac{a}{b}\right)^{-1}\right]^n = \left(\frac{b}{a}\right)^n$$

and

$$\left(\frac{a}{b}\right)^{-n} = \left(\frac{b}{a}\right)^n$$

Problem 6 Simplify.

a. $\left(\dfrac{x^5}{y^{-3}}\right)^{-4}$ **b.** $\left(\dfrac{2x^{-3}y^3}{3y^3}\right)^2$

This means that a fraction raised to the -nth power is equivalent to its *reciprocal* raised to the nth power. In part a, Example 6, we could write

$$\left(\frac{x^4}{y^{-3}}\right)^{-2} = \left(\frac{y^{-3}}{x^4}\right)^2 = \frac{y^{-6}}{x^8} = \frac{1}{x^8 y^6}$$

You can use this method when working the exercises.

F. SCIENTIFIC NOTATION

In science and in other areas, very large or very small numbers frequently occur. For example, a red cell of human blood contains 270,000,000 hemoglobin molecules, and the mass of a single carbon atom is 0.000 000 000 000 000 000 000 019 9 gram. Numbers in this form are difficult to write and to work with, so they are written in scientific notation. We have the following definition.

A number is said to be in **scientific notation** if it is written in the form

$$m \times 10^n$$

where m is a number greater than or equal to 1 and less than 10 and n is an integer.

For any given number, the m is obtained by placing the decimal point so that there is exactly one nonzero digit to its left. n is then the number of places that the decimal point must be moved from its position in m to its original position; it is positive if the point must be moved to the right and negative if the point must be moved to the left. Thus,

$5.3 = 5.3 \times 10^0$	Decimal point in 5.3 must be moved 0 places.
$87 = 8 7 \times 10^1 = 8.7 \times 10$	Decimal point in 8.7 must be moved 1 place to the *right* to get 87.
$68,000 = 6.8 \times 10^4$	Decimal point in 6.8 must be moved 4 places to the *right* to get 68,000.
$0.49 = 4.9 \times 10^{-1}$	Decimal point in 4.9 must be moved 1 place to the *left* to get 0.49.
$0.072 = 7.2 \times 10^{-2}$	Decimal point in 7.2 must be moved 2 places to the *left* to get 0.072.
$0.0003875 = 3.875 \times 10^{-4}$	Decimal point in 3.875 must be moved 4 places to the *left* to get 0.0003875.

EXAMPLE 7 Write in scientific notation.

a. 270,000,000
b. 0.000 000 000 000 000 000 000 019 9

Solution

a. $270,000,000 = 2.7 \times 10^8$
b. $0.000\ 000\ 000\ 000\ 000\ 000\ 000\ 019\ 9 = 1.99 \times 10^{-23}$

Problem 7 Write in scientific notation.
a. 350,000
b. 0.000 000 378

EXAMPLE 8   Write in standard decimal notation.

a. 2.5×10^{10} **b.** 7.4×10^{-6}

Solution

a. $2.5 \times 10^{10} = 25,000,000,000$
b. $7.4 \times 10^{-6} = 0.0000074$ ▲

We can use the laws of exponents when working with numbers in decimal notation. The next example shows you how.

EXAMPLE 9   Do the following calculation, and write the answer in scientific notation: $(5 \times 10^4) \times (9 \times 10^{-7})$.

Solution

$$
\begin{aligned}
(5 \times 10^4) \times (9 \times 10^{-7}) &= (5 \times 9) \times (10^4 \times 10^{-7}) \\
&= 45 \times 10^{4-7} \\
&= 45 \times 10^{-3} \\
&= 4.5 \times 10^1 \times 10^{-3} \\
&= 4.5 \times 10^{1-3} \\
&= 4.5 \times 10^{-2} \qquad \text{▲}
\end{aligned}
$$

Problem 8   Write in decimal notation.
a. 3.5×10^5 **b.** 8.2×10^{-3}

Problem 9   Do the calculation and write the answer in scientific notation: $(3 \times 10^3)(8 \times 10^{-7})$.

NAME

CLASS

SECTION

ANSWERS

A. In Problems 1–10 evaluate.

1. $-4^2 =$ _____

2. $(-4)^2 =$ _____

3. $(-5)^2 =$ _____

4. $-5^2 =$ _____

5. $-5^3 =$ _____

6. $(-5)^3 =$ _____

7. $(-6)^4 =$ _____

8. $-6^4 =$ _____

9. $-2^5 =$ _____

10. $(-2)^5 =$ _____

1. _____
2. _____
3. _____
4. _____
5. _____
6. _____
7. _____
8. _____
9. _____
10. _____

B. In Problems 11–20 write as a fraction in simplified form.

11. 4^{-2}

12. 2^{-3}

13. 5^{-3}

14. 7^{-2}

15. 3^{-4}

16. 6^{-3}

17. x^{-5}

18. y^{-7}

19. a^{-3}

20. b^{-4}

11. _____
12. _____
13. _____
14. _____
15. _____
16. _____
17. _____
18. _____
19. _____
20. _____

C. In Problems 21–50 perform the indicated operations and simplify.

21. $2^{-4} \cdot 2^{-2}$

22. $4^{-1} \cdot 4^{-2}$

23. $(3x^6) \cdot (4x^{-4})$

24. $(4y^7) \cdot (5y^{-3})$

25. $(-3y^{-3}) \cdot (5y^5)$

26. $(-5x^{-7}) \cdot (4x^8)$

27. $(-4a^3) \cdot (-5a^{-8})$

28. $(-2b^4) \cdot (-3b^{-7})$

21. _____
22. _____
23. _____
24. _____
25. _____
26. _____
27. _____
28. _____

29. $(3x^{-5}) \cdot (5x^2y)(-2xy^2)$

30. $(4y^{-6}) \cdot (5xy^4)(-2x^2y)$

31. $(-2x^{-3}y^2)(3x^{-2}y^3)(4xy)$

32. $(-3xy^{-5})(4x^2y)(2x^3y^2)$

33. $(4a^{-2} \cdot b^{-3})(5a^{-1}b^{-1})(-2ab)$

34. $(2a^{-5} \cdot b^{-2})(3a^{-1}b^{-1})(5ab)$

35. $(6a^{-3} \cdot b^3)(5a^2b^2)(-ab^{-5})$

36. $(7a^6 \cdot b^{-6})(2ab^5)(-a^{-7}b)$

37. $\dfrac{8x^7}{4x^3}$

38. $\dfrac{8a^3}{4a^2}$

39. $\dfrac{-8a^4}{-16a^2}$

40. $\dfrac{-9y^5}{-18y^2}$

41. $\dfrac{12x^5y^3}{-6x^2y}$

42. $\dfrac{18x^6y^2}{-9xy}$

43. $\dfrac{-6x^{-4}}{12x^{-5}}$

44. $\dfrac{8x^{-3}}{4x^{-4}}$

45. $\dfrac{-14a^{-5}}{-21a^{-2}}$

46. $\dfrac{-2a^{-6}}{-6a^{-3}}$

47. $\dfrac{-27a^{-4}}{-36a^{-4}}$

48. $\dfrac{-5x^{-3}}{10x^{-3}}$

49. $\dfrac{3a^{-2} \cdot b^5}{2a^4b^2}$

50. $\dfrac{x^{-3} \cdot y^6}{x^4 \cdot y^3}$

D. In Problems 51–60 simplify.

51. $(2x^3y^{-2})^3$

52. $(3x^2y^{-3})^2$

53. $(2x^{-2}y^3)^2$

54. $(3x^{-4}y^4)^3$

55. $(-3x^3y^2)^{-3}$

56. $(-2x^5y^4)^{-4}$

57. $(x^{-6}y^{-3})^2$

58. $(y^{-4}z^{-3})^5$

59. $(x^{-4}y^{-4})^{-3}$

60. $(y^{-5}z^{-3})^{-4}$

29. _____
30. _____
31. _____
32. _____
33. _____
34. _____
35. _____
36. _____
37. _____
38. _____
39. _____
40. _____
41. _____
42. _____
43. _____
44. _____
45. _____
46. _____
47. _____
48. _____
49. _____
50. _____

51. _____
52. _____
53. _____
54. _____
55. _____
56. _____
57. _____
58. _____
59. _____
60. _____

ANSWERS (to problems on page 66)
8. a. 350,000 **b.** 0.0082
9. 2.4×10^{-3}

E. In Problems 61–70 simplify.

61. $\left(\dfrac{a}{b^3}\right)^2$ **62.** $\left(\dfrac{a^2}{b}\right)^3$

63. $\left(\dfrac{-3a}{2b^2}\right)^{-3}$ **64.** $\left(\dfrac{-2a^2}{3b^0}\right)^{-2}$

65. $\left(\dfrac{a^{-4}}{b^2}\right)^{-2}$ **66.** $\left(\dfrac{a^{-2}}{b^3}\right)^{-3}$

67. $\left(\dfrac{x^5}{y^{-2}}\right)^{-3}$ **68.** $\left(\dfrac{x^6}{y^{-3}}\right)^{-2}$

69. $\left(\dfrac{x^{-4}y^3}{x^5y^5}\right)^{-3}$ **70.** $\left(\dfrac{x^{-2}y^0}{x^7y^2}\right)^{-2}$

F. In Problems 71–74 write in scientific notation.

71. 268,000,000   (U.S. population in the year 2000)

72. 1,900,000,000   (dollars spent on waterbeds and accessories in one year)

73. 0.00024   (probability of four of a kind in poker)

74. 0.00000009   (wavelength in centimeters of an X-ray)

In Problems 75–78 write in decimal notation.

75. 8×10^6 (bagels eaten per day in the United States)

76. $\$6.85 \times 10^9$ (estimated wealth of the five wealthiest women)

77. 2.3×10^{-1} (kilowatts per hour used by your TV)

78. 4×10^{-11} J   (energy released by splitting one uranium atom)

In Problems 79–84, write the answer in scientific notation.

79. The width of the asteroid belt is 2.8×10^8 km. The speed of *Pioneer 10* in passing through this belt was 1.4×10^5 km/hr. Thus *Pioneer 10* took

$$\dfrac{2.8 \times 10^8}{1.4 \times 10^5} \text{ hr}$$

to go through the belt. How many hours was that?

80. The mass of Earth is 6×10^{21} tons. The sun is about 300,000 times as massive. Thus the mass of the sun is $(6 \times 10^{21}) \times 300,000$ tons. How many tons is that?

81. The velocity of light can be measured by knowing the distance from the sun to Earth (1.47×10^{11} m) and the time it takes for sunlight to reach Earth (490 sec). Thus, the velocity of light is

$$\dfrac{1.47 \times 10^{11}}{490} \text{ m/sec}$$

How many meters per second is that?

61. _____

62. _____

63. _____

64. _____

65. _____

66. _____

67. _____

68. _____

69. _____

70. _____

71. _____

72. _____

73. _____

74. _____

75. _____

76. _____

77. _____

78. _____

79. _____

80. _____

81. _____

82. United States oil reserves are estimated to be 3.5×10^{10} barrels. Production amounts to 3.2×10^9 barrels per year. At this rate, how long would U.S. oil reserves last? (Give answer to the nearest year.)

83. The world's oil reserves are estimated to be 6.28×10^{11} barrels. Production is 2.0×10^{10} barrels per year. At this rate, how long would the world's oil reserves last? (Give answer to the nearest year.)

84. Scientists have estimated that the total energy received from the sun each minute is 1.02×10^{19} calories. Since the area of Earth is 5.1×10^8 km^2 (square kilometers), the amount of energy received per square centimeter of Earth's surface per minute (the solar constant) is

$$\frac{1.02 \times 10^{19}}{(5.1 \times 10^8) \times 10^{10}} \qquad \textit{Note:} \quad 1 \text{ km}^2 = 10^{10} \text{ cm}^2$$

How many calories per square centimeter is that?

✓ **SKILL CHECKER**

Find.

85. a. $(-3.2)(-1.4)(-2.2)$ **b.** $(-1.1)(-1.2)(-2.1)$

86. a. $\dfrac{-3.2}{1.6}$ **b.** $\dfrac{-4.8}{-1.2}$

1.6 USING YOUR KNOWLEDGE

If you have a scientific calculator, and you multiply 9,800,000 by 4,500,000, the display will show

    4.41   13

This means that the answer is 4.41×10^{13}.

1. The display on a calculator shows

    3.34   5

Write this number in scientific notation.

2. The display on a calculator shows

    -9.97 -6

Write this number in scientific notation.

3. To enter large or small numbers in a calculator with scientific notation, you must write the number using this notation first. Thus, to enter the number 8,700,000,000 in the calculator you must know that 8,700,000,000 is 8.7×10^9, *then* you can key in

    8 · 7 EE↓ 9   or   8 · 7 EXP 9

The calculator displays

    8.7   09

a. What would the display read when you enter the number 73,000,000,000?

b. What would the display read when you enter the number 0.000000123?

82. _____

83. _____

84. _____

85. a. _____ **b.** _____

86. a. _____ **b.** _____

1. _____

2. _____

3. a. _____
    b. _____

1.7 ORDER OF OPERATIONS

OBJECTIVES

REVIEW

Before starting this section, you should know:

1. How to perform addition, subtraction, multiplication, and division with signed numbers.
2. How to find powers of integers.

OBJECTIVES

You should be able to:

A. Evaluate numerical expressions containing grouping symbols.
B. Evaluate expressions using the correct order of operations.

How do you calculate your ideal heart rate when swimming? The way to do this is as follows:

Subtract your age from 205 and multiply by 0.70

This means that if you are *a* years old, you should

Subtract your age from 205 and multiply by 0.70

$205 - a \cdot 0.70$

Should you subtract your age from 205 first and then multiply by 0.70, or multiply your age by 0.70 first and then subtract from 205? In algebra, we can make our meaning clear by using parentheses. The formula should be written as $(205 - a) \cdot 0.70$ to indicate that the subtraction should be done first. (Try it by substituting your age for *a*.)

A. GROUPING SYMBOLS AND OPERATIONS

From the preceding discussion we can conclude that the *parentheses are grouping symbols* that are used to indicate which operations are to be performed first. The square brackets [ ] and the braces { } are also grouping symbols, which can be used in the same manner as the parentheses. Thus

$$4 \cdot (3 + 2), \qquad 4 \cdot [3 + 2], \quad \text{and} \quad 4 \cdot \{3 + 2\}$$

all mean that we must first add 3 and 2 and then multiply this sum by 4.

EXAMPLE 1 Evaluate.

a. $(-4 \cdot 5) + 6$ **b.** $-4 \cdot (5 + 6)$

Solution

a. $(-4 \cdot 5) + 6 = -20 + 6$
$$= -14$$

b. $-4 \cdot (5 + 6) = -4 \cdot 11$
$$= -44 \qquad \blacktriangle$$

Problem 1 Evaluate.
a. $(-6 \cdot 4) + 7$ **b.** $-6(4 + 7)$

The placement of parentheses is equally important in evaluating numerical expressions involving multiplication and division. For example, the expressions $(-32 \div 4) \cdot 2$ and $-32 \div (4 \cdot 2)$ have different meanings. Thus

$$(-32 \div 4) \cdot 2 \qquad -32 \div (4 \cdot 2)$$
$$= -8 \cdot 2 \qquad\qquad = -32 \div 8$$
$$= -16 \qquad\qquad = -4$$

EXAMPLE 2 Evaluate.

a. $-48 \div (4 \cdot 3)$ **b.** $(-48 \div 4) \cdot 3$

Solution

a. $-48 \div (4 \cdot 3) = -48 \div 12$
$$= -4$$
b. $(-48 \div 4) \cdot 3 = -12 \cdot 3$
$$= -36$$

The numerical expressions discussed so far included only three numbers. There are other expressions that may involve more than three numbers. For example, in the accompanying ad, a mattress and box-springs and a bedspread cost $88 + $14. If Ms Perez decides to buy two of each of these and a lamp, her purchase (in dollars) will amount to $[2 \cdot (88 + 14)] + 12$. Since only two numbers may be added at one time, the first step in the evaluation of $[2 \cdot (88 + 14)] + 12$ is to find the sum of 88 and 14. The evaluation is shown next.

$$[2 \cdot (88 + 14)] + 12 = [2 \cdot (102)] + 12$$
$$= 204 + 12$$
$$= 216$$

Thus, her purchase will amount to $216.

EXAMPLE 3 Evaluate $[-5 \cdot (6 + 4)] + 9$.

Solution

$$[-5 \cdot (6 + 4)] + 9 = [-5 \cdot 10] + 9$$
$$= -50 + 9$$
$$= -41 \qquad \blacktriangle$$

Expressions involving subtraction can be evaluated similarly. Thus $[5 \cdot (9 - 2)] - 5$ is evaluated as follows:

$$[5 \cdot (9 - 2)] - 5 = [5 \cdot 7] - 5$$
$$= 35 - 5$$
$$= 30$$

EXAMPLE 4 Evaluate $[-10 \cdot (8 - 3)] - 9$.

Solution

$$[-10 \cdot (8 - 3)] - 9 = [-10 \cdot 5] - 9$$
$$= -50 - 9$$
$$= -59 \qquad \blacktriangle$$

In some cases we have to evaluate expressions in which a bar is used to indicate division. For instance, if we wish to convert a temperature given in degrees Fahrenheit to degrees Celsius, we have to evaluate

Problem 2 Evaluate.
a. $-36 \div (3 \cdot 4)$ **b.** $(-36 \div 3) \cdot 4$

Problem 3 Evaluate $[-8(7 + 2)] + 8$.

Problem 4 Evaluate $[-10(9 - 6)] - 9$.

ANSWER (to problem on page 71)
1. **a.** -17 **b.** -66

the expression

$$\frac{5 \cdot (F - 32)}{9}$$

where F represents the temperature in degrees Fahrenheit. Thus if the temperature is 77°F, the corresponding Celsius temperature is calculated as follows:

$$\frac{5 \cdot (77 - 32)}{9} = \frac{5 \cdot (45)}{9}$$

You can also do this by dividing 45 by 9 first and then multiplying the result, 5, by 5, obtaining 25.

$$= \frac{225}{9}$$
$$= 25$$

Thus the corresponding temperature is 25°C.

EXAMPLE 5 In September 1933, a freak heat flash struck the city of Coimbra, in Portugal. On this day, the temperature rose to 158°F for 120 sec. How many degrees Celsius is that?

Solution In this case $F = 158$; thus the Celsius temperature is given by

$$\frac{5 \cdot (158 - 32)}{9} = \frac{5 \cdot (126)}{9}$$

You can also do this by dividing 126 by 9 first and then multiplying the result, 14, by 5, obtaining 70.

$$= \frac{630}{9}$$
$$= 70$$

Thus the corresponding Celsius temperature is 70°C.

Problem 5 Evaluate $\dfrac{5 \cdot (149 - 32)}{9}$.

B. THE ORDER OF OPERATIONS

If an expression does not contain parentheses or brackets, we must establish the order in which operations are to be performed. For example, the expression $6^2 + 9 \div 3$ can be evaluated in two ways:

$6^2 + 9 \div 3$	Square 6 and	$6^2 + 9 \div 3$	Square 6,
$36 + 3$	divide 9 by 3.	$36 + 9 \div 3$	Add $36 + 9$.
39	Add.	$45 \div 3$	Divide by 3.
		15	

To avoid this ambiguity, we agree to perform any sequence of operations in the following order.

ORDER OF OPERATIONS

1. Do the operations inside parentheses (or other grouping symbols) starting with the innermost grouping symbols and operations above and below division bars.

2. Evaluate all exponential expressions.

3. Perform multiplications and divisions as they occur from left to right.

4. Perform additions and subtractions as they occur from left to right.

You can remember the order of operations if you remember

Please	(Parentheses)
Dad	(Division bar)
Excuse	(Exponent)
My	(Multiplication)
Dear	(Division)
Aunt	(Addition)
Sally	(Subtraction)

With this convention, the expression $6^2 + 9 \div 3$ was evaluated as shown on the left, yielding 39 for the answer.

EXAMPLE 6   Evaluate the expression $-6^2 + \dfrac{(4 - 8)}{2} + 10 \div 5$.

Solution

$$-6^2 + \frac{(4 - 8)}{2} + 10 \div 5$$

$$= -6^2 + \frac{-4}{2} + 10 \div 5 \qquad \text{Perform the operation inside parentheses.}$$

$$= -36 + \frac{-4}{2} + 10 \div 5 \qquad \text{Evaluate } -6^2 = -36.$$

$$= -36 + (-2) + 2 \qquad \text{Perform multiplications and divisions as they occur from left to right.}$$

$$= -38 + 2 \qquad\qquad -36 + (-2) = -38$$

$$= -36 \qquad\qquad\quad -38 + 2 = -36$$

Problem 6   Evaluate

$$-7^2 + \frac{(2 - 8)}{2} + 20 \div 5.$$

A. In Problems 1–30 evaluate the expression.

ANSWERS

1. **a.** $(-10 \cdot 3) + 4$
   b. $-10 \cdot (3 + 4)$

   1. a. _____
      b. _____

2. **a.** $(6 \cdot 4) + 6$
   b. $6 \cdot (4 + 6)$

   2. a. _____
      b. _____

3. **a.** $(36 \div 4) \cdot 3$
   b. $36 \div (4 \cdot 3)$

   3. a. _____
      b. _____

4. **a.** $(-28 \div 7) \cdot 2$
   b. $-28 \div (7 \cdot 2)$

   4. a. _____
      b. _____

5. $[-5 \cdot (8 + 2)] + 3$ 6. $[7 \cdot (4 + 3)] + 1$

   5. _____
   6. _____

7. $-7 + [3 \cdot (4 + 5)]$ 8. $-8 + [3 \cdot (4 + 1)]$

   7. _____
   8. _____

9. $[-6 \cdot (4 - 2)] - 3$ 10. $[-2(7 - 5)] - 8$

   9. _____
   10. _____

11. $3 - [8 \cdot (5 - 3)]$ 12. $7 - [3(4 - 5)]$

   11. _____
   12. _____

13. $-5 \cdot 6 - 6$ 14. $-5 \cdot 2 - 2$

   13. _____
   14. _____

15. $-7 \cdot 3 \div 3 - 3$ 16. $-36 \cdot 2 \div 18 - 4$

   15. _____
   16. _____

17. $(-20 - 5 + 3 \div 3) \div 6$ 18. $(-10 - 2 + 10 \div 5) \cdot 4$

   17. _____
   18. _____

19. $\dfrac{8 + (-3)}{5} - 1$ 20. $\dfrac{7 + (-3)}{2} - 4$

   19. _____
   20. _____

21. $\dfrac{4 \cdot (6 - 2)}{-8} - \dfrac{6}{-2}$ 22. $\dfrac{5 \cdot (6 - 2)}{-4} - \dfrac{16}{-4}$

   21. _____
   22. _____

23. $-8[3 - 2(4 + 1)] + 1$ 24. $6[7 - 2(5 - 7)] - 2$

   23. _____
   24. _____

25. $48 \div \{4(8 - 2[3 - 1])\}$ 26. $-96 \div \{4(8 - 2[1 - 3])\}$

   25. _____
   26. _____

27. $\left[\dfrac{9 - (-3)}{8 - 6}\right]\left[\dfrac{3 + (-8)}{7 - 2}\right]$ 28. $\left[\dfrac{6 + (-2)}{3 + (-7)}\right]\left[\dfrac{8 + (-12)}{2 - 4}\right]$

   27. _____
   28. _____

EXERCISE 1.7

29. $\dfrac{3 - 5\left(\dfrac{4 + 2}{2 + 1}\right) - 2}{-4 + 3\left(\dfrac{4 - 2}{4 - 6}\right) - 2}$

30. $\dfrac{8 + 2\left(\dfrac{9 - 15}{3 - 1}\right) - 2}{-4 + 8\left(\dfrac{6 - 3}{1 - 4}\right) + 12}$

29. _____

30. _____

B. In Problems 31–40 use the correct order of operations and simplify.

31. $4 \div 2 + 3 - 5^2$

32. $8 \div 4 + 7 - 2^2$

33. $4 + 6 \cdot 4 \div 2 - 2^3$

34. $6 + 6 \div 3 - 3^3$

35. $-5^2 + \dfrac{(2 - 10)}{4} + 12 \div 4$

36. $-4^2 + \dfrac{(3 - 7)}{2} + 18 \div 9$

37. $-3^3 + 4 - 6 \cdot 8 \div 4 - \dfrac{8 - 2}{-3}$

38. $-2^3 + 6 - 6 \div 3 \cdot 2 - \dfrac{9 - 3}{-6}$

39. $4 \cdot 9 \div 3 \cdot 10^3 - 2 \cdot 10^2$

40. $5 \cdot 8 \div 4 \cdot 10^3 - 2 \cdot 10^2$

31. _____

32. _____

33. _____

34. _____

35. _____

36. _____

37. _____

38. _____

39. _____

40. _____

✓ **SKILL CHECKER**

Simplify.

41. $3(x - 4) - 5x$

42. $4(x - 5) - 6x$

43. $-3(x - 5) - 6x$

44. $-5(x - 4) - 3x$

41. _____

42. _____

43. _____

44. _____

1.7 USING YOUR KNOWLEDGE

Do you know any computer languages? One of the most popular ones is **BASIC** (**B**eginners **A**ll-purpose **S**ymbolic **I**nstruction **C**ode). In BASIC, the symbols for the mathematical operations are:

SYMBOL	MEANING	EXAMPLE	
+	Plus	$3 + 5$	
−	Minus	$3 - 5$	
*	Times	$3 * 5$	3 times 5
/	Divided by	$3 / 5$	3 divided by 5
^	With exponent	$3 \hat{}\ 5$	3^5 or 3 with exponent 5

(*Note:* Some systems use ** and other systems use the upward-pointing arrow for exponentiation.)

Parentheses are used just as in ordinary mathematical notation. For example, in BASIC, the expression $4 \times 3^2 - (8 + 9)(5 \div 12)^4$ would appear as follows:

$4 * 3 \hat{}\ 2 - (8 + 9) * (5 / 12) \hat{}\ 4$

The order of operations is exactly as in algebra. Use your knowledge to do these problems.

In Problems 1–8 write each expression in BASIC.

1. $(3 + 4) \div (5 + 9)$

2. $(5 - 2)(7 + 3)$

3. $3^2 + 4^2$

4. $2 \times 3 \div 4 \times 6$

5. $2(3^3) - 5(4^2)$

6. $3(2^5) - 2(3^4)$

7. $\dfrac{5 \times 8}{6 \times 9}$

8. $\dfrac{6 - 2}{5 \times 4}$

In Problems 9–18 evaluate the given BASIC expression.

9. $4 + 8\,/\,2 - 24\,/\,6$

10. $3 * 4 - 2$

11. $(4 + 8)\,/\,2 - 24\,/\,6$

12. $3 * 4 - 2 * 3$

13. $2 \char`\^ 3 + 3 * 2 \char`\^ 2$

14. $3 * 5 \char`\^ 2 - 4 * 5$

15. $3 * 4\,/\,6 * 8$

16. $3 * 4\,/\,(6 * 8)$

17. $4 * 5 \char`\^ 3$

18. $(4 * 5) \char`\^ 3$

1. _____
2. _____
3. _____
4. _____
5. _____
6. _____
7. _____
8. _____
9. _____
10. _____
11. _____
12. _____
13. _____
14. _____
15. _____
16. _____
17. _____
18. _____

SUMMARY

REAL NUMBER PROPERTIES

If a, b, and c are real numbers:

NAME	ADDITION	MULTIPLICATION
Commutative law	$a + b = b + a$	$a \cdot b = b \cdot a$
Associative law	$a + (b + c) = (a + b) + c$	$a \cdot (b \cdot c) = (a \cdot b) \cdot c$
Identity	$a + 0 = 0 + a = a$ (0 is the identity)	$1 \cdot a = a \cdot 1 = a$ (1 is the identity)
Inverse	For each real number a, there is a unique inverse $-a$ such that $a + (-a) = -a + a = 0$.	For each nonzero real number a, there is a unique real number $\dfrac{1}{a}$ such that $a \cdot \dfrac{1}{a} = \dfrac{1}{a} \cdot a = 1$ $\left(\dfrac{1}{a} \text{ is called reciprocal of } a\right)$.

Distributive property (over addition) $a(b + c) = ab + ac$
Distributive property (over subtraction) $a(b - c) = ab - ac$

(Continued)

SECTION	ITEM	MEANING	EXAMPLE
1.1A	Empty or null set $\varnothing$	The set containing no elements	The set of natural numbers between 5 and 6 is $\varnothing$.
1.1A	Natural numbers	$N = \{1, 2, 3, \ldots\}$	2, 76, and 308 are natural numbers.
1.1A	Whole numbers	$W = \{0, 1, 2\ldots\}$	0, 8, and 93 are whole numbers.
1.1A	Integers	$I = \{\ldots, -2, -1, 0, 1, 2, \ldots\}$	-7 and 23 are integers.
1.1A	Positive integers	$I^{+} = \{+1, +2, +3, \ldots\}$	7, 19, and 4 are positive integers.
1.1A	Negative integers	$I^{-} = \{\ldots, -3, -2, -1\}$	-3, -8, and -19 are negative integers.
1.1A	Rational numbers	$Q = \left\{ r \mid r = \dfrac{a}{b},\ a \text{ and } b \text{ are integers and } b \neq 0 \right\}$	$\dfrac{1}{5}$, $-\dfrac{2}{3}$, 0, 9, 1.4, and $0.\bar{3}$ are rational numbers.
1.1A	Irrational numbers	$H = \{x \mid x \text{ is not rational}\}$	$\sqrt{2}$ and π are irrational numbers.
1.1A	Real numbers (R)	The set of all rationals and irrationals	$\dfrac{2}{7}$, $-\dfrac{2}{3}$, 0, 9, 1.4, $0.\bar{3}$, $\sqrt{2}$, and π are real numbers.
1.1B	Rational numbers	The set Q of rational numbers is the same as the set of terminating or repeating decimals.	0.345 and $0.\bar{3}$ are rational numbers.
1.1B	$N \subset W \subset I \subset Q \subset R$	N is a subset of W, W is a subset of I, and so on.	Every natural number is a whole number, every whole number is an integer, and so on.
1.2C	Absolute value $|a|$	The distance from 0 to a on the number line.	$|-8| = 8$, $\left\lvert\dfrac{2}{3}\right\rvert = \dfrac{2}{3}$, and $|-0.4| = 0.4$
1.3A	Adding real numbers with the same sign	Add their absolute values and give the sum the common sign.	$-3 + (-7) = -(|-3| + |-7|)$ $= -10$
1.3A	Adding real numbers with different signs	Subtract the smaller absolute value from the greater absolute value and use the sign of the number with the greater absolute value.	$3 + (-5) = -(5 - 3) = -2$ $-7 + 9 = +(9 - 7) = 2$
1.3B	Subtraction	If a and b are real numbers, $a - b = a + (-b)$.	$3 - (-4) = 3 + 4 = 7$
1.4A	$a \cdot b$, ab, $(a)(b)$, $a(b)$, $(a)b$	The product of a and b	
1.4A	Multiplying real numbers with different signs	Multiply their absolute values; the product is negative.	$3 \cdot (-4) = -12$, $-7 \cdot 2 = -14$
1.4A	Multiplying real numbers with the same sign	Multiply their absolute values; the product is positive.	$3 \cdot 8 = 24$ and $(-9)(-2) = 18$
1.4A	Multiplication of fractions	$\dfrac{a}{b} \cdot \dfrac{c}{d} = \dfrac{a \cdot c}{b \cdot d}$	$\left(-\dfrac{3}{4}\right) \cdot \dfrac{2}{7} = -\dfrac{3}{14}$
1.4B	Division	If a, b, c are real numbers $\dfrac{a}{b} = c$ means $a = bc$ ($b \neq 0$).	$\dfrac{6}{3} = 2$ means $6 = 3 \cdot 2$.
1.4B	Dividing signed numbers	The quotient of two real numbers with the same sign is positive, with different signs, negative.	$\dfrac{6}{-3} = -2$, $\dfrac{-6}{3} = -2$, and $\dfrac{-8}{-2} = 4$
1.4B	0 in division problems	$\dfrac{0}{a} = 0$ ($a \neq 0$) $\dfrac{a}{0}$ is not defined	$\dfrac{0}{9} = 0$, $\dfrac{0}{-8} = 0$, and $\dfrac{0}{-2.4} = 0$ $\dfrac{9}{0}$, $\dfrac{-8}{0}$, and $\dfrac{-2.4}{0}$ are not defined.

(Continued)

SECTION	ITEM	MEANING	EXAMPLE
1.4B	Reciprocal	The reciprocal of a is $\frac{1}{a}$, $a \neq 0$.	The reciprocal of -7 is $-\frac{1}{7}$.
1.4B	Division of fractions	$\frac{a}{b} \div \frac{c}{d} = \frac{a}{b} \cdot \frac{d}{c}$	$\frac{3}{4} \div \frac{9}{2} = \frac{3}{4} \cdot \frac{2}{9} = \frac{1}{6}$
1.5A	$-1 \cdot a$	$-1 \cdot a = -a$	$-1 \cdot 4 = -4$
1.5A	$-(a + b)$	$-(a + b) = -a - b$	$-(x + 7) = -x - 7$
1.5A	$-(a - b)$	$-(a - b) = -a + b$	$-(x - 3) = -x + 3$
1.5B	Similar, or like, terms	Two or more terms that differ only in their numerical coefficient	$-3a$ and $7a$ are similar, or like, terms.
1.6A	Exponent	$a^n = a \cdot a \cdot a \cdots a$ (n factors) n is called the exponent	$3^4 = 3 \cdot 3 \cdot 3 \cdot 3$
1.6A	Base	In the expression a^n, a is called the base.	In the expression 2^5, 2 is the base.
1.6B	Negative exponent	$a^{-n} = \frac{1}{a^n}, \quad a \neq 0$	$5^{-2} = \frac{1}{5^2} = \frac{1}{25}$
1.6B	Zero exponent	$a^0 = 1, \quad a \neq 0$	$2^0 = 1$ and $\left(\frac{-1}{4}\right)^0 = 1$
1.6C	First law of exponents	$a^m \cdot a^n = a^{m+n}$	$x^5 \cdot x^4 = x^9$
1.6C	Second law of exponents	$\frac{a^m}{a^n} = a^{m-n}, a \neq 0$	$\frac{x^8}{x^3} = x^5$
1.6D	Third law of exponents	$(a^m)^n = a^{m \cdot n}$	$(x^2)^5 = x^{10}$
1.6D	Raising powers to powers	$(a^m b^n)^k = a^{m \cdot k} b^{n \cdot k}$	$(x^3 y^5)^6 = x^{3 \cdot 6} y^{5 \cdot 6} = x^{18} y^{30}$
1.6E	Raising a quotient to a power	$\left(\frac{a^m}{b^n}\right)^k = \frac{a^{m \cdot k}}{b^{n \cdot k}}$	$\left(\frac{x^5}{y^3}\right)^4 = \frac{x^{5 \cdot 4}}{y^{3 \cdot 4}} = \frac{x^{20}}{y^{12}}$
1.6F	Scientific notation	A number is in scientific notation when it is written in the form $m \times 10^n$, where m is greater than or equal to 1 and less than 10 and n is an integer.	$352 = 3.52 \times 10^2$ is in scientific notation.
1.7	Order of operations (from left to right)	Parentheses Division bars Exponentiation Multiplication Division Addition Subtraction	

ANSWERS

(If you need help with these exercises, look in the section indicated in brackets.)

1. [1.1A] Use set notation to list the natural numbers:
   a. Between 3 and 9
   b. Between 4 and 8

   1. a. _____
      b. _____

2. [1.1A] Classify the given number by making a check mark (✓) in the appropriate place.

   2. _____

Set	0.3	0	$\dfrac{-3}{4}$	-5	$\sqrt{3}$
Natural number					
Whole number					
Integer					
Rational numbers					
Irrational number					
Real number					

3. [1.1B] Write as a decimal.

   a. $\dfrac{1}{5}$

   b. $\dfrac{2}{5}$

   3. a. _____
      b. _____

4. [1.1B] Write as a decimal.

   a. $\dfrac{1}{9}$

   b. $\dfrac{2}{9}$

   4. a. _____
      b. _____

5. [1.1B] Write as a fraction.
   a. 0.31
   b. 0.41

   5. a. _____
      b. _____

6. [1.2A] Graph on the number line.

   a. $\dfrac{1}{3}$

   b. $\dfrac{3}{4}$

   6. a. ⟵―|―――|―――|―――|―――|―⟶
              -2 -1 0   1   2

      b. ⟵―|―――|―――|―――|―――|―⟶
              -2 -1 0   1   2

7. [1.2B] Find the additive inverse.
   a. 5
   b. -8

   7. a. _____
      b. _____

8. [1.2B] Find the additive inverse.
    a. -3.5

    b. $\dfrac{3}{4}$

    8. a. _____
       b. _____

9. [1.2C] Find.
    a. $|-9|$
    b. $|4.2|$

    9. a. _____
       b. _____

10. [1.2C] Find.

    a. $\left|-\dfrac{1}{8}\right|$

    b. $|0.\overline{4}|$

    10. a. _____
        b. _____

11. [1.3A] Find.
    a. $-3 + (-8)$
    b. $-5 + 2$

    11. a. _____
        b. _____

12. [1.3B] Find.

    a. $\dfrac{1}{7} - \dfrac{3}{7}$

    b. $-0.2 - 0.4$

    12. a. _____
        b. _____

13. [1.3B] Find.
    a. $8 - (-4)$
    b. $-3 - (-7)$

    13. a. _____
        b. _____

14. [1.3C] Which law is illustrated in the statement?
    a. $(4 + 9) + 5 = 5 + (4 + 9)$
    b. $(3 + 5) + 8 = 3 + (5 + 8)$

    14. a. _____
        b. _____

15. [1.4A] Find.
    a. $9 \cdot (-4)$
    b. $-2.4 \cdot 6$

    15. a. _____
        b. _____

16. [1.4A] Find.

    a. $\left(-\dfrac{3}{4}\right) \cdot \dfrac{7}{8}$

    b. $\left(-\dfrac{5}{6}\right) \cdot \left(-\dfrac{2}{7}\right)$

    16. a. _____
        b. _____

17. [1.4B] Find.

    a. $\dfrac{0}{7}$

    b. $\dfrac{8}{0}$

    17. a. _____
        b. _____

18. [1.4B] Find the reciprocal.

    a. $-\dfrac{3}{5}$

    b. 0.3

    18. a. _____
        b. _____

19. [1.4B] Find.

a. $-\dfrac{3}{5} \div \dfrac{4}{15}$

b. $\dfrac{3.6}{-1.2}$

19. a. _____
    b. _____

20. [1.5A] Remove parentheses (simplify).
a. $-3(x - 7)$
b. $3(x + 8) - (x + 7)$

20. a. _____
    b. _____

21. [1.5B] Simplify.
a. $3(x + 2y - 2) - 2(2x - 2y + 5)$
b. $[(5x^2 - 3) + (4x + 5)] - [(x - 4) + (2x^2 - 2)]$

21. a. _____
    b. _____

22. [1.6A] Evaluate.
a. $(-3)^4$
b. -3^4

22. a. _____
    b. _____

23. [1.6B] Write as a fraction.
a. $(-8)^{-3}$
b. x^{-10}

23. a. _____
    b. _____

24. [1.6C] Multiply and simplify.
a. $(3x^4y)(-5x^{-8}y^9)$
b. $(4x^{-3}y^{-1})(-6x^{-8}y^{-7})$

24. a. _____
    b. _____

25. [1.6C] Divide and simplify.

a. $\dfrac{48x^4}{16x^6}$

b. $\dfrac{8x^5}{-2x^{-6}}$

25. a. _____
    b. _____

26. [1.6C] Divide and simplify.

a. $\dfrac{-5x^{-3}}{15x^{-4}}$

b. $\dfrac{8x^{-4}}{-4x^7}$

26. a. _____
    b. _____

27. [1.6D] Simplify.
a. $(-2x^7y^{-6})^3$
b. $(-2x^{-6}y^{-6})^4$

27. a. _____
    b. _____

28. [1.6E) Simplify.

a. $\left(\dfrac{x^6}{y^{-3}}\right)^{-4}$

b. $\left(\dfrac{x^{-5}}{y^3}\right)^{-5}$

28. a. _____
    b. _____

29. [1.6F] Write in scientific notation.
a. 340,000
b. 0.000047

29. a. _____
    b. _____

30. [1.6F] Write in decimal notation.
   a. 3.7×10^4
   b. 7.8×10^{-3}

31. [1.7A] Evaluate.
   a. $[-8 \cdot (9 + 2)] + 13$
   b. $[-7(3 - 8)] + 15$

32. [1.7A] Evaluate.
   a. $6^2 \div 3 - 9 \cdot 2 \div 3 + 3$

   b. $\dfrac{5 \cdot (68 - 32)}{9}$

33. [1.7B] Evaluate.

   a. $-3^2 + \dfrac{(4 - 10)}{2} + 15 \div 3$

   b. $-4^3 + \dfrac{(2 - 10)}{2} - 25 \div 5$

30. a. _____
    b. _____

31. a. _____
    b. _____

32. a. _____
    b. _____

33. a. _____
    b. _____

_____
_____
_____
_____
_____
_____

(Answers on pages 88–89)

ANSWERS

1. Use set notation to list the natural numbers between 5 and 9.

   1. _____

2. Classify the given number by making a check mark ($\checkmark$) in the appropriate place.

   2. _____

Set	0.5	0	-6	$\dfrac{-2}{7}$	$\sqrt{5}$
Natural number					
Whole number					
Integer					
Rational numbers					
Irrational number					
Real number					

3. Write $\dfrac{1}{3}$ as a decimal.

   3. _____

4. Write 0.33 as a fraction.

   4. _____

5. Graph the additive inverse of $\dfrac{1}{3}$ on the number line.

   5.

6. Find.
   a. $|-9|$
   b. $|0.5|$

   6. a. _____
   b. _____

7. Find.
   a. $-9 + 5$
   b. $-0.8 + (-0.7)$

   7. a. _____
   b. _____

8. Find.
   a. $-16 - 7$
   b. $-0.6 - (-0.4)$

   8. a. _____
   b. _____

9. Which law is being illustrated?
   a. $(7 + 3) + 6 = (3 + 7) + 6$
   b. $(2 + 9) + 4 = 2 + (9 + 4)$

9. a. _____
   b. _____

10. Find.
   a. $6 \cdot (-9)$
   b. $-4 \cdot (-1.2)$

10. a. _____
    b. _____

11. Find.

   a. $-\dfrac{1}{2} \cdot \dfrac{2}{9}$

   b. $-\dfrac{3}{2} \div \dfrac{9}{8}$

11. a. _____
    b. _____

12. Simplify.
   a. $-5(x + 7)$
   b. $7x - (3x + 1) + (2x + 2)$

12. a. _____
    b. _____

13. Simplify $[(5x^2 - 3) + (3x + 7)] - [(x - 3) + (2x^2 - 2)]$

13. _____

14. Evaluate.
   a. $(-3)^4$
   b. -3^4

14. a. _____
    b. _____

15. Write as a fraction.
   a. 7^{-2}
   b. x^{-8}

15. a. _____
    b. _____

16. Perform the indicated operation and simplify.
   a. $(3x^4 y)(-4x^{-8} y^8)$

   b. $\dfrac{48x^4}{16x^{-8}}$

16. a. _____
    b. _____

17. Simplify.
   a. $(-2x^8 y^{-2})^3$

   b. $\left(\dfrac{x^5}{y^{-3}}\right)^{-3}$

17. a. _____
    b. _____

18. Do the calculation and write the answer in scientific notation.

    $(7.1 \times 10^5) \times (4 \times 10^{-7})$

18. _____

19. Evaluate.
    a. $[-7(4 + 3)] + 9$

    b. $\dfrac{5 \cdot (131 - 32)}{9}$

19. a. _____
    b. _____

20. Evaluate $-4^3 + \dfrac{(6 - 12)}{2} + 15 \div 3$.

20. _____

	REVIEW			
IF YOU MISSED QUESTION	**SECTION**	**EXAMPLES**	**PAGE**	**ANSWERS**
1	1.1	1	10	1. $\{6, 7, 8\}$

Set	0.5	0	-6	$\dfrac{-2}{7}$	$\sqrt{5}$
N					
W		✓			
I		✓	✓		
Rat.	✓	✓	✓	✓	
Irr.					✓
R	✓	✓	✓	✓	✓

2	1.1	2	11	2. R
3	1.1	3	11	3. $0.\overline{3}$
4	1.1	4	12	4. $\dfrac{33}{100}$
5	1.2	1, 2, 3	17–19	5. $\xleftarrow{\quad}\overset{\displaystyle -\frac{1}{3}}{\underset{-2\ -1\ \ 0\ \ 1\ \ 2}{\bullet}}\xrightarrow{\quad}$
6a	1.2	4	19	6. a. 9
6b	1.2	4	19	b. 0.5
7a	1.3	1, 2	26–27	7. a. -4
7b	1.3	1, 2	26–27	b. -1.5
8a	1.3	3	28	8. a. -23
8b	1.3	3	28	b. -0.2
9a	1.3	4	29	9. a. Commutative law of addition
9b	1.3	4	29	b. Associative law of addition
10a	1.4	1, 2	36–37	10. a. -54
10b	1.4	1, 2	36–37	b. 4.8
11a	1.4	3	38	11. a. $-\dfrac{1}{9}$
11b	1.4	4, 5, 6	39–40	b. $-\dfrac{4}{3}$
12a	1.5	1	48	12. a. $-5x - 35$
12b	1.5	2, 3, 4	49–51	b. $6x + 1$
13	1.5	5	51	13. $3x^2 + 2x + 9$
14a	1.6	1	58	14. a. 81
14b	1.6	1	58	b. -81
15a	1.6	2	59	15. a. $\dfrac{1}{49}$
15b	1.6	2	59	b. $\dfrac{1}{x^8}$

IF YOU MISSED QUESTION	SECTION	EXAMPLES	PAGE	ANSWERS
16a	1.6	3	61	**16. a.** $-\dfrac{12y^9}{x^4}$
16b	1.6	4	62	**b.** $3x^{12}$
17a	1.6	5	63	**17. a.** $-\dfrac{8x^{24}}{y^6}$
17b	1.6	6	64	**b.** $\dfrac{1}{x^{15}y^9}$
18	1.6	7, 8, 9	65–66	**18.** 2.84×10^{-1}
19a	1.7	1, 2, 3, 4	71–72	**19. a.** -40
19b	1.7	5	73	**b.** 55
20	1.7	6	74	**20.** -62

LINEAR EQUATIONS AND INEQUALITIES IN ONE VARIABLE

In this chapter we use the principles learned about the real numbers to solve linear equations and inequalities in one variable. We also discuss equations and inequalities involving absolute values. We then learn how to solve for a particular variable in a formula and end the chapter discussing integer, geometry, percent and investment, distance, and mixture problems.

NAME

CLASS

SECTION

(Answers on page 94)

ANSWERS

1. Does 3 satisfy the equation $6 = 9 - x$?

1. _____

2. Solve $\dfrac{4}{5}y - 2 = 6$.

2. _____

3. Solve $x + 5 = 2(3x - 1)$.

3. _____

4. Solve $\dfrac{x + 2}{3} - \dfrac{x - 2}{5} = 2$.

4. _____

5. Solve $\dfrac{7}{5} - \dfrac{x}{15} = \dfrac{2(x + 7)}{30}$.

5. _____

6. Solve $0.05P + 0.07(1500 - P) = 87$.

6. _____

7. Solve $\left|\dfrac{3}{4}x + 2\right| + 4 = 12$.

7. _____

8. Solve $|x - 2| = |x - 4|$.

8. _____

9. Graph $x \geq -2$.

9. _____

10. Solve and graph $3(x - 1) \leq 6x + 3$.

10. _____

11. Solve and graph $\dfrac{x}{5} - \dfrac{x}{3} < \dfrac{x - 5}{5}$.

11. _____

12. Solve and graph $\{x \mid x > -3 \text{ and } x < 2\}$.

12. _____

13. Solve and graph $\{x \mid x < -2 \text{ or } x \geq 3\}$.

13. _____

14. Solve and graph $x + 1 \leq 3$ and $-2x < 6$.

14. _____

15. Solve and graph $-3 < -3x + 6 \leq 3$.

15. _____

16. Solve and graph $|2x - 1| \leq 3$.

16. _____

17. Solve and graph $|3x + 1| > 4$.

17. _____

18. Given $H = 2.75h + 71.48$.
   a. Solve for h.
   b. Find h if $H = 137.48$.

18. **a.** _____
    b. _____

19. Solve for A in $B = \frac{3}{4}(A - 10)$.

19. _____

20. The perimeter of a rectangle is $P = 2L + 2W$, where L is the length and W is the width.
   a. Solve for L.
   b. If the perimeter is 100 ft and the length is 30 ft more than the width, what are the dimensions?

20. **a.** _____
    b. _____

21. The sum of three consecutive odd integers is 99. What are the three integers?

21. _____

22. A woman's salary was increased by 20% to $18,000. What was her salary before the increase?

22. _____

23. An investor bought some municipal bonds yielding 5% annually and some certificates of deposit yielding 7%. If his total investment amounts to $20,000 and his annual interest is $1160, how much money is invested in bonds and how much in certificates of deposit?

24. A freight train leaves a station traveling at 30 mi/hr. Two hours later, a passenger train leaves the same station in the same direction at 40 mi/hr. How far from the station does the passenger train overtake the freight train?

25. How many gallons of a 30% salt solution must be mixed with 40 gal of a 12% salt solution to obtain a 20% solution?

23. _____

24. _____

25. _____

IF YOU MISSED QUESTION	SECTION	EXAMPLES	PAGE	ANSWERS
1	2.1	1	96	**1.** Yes
2	2.1	2	97	**2.** 10
3	2.1	3, 4	98	**3.** $\dfrac{7}{5}$
4	2.1	5	99	**4.** 7
5	2.1	6	100	**5.** 7
6	2.2	1, 2	108	**6.** 900
7	2.2	3, 4	109	**7.** $8, -\dfrac{40}{3}$
8	2.2	5	110	**8.** 3
9	2.3	1	116	**9.**
10	2.3	3	118	**10.**
11	2.3	4	120	**11.**
12	2.4	1	125	**12.**
13	2.4	2	126	**13.**
14	2.4	3	127	**14.**
15	2.4	4	128	**15.**
16	2.4	5	129	**16.**
17	2.4	6	130	**17.**
18a	2.5	1	135–136	**18. a.** $h = \dfrac{H - 71.48}{2.75}$
18b	2.5	1	135–136	**b.** 24
19	2.5	2	136	**19.** $A = \dfrac{4B + 30}{3}$
20a	2.5	3	137	**20. a.** $L = \dfrac{P - 2W}{2}$
20b	2.5	3	137	**b.** 10 ft by 40 ft
21	2.6	1	145	**21.** 31, 33, 35
22	2.6	3	146	**22.** $15,000
23	2.6	4	147	**23.** $12,000 bonds; $8000 CD
24	2.6	5	148	**24.** 240 mi
25	2.6	6	149	**25.** 32

2.1   LINEAR EQUATIONS IN ONE VARIABLE

Courtesy of Princeton University Library.

OBJECTIVES

REVIEW

Before starting this section, you should know:

1. How to add, subtract, multiply, and divide positive and negative numbers.
2. How to use the commutative and associative laws of addition and multiplication.
3. That opposites add to 0 and that the product of reciprocals is 1.

OBJECTIVES

You should be able to:

A. Determine if a number is a solution of an equation.
B. Solve equations using the properties of equality.
C. Solve linear equations in one variable using the procedure given in the text.

Do you know the name of the man in the photo? It is Albert Einstein. Einstein discovered one of the most important *equations* of our time, $E = mc^2$. (E is the energy, m is the mass, and c is the velocity of light).

An **equation** is a sentence using "$=$" for its verb.

Some equations are *true* ($2 + 2 = 4$, $5 - 3 = 2$), some are *false* ($2 + 2 = 22$, $5 - 3 = -2$), and some are neither true nor false. For example, the equation $x + 1 = 5$ is neither true nor false.

A.   SOLUTIONS OF AN EQUATION

In the equation $x + 1 = 5$, the **variable** x can be replaced by many numbers, but only one number will make the resulting statement true. This number is called the *solution* of the equation.

The **solutions** of an equation are the replacements of the variable that make the equation a *true* statement. To **solve** an equation is to find all its solutions.

To determine if a number is a solution of an equation, we replace the variable by the number. For example, 4 is a solution of $x + 1 = 5$ because replacing x by 4 in the equation yields $4 + 1 = 5$, a true statement. Since 4 is the only number that yields a true statement, the solution set of $x + 1 = 5$ is $\{4\}$.

EXAMPLE 1 Determine if

a. 8 is a solution of $x - 5 = 3$.
b. 5 is a solution of $3 = 2 - y$.
c. 6 is a solution of $\frac{1}{3}z - 4 = 2z - 14$.

Problem 1 Determine if
a. 6 is a solution of $x - 2 = 4$.
b. 7 is a solution of $4 = 11 - y$.
c. 4 is a solution of $\frac{1}{2}z - 2 = x - 2$.

Solution

a. Substituting 8 for x in $x - 5 = 3$, we have $8 - 5 = 3$, a true statement. Thus 8 is a solution of $x - 5 = 3$.

b. Substituting 5 for y in $3 = 2 - y$, we obtain $3 = 2 - y$, which is false. Hence, 5 is not a solution of $3 = 2 - y$.

c. If we replace z by 6, we obtain:

$$\frac{1}{3}(6) - 4 = 2(6) - 14$$
$$2 - 4 = 12 - 14$$
$$-2 = -2$$

a true statement. Thus, 6 is a solution of $\frac{1}{3}z - 4 = 2z - 14$.

B. SOLVING EQUATIONS USING THE PROPERTIES OF EQUALITY

We have learned how to determine if a number is a solution of an equation; now we will learn how to find these solutions. The procedure is to find an *equivalent* equation whose solution is obvious. For example, the equations $x = 2$ and $x + 3 = 5$ are equivalent because $x = 2$, with the obvious solution 2, is the only solution of the equation $x + 3 = 5$.

> Two or more equations are **equivalent** if they have the same solution set.

To solve equations, we use the idea that adding or subtracting the same number on both sides and multiplying or dividing both sides of an equation by the same quantity preserves the equality. Here is the principle.

PROPERTIES OF EQUALITY

If C is a real number, then the following equations are all equivalent.

$$A = B$$

Add C.	$A + C = B + C$	
Subtract C.	$A - C = B - C$	
Multiply by C.	$A \cdot C = B \cdot C$	$(C \neq 0)$
Divide by C.	$\dfrac{A}{C} = \dfrac{B}{C}$	$(C \neq 0)$

Now, suppose we want to solve $x + 5 = 7$. Since we are trying to find a value of x that will satisfy the equation, we try to have x by itself on one side. To do this, we "undo" the addition of 5 by *adding* the inverse of 5 on both sides. Thus

$$x + 5 = 7 \qquad \text{Given.}$$
$$x + 5 + (-5) = 7 + (-5) \qquad \text{Add } (-5) \text{ to both sides.}$$
$$x + 0 = 2 \qquad 5 + (-5) = 0$$
$$x = 2$$

The solution is 2, and the solution set is {2}. Note that you can also solve the equation by subtracting 5 from both sides:

$$x + 5 = 7 \qquad \text{Given.}$$
$$x + 5 - 5 = 7 - 5 \qquad \text{Subtract 5 from both sides.}$$
$$x + 0 = 2 \qquad 5 - 5 = 0$$
$$x = 2$$

Using the same idea, we can solve $8 = x - 7$ by adding 7 to both sides. Thus

$$8 = x - 7 \qquad \text{Given.}$$
$$8 + 7 = x - 7 + 7 \qquad \text{Add 7 to both sides.}$$
$$15 = x$$

The solution is 15, and the solution set is {15}.

EXAMPLE 2 Solve.

a. $2x - 4 = 6$ **b.** $\frac{2}{3}y - 3 = 9$

Solution

a. To solve this equation, we want the variable x by itself on one side of the equation. We start by adding 4 to both sides.

$$2x - 4 = 6 \qquad \text{Given.}$$
$$2x - 4 + 4 = 6 + 4 \qquad \text{Add 4 to both sides.}$$
$$2x = 10$$
$$\frac{1}{2} \cdot 2x = \frac{1}{2} \cdot 10 \qquad \text{Multiply both sides by } \frac{1}{2}.$$
$$x = 5 \qquad \frac{1}{2} \cdot 2 = 1 \text{ and } \frac{1}{2} \cdot 10 = 5$$

Note that to solve $2x = 10$, we could divide both sides by 2:

$$\frac{2x}{2} = \frac{10}{2}$$

obtaining the same result:

$$x = 5$$

To check this solution, we substitute 5 for x in the original equation and use the following diagram:

$$2x - 4 \overset{?}{=} 6$$

$2(5) - 4$	6
$10 - 4$	
6	

Since both sides yield 6, our result is correct.

Problem 2 Solve.
a. $3x - 8 = 4$ **b.** $\frac{3}{4}x - 5 = 10$

b. $\frac{2}{3}y - 3 = 9$ Given.

$\frac{2}{3}y - 3 + 3 = 9 + 3$ Add 3 to both sides.

$\frac{2}{3}y = 12$

$\frac{3}{2} \cdot \frac{2}{3}y = \frac{3}{2} \cdot 12$ Multiply both sides by $\frac{3}{2}$.

$y = 18$ $\frac{3}{2} \cdot 12 = \frac{3}{2} \cdot \frac{12}{1} = 18$

You can check this answer by substituting 18 for y in $\frac{2}{3}y - 3 = 9$.

EXAMPLE 3   Solve $4a - 7 = a + 4$.

Solution   We start by adding 7 to both sides. We then add the inverse of a on both sides, so that only variables are on the left. Here are the steps:

$4a - 7 = a + 4$ Given.

$4a - 7 + 7 = a + 4 + 7$ Add 7 to both sides.

$4a = a + 11$

$4a + (-a) = a + (-a) + 11$ Add $(-a)$ to both sides so that all

$3a = 11$ variables are on the left.

$a = \frac{11}{3}$ Divide by 3 $\left(\text{or multiply by } \frac{1}{3}\right)$.

Thus, the solution is $\frac{11}{3}$, as can be checked by substituting $\frac{11}{3}$ for a in $4a - 7 = a + 4$.   ▲

Sometimes we need to simplify an equation before solving it. Thus, to solve $x + 6 = 2(2x - 3)$ we use the distributive law to simplify the right side of the equation. We do this next.

EXAMPLE 4   Solve $x + 6 = 2(2x - 3)$.

Solution

$x + 6 = 2(2x - 3)$ Given.

$x + 6 = 6x - 6$ Simplify the right side.

Now, we have two choices; we can have the variables on the right or on the left. To avoid negative expressions, we keep them on the right by adding (6) to both sides. We have:

$x + 6 + (6) = 6x - 6 + (6)$

$x + 12 = 6x$

$x + (-x) + 12 = 6x + (-x)$ Add $(-x)$ so all variables are

$12 = 5x$ on the right.

$\frac{12}{5} = x$ Divide by 5 $\left(\text{or multiply by } \frac{1}{5}\right)$.

Thus, the solution is $\frac{12}{5}$.   ▲

Problem 3   Solve $5a - 8 = a + 5$.

Problem 4   Solve $3 + x = 2(2x - 3)$.

If an equation involves fractions, we "clear" them by multiplying both sides by the smallest number that is a multiple of each denominator, that is, the lowest common denominator (LCD). (See Using Your Knowledge for a shortcut in finding LCD's.) Thus to solve $\dfrac{x}{6} + \dfrac{x}{4} = 10$, we have to find the LCD of 6 and 4. One way is to pick the larger of the two numbers and double it, triple it, etc., until the other number divides into the result. Using this idea, we can find that 12 is the LCD. Multiplying both sides by 12, we have:

$$12 \cdot \left(\frac{x}{6} + \frac{x}{4}\right) = 12 \cdot 10$$

$$12 \cdot \frac{x}{6} + 12 \cdot \frac{x}{4} = 12 \cdot 10 \quad \text{Use the distributive law.}$$

$$2x + 3x = 120$$

$$5x = 120 \qquad \text{Combine like terms.}$$

$$x = 24 \qquad \text{Divide both sides by 5.}$$

Thus, the solution of $\dfrac{x}{6} + \dfrac{x}{4} = 10$ is 24. Here is the check:

$$\frac{x}{6} + \frac{x}{4} \overset{?}{=} 10$$

$\dfrac{24}{6} + \dfrac{24}{4}$	10
$4 + 6$	
10	

EXAMPLE 5 Solve.

a. $\dfrac{x+1}{3} + \dfrac{x-1}{10} = 5$ **b.** $\dfrac{x+1}{3} - \dfrac{x-1}{8} = 4$

Solution

a. The LCD of 3 and 10 is $3 \cdot 10 = 30$, since 3 and 10 do not have any common factors. Multiplying both sides by 30 we have

$$30\left[\frac{x+1}{3} + \frac{x-1}{10}\right] = 30 \cdot 5$$

$$30\left(\frac{x+1}{3}\right) + 30\left(\frac{x-1}{10}\right) = 30 \cdot 5$$

$$10(x+1) + 3(x-1) = 150$$

$$10x + 10 + 3x - 3 = 150 \quad \text{Use the distributive property.}$$

$$13x + 7 = 150 \quad \text{Add like terms.}$$

$$13x = 143 \quad \text{Subtract 7.}$$

$$x = 11 \quad \text{Divide by 13.}$$

The check is left to the student.

Problem 5 Solve.

a. $\dfrac{x+2}{4} + \dfrac{x-1}{5} = 3$

b. $\dfrac{x+3}{2} - \dfrac{x-2}{3} = 5$

b. Here the LCD is $3 \cdot 8 = 24$. Multiplying both sides by 24, we obtain

$$24\left[\frac{x+1}{3} - \frac{x-1}{8}\right] = 24 \cdot 4$$

$$24\left(\frac{x+1}{3}\right) - 24\left(\frac{x-1}{8}\right) = 24 \cdot 4$$

$$8(x+1) - 3(x-1) = 96$$

$8x + 8 - 3x + 3 = 96$ Use the distributive property.

$5x + 11 = 96$ Add like terms.

$5x = 85$ Subtract 11.

$x = 17$ Divide by 5.

Be sure you check this answer in the original equation.

C. SOLVING LINEAR EQUATIONS

All the equations we have solved in this section can be written in the form $ax + b = c$, where a, b, and c are real numbers. Equations of this type are called **linear equations.** (You will see in later chapters that the graph of $ax + b = c$ is a *straight line.*) Thus $2x - 4 = 6$ and $\frac{x}{10} + \frac{x}{8} = 9$ are linear equations. In the equation $2x - 4 = 6$, $2x$ and 4 are called **terms.** $2x$ is a variable term and 4 is a constant term. We will use this terminology to give you a general procedure used to solve linear equations.

We have done an example in the margin.

> **PROCEDURE FOR SOLVING LINEAR EQUATIONS**
>
> 1. If there are fractions, multiply both sides of the equation by the LCD of the fractions.
>
> 2. Remove parentheses and collect like terms (simplify) if necessary.
>
> 3. Add or subtract the same number on both sides of the equation so that one side has only variables.
>
> 4. Add or subtract the same expression on both sides of the equation so that the other side has only numbers.
>
> 5. If the coefficient of the variable is not 1, divide both sides of the equation by this coefficient (or, equivalently, multiply by the reciprocal of the coefficient of the variable).
>
> 6. Be sure to check your answer in the original equation.

EXAMPLE 6 Solve.

a. $\dfrac{7}{24} = \dfrac{x}{8} + \dfrac{1}{6}$ **b.** $\dfrac{1}{5} - \dfrac{x}{4} = \dfrac{7(x+3)}{10}$

Solution

a. We use the six steps given above.

Given: $\dfrac{x}{4} - \dfrac{1}{6} = \dfrac{7}{12}(x - 2)$

1. $12\left(\dfrac{x}{4} - \dfrac{1}{6}\right) = 12 \cdot \dfrac{7}{12}(x - 2)$

2. $12 \cdot \dfrac{x}{4} - 12 \cdot \dfrac{1}{6} = 12 \cdot \dfrac{7}{12}(x - 2)$

   or

$$3x - 2 = 7(x - 2)$$
$$= 7x - 14$$

3. $3x - 2 + 14 = 7x - 14 + 14$

    $3x + 12 = 7x$

Note that we have more x's on the right, so subtract $3x$ from both sides.

4. $3x - 3x + 12 = 7x - 3x$

        $12 = 4x$

5. $\dfrac{12}{4} = \dfrac{4x}{4}$

      $3 = x$

The solution is 3.

6. *Check.*

$$\frac{x}{4} - \frac{1}{6} \stackrel{?}{=} \frac{7}{12}(x - 2)$$

$$\begin{array}{c|c} \dfrac{3}{4} - \dfrac{1}{6} & \dfrac{7}{12}(3 - 2) \\[2mm] \dfrac{9}{12} - \dfrac{2}{12} & \dfrac{7}{12} \cdot 1 \\[2mm] \dfrac{7}{12} & \dfrac{7}{12} \end{array}$$

Since both sides are equal, the solution is correct.

Problem 6 Solve.

a. $\dfrac{7}{12} = \dfrac{x}{4} + \dfrac{1}{3}$ **b.** $\dfrac{1}{3} - \dfrac{x}{5} = \dfrac{8(x+2)}{15}$

---

1. Multiply by 24, the LCD:

$$24 \cdot \frac{7}{24} = 24\left(\frac{x}{8} + \frac{1}{6}\right)$$

$$24 \cdot \frac{7}{24} = 24 \cdot \frac{x}{8} + 24 \cdot \frac{1}{6}$$

2. Simplify:

$$7 = 3x + 4$$

3. Subtract 4 (or add -4):

$$7 - 4 = 3x + 4 - 4$$

4. The left side has numbers only:

$$3 = 3x$$

5. Divide by 3 $\left(\text{or multiply by } \frac{1}{3}\right)$:

$$\frac{3}{3} = \frac{3x}{3}$$

$$1 = x$$

$$x = 1$$

Check. Since $\dfrac{1}{8} = \dfrac{3}{24}$ and $\dfrac{1}{6} = \dfrac{4}{24}$,

$$\frac{7}{24} = \frac{1}{8} + \frac{1}{6} = \frac{3}{24} + \frac{4}{24}$$

which is true.

b. Given:

$$\frac{1}{5} - \frac{x}{4} = \frac{7(x + 3)}{10}$$

1. Multiply by 20, the LCD:

$$20\left(\frac{1}{5} - \frac{x}{4}\right) = 20 \cdot \left[\frac{7(x + 3)}{10}\right]$$

$$20 \cdot \frac{1}{5} - 20 \cdot \frac{x}{4} = 20 \cdot \frac{7(x + 3)}{10}$$

2. Simplify:

$$4 - 5x = 14(x + 3)$$
$$= 14x + 42$$

3. Subtract 42 (or add -42):

$$4 - 5x - 42 = 14x + 42 - 42$$
$$-38 - 5x = 14x$$

4. Add $5x$:

$$-38 - 5x + 5x = 14x + 5x$$
$$-38 = 19x$$

5. Divide by 19 $\left(\text{or multiply by } \frac{1}{19}\right)$:

$$\frac{-38}{19} = \frac{19x}{19}$$
$$-2 = x$$

Check. $\quad \dfrac{1}{5} - \dfrac{x}{4} \overset{?}{=} \dfrac{7(x + 3)}{10} \quad$ for $x = -2$

$\dfrac{1}{5} - \dfrac{(-2)}{4}$	$\dfrac{7(-2 + 3)}{10}$
$\dfrac{1}{5} + \dfrac{1}{2}$	$\dfrac{7(1)}{10}$
$\dfrac{7}{10}$	$\dfrac{7}{10}$

▲

So far all our equations have had one solution. There are two other possibilities:

1. Equations with *no* solution
2. Equations with *infinitely many* solutions

Consider the following example.

$$-4x + 6 = 2(1 - 2x) + 3$$
$$-4x + 6 = 2 - 4x + 3 \qquad \text{Clear parentheses.}$$
$$-4x + 6 = -4x + 5 \qquad \text{Collect like terms.}$$
$$-4x + 6 + (-6) = -4x + 5 + (-6) \qquad \text{Add } -6.$$
$$-4x = -4x + (-1)$$
$$-4x + 4x = -4x + 4x + (-1) \qquad \text{Add } 4x.$$
$$0 = (-1) \qquad \text{False!} \qquad \text{Simplify.}$$

We cannot find an x to satisfy this equation. This equation has no solution; its solution set is $\varnothing$ (the empty set). On the other hand, consider the following equation:

$$-4x + 6 = 2(1 - 2x) + 4$$
$$-4x + 6 = 2 - 4x + 4 \qquad \text{Clear parentheses.}$$
$$-4x + 6 = -4x + 6 \qquad \text{Collect like terms.}$$
$$6 = 6 \qquad \text{True!} \qquad \text{Add } 4x.$$

Here, any x will be a solution (try 0 or 1 in the original equation). This equation is called an **identity.** Any real number is a solution. Its solution set is R, the set of real numbers.

EXAMPLE 7   Solve.

a. $3x + 8 = 3(x + 1) + 5$ **b.** $3x + 8 = 3(x + 1) + 2$

Solution

a. We use the six-step procedure we studied.

Given: $\qquad\qquad\qquad 3x + 8 = 3(x + 1) + 5$

1. There are no fractions:

2. Remove parentheses: $\qquad 3x + 8 = 3x + 3 + 5$

Collect like terms: $\qquad\ 3x + 8 = 3x + 8$

We can stop here. This equation is always true. It is an identity. Its solution set is R.

b. Given. $\qquad\qquad\qquad\ 3x + 8 = 3(x + 1) + 2$

1. There are no fractions:

2. Remove parentheses: $\qquad 3x + 8 = 3x + 3 + 2$

Collect like terms: $\qquad\ 3x + 8 = 3x + 5$

We can stop here. This equation is always false. (Try subtracting $3x$ on both sides.) It has no solution. Its solution set is $\varnothing$.

Problem 7   Solve.
a. $4(x + 1) + 3 = 4x + 7$
b. $5(x + 2) + 1 = 3 + 5x$

ANSWER   (to problem on page 100)
6. a. 1   **b.** -1

NAME

CLASS

SECTION

ANSWERS

A. In Problems 1–10 determine if the number in the box is a solution of the equation.

1. $2x + 8 = 14;$ $\boxed{3}$

2. $5x + 5 = 10;$ $\boxed{2}$

3. $-2x + 1 = 3;$ $\boxed{-1}$

4. $-3x + 4 = 10;$ $\boxed{-2}$

5. $2y - 5 = y - 2;$ $\boxed{3}$

6. $3y - 7 = y + 1;$ $\boxed{4}$

7. $\dfrac{4}{5}t - 1 = 5t;$ $\boxed{\dfrac{-1}{5}}$

8. $6t + 1 = t + \dfrac{2}{3};$ $\boxed{\dfrac{-1}{3}}$

9. $\dfrac{1}{2}x + 5 = 5 - \dfrac{1}{3}x;$ $\boxed{3}$

10. $\dfrac{-1}{3}x + 1 = x - 3;$ $\boxed{3}$

1. _____

2. _____

3. _____

4. _____

5. _____

6. _____

7. _____

8. _____

9. _____

10. _____

B. In Problems 11–36 solve the equation.

11. $3x - 4 = 8$

12. $5a + 16 = 6$

13. $2y + 8 = 10$

14. $4b - 6 = 2$

15. $-3z - 6 = -12$

16. $-4r - 3 = 5$

17. $-5y + 2 = -8$

18. $-3x + 2 = -10$

19. $3x + 5 = x + 19$

20. $4x + 6 = x + 9$

21. $7(x - 1) - 3 + 5x = 3(4x - 3) + x$

22. $5(x + 1) + 3x + 2 = 8(x + 2) + 2x + 1$

23. $6v - 8 = 8v + 8$

24. $8t + 3 = 15t - 11$

25. $7m - 4m + 12 = 0$

26. $10k + 25 - 5k = 35$

27. $4(2 - z) + 8 = 8(2 - z)$

28. $4(3 - y) + 3 = 12(3 - y)$

11. _____

12. _____

13. _____

14. _____

15. _____

16. _____

17. _____

18. _____

19. _____

20. _____

21. _____

22. _____

23. _____

24. _____

25. _____

26. _____

27. _____

28. _____

29. $5(x + 3) = 3(x + 3) + 6$

30. $y - (5 - 2y) = 7(y - 1) - 2$

31. $5(4 - 3a) = 7(3 - 4a)$

32. $\dfrac{3}{4}y - 4 = \dfrac{1}{4}y - 2$

33. $-\dfrac{7}{8}c + 5 = -\dfrac{5}{8}c + 3$

34. $x + \dfrac{2}{3}x = 10$

35. $-2x + \dfrac{1}{4} = 2x + \dfrac{4}{5}$

36. $6x + \dfrac{1}{7} = 2x - \dfrac{2}{7}$

29. _____

30. _____

31. _____

32. _____

33. _____

34. _____

35. _____

36. _____

C. In Problems 37–62 use the six-step procedure given in the text to solve.

37. $\dfrac{t}{6} + \dfrac{t}{8} = 7$

38. $\dfrac{f}{9} + \dfrac{f}{12} = 14$

39. $\dfrac{x}{2} + \dfrac{x}{5} = \dfrac{7}{10}$

40. $\dfrac{a}{3} + \dfrac{a}{7} = \dfrac{20}{21}$

41. $\dfrac{c}{3} - \dfrac{c}{5} = 2$

42. $\dfrac{F}{4} - \dfrac{F}{7} = 3$

43. $\dfrac{W}{6} - \dfrac{W}{8} = \dfrac{5}{12}$

44. $\dfrac{m}{6} - \dfrac{m}{10} = \dfrac{4}{3}$

45. $\dfrac{x}{5} - \dfrac{3}{10} = \dfrac{1}{2}$

46. $\dfrac{3y}{7} - \dfrac{1}{14} = \dfrac{1}{14}$

47. $\dfrac{x + 4}{4} - \dfrac{x + 2}{3} = -\dfrac{1}{2}$

48. $\dfrac{w - 1}{2} + \dfrac{w}{8} = \dfrac{7w + 1}{16}$

49. $\dfrac{x + 1}{4} - \dfrac{2x - 2}{3} = 3$

50. $\dfrac{z + 4}{3} = \dfrac{z + 6}{4}$

51. $\dfrac{2h - 1}{3} = \dfrac{h - 4}{12}$

52. $\dfrac{5 - 6y}{7} - \dfrac{-7 - 4y}{3} = 2$

53. $\dfrac{2w + 3}{2} - \dfrac{3w + 1}{4} = 1$

54. $\dfrac{8x - 23}{6} + \dfrac{1}{3} = \dfrac{5}{2}x$

55. $\dfrac{7r + 2}{6} + \dfrac{1}{2} = \dfrac{r}{4}$

56. $\dfrac{x + 1}{2} + \dfrac{x + 2}{3} + \dfrac{x + 4}{4} = -8$

57. $\dfrac{x - 5}{2} - \dfrac{x - 4}{3} = \dfrac{x - 3}{2} - (x - 2)$

58. $\dfrac{x + 1}{2} + \dfrac{x + 2}{3} + \dfrac{x + 3}{4} = 16$

59. $4(x - 2) + 4 = 4x - 4$

60. $8 - (2 - 3x) = 3(x + 2)$

37. _____

38. _____

39. _____

40. _____

41. _____

42. _____

43. _____

44. _____

45. _____

46. _____

47. _____

48. _____

49. _____

50. _____

51. _____

52. _____

53. _____

54. _____

55. _____

56. _____

57. _____

58. _____

59. _____

60. _____

61. $4(x - 2) + 3 = 4x - 4$

62. $8 - (2 - 3x) = 3(x + 4)$

61. _____

62. _____

✓ **SKILL CHECKER**

Find.

63. $|-5|$

64. $\left|-\dfrac{2}{3}\right|$

65. $|0.3|$

66. $|-0.8|$

67. $-|-0.9|$

68. $-|\sqrt{3}|$

69. $-|-\sqrt{5}|$

70. $-|\sqrt{7}|$

63. _____

64. _____

65. _____

66. _____

67. _____

68. _____

69. _____

70. _____

2.1 USING YOUR KNOWLEDGE

We already learned that the LCD of $\dfrac{x}{6} + \dfrac{x}{4} = 10$ is found by selecting the larger of the two numbers (6) and doubling it, tripling it, etc., until the other denominator (4) divides into the number. We can write the procedure like this:

6   12   18   24 . . .

4

The LCD is 12, the smallest multiple of 6 and 4. But there is a shorter way of finding the least common multiple of two numbers. Here is the way it works. Write the numbers in a row, as shown.

$\underline{|\;6\quad 4}$

Divide by the largest number that will divide both numbers, which is 2.

$\begin{array}{c|cc} 2 & 6 & 4 \\ \hline & 3 & 2 \end{array}\;\; 2 \cdot 3 \cdot 2 = 12$

Write the answers under 6 and 4. Since no other number divides the results, 3 and 2, the LCM is the product of the divisor 2 and the final quotients, 3 and 2. Thus, the LCM of 6 and 4 is $2 \cdot 3 \cdot 2 = 12$. Now, let us say that you want the LCD of $\frac{1}{12}$ and $\frac{1}{64}$. We proceed in the same way:

1. Write 12 and 64 in a row as shown.

$\underline{|\;12\quad 64}$

2. Divide by the largest number that will divide 12 and 64, such as 4.

$\begin{array}{c|cc} 4 & 12 & 64 \\ \hline & 3 & 16 \end{array}$

(Do not worry if you do not divide by the largest divisor. If you divided by 2, you would divide by 2 again in the next step. The final result would still be the same.)

ANSWER (to problem on page 102)
7. a. Identity   **b.** No solution

EXERCISE 2.1

Since no other number divides 3 and 16, the LCD is the product of the divisor 4 and the final quotients 3 and 16, that is, $4 \cdot 3 \cdot 16 = 192$.

Use this procedure to find the LCD of each pair of fractions.

1. $\dfrac{1}{40}$ and $\dfrac{1}{18}$ 2. $\dfrac{1}{24}$ and $\dfrac{1}{30}$

3. $\dfrac{1}{65}$ and $\dfrac{1}{26}$ 4. $\dfrac{1}{40}$ and $\dfrac{1}{32}$

1. _____

2. _____

3. _____

4. _____

2.2 EQUATIONS INVOLVING DECIMALS AND ABSOLUTE VALUE

OBJECTIVES

REVIEW

Before starting this section, you should know:

1. How to collect like terms.
2. How to remove parentheses by using the distributive law.
3. How to solve linear equations.

OBJECTIVES

You should be able to:

A. Solve linear equations involving decimals by clearing the decimals.
B. Solve equations involving absolute values.

Do you have a toll road where you live? How much is the toll? Here, it is \$0.35 for each car. At the end of one hour, if \$45.85 was collected, how many cars paid the toll? If we let n be the number of cars, the amount collected was $0.35n$. Since the total paid was \$45.85, we have:

$$\text{amount per car} \times \text{number of cars} = \text{total}$$
$$0.35n = 45.85$$

We can solve this equation by dividing both sides by 0.35, but we can also solve by clearing decimals if we multiply both sides by 100 (or 10^2). (When we multiply a decimal by 100, we move the decimal point two places to the right.)

$0.35n = 45.85$	Given.
$(100)0.35n = (100)45.85$	Multiply by 100.
$35n = 4585$	Move the decimal point.
$\dfrac{35n}{35} = \dfrac{4585}{35}$	Divide by 35.
$n = 131$	

Thus 131 cars paid the toll.

A. EQUATIONS INVOLVING DECIMALS

Equations involving decimals can be solved by clearing decimals and following the procedure given in the preceding section. To clear decimals, look at the coefficient of the variable and multiply by a power of ten having as many zeros as the coefficient of the variable has decimal places. We do this in the next example.

EXAMPLE 1 Solve $14.5 - 3.15x = 5.5$.

Solution Since the coefficient of $x(3.15)$ has 2 decimal places we multiply both sides by 100 and proceed as usual.

$14.5 - 3.15x = 5.5$	Given.
$100(14.5 - 3.15x) = 100 \cdot 5.5$	Multiply both sides by 100.
$100 \cdot 14.5 - 100 \cdot 3.15x = 100 \cdot 5.5$	Use the distributive law.
$1450 - 315x = 550$	Multiply.
$(-1450) + 1450 - 315x = 550 + (-1450)$	Add (-1450).
$-315x = -900$	Simplify.
$\dfrac{-315x}{-315} = \dfrac{-900}{-315}$ or $\dfrac{20}{7}$	Divide by -315.

The solution is $\frac{20}{7}$, which you can check by substituting in the original equation. ▲

Later in this chapter we solve word problems that involve equations such as $1.30P + 1.50(50 - P) = 72.50$. We show you how to solve this equation next.

EXAMPLE 2 Solve $1.30P + 1.50(50 - P) = 72.50$.

Solution

$1.30P + 1.50(50 - P) = 72.50$	Given.
$100 \cdot [1.30P + 1.50(50 - P)] = 100 \cdot 72.50$	Multiply both sides by 100.
$100 \cdot 1.30P + 100 \cdot 1.50(50 - P) = 100 \cdot 72.50$	Use the distributive law
$130P + 150(50 - P) = 7250$	Multiply.
$130P + 7500 - 150P = 7250$	Use the distributive law
$7500 - 20P = 7250$	Collect like terms.
$(-7500) + 7500 - 20P = 7250 + (-7500)$	Add (-7500).
$-20P = -250$	Simplify.
$\dfrac{-20P}{-20} = \dfrac{-250}{-20}$	Divide by -20.
$P = \dfrac{25}{2}$	

The solution is $\frac{25}{2}$. Make sure you check this!

B. ABSOLUTE VALUE EQUATIONS

The absolute value of a, denoted by $|a|$, is the distance between a and 0 on the number line. Thus, $|3| = 3$, $|-5| = 5$, and $|-0.7| = 0.7$. The equation $|x| = 2$ is read as "the distance between x and 0 on the number line is 2." Since x is 2 units from 0, x can only be $+2$ or -2, that is, $x = \pm 2$. In general, we have the following.

> If $a \geq 0$, the solutions of $|x| = a$ are $x = a$ and $x = -a$.

Problem 1 Solve $13.5 - 2.85x = 4.5$.

Problem 2 Solve $0.50x + 0.80(90 - x) = 60.30$.

EXAMPLE 3 Solve.

a. $|x| = 8$ **b.** $|y| = 2.5$ **c.** $|z| = -6$

Problem 3 Solve.
a. $|x| = 7$ **b.** $|y| = 3.4$
c. $|z| = -2$

Solution

a. Since the numbers 8 and -8 are 8 units from 0, the solutions of $|x| = 8$ are 8 and -8. The solution set of $|x| = 8$ is $\{-8, 8\}$.

b. The solutions of $|y| = 2.5$ are $y = 2.5$ and $y = -2.5$. Thus, the solution set of $|y| = 2.5$ is $\{-2.5, 2.5\}$.

c. The absolute value of a number is never negative (it represents a distance). Thus, the equation $|z| = -6$ has no solution. Its solution set is $\varnothing$ (the empty set). ▲

The ideas of Example 3 can be generalized to more complicated equations. Look at the pattern:

$|x| = 8$ has solutions	$x = 8$	and	$x = -8$
$|y| = 2.5$ has solutions	$y = 2.5$	and	$y = -2.5$
$|x + 1| = 8$ has solutions	$x + 1 = 8$	and	$x + 1 = -8$
(Solve each equation by	$x = 7$	and	$x = -9$
adding (-1) on each side.)			

We can check the solutions by substituting them into the equation $|x + 1| = 8$. Substituting 7 for x, we obtain $|7 + 1| = |8| = 8$, a true statement. If we substitute -9 for x, we have $|-9 + 1| = |-8| = 8$, also true.

EXAMPLE 4 Solve.

a. $|2x + 1| = 6$ **b.** $|\frac{3}{4}x + 1| + 5 = 11$

Problem 4 Solve.
a. $|3x + 1| = 5$ **b.** $|\frac{2}{3}x + 1| + 3 = 9$

Solution

a. The solutions of $|2x + 1| = 6$ are:

$$2x + 1 = 6 \quad \text{and} \quad 2x + 1 = -6$$
$$2x = 5 \quad \text{and} \quad 2x = -7 \quad \text{Add } (-1) \text{ to both sides.}$$
$$x = \frac{5}{2} \quad \text{and} \quad x = -\frac{7}{2} \quad \text{Multiply both sides by } \frac{1}{2}.$$

Thus, the solution set of $|2x + 1| = 6$ is $\{-\frac{7}{2}, \frac{5}{2}\}$.

b. In order to use the definition of absolute value, we must have $|\frac{3}{4}x + 1|$ by itself. Thus, we first add (-5) to both sides.

$$\left|\frac{3}{4}x + 1\right| + 5 = 11 \qquad \text{Given.}$$

$$\left|\frac{3}{4}x + 1\right| + 5 + (-5) = 11 + (-5) \qquad \text{Add } -5.$$

$$\left|\frac{3}{4}x + 1\right| = 6 \qquad \text{Simplify.}$$

The solutions of this equation are:

$$\frac{3}{4}x + 1 = 6 \qquad \text{and} \qquad \frac{3}{4}x + 1 = -6$$

$$\frac{3}{4}x = 5 \qquad \text{and} \qquad \frac{3}{4}x = -7 \qquad \text{Add } -1.$$

$$\frac{4}{3} \cdot \frac{3}{4}x = \frac{4}{3} \cdot 5 \quad \text{and} \quad \frac{4}{3} \cdot \frac{3}{4}x = \frac{4}{3} \cdot (-7) \quad \text{Multiply by } \frac{4}{3}.$$

$$x = \frac{20}{3} \qquad \text{and} \qquad x = -\frac{28}{3} \qquad \text{Simplify.}$$

Thus the solution set of $\left|\frac{3}{4}x + 1\right| + 5 = 11$ is $\left\{-\frac{28}{3}, \frac{20}{3}\right\}$.

EXAMPLE 5   Solve $|x - 3| = |x - 5|$.

The equation $|x - 3| = |x - 5|$ is true if $x - 3$ and $x - 5$ are equal to each other, or if they are opposites. Now, the opposite of $(x - 5)$ is $-(x - 5) = -1 \cdot (x - 5) = -x + 5$. Here are the two cases:

EQUAL	**OPPOSITES**
$x - 3 = x - 5$	$x - 3 = -(x - 5)$
No solution here.	$x - 3 = -x + 5$
(If we add $(-x)$ we obtain $-3 = -5$)	$x = -x + 8$ Add 3.
	$2x = 8$ Add x.
	$x = 4$ Divide by 2.

Thus, the solution set is $\{4\}$. We can verify this by replacing x by 4 in $|x - 3| = |x - 5|$, obtaining:

$$|4 - 3| = |4 - 5|$$
$$|1| = |-1|$$
$$1 = 1$$

This is a true statement. Thus our solution is correct.

Problem 5   Solve $|x - 5| = |x - 8|$.

NAME

CLASS

SECTION

ANSWERS

A. In Problems 1–16, solve.

1. $6.3x - 8.4 = 16.8$

2. $15.5a + 49.6 = 18.6$

3. $-12.6y - 25.2 = 50.4$

4. $6.4y - 19.2 = 32$

5. $2.1y + 3.5 = 0.7y + 83.3$

6. $2.4x + 3.6 = 0.6x + 5.4$

7. $3.5(x + 3) = 2.1(x + 3) + 4.2$

8. $7.2(3 - t) = 2.4(3 - t) + 4.8$

9. $0.40y + 0.20(32 - y) = 9.60$

10. $0.30x + 0.35(50 - x) = 16$

11. $0.65x + 0.40(50 - x) = 25.375$

12. $0.09y + 0.12(200 - y) = 19.20$

13. $0.06P + 0.08(2000 - P) = 130$

14. $0.15P + 0.10(6000 - P) = 660$

15. $0.30y + 1.80 = 0.20(y + 12)$

16. $0.10x + 4 = 0.30(x + 10)$

B. In Problems 17–46 solve.

17. $|x| = 13$

18. $|y| = 17$

19. $|y| = 2.3$

20. $|x| = 3.7$

21. $|x| = 0$

22. $|y| = -3$

23. $|z| = -4$

24. $|x + 1| = 10$

25. $|x + 7| = 2$

26. $|x + 9| = 3$

27. $|2x - 4| = 8$

28. $|3x - 6| = 9$

1. _____
2. _____
3. _____
4. _____
5. _____
6. _____
7. _____
8. _____
9. _____
10. _____
11. _____
12. _____
13. _____
14. _____
15. _____
16. _____

17. _____
18. _____
19. _____
20. _____
21. _____
22. _____
23. _____
24. _____
25. _____
26. _____
27. _____
28. _____

29. $|5a - 2| = 8$

30. $|6b - 3| = 9$

31. $\left|\dfrac{1}{2}x + 4\right| = 6$

32. $\left|\dfrac{1}{3}x + 2\right| = 7$

33. $\left|\dfrac{2}{3}z - 3\right| = 9$

34. $\left|\dfrac{2}{5}x - 6\right| = 4$

35. $|x + 2| = |x + 4|$

36. $|y + 6| = |y + 2|$

37. $|2y - 4| = |4y + 6|$

38. $|3x - 2| = |6x + 4|$

39. $2|a + 1| - 3 = 9$

40. $2|3a + 1| + 5 = 13$

41. $3|2x + 1| - 4 = -6$

42. $5|x - 1| - 6 = -8$

43. $|x - 4| = |4 - x|$

44. $|2x - 2| = |2 - 2x|$

45. $|5x - 10| = |10 - 5x|$

46. $|6x - 3| = |3 - 6x|$

47. Problems 43–46 were of the form $|a - b| = |b - a|$.
   a. What is the solution of $|a - b| = |b - a|$?
   b. Show that the statement is true when $a = 9$ and $b = 5$.
   c. Show that the statement is true when $a = 0$ and $b = -4$.

48. The statement $|a + b| \leq |a| + |b|$ is called the **triangle inequality.** Show that $|a + b| = |a| + |b|$ when
   a. $a = 7$ and $b = 3$
   b. $a = -5$ and $b = -4$

✓ SKILL CHECKER

Fill in the blank with $<$ or $>$ to make the resulting statement true.

49. **a.** -3 _____ -1
   b. -1.3 _____ -1.4
   c. $\dfrac{1}{3}$ _____ $\dfrac{1}{2}$

50. **a.** $-\dfrac{1}{3}$ _____ $-\dfrac{1}{2}$
   b. $-\dfrac{1}{5}$ _____ $-\dfrac{1}{2}$
   c. $-\dfrac{1}{4}$ _____ $-\dfrac{1}{2}$

29. _____
30. _____
31. _____
32. _____
33. _____
34. _____
35. _____
36. _____
37. _____
38. _____
39. _____
40. _____
41. _____
42. _____
43. _____
44. _____
45. _____
46. _____
47. **a.** _____
   b. _____
   c. _____
48. **a.** _____
   b. _____
49. **a.** _____
   b. _____
   c. _____
50. **a.** _____
   b. _____
   c. _____

2.2 USING YOUR KNOWLEDGE

Do you know what *percent* means? It means "by the hundred." Thus 40 percent, written as 40%, means $\frac{40}{100}$, or 0.40. When we say percent, we can simply write the number in hundredths. For example, 80% = 0.80, 30% = 0.30, 5% = 0.05, and 183% = 1.83. We can use this knowledge and the facts we have learned in this section to solve problems like this one: 60% of the calories in a McDonald's Biscuit with sausage and eggs are fat calories (calories derived from the fat in the food). If there are 351 fat calories in this product, what is the total number of calories in a McDonald's biscuit with sausage and eggs? Let this number be c. Since 60% of c is 351, we have:

$$60\% \cdot c = 351$$
$$0.60c = 351$$
$$60c = 35,100 \quad \text{Multiplying both sides by 100.}$$
$$c = \frac{35,100}{60} = 585 \quad \text{Dividing by 60.}$$

Thus, there are 585 total calories in a McDonald's biscuit with sausage and eggs.

1. Of the total calories in a McDonald's apple pie, 50% are fat calories. If there are 125 fat calories in a McDonald's apple pie, how many total calories are there in the apple pie?

    1. _____

2. Of the total calories in a Big Mac, 55% are fat calories. If 313.5 of the calories in a Big Mac are fat calories, how many total calories are there in a Big Mac?

    2. _____

3. The Burger King Whopper also contains 55% fat calories. If 343.75 of the calories in a Whopper are fat calories, how many total calories are there in a Whopper?

    3. _____

ANSWER (to problem on page 110)

5. $\frac{13}{2}$

2.3 LINEAR INEQUALITIES

REVIEW

Before starting this section, you should know:

1. The procedure used to solve linear equations.
2. How to add, subtract, multiply and divide integers.

OBJECTIVES

You should be able to:

A. Graph linear inequalities.
B. Solve and graph linear inequalities.
C. Use inequalities to write sentences given in words.

Suppose a rental car costs $25 each day and $0.20 per mile. If you drive m miles, the cost C for the day is

$$C = 0.20m + 25$$

Now, suppose your daily cost must be under $50. This means that: $0.20m + 25 < 50$. How many miles can you drive? The expression $0.20m + 25 < 50$ is an example of a linear inequality.

> A **linear inequality** in one variable is an inequality that can be written in the form
>
> $$ax + b < c$$
>
> where a, b, and c are real numbers, and $a \neq 0$.

The inequality $ax + b < c$ is still a linear inequality if the $<$ symbol is replaced by $>$, $\leq$ (less than or equal to), or $\geq$ (greater than or equal to). Thus, $2x + 5 < 8$, $-3x - 7 \geq 9$, $-\frac{1}{2}x + 8 \geq -\frac{3}{4}$, and $0.20m + 25 < 50$ are all linear inequalities. As with linear equations, we *solve* an inequality by finding all the replacements of the variable that make the inequality a true statement. This is done by finding an *equivalent* inequality whose solution is obvious. For example, the inequalities $x + 2 > 5$ and $x > 3$ are equivalent. They are both satisfied by all real numbers greater than 3. We write the *solution set* of $x + 2 > 5$ as

$$\{x \mid x > 3\}$$

Read "the set of all x's such that x is greater than 3."

A. GRAPHING INEQUALITIES

There are infinitely many numbers in the solution set of the inequality $x > 3$. (4, 7.5, $\sqrt{10}$, and $\frac{10}{3}$ are a few of them.) Since we cannot list all these numbers, we show all solutions of $x > 3$ **graphically** by using the number line. This type of representation is called the **graph** of the inequality. The heavy line in the figure is the graph of $x > 3$.

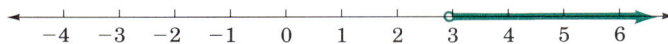

Note that the number 3 is *excluded* from the graph. This fact is shown by drawing a small open circle around the point 3.

EXAMPLE 1 Graph.

a. $x \geq -1$ **b.** $x < -2$

Solution

a. The numbers that satisfy the inequality $x \geq -1$ are the numbers that are *greater than or equal to* -1, that is, the number -1 and all the numbers to the right of -1. (Note that $\geq$ points to the right.) The graph is shown in the following figure.

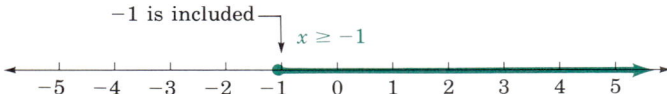

b. The numbers that satisfy the inequality $x < -2$ are the numbers that are *less than* -2, that is, the numbers to the *left* but *not including* -2 (note that $<$ points to the left). The graph of these points is shown in the following figure.

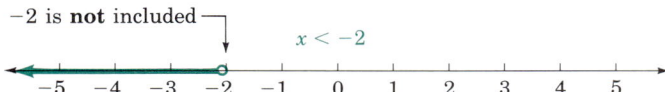

B. SOLVING LINEAR INEQUALITIES

We mentioned that the inequalities $x + 2 > 5$ and $x > 3$ are equivalent. This is because we added (-2) to both sides of $x + 2 > 5$, obtaining the equivalent inequality

$$x + 2 + (-2) > 5 + (-2)$$

or

$$x > 3$$

We used the first of the following properties:

PROPERTIES OF INEQUALITIES
If C is a real number, then the following inequalities are all *equivalent*:

	$A < B$
Add C.	$A + C < B + C$
Subtract C.	$A - C < B - C$

Problem 1 Graph.
a. $x \geq -2$

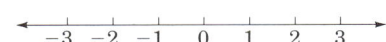

b. $x < 1$

Now, consider the inequality $2x - 3 < x + 1$. To solve this inequality, we first put all the variables by themselves on one side. Thus we proceed as follows:

$$2x - 3 < x + 1 \qquad \text{Given.}$$
$$2x - 3 + 3 < x + 1 + 3 \qquad \text{Add 3.}$$
$$2x < x + 4 \qquad \text{Simplify.}$$
$$2x - x < x - x + 4 \qquad \text{Add } -x \text{ (or subtract } x).$$
$$x < 4 \qquad \text{Simplify.}$$

The solution set is $\{x \mid x < 4\}$.

The graph of this inequality is shown in the figure in the margin. You can check that this solution is correct, by selecting a number from the graph (say 0) and replacing x with the number 0 in the original inequality obtaining $2(0) - 3 < 0 + 1$ or $-1 < 3$, a true statement. Of course, this is only a "partial" check since we did not try *all* the numbers in the graph.

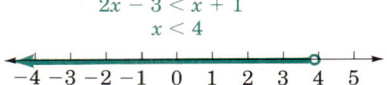

EXAMPLE 2   Solve and graph.

a. $3x - 2 < 2(x - 2)$ **b.** $4(x + 1) \geq 3x + 7$

Solution

a.
$$3x - 2 < 2(x - 2) \qquad \text{Given.}$$
$$3x - 2 < 2x - 4 \qquad \text{Simplify.}$$
$$3x - 2 + 2 < 2x - 4 + 2 \qquad \text{Add 2.}$$
$$3x < 2x - 2 \qquad \text{Simplify.}$$
$$3x - 2x < 2x - 2x - 2 \qquad \text{Add } -2x \text{ (or subtract } 2x).$$
$$x < -2 \qquad \text{Simplify.}$$

The solution set is $\{x \mid x < 2\}$ and the graph is shown in the following figure.

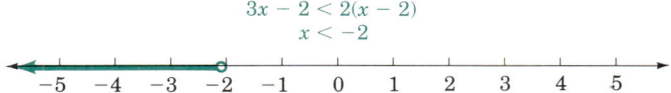

b.
$$4(x + 1) \geq 3x + 7 \qquad \text{Given.}$$
$$4x + 4 \geq 3x + 7 \qquad \text{Simplify.}$$
$$4x + 4 - 4 \geq 3x + 7 - 4 \qquad \text{Add } -4 \text{ (or subtract 4).}$$
$$4x \geq 3x + 3 \qquad \text{Simplify.}$$
$$4x - 3x \geq 3x - 3x + 3 \qquad \text{Add } -3x \text{ (or subtract } 3x).$$
$$x \geq 3 \qquad \text{Simplify.}$$

The solution set is $\{x \mid x \geq 3\}$. The graph of this inequality appears in the following figure.

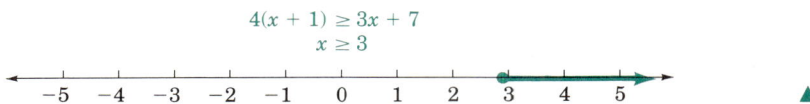

Before we state the multiplication and division properties of inequalities, let us see what happens if we multiply an inequality by a

Problem 2   Solve and graph.
a. $4x - 7 < 3(x - 2)$
b. $3(x + 1) \geq 2x + 5$

positive number. Consider these true inequalities:

$$1 < 3 \qquad\qquad -1 < 3 \qquad\qquad -3 < -1$$
$$4 \cdot 1 < 4 \cdot 3 \qquad 4 \cdot (-1) < 4 \cdot 3 \qquad 4 \cdot (-3) < 4 \cdot (-1) \quad \text{Multiply by 4.}$$
$$4 < 12 \qquad\qquad -4 < 12 \qquad\qquad -12 < -4$$

Note that the resulting inequalities are all *true* and that the inequality symbol points in the *same* direction as the original. We say that multiplication by a *positive* number *preserves the sense*, or direction, of the inequality.

Now, let us multiply the original inequality by a *negative* number, say -4. We have

$$1 < 3 \qquad\qquad -1 < 3 \qquad\qquad -3 < -1$$
$$-4 \cdot 1 \ ? \ -4 \cdot 3 \qquad -4 \cdot (-1) \ ? \ -4 \cdot 3 \qquad -4 \cdot (-3) \ ? \ -4 \cdot (-1) \quad \text{Multiply}$$
$$-4 > -12 \qquad\qquad 4 > -12 \qquad\qquad 12 > 4 \qquad\qquad \text{by } -4$$

This time, however, we had to reverse the direction of the inequalities. Thus, multiplying both sides of an inequality by a *negative* number *reverses the sense* of the inequality. Note that since division is defined in terms of multiplication, these two properties apply to division as well. They are stated as follows.

PROPERTIES OF INEQUALITIES

If C is a real number, then the following inequalities are all equivalent:

$$A < B$$
$$A \cdot C < B \cdot C, \quad \text{if } C \text{ is positive } (C > 0)$$
$$\frac{A}{C} < \frac{B}{C}, \quad \text{if } C \text{ is positive } (C > 0)$$
$$A \cdot C > B \cdot C, \quad \text{if } C \text{ is negative } (C < 0)$$
$$\frac{A}{C} > \frac{B}{C}, \quad \text{if } C \text{ is negative } (C < 0)$$

Note that we can still multiply or divide both sides of an inequality by any nonzero number *as long as we remember to reverse the sense of the inequality if the number is negative.* Thus to solve $-2x > 4$, we must divide both sides by -2 (or, equivalently, multiply by $-\frac{1}{2}$). When doing this, we must remember to reverse the sense (direction) of the inequality.

$$-2x > 4 \qquad \text{Given.}$$
$$\frac{-2x}{-2} < \frac{4}{-2} \quad \text{Multiply by } -\tfrac{1}{2} \text{ (or divide by } -2\text{).}$$
$$x < -2 \qquad \text{Direction reversed}$$

EXAMPLE 3 Solve and graph.

a. $4(x - 1) \leq 6x + 2$ **b.** $2x + 9 \geq 5x + 3$

Problem 3 Solve and graph.
a. $3(x - 1) \leq 5x + 1$
b. $3x + 11 \geq 5x + 5$

Solution

a. We follow the six-step procedure given on page 100.

Given.	$4(x - 1) \leq 6x + 2$

1. There are no fractions.

2. Remove parentheses. $\qquad 4x - 4 \leq 6x + 2$

3. Add 4 to both sides. $\qquad 4x - 4 + 4 \leq 6x + 2 + 4$

$$4x \leq 6x + 6$$

4. Add $-6x$ (or subtract $6x$). $\qquad 4x + (-6x) \leq 6x + (-6x) + 6$

$$-2x \leq 6$$

5. Multiply by $-\dfrac{1}{2}$ (or divide by -2) and *reverse* the sense of the inequality.

$$\frac{-2x}{-2} \geq \frac{6}{-2}$$

$$x \geq -3$$

Thus, the solution set is $\{x \mid x \geq -3\}$. The graph is shown.

If we wish to avoid multiplying (or dividing) by negative numbers, we can add -6 in step 3 and $-4x$ in step 4, obtaining $-6 \leq 2x$. Then, dividing by 2, we get $-3 \leq x$, which is equivalent to $x \geq -3$.

b. Again, we isolate the x's on one side. This time, let us avoid multiplying or dividing by negative numbers. We do this by noting that there are more x's on the right side of the inequality. Thus, we isolate the x's on the right.

Given.	$2x + 9 \geq 5x + 3$

1. There are no fractions.

2. The inequality is simplified.

3. Add -3. $\qquad 2x + 9 + (-3) \geq 5x + 3 + (-3)$

$$2x + 6 \geq 5x$$

4. Add $-2x$. $\qquad (-2x) + 2x + 6 \geq 5x + (-2x)$

$$6 \geq 3x$$

5. Multiply by $\dfrac{1}{3}$ (or divide by 3).

$$\frac{6}{3} \geq \frac{3x}{3}$$

$$2 \geq x$$

We can write this as $x \leq 2$, so the solution set is $\{x \mid x \leq 2\}$. The graph is shown next.

 ▲

The last two inequalities contained no fractions. If fractions are present, we clear them by multiplying both sides of the inequality by the LCD of the fractions involved, as shown next.

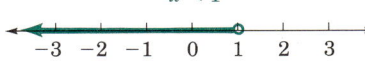
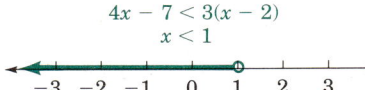
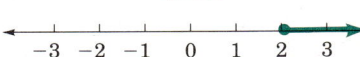

EXAMPLE 4   Solve and graph $\dfrac{x}{4} - \dfrac{x}{6} > \dfrac{x-3}{6}$

Solution

Given.

$$\frac{x}{4} - \frac{x}{6} > \frac{x-3}{6}$$

1. Multiply both sides by 12 the LCD of 4 and 6.

$$12\left(\frac{x}{4} - \frac{x}{6}\right) > 12\left(\frac{x-3}{6}\right)$$

$$12\left(\frac{x}{4}\right) - 12\frac{x}{6} > 12\left(\frac{x-3}{6}\right)$$

2. Simplify.

$$3x - 2x > 2(x-3)$$
$$x > 2x - 6$$

3. There are no numbers on the left.

4. Add $-2x$ (or subtract $2x$).

$$x + (-2x) > 2x + (-2x) - 6$$
$$-x > -6$$

5. Multiply by -1 (or divide by the coefficient of x, -1), and reverse the sense of the inequality.

$$\frac{-x}{-1} < \frac{-6}{-1}$$
$$x < 6$$

The solution set is $\{x \mid x < 6\}$, and the graph is shown.

C. APPLICATIONS

At the beginning of this section, we mentioned that the cost of renting a car must be under $50. We then translated this phrase by writing "< 50."

Here are some other words and their translation using inequalities.

WORDS	TRANSLATION	IN SYMBOLS
x is at least 10	x is 10 or more	$x \ge 10$
x is at most 20	x is 20 or less	$x \le 20$
x is no more than 30	x is 30 or less	$x \le 30$
x is no less than 40	x is 40 or more	$x \ge 40$

EXAMPLE 5   Translate into an inequality.

a. The height h of a human (in feet) has never been known to exceed 9 ft.

b. The weight w of a human is at most 1400 lb.

c. The number n of puppies born in a single litter is no more than 23.

d. The cat population p in the United States (1989) is at least 58 million.

Solution

a. $h \le 9$ (Robert Wadlow was the tallest at 8 ft 11.1 in.)
b. $w \le 1400$ (Jon Browner Minnoch weighed 1400 lb.)
c. $n \le 23$ (Lena, a foxhound, had 23 live puppies June 9, 1944.)
d. $p \ge 58$ (million)

Problem 4   Solve and graph
$$\frac{x}{3} - \frac{x}{4} > \frac{x-4}{4}$$

Problem 5   Translate into an inequality.
a. x does not exceed 23.
b. y is at most 180.
c. z is no more than 10.
d. p is at least 45.

EXERCISE 2.3

A. In Problems 1–10 graph the inequality.

1. $x > 3$

2. $x > 4$

3. $x \leq -3$

4. $x \leq -4$

5. $2x \geq 6$

6. $3x \geq 8$

7. $-3x \leq 3$

8. $-5x \leq 10$

9. $-4x \geq -8$

10. $-6x \geq -12$

B. Solve and graph.

11. $3x + 6 \leq 9$

12. $4y - 9 \leq 3$

13. $-2y - 4 \geq -10$

14. $-2z - 2 \geq -10$

15. $-3x + 1 \leq -14$

16. $-3x + 4 \leq -8$

17. $3a + 6 \leq a + 10$

18. $4b + 4 \leq b + 7$

19. $7z - 12 \geq 8z - 8$

20. $3z + 7 \geq 5z + 19$

21. $10 - 5x \leq 7 - 8x$

22. $6 - 4y \leq -14 + 6y$

ANSWERS

1. _____

2. _____

3. _____

4. _____

5. _____

6. _____

7. _____

8. _____

9. _____

10. _____

11. _____

12. _____

13. _____

14. _____

15. _____

16. _____

17. _____

18. _____

19. _____

20. _____

21. _____

22. _____

ANSWER (to problem on page 118)

3. **a.**
$$3(x - 1) \leq 5x + 1$$
$$x \geq -2$$

   b.
$$3x + 11 \geq 5x + 5$$
$$x \leq 3$$

23. $5(x + 2) \leq 3(x + 3) + 1$

24. $5(4 - 3x) < 7(3 - 4x) + 12$

25. $-4x + \dfrac{1}{2} \geq 4x + \dfrac{8}{5}$

26. $12x + \dfrac{2}{7} \geq 4x - \dfrac{4}{7}$

27. $\dfrac{x}{5} - \dfrac{x}{4} \leq 1$

28. $\dfrac{x}{3} - \dfrac{x}{2} \leq 1$

29. $\dfrac{7x + 2}{6} + \dfrac{1}{2} \geq \dfrac{3}{4}x$

30. $\dfrac{8x - 23}{6} + \dfrac{1}{3} \geq \dfrac{5}{2}x$

C. In Problems 31–35, translate into an inequality.

31. The height h (in feet) of any mountain does not exceed that of Mt. Everest, 29,028 ft.

32. The number e of possible eclipses in a year is at most 7.

33. The number e of possible eclipses in a year is at least 2.

34. The altitude h (in feet) attained by the first liquid-fueled rocket was no more than 41 ft.

35. There are no less than 4×10^{25} nematode sea worms in the world. (Let n be the number of nematodes.)

36. There are at least 713 million people (p) that speak Mandarin Chinese.

✓ **SKILL CHECKER**

Fill in the blank with $<$ or $>$ to make the result a true statement.

37. $-7 \underline{\hspace{1cm}} -8$

38. $\dfrac{1}{3} \underline{\hspace{1cm}} \dfrac{1}{2}$

39. $0.34 \underline{\hspace{1cm}} 0.342$

40. $-0.234 \underline{\hspace{1cm}} -0.233$

Find.

41. $|-9|$

42. $\left| -\dfrac{1}{5} \right|$

43. $|-0.34|$

44. $|\sqrt{2}|$

23. _____

24. _____

25. _____

26. _____

27. _____

28. _____

29. _____

30. _____

31. _____

32. _____

33. _____

34. _____

35. _____

36. _____

37. _____

38. _____

39. _____

40. _____

41. _____

42. _____

43. _____

44. _____

2.3 USING YOUR KNOWLEDGE

The graphs of the inequalities $x \geq -1$ and $x < -2$ of Example 1 are examples of **intervals** on the number line. A notation called **interval notation** is used for writing these intervals. Thus, the interval $\{x \mid x \geq -1\}$ is written as $[-1, \infty)$. The symbol ∞ does not represent a number; it is merely used to indicate that the interval

ANSWERS (to problems on page 120)

$x < 6$

4.

$$-6 \quad -4 \quad -2 \quad 0 \quad 2 \quad 4 \quad 6$$

5. a. $x \leq 23$ **b.** $y \leq 180$ **c.** $z \leq 10$

d. $p \geq 45$

includes all numbers greater than or equal to -1. The square bracket indicates that the -1 is part of the interval. The interval is called a **closed** interval (if -1 were excluded, we would write $(-1, \infty)$, an **open** interval). The interval notation for $\{x \mid x < -2\}$ is $(-\infty, -2)$. Here the $-\infty$ indicates that the interval includes all numbers less than -2. Note that -2 is *not* included in the interval; it is an open interval. Here are some types of open intervals, their notation, and their graphs.

SET NOTATION	INTERVAL NOTATION	GRAPH
$\{x \mid a < x\}$	$(a, +\infty)$	a
$\{x \mid x < b\}$	$(-\infty, b)$	b
$\{x \mid a \le x\}$	$[a, +\infty)$	a
$\{x \mid x \le b\}$	$(-\infty, b]$	b

Use interval notation to write.

1. $\{x \mid x \ge -4\}$

2. $\{x \mid x < 5\}$

3. $\{x \mid x \le -6\}$

4. $\{x \mid x > 9\}$

1. _____

2. _____

3. _____

4. _____

2.4 COMPOUND AND ABSOLUTE VALUE INEQUALITIES

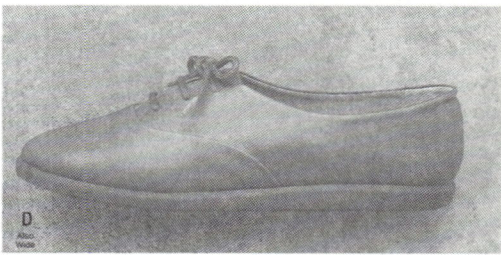

Cut $3 to $5

OBJECTIVES

REVIEW

Before starting this section, you should know:

1. How to graph an inequality.
2. How to solve linear inequalities.
3. How to find the absolute value of a number.

OBJECTIVES

You should know:

A. How to solve compound inequalities using the words *and* or *or*.
B. How to solve inequalities involving absolute values.

Do you want to buy some shoes? You can get them from this catalog and cut (save) $3 to $5. If s is the amount saved, then we can write $3 \le s \le 5$. Read "3 is less than or equal to s and s is less than or equal to 5." If s is a real number, the graph of $3 \le s \le 5$ is

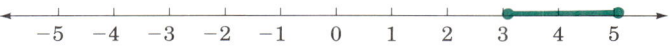

A. COMPOUND INEQUALITIES

The inequality $3 \le s \le 5$ is a **compound** inequality because it is equivalent to two other inequalities:

$$3 \le s \quad and \quad s \le 5$$

Another way of graphing this inequality is to use the idea of *intersection*.

> If A and B are two sets, the **intersection** of A and B, denoted by $A \cap B$, is the set of elements that are in both A *and* B.

The graph of $3 \le s$ is

The graph of $s \le 5$ is

The intersection $A \cap B$ is

Note that the intersection is the part the two graphs have in common. It is the graph of $\{s \mid 3 \le s \text{ and } s \le 5\}$.

EXAMPLE 1 Graph $\{x \mid x > -2 \text{ and } x < 1\}$.

Solution The graph of $x > -2$ is

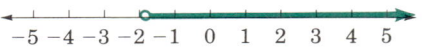

Problem 1 Graph $\{x \mid x > -1 \text{ and } x < 2\}$.

The graph of $x < 1$ is

The *intersection* is

$$\text{number line graph: } -5 \ -4 \ -3 \ -2 \ -1 \ 0 \ 1 \ 2 \ 3 \ 4 \ 5$$

▲

There is another type of compound inequality, which uses the word *or* (instead of *and*) for its connective. The graph of such an inequality is based on the idea of *union*.

> If A and B are sets, the **union** of A and B, denoted by $A \cup B$, is the set of elements that are in either A *or* B.

Now, suppose we wish to graph $\{x \mid x < -2 \text{ or } x > 1\}$. We proceed as before but graph the *union*. The graph of $x < -2$ is

$$\text{number line graph: } -5 \ -4 \ -3 \ -2 \ -1 \ 0 \ 1 \ 2 \ 3 \ 4 \ 5$$

The graph of $x > 1$ is

$$\text{number line graph: } -5 \ -4 \ -3 \ -2 \ -1 \ 0 \ 1 \ 2 \ 3 \ 4 \ 5$$

The union is

$$\text{number line graph: } -5 \ -4 \ -3 \ -2 \ -1 \ 0 \ 1 \ 2 \ 3 \ 4 \ 5$$

EXAMPLE 2 Graph $\{x \mid x < -3 \text{ or } x \geq 1\}$.

Solution We do each of the graphs separately and then take the *union* of the two graphs. The graph of $x < -3$ is

$$\text{number line graph: } -5 \ -4 \ -3 \ -2 \ -1 \ 0 \ 1 \ 2 \ 3 \ 4 \ 5$$

The graph of $x \geq 1$ is

$$\text{number line graph: } -5 \ -4 \ -3 \ -2 \ -1 \ 0 \ 1 \ 2 \ 3 \ 4 \ 5$$

The union is

$$\text{number line graph: } -5 \ -4 \ -3 \ -2 \ -1 \ 0 \ 1 \ 2 \ 3 \ 4 \ 5$$

▲

Now, suppose you wish to solve the inequality $2x + 7 > 9$ *and* $x + 3 \leq -1$. We can solve each inequality, obtaining:

$2x + 7 < 9$	Given.	and	$x + 3 \geq -1$	
$2x < 2$	Add -7.		$x \geq -4$	Add -3.
$x < 1$	Divide by 2.			

Problem 2 Graph $\{x \mid x < -2 \text{ or } x \geq 3\}$.

The graph of $x < 1$ is

The graph of $x \geq -4$ is

The intersection is

Note that the intersection consists of all points such that $-4 \leq x$ *and* $x < 1$, which we can write as $-4 \leq x < 1$. A compound inequality that can be written in the form $a < x$ *and* $x < b$ (a and b real numbers) can be expressed more concisely as

   $a < x < b$

The graph of $a < x < b$ is the solution set of $a < x$ and $x < b$. We use this idea to solve compound inequalities in the next example.

EXAMPLE 3   Solve and graph.

a. $1 < x$ and $x < 3$ **b.** $5 \geq -x$ and $x \leq -3$
c. $x + 1 \leq 5$ and $-2x < 6$

Solution

a. $1 < x$ and $x < 3$ is written as $1 < x < 3$. The graph is

b. Since we wish to write this inequality in the form $a < x < b$, we multiply the first inequality by -1, obtaining

   $-5 \leq x$ and $x \leq -3$

that is

   $-5 \leq x \leq -3$

The graph is

c. We solve the first inequality by subtracting 1 from both sides. We then have

   $x \leq 4$ and $-2x < 6$

We then divide the second inequality by -2:

   $x \leq 4$ and $x > -3$

Rearranging these inequalities gives

   $-3 < x$ and $x \leq 4$

   $-3 < x \leq 4$

Problem 3   Solve and graph.
a. $2 < x$ and $x < 4$
b. $3 \geq -x$ and $x \leq -1$
c. $x + 2 \leq 6$ and $-3x < 9$

The graph is

▲

Now, suppose we want to solve $-5 < 2x + 3 \leq 9$. We can solve this inequality by using the inequality properties we studied, keeping in mind that if we do any operations on the center expression $(2x + 3)$ we must do the same operation on the outside expressions. As before, we wish to have the x by itself (in the middle). We start by adding (-3) to all three parts. We then have:

$-5 <$	$2x + 3$	≤ 9	Given.
$-5 + (-3) <$	$2x + 3 + (-3)$	$\leq 9 + (-3)$	Add (-3).
$-8 <$	$2x$	≤ 6	Simplify.
$\dfrac{-8}{2} <$	$\dfrac{2x}{2}$	$\leq \dfrac{6}{2}$	Multiply by $\dfrac{1}{2}$ (or divide by 2).
$-4 <$	x	≤ 3	Simplify.

The graph is

EXAMPLE 4 Solve and graph $-2 \leq -3x - 5 < 4$.

Solution We start by adding 5 to each part.

$-2 + 5 \leq$	$-3x - 5 + 5$	$< 4 + 5$	
$3 \leq$	$-3x$	< 9	Simplify.
$\dfrac{3}{-3} \geq$	$\dfrac{-3x}{-3}$	$> \dfrac{9}{-3}$	Divide by -3 and *reverse* the sense (direction).
$-1 \geq$	x	> -3	Simplify.
$-3 <$	x	≤ -1	Rewrite.

The graph is

B. ABSOLUTE VALUE INEQUALITIES

As you recall, $|x| = 2$ means that the distance from 0 to x on the number line is 2 units. What would $|x| < 2$ and $|x| > 2$ mean? Here are the statements, their translations, their graphs, and the meaning in terms of the compound inequalities we just studied.

STATEMENT TRANSLATION

$|x| = 2$ x is exactly 2 units from 0.

$|x| < 2$ x is less than 2 units from 0.

Problem 4 Solve and graph
$-4 \leq -2x - 2 < 6$.

$|x| > 2$ x is more than 2 units from 0.

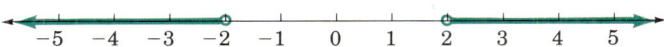

As you can see, the absolute value inequality $|x| < 2$ is equivalent to $-2 < x < 2$. In general, the following holds.

$|x| < a$ is equivalent to $-a < x < a$, $a > 0$.
The same relationship applies if $\leq$ replaces $<$.

$|x| < 4$ is equivalent to $-4 < x < 4$
$|y| \leq 2.5$ is equivalent to $-2.5 \leq y \leq 2.5$
$|z| < \dfrac{3}{4}$ is equivalent to $-\dfrac{3}{4} < z < \dfrac{3}{4}$

What does $|x + 1| < 2$ mean? If we use $x + 1$ instead of x and 2 instead of a, then $|x| < a$ is equivalent to $-a < x < a$ becomes $|x + 1| < 2$ is equivalent to $-2 < x + 1 < 2$.

$-2 + (-1) < x + 1 + (-1) < 2 + (-1)$ Add -1.
$\qquad -3 < \qquad x \qquad < 1$ Simplify.

The graph is shown.

EXAMPLE 5   Solve and graph $|3x - 4| \leq 8$.

Solution $|3x + 4| \leq 8$ is equivalent to $-8 \leq 3x + 4 \leq 8$.

$-8 + (-4) \leq 3x + 4 + (-4) \leq 8 + (-4)$ Add -4.
$\qquad -12 \leq \qquad 3x \qquad \leq 4$ Simplify.
$\qquad \dfrac{-12}{3} \leq \qquad \dfrac{3x}{3} \qquad \leq \dfrac{4}{3}$ Divide by 3.
$\qquad -4 \leq \qquad x \qquad \leq \dfrac{4}{3}$ Simplify.

The graph is shown.

Problem 5   Solve and graph $|3x - 2| \leq 5$.

Let us look at $|x| > 2$. Its graph consists of all points such that $x < -2$ or $x > 2$. Thus $|x| > 2$ is equivalent to $x < -2$ or $x > 2$. If we generalize this result, we obtain:

$|x| > a$ is equivalent to $x < -a$ or $x > a$, $a > 0$.
The same relationship applies if $\geq$ replaces $>$.

$|x| > 5 \quad$ is equivalent to $\quad x < -5$ or $x > 5$

$|x| > 3.4 \quad$ is equivalent to $\quad x < -3.4$ or $x > 3.4$

$|x| \geq \dfrac{1}{5} \quad$ is equivalent to $\quad x \leq -\dfrac{1}{5}$ or $x \geq \dfrac{1}{5}$

What does $|x - 1| > 3$ mean? It means $x - 1 < -3$ or $x - 1 > 3$.

$x < -2 \quad$ or $\quad x > 4 \quad$ Solve each inequality by adding 1.

The graph is shown.

EXAMPLE 6 Solve and graph $|-2x + 3| > 1$.

Solution The inequality $|-2x + 3| > 1$ is equivalent to

$$-2x + 3 < -1 \qquad \text{or} \qquad -2x + 3 > 1$$

Now add (-3) on both sides

$$-2x + 3 + (-3) < -1 + (-3) \quad \text{or} \quad -2x + 3 + (-3) > 1 + (-3)$$
$$-2x < -4 \qquad \text{or} \qquad -2x > -2 \qquad \text{Simplify}$$

Next divide by -2 and reverse the inequality signs

$$\dfrac{-2x}{-2} > \dfrac{-4}{-2} \qquad \text{or} \qquad \dfrac{-2x}{-2} < \dfrac{-2}{-2}$$
$$x > 2 \qquad \text{or} \qquad x < 1 \qquad \text{Simplify}$$

The graph is shown.

▲

Since the absolute value of a number is always positive, inequalities of the form $|x| < a$, where a is negative, will have no solution. Thus $|x| < -2$ and $|-x - 7| < -3$ have no solution. On the other hand, inequalities such as $|x| > a$, where a is negative, are *always* true because $|x|$ is always positive. Any real number will satisfy this inequality; the solution set is the set of real numbers. Thus $|x| > -9$ and $|-3x - 8| > -10$ are always true and have the real numbers as their solution set.

Problem 6 Solve and graph $|-2x + 1| > 5$.

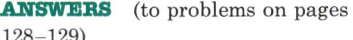

EXERCISE 2.4

A. In Problems 1–30 solve and graph.

1. $\{x \mid x \le 4 \quad \text{and} \quad x \ge -2\}$
2. $\{x \mid x > 0 \quad \text{and} \quad x \le 5\}$
3. $\{x \mid x + 1 \le 7 \quad \text{and} \quad x > 2\}$
4. $\{x \mid x > -5 \quad \text{and} \quad x - 1 < 0\}$
5. $\{x \mid 2x - 1 > 1 \quad \text{and} \quad x + 1 < 4\}$
6. $\{x \mid x - 1 < 1 \quad \text{or} \quad 3x - 1 > 11\}$
7. $\{x \mid x < -5 \quad \text{or} \quad x > 5\}$
8. $\{x \mid x \ge 0 \quad \text{or} \quad x < -2\}$
9. $\{x \mid x + 1 \ge 2 \quad \text{and} \quad x \le 4\}$
10. $\{x \mid x \le 5 \quad \text{and} \quad x > -1\}$
11. $\{x \mid x < 3 \quad \text{and} \quad -x < -2\}$
12. $\{x \mid -x < 5 \quad \text{and} \quad x < 2\}$
13. $\{x \mid x + 1 < 4 \quad \text{and} \quad -x < -1\}$
14. $\{x \mid x - 2 < 1 \quad \text{and} \quad -x < 2\}$
15. $\{x \mid x - 2 < 3 \quad \text{and} \quad 2 > -x\}$
16. $\{x \mid x - 3 < 1 \quad \text{and} \quad 1 > -x\}$
17. $\{x \mid x + 2 < 3 \quad \text{and} \quad -4 < x + 1\}$
18. $\{x \mid x + 4 < 5 \quad \text{and} \quad -1 < x + 2\}$
19. $\{x \mid x - 1 > 2 \quad \text{and} \quad x + 7 < 11\}$
20. $\{x \mid x - 2 > 1 \quad \text{and} \quad -x > -5\}$
21. $-3 < x - 1 < 3$
22. $-4 < x + 1 < 4$
23. $-8 < 2y + 4 < 6$
24. $-2 \le 3x + 1 \le 7$
25. $4 \le 3y - 8 \le 10$
26. $3 \le 4z + 3 \le 5$
27. $-1 < \dfrac{x}{2} < 2$
28. $-2 < \dfrac{y}{2} < 1$
29. $2 < 4 + \dfrac{2}{3}a < 6$
30. $1 < 5 + \dfrac{4}{5}b < 9$

B. In Problems 31–62 solve and graph.

31. $|x| < 4$

32. $|y| \leq 1.5$

33. $|z| \leq 2.4$

34. $|x| - 3 < 1$

35. $|a| - 2 \leq 2$

36. $|b| + 2 \leq 5$

37. $|x - 1| < 2$

38. $|x - 3| \leq 1$

39. $|x + 3| < -2$

40. $|2x - 3| < -4$

41. $|2x + 3| \leq 1$

42. $|3x + 2| < 5$

43. $|4x + 2| - 4 < 2$

44. $|3x + 3| - 2 < 4$

45. $|x| > 2$

46. $|y| \geq 2.5$

47. $|z| \geq 1.4$

48. $|x| - 2 > 1$

49. $|a| - 1 \geq 2$

50. $|b| + 1 \geq 5$

51. $|x - 1| > 1$

52. $|x - 3| \geq 2$

53. $|x + 3| > -1$

54. $|2x - 3| > -1$

55. $|2x + 3| \geq 1$

56. $|3x + 2| > 5$

57. $|3 - 4x| > 7$

58. $|5 - 3x| > 11$

59. $|4x + 2| - 4 \geq 2$

60. $|3x + 3| - 2 \geq 4$

61. $\left|2 - \dfrac{1}{2}a\right| > 1$

62. $\left|1 - \dfrac{2}{3}b\right| > 2$

31.
32.
33.
34.
35.
36.
37.
38.
39.
40.
41.
42.
43.
44.
45.
46.
47.
48.
49.
50.
51.
52.
53.
54.
55.
56.
57.
58.
59.
60.
61.
62.

ANSWER (to problem on page 130)

6. $x < -2$ or $x > 3$

✓ **SKILL CHECKER**

Solve.

63. $3x + 2(x - 1) = 8$

64. $2x + 3(x - 1) = 12$

65. $-3(x + 1) + 5x = 2$

66. $-4(x + 1) - 6x = 8$

63. _____

64. _____

65. _____

66. _____

2.4 USING YOUR KNOWLEDGE

So far, all the inequalities we have considered have been written symbolically. Use your knowledge of the words we used at the end of Section 2.3 and translate absolute value inequalities from words to symbols; then graph the result.

1. The absolute value of a number x is less than 5.

2. The absolute value of a number y is more than 3.

3. If the absolute value of a number z is found, the result is at least 3.

4. The absolute value of the sum x plus 1 is at most 3.

1. _____

2. _____

3. _____

4. _____

2.5 FORMULAS

TOLLS

2	AXLES	.35
3	AXLES	.50
	EACH ADDITIONAL AXLE	.25

FLORIDA DEPARTMENT OF TRANSPORTATION

OBJECTIVES

REVIEW

Before starting this section, you should know:

1. How to solve linear equations.
2. How to evaluate an expression.

OBJECTIVE

You should be able to solve a formula for a specified variable.

If your truck has 5 axles, how much do you have to pay? The cost C is given by

$$C = 0.50 + 0.25(n - 3)$$

For a truck with 5 axles, $n = 5$ and the answer is

$$C = 0.50 + 0.25(5 - 3)$$
$$= 0.50 + 0.25(2)$$
$$= 0.50 + 0.50$$
$$= \$1.00$$

Many problems in algebra can be solved if the correct formula is used. Now, suppose a trucker pays \$1.25 for the toll. How many axles did the truck have? Here, we are interested in n in the equation $1.25 = 0.50 + 0.25(n - 3)$. Thus we are *solving for a specified variable*. The steps that we use are similar to those we used in solving linear equations. To remind you that we are solving for n, the n is in color. Here is the procedure.

$1.25 = 0.50 + 0.25(n - 3)$	Given.
$1.25 = 0.50 + 0.25n - 0.75$	Simplify.
$1.25 = -0.25 + 0.25n$	Combine like terms.
$0.25 + 1.25 = -0.25 + 0.25 + 0.25n$	Add 0.25.
$1.50 = 0.25n$	Simplify.
$\dfrac{1.50}{0.25} = n$	Divide by 0.25.
$6 = n$	

Thus the truck had 6 axles.

EXAMPLE 1 Anthropologists know how to estimate the height of a man (in centimeters) by using only a bone as a clue. They use the formula

$$H = 2.89h + 70.64$$

Height of the man   Length of the humerus

Problem 1 The comparable formula for a woman is $H = 2.75h + 71.48$.
a. Solve for h.
b. A woman is 126.48 cm tall. How long is her humerus?

a. Solve for h.

b. A man is 157.34 cm tall. How long is his humerus?

Solution

a. We have to solve for h, that is, isolate the h on one side of the equation. Given

$$H = 2.89h + 70.64$$
$$H + (-70.64) = 2.89h + 70.64 + (-70.64) \quad \text{Add } -70.64.$$
$$H - 70.64 = 2.89h \quad \text{Simplify.}$$
$$\frac{H - 70.64}{2.89} = h \quad \text{Divide by 2.89}$$

b. Substitute 157.34 for H to get

$$h = \frac{157.34 - 70.64}{2.89} = \frac{86.7}{2.89} = 30$$

Thus the man's humerus is 30 cm long.

EXAMPLE 2 The formula for converting degrees Fahrenheit (°F) to degrees Celsius (°C) is

$$C = \frac{5}{9}(F - 32)$$

a. Solve for F.

b. If the temperature is 35°C, what is the equivalent Fahrenheit temperature?

Solution

a. We use the procedure for solving linear equations given on page 100.

Given. $\qquad\qquad\qquad C = \dfrac{5}{9}(F - 32)$

1. Clear fractions by multiplying by 9 and simplify.
$$9 \cdot C = 9 \cdot \frac{5}{9}(F - 32)$$
$$9C = 5(F - 32)$$

2. Remove parentheses.
$$9C = 5F - 180$$

3. Add 180.
$$9C + 180 = 5F$$

4. Divide by 5.
$$\frac{9C + 180}{5} = F$$

b. Substitute 35 for C.
$$\frac{9 \cdot 35 + 180}{5} = \frac{495}{5} = 99$$

Thus the temperature was 99°F. ▲

Many of the formulas we encounter in algebra come from geometry. In geometry, the distance around a polygon (a figure) is called the **perimeter** and the number of square units occupied by the figure is its **area**. Here are the perimeters and areas of some geometric figures.

Problem 2

a. Solve for a in the formula $A = \frac{1}{2}h(a + b)$.

b. If $A = 40$, $h = 10$, and $b = 5$, what is a?

NAME	SHAPE	PERIMETER	AREA
Square		$P = 4S$	$A = S^2$
Rectangle		$P = 2L + 2W$	$A = LW$
Triangle		$P = s_1 + s_2 + b$	$A = \dfrac{1}{2}bh$
Circle		$C = 2\pi r$	$A = \pi r^2$

Note that $2r = d$ and $\pi \approx 3.14$. C is the circumference.

EXAMPLE 3 The perimeter P of a rectangle is given by $P = 2L + 2W$.

a. Solve for W.

b. The largest poster ever made was a rectangular greeting card 166 ft long and with a perimeter of 458.50 ft. How wide was this poster?

Solution

a.
$$P = 2L + 2W \qquad \text{Given.}$$
$$P + (-2L) = 2L + (-2L) + 2W \qquad \text{Add } -2L.$$
$$P - 2L = 2W \qquad \text{Simplify.}$$
$$\frac{P - 2L}{2} = W \qquad \text{Divide by 2.}$$

b. Since the perimeter P was 458.50 ft and the length L was 166 ft, we substitute 458.50 for P and 166 for L, obtaining:

$$W = \frac{458.50 - 2 \cdot 166}{2}$$

$$= \frac{458.50 - 332}{2} = 63.25 \text{ ft}$$

Thus the poster was 63.25 ft wide.

Problem 3
a. Solve for L in the formula $P = 2L + 2W$.
b. The Mona Lisa, a painting by Leonardo da Vinci, has a perimeter of 102.8 in. and is 20.9 in. wide. What is the length of this painting?

EXAMPLE 4 If you invest P dollars at the simple interest rate r, the amount of money A you will receive at the end of t years is $A = P + Prt$.

a. Solve for P.

b. At the end of 5 yr an investor receives \$2250 on her investment at 10% simple interest. How much money did she invest?

Solution

a. This time, P appears twice on the right side of the equation. Using the distributive law, we write: $A = P(1 + rt)$. To have P by itself on the right, we need only divide by $(1 + rt)$. Here are all the steps:

$$A = P + Prt \quad \text{Given.}$$
$$A = P(1 + rt) \quad \text{Use the distributive law.}$$
$$\frac{A}{1 + rt} = \frac{P(1 + rt)}{1 + rt} \quad \text{Divide by } 1 + rt.$$
$$P = \frac{A}{1 + rt}$$

b. We substitute 2250 for A, $10\% = 0.10$ for r, and 5 for t, obtaining:

$$P = \frac{2250}{1 + 0.10 \cdot 5} = \frac{2250}{1.5} = 1500$$

Thus \$1500 was invested.

EXAMPLE 5 Do you know how to add $a_1 + a_2 + a_3 + \cdots + a_n$? If the difference between successive terms is a constant, the sum is:

$$S_n = \frac{n(a_1 + a_n)}{2}$$

(We show you why in Chapter 12.)

a. Solve for n.

b. $S_n = 2 + 4 + 6 + \cdots + 100 = 2550$. Thus, $a_1 = 2$, $a_n = 100$, and $S_n = 2550$. Verify that we have added 50 terms.

Solution

a. Since we want to have n by itself on the right, we multiply by 2 and divide by $a_1 + a_n$. Here are the steps:

$$S_n = \frac{n(a_1 + a_n)}{2} \quad \text{Given.}$$
$$2 \cdot S_n = 2 \cdot \frac{n(a_1 + a_n)}{2} \quad \text{Multiply by 2.}$$
$$2S_n = n(a_1 + a_n) \quad \text{Simplify.}$$
$$\frac{2S_n}{a_1 + a_n} = \frac{n(a_1 + a_n)}{a_1 + a_n} \quad \text{Divide by } a_1 + a_n.$$
$$\frac{2S_n}{a_1 + a_n} = n \quad \text{Simplify.}$$

b. We substitute $a_1 = 2$, $a_n = 100$, and $S_n = 2550$ in the formula $n = \dfrac{2S_n}{a_1 + a_n}$ to get $n = \dfrac{2 \cdot 2550}{2 + 100} = \dfrac{5100}{102} = 50$.

Problem 4 If $2A = ah + bh$
a. Solve for h.
b. If A is 150, $a = 10$, and $b = 20$, what is h?

Problem 5 If $A = \dfrac{n(P_1 + P_n)}{2}$
a. Solve for n.
b. If $A = 50$, $P_1 = 32$, $P_n = 18$, find n.

ANSWERS (to problems on pages 136–137)

2. a. $a = \dfrac{2A}{h} - b$, or $a = \dfrac{2A - bh}{h}$ **b.** 3

3. a. $L = \dfrac{P - 2W}{2}$ **b.** 30.5 in.

NAME

CLASS

SECTION

ANSWERS

In Problems 1–12 solve each formula for the indicated letter.

1. $V = \pi r^2 h$ for h

2. $V = \frac{1}{3}\pi r^2 h$ for h

3. $V = LWH$ for W

4. $V = LWH$ for H

5. $P = s_1 + s_2 + b$ for b

6. $P = s_1 + s_2 + b$ for s_2

7. $A = \pi(r^2 + rs)$ for s

8. $T = 2\pi(r^2 + rh)$ for h

9. $\dfrac{V_2}{V_1} = \dfrac{P_1}{P_2}$ for V_2

10. $\dfrac{V_2}{V_1} = \dfrac{P_1}{P_2}$ for P_1

11. $\dfrac{V_2}{V_1} = \dfrac{P_1}{P_2}$ for P_2

12. $\dfrac{V_2}{V_1} = \dfrac{P_1}{P_2}$ for V_1

1. _____

2. _____

3. _____

4. _____

5. _____

6. _____

7. _____

8. _____

9. _____

10. _____

11. _____

12. _____

13. The distance D traveled in time t by an object moving at rate R is given by $D = RT$.
   a. Solve for T.
   b. The distance between two cities is 220 mi. How long would it take a driver traveling at 55 mi/hr to go from A to B?

13. a. _____
   b. _____

14. The ideal height H (in inches) of a man is related to his weight W (in pounds) by the formula $W = 5H - 190$.
   a. Solve for H.
   b. If a man weighs 160 pounds, how tall should he be?

14. a. _____
   b. _____

15. The number of hours H a growing child should sleep is $H = 17 - \dfrac{A}{2}$, where A is the age of the child.
   a. Solve for A.
   b. The parents of an infant cannot wait until the child sleeps just 8 hr a day. At what age will that happen?

15. a. _____
   b. _____

16. The efficiency energy rating EER of an air conditioner is given by $EER = \dfrac{BTU}{W}$, where BTU is the cooling capacity (per hour) and W is the watts of energy consumed.
   a. Solve for W.
   b. How many watts of electricity would an air conditioner with EER 9 and rated at 9000 BTU consume in 1 hr?

16. a. _____
   b. _____

17. The operating profit margin for a business is

$$OPM = \frac{CGS + OE}{NS},$$ where CGS is the cost of goods sold, OE is

the operating expense, and NS is the net sales.
  a. Solve for CGS.
  b. If the operating expenses of a business amounted to
     $18,500, the operating profit margin was 96% = 0.96, and the
     net sales were $50,000, what was the cost of the goods sold?

17. a. _____
  b. _____

18. The acid-test (AT) ratio for a business is given by: $AT = \dfrac{C + R}{CL}$,

where C is the cash, R is the amount of receivables, and CL is
the current liability.
  a. Solve for R.
  b. If the AT ratio for a business is 1, its current liability is
     $7800, and the business has $1200 cash on hand, what
     should the accounts receivable be?

18. a. _____
  b. _____

19. The probability P of an event is $P = \dfrac{F}{F + U}$, where F is the

number of favorable outcomes for the event and U is the
number of unfavorable outcomes for the event.
  a. Solve for U.
  b. The probability of throwing a 4 with a die (plural dice) is $\frac{1}{6}$.
     If a die is thrown a number of times and 200 of the throws
     are 4's (favorable), how many unfavorable throws would
     you expect?

19. a. _____
  b. _____

20. The area A of a trapezoid is given by $A = \frac{1}{2}h(a + b)$.
  a. Solve for b.
  b. If the area of a trapezoid is 60 square units, its height h is
     10 units, and side a is 7 units, what is the length of side b?

20. a. _____
  b. _____

✓ **SKILL CHECKER**

Solve.

21. $0.06P + 0.08(10,000 - P) = 730$

21. _____

22. $0.05s + 0.06(10,000 - s) = 2200$

22. _____

23. $0.05s + 0.10(6000 - s) = 500$

23. _____

24. $0.05P + 0.07(10,000 - P) = 18,000$

24. _____

2.5 USING YOUR KNOWLEDGE

In this section, you learned how to solve for a specified variable
in a given equation. This idea can sometimes be used to do some
detective work. Suppose that a femur bone measuring 20 in. in
length is found. The relationship between the length L of the
femur bone of a man and his height H is given by

$$H = 1.88L + 32 \tag{1}$$

where H and L are both measured in inches. The corresponding
equation for a woman is

$$H = 1.95L + 29 \tag{2}$$

1. **a.** Solve Equation (1) for L.
   b. Can the femur belong to a man whose height was 6 ft?
   c. Can the bone belong to a man whose height was 69.6 in.?

2. **a.** Solve Equation (2) for L.
   b. Can a femur measuring 20 in. in length belong to a woman whose height was 5 ft 8 in?

3. A police pathologist wants to check the accuracy of Equations (1) and (2). Use your calculator to find L for:
   a. A man 5 ft 6 in. in height.
   b. A man 5 ft 8 in. in height.
   c. A woman 5 ft 6 in. in height.
   d. A woman 5 ft 10 in. in height.

4. Find the height to the nearest tenth of an inch of a woman whose femur is the same length as the femur of a 5-ft, 10-in. man.

1. a. _____
   b. _____
   c. _____

2. a. _____
   b. _____

3. a. _____
   b. _____
   c. _____
   d. _____

4. _____

2.6 APPLICATIONS: WORD PROBLEMS

Courtesy of NASA.

In the preceding sections, we learned how to solve certain kinds of equations. Now, we are ready to apply this knowledge to solve problems. These problems will be stated in words, and are consequently called **word,** or **story, problems.** Word problems frighten many students, but do not panic. We have a surefire method for tackling such problems.

Let us start with a problem that you might have heard about. Do you know the name of the heaviest glider in the world? It is the *Columbia*! When fully loaded, the glider and its payload weigh 215,000 lb. The glider itself weighs 85,000 lb more than the payload. What is the weight of each? There you have it, a word problem. Here is the way to solve such problems.

PROCEDURE FOR SOLVING WORD PROBLEMS

1. Read the problem carefully, and decide what it asks for (the unknown).

2. Select a variable to represent this unknown.

3. Translate the problem into the language of algebra.

4. Use the rules of algebra to solve for the unknown.

5. Verify the solution.

How can you remember these five steps? It's easy—just look at the first letter in each sentence. We call this method the **RSTUV** method. Here is how we use this method to solve the glider problem.

1. Read the problem slowly, not once but two or three times.
2. Select the variable p to be the number of pounds in the weight of the payload. Since the glider weighs 85,000 lb more than the payload, the glider weighs $p + 85,000$ lb.

3. Translate the problem into the language of algebra.

The glider and its payload weigh 215,000 lb.

$$(p + 85{,}000) + p = 215{,}000$$

4. Use algebra to solve the equation.

$(p + 85{,}000) + p = 215{,}000$	Given.
$p + 85{,}000 + p = 215{,}000$	Remove parentheses.
$2p + 85{,}000 = 215{,}000$	Combine the p's.
$2p = 215{,}000 - 85{,}000$	Subtract 85,000.
$2p = 130{,}000$	
$p = 65{,}000$	Divide by 2.

Thus, the payload weighs 65,000 lb and the glider weighs $65{,}000 + 85{,}000 = 150{,}000$ lb.

You have probably noticed that certain words occur quite frequently in word problems. To help you translate these words into the language of algebra, here is a small mathematics dictionary.

TABLE 1
Mathematics Dictionary

WORDS	TRANSLATION	EXAMPLE	TRANSLATION
Add, more than, sum, increased by, added to	$+$	Add 7 to n 7 more than n The sum of n and 7 n increased by 7 7 added to n	$n + 7$
Subtract, less than, minus, difference, decreased by, subtracted from	$-$	Subtract 9 from x 9 less than x x minus 9 Difference of x and 9 x decreased by 9 9 subtracted from x	$x - 9$
Of, the product, times, multiply by	$\times$	$\frac{1}{2}$ of a number x The product of $\frac{1}{2}$ and x $\frac{1}{2}$ times a number x Multiply $\frac{1}{2}$ by x	$\frac{1}{2}x$
Divide, divided by, the quotient of	$\div$	Divide 10 by x 10 divided by x The quotient of 10 and x	$\dfrac{10}{x}$

Of course, most of these words are used in conjunction with the word *equals*. Here are some words that mean equals:

The same, yields, gives, is $=$

A. INTEGER PROBLEMS

A popular algebra problem deals with integers. If you are given an integer, can you find the integer that comes right after it? For example, the integer that comes after 7 is $7 + 1 = 8$, and the one that comes after -6 is $-6 + 1$.

> If n is any integer, the next **consecutive** integer is $n + 1$

On the other hand, if you are given an *even* integer such as 6, the next *even* integer is 8 (add 2 this time). The next *even* integer after 34 is $34 + 2 = 36$.

> If n is an *even* (or *odd*) integer, the next *even* (or *odd*) integer is $n + 2$

We use this idea in the next example. (Don't forget to consult the dictionary when necessary.)

EXAMPLE 1 The sum of three consecutive odd integers is 129. Find the integers.

Problem 1 The sum of three consecutive odd integers is 249. Find the integers.

Solution We use the RSTUV method.

1. Read the problem and note that we are asking for three consecutive odd *integers*.
2. Select n to be the first of the integers. Since we want three consecutive odd integers, we need to find the next two consecutive odd integers. The next odd integer after n is $n + 2$ and the one after $n + 2$ is $n + 4$. Thus the three consecutive odd integers are n, $n + 2$, and $n + 4$.
3. Translate the problem into the language of algebra.

The sum of 3 consecutive odd integers is 129

$$n + (n + 2) + (n + 4) = 129$$

4. Use algebra to solve this equation.

$$\begin{aligned}
n + (n + 2) + (n + 4) &= 129 && \text{Given.} \\
n + n + 2 + n + 4 &= 129 && \text{Remove parentheses.} \\
3n + 6 &= 129 && \text{Combine like terms.} \\
3n + 6 - 6 &= 129 - 6 && \text{Subtract 6.} \\
3n &= 123 && \text{Simplify.} \\
\frac{3n}{3} &= \frac{123}{3} && \text{Divide by 3.} \\
n &= \mathbf{41}
\end{aligned}$$

Thus the three consecutive odd integers are **41**, **43**, and **45**.

5. Verify that the sum of the three integers is 129. Since $41 + 43 + 45 = 129$ our result is correct.

B. GEOMETRY PROBLEMS

As you recall, the *perimeter* of (distance around) a rectangle is $P = 2L + 2W$, where L is the length and W is the width of the rectangle. Here is a problem involving this formula.

EXAMPLE 2 The students at Osaka Gakun University made a rect-angular poster whose length was 130 ft more than its width. If the pe-rimeter of the poster was 416 ft, give its dimensions.

Solution

1. Read the problem.
2. Select a variable to represent the unknown. We let L be the length and W the width.
3. Translate the problem. This time, we make a sketch to help us visu-alize the situation.

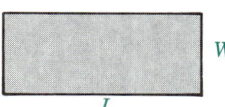

The length is 130 more than the width

$$L = W + 130$$

Now, the perimeter is:

$$P = 2L + 2W$$

Substituting 416 for P and $W + 130$ for L,

$$P = 2L + 2W$$

becomes

$$416 = 2(W + 130) + 2W$$

4. Use algebra to solve.

$416 = 2(W + 130) + 2W$	Given.
$416 = 2W + 260 + 2W$	Remove parentheses.
$416 = 260 + 4W$	Collect like terms.
$156 = 4W$	Subtract 260.
$39 = W$	Divide by 4.

The length $L = W + 130$; thus $L = 39 + 130 = 169$. The dimensions are 169 ft long and 39 ft wide.

5. *Verification:* Is $P = 2L + 2W$? Is $416 = 2 \cdot 169 + 2 \cdot 39$?

$$416 = 338 + 78 \qquad \text{Yes.}$$

C. PERCENT AND INVESTMENT PROBLEMS

EXAMPLE 3 It is estimated that by the year 2000 the annual number of airline passengers at O'Hare airport in Chicago will reach 42 million, a 50% increase over this year's figures. How many passengers will use O'Hare this year?

Solution

1. Read the problem.
2. Let n millions be the number of passengers this year.
3. The number of passengers will increase by 50%. The increase is 50% of n, that is, $0.50n$.

Problem 2 A poster is 100 in. longer than it is wide. If its perimeter is 360 in., give its dimensions.

Problem 3 The estimated annual number of passengers at Atlanta airport in the year 2000 will reach 31.2 million, a 30% increase over this year's figures. How many passengers will use Atlanta airport this year?

ANSWER (to problem on page 145)
1. 81, 83, 85

A 50%   this year's   will   42
increase over   figure   reach (million)

$$0.50n + n = 42$$

4. We now solve this equation.

$$n + 0.50n = 42 \qquad \text{Given.}$$

$$1.50n = 42 \qquad \text{Collect like terms.}$$

$$n = \frac{42}{1.5} = 28 \qquad \text{Divide by 1.5}$$

Thus the number of passengers this year will be 28 million.

5. $28 + 0.50(28) = 28 + 14 = 42$. Thus the answer is correct.   ▲

Investment problems also use percents. If you invest P dollars at a rate r, your annual interest is

$$I = Pr$$

When working this type of problem, it is helpful to enter all the information in a table. We do this in the next example.

EXAMPLE 4 A woman has some stocks yielding 5% annually and some bonds that yield 10%. If her investment totals $6000 and her annual income from the investment is $500, how much does she have invested in stocks and how much in bonds?

Solution

1. Read the problem. We want to find how much she has invested in stocks and how much in bonds.

2. Let s be the amount she has invested in stocks. This makes the amount invested in bonds $(6000 - s)$.

3. We enter the information in the chart below:

	P ×	**r** =	**I**
Stocks	s	0.05	$0.05s$
Bonds	$6000 - s$	0.10	$0.10(6000 - s)$

The total interest is the sum of the entries in the last column, that is, $0.05s$ and $0.10(6000 - s)$. This amount must be $500, that is,

$$0.05s + 0.10(6000 - s) = 500$$

4. Solve the equation.

$$0.05s + 0.10(6000 - s) = 500 \qquad \text{Given.}$$

$$0.05s + 600 - 0.10s = 500 \qquad \text{Simplify.}$$

$$600 - 0.05s = 500 \qquad \text{Multiply.}$$

$$600 - 600 - 0.05s = 500 - 600 \qquad \text{Subtract 600.}$$

$$-0.05s = -100 \qquad \text{Simplify.}$$

$$\frac{-0.05s}{-0.05} = \frac{-100}{-0.05} \qquad \text{Divide by } -0.05.$$

$$s = 2000$$

Problem 4 A man has two investments totaling $8000. One investment yields 5% and the other 10%. If the total annual interest is $650, how much is invested at each rate?

Thus the woman has $2000 in stocks and the rest, that is, $4000, in bonds.

5. To verify the answer, note that 5% of 2000 = 100 and 10% of 4000 = 400. Thus the total interest is indeed $500.

D. DISTANCE PROBLEMS

When traveling at a constant rate R, the distance D traveled in time T is given by

$$D = RT$$

Since this formula is very similar to the one used for interest problems, we also use a table to enter the information. Here is the way we do it.

EXAMPLE 5   A Supercruiser bus leaves Miami traveling at an average rate of 40 mi/hr. Three hours later a car leaves Miami for San Francisco traveling at 55 mi/hr. How far from Miami does the car overtake the bus?

Solution

1. Read the problem carefully. We need to know the time it takes the car to overtake the bus.
2. Select the variable T to represent this time.
3. Translate the given information and write it in a chart. Note that if the car goes for T hours the bus goes for $T + 3$ hours (since it left 3 hr earlier).

	R $\times$	T $=$	D
Car	55	T	$55T$
Bus	40	$T + 3$	$40(T + 3)$

When the car overtakes the bus, they will have traveled the same distance. According to the chart, the car has traveled $55T$ and the bus $40(T + 3)$ miles. Thus

$$55T = 40(T + 3)$$

4. Solve the equation.

$55T = 40(T + 3)$	Given.
$55T = 40T + 120$	Simplify.
$55T - 40T = 40T - 40T + 120$	Subtract $40T$.
$15T = 120$	Simplify.
$\dfrac{15T}{15} = \dfrac{120}{15}$	Divide by 15.
$T = 8$	

Thus it takes the car 8 hr to overtake the bus. In 8 hr, the car travels $8 \cdot 55 = 440$ mi. Thus, the car overtakes the bus 440 mi from Miami.

5. Verify the answer. The car travels for 8 hr at 55 mi/hr; thus it travels $55 \times 8 = 440$ mi, whereas the bus went at 40 mi/hr for 11 hr, a total of $40 \times 11 = 440$ mi.

Problem 5   A bus leaves Los Angeles traveling at 50 mi/hr. An hour later a car leaves at 60 mi/hr to try to catch the bus. How far from Los Angeles does the car catch the bus?

E. MIXTURE PROBLEMS

The last type of problem we discuss is the **mixture problem,** a type of problem in which two or more things are put together to form a mixture. Again, we use a chart to enter the information in this type of problem.

EXAMPLE 6   How many ounces of a 50% acetic acid solution should a photographer add to 32 oz of a 5% acetic acid solution to obtain a 10% solution?

Solution   We use the RSTUV method.

1. Read the problem and note that we are asking for the number of ounces of the 50% solution that should be added.
2. Select x to stand for the number of ounces of 50% solution to be added.
3. To translate the problem, we first use a chart. In this case, the heading for the chart should contain the percent of acetic acid and the amount to be mixed. The product of these two numbers would give us the amount of pure acetic acid. This information does indeed appear as the heading of the chart below.

	%	×	AMOUNT	=	AMOUNT OF PURE ACID
50% solution	0.50		x		$0.50x$
5% solution	0.05		32		1.60
10% solution	0.10		$x + 32$		$0.10(x + 32)$

The percents have been converted to decimals

Since we have x oz of one and 32 oz of the other, we have $(x - 32)$ ounces of the mixture

Since the sum of the total amounts of pure acetic acid should be the same as the amount of pure acetic acid in the final mixture, we have

$$0.50x + 1.60 = 0.10(x + 32)$$

4. We now solve this equation.

$5x + 16 = x + 32$	Multiply by 10 to get rid of the decimals.
$5x + 16 - 16 = x + 32 - 16$	Subtract 16.
$5x = x + 16$	Simplify.
$5x - x = x - x + 16$	Subtract x.
$4x = 16$	Simplify.
$\dfrac{4x}{4} = \dfrac{16}{4}$	Divide by 4.
$x = 4$	

Thus the photographer must add 4 oz. of the 50% solution.

5. The verification of this fact is left to the student.

Problem 6   How many gallons of a 10% salt solution should be added to 15 gal of a 20% salt solution to obtain a 16% solution?

A. Solve the integer problems (1–12). Consult your mathematics dictionary if necessary.

ANSWERS

1. The sum of three consecutive even integers is 138. Find the integers.

   1. _____

2. The sum of three consecutive odd integers is 135. Find the integers.

   2. _____

3. The sum of three consecutive odd integers is −27. Find the integers.

   3. _____

4. The sum of three consecutive even integers is −24. Find the integers.

   4. _____

5. The sum of two numbers is 179 and one of them is 5 more than the other. Find the numbers.

   5. _____

6. The larger of two numbers is six times the smaller. Their sum is 147. Find the numbers.

   6. _____

7. The Beatles have 20 more Recording Industry Association of America (RIAA) awards than Paul McCartney. If the number of awards received by the Beatles and McCartney total 74, how many awards did the Beatles have?

   7. _____

8. The number of rooms in the Detroit Plaza Hotel exceeds the number of rooms in the Peachtree Plaza by 300. If the combined number of rooms in these hotels is 2500, how many rooms are there in each hotel?

   8. _____

9. Norway has 2811 merchant ships, 3 times as many as Sweden. How many merchant ships does Sweden have?

   9. _____

10. To find the weight of an object on the moon, you can divide its weight on earth by 6. The crew of *Apollo 16* collected lunar rocks and soil weighing 35.5 lb on the moon. What is their weight on earth?

    10. _____

11. The weight of an object on the moon is obtained by dividing its earth weight by 6. If an astronaut weighs 28 lb on the moon, what is the corresponding earth weight?

    11. _____

12. The greatest weight difference ever recorded in a major boxing bout is 140 lb in a match between John Fitzsimmons and Ed Purkhorst. If the combined weight of the contestants was 484 lb, find Fitzsimmons' weight. (He was the lighter of the two.)

    12. _____

B. Solve the geometry problems (13–16).

13. The largest painting in the world used to be the Panorama of the Mississippi, by John Banvard. If the length of this painting was 4988 ft more than its width and its perimeter was 10,024 ft, find the dimensions of this rectangular painting.

    13. _____

14. The largest painting now in existence is probably the Battle of Gettysburg. The length of this painting exceeds its width by 340 ft. If the perimeter of this rectangular painting is 960 ft, find its dimensions.

14. _____

15. The scientific building with the greatest capacity is the Vehicle Assembly Building (VAB) at the John F. Kennedy Space Center. The width of this building is 198 feet less than its length. If the perimeter of the rectangular building is 2468 ft, find its dimensions.

15. _____

16. The largest Fair Hall is in Hanover, Germany. The length of this hall exceeds its width by 295 ft. If the perimeter of this rectangular hall is 4130 ft, find its dimensions.

16. _____

C. Solve the percent and investment problems (17–24).

17. In the first four months of this year the Dow Jones Industrial Average (DJIA) rose 25% to 2400. What was it at the beginning of the year?

17. _____

18. The number of beverages introduced in the market during two consecutive years decreased by 6% to 611 new beverages. How many beverages were introduced the year before?

18. _____

19. The total amount spent in corporate travel and entertainment will increase by 21% from 1988 to 1990. If the amount spent will reach $115 billion in 1990, how much was spent in 1988?

19. _____

20. The annual amount spent on entertainment by the average person will increase by 33% to $1463. How much is the average person spending now on entertainment?

20. _____

21. Two sums of money totaling $15,000 earn, respectively, 5% and 7% annual interest. If the interest from both investments amounts to $870, how much is invested at each rate?

21. _____

22. An investor invested $20,000, part at 6% and the rest at 8%. Find the amount invested at each rate if the annual income from the two investments is $1500.

22. _____

23. A woman invested $25,000, part at 7.5% and the rest at 6%. If her annual interest from these two investments amounted to $1620, how much money did she have invested at each rate?

23. _____

24. A man has a savings account that pays 5% annual interest and some certificates of deposit paying 7% annually. His total interest from the two investments is $1100, and the total amount of money in the two investments is $18,000. How much money does he have in the savings account?

24. _____

D. Solve the distance problems (25–30).

25. Two hours after a car leaves a certain town traveling at an average speed of 60 km/hr, a highway patrol officer leaves from the same starting point to overtake the car. If the average speed of the patrolman is 90 km/hr, how far from town does the officer overtake the car?

25. _____

26. A group of smugglers in a car crosses the border traveling in a straight line at 96 km/hr. An hour later, the border patrol starts after them in a light plane flying 144 km/hr.
   a. How long will it be before the border patrol reaches the smugglers?
   b. At what distance from the border will the smugglers be reached?

26. _____

27. A freight train leaves the station traveling at 30 mi/hr. One hour later, a passenger train leaves the same station on a parallel track traveling at 60 mi/hr. How far from the station does the passenger train overtake the freight train?

27. _____

28. A bus leaves the station traveling at 60 km/hr. Two hours later, the wife of one of the passengers shows up at the station with a briefcase belonging to an absentminded professor riding the bus. If she immediately starts after the bus at 90 km/hr, how far from the station is the briefcase reunited with the professor?

28. _____

29. An accountant and her boss have to travel to a nearby town. The accountant catches a train traveling at 50 mi/hr while the boss leaves 1 hr later in a car traveling at 60 mi/hr. They have decided to meet at the train station and, strangely enough, they get there at exactly the same time! If the train and the car traveled in a straight line on parallel paths, how far is it from one town to the other?

29. _____

30. The basketball coach at a local high school left for work on her bicycle, traveling at 15 mi/hr. Half an hour later, her husband noticed that she had left her lunch. He got in his car and took her lunch to her traveling at 60 mi/hr. Luckily, he got to school at exactly the same time as his wife. How far is it from the house to the school?

30. _____

E. Solve the mixture problems (31–40).

31. How many liters of a 40% glycerin solution must be mixed with 10 L of an 80% glycerin solution to obtain a 65% solution?

31. _____

32. How many parts of glacial acetic acid (99.5%) must be added to 100 parts of a 10% solution of acetic acid to give a 28% solution?

32. _____

33. If the price of copper is 65¢ per pound and the price of zinc is 30¢ per pound, how many pounds of copper and zinc should be mixed to make 70 lb of brass selling for 45¢ per pound?

33. _____

34. Oolong tea sells for $19 per pound. How many pounds of Oolong should be mixed with regular tea selling at $4 per pound to produce 50 lb of tea selling for $7 per pound?

34. _____

35. You think the prices of coffee are high? You have seen nothing yet! Blue Jamaican coffee sells for about $5 per pound! How many pounds of Blue Jamaican should be mixed with 80 lb of regular coffee selling at $2 per pound so that the result is a mixture selling for $2.60 per pound? (You can cleverly advertise this as "Containing the incomparable Blue Jamaican coffee.")

35. _____

36. How many ounces of vermouth containing 10% alcohol should be added to 20 oz of gin containing 60% alcohol so that the resulting pitcher of martinis will contain 30% alcohol?

37. How many gallons of a 10% salt solution should be added to 15 gal of a 20% salt solution to obtain a 16% solution?

38. How many gallons of a 30% salt solution must be added to 12 gal of a 15% salt solution to obtain a 20% salt solution?

39. A dietician wishes to prepare a special meal containing 50% carbohydrates. How many grams of a certain food containing 20% carbohydrates should be added to 100 g of another food containing 60% carbohydrates to obtain the desired 50% mixture?

40. How many gallons of rocket fuel containing 90% hydrogen peroxide should be mixed with rocket fuel containing 98% hydrogen peroxide to obtain 3000 gal of fuel containing 96% hydrogen peroxide?

36. _____

37. _____

38. _____

39. _____

40. _____

✓ **SKILL CHECKER**

Simplify.

41. $a^8 \cdot a^{-5}$

42. $a^{-3} \cdot a^7$

43. $\dfrac{a^5}{a^2}$

44. $\dfrac{a^5}{a^{-2}}$

45. $(a^3 b^{-5})^4$

46. $(a^{-3} b^5)^5$

47. $(a^3 b^{-2})^{-4}$

48. $(a^{-3} b^4)^{-5}$

49. $\left(\dfrac{a^3}{b^{-3}}\right)^{-4}$

50. $\left(\dfrac{a^{-4}}{b^3}\right)^{-5}$

41. _____

42. _____

43. _____

44. _____

45. _____

46. _____

47. _____

48. _____

49. _____

50. _____

2.6 USING YOUR KNOWLEDGE

Some of the mixture problems given in this section can be solved using the *guess-and-correct* procedure developed by Dr. Harrison A. Gesselmann of Cornell University. The procedure depends on taking a guess at the answer and then using the calculator to *correct* this guess. For example, suppose we have to mix gold and platinum to obtain 10 oz of a mixture selling for $415 per ounce. Our first guess is to use *equal amounts* (5 oz each) of gold and platinum. This gives a mixture with a price per pound equal to the average price of gold ($400) and platinum ($475), that is,

$$\frac{400 + 475}{2} = \$437.50 \text{ per pound}$$

As you can see from the figure

5 oz	Desired	Average	5 oz
$400	$415	$437.50	$475

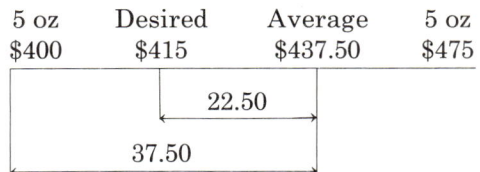

more gold must be used in order to bring the $437.50 average down to the desired $415. Thus the correction factor for the additional amount of gold that must be used is

$$\frac{22.50}{37.50} \times 5 \text{ oz}$$

This expression can be obtained by the keystroke sequence

| 22.50 | ÷ | 37.50 | × | 5 | = |

which gives the correction factor 3. The correct amount is:

first guess	5 oz of gold
+ correction	3 oz of gold
total =	8 oz of gold

and the remaining 2 oz of platinum.

If your instructor wishes, use this method to work Problem 33.

The guess-and-correct method also works for *investment problems* in which we must find how much is invested at certain rates when the *final* amount of interest is known. For example, suppose that $10,000 has been invested, part at 6% and the rest at 8%. We also know the final interest for the year, $730, and we must find how much was invested at each rate. Our first guess is that equal amounts ($5000 each) were invested at each rate. This would give a rate of return of

$$\frac{6 + 8}{2} = 7\%$$

The actual rate of return is

$$\frac{730}{10,000} = 7.3\%$$

so we must invest more of the money at 8% to bring the average up to the desired amount (see the figure).

$5000	Average	Desired	$5000
6%	7%	7.3%	8%

Thus the correction factor would be

$$\frac{0.3}{1} \times 5000$$

more dollars invested at 8%, that is,

or $1500 more dollars than the original $5000 guess must be invested at 8%. Thus we must invest $5000 + $1500, or $6500, at 8% and the rest ($3500) at 6%.

If your instructor approves, try Problems 22 and 24.

SUMMARY

SECTION	ITEM	MEANING	EXAMPLE
2.1	Equation	A sentence using = as its verb	$x + 1 = 5$ and $3 - y = 7$ are equations.
2.1A	Solutions of an equation	The replacements of the variable that make the equation a true statement.	5 is a solution of $x + 3 = 8$.
2.1B	Equivalent	Two or more equations are equivalent if they have the same solution set.	$x + 2 = 5$ and $x = 3$ are equivalent.
2.1B	Properties of equality	You can *add* or *subtract* the same number on both sides and *multiply* or *divide* both sides of an equation by the same nonzero quantity and the result will be an equivalent equation.	
2.1C	Linear equation	An equation that can be written in the form $ax + b = c$, a, b, and c are real numbers, $a \neq 0$	$3x + 7 = 9$ and $-x - 3 = \frac{2}{3}$ are linear equations.
2.2A	Absolute value equation	The solutions of $|x| = a$ are $x = a$ or $x = -a$ for $a \geq 0$.	The solutions of $|x| = 2$ are 2 or -2.
2.2A	Linear inequality	An inequality that can be written in the form $ax + b > c$, where a, b, and c are real numbers	$-2x + 3 > 5$ and $3 - 2x > 7$ are linear inequalities.
2.2A	Properties of inequalities	If $A < B$, then $A + C < B + C$. If $A < B$, then $A - C < B - C$. If $A < B$, then $A \cdot C < B \cdot C$ when $C > 0$.	

(Continued)

SECTION	ITEM	MEANING	EXAMPLE
		If $A < B$, then $A \cdot C > B \cdot C$ when $C < 0$. If $A < B$, then $\dfrac{A}{C} < \dfrac{B}{C}$ when $C > 0$. If $A < B$, then $\dfrac{A}{C} > \dfrac{B}{C}$ when $C < 0$.	
2.3A	Intersection of sets	If A and B are sets, the intersection of A and B, denoted by $A \cap B$, is the set of elements that are in both A and B.	
2.3A	Union of sets	If A and B are sets, the union of A and B, denoted by $A \cup B$, is the set of elements that are in either A or B.	
2.4B	$|x| < a$	$|x| < a$ is equivalent to $-a < x < a$.	$|x| < 3$ is equivalent to $-3 < x < 3$.
2.4B	$|x| > a$	$|x| > a$ is equivalent to $x > a$ or $x < -a$.	$|x| > 3$ is equivalent to $x > 3$ or $x < -3$.
2.5	Perimeter	The distance around a polygon (figure)	The perimeter P of a rectangle is $P = 2L + 2W$, where L is the length and W the width of the rectangle.
2.6	Word problem	The RSTUV procedure to solve a word problem is: Read Select a variable Translate Use algebra Verify	

NAME

CLASS

SECTION

ANSWERS

(If you need help with these exercises, look in the section indicated in brackets.)

1. [2.1A] Does -3 satisfy the equation?
   a. $7 = 8 - x$
   b. $9 = 8 + x$
   c. $4 = 1 - x$

   1. a. _____
      b. _____
      c. _____

2. [2.1B] Solve.

   a. $\dfrac{2}{3}y - 3 = 5$

   b. $\dfrac{2}{3}y - 5 = 5$

   c. $\dfrac{2}{3}y - 7 = 5$

   2. a. _____
      b. _____
      c. _____

3. [2.1B] Solve.
   a. $x + 2 = 2(2x - 2)$
   b. $x + 3 = 3(2x - 4)$
   c. $x + 4 = 4(2x - 6)$

   3. a. _____
      b. _____
      c. _____

4. [2.1B] Solve.

   a. $\dfrac{x + 4}{3} - \dfrac{x - 4}{5} = 4$

   b. $\dfrac{x + 6}{3} - \dfrac{x - 6}{5} = 6$

   c. $\dfrac{x + 8}{3} - \dfrac{x - 8}{5} = 8$

   4. a. _____
      b. _____
      c. _____

5. [2.1C] Solve.

   a. $\dfrac{x}{4} - \dfrac{x}{3} = \dfrac{x - 4}{4}$

   b. $\dfrac{x}{5} - \dfrac{x}{3} = \dfrac{x - 5}{5}$

   c. $\dfrac{x}{7} - \dfrac{x}{3} = \dfrac{x - 7}{7}$

   5. a. _____
      b. _____
      c. _____

6. [2.2A] Solve.
   a. $0.05P + 0.10(2000 - P) = 175$
   b. $0.08P + 0.10(5000 - P) = 460$
   c. $0.06P + 0.10(10,000 - P) = 840$

   6. a. _____
      b. _____
      c. _____

7. [2.2B] Solve.

    a. $\left|\dfrac{2}{7}x + 2\right| + 5 = 9$

    b. $\left|\dfrac{2}{7}x + 2\right| + 3 = 9$

    c. $\left|\dfrac{2}{7}x + 2\right| + 1 = 9$

7. a. _____
    b. _____
    c. _____

8. [2.2B] Solve.
    a. $|x - 1| = |x - 3|$
    b. $|x - 3| = |x - 5|$
    c. $|x - 5| = |x - 7|$

8. a. _____
    b. _____
    c. _____

9. [2.3A] Graph.
    a. $x \geq -2$
    b. $x \geq -3$
    c. $x \geq -4$

9. a. _____
    b. _____
    c. _____

10. [2.3B] Solve and graph.
    a. $2(x - 1) \leq 4x + 4$
    b. $3(x - 1) \leq 6x + 3$
    c. $4(x - 1) \leq 8x + 4$

10. a. _____
    b. _____
    c. _____

11. [2.3B] Solve and graph.

    a. $\dfrac{x}{4} - \dfrac{x}{3} < \dfrac{x - 4}{4}$

    b. $\dfrac{x}{5} - \dfrac{x}{3} < \dfrac{x - 5}{5}$

    c. $\dfrac{x}{7} - \dfrac{x}{3} < \dfrac{x - 7}{7}$

11. a. _____
    b. _____
    c. _____

12. [2.4A] Solve and graph.
    a. $\{x \,|\, x > -1 \text{ and } x < 2\}$
    b. $\{x \,|\, x > -2 \text{ and } x < 3\}$
    c. $\{x \,|\, x > -3 \text{ and } x < 4\}$

12. a. _____
    b. _____
    c. _____

13. [2.4A] Solve and graph.
    a. $\{x \,|\, x < -2 \text{ or } x \geq 3\}$
    b. $\{x \,|\, x < -3 \text{ or } x \geq 2\}$
    c. $\{x \,|\, x < -4 \text{ or } x \geq 1\}$

13. a. _____
    b. _____
    c. _____

14. [2.4A] Solve and graph.
    a. $x + 1 \leq 3 \text{ and } -4x < 8$
    b. $x + 1 \leq 4 \text{ and } -3x < 9$
    c. $x + 1 \leq 5 \text{ and } -2x < 4$

14. a. _____
    b. _____
    c. _____

15. [2.4A] Solve and graph.
    a. $-4 \leq -2x - 6 < 4$
    b. $-3 \leq -2x - 5 < 3$
    c. $-6 \leq -2x - 4 < 2$

15. a. _____
    b. _____
    c. _____

16. [2.4B] Solve and graph.
   a. $|3x - 1| \leq 2$
   b. $|4x - 1| \leq 3$
   c. $|5x - 1| \leq 4$

16. a. _____
   b. _____
   c. _____

17. [2.4B] Solve and graph.
   a. $|3x - 1| \geq 2$
   b. $|4x - 1| \geq 3$
   c. $|5x - 1| \geq 4$

17. a. _____
   b. _____
   c. _____

18. [2.5] Solve for h and evaluate when $H = 82.48$.
   a. $H = 2.5h + 72.48$
   b. $H = 2.5h + 77.48$
   c. $H = 2.5h + 84.98$

18. a. _____
   b. _____
   c. _____

19. [2.5] Solve for A.

   a. $B = \dfrac{2}{7}(A - 7)$

   b. $B = \dfrac{3}{7}(A - 7)$

   c. $B = \dfrac{4}{7}(A - 7)$

19. a. _____
   b. _____
   c. _____

20. [2.5] The perimeter of a rectangle is $P = 2L + 2W$, where L is the length and W is the width. Solve for L and give the dimensions when
   a. The perimeter is 180 ft and the length is 10 ft more than the width.
   b. The perimeter is 220 ft and the length is 10 ft more than the width.
   c. The perimeter is 260 ft and the length is 10 ft more than the width.

20. a. _____
   b. _____
   c. _____

21. [2.6A] Find three consecutive odd integers whose sum is
   a. 153
   b. 159
   c. 207

21. a. _____
   b. _____
   c. _____

22. [2.6B] A woman's salary was increased by 20%. What was her salary before the increase if her new salary was
   a. $24,000
   b. $36,000
   c. $18,000

22. a. _____
   b. _____
   c. _____

23. [2.6C] An investor bought some municipal bonds yielding 5% annually and some certificates of deposit yielding 10% annually. If his total investment amounts to $20,000, find how much money is invested in bonds and how much in certificates of deposit if his annual interest is
   a. $1750
   b. $1150
   c. $1500

23. a. _____
   b. _____
   c. _____

24. [2.6D] A car leaves a town traveling at 40 mi/hr. An hour later another car leaves the same town in the same direction traveling at 50 mi/hr.
 a. How far from town does the second car overtake the first?
 b. Repeat the problem if the first car is traveling at 50 mi/hr and the second one, 60 mi/hr.
 c. Repeat the problem if the first car is traveling at 40 mi/hr and the second one, 60 mi/hr.

24. a. _____
    b. _____
    c. _____

25. [2.6E] How many liters of a 40% salt solution must be mixed with 50 L of a 10% salt solution to obtain
 a. A 30% solution
 b. A 20% solution
 c. A 10% solution

25. a. _____
    b. _____
    c. _____

NAME

CLASS

SECTION

(Answers on page 166)

ANSWERS

1. Does 4 satisfy the equation $5 = 9 - x$?

1. _____

2. Solve $\frac{4}{5}y - 3 = 9$

2. _____

3. Solve $x + 1 = 2(3x - 3)$.

3. _____

4. Solve $\dfrac{x + 4}{3} - \dfrac{x - 4}{5} = 4$.

4. _____

5. Solve $\dfrac{6}{5} - \dfrac{x}{15} = \dfrac{2(x + 4)}{25}$.

5. _____

6. Solve $0.06P + 0.07(1500 - P) = 96$.

6. _____

7. Solve $\left|\dfrac{3}{4}x + 2\right| + 4 = 9$.

7. _____

8. Solve $|x - 3| = |x - 7|$.

8. _____

9. Graph $x \geq -1$.

9. _____

10. Solve and graph $4(x - 1) \leq 8x + 4$.

10. _____

11. Solve and graph $\dfrac{x}{7} - \dfrac{x}{3} < \dfrac{x - 7}{7}$.

11. _____

12. Solve and graph $\{x | x > -2 \text{ and } x < 3\}$.

12. _____

13. Solve and graph $\{x | x < -1 \text{ or } x \geq 2\}$.

13. _____

14. Solve and graph $x + 1 \leq 4$ and $-2x < 6$.

14. _____

15. Solve and graph $-4 \leq -2x - 6 < 0$.

15. _____

16. Solve and graph $|2x - 1| \leq 5$.

16. _____

17. Solve and graph $|2x + 1| > 3$.

17. _____

18. $H = 2.75h + 71.48$
   a. Solve for h.
   b. Find h if $H = 140.23$.

18. a. _____
    b. _____

19. Solve for A in $B = \frac{3}{4}(A - 8)$.

19. _____

20. The perimeter of a rectangle is $P = 2L + 2W$, where L is the length and W is the width.
   a. Solve for L.
   b. If the perimeter is 100 ft and the length is 20 ft more than the width, what are the dimensions?

20. a. _____
    b. _____

21. The sum of three consecutive odd integers is 117. What are the three integers?

21. _____

22. A woman's salary was increased by 20% to $24,000. What was her salary before the increase?

22. _____

23. An investor bought some municipal bonds yielding 5% annually and some certificates of deposit yielding 7%. If his total investment amounts to $20,000 and his annual interest is $1100, how much money is invested in bonds and how much in certificates of deposit?

23. _____

24. A freight train leaves a station traveling at 40 mi/hr. Two hours later, a passenger train leaves the same station in the same direction at 60 mi/hr. How far from the station does the passenger train catch the freight train?

24. _____

25. How many gallons of a 30% salt solution must be mixed with 40 gal of a 12% salt solution to obtain a 20% solution?

25. _____

IF YOU MISSED QUESTION	SECTION	EXAMPLES	PAGE	ANSWERS
1	2.1	1	96	**1.** Yes
2	2.1	2	97	**2.** 15
3	2.1	3, 4	98	**3.** $\frac{7}{5}$
4	2.1	5	99	**4.** 14
5	2.1	6	100	**5.** 6
6	2.2	1, 2	108	**6.** 900
7	2.2	3, 4	109	**7.** $4, -\frac{28}{3}$
8	2.2	5	110	**8.** 5
9	2.3	1	116	**9.** number line graph, $-3\;-2\;-1\;0\;1\;2\;3$
10	2.3	3	118	**10.** number line graph, $-3\;-2\;-1\;0\;1\;2\;3$
11	2.3	4	120	**11.** number line graph, $-3\;-2\;-1\;0\;1\;2\;3$
12	2.4	1	125	**12.** number line graph, $-3\;-2\;-1\;0\;1\;2\;3$
13	2.4	2	126	**13.** number line graph, $-3\;-2\;-1\;0\;1\;2\;3$
14	2.4	3	127	**14.** number line graph, $-3\;-2\;-1\;0\;1\;2\;3$
15	2.4	4	128	**15.** number line graph, $-3\;-2\;-1\;0\;1\;2\;3$
16	2.4	5	129	**16.** number line graph, $-3\;-2\;-1\;0\;1\;2\;3$
17	2.4	6	130	**17.** number line graph, $-3\;-2\;-1\;0\;1\;2\;3$
18a	2.5	1	135–136	**18. a.** $h = \dfrac{H - 71.48}{2.75}$
18b	2.5	1	135–136	**b.** 25
19	2.5	2	136	**19.** $A = \dfrac{4B + 24}{3}$
20a	2.5	3	137	**20. a.** $L = \dfrac{P - 2W}{2}$
20b	2.5	3	137	**b.** 15 ft by 35 ft
21	2.6	1	145	**21.** 37, 39, 41
22	2.6	3	146	**22.** $20,000
23	2.6	4	147	**23.** $15,000 bonds; $5000 CD
24	2.6	5	148	**24.** 240 mi
25	2.6	6	149	**25.** 32

CHAPTER 3

EXPONENTS AND POLYNOMIALS

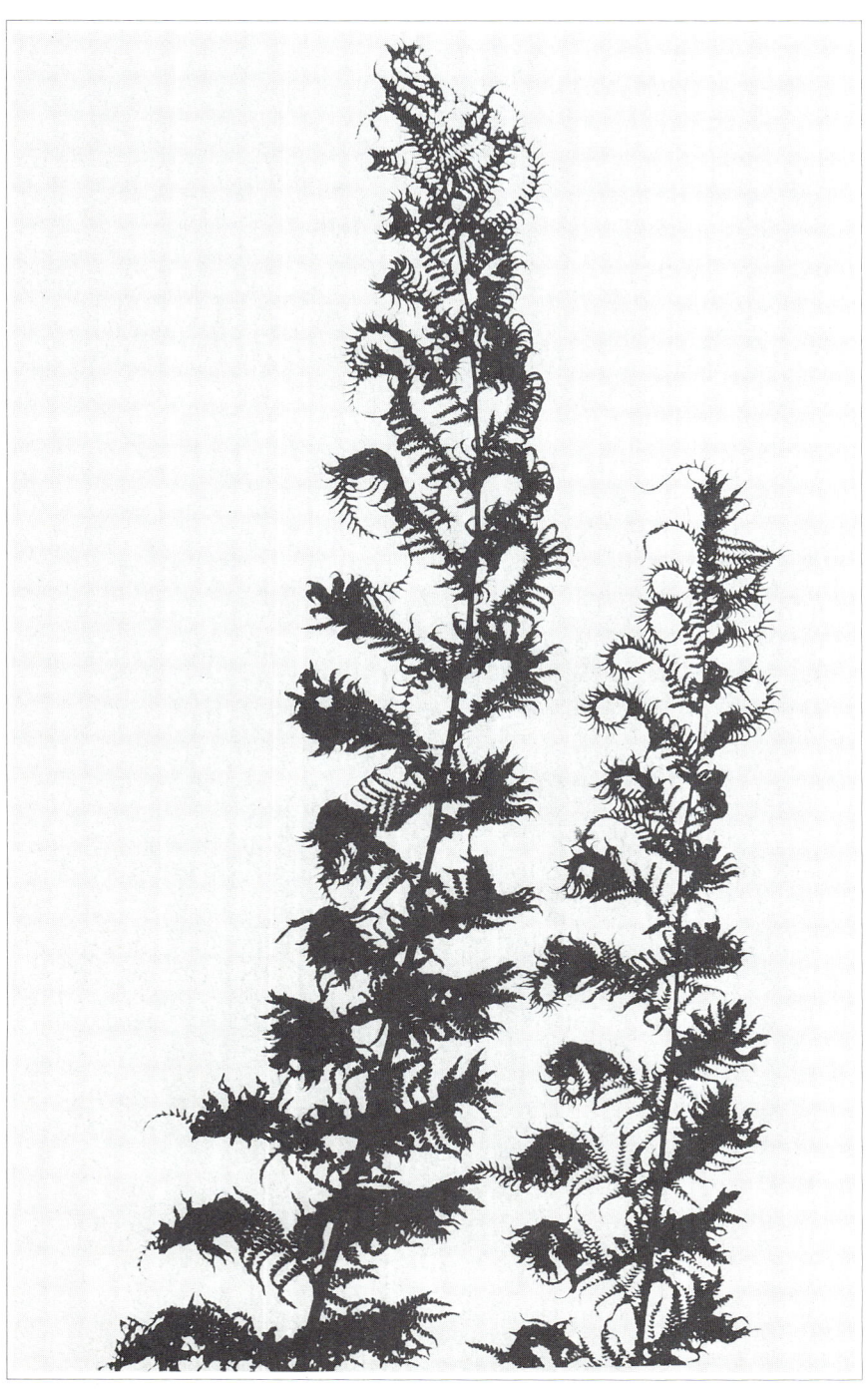

In this chapter we study how to add, subtract, multiply and divide polynomials. We make extensive use of the properties of exponents studied in chapter 1 as we multiply and divide polynomials. We then learn how to "undo" the multiplication of certain kinds of polynomials, a process called factoring. We end the chapter by learning how to use factoring to solve equations and applying this knowledge to problems related to the Pythagorean theorem.

NAME

CLASS

SECTION

ANSWERS

(Answers on page 170)

1. Classify as a monomial, binomial, or trinomial and give the degree of $-x^2 + xy^2z^4 - x^6$.

2. Rewrite $3xxxyyy - 8xxyyy - 4zzzz$ using exponents and give its degree.

3. Let $P(x) = x^2 - 3x + 2$. Find $P(-1)$.

4. Add $5x^3 + 8x^2 - 6x - 4$ and $6 - 3x + x^2 - 2x^3$.

5. Subtract $8x^3 - 6x^2 + 4x - 2$ from $5x^3 + 3x^2 + 2$.

6. Multiply $-2x^2y(x^2 + 5xy - 3y^3)$.

7. Multiply $(x - 3)(x^2 - 4x - 5)$.

8. Multiply $(3x + 4y)(4x - 7y)$.

9. Multiply
   a. $(2x + 5y)^2$ b. $(3x - 4y)^2$

10. Multiply $(5x + 4y)(5x - 4y)$.

11. Factor $12x^5 - 16x^4 + 8x^3 + 20x^2$.

12. Factor $6x^6 - 4x^4 + 15x^3 - 10x$.

13. Factor
    a. $x^2 - 2xy - 24y^2$ b. $2x^2 - xy - 15y^2$

14. Factor $24x^4y + 4x^3y^2 - 8x^2y^3$.

15. Factor
    a. $16x^2 - 24xy + 9y^2$ b. $9x^2 + 12xy + 4y^2$

16. Factor $16x^4 - y^4$.

17. Factor $x^2 - 8x + 16 - y^2$.

18. Factor
    a. $8x^3 + 27y^3$ b. $27y^3 - 8x^3$

19. Factor
    a. $8x^6 - x^3y^3$ b. $6x^6 + 24x^4$

20. Factor
    a. $18x^4 + 24x^3y + 8x^2y^2$ b. $12x^2y^2 - 36xy^3 + 27y^4$

21. Factor $9x^3y - 33x^2y^2 - 12xy^3$.

22. Factor $12x^3 - 8x^2 - 3x + 2$.

23. Solve
    a. $x^2 = -2x + 8$ b. $6x^2 - x = 2$

24. Solve $x^3 + 4x^2 - x - 4 = 0$.

25. The sides of a right triangle are x, $x + 7$, and $x - 8$ units long. Find the dimensions of the triangle.

ANSWERS

1. _____
2. _____
3. _____
4. _____
5. _____
6. _____
7. _____
8. _____
9. a. _____
   b. _____
10. _____
11. _____
12. _____
13. a. _____
    b. _____
14. _____
15. a. _____
    b. _____
16. _____
17. _____
18. a. _____
    b. _____
19. a. _____
    b. _____
20. a. _____
    b. _____
21. _____
22. _____
23. a. _____
    b. _____
24. _____
25. _____

IF YOU MISSED QUESTION	SECTION	EXAMPLES	PAGE	ANSWERS
1	3.1	1, 2	172	**1.** Trinomial: 7
2	3.1	3	173	**2.** $3x^3y^3 - 8x^2y^3 - 4z^4$; 6
3	3.1	4, 5	174	**3.** 6
4	3.2	1, 2	180	**4.** $3x^3 + 9x^2 - 9x + 2$
5	3.2	3	181	**5.** $-3x^3 + 9x^2 - 4x + 4$
6	3.3	1	188	**6.** $-2x^4y - 10x^3y^2 + 6x^2y^4$
7	3.3	2	189	**7.** $x^3 - 7x^2 + 7x + 15$
8	3.3	3	190	**8.** $12x^2 - 5xy - 28y^2$
9a	3.3	4	191	**9. a.** $4x^2 + 20xy + 25y^2$
9b	3.3	4	191	**b.** $9x^2 - 24xy + 16y^2$
10	3.3	5	192	**10.** $25x^2 - 16y^2$
11	3.4	1, 2, 3	198–199	**11.** $4x^2(3x^3 - 4x^2 + 2x + 5)$
12	3.4	4, 5	199–200	**12.** $x(3x^2 - 2)(2x^3 + 5)$
13a	3.5	1, 2	206	**13. a.** $(x + 4y)(x - 6y)$
13b	3.5	3	207	**b.** $(2x + 5y)(x - 3y)$
14	3.5	6	209–210	**14.** $4x^2y(3x + 2y)(2x - y)$
15a	3.6	1, 2	216	**15. a.** $(4x - 3y)^2$
15b	3.6	1, 2	216	**15. b.** $(3x + 2y)^2$
16	3.6	3	217	**16.** $(4x^2 + y^2)(2x + y)(2x - y)$
17	3.6	4	217	**17.** $(x - 4 + y)(x - 4 - y)$
18a	3.6	5	218	**18. a.** $(2x + 3y)(4x^2 - 6xy + 9y^2)$
18b	3.6	5	218	**b.** $(3y - 2x)(9y^2 + 6xy + 4y^2)$
19a	3.7	1	226	**19. a.** $x^3(2x - y)(4x^2 + 2xy + y^2)$
19b	3.7	1	226	**b.** $6x^4(x^2 + 4)$
20a	3.7	2	226	**20. a.** $2x^2(3x + 2y)^2$
20b	3.7	2	226	**b.** $3y^2(2x - 3y)^2$
21	3.7	3	226–227	**21.** $3xy(3x + y)(x - 4y)$
22	3.7	5	227	**22.** $(3x - 2)(2x + 1)(2x - 1)$
23a	3.8	1, 2, 3	234–237	**23. a.** $2, -4$
23b	3.8	1, 2, 3	234–237	**b.** $\frac{2}{3}, -\frac{1}{2}$
24	3.8	4	237	**24.** $1, -1, -4$
25	3.8	5	238	**25.** 5, 12, 13 units

3.1 POLYNOMIALS

Courtesy of the *Guinness Book of World Records.*

OBJECTIVES

REVIEW

Before starting this section, you should know:

1. The definition of base and exponent (Section 1.6.)
2. How to evaluate an expression.

OBJECTIVES

You should know how to:

A. Classify a polynomial.
B. Find the degree of a polynomial.
C. Rewrite a polynomial using exponents.
D. Evaluate a polynomial.

The photograph shows a man jumping from an altitude of 118 feet. Do you know how high above sea level he will be after falling downward for t seconds? This height is given by

$$H = -16t^2 + 118$$

The right side of this formula is an algebraic expression called a *polynomial.*

A. MONOMIALS, BINOMIALS, AND TRINOMIALS

An expression consisting of a constant times a product of variables with whole number exponents is called a **monomial.** For example,

$$2x, \quad -5x^2y, \quad 10x^5yz^2, \quad \text{and} \quad -\frac{2}{3}x^4$$

are all monomials. A **polynomial** is just a sum or difference of monomials. Thus

$$2x^2 - 4xy - 7y^3$$

is a polynomial. The individual monomials in a polynomial are called the **terms** of the polynomial. The numerical factor of a term is called the **coefficient.** In the polynomial $2x^2 + xy - 7y^3$, the terms are $2x^2$, with coefficient 2; xy, with coefficient 1 (since $xy = 1xy$); and $-7y^3$, with coefficient -7. Note that $\frac{5}{x}$, $\frac{(x+y)}{z}$, and $2 + \sqrt{x}$ are *not* polynomials. (In the first two expressions we are dividing by a variable. The second expression involves taking the square root of the variable.)

A polynomial with two terms is called a **binomial.** Thus, $x - y$, $-3x^2 + xy$, and $16t^2 + 118$ are binomials. Similarly, a polynomial with three terms is called a **trinomial.** For example, $x + y - z^5$, $2x^2 + xy - 7y^3$, and $z^2 - 2z - \frac{1}{5}$ are trinomials.

EXAMPLE 1   Classify the following polynomials as monomials, binomials, or trinomials.

a. $x + x^2$ **b.** $-9x$ **c.** $3x^2 - y + 3xyz$

Solution

a. Since $x + x^2$ has two terms, $x + x^2$ is a *binomial*.
b. Since $-9x$ has one term, $-9x$ is a *monomial*.
c. Since $3x^2 - y + 3xyz$ has three terms, $3x^2 - y + 3xyz$ is a *trinomial*.

B.   THE DEGREE OF A POLYNOMIAL

In Example 1, the polynomials $x + x^2$ and $-9x$ contain only one variable, x. Polynomials can also be classified according to the *greatest exponent* of the variable. This number is called the *degree* of the polynomial.

> The **degree** of a polynomial in one variable is the *greatest* exponent of the variable it contains.

Thus, $-8x^5$ is of the *fifth* degree. $-3x^2 + 8x^4 - 2x$ is of the *fourth* degree and $0.5x$ is of the *first* degree. (Note that $x = x^1$.) Since $x^0 = 1$, $-3 = -3 \cdot 1 = -3x^0$. Similarly, $9 = 9x^0$. Thus, the degree of nonzero numbers such as -3 and 9 is 0. The number 0 itself is called the **zero polynomial** and is not assigned a degree. These ideas can be extended to include polynomials in more than one variable. For example, the expression $3x^2 - y$ is a polynomial in *two* variables (x and y) and $3x^2 - 2xy - 3xyz^2$ is a polynomial in *three* variables (x, y, and z.) To find the degree of these polynomials, we look at the degree of each term. Here is the definition:

> The degree of a polynomial in several variables is the greatest sum of the exponents in any one term of the polynomial.

Thus, the degree of $3x^2 - y$ is 2 (the degree of the first term, $3x^2$). To find the degree of $3x^2 - 2xy - 3xyz^2$, find the degree of each term:

$$3x^2 \quad - \quad 2x^1y^1 \quad - \quad 3x^1y^1z^2$$

degree 2   degree $1 + 1 = 2$ degree $1 + 1 + 2 = 4$

Since the greatest sum of exponents is 4, the degree of $3x^2 - 2xy - 3xyz^2$ is 4.

EXAMPLE 2   Find the degree of the given polynomials.

a. $-5x^2 + 3x^5 + 9$ **b.** $-x^2 + xy^2z^3 - x^5$ **c.** 3 **d.** 0

Solution

a. The degree of $-5x^2 + 3x^5 + 9$ is 5.
b. The degree of $-x^2 + x^1 y^2 z^3 - x^5$ is $1 + 2 + 3 = 6$.

Problem 1   Classify the following polynomials as monomials, binomials, or trinomials.
a. $x^2 - x$ **b.** $-3xy$
c. $\frac{2}{3}x^2 - 8 - 3x$

Problem 2   Find the degree of
a. $-2x^2 + 9 - 7x^4$
b. $-3ab^2c + x^2 - x^3$
c. 8

c. The degree of 3 is 0.

d. 0 has no degree.

C. REWRITING POLYNOMIALS USING EXPONENTS

If we have a polynomial in several variables, it is easier to determine its degree if the multiplications indicated by exponents are written out. Thus, if $3x^2 - 2xy - 3xyz^2$ is written as $3 \cdot x \cdot x - 2 \cdot x \cdot y - 3 \cdot x \cdot y \cdot z \cdot z$ you can see that the last term involves variables as factors $1 + 1 + 2 = 4$ times; thus the degree of the polynomial is 4. We use this idea in the next example.

EXAMPLE 3   Rewrite each of the given polynomials using exponents and state the degree of the polynomial.

a. $(x \cdot x) + (x \cdot y \cdot y) + (x \cdot x \cdot y \cdot y)$ **b.** $8xxxyy - 2xxyyzzz$

Solution

a. $(x \cdot x) + (x \cdot y \cdot y) + (x \cdot x \cdot y \cdot y)$ can be written as $x^2 + xy^2 + x^2y^2$. The degree of this polynomial is $2 + 2 = 4$.

b. $8xxxyy - 2xxyyzzz$ can be written as $8x^3y^2 - 2x^2y^2z^3$. The degree of this polynomial is $2 + 2 + 3 = 7$.

Problem 3   Rewrite each polynomial using exponents and give its degree.
a. $(y \cdot y \cdot y) + 5 \cdot x \cdot y - x \cdot y \cdot z \cdot z$
b. $7 \cdot x \cdot x \cdot x \cdot y - 4 \cdot x \cdot x \cdot y \cdot y \cdot y$

D. EVALUATING POLYNOMIALS

In mathematics, polynomials in one variable are sometimes represented by using symbols such as $P(t)$ (read "P of t"), $Q(x)$, and $D(y)$, where the symbol in parentheses indicates the variable being used. For example, we may have:

$$P(t) = -16t^2 + 10t - 15$$
$$Q(x) = x^2 - 3x$$
$$D(y) = -3y - 9$$

With this notation, it is easy to indicate the value of a polynomial for specific values of the variable. Thus, $P(1)$ represents the value of the polynomial $P(t)$ when 1 is substituted for t in the polynomial. Similarly, $Q(3)$ represents the value of $Q(x)$ for $x = 3$. Thus

$$P(t) = -16t^2 + 10t - 15$$
$$P(1) = -16(1)^2 + 10(1) - 15$$
$$= -16 + 10 - 15 = -21$$

$P(1)$ represents the value of the polynomial $P(t)$ when $t = 1$. We also say that we are *evaluating* $P(t)$ at $t = 1$.

Similarly,

$$Q(x) = x^2 - 3x$$
$$Q(3) = 3^2 - 3(3) = 0$$

Using these ideas, we can find the height above sea level of the man mentioned at the beginning of this section. As you recall, the altitude of the man after t sec was given as $H(t) = -16t^2 + 118$. Thus, after 1 sec his altitude will be

$$H(1) = -16(1)^2 + 118 = 102 \text{ ft}$$

Here we are evaluating $H(t)$ when $t = 1$.

After 2 sec it will be

$$H(2) = -16(2)^2 + 118 = -64 + 118 = 54 \text{ ft} \quad \text{Here we are evaluating } H(t) \text{ when } t = 2.$$

EXAMPLE 4   Find the altitude of the man mentioned at the beginning of this section after 3 sec.

Solution   After 3 sec the formula tells us that the altitude of the man will be $P(3) = -16(3)^2 + 118 = -26$ ft, that is, 26 ft below sea level. Since the water is only 12 ft deep at this point, divers cannot continue to free fall after they hit the surface. In other words, the formula does not apply.

EXAMPLE 5   Let $P(x) = x^2 - 2x + 3$ and $Q(x) = x^2 + 3x - 5$.

a. Find $P(0)$.   **b.** Find $Q(-1)$.   **c.** Find $P(0) + Q(-1)$.

Solution

a. To find $P(0)$, we substitute 0 for x in $P(x)$.

$$P(x) = x^2 - 2x + 3$$
$$P(0) = 0^2 - 2 \cdot 0 + 3$$
$$= 0 - 0 + 3$$
$$= 3$$

Hence, $P(0) = 3$.

b. $Q(x) = x^2 + 3x - 5$.

$$Q(-1) = (-1)^2 + 3(-1) - 5$$
$$= 1 - 3 - 5$$
$$= -7$$

Hence, $Q(-1) = -7$.

c. Since $P(0) = 3$ and $Q(-1) = -7$,

$$P(0) + Q(-1) = 3 + (-7) = -4$$

Problem 4   Find the altitude of the man after 2.5 sec.

Problem 5   If $P(x)$ and $Q(x)$ are as in Example 5, find
a. $P(1)$ **b.** $Q(-2)$
c. $P(1) + Q(-2)$

___
___
___
___
___
___

NAME _____

CLASS _____

SECTION _____

ANSWERS

A, B. In Problems 1–10, classify the given polynomial as a monomial, binomial, or trinomial and give its degree.

1. xyz^2

2. u^2vw^3

3. $x^2 + yz^2$

4. $x^2y + z^3 - x^6$

5. $x + y^2 + z^3$

6. $xy + y^3$

7. $x^2yz - xy^3 - u^2v^3$

8. 8

9. 0

10. $3xyz - uv^2 + v^7$

1. _____
2. _____
3. _____
4. _____
5. _____
6. _____
7. _____
8. _____
9. _____
10. _____

B, C. In Problems 11–16 use exponents to rewrite each expression as a polynomial and give its degree.

11. $x \cdot x \cdot x + y \cdot y \cdot y \cdot y$

12. $(ab)(ab) - (xy)(xy)$

13. $xxxx + yyyy + zzzz$

14. $2aabb - 4(xy)(xy)$

15. $3xx - 4yyy + 5(xy)(xy)$

16. $9x^2 - 6(xyz)(xyz)$

11. _____
12. _____
13. _____
14. _____
15. _____
16. _____

D. In Problems 17–26, evaluate the polynomial for the specified values of the variables.

17. z^3 for $z = -2$

18. $(xy)^3$ for $x = 2, y = -1$

19. $(x - 2y + z)^2$ for $x = 2, y = -1, z = -2$

20. $(x - y)(x - z)$ for $x = 2, y = -1, z = -2$

21. If $P(x) = 4x^2 + 4x - 1$, find $P(-1)$.

22. If $P(x) = -3x^2 + 3x + 2$, find $P(0)$.

23. If $Q(y) = y^2 - 7y - 2$, find $Q(-2)$.

24. If $R(t) = t^2 - 2t + 7$, find $R(-2)$.

25. If $S(u) = -16u^2 + 120$, find $S(4)$.

26. If $V(t) = -16t^2 + 80t$, find $V(3)$.

17. _____
18. _____
19. _____
20. _____
21. _____
22. _____
23. _____
24. _____
25. _____
26. _____

27. $P(x) = 2x^2 + 3x$ and $Q(y) = -3y^2 - 7y + 1$.
  a. Find $P(0)$.
  b. Find $Q(1)$.
  c. Find $P(0) + Q(1)$.

27. a. _____
  b. _____
  c. _____

28. $P(x) = 3x^2 - 2x + 5$ and $Q(y) = -2y^2 + 3y - 1$.
  a. Find $P(-2)$.
  b. Find $Q(0)$.
  c. Find $P(-2) + Q(0)$.

28. a. _____
  b. _____
  c. _____

29. $P(x) = x^2 - 2x + 5$ and $Q(y) = -2y^2 + 5y - 1$.
  a. Find $P(-1)$.
  b. Find $Q(1)$.
  c. Find $P(-1) - Q(1)$.

29. a. _____
  b. _____
  c. _____

30. $P(x) = x^2 - 3x + 5$ and $Q(y) = -2y^2 + 5y - 1$.
  a. Find $P(-2)$.
  b. Find $Q(2)$.
  c. Find $P(-2) - Q(2)$.

30. a. _____
  b. _____
  c. _____

✓ **SKILL CHECKER**

Simplify.

31. $[(5x^2 - 3) + (4x + 5)] + [(x - 2) + (2x^2 - 3)]$

31. _____

32. $[(6x^2 - 2) + (5x + 2)] + [(x - 3) + (3x^2 - 1)]$

32. _____

33. $[(5x^2 - 3) - (4x + 5)] - [(x - 2) + (2x^2 - 3)]$

33. _____

34. $[(6x^2 - 2) - (5x + 2)] - [(x - 3) + (3x^2 - 1)]$

34. _____

3.1 USING YOUR KNOWLEDGE

Polynomials are used to represent different quantities in a variety of fields. For example, the height $H(t)$ (in feet) of an object thrown straight up with an initial velocity of 64 ft/sec after t sec is given by

$$H(t) = -16t^2 + 64t$$

1. **a.** Find the height of the object after 1 sec.
  b. Find the height of the object after 2 sec.

1. a. _____
  b. _____

2. The total dollar cost $C(x)$ of manufacturing x units of a certain product each week is given by

$$C(x) = 10x + 400$$

  a. Find the cost of manufacturing 500 units.
  b. Find the cost of manufacturing 1000 units.

2. a. _____
  b. _____

3. The dollar revenue obtained by selling x units of a certain product each week is given by

$$R(x) = 100x - 0.03x^2$$

  a. Find the revenue when 500 units are sold
  b. Find the revenue when 1000 units are sold

3. a. _____
  b. _____

4. In business, you can estimate your gross profits by subtracting the cost from the revenue. If the cost C and revenue R are given by

$$C(x) = 10x + 400 \text{ and } R(x) = 100x - 0.03x^2$$

find the gross profit when
a. 500 units are sold
b. 1000 units are sold

4. a. _____
   b. _____

5. If a \$100,000 computer depreciates 10% each year its value $V(t)$ after t years is given by:

$$V(t) = 100{,}000 - 0.10t(100{,}000)$$
$$= 100{,}000(1 - 0.10t)$$

a. Find the value of the computer after 5 yr.
b. Find the value of the computer after 10 yr.

5. a. _____
   b. _____

As you will see in the work of some later chapters, it is frequently necessary to evaluate a polynomial in x for some given values of x. If these values of x are not small, simple numbers, then the arithmetic can become quite tedious. However, if you have a small hand calculator that can do addition, subtraction, and multiplication, all of the tiresome arithmetic can easily be done on your calculator. Let's illustrate this with an example.

Suppose that $P(x) = 2x^3 - 7x^2 + 5x - 10$ and we wish to calculate $P(3.1)$. First, we use the distributive and associative laws to rewrite $P(x)$ as follows:

$$P(x) = (2x^3 - 7x^2 + 5x) - 10 = x(2x^2 - 7x + 5) - 10$$
$$= x[x(2x - 7) + 5] - 10$$

Although it looks as if we have complicated matters, we have actually broken the computation down into the individual, small steps that we need in order to use the calculator. Follow these steps as we do the calculation starting from the innermost parentheses and working out.

1. Multiply $\underline{2}$ by the value of x, 3.1, to get 6.2.
2. Subtract $\underline{7}$ to get -0.8.
3. Multiply -0.8 by the value of x, 3.1, to get -2.48.
4. Add $\underline{5}$ to -2.48 to get 2.52.
5. Multiply 2.52 by the value of x, 3.1, to get 7.812.
6. Subtract $\underline{10}$ from 7.812 to get -2.188. This is the value of $P(3.1)$.

The underlined numbers are obtained from the coefficients of $P(x)$. In the even-numbered steps, these numbers are added or subtracted according to the sign of the coefficient.

If you have storage and recall keys on your calculator, you can store the value of x and recall it as needed. Here are the procedures for finding $P(3.1)$ following the steps just outlined.

Algebraic logic

| 3.1 | STO | 2 | × | RCL | − | 7 |

| = | × | RCL | + | 5 | = | × | RCL | − | 10 | = |

EXERCISE 3.1

RPN logic

| 3.1 | STO | 2 | ENTER |

| RCL | × | 7 | − | RCL | × | 5 | + | RCL | × | 10 | − |

Try it. You should get the answer -2.188. A little practice will enable you to do calculations such as the preceding one in just a few seconds. Use the procedure to find $P(3.21)$ for the same polynomial. You should come out with 0.073622.

Notice that the important numbers in the calculation are the numerical coefficients of the polynomial. For the given $P(x)$, we can list the coefficients as

$$2, \quad -7, \quad +5, \quad -10$$

The first of these is the one we multiply by the value of x in the first step. The others are added or subtracted according to their signs in every even-numbered step before multiplying by the value of x in the succeeding step. If a power of x is missing, be sure to supply the coefficient zero for this term.

Use your calculator to find the indicated values of the polynomials in the following problems.

1. $P(x) = x^3 - 4x^2 + 2x - 5; \quad P(2.56)$

2. $Q(x) = 3x^3 - x^2 + x + 2; \quad Q(-0.24)$

3. $H(t) = 8t^3 + 7t^2 + 5t + 30; \quad H(-1.68)$

4. $G(y) = 8y^3 - 7y - 80; \quad G(2.35)$

5. $R(x) = x^4 + 5x^3 + 3x - 20; \quad R(1.38)$

6. $S(t) = 5t^4 + 8t^3 - t^2 + t - 52; \quad S(-2.44)$

1. _____

2. _____

3. _____

4. _____

5. _____

6. _____

3.2 ADDITION AND SUBTRACTION OF POLYNOMIALS

OBJECTIVES

REVIEW

Before starting this section, you should know:

1. The properties of the real numbers.
2. How to collect like terms.
3. How to remove parentheses in an expression preceded by a minus sign.

OBJECTIVES

You should be able to:

A. Add two or more polynomials.
B. Subtract two or more polynomials.
C. Solve applications involving sums or differences of polynomials.

The woman is selling jewelry at 10% over cost. Her profit P is the amount R (revenue) of money she gets for the jewelry minus the cost C—that is,

$$P = R - C$$

In many instances, R and C are polynomials. Thus, suppose $R(p) = 60p - 0.3p^2$, where p is the price per unit, and $C(p) = 4000 - 20p$. To find the profit, we need to find $R(p) - C(p)$. We will discuss how to do that in this section.

A. ADDING POLYNOMIALS

Suppose we wish to add $(5x^2 + 3x + 9) + (7x^2 + 2x + 1)$. For our solution, we use the commutative, associative, and distributive properties studied in Chapter 1. With these facts, the terms in the expression $(5x^2 + 3x + 9) + (7x^2 + 2x + 1)$ can be added as follows:

$$(5x^2 + 3x + 9) + (7x^2 + 2x + 1)$$
$$= (5x^2 + 7x^2) + (3x + 2x) + (9 + 1) \qquad \text{Group like terms.}$$
$$= (5 + 7)x^2 + (3 + 2)x + (9 + 1) \qquad \text{Use the distributive law.}$$
$$= 12x^2 + 5x + 10$$

This addition is sometimes done by writing the terms of the polynomials in order of descending (or ascending) degree and then placing like terms in a column, as shown next.

$$
\begin{array}{r}
5x^2 + 3x + 9 \\
+\ \ 7x^2 + 2x + 1 \\
\hline
12x^2 + 5x + 10
\end{array}
$$

As we have already seen, $a - b = a + (-b)$. Thus the signs in a polynomial are always taken to indicate positive or negative coefficients, and the operation involved is assumed to be addition. Thus

$$(6x^2 - 9x - 8) + (x^2 + 3x - 1)$$
$$= [6x^2 + (-9x) + (-8)] + [x^2 + 3x + (-1)]$$
$$= (6x^2 + x^2) + (-9x + 3x) + [-8 + (-1)]$$
$$= (6 + 1)x^2 + (-9 + 3)x + [-8 + (-1)]$$
$$= 7x^2 + (-6)x + (-9)$$
$$= 7x^2 - 6x - 9$$

Using the column method, this problem can be shortened to

$$+\begin{array}{r} 6x^2 - 9x - 8 \\ x^2 + 3x - 1 \\ \hline 7x^2 - 6x - 9 \end{array}$$

It is important to note that the scheme used to add polynomials is dependent on the fact that the same laws used in the addition of numbers also apply to polynomials. We list these laws here for your convenience.

If P, Q, and R are polynomials:

$P + Q = Q + P$	Commutative law of addition
$P + (Q + R) = (P + Q) + R$	Associative law of addition
$P(Q + R) = PQ + PR$	Distributive laws
$(Q + R)P = QP + RP$	

EXAMPLE 1 Add $10x^3 + 8x^2 - 7x - 3$ and $9 - 4x + x^2 - 5x^3$.

Solution We write $9 - 4x + x^2 - 5x^3$ in descending order: $-5x^3 + x^2 - 4x + 9$. We then place like terms in a column and add as shown next.

$$+\begin{array}{r} 10x^3 + 8x^2 - 7x - 3 \\ -5x^3 + x^2 - 4x + 9 \\ \hline 5x^3 + 9x^2 - 11x + 6 \end{array}$$

EXAMPLE 2 Add $-8x^2 + 7x - 8$ and $3x^2 + 5$.

Solution The polynomials are already in descending order, so we place like terms in a column, leaving space for any missing terms, and add as follows:

$$+\begin{array}{r} -8x^2 + 7x - 8 \\ 3x^2 \quad + 5 \\ \hline -5x^2 + 7x - 3 \end{array}$$ x term is missing

B. SUBTRACTING POLYNOMIALS

To subtract polynomials, we first recall that

$$a - (b + c) = a - b - c$$

For example, the difference between the revenue $R(p) = 60 - 0.3p^2$ and the cost $C(p) = 4000 - 20p$ is:

$$(60p - 0.3p^2) - (4000 - 20p) = 60p - 0.3p^2 - 4000 + 20p$$
$$= -0.3p^2 + 80p - 4000$$

Problem 1 Add:
$9x^3 + 7x^2 - 3x + 5$ and
$8 - 3x + x^2 - 6x^3$

Problem 2 Add $5x^2 + 3$ and $-7x^2 + 5x - 4$.

Similarly,

$$(3x^2 + 4x - 5) - (5x^2 + 2x + 3) = 3x^2 + 4x - 5 - 5x^2 - 2x - 3$$
$$= -2x^2 + 2x - 8$$

Note that to subtract $(5x^2 + 2x + 3)$ from $(3x^2 + 4x - 5)$, we changed the sign of each term in $(5x^2 + 2x + 3)$ and then added. This procedure can also be done in columns as shown:

$$
\begin{array}{r}
3x^2 + 4x - 5 \\
(-)\ 5x^2 + 2x + 3 \\
\hline
\end{array}
\quad \xrightarrow{\text{Is written}} \quad
\begin{array}{r}
3x^2 + 4x - 5 \\
(+)\ -5x^2 - 2x - 3 \\
\hline
-2x^2 + 2x - 8
\end{array}
$$

EXAMPLE 3 Subtract $9x^3 - 7x^2 + 3x - 5$ from $6x^3 + 2x^2 + 5$

$$
\begin{array}{r}
6x^3 + 2x^2 \quad\ + 5 \\
-\ 9x^3 - 7x^2 + 3x - 5 \\
\hline
\end{array}
\quad \xrightarrow{\text{Is written}} \quad
\begin{array}{r}
6x^3 + 2x^2 \qquad + \ 5 \\
-9x^3 + 7x^2 - 3x + \ 5 \\
\hline
-3x^3 + 9x^2 - 3x + 10
\end{array}
$$

Of course, the same result can be obtained by combining like terms, using the usual rules of signs. Thus

$$(6x^3 + 2x^2 + 5) - (9x^3 - 7x^2 + 3x - 5)$$
$$= (6 - 9)x^3 + [2 - (-7)]x^2 - 3x + [5 - (-5)]$$
$$= -3x^3 + 9x^2 - 3x + 10$$

C. APPLICATIONS

As we mentioned at the beginning of this section, the profit P derived from selling x units of a product is related to the cost C and the revenue R, and is given by $P = R - C$.

EXAMPLE 4 A company produces video cassettes at a weekly cost $C = 2x + 1000$ (dollars). What is their weekly profit P if their revenue R is given by $R = 50x - 0.1x^2$ and they produce and sell 300 cassettes a week?

Solution We need to find

$$
\begin{aligned}
P = \qquad R \qquad - \qquad C \\
= (50x - 0.1x^2) - (2x + 1000) \\
= 50x - 0.1x^2 - 2x - 1000 \\
= -0.1x^2 + 48x - 1000
\end{aligned}
$$

If they produced 300 cassettes, $x = 300$ and

$$
\begin{aligned}
P &= -0.1(300)^2 + 48(300) - 1000 \\
&= -9000 + 14{,}400 - 1000 \\
&= 4400 \text{ (dollars)}
\end{aligned}
$$

Thus their profit P when they sell 300 cassettes is $4400.

Problem 3 Subtract $8x^3 - 6x^2 + 5x - 3$ from $4x^3 + 4x^2 + 3$.

Problem 4 Repeat Example 4 if the revenue is $R = 60x - 0.1x^2$.

A, B.   In Problems 1–25, perform the indicated operations.

ANSWERS

1. $(x^2 + 4x - 8) + (5x^2 - 4x + 3)$

2. $(3x^2 + 2x + 1) + (8x^2 - 7x + 5)$

3. $(5x^2 + 4 + 3x) + (-4x^2 - 5x - 8)$

4. $(-5x^2 + 4x - 3) + (4x + 6x^2 - 7)$

5. $(4x^2 + 7x - 5) - (3x + x^2 + 4)$

6. $(8x^2 - 6x + 3) - (4x + 2x^2 - 6)$

7. $(6y - 3y^2 - 5) - (8y^2 + 7y - 2)$

8. $(5y - 4y^2 - 2) - (5y^2 - 3y + 6)$

9. $(x^3 - 6x^2 + 4x - 2) + (3x^3 - 6x^2 + 5x - 4)$

10. $(-6x^3 - 3x + 2x^2 + 2) + (2x^3 - 6x^2 + 8x - 4)$

11. $(-8y^3 + 5y + 7y^2 - 5) + (8y^3 + 7y - 6)$

12. $(5y^3 + 3y - 8) + (-9y^3 - 6y + 2y^2 + 3)$

13. $(6v^3 - 3v^2 + 2v - 5) - (3v^3 + v - v^2 + 2)$

14. $(3v^3 - 7v^2 + 3v - 1) - (5v^3 + 3v^2 - 6v + 4)$

15. $(4u^3 - 5u^2 - u + 3) - (2u + 9u^3 - 7)$

16. $(x^3 + y^3 - 8xy + 3) + (10xy - y^3 + 2x^3 - 6)$

17. $(x^3 + y^3 - 6xy + 7) + (3x^3 - y^3 + 8xy - 8)$

18. $(2x^3 - y^3 + 3xy - 5) - (x^3 + y^3 - 3xy + 9)$

19. $(x^3 - y^3 + 5xy - 2) - (x^2 - y^3 + 5xy + 2)$

20. $(4x^2 + y^2 - 3x^2y^2) - (x^3 + 3y^2 - 3x^2y^2)$

21. $(a + a^2) + (9a - 4a^2) + (a^2 - 5a)$

22. $(2x - 5x^2) + (7x - x^2) - (x + x^2)$

23. $2y + (x + 3y) - (x + y)$

24. $8y - (y + 3x) + 7y$

25. $(3x^2 + y) - (x^2 - 3y) + (3y + x^2)$

1. _____

2. _____

3. _____

4. _____

5. _____

6. _____

7. _____

8. _____

9. _____

10. _____

11. _____

12. _____

13. _____

14. _____

15. _____

16. _____

17. _____

18. _____

19. _____

20. _____

21. _____

22. _____

23. _____

24. _____

25. _____

In Problems 26–30, let $P(x) = x^2 - 2x + 3$ and $Q(x) = 2x^2 + 3x - 1$.
Find.

26. $P(x) - Q(x)$ 27. $P(0) + Q(0)$

28. $P(1) - Q(-1)$ 29. $P(x) - P(x)$

26. _____

27. _____

28. _____

29. _____

30. $[P(x) + Q(x)] + P(x)$

30. _____

In Problems 31–40, justify each of the equalities by using one of the three laws given in the text.

31. $3x^2 + 9x = 9x + 3x^2$

31. _____

32. $-y^3 + 7y = 7y + (-y)^3$

32. _____

33. $(8x + 9)4 = 32x + 36$

33. _____

34. $(7 - 2x)(-3) = -21 + 6x$

34. _____

35. $x^2 + (x + 5) = (x^2 + x) + 5$

35. _____

36. $(y^3 + y^2) + 4 = y^3 + (y^2 + 4)$

36. _____

37. $x^7 + (3 + x) = x^7 + (x + 3)$

37. _____

38. $(y + 7) + y^3 = y^3 + (y + 7)$

38. _____

39. $3(x^2 + 5) = 3x^2 + 15$

39. _____

40. $8(x^3 - 5x) = 8x^3 + 40x$

40. _____

C. Applications.

In Problems 41–45 $P = R - C$, where P is the profit, R is the revenue, and C is the cost.

41. The cost C in dollars of producing x items is given by $C = 100 + 0.3x$. If the revenue $R = 1.50x$, find the profit P when 100 items are produced and sold.

41. _____

42. The cost C in dollars of producing x pairs of jogging shoes is given by $C = 2000 + 60x$. If the revenue $R = 90x$, find the profit when 200 pairs are produced and sold.

42. _____

43. The cost C in dollars of producing x pairs of jeans is given by $C = 1500 + 20x$. If the revenue $R = 50x - x^2/20$, find the profit when 100 pairs of jeans are produced and sold.

43. _____

44. The cost C in dollars of producing x pairs of sunglasses is given by $C = 30,000 + 60x$. Find the profit when 300 pairs of sunglasses are manufactured and sold if the revenue $R = 200x - x^2/30$.

44. _____

45. The cost C in dollars of producing x pairs of shoes is given by $C = 100,000 + 30x$. Find the profit when 500 pairs of shoes are manufactured and sold if the revenue $R = 300x - x^2/50$.

45. _____

✓ **SKILL CHECKER**

Multiply.

46. $6x^2y \cdot 2x$

47. $6x^2y \cdot 3xy$

46. _____

47. _____

48. $-2x^3y \cdot (-2y^2)$

49. $-2x^3y \cdot (-7xy)$

48. _____

49. _____

50. $(2x)^2$

50. _____

3.2 USING YOUR KNOWLEDGE

The addition of polynomials can be used to find the sum of the areas of several rectangles. Thus, to find the total area of the rectangles, add the individual areas as shown.

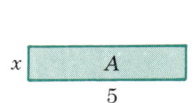 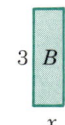 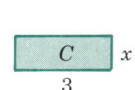 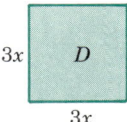

The total area is:

$$\underbrace{\underbrace{\text{Area of A}}_{} + \underbrace{\text{area of B}}_{} + \underbrace{\text{area of C}}_{} + \underbrace{\text{area of D}}_{}}$$

$$\underbrace{5x \quad + \quad 3x \quad + \quad 3x}_{11x} \quad + \quad (3x)^2$$
$$ \quad + \quad 9x^2$$

or

$9x^2 + 11x$ in descending order

Find the sum of the areas of the shaded rectangles.

1.

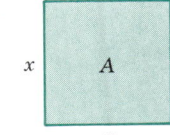 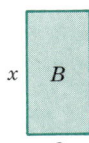 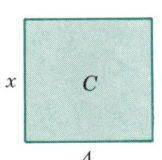 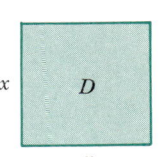

1. _____

2.

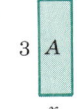 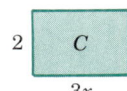 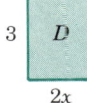

2. _____

3.

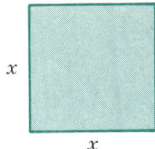 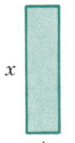 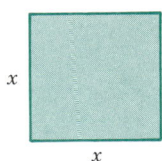

3. _____

4.

 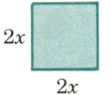 

4. _____

5.

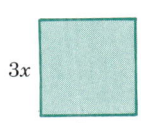 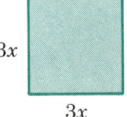

5. _____

3.3 MULTIPLICATION OF POLYNOMIALS

OBJECTIVES

REVIEW

Before starting this section, you should know:

1. How to use the distributive law to simplify an expression.
2. How to use the laws of exponents (Sections 1.5, 1.6).

OBJECTIVES

You should be able to:

A. Multiply a monomial by a polynomial.
B. Multiply a binomial by a trinomial.
C. Use the FOIL method to multiply two binomials.
D. Square a binomial sum or difference.
E. Find the product of the sum and difference of two squares.
F. Use the ideas discussed to solve applications.

How much does the beam bend (deflect) when a car or truck goes over the bridge? There is a formula that can tell us! For a certain beam of length L at a distance x from one end, the deflection is given by

$$(x - L)(x - 2L)$$

To multiply these two binomials we first learn how to do several types of related multiplications.

A. MULTIPLYING MONOMIALS BY POLYNOMIALS

When multiplying polynomials such as $6x^2$ and $2x + 3xy$, we use the commutative, associative, and distributive laws of multiplication for polynomials. These laws are generalizations of the laws discussed in Chapter 1. We state them here for polynomials.

If P, Q, and R are polynomials	
$P \cdot Q = Q \cdot P$	Commutative law
$P \cdot (Q \cdot R) = (P \cdot Q) \cdot R$	Associative law
$P(Q + R) = PQ + PR$	Distributive laws
$(Q + R)P = QP + RP$	

To multiply $6x^2y(2x + 3xy)$, we proceed as follows:

$$6x^2y(2x + 3xy) = 6x^2y(2x) + 6x^2y(3xy) \quad \text{Use the distributive law.}$$
$$= 12x^3y + 18x^3y^2 \quad \text{Multiply.}$$

Note that we multiplied the coefficients but added the exponents. We can multiply $4x^2(3x^3 + 7x^2 - 4x + 8)$ in a similar way. Thus

$$4x^2(3x^3 + 7x^2 - 4x + 8) = 4x^2(3x^3) + 4x^2(7x^2) + 4x^2(-4x) + 4x^2(8)$$
$$= 12x^5 + 28x^4 - 16x^3 + 32x^2$$

EXAMPLE 1 Multiply.

a. $5x^2(3x^3 + 3x^2 - 2x - 3)$ **b.** $-2x^3y(x^2 + 7xy - 2y^2)$

Solution

a. $5x^2(3x^3 + 3x^2 - 2x - 3) = 15x^5 + 15x^4 - 10x^3 - 15x^2$

b. $-2x^3y(x^2 + 7xy - 2y^2) = -2x^5y - 14x^4y^2 + 4x^3y^3$

Problem 1 Multiply.
a. $4x^3(5x^3 + 3x^2 - 2x - 5)$
b. $-3xy^2(x^3 - 7xy - 2x^2)$

B. MULTIPLYING TWO POLYNOMIALS

To multiply $(3x + 2)(x + 3)$, we use the fact that

$$a(b + c) = ab + ac$$

and multiply each term of $x + 3$ by $3x + 2$, obtaining

$$(3x + 2)(x + 3) = (3x + 2)(x) + (3x + 2)(3)$$
$$= 3x^2 + 2x + 9x + 6$$
$$= 3x^2 + 11x + 6$$

This multiplication can also be done by arranging the work as in ordinary multiplication and placing like terms in the same column. The procedure looks like this:

$$
\begin{array}{r}
x + 3 \\
3x + 2 \\
\hline
2x + 6 \quad \text{Multiply by 2.} \\
3x^2 + 9x \quad\quad \text{Multiply by } 3x. \\
\hline
3x^2 + 11x + 6 \quad \text{Add like terms.}
\end{array}
$$

Now let us use this same technique to multiply two polynomials like $(x + 5)$ and $(x^2 + x - 2)$.

STEP 1	**STEP 2**	**STEP 3**
$x^2 + x - 2$	$x^2 + x - 2$	$x^2 + x - 2$
$x + 5$	$x + 5$	$x + 5$
$\overline{5x^2 + 5x - 10}$	$\overline{5x^2 + 5x - 10}$	$\overline{5x^2 + 5x - 10}$
	$x^3 + x^2 - 2x$	$x^3 + x^2 - 2x$
		$\overline{x^3 + 6x^2 + 3x - 10}$
$5(x^2 + x - 2)$	$x(x^2 + x - 2)$	
$= 5x^2 + 5x - 10$	$= x^3 + x^2 - 2x$	

For obvious reasons this method is called the **vertical scheme** and can be used when one of the polynomials to be multiplied has three or more terms. Of course, we could have obtained the same result by using the distributive law and multiplying $(x + 5)$ by x^2, $(x + 5)$ by x, and $(x + 5)$ by -2. The result would look like this:

$$(x + 5)(x^2 + x - 2) = (x + 5)(x^2) + (x + 5)x + (x + 5)(-2)$$
$$= x^3 + 5x^2 + x^2 + 5x - 2x - 10$$
$$= x^3 + (5x^2 + x^2) + (5x - 2x) - 10$$
$$= x^3 + 6x^2 + 3x - 10$$

EXAMPLE 2   Multiply $(x - 3)(x^2 - 2x - 4)$.

Solution   Using the vertical scheme:

$$
\begin{array}{r}
x^2 - 2x - \ 4 \\
x - \ 3 \\
\hline
-3x^2 + 6x + 12 \\
x^3 - 2x^2 - 4x \\
\hline
x^3 - 5x^2 + 2x + 12
\end{array}
$$

Multiply by -3.
Multiply by x.
Add like terms.   ▲

The result of multiplying $(x - 3)(x^2 - 2x - 4)$ is $x^3 - 5x^2 + 2x + 12$. You can also do this problem by multiplying $x(x^2 - 2x - 4)$ first and then multiplying $-3(x^2 - 2x - 4)$, obtaining the result:

$$
\begin{aligned}
(x - 3)(x^2 - 2x - 4) &= x(x^2 - 2x - 4) - 3(x^2 - 2x - 4) \\
&= x^3 - 2x^2 - 4x - 3x^2 + 6x + 12 \\
&= x^3 + (-2x^2 - 3x^2) + (-4x + 6x) + 12 \\
&= x^3 - 5x^2 + 2x + 12
\end{aligned}
$$

Note that the same result is obtained in both cases. Here is the idea we used:

RULE TO MULTIPLY ANY TWO POLYNOMIALS

To multiply two polynomials, multiply each term of one by every term of the other and add the results.

C.   MULTIPLYING TWO BINOMIALS

Let us use the rule we just mentioned to multiply $(x + 4)$ and $(x - 7)$. We have to multiply each term of $x - 7$ by every term of $x + 4$. We write:

$$
\begin{aligned}
(x + 4)(x - 7) &= x \cdot x + x \cdot (-7) + 4 \cdot x + 4 \cdot (-7) \\
&= x^2 - 7x + 4x - 28 \\
&= x^2 - 3x - 28
\end{aligned}
$$

If you look at the second line, you can see what the procedure for multiplying two binomials is.

The first term, x^2, is the product of the first terms.

$(x + 4)(x - 7)$ x^2 First terms

The second term, $-7x$, is the product of the outside terms.

$(x + 4)(x - 7)$ $-7x$ Outside terms

The third term, $+4x$, is the product of the inside terms.

$(x + 4)(x - 7)$ $4x$ Inside terms

Problem 2   Multiply $(x - 2)(x^2 - 4x - 3)$.

The last term, -28, is the product of the last two terms.

$(x + 4)(x - 7) \qquad -28 \quad$ Last terms

Thus, to multiply $(x + 4)(x - 7)$, we simply write:

$$(x + 4)(x - 7) = x^2 \quad - 7x \quad + 4x \quad - 28$$

$$\qquad\qquad\qquad \underset{\text{F}}{\text{First}} \ \ \underset{\text{O}}{\text{Outside}} \ \ \underset{\text{I}}{\text{Inside}} \ \ \underset{\text{L}}{\text{Last}}$$

$$= x^2 - 3x - 28$$

Here is the general rule for multiplying two binomials using FOIL.

PRODUCT OF TWO BINOMIALS

$$\qquad\qquad\qquad\qquad\qquad \text{F} \quad\ \text{O} \quad\ \text{I} \quad\ \ \text{L}$$

Special Product (1) $\qquad (x + a)(x + b) = x^2 + bx + ax + ab$

$$= x^2 + (b + a)x + ab$$

Of course, we call this the FOIL method. We shall do one more example, step by step, so you can have more practice.

F $\qquad (x + 7)(x - 4) = x^2$

O $\qquad (x + 7)(x - 4) = x^2 - 4x$

I $\qquad (x + 7)(x - 4) = x^2 - 4x + 7x$

L $\qquad (x + 7)(x - 4) = x^2 - 4x + 7x - 28$

Thus

$$(x + 7)(x - 4) = x^2 + 3x - 28$$

EXAMPLE 3 Multiply.

a. $(5x + 2y)(2x + 3y)$
b. $(3x - y)(4x - 3y)$

Solution

$$\qquad\qquad\qquad\qquad \underset{\text{F}}{\ } \qquad\quad \underset{\text{O}}{\ } \qquad\quad \underset{\text{I}}{\ } \qquad\quad \underset{\text{L}}{\ }$$

a. $(5x + 2y)(2x + 3y) = (5x)(2x) + (5x)(3y) + (2y)(2x) + (2y)(3y)$

$$= 10x^2 + 15xy + 4xy + 6y^2$$

$$= 10x^2 + 19xy + 6y^2$$

$$\qquad\qquad\qquad\qquad \underset{\text{F}}{\ } \qquad\quad \underset{\text{O}}{\ } \qquad\quad \underset{\text{I}}{\ } \qquad\quad \underset{\text{L}}{\ }$$

b. $(3x - y)(4x - 3y) = (3x)(4x) + (3x)(-3y) + (-y)(4x) + (-y)(-3y)$

$$= 12x^2 - 9xy - 4xy + 3y^2$$

$$= 12x^2 - 13xy + 3y^2$$

D. SQUARING SUMS OR DIFFERENCES OF BINOMIALS

Now, suppose we want to find $(x + 7)^2$. The exponent 2 means that we must multiply $(x + 7)(x + 7)$. Using FOIL we write:

$$(x + 7)(x + 7) = x^2 + 7x + 7x + 7 \cdot 7$$

$$= x^2 + 14x + 49$$

Problem 3 Multiply.
a. $(4x + 3y)(3x + 2y)$
b. $(5x - y)(2x - 3y)$

Did you see how the middle term was calculated? It is the sum of $7x$ and $7x$, that is, $2 \cdot 7x$. Also, the last term is 7^2. In general, we have the following.

SQUARE OF A BINOMIAL SUM

Special product (2) $(x + a)(x + a) = x^2 + ax + ax + a \cdot a$
$$= x^2 + 2ax + a^2$$

The same pattern applies to the difference of two binomials.

SQUARE OF A BINOMIAL DIFFERENCE

Special product (3) $(x - a)(x - a) = x^2 - ax - ax + a^2$
$$= x^2 - 2ax + a^2$$

Thus, to find the square of a binomial, add the square of the first term, twice the product of the two terms, and the square of the second term. The sign of the middle term is $+$ for binomial sums and $-$ for binomial differences.

EXAMPLE 4   Multiply.

a. $(2x + 3y)^2$ **b.** $(3x - 2y)^2$

Solution

a. $(2x + 3y)^2 = (2x)^2 + 2 \cdot 2x \cdot 3y + (3y)^2$
$$= 4x^2 + 12xy + 9y^2$$
b. $(3x - 2y) = (3x)^2 + 2 \cdot 3x \cdot (-2y) + (-2y)^2$
$$= 9x^2 - 12xy + 4y^2$$

Problem 4   Multiply.
a. $(3x + 2y)^2$ **b.** $(2x - 3y)^2$

E.   PRODUCT OF A SUM AND DIFFERENCE

We have one more special product, and this is really special. Suppose we multiply the sum of two terms by the difference of the same two terms, that is, suppose we want the product

$(x - 7)(x + 7)$

Using FOIL, we get

$(x - 7)(x + 7) = x^2 + 7x - 7x + 7^2$
$$= x^2 + 0x - 7^2$$
$$= x^2 - 49$$

Since multiplication is commutative,

$(x + 7)(x - 7) = x^2 - 49$

In general, we have the following.

PRODUCT OF THE SUM AND DIFFERENCE OF TWO MONOMIALS

Special Product (4) $(x - a)(x + a) = x^2 - a^2$

$(x + a)(x - a) = x^2 - a^2$

EXAMPLE 5 Multiply.

a. $(x + 10)(x - 10)$ **b.** $(2x + y)(2x - y)$ **c.** $(3x - 5y)(3x + 5y)$

Solution

a. $(x + 10)(x - 10) = x^2 - 10^2$

$= x^2 - 100$

b. $(2x + y)(2x - y) = (2x)^2 - y^2$

$= 4x^2 - y^2$

c. $(3x - 5y)(3x + 5y) = (3x)^2 - (5y)^2$

$= 9x^2 - 25y^2$

F. APPLICATIONS

We have mentioned that the profit P is the revenue R minus the cost C, that is, $P = R - C$. Do you know how revenue is calculated? Suppose you have 10 skateboards and sell them for \$50 each. Your revenue is $R = 10 \cdot 50 = \$500$. In general,

$$R = \begin{pmatrix} \text{number of} \\ \text{items sold} \end{pmatrix} \cdot \begin{pmatrix} \text{price of} \\ \text{each item} \end{pmatrix}$$

or

$$R = xp$$

EXAMPLE 6 The research department of a company determines that the demand x for skateboards is given by $x = 1000 - 10p$, where p is the price of each skateboard.

a. Write a formula for the revenue R.

b. Find the revenue obtained by selling the skateboards for \$50 each.

Solution

a. The revenue is

$R = xp$

$R = (1000 - 10p)p$ Substitute $x = 1000 - 100p$.

$= 1000p - 10p^2$

b. When $p = 50$,

$R = 1000(50) - 10(50)^2$

$= 50{,}000 - 10(2500)$

$= 50{,}000 - 25{,}000$

$= 25{,}000$

Thus the revenue is \$25,000.

Problem 5 Multiply.
a. $(x + 5)(x - 5)$
b. $(3x + y)(3x - y)$
c. $(5x - 3y)(5x + 3y)$

Problem 6 Repeat Example 6 if the demand is $x = 1000 - 20p$.

NAME

CLASS

SECTION

ANSWERS

A. In Problems 1–10 do the indicated multiplications.

1. $3x(4x - 2)$ **2.** $4x(x - 6)$

3. $-3x^2(x - 3)$ **4.** $-5x^3(x^2 - 8)$

5. $-8x(3x^2 - 2x + 1)$ **6.** $-4x^2(3x^2 - 5x - 1)$

7. $-3xy^2(6x^2 + 3y^2 - 7)$ **8.** $-2x^2y^3(6xy^3 - 2x^2y + 9)$

9. $2xy^3(3x^2y^3 - 5xy^2 + xy)$ **10.** $3x^4y(6x^3y^2 - 10x^2y + xy)$

1. _____

2. _____

3. _____

4. _____

5. _____

6. _____

7. _____

8. _____

9. _____

10. _____

B. Multiply.

11. $(x + 3)(x^2 + x + 5)$ **12.** $(x + 2)(x^2 + 5x + 6)$

13. $(x + 4)(x^2 - x + 3)$ **14.** $(x + 5)(x^2 - x + 2)$

15. $(x + 3)(x^2 - x - 2)$ **16.** $(x + 4)(x^2 - x - 3)$

17. $(x - 2)(x^2 + 2x + 4)$ **18.** $(x - 3)(x^2 + x + 1)$

19. $(x^2 - 1)(x^2 - x + 2)$ **20.** $(x^2 - 2)(x^2 - 2x + 1)$

11. _____

12. _____

13. _____

14. _____

15. _____

16. _____

17. _____

18. _____

19. _____

20. _____

C. Multiply.

21. $(3x + 2)(3x + 1)$ **22.** $(x + 5)(2x + 7)$

23. $(5x - 4)(x + 3)$ **24.** $(2x - 1)(x + 5)$

25. $(3a - 1)(a + 5)$ **26.** $(3a - 2)(a + 7)$

27. $(y + 5)(2y - 3)$ **28.** $(y + 1)(5y - 1)$

21. _____

22. _____

23. _____

24. _____

25. _____

26. _____

27. _____

28. _____

29. $(x - 3)(x - 5)$

30. $(x - 6)(x - 1)$

31. $(2x - 1)(3x - 2)$

32. $(3x - 5)(x - 1)$

33. $(2x - 3a)(2x + 5a)$

34. $(5x - 2a)(x + 5a)$

35. $(x + 7)(x + 8)$

36. $(x + 1)(x + 9)$

37. $(2a + b)(2a + 4b)$

38. $(3a + 2b)(3a + 5b)$

D, E. Use the special products to multiply.

39. $(4u + v)^2$

40. $(3u + 2v)^2$

41. $(2y + z)^2$

42. $(4y + 3z)^2$

43. $(3a - b)^2$

44. $(4a - 3b)^2$

45. $(a + b)(a - b)$

46. $(a + 4)(a - 4)$

47. $(5x - 2y)(5x + 2y)$

48. $(2x - 7y)(2x + 7y)$

49. $-(3a - b)(3a + b)$

50. $-(2a - 5b)(2a + 5b)$

51. $3x(x + 1)(x + 2)$

52. $3x(x + 2)(x + 3)$

53. $-3x(x - 1)(x - 3)$

54. $-2x(x - 5)(x - 1)$

55. $x(x + 3)^2$

56. $3x(x + 7)^2$

57. $-2x(x - 1)^2$

58. $-5x(x - 3)^2$

29. _____
30. _____
31. _____
32. _____
33. _____
34. _____
35. _____
36. _____
37. _____
38. _____

39. _____
40. _____
41. _____
42. _____
43. _____
44. _____
45. _____
46. _____
47. _____
48. _____
49. _____
50. _____
51. _____
52. _____
53. _____
54. _____
55. _____
56. _____
57. _____
58. _____

ANSWERS (to problems on page 192)
5. a. $x^2 - 25$
  b. $9x^2 - y^2$
  c. $25x^2 - 9y^2$
6. $0 (no profit)

59. $(2x + y)(2x - y)y^2$ **60.** $(3x + y)(3x - y)x^2$

59. _____

60. _____

F. Applications.

In Problems 61–62 $R = xp$, where x is the number of items sold and p is the price of the item.

61. The demand x for a certain product is given by $x = 1000 - 30p$.
  a. Write a formula for R.
  b. What is the revenue when the price is \$20?

61. a. _____
    b. _____

62. A company manufactures and sells x jogging suits at p dollars every day. If $x = 3000 - 30p$, write a formula for the daily revenue R and use it to find the revenue on a day in which the suits were selling for \$40.

62. _____

63. The heat transmission between two objects of temperature T_2 and T_1 involves the expression

$$(T_1^2 + T_2^2)(T_1^2 - T_2^2)$$

Multiply this expression.

63. _____

64. The deflection of a certain beam involves the expression $w(l^2 - x^2)^2$. Multiply this expression.

64. _____

65. The heat output from a natural draught convector is given by $K(t_n - t_a)^2$. Multiply this expression.

65. _____

✓ **SKILL CHECKER**

The distributive law can be written as $ab + ac = a(b + c)$. Use this idea to rewrite each expression.

66. $3x + 3y$ **67.** $5x + 5y$

66. _____

67. _____

68. $2xz + 2xy$ **69.** $3ab + 3ac$

68. _____

69. _____

70. $6bc + 6bd$

70. _____

3.3 USING YOUR KNOWLEDGE

A common fallacy (mistake) when multiplying binomials is to assume that

$$(x + y)^2 = x^2 + y^2$$

Here are some arguments that should convince you otherwise.

1. Let $x = 1$, $y = 2$.
  a. What is $(x + y)^2$?
  b. What is $x^2 + y^2$?
  c. Is $(x + y)^2 = x^2 + y^2$?

1. a. _____
   b. _____
   c. _____

2. Let $x = 2$, $y = 1$.
  a. What is $(x - y)^2$?
  b. What is $x^2 - y^2$?
  c. Is $(x - y)^2 = x^2 - y^2$?

2. a. _____
   b. _____
   c. _____

3. Look at the large square in the margin. Its area is $(x + y)^2$. The square is divided into four smaller areas numbered 1, 2, 3, and 4.

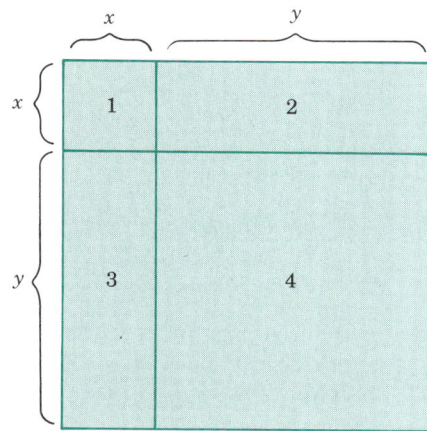

    a. What is the area of square 1?
    b. What is the area of rectangle 2?
    c. What is the area of square 4?
    d. What is the area of rectangle 3?

4. The total area of the square is $(x + y)^2$. It is also the sum of the four areas numbered 1, 2, 3, and 4. What is the sum of these four areas? (Simplify your answer.)

5. From your answer to Problem 4, what can you say about $x^2 + 2xy + y^2$ and $(x + y)^2$?

6. $x^2 + y^2$ is the sum of the areas of squares 1 and 3. Is $x^2 + y^2 = (x + y)^2$?

3. a. _____
   b. _____
   c. _____
   d. _____

4. _____

5. _____

6. _____

3.4 THE GREATEST COMMON FACTOR AND FACTORING BY GROUPING

BEST CD YIELD RATE
6 MOS 8.34% 8.01%
30 MOS 8.34% 8.01%

If you buy a 1-yr, $1000 certificate, what amount A would you get at the end of the year? You will get $1000 plus interest. Since the annual interest is $1000 \cdot 0.0834$, the amount will be

$A = 1000 + 1000 \cdot (0.0834)$

In general, if you invest P dollars at an annual rate r, the amount A dollars that you get is

$A = P + Pr$

The expression $P + Pr$ can be written in a simpler way if we **factor** it, that is, if we write it as the product of its factors. Factoring is the reverse of multiplication. Thus if we tell you to multiply P by $1 + r$, you will write $P(1 + r) = P + Pr$. If we tell you to *factor* $P + Pr$, you should write

$P + Pr = P(1 + r)$ Use the distributive law.

Here are some more examples.

FINDING THE PRODUCT	FINDING THE FACTORS
$5(x + 3) \qquad = 5x + 15$	$5x + 15 \qquad = 5(x + 3)$
$6x(x + y) \qquad = 6x^2 + 6xy$	$6x^2 + 6xy \qquad = 6x(x + y)$
$-6x^2(2x + 7y) = -12x^3 - 42x^2y$	$-12x^3 - 42x^2y = -6x^2(2x + 7y)$

How do we know if we are through with the factorization, that is, how do we know if an expression is **completely factored?** We can completely factor an expression by using the *greatest common factor* (GCF).

A. THE GREATEST COMMON FACTOR

When factoring polynomials with integer coefficients, the GCF is defined as follows.

The **greatest common factor (GCF)** of a polynomial is the monomial with (1) the highest degree and (2) the coefficient of greatest absolute value and that is a factor of each term.

OBJECTIVES

REVIEW

Before starting this section, you should know:

1. The distributive law.
2. The laws of exponents.

OBJECTIVES

You should be able to:

A. Factor out the greatest common factor in a polynomial.
B. Factor a polynomial with four terms by grouping.

Thus to factor $12x^3 + 18x^5$, we could write

$$12x^3 + 18x^5 = 2x(6x^2 + 9x^4) \quad \text{Check by multiplying!}$$

But this is not completely factored. The greatest number that is a factor of 12 and 18 is 6 and the highest power of x that is a factor of x^3 and x^5 is x^3. Thus, the GCF is $6x^3$. The factorization is

$$12x^3 + 18x^5 = 6x^3(2 + 3x^2)$$

It might help your accuracy and understanding if you write an intermediate step showing the greatest common factor present in each term. Thus when factoring $12x^3 + 18x^5$ you can write

$$12x^3 + 18x^5 = 6x^3 \cdot 2 + 6x^3 \cdot 3x^2$$
$$= 6x^3(2 + 3x^2)$$

One more convention: when factoring expressions such as $-6x + 18$, we have two choices:

$$-6(x - 3) \quad \text{and} \quad 6(-x + 3)$$

The factorization $-6(x - 3)$ is the *preferred* choice, since the first term of the binomial $x - 3$ has a positive sign.

EXAMPLE 1   Factor.

a. $8x - 24$ **b.** $-6y^2 + 12y$ **c.** $10x^2 - 25x^3$

Solution

a. $8x - 24 = 8 \cdot x - 8 \cdot 3$
$$= 8(x - 3)$$
b. $-6y^2 + 12y = -6y \cdot y - 6y \cdot (-2)$
$$= -6y(y - 2)$$
c. $10x^2 - 25x^3 = 5x^2 \cdot 2 - 5x^2 \cdot 5x$
$$= 5x^2(2 - 5x)$$

Check your answers by multiplying out the expressions on the right of the equal sign.   ▲

We can also factor polynomials with more than two terms, as shown next.

EXAMPLE 2   Factor.

a. $6x^3 + 12x^4 + 18x^2$ **b.** $10x^6 - 15x^5 + 20x^7 + 30x^2$

Solution

a. $6x^3 + 12x^4 + 18x^2 = 6x^2 \cdot x + 6x^2 \cdot 2x^2 + 6x^2 \cdot 3$
$$= 6x(x + 2x^2 + 3)$$
b. $10x^6 - 15x^5 + 20x^7 + 30x^2 = 5x^2 \cdot 2x^4 - 5x^2 \cdot 3x^3$
$$+ 5x^2 \cdot 4x^5 + 5x^2 \cdot 6$$
$$= 5x^2(2x^4 - 3x^3 + 4x^5 + 6)$$

Check by multiplying.

EXAMPLE 3   Factor $\frac{3}{4}x^2 - \frac{1}{4}x^4 + \frac{5}{4}x^5$.

Problem 1   Factor.
a. $6x - 48$ **b.** $-3y^2 + 21y$
c. $4x^2 - 32x^3$

Problem 2   Factor.
a. $7x^3 + 14x^4 - 49x^2$
b. $3x^6 - 6x^5 + 12x^7 + 27x^2$

Problem 3   Factor $\frac{2}{5}x^2 - \frac{3}{5}x^4 - \frac{1}{5}x^5$.

Solution Here, we can see that the greatest common factor is $\frac{1}{4}x^2$. Thus

$$\frac{3}{4}x^2 - \frac{1}{4}x^4 + \frac{5}{4}x^5 = \frac{1}{4}x^2(3 - x^2 + 5x^3)$$

Check by multiplying.

B. FACTORING BY GROUPING

Can we factor $x^3 + 2x^2 + 3x + 6$? It seems that there is no common factor except 1. However, we can group and factor the first two terms and also the last two terms and then use the distributive law as shown next.

STEP 1. Group terms with common factors using the associative law.

$$x^3 + 2x^2 + 3x + 6 = (x^3 + 2x^2) + (3x + 6)$$

STEP 2. Factor each resulting binomial.

$$= x^2(x + 2) + 3(x + 2)$$

STEP 3. Factor out the GCF, $(x + 2)$.

$$= (x + 2)(x^2 + 3)$$

Thus $x^3 + 2x^2 + 3x + 6 = (x + 2)(x^2 + 3)$. Note that $x^2(x + 2) + 3(x + 2) = (x^2 + 3)(x + 2)$ since $QP + RP = (Q + R)P$. Hence $x^3 + 2x^2 + 3x + 6 = (x^2 + 3)(x + 2)$. Either $(x + 2)(x^2 + 3)$ or $(x^2 + 3)(x + 2)$ is a correct factorization for $x^3 + 2x^2 + 3x + 6$. Check by multiplying.

EXAMPLE 4 Factor.

a. $3x^3 + 6x^2 + 4x + 8$ **b.** $6x^3 - 3x^2 - 4x + 2$

Problem 4 Factor.
a. $2x^3 + 2x^2 + 3x + 3$
b. $6x^3 - 9x^2 - 2x + 3$

Solution

a. We proceed by steps.

STEP 1. Group terms with common factors using the associative law.

$$3x^3 + 6x^2 + 4x + 8 = (3x^3 - 6x^2) + (4x + 8)$$

STEP 2. Factor each resulting binomial.

$$= 3x^2(x + 2) + 4(x + 2)$$

STEP 3. Factor out the GCF, $(x + 2)$, using the distributive law.

$$= (x + 2)(3x^2 + 4)$$

Note that you can write $3x^3 + 6x^2 + 4x + 8$ as $(3x^2 + 4)(x + 2)$ in Step 3. Since $(x + 2)(3x^2 + 4) = (3x^2 + 4)(x + 2)$, both answers are correct!

b. **STEP 1.** Use the associative law.

$$6x^3 - 3x^2 - 4x + 2 = (6x^3 - 3x^2) - (4x - 2)$$

Note that
$-(4x - 2) = -4x + 2$

STEP 2. Factor each resulting binomial.

$$= 3x^2(2x - 1) - 2(2x - 1)$$

STEP 3. Factor out the GCF, $(2x - 1)$.

$$= (2x - 1)(3x^2 - 2)$$

Thus $6x^3 - 3x^2 - 4x + 2 = (2x - 1)(3x^2 - 2)$. Check this.

EXAMPLE 5 Factor.

a. $2x^4 - 4x^3 - x^2 + 2x$ **b.** $6x^6 - 9x^4 + 4x^3 - 6x$

Solution

a. First, factor out the common factor x and write

$$2x^4 - 4x^3 - x^2 + 2x = x(2x^3 - 4x^2 - x + 2)$$

Now, factor $2x^3 - 4x^2 - x + 2$ as follows:

STEP 1. $2x^3 - 4x^2 - x + 2 = (2x^3 - 4x^2) - (x - 2)$

STEP 2. $\qquad\qquad\qquad = 2x^2(x - 2) - 1(x - 2)$

STEP 3. $\qquad\qquad\qquad = (x - 2)(2x^2 - 1)$

Thus $2x^4 - 4x^3 - x^2 + 2x = x(x - 2)(2x^2 - 1)$.

b. Factor out x, obtaining

$$6x^6 - 9x^4 + 4x^3 - 6x = x(6x^5 - 9x^3 + 4x^2 - 6)$$

Then, factor $6x^5 - 9x^3 + 4x^2 - 6$.

STEP 1. $6x^5 - 9x^3 + 4x^2 - 6 = (6x^5 - 9x^3) + (4x^2 - 6)$

STEP 2. $\qquad\qquad\qquad = 3x^3(2x^2 - 3) + 2(2x^2 - 3)$

STEP 3. $\qquad\qquad\qquad = (2x^2 - 3)(3x^3 + 2)$

Thus

$$6x^6 - 9x^4 + 4x^3 - 6x = x(2x^2 - 3)(3x^3 + 2)$$

Don't forget the x!

Check this.

Problem 5 Factor.
a. $3x^4 - 6x^3 - x^2 + 2x$
b. $6x^6 - 9x^4 + 2x^3 - 3x$

ANSWERS   (to problems on page 198–199)

1. a. $6(x - 8)$ **b.** $-3y(y - 7)$
   c. $4x^2(1 - 8x)$
2. a. $7x^2(x + 2x^2 - 7)$
   b. $3x^2(x^4 - 2x^3 + 4x^5 + 9)$
3. $\dfrac{1}{5}x^2(2 - 3x^2 - x^3)$
4. a. $(x + 1)(2x^2 + 3)$
   b. $(2x - 3)(3x^2 - 1)$

NAME

CLASS

SECTION

ANSWERS

A. Factor.

1. $8x + 16$

2. $15x + 45$

3. $9y - 18$

4. $11y - 88$

5. $-5y + 25$

6. $-4y + 28$

7. $-8x - 24$

8. $-6x - 36$

9. $4x^2 + 36x$

10. $5x^3 + 20x$

11. $6x - 42x^3$

12. $7x - 14x^5$

13. $-5x^2 - 35x^4$

14. $-8x^3 - 16x^6$

15. $3x^3 + 6x^2 + 39x$

16. $8x^3 + 4x^2 - 36x$

17. $63y^3 - 18y^2 + 27y$

18. $10y^3 - 5y^2 + 20y$

19. $36x^6 + 12x^5 - 18x^4 + 30x^2$

20. $15x^7 - 15x^6 + 30x^3 - 20x^2$

21. $48y^8 + 16y^5 - 24y^4 + 8y^3$

22. $12y^9 - 4y^6 + 6y^5 + 8y^4$

23. $\frac{4}{7}x^3 + \frac{3}{7}x^2 - \frac{9}{7}x + \frac{3}{7}$

24. $\frac{2}{5}x^3 + \frac{3}{5}x^2 - \frac{2}{5}x + \frac{4}{5}$

25. $\frac{7}{8}y^9 + \frac{3}{8}y^6 - \frac{5}{8}y^4 + \frac{5}{8}y^2$

26. $\frac{4}{3}y^7 - \frac{1}{3}y^5 + \frac{2}{3}y^4 - \frac{5}{3}y^3$

B. Factor by grouping.

27. $x^3 + 2x^2 + x + 2$

28. $x^3 + 3x^2 + x + 3$

ANSWERS

1. _____
2. _____
3. _____
4. _____
5. _____
6. _____
7. _____
8. _____
9. _____
10. _____
11. _____
12. _____
13. _____
14. _____
15. _____
16. _____
17. _____
18. _____
19. _____
20. _____
21. _____
22. _____
23. _____
24. _____
25. _____
26. _____

27. _____
28. _____

29. $y^3 - 3y^2 + y - 3$ **30.** $y^3 - 5y^2 + y - 5$

31. $4x^3 + 6x^2 + 2x + 3$ **32.** $6x^3 + 3x^2 + 2x + 1$

33. $6x^3 - 2x^2 + 3x - 1$ **34.** $6x^3 - 9x^2 + 2x - 3$

35. $4y^3 + 8y^2 + y + 2$ **36.** $2y^3 + 6y^2 + y + 3$

37. $2a^6 + 3a^4 + 2a^2 + 3$ **38.** $3a^6 + 2a^4 + 3a^2 + 2$

39. $3x^5 + 12x^3 + x^2 + 4$ **40.** $2x^5 + 2x^3 + x^2 + 1$

41. $6y^5 + 9y^3 + 2y^2 + 3$ **42.** $12y^5 + 8y^3 + 3y^2 + 2$

43. $4y^7 + 12y^5 + y^4 + 3y^2$ **44.** $2y^7 + 2y^5 + y^4 + y^2$

45. $3a^7 - 6a^5 - 2a^4 + 4a^2$ **46.** $4a^7 - 12a^5 - 3a^4 + 9a^2$

47. Factor $\alpha L t_2 - \alpha L t_1$, where α is the coefficient of linear expansion, L is the length of the material, and t_2 and t_1 are the temperatures in degrees Celsius.

48. Factor the expression $-kx - kl$, which represents the restoring force of a spring stretched an amount l to its equilibrium position and then an additional x units.

49. When solving for the equivalent resistance of two circuits, we have to factor the expression $R^2 - R - R + 1$. Factor this expression by grouping.

50. The bending moment of a cantilever beam of length L, at x inches from its support, involves the expression $L^2 - Lx - Lx + x^2$. Factor this expression by grouping.

✓ SKILL CHECKER

Multiply.

51. $(x + 3)(x + 4)$ **52.** $(x + 7)(x + 2)$

29. _____
30. _____
31. _____
32. _____
33. _____
34. _____
35. _____
36. _____
37. _____
38. _____
39. _____
40. _____
41. _____
42. _____
43. _____
44. _____
45. _____
46. _____
47. _____
48. _____
49. _____
50. _____
51. _____
52. _____

ANSWER (to problem on page 200)
5. a. $x(x - 2)(3x^2 - 1)$
  b. $x(2x^2 - 3)(3x^3 + 1)$

53. $(x + 5)(x - 2)$ **54.** $(x - 5)(x + 3)$

55. $(5x + 2y)^2$ **56.** $(3x + 4y)^2$

57. $(5x - 2y)^2$ **58.** $(3x - 4y)^2$

59. $(u + 6)(u - 6)$ **60.** $(2a + 7b)(2a - 7b)$

53. _____

54. _____

55. _____

56. _____

57. _____

58. _____

59. _____

60. _____

3.4 USING YOUR KNOWLEDGE

There are many formulas that can be simplified by factoring. Here are a few.

1. The vertical shear at any section of a cantilever beam of uniform cross section is

$$-wl + wz$$

Factor this expression.

2. The bending moment of any section of a cantilever beam of uniform cross section is

$$-Pl + Px$$

Factor this expression.

3. The surface area of a square pyramid is

$$a^2 + 2as$$

Factor this expression.

4. The energy of a moving object is given by

$$800m - mv^2$$

Factor this expression.

5. The height (in feet after t seconds) of a rock thrown from the roof of a certain building is given by

$$-16t^2 + 80t + 240$$

Factor this expression. *Hint:* -16 is a common factor.

1. _____

2. _____

3. _____

4. _____

5. _____

CALCULATOR CORNER

The factoring techniques studied in this section can be used to evaluate higher-degree polynomials. Moreover, the evaluation can be done with a calculator performing only the four basic operations of addition, subtraction, multiplication, and division. We illustrate the five-step procedure below for the polynomial $2a^3 + 3a^2 + 4a + 5$.

STEP 1. Given. $2a^3 + 3a^2 + 4a + 5$

STEP 2. Group the terms involving a, and factor a. $(2a^2 + 3a + 4)a + 5$

STEP 3. Repeat this process for the expression in parentheses. $[(2a + 3)a + 4]a + 5$

STEP 4. Repeat the process within the innermost grouping symbol. $\{[(2)a + 3]a + 4\}a + 5$

STEP 5. You can stop when the innermost expression (the 2) is a constant.

The keystroke sequence to evaluate this polynomial for any number a would then be:

$$\boxed{2}\;\boxed{\times}\;\boxed{a}\;\boxed{+}\;\boxed{3}\;\boxed{=}\;\boxed{\times}\;\boxed{a}\;\boxed{+}\;\boxed{4}\;\boxed{=}\;\boxed{\times}\;\boxed{a}\;\boxed{+}\;\boxed{5}\;\boxed{=}$$

What is so great about that? Well, notice that after you enter the first 2, you simply repeat over and over the keystrokes

$$\boxed{\times}\;\boxed{a}\;\boxed{+}\;\boxed{\text{a number}}\;\boxed{=}$$

For example, when $a = 3$, we obtain

$$\boxed{2}\;\boxed{\times}\;\boxed{3}\;\boxed{+}\;\boxed{3}\;\boxed{=}$$
$$\boxed{\times}\;\boxed{3}\;\boxed{+}\;\boxed{4}\;\boxed{=}\;\boxed{\times}\;\boxed{3}\;\boxed{+}\;\boxed{5}\;\boxed{=}$$

The result would be 98. In case some terms are "missing" in a given polynomial, they are inserted with a zero coefficient. Thus, to write $3a^3 + 5a + 6$ using this procedure, we first insert the "missing" a^2 term and write

$$3a^3 + 0a^2 + 5a + 6 = \{[(3)a + 0]a + 5\}a + 6$$

Use this procedure to evaluate the polynomials in Problems 31 through 34 for x (or y) equal 2.

3.5 FACTORING TRINOMIALS

OBJECTIVES

REVIEW

Before starting this section, you should know:

1. How to multiply two binomials using the FOIL method.
2. How to add and multiply signed numbers (Sections 1.3, 1.4).

OBJECTIVES

You should be able to:

A. Factor a trinomial of form $x^2 + bx + c$.
B. Factor a trinomial of the form $ax^2 + bx + c$ using trial and error.
C. Factor a trinomial of the form $ax^2 + bx + c$ using the ac method.

How many hundred gallons of water per minute can the engine pump? If the hose is 100 ft long, it can pump $2g^2 + g - 36$ gal/min. The expression $2g^2 + g - 36$ is factorable. How do we factor it? By using reverse multiplication. Let us start with a simpler problem.

A. FACTORING TRINOMIALS OF THE FORM $x^2 + bx + c$ (b and c integers)

In Section 3.4, we multiplied two binomials using FOIL:

$$\overset{\text{F\quad O\quad I\quad L}}{(x + 4)(x - 7) = x^2 - 7x + 4x - 28}$$

and

$$(x + 7)(x - 4) = x^2 - 4x + 7x - 28$$

To factor $x^2 - 3x - 28$, we recall that

$$(x + a)(x + b) = x^2 + bx + ax + ab$$
$$= x^2 + (b + a)x + ab$$

Rewriting this trinomial, we have the following factoring form.

> F-1 $\quad x^2 + (b + a)x + ab = (x + a)(x + b)$

This means that to factor a trinomial with a leading coefficient of 1, we need two integers a and b whose product is the last term and whose sum is the coefficient of the middle term.
　　Since

$$x^2 + \underbrace{(b + a)}x + ab = (x + a)(x + b)$$

write

$$x^2 - \quad 3x \quad - 28 = (x + a)(x + b)$$

We need two integers a and b whose product ab is -28 and whose sum is -3. Since the product is negative, one number must be positive and the other negative, with the larger number being negative, since the sum is -3. The numbers are -7 and 4. (*Check:* $-7 \cdot 4 = -28$ and $-7 + 4 = -3$, the coefficient of the middle term.) Thus

$$x^2 - 3x - 28 = (x + 4)(x - 7)$$

Since the multiplication of polynomials is commutative,

$$x^2 - 3x - 28 = (x - 7)(x + 4).$$

Now, suppose we want to factor $x^2 - 8x + 12$. This time, we need two integers whose product is 12 and whose sum is -8 (the coefficient of the middle term). The numbers are -6 and -2. (*Check:* $(-6)(-2) = 12$ and $(-6) + (-2) = -8$.) Thus

$$x^2 - 8x + 12 = (x - 6)(x - 2)$$

You can check this by multiplying $(x - 6)$ by $(x - 2)$.

EXAMPLE 1 Factor.

a. $x^2 + 7x + 12$ **b.** $x^2 - 6x + 8$ **c.** $x^2 - 3x - 4$

Solution

a. To factor $x^2 + 7x + 12$, we need two integers whose product is 12 and whose sum is 7. The numbers are 3 and 4. Thus

$$x^2 + 7x + 12 = (x + 3)(x + 4)$$

b. To factor $x^2 - 6x + 8$, we need two integers whose product is 8 and whose sum is -6. Since the product is positive, both numbers must be negative. They are -4 and -2. (*Check:* $(-4)(-2) = 8$ and $(-4) + (-2) = -6$.) Thus

$$x^2 - 6x + 8 = (x - 2)(x - 4)$$

c. This time, we need integers with product -4 and sum -3. The numbers are -4 and 1. Thus

$$x^2 - 3x - 4 = (x + 1)(x - 4)$$

You can check all these results by multiplying. For example, $(x + 1)(x - 4) = x^2 - 4x + x - 4 = x^2 - 3x - 4$.

EXAMPLE 2 Factor.

a. $x^2 - 8xy + 7y^2$ **b.** $x^2 + 3xy + 7y^2$

Solution

a. We need two integers whose product is 7 and whose sum is -8. Since the product is positive, both numbers must be negative. The numbers are -1 and -7. Thus

$$\begin{aligned} x^2 - 8xy + 7y^2 &= (x - 1y)(x - 7y) \\ &= (x - y)(x - 7y) \quad \text{Check by multiplication.} \end{aligned}$$

b. We need two integers whose product is 7 and whose sum is 3. There are no such numbers. The polynomial $x^2 + 3xy + 7y^2$ is not factorable. A polynomial that is not factorable using integer coefficients is a **prime polynomial.**

Problem 1 Factor.
a. $x^2 + 7x + 10$ **b.** $x^2 - 3x - 10$
c. $x^2 - 5x - 6$

Problem 2 Factor.
a. $x^2 - 2xy + 5y^2$
b. $x^2 - 7xy + 10y^2$

B. FACTORING TRINOMIALS OF THE FORM $ax^2 + bx + c$

To factor the polynomial $2g^2 + g - 36$ mentioned at the beginning of this section, we can rely on our experience with FOIL. To obtain $2g^2 + 1g - 36$, we must multiply

$$(2g + \underline{\quad})(g + \underline{\quad})$$

We need to fill the blanks with two integers that have a product of -36 and that give a middle term of $1g$. We can list the possibilities as follows.

TRIAL FACTORS	MIDDLE TERM
$(2g + \underline{1})(g - \underline{36})$	$-71g$
$(2g + \underline{2})(g - \underline{18})$	$-34g$

Note that $(2g + 2) = 2(g + 1)$, so we will not use even numbers in both terms of a factor.

$(2g + \underline{3})(g - \underline{12})$	$-21g$
$(2g + \underline{9})(g - \underline{4})$	$1g$

Thus $2g^2 + g - 36 = (2g + 9)(g - 4)$.

EXAMPLE 3   Factor $6x^2 + 17x + 12$.

Solution   The factors of 6 are 6 and 1 or 3 and 2. The possible combinations are

$$(6x + \underline{\quad})(x + \underline{\quad}) \quad \text{or} \quad (3x + \underline{\quad})(2x + \underline{\quad})$$

TRIAL FACTORS	
$(6x + 12)(x + 1)$	$(6x + 1)(x + 12)$
$(6x + 6)(x + 2)$	$(6x + 2)(x + 6)$
$(6x + 4)(x + 3)$	$(6x + 3)(x + 4)$
$(3x + 12)(2x + 1)$	$(3x + 1)(2x + 12)$
$(3x + 6)(2x + 2)$	$(3x + 2)(2x + 6)$
$(3x + 4)(2x + 3)$	$(3x + 3)(2x + 4)$
$(3x + 3)(2x + 4)$	$(3x + 4)(2x + 3)$

The only combination yielding the correct middle term is

$$(3x + 4)(2x + 3) = 6x^2 + 9x + 8x + 12. \qquad \blacktriangle$$

Problem 3   Factor $6x^2 + 13x + 6$.

C. THE ac TEST

The method used in Example 3 is not efficient when the coefficients are large. Moreover, we do not even know whether the polynomial is factorable. We remedy this situation with the following test.

> **THE *ac* TEST**
>
> The polynomial $ax^2 + bx + c$ is factorable only if there are two integers whose product is ac and whose sum is b.

Thus, to find if $3x^2 + 2x + 5$ is factorable, we first look at $3 \cdot 5 = 15$. If we can find two integers whose product is 15 and whose sum is 2, we can factor the polynomial. No such integers exist, so $3x^2 + 2x + 5$ is *prime*. On the other hand, the polynomial $5x^2 + 11x + 2$ is factorable, since we can find two integers (10 and 1) whose product is $5 \cdot 2 = 10$ and whose sum is 11. The number ac plays such an important part in the procedure used to factor $ax^2 + bx + c$ that we call it the **key number.**

We now give you a procedure that uses the key number to factor polynomials. For example, suppose we want to factor $2x^2 - 7x - 4$. We proceed as follows.

1. Find the key number $(2 \cdot (-4) = -8)$. $2x^2 - 7x - 4$ -8

2. Find the factors of the key number and use the appropriate ones to rewrite the middle term. $2x^2 - 8x + 1x - 4$ $-8, 1$

3. Group the terms into pairs (as in Section 3.4). $(2x^2 - 8x) + (1x - 4)$

4. Factor each pair. $2x(x - 4) + 1(x - 4)$

5. Note that $(x - 4)$ is the GCF. $(x - 4)(2x + 1)$

Thus, $2x^2 - 7x - 4 = (x - 4)(2x + 1)$. You should check that this is the correct factorization by multiplying $(x - 4)$ by $(2x + 1)$.

Suppose you want to factor the trinomial $5x^2 + 7x + 2$. Here is one way of doing it.

1. Find the key number $(5 \cdot 2 = 10)$. $5x^2 + 7x + 2$ 10

2. Find the factors of the key number and use them to rewrite the middle term. $5x^2 + 5x + 2x + 2$ $5, 2$

3. Group the terms into pairs. $(5x^2 + 5x) + (2x + 2)$

4. Factor each pair. $5x(x + 1) + 2(x + 1)$

5. Note that $(x + 1)$ is the GCF. $(x + 1)(5x + 2)$

Thus, $5x^2 + 7x + 2 = (5x + 2)(x + 1)$.

Another way of proceeding is as follows.

1. Find the key number. $5x^2 + 7x + 2$ 10

2. Find the factors of the key number and use them to rewrite the middle term. $5x^2 + 2x + 5x + 2$ $2, 5$

3. Group the terms into pairs. $(5x^2 + 2x) + (5x + 2)$

4. Factor each pair. $x(5x + 2) + 1(5x + 2)$

5. Note that $(5x + 2)$ is the GCF. $(5x + 2)(x + 1)$

In this case, we found that

$$5x^2 + 7x + 2 = (x + 1)(5x + 2)$$

Which is the correct factorization, $(x + 1)(5x + 2)$ or $(5x + 2)(x + 1)$? Both are correct! The multiplication of polynomials is commutative, and the order in which the product is written makes no difference. You can write the factorization of $ax^2 + bx + c$ in *two* ways.

EXAMPLE 4 Factor.

a. $6x^2 - 3x + 4$ **b.** $4x^2 - 3 - 4x$

Solution

a. We proceed by steps:

   1. Find the key number $(6 \cdot 4 = 24)$. $6x^2 - 3x + 4$ ㉔
   2. Find the factors of the key numbers and use them to rewrite the middle term. Unfortunately, it is impossible to find two integers with product 24 and sum -3. This trinomial is *not* factorable.

b. We first rewrite the polynomial (in descending order) as $4x^2 - 4x - 3$ and then proceed by steps.

   1. Find the key number $4x^2 - \underline{4x} - 3$ $\boxed{-12}$
   $(4 \cdot (-3) = -12)$.

   2. Find the factors of the key $4x^2 - \underline{6x + 2x} - 3$ $-6, 2$
   number and use them to rewrite
   the middle term.

   3. Group the terms into pairs. $(4x^2 - 6x) + (2x - 3)$

   4. Factor each pair. $2x(2x - 3) + 1(2x - 3)$

   5. Note that $(2x - 3)$ is the GCF. $(2x - 3)(2x + 1)$

Thus, $4x^2 - 4x - 3 = (2x - 3)(2x + 1)$, as can easily be verified by multiplication.

EXAMPLE 5   Factor $6x^2 + xy - y^2$.

Solution

   1. Find the key number $6x^2 + \underline{xy} - y^2$ $\boxed{-6}$
   $(6 \cdot (-1) = -6)$.

   2. Find the factors of $6x^2 + \underline{3xy - 2xy} - y^2$ $3, -2$
   the key number and
   use them to rewrite
   the middle term.

   3. Group the terms into $(6x^2 + 3xy) - (2xy + y^2)$ Note that
   pairs. $-(2xy + y^2) =$
                                             $-2xy - y^2$.

   4. Factor each pair. $3x(2x + y) - y(2x + y)$

   5. Note that $(2x + y)$ is $(2x + y)(3x - y)$
   the GCF.

Thus $6x^2 + xy - y^2 = (2x + y)(3x - y)$.   ▲

If the terms of the trinomial have a common factor, we factor it out first, as in the next example.

EXAMPLE 6   Factor $12x^3y^2 + 14x^2y^3 - 6xy^4$.

Solution   The greatest common factor of these three terms is $2xy^2$. Thus $12x^3y^2 + 14x^2y^3 - 6xy^4 = 2xy^2(6x^2 + 7xy - 3y^2)$. We then factor $6x^2 + 7xy - 3y^2$.

   1. The key number is -18. $6x^2 + \underline{7xy} - 3y^2$

   2. The factors of -18 with a sum of 7 $6x^2 + \underline{9xy - 2xy} - 3y^2$
   are 9 and -2. Rewrite the middle
   term.

Problem 4   Factor.
a. $5x^2 - 2x + 2$ **b.** $3x^2 - 4 - 4x$

Problem 5   Factor $2x^2 + xy - 3y^2$.

Problem 6   Factor
$12x^4y + 2x^3y^2 - 4x^2y^3$.

3. Group the terms into pairs. $\qquad (6x^2 + 9xy) - (2xy + 3y^2)$

4. Factor each pair. $\qquad 3x(2x + 3y) - y(2x + 3y)$

5. Note that $(2x + 3y)$ is the GCF. $\qquad (2x + 3y)(3x - y)$

Thus

$$12x^3y^2 + 14x^2y^3 - 6xy^4 = 2xy^2(6x^2 + 7xy - 3y^2)$$
$$= 2xy^2(2x + 3y)(3x - y)$$

You can check this by multiplying all the factors on the right side of the equation.

NAME

CLASS

SECTION

ANSWERS

A. In Problems 1–16, factor.

1. $x^2 + 5x + 6$

2. $x^2 + 15x + 56$

3. $a^2 + 7a + 10$

4. $a^2 + 10a + 24$

5. $x^2 + x - 12$

6. $x^2 + 5x - 6$

7. $x^2 - 2 + x$

8. $x^2 - 18 - 7x$

9. $x^2 - x - 2$

10. $x^2 - 5x - 14$

11. $x^2 - 3x - 10$

12. $x^2 - 4x - 21$

13. $a^2 - 16a + 63$

14. $a^2 - 4a + 3$

15. $y^2 + 22 - 13y$

16. $y^2 + 11 - 12y$

B, C. In Problems 17–36, factor if possible.

17. $9x^2 + 37x + 4$

18. $2x^2 + 5x + 2$

19. $3a^2 - 5a - 2$

20. $8a^2 - 2a - 21$

21. $2y^2 - 3y - 20$

22. $6y^2 - 13y - 5$

23. $4x^2 - 11x + 6$

24. $16x^2 - 16x + 3$

25. $6x^2 + x - 12$

26. $20y^2 + y - 1$

27. $21a^2 + 11a - 2$

28. $18x^2 - 3x - 10$

1. _____
2. _____
3. _____
4. _____
5. _____
6. _____
7. _____
8. _____
9. _____
10. _____
11. _____
12. _____
13. _____
14. _____
15. _____
16. _____

17. _____
18. _____
19. _____
20. _____
21. _____
22. _____
23. _____
24. _____
25. _____
26. _____
27. _____
28. _____

29. $6x^2 + 7xy - 3y^2$ **30.** $3x^2 + 13xy - 10y^2$

31. $7x^4 - 10x^3y + 3x^2y^2$ **32.** $6x^4 - 17x^3y + 5x^2y^2$

33. $15x^2y^3 - xy^4 - 2y^5$ **34.** $5x^2y^3 - 6xy^4 - 8y^5$

35. $15x^3y^2 - 2x^2y^3 - 2xy^4$ **36.** $4x^3y^2 - 13x^2y^3 - 3xy^4$

37. To find the flow g (in hundreds of gallons per minute) in 100 ft of $2\frac{1}{2}$-in. rubber-lined hose when the friction loss is 21 lb/in.^{2} we need to factor the expression

$2g^2 + g - 21$

Factor this expression.

38. To find the flow g (in hundreds of gallons per minute) in 100 ft of $2\frac{1}{2}$-in. rubber-lined hose when the friction loss is 55 lb/in.^{2}, we must factor the expression

$2g^2 + g - 55$

Factor this expression.

39. When solving for the equivalent resistance R of two electric circuits we find the expression

$2R^2 - 3R + 1$

Factor this expression.

40. To find the time t at which an object thrown upward at 12 m/sec will be 4 m above the ground, we must factor the expression

$5t^2 - 12t + 4$

Factor this expression.

✓ **SKILL CHECKER**

Multiply.

41. $(2a + b)^2$ **42.** $(3a + 2b)^2$

43. $(a - 2b)^2$ **44.** $(2a - 3b)^2$

45. $(a + b)(a - b)$ **46.** $(a - 2b)(a + 2b)$

47. $(2x - 3y)(2x + 3y)$ **48.** $(5x + 7y)(5x - 7y)$

29. _____
30. _____
31. _____
32. _____
33. _____
34. _____
35. _____
36. _____
37. _____

38. _____

39. _____

40. _____

41. _____
42. _____
43. _____
44. _____
45. _____
46. _____
47. _____
48. _____

3.5 USING YOUR KNOWLEDGE

The ideas presented in this section are very important in many other fields. Use your knowledge to factor the following problems.

1. To find the deflection of a beam of length L at a distance of 3 ft from its end, we must evaluate the expression

    $2L^2 - 9L + 9$

    Factor this expression.

2. In Problem 1, if the distance from the end is x feet, then we must factor the expression

    $2L^2 - 3xL + x^2$

    Factor this expression.

3. The distance (in meters) traveled in t sec by an object thrown upward at 12 m/sec is

    $-5t^2 + 12t$

    To determine the time at which the object will be 7 m above ground, we must solve the equation

    $5t^2 - 12t + 7 = 0$

    Factor the trinomial on the left side of this equation.

1. _____

2. _____

3. _____

OBJECTIVES

REVIEW

Before starting this section, you should know:

1. How to square a binomial.
2. How to multiply a binomial sum by a binomial difference.
3. How to multiply a binomial and a trinomial.

OBJECTIVES

You should be able to:

A. Factor a perfect square trinomial.
B. Factor the difference of two squares.
C. Factor the sum or the difference of two cubes.

How do you know the crane can support the weight of the car? It depends on the bending moment for the crane. At x feet from its support, the bending moment involves the expression

$$\frac{w}{2}(x^2 - 20x + 100)$$

where w is the weight of the crane in pounds per foot. The expression $x^2 - 20x + 100$ is the result of expanding the binomial $(x - 10)^2$ and is called a **perfect square trinomial.** We learned that

$$(a + b)^2 = a^2 + 2ab + b^2$$

and

$$(a - b)^2 = a^2 - 2ab + b^2$$

The trinomials on the right side of the equations are perfect square trinomials.

A. FACTORING PERFECT SQUARE TRINOMIALS

To be able to factor perfect square trinomials, we rewrite the preceding formulas as shown.

F-2	$x^2 + 2ax + a^2 = (x + a)^2$
F-3	$x^2 - 2ax + a^2 = (x - a)^2$

In a perfect square trinomial

1. The first and last terms (x^2 and a^2) are perfect squares.
2. There are no minus signs before x^2 or a^2.

3. The middle term is twice the product of the square root of the first and last terms ($2ax$) or is the additive inverse of this product ($-2ax$).

If you can write the trinomial in the form shown on the left of F-2, then you can factor it as shown on the right. Thus to factor $x^2 + 6x + 9$, note that the first and last terms are perfect squares ($(x)^2 = x^2$ and $3^2 = 9$), there are no minus signs before x^2 or 9, and the middle term is $2 \cdot 3 \cdot x = 6x$. Hence,

$$x^2 + 6x + 9 = x^2 + 2 \cdot 3x + 3^2 = (x + 3)^2.$$

Similarly, $x^2 - 8x + 16$ is a perfect square trinomial whose middle term is the negative of twice the product of the square root of the first and last terms: that is, $-2 \cdot 4 \cdot x = -8x$. Thus

$$x^2 - 8x + 4^2 = x^2 - 2 \cdot 4 \cdot x + 4^2 = (x - 4)^2$$

Note that you can also factor $x^2 - 8x + 16$ by finding two factors whose product is 16 and whose sum is -8, as we did in the previous section. Of course, the factors are -4 and -4. Thus $x^2 - 8x + 16 = (x - 4)(x - 4) = (x - 4)^2$.

EXAMPLE 1 Factor (if possible).

a. $x^2 - 10x + 25$ **b.** $x^2 + 12x + 36$ **c.** $x^2 + 7x + 49$

Solution

a. $x^2 - 10x + 25 = x^2 - 2 \cdot 5 \cdot x + 5^2 = (x - 5)^2$
b. $x^2 + 12x + 36 = x^2 + 2 \cdot 6 \cdot x + 6^2 = (x + 6)^2$

You can verify that a and b are correct by finding $(x - 5)^2$ and $(x + 6)^2$.

c. $x^2 + 7x + 49$ has perfect squares for first (x^2) and last (7^2) terms. However, the middle term is *not* $2 \cdot 7 \cdot x$. Thus, $x^2 + 7x + 49$ is not a perfect square; it is not even factorable. (We cannot find two integers whose product is 49 and whose sum is 7.) ▲

We can use the same idea to factor trinomials in two variables. Thus, to factor $25x^2 + 20xy + 4y^2$, we write

$$25x^2 + 20xy + 4y^2 = (5x)^2 + 2 \cdot (5x)(2y) + (2y)^2$$
$$= (5x + 2y)^2$$

EXAMPLE 2 Factor (if possible).

a. $9x^2 - 12xy + 4y^2$ **b.** $4x^2 - 10xy + 25y^2$

Solution

a. $9x^2 - 12xy + 4y^2 = (3x)^2 - 2 \cdot 3x \cdot 2y + (2y)^2$
$$= (3x - 2y)^2$$
b. Even though the first term, $(2x)^2$, and last term, $(5y)^2$, are perfect squares, the middle term is *not* $2 \cdot 2x \cdot 5y$. Thus $4x^2 - 10xy + 25y^2$ is not a perfect square. The *ac* test shows that it is prime.

B. FACTORING THE DIFFERENCE OF TWO SQUARES

As you recall from Section 3.3,

$$(x + a)(x - a) = x^2 - a^2$$

Problem 1 Factor (if possible).
a. $x^2 - 16x + 64$ **b.** $x^2 + 8x + 64$
c. $x^2 + 18x + 81$

Problem 2 Factor (if possible).
a. $4x^2 - 4xy + 9y^2$
b. $4x^2 - 12xy + 9y^2$

$x^2 - a^2$ is called the **difference of two squares.** The corresponding factoring formula is as follows.

$$F\text{-}4 \qquad x^2 - a^2 = (x + a)(x - a)$$

Thus to factor $x^2 - 9$, we write

$$x^2 - 9 = x^2 - 3^2 = (x + 3)(x - 3)$$

Similarly,

$$25x^2 - 16y^2 = (5x)^2 - (4y)^2 = (5x + 4y)(5x - 4y)$$

EXAMPLE 3 Factor.

a. $9x^2 - 1$ **b.** $81x^4 - 16y^4$

Solution

a. $9x^2 - 1 = (3x)^2 - 1^2 = (3x + 1)(3x - 1)$

b. $81x^4 - 16y^4 = (9x^2)^2 - (4y^2)^2$ ⎯⎯ Difference of two squares

$$= (9x^2 + 4y^2)(9x^2 - 4y^2)$$
$$= \underbrace{(9x^2 + 4y^2)}_{\text{Not factorable}}\underbrace{(3x + 2y)(3x - 2y)}_{\text{Factored}}$$ ▲

The polynomial $(x + y)^2 - 4$ is also the difference of two squares. If you think of $(x + y)$ as A and 4 as 2^2, you are factoring

$$A^2 - 2^2 = (A + 2)(A - 2)$$

or equivalently

$$(x + y)^2 - 2^2 = (x + y + 2)(x + y - 2)$$

If $(x + y)^2$ appears as $x^2 + 2xy + y^2$, factor it first. The procedure is

$x^2 + 2xy + y^2 - 4$	Given.
$= (x + y)^2 - 2^2$	Factor the trinomial.
$= (x + y + 2)(x + y - 2)$	Factor the difference.

EXAMPLE 4 Factor.

a. $x^2 - 6x + 9 - y^2$ **b.** $x^2 + 6xy + 9y^2 - 4$

Solution

a. The first three terms are a perfect square trinomial. Thus $x^2 - 6x + 9 = (x - 3)^2$ and

$$x^2 - 6x + 9 - y^2 = (x^2 - 6x + 9) - y^2$$
$$= (x - 3)^2 - y^2 \leftarrow \text{This is the difference of two squares.}$$
$$= (x - 3 + y)(x - 3 - y)$$
$$= (x + y - 3)(x - y - 3)$$

b. Since $x^2 + 6xy + 9y^2 = (x + 3y)^2$,

$$x^2 + 6xy + 9y^2 - 4 = (x^2 + 6xy + 9y^2) - 4$$
$$= (x + 3y)^2 - 4 \leftarrow \text{This is the difference of two squares.}$$
$$= (x + 3y + 2)(x + 3y - 2)$$

Problem 3 Factor.
a. $16x^2 - 1$ **b.** $16x^4 - 81y^4$

Problem 4 Factor.
a. $x^2 + 2xy + y^2 - z^2$
b. $x^2 + 10xy + 25y^2 - 16$

C. THE SUM AND DIFFERENCE OF TWO CUBES

We know how to factor the difference of two squares. Can we factor the difference of two cubes—that is, $x^3 - a^3$? Yes! As a matter of fact, we can also factor $x^3 + a^3$, the sum of two cubes. Here are the formulas.

> F-5 $x^3 + a^3 = (x + a)(x^2 - ax + a^2)$
>
> F-6 $x^3 - a^3 = (x - a)(x^2 + ax + a^2)$

Since we did not give corresponding product formulas, we verify these results:

$$
\begin{array}{l}
\quad x^2 - ax + a^2 \\
\underline{\quad\quad\quad\quad x + a} \\
\quad ax^2 - a^2x + a^3 \quad \leftarrow \text{Multiply } a(x^2 - ax + a^2). \\
\underline{x^3 - ax^2 + a^2x \quad\quad\ \leftarrow \text{Multiply } x(x^2 - ax + a^2).} \\
x^3 \quad\quad\quad\quad\quad + a^3
\end{array}
$$

Similarly,

$$
\begin{array}{l}
\quad x^2 + ax + a^2 \\
\underline{\quad\quad\quad\quad x - a} \\
\ - ax^2 - a^2x - a^3 \quad \leftarrow \text{Multiply } -a(x^2 + ax + a^2). \\
\underline{x^3 + ax^2 + a^2x \quad\quad\ \leftarrow \text{Multiply } x(x^2 + ax + a^2).} \\
x^3 \quad\quad\quad\quad\quad - a^3
\end{array}
$$

Now that we have verified the formulas, we can factor sums and differences of cubes. For example,

$$
\begin{aligned}
8 - x^3 &= 2^3 - x^3 \\
&= (2 - x)(2^2 + 2x + x^2) \\
&= (2 - x)(4 + 2x + x^2)
\end{aligned}
$$

and

$$
\begin{aligned}
x^3 + 125 &= x^3 + 5^3 \\
&= (x + 5)(x^2 - 5x + 5^2) \\
&= (x + 5)(x^2 - 5x + 25)
\end{aligned}
$$

EXAMPLE 5 Factor.

a. $27 - \frac{1}{64}x^3$ **b.** $125x^3 + 64y^3$

Solution

a. $27 - \frac{1}{64}x^3 = 3^3 - \left(\frac{1}{4}x\right)^3$

$$
\begin{aligned}
&= \left(3 - \frac{1}{4}x\right)\left[3^2 + 3 \cdot \frac{1}{4}x + \left(\frac{1}{4}x\right)^2\right] \\
&= \left(3 - \frac{1}{4}x\right)\left(9 + \frac{3}{4}x + \frac{1}{16}x^2\right)
\end{aligned}
$$

b. $125x^3 + 64y^3 = (5x)^3 + (4y)^3$

$$
\begin{aligned}
&= (5x + 4y)[(5x)^2 - 5x \cdot 4y + (4y)^2] \\
&= (5x + 4y)(25x^2 - 20xy + 16y^2)
\end{aligned}
$$

Problem 5 Factor.

a. $64 - \frac{1}{27}x^3$ **b.** $8x^3 + 27y^3$

EXAMPLE 6 Factor $x^6 - 64$ completely.

Solution

a. $x^6 - 64 = x^6 - 2^6$

$\qquad = [x^3]^2 - [2^3]^2$

$\qquad = (x^3 + 2^3)(x^3 - 2^3)$

$\qquad = [(x + 2)(x^2 - 2x + 4)][(x - 2)(x^2 + 2x + 4)]$

Note that you can also factor $x^6 - 64$ by writing it as the difference of two cubes, that is, $x^6 - 64 = (x^2)^3 - (2^2)^3$. Try it!

Problem 6 Factor $x^6 - 1$ completely.

NAME

CLASS

SECTION

ANSWERS

A. Factor.

1. $x^2 + 2x + 1$

2. $x^2 + 20x + 100$

3. $y^2 + 22y + 121$

4. $y^2 + 14x + 49$

5. $1 + 4x + 4x^2$

6. $1 + 6x + 9x^2$

7. $9x^2 + 30xy + 25y^2$

8. $25x^2 + 30xy + 9y^2$

9. $36a^2 + 48a + 16$

10. $9a^2 + 60c + 100$

11. $y^2 - 2y + 1$

12. $25 - 10y + y^2$

13. $49 - 14x + x^2$

14. $x^2 - 100x + 2500$

15. $49a^2 - 28ax + 4x^2$

16. $4a^2 - 12ax + 9x^2$

17. $16x^2 - 24xy + 9y^2$

18. $9x^2 - 42xy + 49y^2$

B. Factor.

19. $9x^4 + 12x^2 + 4$

20. $25y^4 + 20y^2 + 4$

21. $16x^4 - 24x^2 + 9$

22. $4y^4 - 20y^2 + 25$

23. $1 + 2x^2 + x^4$

24. $4 + 12x^2 + 9x^4$

25. $y^2 - 64$

26. $y^2 - 121$

27. $a^2 - \dfrac{1}{9}$

28. $x^2 - \dfrac{1}{16}$

1. _____
2. _____
3. _____
4. _____
5. _____
6. _____
7. _____
8. _____
9. _____
10. _____
11. _____
12. _____
13. _____
14. _____
15. _____
16. _____
17. _____
18. _____

19. _____
20. _____
21. _____
22. _____
23. _____
24. _____
25. _____
26. _____
27. _____
28. _____

29. $64 - b^2$

30. $81 - b^2$

31. $36a^2 - 49b^2$

32. $36a^2 - 25b^2$

33. $\dfrac{x^2}{9} - \dfrac{y^2}{16}$

34. $\dfrac{x^2}{16} - \dfrac{9y^2}{25}$

35. $a^2 + 4ab + 4b^2 - c^2$

36. $9a^2 + 6ab + b^2 - 1$

37. $4x^2 - 4xy + y^2 - 1$

38. $9x^2 - 30xy + 25y^2 - 9$

39. $9y^2 - 12xy + 4x^2 - 25$

40. $16y^2 - 40xy + 25x^2 - 36$

41. $16a^2 - (x^2 + 6xy + 9y^2)$

42. $25a^2 - (4x^2 - 4xy + y^2)$

43. $y^2 - a^2 + 2ab - b^2$

44. $9y^2 - 9x^2 + 6xz - z^2$

29. _____

30. _____

31. _____

32. _____

33. _____

34. _____

35. _____

36. _____

37. _____

38. _____

39. _____

40. _____

41. _____

42. _____

43. _____

44. _____

C. Factor.

45. $x^3 + 125$

46. $x^3 + 64$

47. $1 + a^3$

48. $343 + a^3$

49. $8x^3 + y^3$

50. $125x^3 + 8y^3$

51. $x^3 - 1$

52. $x^3 - 216$

53. $125a^3 - 8b^3$

54. $216a^3 - 125b^3$

55. $x^6 - 64$

56. $y^6 - 1$

57. $x^6 - \dfrac{1}{64}$

58. $y^6 - 729$

59. $\dfrac{x^6}{64} - 1$

60. $\dfrac{y^6}{729} - 1$

45. _____

46. _____

47. _____

48. _____

49. _____

50. _____

51. _____

52. _____

53. _____

54. _____

55. _____

56. _____

57. _____

58. _____

59. _____

60. _____

✓ **SKILL CHECKER**

Multiply.

61. $(x + 3)(x - 5)$

62. $(x - 5)(x + 7)$

63. $(x - 8)(x + 2)$

64. $(x + 2y)^2$

65. $(2x - 3y)^2$

61. _____

62. _____

63. _____

64. _____

65. _____

3.6 USING YOUR KNOWLEDGE

Have you heard of supply and demand? In business the supply and demand of a product can be expressed by using a polynomial.

1. When x units of an item are demanded by consumers, the price per unit is given by

$D(x) = 100 - x^2$

Factor $100 - x^2$.

2. When x units are supplied by sellers, the price per unit of an item is given by

$S(x) = x^3 + 216$

Factor $x^3 + 216$.

3. When x units of a certain item are produced, the cost is given by

$C(x) = 8x^3 + 1$

Factor $8x^3 + 1$.

1. _____

2. _____

3. _____

REVIEW

Before starting this section, you should know factoring formulas F-1 through F-6.

OBJECTIVE

You should be able to factor a given polynomial by using the procedure explained in the text.

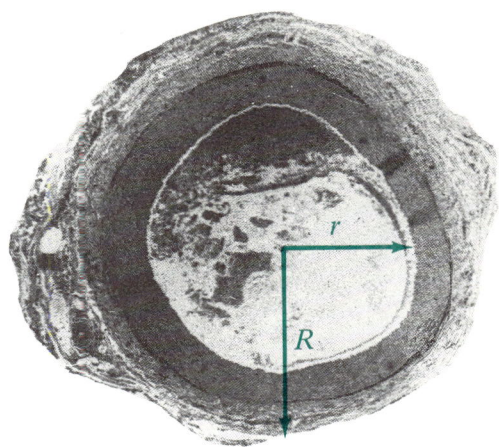

Courtesy of Bil Davis and Eldra Solomon

Have you heard of blocked arteries? The velocity of the blood inside a blocked artery (see photo) depends on the diameter of the inside wall (r) and outside wall (R) and is given by

$$CR^2 - Cr^2, \qquad C \text{ a constant}$$

How do we factor this expression? We follow a general pattern that may use one or more of the techniques we have learned.

To factor a given polynomial, follow these steps.

A GENERAL FACTORING STRATEGY

1. Factor out the GCF, if there is one.

2. Look at the number of terms in the given polynomial (or inside the parentheses if the GCF was factored out).

   • If there are *two terms,* check for:
     Difference of two squares

     $$x^2 - a^2 = (x + a)(x - a)$$

     Difference of two cubes

     $$x^3 - a^3 = (x - a)(x^2 + ax + a^2)$$

     Sum of two cubes

     $$x^3 + a^3 = (x + a)(x^2 - ax + a^2)$$

     Note that the sum of two squares, $x^2 + a^2$, is not factorable.

   • If there are *three terms,* check for:
     Perfect square trinomial.

     $$x^2 + 2ab + b^2 = (x + a)^2$$
     $$x^2 - 2ab + b^2 = (x - a)^2$$

     Trinomials of the form

     $$ax^2 + bx + c$$

     Use the *ac* method or trial and error.

   • If there are *four terms:*
     Factor by grouping.

3. Check the result by multiplying the factors.

Thus, to factor $CR^2 - Cr^2$, we follow the steps.

STEP 1. Factor out the GCF, C. $\quad\quad CR^2 - Cr^2 = C(R^2 - r^2)$

STEP 2. Factor the difference of two $\quad\quad\quad\quad\quad = C(R + r)(R - r)$
squares inside the parentheses.

EXAMPLE 1   Factor.

a. $8x^5 - x^2y^3$ **b.** $6x^5 + 24x^3$

Solution

a. We use the steps in our procedure.

    STEP 1. Factor out the $\quad\quad 8x^5 - x^2y^3 = x^2(8x^3 - y^3)$
              GCF, x^2.

    STEP 2. Factor the $\quad\quad\quad\quad = x^2(2x - y)(4x^2 + 2xy + y^2)$
              difference of two
              cubes inside the
              parentheses.

b. We first factor out $6x^3$ the GCF: $6x^5 + 24x^3 = 6x^3(x^2 + 4)$. Since
$x^2 + 4$ is the sum of two squares, it is not factorable. Thus, the
complete factorization of $6x^5 + 24x^3$ is $6x^3(x^2 + 4)$.

EXAMPLE 2   Factor.

a. $12x^5 + 12x^4y + 3x^3y^2$ **b.** $36x^2y^2 - 24xy^3 + 4y^4$

Solution

a. As usual, factor out the GCF, $3x^3$, first. Then note that we have a
perfect trinomial square inside the parentheses. Here are the steps
we need.

    STEP 1. Factor out $\quad\quad 12x^5 + 12x^4y + 3x^3y^2 = 3x^3(\underbrace{4x^2 + 4xy + y^2})$
              the GCF, $3x^3$.

    STEP 2. Factor the $\quad\quad\quad\quad\quad = 3x^3(2x + y)^2$
              perfect
              square
              trinomial.

b. **STEP 1.** Factor out $\quad 36x^2y^2 - 24xy^3 + 4y^4 = 4y^2(\underbrace{9x^2 - 6xy + y^2})$
              the GCF, $4y^2$.

    STEP 2. Factor the $\quad\quad\quad\quad = 4y^2(3x - y)^2$
              perfect
              square
              trinomial
              inside the
              parentheses.

EXAMPLE 3   Factor.

a. $4x^3y - 10x^2y^2 - 6xy^3$ **b.** $2x^5 + x^4y + x^3y^2$

Problem 1   Factor.
a. $27x^5 - x^2y^3$ **b.** $8x^5 + 72x^3$

Problem 2   Factor.
a. $36x^5 + 24x^4y + 4x^3y^2$
b. $12x^2y^2 - 12xy^3 + 3y^4$

Problem 3   Factor.
a. $9x^3y - 15x^2y^2 - 6xy^3$
b. $24x^4 + x^3y + x^2y^2$

Solution

a. The GCF is $2xy$. After factoring out this GCF, we have three terms inside the parentheses. We can use the *ac* method or trial and error to finish the problem. The steps are as follows.

STEP 1. Factor out the GCF, $2xy$.

$$4x^3y - 10x^2y^2 - 6xy^3 = 2xy(2x^2 - 5xy - 3y^2)$$

STEP 2. Use the *ac* method or trial and error to factor $2x^2 - 5xy - 3y^2$.

$$= 2xy(2x + y)(x - 3y)$$

STEP 3. Check the answer by multiplying the factors.

b. Factor out the GCF, x^3.

$$2x^5 + x^4y + x^3y^2 = x^3(2x^2 + xy + y^2)$$

Note that the expression inside the parentheses is *not* factorable. According to the *ac* method, the key number is 2. We need two integers whose product is 2 and whose sum is 1; no such integers exist.

EXAMPLE 4 Factor $2x^5 + x^4y + x^3y^2$.

Solution We start by factoring out the GCF, which is x^3.

$$2x^5 + x^4y + x^3y^2 = x^3(2x^2 + xy + y^2)$$

$2x^2 + xy + y^2$ is a trinomial, but it is *not* factorable. (The *ac* is $2 \cdot 1 = 2$, and there are no factors whose product is 2 and whose sum is 1.) Thus, the factorization shown is the complete factorization.

EXAMPLE 5 Factor $4x^3 - 12x^2 - x + 3$.

Solution In this case, there is no common factor. Since the polynomial has four terms, we try to factor by grouping.

$$4x^3 - 12x^2 - x + 3 = 4x^2(x - 3) - (x - 3) \qquad \text{Group into pairs.}$$
$$= (x - 3)(4x^2 - 1) \qquad \text{Factor out the GCF, } (x - 3).$$
$$= (x - 3)(2x + 1)(2x - 1) \qquad \text{Factor the difference of two squares, } (4x^2 - 1).$$

Problem 4 Factor.
$2x^4 + x^3y + 2x^2y^2$.

Problem 5 Factor $9x^3 - 18x^2 - x + 2$.

NAME

CLASS

SECTION

ANSWERS

Factor.

1. $3x^4 - 3x^3 - 18x^2$

2. $4x^5 - 12x^4 - 16x^3$

3. $5x^4 + 10x^3y - 40x^2y^2$

4. $6x^7 + 18x^6y - 60x^5y^2$

5. $-3x^6 - 6x^5 - 21x^4$

6. $-6x^5 - 18x^4 - 12x^3$

7. $2x^6y - 4x^5y^2 - 10x^4y^3$

8. $3x^8y - 12x^7y^2 - 9x^6y^3$

9. $-4x^6 - 12x^5y - 18x^4y^2$

10. $-5x^6 - 25x^5 - 30x^4$

11. $6x^3y^2 + 12x^2y^2 + 2xy^2 + 4y^2$

12. $6x^3y^2 + 24x^2y^2 + 3xy^2 + 12y^2$

13. $-9x^4y - 9x^3y - 6x^2y - 6xy$

14. $-8x^4y - 16x^3y - 6x^2y - 12xy$

15. $-4x^4 - 4x^3y + 2x^2y + 2xy^2$

16. $-9x^4 - 18x^3y + 3x^2y + 6xy^2$

17. $3x^2y^2 + 24xy^3 + 48y^4$

18. $8x^2y^2 + 24xy^3 + 18y^4$

19. $-18kx^2 - 24kxy - 8ky^2$

20. $-12kx^2 - 60kxy - 75ky^2$

21. $16x^3y^2 - 48x^2y^3 + 36xy^4$

22. $45x^3y^2 - 60x^2y^3 + 20xy^4$

23. $kx^2 - 12kx + 36$

24. $kx^2 - 20kx + 25$

25. $3x^5 + 12x^4y + 12x^3y^2$

26. $2x^5 + 16x^4y + 32x^3y^2$

27. $18x^6 + 12x^5y + 2x^4y^2$

28. $12x^6 + 12x^5y + 3x^4y^2$

29. $12x^4y^2 - 36x^3y^3 + 27x^2y^4$

30. $18x^4y^2 - 24x^3y^3 + 8x^2y^4$

ANSWERS

1. _____
2. _____
3. _____
4. _____
5. _____
6. _____
7. _____
8. _____
9. _____
10. _____
11. _____
12. _____
13. _____
14. _____
15. _____
16. _____
17. _____
18. _____
19. _____
20. _____
21. _____
22. _____
23. _____
24. _____
25. _____
26. _____
27. _____
28. _____
29. _____
30. _____

31. $6x^3 + 12x^2 - 6x - 12$

32. $4x^3 + 16x^2 - 16x - 64$

33. $7x^4 - 7y^4$

34. $9x^4 - 9z^4$

35. $2x^6 - 32x^2y^4$

36. $x^7 - 81x^3y^4$

37. $-2x^2 - 12x - 18$

38. $-2x^2 - 20x - 50$

39. $-3x^2 - 12x - 12$

40. $-4x^2 - 24x - 36$

41. $-4x^4 - 4x^3y - x^2y^2$

42. $-9x^4 - 6x^3y - x^2y^2$

43. $-9x^2y^2 - 12xy^3 - 4y^4$

44. $-4x^2y^2 - 12xy^3 - 9y^4$

45. $-8x^2y^2 + 24xy^3 - 18y^4$

46. $-18x^4 + 24x^3y - 8x^2y^2$

47. $-18x^3 - 24x^2y - 8xy^2$

48. $-12x^3 - 36x^2y - 27xy^2$

49. $-18x^3 - 60x^2y - 50xy^2$

50. $-12x^3 - 60x^2y - 75xy^2$

51. $-x^3 + xy^2$

52. $-x^3 + 9xy^2$

53. $-x^4 + 4x^2y^2$

54. $-x^4 + 16x^2y^2$

55. $-4x^4 + 9x^2y^2$

56. $-9x^4 + 4x^2y^2$

57. $-8x^3 + 18xy^2$

58. $-12x^3 + 3x$

59. $-18x^4 + 8x^2y^2$

60. $-12x^4 + 27x^2y^2$

61. $27x^2 - x^5$

62. $64x^3 - x^6$

63. $x^7 - 8x^4$

64. $8x^{10} - \dfrac{1}{27}x^7$

65. $27x^4 + 8x^7$

66. $8x^5 + 27x^8$

31. _____
32. _____
33. _____
34. _____
35. _____
36. _____
37. _____
38. _____
39. _____
40. _____
41. _____
42. _____
43. _____
44. _____
45. _____
46. _____
47. _____
48. _____
49. _____
50. _____
51. _____
52. _____
53. _____
54. _____
55. _____
56. _____
57. _____
58. _____
59. _____
60. _____
61. _____
62. _____
63. _____
64. _____
65. _____
66. _____

67. $27x^7 + 64x^4y^3$ **68.** $8x^8 + 27x^5y^3$

✓ **SKILL CHECKER**

Factor.

69. $6x^2 - x - 2$ **70.** $6x^2 - 7x - 3$

71. $12x^2 - x - 1$ **72.** $10x^2 - 17x + 3$

3.7   USING YOUR KNOWLEDGE

Many of the ideas presented in this section are used by engineers and technicians. Use your knowledge to solve the following problems.

1. The bend allowance needed to bend a piece of metal of thickness t through an angle A when the inside radius of the bend is IR is given by the expression

$$\frac{2\pi A}{360} IR + \frac{2\pi A}{360} Kt \quad K \text{ a constant}$$

Factor this expression.

2. The change in kinetic energy of a moving object of mass m with initial velocity v_1 and terminal velocity v_2 is given by

$$\frac{1}{2}mv_1^2 - \frac{1}{2}mv_2^2$$

Factor this expression.

3. The parabolic distribution of shear stress on the cross section of a certain beam is given by

$$\frac{3Sd^2}{2bd^3} - \frac{12Sz^2}{2bd^3}$$

Factor this expression.

4. The polar moment of inertia J of a hollow round shaft of inner diameter d_1 and outer diameter d is given by

$$\frac{\pi d^4}{32} - \frac{\pi d_1^4}{32}$$

Factor this expression.

OBJECTIVES

REVIEW

Before starting this section, you should know how to:

1. Factor polynomials.
2. Solve linear equations.
3. Evaluate expressions.

OBJECTIVES

You should be able to:

A. Solve quadratic equations by factoring.
B. Use the Pythagorean theorem to find the length of one side of a right triangle when the length of the two other sides are given.

How much water is the engine pumping if the friction loss is 36 lb/in.^{2}? You can find out by solving the equation

$$2g^2 + g - 36 = 0 \quad (g \text{ in hundreds of gal/min})$$

This equation is a *quadratic equation in standard form.* Here is the definition.

An equation that can be written in the **standard form**

$$ax^2 + bx + c = 0$$

where a, b, and c are constants and a is not 0, $(a \neq 0)$ is a **quadratic equation.**

Here are some other quadratic equations:

$$x^2 = 5, \quad 3x^2 - 8x + 7 = 0, \quad \text{and} \quad x^2 - 2x = 4$$

Of these, only $3x^2 - 8x + 7 = 0$ is in standard form, with $a = 3$, $b = -8$, and $c = 7$.

A. SOLVING EQUATIONS BY FACTORING

The equation $2g^2 + g - 36 = 0$ can be solved by factoring. As you recall from Section 3.5, $2g^2 + g - 36$ can be factored as $(2g + 9)(g - 4)$. We then write

$$2g^2 + g - 36 = 0 \quad \text{Given.}$$
$$(2g + 9)(g - 4) = 0 \quad \text{Factor.}$$

Note that the product of the factors is 0. The only way this can happen is if at least one of the factors is 0. (Try getting 0 for an answer without having any 0 factors!) Here is the property we need.

ZERO-FACTOR PROPERTY

For all real numbers a and b, $a \cdot b = 0$ means that $a = 0$ or $b = 0$ (or both).

Thus

$$(2g + 9)(g - 4) = 0$$

means that

$$
\begin{array}{lll}
2g + 9 = 0 & \text{or} & g - 4 = 0 \\
2g = -9 & \text{or} & g = 4 \quad \text{Solve the linear equations.} \\
g = -\dfrac{9}{2} & \text{or} & g = 4
\end{array}
$$

The two possible solutions are $g = -\frac{9}{2}$ and $g = 4$. Since g is the flow of water, g must be positive, so we discard the negative solution $g = -\frac{9}{2}$. Thus, the engine can pump 400 gallons per minute.

EXAMPLE 1 Solve by factoring.

a. $x^2 - 9 = 0$ **b.** $x^2 + 8x = 0$

Solution

a.
$$
\begin{array}{ll}
x^2 - 9 = 0 & \\
(x + 3)(x - 3) = 0 & \text{Factor.} \\
x + 3 = 0 \quad \text{or} \quad x - 3 = 0 & \text{Use the zero-factor property, with} \\
& a = x + 3 \text{ and } b = x - 3. \\
x = -3 \quad \text{or} \quad x = 3 & \text{Solve the equations } x + 3 = 0 \text{ and} \\
& x - 3 = 0.
\end{array}
$$

We can check the solutions by substituting in the original equation, $x^2 - 9 = 0$.

Check:
$$
\begin{array}{c|c}
x^2 - 9 = 0 & x^2 - 9 = 0 \\
\hline
(-3)^2 - 9 \;\big|\; 0 & 3^2 - 9 \;\big|\; 0 \\
9 - 9 & 9 - 9 \\
0 & 0
\end{array}
$$

In both cases the result is true. The solution set is $\{3, -3\}$.

b.
$$
\begin{array}{ll}
x^2 + 8x = 0 & \\
x(x + 8) = 0 & \text{Factor.} \\
x = 0 \quad \text{or} \quad x + 8 = 0 & \text{Use the zero-factor property, with} \\
& a = x \text{ and } b = x + 8. \\
x = 0 \quad \text{or} \quad x = -8 & \text{Solve the equation } x + 8 = 0.
\end{array}
$$

Problem 1 Solve by factoring.
a. $x^2 - 16 = 0$ **b.** $x^2 + 9x = 0$

Check:

$$\begin{array}{c|c}
x^2 + 8x = 0 \\
\hline
(0)^2 + 8(0) & 0 \\
0 + 0 \\
0
\end{array}
\qquad
\begin{array}{c|c}
x^2 + 8x = 0 \\
\hline
(-8)^2 + 8(-8) & 0 \\
64 - 64 \\
0
\end{array}$$

Since both results check, the solution set is $\{0, -8\}$. ▲

The equation $x^2 = -x + 6$ is *not* in standard form. To write it in standard form, we add x and subtract 6 on both sides, obtaining $x^2 + x - 6 = 0$, which can be solved by factoring. As you recall, to factor $x^2 + x - 6$, we need to find integers whose sum is 1 and whose product is -6. These integers are 3 and -2. Thus we have

$$\begin{array}{ll}
x^2 = -x + 6 & \text{Given.} \\
x^2 + x - 6 = 0 & \text{Write in standard form} \\
& \text{(add } x \text{ and subtract 6).} \\
(x + 3)(x - 2) = 0 & \text{Factor.} \\
x + 3 = 0 \quad \text{or} \quad x - 2 = 0 & \text{Use the zero-factor property.} \\
x = -3 \quad \text{or} \qquad x = 2 & \text{Solve } x + 3 = 0 \text{ and } x - 2 = 0.
\end{array}$$

Check:

$$\begin{array}{c|c}
x^2 = -x + 6 \\
\hline
(2)^2 & -(2) + 6 \\
4 & 4
\end{array}
\qquad
\begin{array}{c|c}
x^2 = -x + 6 \\
\hline
(-3)^2 & -(-3) + 6 \\
9 & 3 + 6 \\
& 9
\end{array}$$

The solution set is $\{2, -3\}$.

EXAMPLE 2   Solve by factoring.

a. $x^2 = 6x - 8$ **b.** $x^2 + x = 2$

Solution

a. We first write the equation in standard form by subtracting $6x$ and adding 8, obtaining $x^2 - 6x + 8 = 0$. To factor $x^2 - 6x + 8$, we must find two integers whose sum is -6 and whose product is 8. These numbers are -4 and -2. Here is the procedure,

$$\begin{array}{ll}
x^2 = 6x - 8 & \text{Given.} \\
x^2 - 6x + 8 = 0 & \text{Subtract } 6x \text{ and add 8.} \\
(x - 4)(x - 2) = 0 & \text{Factor.} \\
x - 4 = 0 \quad \text{or} \quad x - 2 = 0 & \text{Use the zero-factor property.} \\
x = 4 \quad \text{or} \qquad x = 2 & \text{Solve } x - 4 = 0 \text{ and } x - 2 = 0.
\end{array}$$

The solution set is $\{2, 4\}$. Check this!

b. The equation $x^2 + x = 2$ is *not* in standard form. To solve an equation by factoring, we must first write the equation in the standard form $ax^2 + bx + c = 0$. Thus, subtracting 2 from both members of $x^2 + x = 2$, we have

$$\begin{array}{ll}
x^2 + x - 2 = 0 & \\
(x + 2)(x - 1) = 0 & \text{Factor.} \\
x + 2 = 0 \quad \text{or} \quad x - 1 = 0 & \text{Use the zero-factor property.} \\
x = -2 \quad \text{or} \qquad x = 1 & \text{Solve } x + 2 = 0 \text{ and } x - 1 = 0.
\end{array}$$

The solution set is $\{1, -2\}$. The check is left for you. ▲

Problem 2   Solve by factoring.
a. $x^2 = 6x - 5$ **b.** $x^2 + x = 6$

To solve the equation $6x^2 - x - 2 = 0$, we first factor $6x^2 - x - 2$ by trial and error or by the *ac* method (shown). For $6x^2 - x - 2 = 0$, the key number is $(6) \cdot (-2) = -12$. Thus, we have to find integers whose product is -12 and whose sum is -1 (that is, -4 and 3) and use these numbers to rewrite the middle term x. We have

$$6x^2 - x - 2 = 0 \qquad \text{Given.}$$
$$6x^2 - 4x + 3x - 2 = 0 \qquad \text{Write the middle term } -x \text{ as } -4x + 3x.$$
$$2x(3x - 2) + (3x - 2) = 0 \qquad \text{Factor the first pair of terms.}$$
$$(3x - 2)(2x + 1) = 0 \qquad \text{Factor out the common factor, } 3x - 2.$$
$$3x - 2 = 0 \quad \text{or} \quad 2x + 1 = 0 \qquad \text{Use the zero-factor property.}$$
$$3x = 2 \quad \text{or} \quad 2x = -1 \qquad \text{Solve } 3x - 2 = 0 \text{ and } 2x + 1 = 0.$$
$$x = \frac{2}{3} \quad \text{or} \quad x = \frac{-1}{2}$$

The solution set is $\{\frac{2}{3}, -\frac{1}{2}\}$. We check this for $x = \frac{2}{3}$.

$$
\begin{array}{c|c}
6x^2 - x - 2 = 0 & \\
\hline
6\left(\frac{2}{3}\right)^2 - \frac{2}{3} - 2 & 0 \\
6\left(\frac{4}{9}\right) - \frac{2}{3} - 2 & \\
\frac{8}{3} - \frac{2}{3} - 2 & \\
2 - 2 & \\
0 &
\end{array}
$$

EXAMPLE 3   Solve by factoring.

a. $12x^2 + 5x - 3 = 0$ **b.** $6x^2 - x = 1$

Solution

a. You can factor $12x^2 + 5x - 3$ by trial and error or note that the key number for $12x^2 + 5x - 3 = 0$ is $(12)(-3) = -36$. Thus

$$12x^2 + 5x - 3 = 0 \qquad \text{Given.}$$
$$12x^2 + 9x - 4x - 3 = 0 \qquad \text{Write } 5x \text{ using coefficients whose sum is } 5 \text{ and whose product is } -36\text{—that is, } 5x = 9x - 4x.$$
$$3x(4x + 3) - (4x + 3) = 0 \qquad \text{Factor the first and last pairs of terms.}$$
$$(4x + 3)(3x - 1) = 0 \qquad \text{Factor out the common factor, } 4x + 3.$$
$$4x + 3 = 0 \quad \text{or} \quad 3x - 1 = 0 \qquad \text{Use the zero-factor property.}$$
$$4x = -3 \quad \text{or} \quad 3x = 1 \qquad \text{Solve } 4x + 3 = 0 \text{ and } 3x - 1 = 0.$$
$$x = \frac{-3}{4} \quad \text{or} \quad x = \frac{1}{3}$$

The solution set is $\{\frac{1}{3}, -\frac{3}{4}\}$. Check this!

Problem 3   Solve by factoring.
a. $12x^2 + 13x - 4 = 0$
b. $8x^2 - 2x = 1$

b. $6x^2 - x = 1$ is *not* in standard form. Subtracting 1 from both members, we have

$$6x^2 - x - 1 = 0 \qquad \text{The key number is } -6.$$

$$6x^2 - 3x + 2x - 1 = 0 \qquad \text{The integers whose product is } -6$$
$$3x(2x - 1) + (2x - 1) = 0 \qquad \text{with sum } -1 \text{ are } -3 \text{ and } 2.$$

$$(2x - 1)(3x + 1) = 0 \qquad \text{Factor } (2x - 1).$$

$$2x - 1 = 0 \quad \text{or} \quad 3x + 1 = 0 \qquad \text{Use the zero-factor property.}$$

$$2x = 1 \quad \text{or} \quad 3x = -1 \qquad \text{Solve the equations } 2x - 1 = 0 \text{ and}$$
$$\qquad\qquad\qquad\qquad\qquad\qquad 3x + 1 = 0.$$

$$x = \frac{1}{2} \quad \text{or} \quad x = \frac{-1}{3}$$

The solution set is $\{\frac{1}{2}, -\frac{1}{3}\}$. Check this!

EXAMPLE 4   Solve $x^3 + 3x^2 - 4x - 12 = 0$.

Problem 4   Solve
$x^3 + 2x^2 - 9x - 18 = 0$.

Solution   Since the polynomial has four terms, we factor it by grouping. Here are the steps.

$$x^3 + 3x^2 - 4x - 12 = 0 \qquad\qquad \text{Given.}$$

$$x^2(x + 3) - 4(x + 3) = 0 \qquad\qquad \text{Group for factoring.}$$

$$(x + 3)(x^2 - 4) = 0 \qquad\qquad \text{Factor out the GCF.}$$

$$(x + 3)(x + 2)(x - 2) = 0 \qquad\qquad \text{Factor } x^2 - 4.$$

$$x + 3 = 0 \quad \text{or} \quad x + 2 = 0 \quad \text{or} \quad x - 2 = 0$$

$$x = -3 \quad \text{or} \quad x = -2 \quad \text{or} \quad x = 2$$

The solution set is $\{2, -2, -3\}$. Check this!

B.   THE PYTHAGOREAN THEOREM

Quadratic equations can be used to find the lengths of the sides of right triangles using the **Pythagorean theorem,** which we state next.

PYTHAGOREAN THEOREM

In any right triangle, the square of the longest side (hypotenuse) is equal to the sum of the squares of the other two sides (the legs). In symbols:

$$c^2 = a^2 + b^2$$

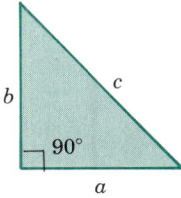

It is interesting to note that there are infinitely many triples of whole numbers (a, b, c) that satisfy the equation $c^2 = a^2 + b^2$. Such triples are called *Pythagorean triples,* and we will find some of them next.

EXAMPLE 5 The length of the three sides of a right triangle are consecutive integers. What are these lengths?

Solution Let the length of the shortest side be x. Since the lengths of the sides are consecutive integers, we have:

Length of the shortest side x
Length of the next side $x + 1$
Length of the hypotenuse $x + 2$

See the diagram. By the Pythagorean theorem

$$(x + 2)^2 = (x + 1)^2 + x^2$$

$x^2 + 4x + 4 = x^2 + 2x + 1 + x^2$	Multiply.
$x^2 + 4x + 4 = 2x^2 + 2x + 1$	Simplify.
$0 = x^2 - 2x - 3$	Subtract x^2, $4x$, and 4 from both sides.
$x^2 - 2x - 3 = 0$	Write in standard form.
$(x - 3)(x + 1) = 0$	Factor.
$x - 3 = 0$ or $x + 1 = 0$	Use the zero-factor property.
$x = 3$ or $x = -1$	Solve $x - 3 = 0$ and $x + 1 = 0$.

Since the lengths of the sides must be positive, we discard the negative answer, -1. Thus the shortest side is 3 units, so that the other two sides are 4 and 5 units.   ▲

By the way, after you find the first Pythagorean triple of the form $(a, a + 1, c) = (3, 4, 5)$, you can find infinitely many more of the same form by using $(3a + 2c + 1, 3a + 2c + 2, 4a + 3c + 2)$. If we let $a = 3$, $b = 4$, and $c = 5$, one such triple is $(3 \cdot 3 + 2 \cdot 5 + 1, 3 \cdot 3 + 2 \cdot 5 + 2, 4 \cdot 3 + 3 \cdot 5 + 2)$ or $(20, 21, 29)$.

Problem 5 The lengths of one side of a right triangle and the hypotenuse are consecutive integers. If the shortest side is 7 units smaller than the longest side, find the length of each of the sides.

A. Solve.

1. $(x + 1)(x + 2) = 0$

2. $(x + 3)(x + 4) = 0$

3. $(x - 1)(x + 4)(x + 3) = 0$

4. $(x + 5)(x - 3)(x + 2) = 0$

5. $\left(x - \dfrac{1}{2}\right)\left(x - \dfrac{1}{3}\right) = 0$

6. $\left(x - \dfrac{1}{4}\right)\left(x - \dfrac{1}{7}\right) = 0$

7. $y(y - 3) = 0$

8. $y(y - 4) = 0$

9. $y^2 - 64 = 0$

10. $y^2 - 1 = 0$

11. $y^2 - 81 = 0$

12. $y^2 - 100 = 0$

13. $x^2 + 6x = 0$

14. $x^2 + 2x = 0$

15. $x^2 - 3x = 0$

16. $x^2 - 8x = 0$

17. $y^2 - 12y = -27$

18. $y^2 - 10y = -21$

19. $y^2 = -6y - 5$

20. $y^2 = -3y - 2$

21. $x^2 = 2x + 15$

22. $x^2 = 4x + 12$

23. $3y^2 + 5y + 2 = 0$

24. $3y^2 + 7y + 2 = 0$

25. $2y^2 - 3y + 1 = 0$

26. $2y^2 - 3y - 20 = 0$

27. $2y^2 - y - 1 = 0$

28. $2y^2 - y - 15 = 0$

1. _____
2. _____
3. _____
4. _____
5. _____
6. _____
7. _____
8. _____
9. _____
10. _____
11. _____
12. _____
13. _____
14. _____
15. _____
16. _____
17. _____
18. _____
19. _____
20. _____
21. _____
22. _____
23. _____
24. _____
25. _____
26. _____
27. _____
28. _____

(Hint for Problems 29–34: Multiply each term by the LCD *first*.)

29. $\dfrac{x^2}{12} + \dfrac{x}{3} - 1 = 0$ **30.** $\dfrac{x^2}{2} - \dfrac{x}{12} - 1 = 0$

31. $\dfrac{x^2}{3} - \dfrac{x}{2} = \dfrac{-1}{6}$ **32.** $\dfrac{x^2}{6} + \dfrac{x}{3} = \dfrac{1}{2}$

33. $\dfrac{x^2}{12} + \dfrac{x}{2} = \dfrac{-2}{3}$ **34.** $\dfrac{x^2}{3} + \dfrac{x}{3} = \dfrac{1}{4}$

35. $(2x - 1)(x - 3) = 3x - 5$ **36.** $(3x + 1)(x - 2) = x + 7$

37. $(2x + 3)(x + 4) = 2(x - 1) + 4$

38. $(5x - 2)(x + 2) = 3(x + 1) - 7$

39. $(2x - 1)(x - 1) = x - 1$ **40.** $(3x - 2)(3x - 1) = 1 - 3x$

41. $x^3 + 4x^2 - 4x - 16 = 0$ **42.** $x^3 - 4x^2 - 4x + 16 = 0$

43. $x^3 - 5x^2 - 9x + 45 = 0$ **44.** $x^3 + 5x^2 - 9x - 45 = 0$

45. $3x^3 + 3x^2 = 12x + 12$ **46.** $2x^3 - 2x^2 - 18x + 18 = 0$

B. Solve the problems using the Pythagorean theorem.

47. The sides of a right triangle are consecutive even integers. Find their lengths.

48. The hypotenuse of a right triangle is 4 cm longer than the shortest side and 2 cm longer than the remaining side. Find the dimensions of the triangle.

49. The hypotenuse of a right triangle is 16 in. longer than the shortest side and 2 in. longer than the remaining side. Find the dimensions of the triangle.

50. The hypotenuse of a right triangle is 8 in. longer than the shortest side and 1 in. longer than the remaining side. Find the dimensions of the triangle.

C. Applications

In Problems 51–54 use

$$h = 5t^2 + V_0 t \quad \text{(meters)}$$

where h is the distance traveled in t sec by an object thrown downward with an initial velocity V_0.

51. An object is thrown downward at 5 m/sec from a height of 10 m. How long would it take the object to hit the ground?

29. _____
30. _____
31. _____
32. _____
33. _____
34. _____
35. _____
36. _____
37. _____
38. _____
39. _____
40. _____
41. _____
42. _____
43. _____
44. _____
45. _____
46. _____

47. _____

48. _____

49. _____

50. _____

51. _____

52. An object is thrown downward from a height of 28 m with an initial velocity of 4 m/sec. How long would it take the object to reach the ground?

52. _____

53. An object is thrown down from a building 15 m high at 10 m/sec. How long would it take the object to hit the ground?

53. _____

54. How long would it take a package thrown downward from a plane at 10 m/sec to hit the ground 175 m below?

54. _____

✓ **SKILL CHECKER**

Simplify.

55. $\dfrac{3x^5}{15x^7}$

56. $\dfrac{10x^{-3}}{5x^6}$

55. _____

56. _____

57. $\dfrac{20x^7}{10x^{-3}}$

58. $\dfrac{8x^{-4}}{16x^{-5}}$

57. _____

58. _____

59. $\dfrac{18x^{-10}}{9x^2}$

60. $\dfrac{4x^{-5}}{2x^{-5}}$

59. _____

60. _____

3.8 USING YOUR KNOWLEDGE

We have mentioned that there are infinitely many Pythagorean triples of the form $(a, a + 1, c)$. One of them is $(3, 4, 5)$ (see Example 5). Another Pythagorean triple is $(20, 21, 29)$. How can we verify this? In order for these three numbers to be the sides of a right triangle, they must satisfy the equation $c^2 = a^2 + b^2$. If we let $a = 20$, $b = 21$, and $c = 29$, we must verify that

$$29^2 = 20^2 + 21^2$$

or

$$841 = 400 + 441$$

which is true. The knowledge of the formula for finding triples,

$$(3a + 2c + 1, 3a + 2c + 2, 4a + 3c + 2)$$

enabled us to find the next triple $(20, 21, 29)$ by substituting $a = 3$, $b = 4$, and $c = 5$ in the formula.

1. Use $(20, 21, 29)$ to find the next Pythagorean triple of the form $(a, a + 1, c)$.

1. _____

2. Use the answer found in 1 to find another Pythagorean triple of the form $(a, a + 1, c)$.

2. _____

CALCULATOR CORNER

A calculator is ideal to check the solutions of quadratic equations. This time we will use the memory key denoted by $\boxed{\text{STO}}$ (or $\boxed{\text{Min}}$ or $\boxed{\text{X} \rightarrow \text{M}}$) and the memory recall key $\boxed{\text{RCL}}$ (or $\boxed{\text{MR}}$). The $\boxed{\text{STO}}$ key will enable you to store the solution and then recall it by using the

ANSWER (to problem on page 238)
5. 5, 12, and 13 units.

$\boxed{\text{RCL}}$ key. Thus, to check the solution $-\frac{3}{4}$ for the equation $12x^2 + 5x - 3 = 0$ of Example 3, store the solution $-\frac{3}{4}$ by using the keystrokes.

$\boxed{3}\;\boxed{+/-}\;\boxed{\div}\;\boxed{4}\;\boxed{=}\;\boxed{\text{STO}}$

Then check the answer by pressing

$\boxed{1}\;\boxed{2}\;\boxed{\times}\;\boxed{\text{RCL}}\;\boxed{x^2}\;\boxed{+}\;\boxed{5}\;\boxed{\times}\;\boxed{\text{RCL}}\;\boxed{=}\;\boxed{-}\;\boxed{3}\;\boxed{=}$

The answer should be 0. To check the other answer $(\frac{1}{3})$, enter

$\boxed{1}\;\boxed{\div}\;\boxed{3}\;\boxed{=}\;\boxed{\text{STO}}$

and then repeat the keystrokes used to check $-\frac{3}{4}$.

If you are allowed to use calculators, check your answers to Problems 23 through 28.

SUMMARY

SECTION	ITEM	MEANING	EXAMPLE
3.1A	Monomial	A constant times a product of variables with whole-number exponents	$3x^2y$, $-7x$, $0.5x^3$, $\frac{2}{3}x^2yz^4$
3.1A	Polynomial	A sum or difference of monomials	$3x^2 - 7x + 8$, $x^2y + y^3$
3.1A	Terms	The individual monomials in a polynomial	The terms of $3x^2 - 7x + 8$ are: $3x^2$, $-7x$, and 8.
3.1A	Coefficient	The numerical factor of a term	The coefficient of $3x^2$ is 3.
3.1A	Binomial	A polynomial with two terms	$5x^2 - 7$ is a binomial.
3.1A	Trinomial	A polynomial with three terms	$-3 + x^2 + x$ is a trinomial.
3.1B	Degree of a polynomial	Largest sum of the exponents in any term	The degree of $x^3 + 7x$ is 3. The degree of $-2x^3yz^2$ is 6.
3.2A	Commutative law of addition	If P and Q are polynomials, $P + Q = Q + P$.	$x + 3x^2 = 3x^2 + x$
3.2A	Associative law of addition	If P, Q, and R are polynomials, $P + (Q + R) = (P + Q) + R$.	$x^2 + (3 + 7x) = (x^2 + 3) + 7x$
3.2A	Distributive laws	If P, Q, and R are polynomials: $P(Q + R) = PQ + PR$ $(Q + R)P = QP + RP$	$2x(x^2 + 3) = 2x^3 + 6x$
3.2A	Commutative law of multiplication	If P and Q are polynomials, $P \cdot Q = Q \cdot P$.	$x \cdot 3x^2 = 3x^2 \cdot x$
3.2A	Associative law of multiplication	If P, Q, and R are polynomials, $P \cdot (Q \cdot R) = (P \cdot Q) \cdot R$.	$x^2 \cdot (3 \cdot 7x) = (x^2 \cdot 3) \cdot 7x$
3.2C	Product of two binomials	$(x + a)(x + b) = x^2 + (b + a)x + ab$	$(x + 5)(x - 7) = x^2 - 2x - 35$
3.2D	Square of a binomial sum	$(x + a)^2 = x^2 + 2ax + a^2$	$(x + 5y)^2 = x^2 + 10xy + 25y^2$
3.2D	Square of a binomial difference	$(x - a)^2 = x^2 - 2ax + a^2$	$(x - 5y)^2 = x^2 - 10xy + 25y^2$
3.2E	Product of a sum and difference	$(x + a)(x - a) = x^2 - a^2$	$(x + 2y)(x - 2y) = x^2 - 4y^2$
3.4A	Greatest common factor (GCF) of a polynomial	The monomial that is a factor of each term with: (1) the highest degree and (2) the coefficient of greatest absolute value	The GCF of $3x^2 + 6x$ is $3x$.

(Continued)

(continued)

SECTION	ITEM	MEANING	EXAMPLE
3.5A	Factoring $x^2 + (b + a)x + ab$	$x^2 + (b + a)x + ab = (x + a)(x + b)$	$x^2 + 5x + 4 = (x + 1)(x + 4)$
3.5B	The ac test	$ax^2 + bx + c$ is factorable only if there are two integers whose product is ac and whose sum is b.	$3x^2 + 5x + 2$ is factorable, since there are two integers with product 6 and sum 5.
3.6A	Factoring perfect square trinomials	$x^2 + 2ax + a^2 = (x + a)^2$ $x^2 - 2ax + a^2 = (x - a)^2$	$x^2 + 10x + 25 = (x + 5)^2$ $x^2 - 14x + 49 = (x - 7)^2$
3.6B	Factoring the difference of two squares	$x^2 - a^2 = (x + a)(x - a)$	$16x^2 - 9y^2 = (4x + 3y)(4x - 3y)$
3.6C	Factoring the sum or difference of two cubes	$x^3 + a^3 = (x + a)(x^2 - ax + a^2)$ $x^3 - a^3 = (x - a)(x^2 + ax + a^2)$	$8x^3 + y^3 = (2x + 3)(4x^2 - 6x + 9)$ $8x^3 - y^3 = (2x - 3)(4x^2 + 6x + 9)$
3.7	General factoring	**1.** Factor out the GCF. **2.** Check for: Difference of two squares Difference of two cubes Sum of two cubes Perfect square trinomials Trinomials of the form $ax^2 + bx + c$ Four terms (grouping)	
3.8	Quadratic equation	An equation that can be written in the form $ax^2 + bx + c = 0$ $(a \neq 0)$	$3x^2 + 5x = -6$ is a quadratic equation.
3.8A	Zero-factor property	For all real numbers a and b, $a \cdot b = 0$ means that $a = 0$, $b = 0$, or both.	$(x + 1)(x + 2) = 0$ means $x + 1 = 0$ or $x + 2 = 0$.
3.8B	Pythagorean theorem	In any right triangle, the square of the longest side is equal to the sum of the squares of the other two sides: $c^2 = a^2 + b^2$.	If the sides of a right triangle are of lengths 3 and 4 and the hypotenuse is 5, $3^2 + 4^2 = 5^2$.

NAME

CLASS

SECTION

ANSWERS

(If you need help with these exercises, look in the section indicated in brackets.)

1. [3.1A,B] Classify as a monomial, binomial, or trinomial and give the degree.
   a. $x^3 + x^2y^3z$
   b. $x^3y^2z^3$
   c. $x^4 - 5x^2y^3 + xyz$

2. [3.1C] Rewrite using exponents and give the degree.
   a. $8xxxyyy - 3xxxyy + 7zzz$
   b. $3xxyyzzz + 4xxyy - 4zzzzz$
   c. $6xxyyzz + 3xyzzz - 4xxyyyzzz$

3. [3.1D] Let $P(x) = x^2 - 3x + 5$. Find.
   a. $P(-1)$
   b. $P(2)$
   c. $P(-3)$

4. [3.2A] Add $2x^3 + 5x^2 - 3x - 1$ and
   a. $8 - 7x + 3x^2 - x^3$
   b. $9 - 8x^2 + 3x^3$
   c. $7 - 4x + x^3$

5. [3.2B] Subtract $7x^3 - 5x^2 + 3x - 1$ from
   a. $4x^3 + 2x^2 + 2$
   b. $7x^3 + 5x^2 + 4x - 7$
   c. $8x^2 - 9x + 3$

6. [3.3A] Multiply.
   a. $-2x^2y(x^2 + 3xy - 2y^3)$
   b. $-3x^2y^2(x^2 + 3xy - 2y^3)$
   c. $-4xy^2(x^2 + 3xy - 2y^3)$

7. [3.3B] Multiply.
   a. $(x - 1)(x^2 - 3x - 2)$
   b. $(x - 2)(x^2 - 3x - 2)$
   c. $(x - 3)(x^2 - 3x - 2)$

8. [3.3C] Multiply.
   a. $(2x + 3y)(4x - 5y)$
   b. $(2x + 3y)(3x - 2y)$
   c. $(2x + 3y)(5x - 3y)$

9. [3.3D] Multiply.
   a. $(2x + 7y)^2$
   b. $(3x + 7y)^2$
   c. $(4x + 7y)^2$

1. a. _____
   b. _____
   c. _____

2. a. _____
   b. _____
   c. _____

3. a. _____
   b. _____
   c. _____

4. a. _____
   b. _____
   c. _____

5. a. _____
   b. _____
   c. _____

6. a. _____
   b. _____
   c. _____

7. a. _____
   b. _____
   c. _____

8. a. _____
   b. _____
   c. _____

9. a. _____
   b. _____
   c. _____

10. [3.3D] Multiply.
   a. $(3x - 7y)^2$
   b. $(4x - 7y)^2$
   c. $(5x - 7y)^2$

10. a. _____
    b. _____
    c. _____

11. [3.3E] Multiply.
   a. $(3x + 2y)(3x - 2y)$
   b. $(4x + 3y)(4x - 3y)$
   c. $(5x + 3y)(5x - 3y)$

11. a. _____
    b. _____
    c. _____

12. [3.4A] Factor.
   a. $15x^5 - 20x^4 + 10x^3 + 25x^2$
   b. $9x^5 - 12x^4 + 6x^3 + 15x^2$
   c. $6x^5 - 8x^4 + 4x^3 + 10x^2$

12. a. _____
    b. _____
    c. _____

13. [3.4B] Factor.
   a. $6x^6 - 2x^4 + 15x^3 - 5x$
   b. $6x^6 - 8x^4 + 15x^3 - 20x$
   c. $6x^6 - 4x^4 + 9x^3 - 6x$

13. a. _____
    b. _____
    c. _____

14. [3.5A] Factor.
   a. $x^2 - 3xy - 18y^2$
   b. $x^2 - 4xy - 12y^2$
   c. $x^2 - 5xy - 6y^2$

14. a. _____
    b. _____
    c. _____

15. [3.5B] Factor.
   a. $2x^2 - 7xy - 30y^2$
   b. $2x^2 - 3xy - 20y^2$
   c. $2x^2 - 5xy - 25y^2$

15. a. _____
    b. _____
    c. _____

16. [3.5B] Factor.
   a. $18x^4y + 3x^3y^2 - 6x^2y^3$
   b. $30x^4y + 5x^3y^2 - 10x^2y^3$
   c. $36x^4y + 6x^3y^2 - 12x^2y^3$

16. a. _____
    b. _____
    c. _____

17. [3.6A] Factor.
   a. $4x^2 - 28xy + 49y^2$
   b. $9x^2 - 42xy + 49y^2$
   c. $16x^2 - 56xy + 49y^2$

17. a. _____
    b. _____
    c. _____

18. [3.6A] Factor.
   a. $9x^2 + 24xy + 16y^2$
   b. $9x^2 + 30xy + 25y^2$
   c. $9x^2 + 36xy + 36y^2$

18. a. _____
    b. _____
    c. _____

19. [3.6B] Factor.
   a. $81x^4 - y^4$
   b. $x^4 - 16y^4$
   c. $81x^4 - 16y^4$

19. a. _____
    b. _____
    c. _____

20. [3.6B] Factor.
   a. $x^2 - 4x + 4 - y^2$
   b. $x^2 - 6x + 9 - y^2$
   c. $x^2 - 8x + 16 - y^2$

20. a. _____
    b. _____
    c. _____

21. [3.6C] Factor.
   a. $27x^3 + 8y^3$
   b. $27x^3 + 64y^3$
   c. $64x^3 + 27y^3$

21. a. _____
    b. _____
    c. _____

22. [3.6C] Factor.
    a. $27x^3 - 8y^3$
    b. $27x^3 - 64y^3$
    c. $64x^3 - 27y^3$

23. [3.7] Factor.
    a. $27x^6 - 8x^3y^3$
    b. $27x^7 - 64x^4y^3$
    c. $64x^8 - 27x^5y^3$

24. [3.7] Factor.
    a. $27x^6 + 3x^4$
    b. $4x^6 + 64x^4$
    c. $2x^6 + 18x^4$

25. [3.7] Factor.
    a. $27x^4 + 36x^3y + 12x^2y^2$
    b. $36x^4 + 48x^3y + 16x^2y^2$
    c. $45x^4 + 60x^3y + 20x^2y^2$

26. [3.7] Factor.
    a. $27x^4 - 36x^3y + 12x^2y^2$
    b. $36x^4 - 48x^3y + 16x^2y^2$
    c. $45x^4 - 60x^3y + 20x^2y^2$

27. [3.7] Factor.
    a. $12x^3y - 44x^2y^2 - 16xy^3$
    b. $15x^3y - 55x^2y^2 - 20xy^3$
    c. $18x^3y - 66x^2y^2 - 24xy^3$

28. [3.7] Factor.
    a. $2x^3 - x^2 - 2x + 1$
    b. $18x^3 - 9x^2 - 2x + 1$
    c. $32x^3 - 16x^2 - 2x + 1$

29. [3.8A] Solve.
    a. $x^2 = -x + 12$
    b. $x^2 = -x + 20$
    c. $x^2 = -2x + 24$

30. [3.8A] Solve.
    a. $6x^2 + x = 1$
    b. $8x^2 + 2x = 1$
    c. $10x^2 + 3x = 1$

31. [3.8A] Solve.
    a. $x^3 + 2x^2 - x - 2 = 0$
    b. $x^3 + 4x^2 - x - 4 = 0$
    c. $x^3 + 2x^2 - 9x - 18 = 0$

32. [3.8B] Find the dimensions of a right triangle whose sides are
    a. x, $x + 2$, and $x + 4$ units long
    b. x, $x + 3$, and $x + 6$ units long
    c. x, $x + 4$, and $x + 8$ units long

22. a. _____
    b. _____
    c. _____

23. a. _____
    b. _____
    c. _____

24. a. _____
    b. _____
    c. _____

25. a. _____
    b. _____
    c. _____

26. a. _____
    b. _____
    c. _____

27. a. _____
    b. _____
    c. _____

28. a. _____
    b. _____
    c. _____

29. a. _____
    b. _____
    c. _____

30. a. _____
    b. _____
    c. _____

31. a. _____
    b. _____
    c. _____

32. a. _____
    b. _____
    c. _____

(*Answers on page* 250)

1. Classify as a monomial, binomial, or trinomial and give the degree of $xy^3z^4 - x^7$.

2. Rewrite $3xxxyyyy - 8xxyyyy - 4zzzzz$ using exponents and give its degree.

3. Let $P(x) = x^2 - 3x + 2$. Find $P(-2)$.

4. Add $6x^3 + 8x^2 - 6x - 4$ and $6 - 3x + x^2 - 3x^3$.

5. Subtract $8x^3 - 6x^2 + 5x - 3$ from $5x^3 + 3x^2 + 3$.

6. Multiply $-3x^2y(x^2 + 5xy - 3y^3)$.

7. Multiply $(x - 2)(x^2 - 4x - 5)$.

8. Multiply $(3x + 5y)(4x - 7y)$.

9. Multiply.
   a. $(2x + 3y)^2$ b. $(3x - 4y)^2$

10. Multiply $(3x + 4y)(3x - 4y)$.

11. Factor $12x^6 - 16x^5 + 8x^4 + 20x^3$.

12. Factor $6x^7 + 6x^5 + 15x^4 + 15x^2$.

13. Factor.
    a. $x^2 - 3xy - 18y^2$ b. $2x^2 + xy - 10y^2$

14. Factor $36x^4y + 12x^3y^2 - 8x^2y^3$.

15. Factor.
    a. $16x^2 - 24xy + 9y^2$ b. $9x^2 + 30xy + 25y^2$

16. Factor $x^4 - 16y^4$. 17. Factor $x^2 - 10x + 25 - y^2$.

18. Factor.
    a. $27x^3 + 8y^3$ b. $8y^3 - 27x^3$

19. Factor.
    a. $8x^7 - x^4y^3$ b. $6x^8 + 24x^6$

20. Factor.
    a. $8x^4 + 24x^3y + 18x^2y^2$ b. $48x^2y^2 - 72xy^3 + 27y^4$

21. Factor $9x^3y - 33x^2y^2 - 12xy^3$.

22. Factor $16x^3 - 12x^2 - 4x + 3$.

23. Solve.
    a. $x^2 = -3x + 10$ b. $6x^2 + 7x = 3$

24. Solve $x^3 - x^2 - 4x + 4 = 0$.

25. The sides of a right triangle are x, $x + 2$, and $x + 4$ units long. Find the dimensions of the triangle.

IF YOU MISSED QUESTION	SECTION	EXAMPLES	PAGE	ANSWERS
1	3.1	1, 2	172	1. Binomial: 8
2	3.1	3	173	2. $3x^3y^4 - 8x^2y^4 - 4z^5$; 7
3	3.1	4, 5	174	3. 12
4	3.2	1, 2	180	4. $3x^3 + 9x^2 - 9x + 2$
5	3.2	3	181	5. $-3x^3 + 9x^2 - 5x + 6$
6	3.3	1	188	6. $-3x^4y - 15x^3y^2 + 9x^2y^4$
7	3.3	2	189	7. $x^3 - 6x^2 + 3x + 10$
8	3.3	3	190	8. $12x^2 - xy - 35y^2$
9a	3.3	4	191	9. a. $4x^2 + 12xy + 9y^2$
9b	3.3	4	191	b. $9x^2 - 24xy + 16y^2$
10	3.3	5	192	10. $9x^2 - 16y^2$
11	3.4	1, 2, 3	198–199	11. $4x^3(3x^3 - 4x^2 + 2x + 5)$
12	3.4	4, 5	199–200	12. $3x^2(x^2 + 1)(2x^3 + 5)$
13a	3.5	1, 2	206	13. a. $(x + 3y)(x - 6y)$
13b	3.5	3	207	b. $(x - 2y)(2x + 5y)$
14	3.5	6	209–210	14. $4x^2y(3x + 2y)(3x - y)$
15a	3.6	1, 2	216	15. a. $(4x - 3y)^2$
15b	3.6	1, 2	216	b. $(3x + 5y)^2$
16	3.6	3	217	16. $(x^2 + 4y^2)(x + 2y)(x - 2y)$
17	3.6	4	217	17. $(x - 5 + y)(x - 5 - y)$
18a	3.6	5	218	18. a. $(3x + 2y)(9x^2 - 6xy + 4y^2)$
18b	3.6	5	218	b. $(2y - 3x)(4y^2 + 6xy + 9y^2)$
19a	3.7	1	226	19. a. $x^4(2x - y)(4x^2 + 2xy + y^2)$
19b	3.7	1	226	b. $6x^6(x^2 + 4)$
20a	3.7	2	226	20. a. $2x^2(2x + 3y)^2$
20b	3.7	2	226	b. $3y^2(4x - 3y)^2$
21	3.7	3	226–227	21. $3xy(3x + y)(x - 4y)$
22	3.7	5	227	22. $(4x - 3)(2x + 1)(2x - 1)$
23a	3.8	1, 2, 3	234–237	23. a. $2, -5$
23b	3.8	1, 2, 3	234–237	b. $\dfrac{1}{3}, -\dfrac{3}{2}$
24	3.8	4	237	24. $1, 2, -2$
25	3.8	5	238	25. $6, 8, 10$

CHAPTER 4

RATIONAL EXPRESSIONS

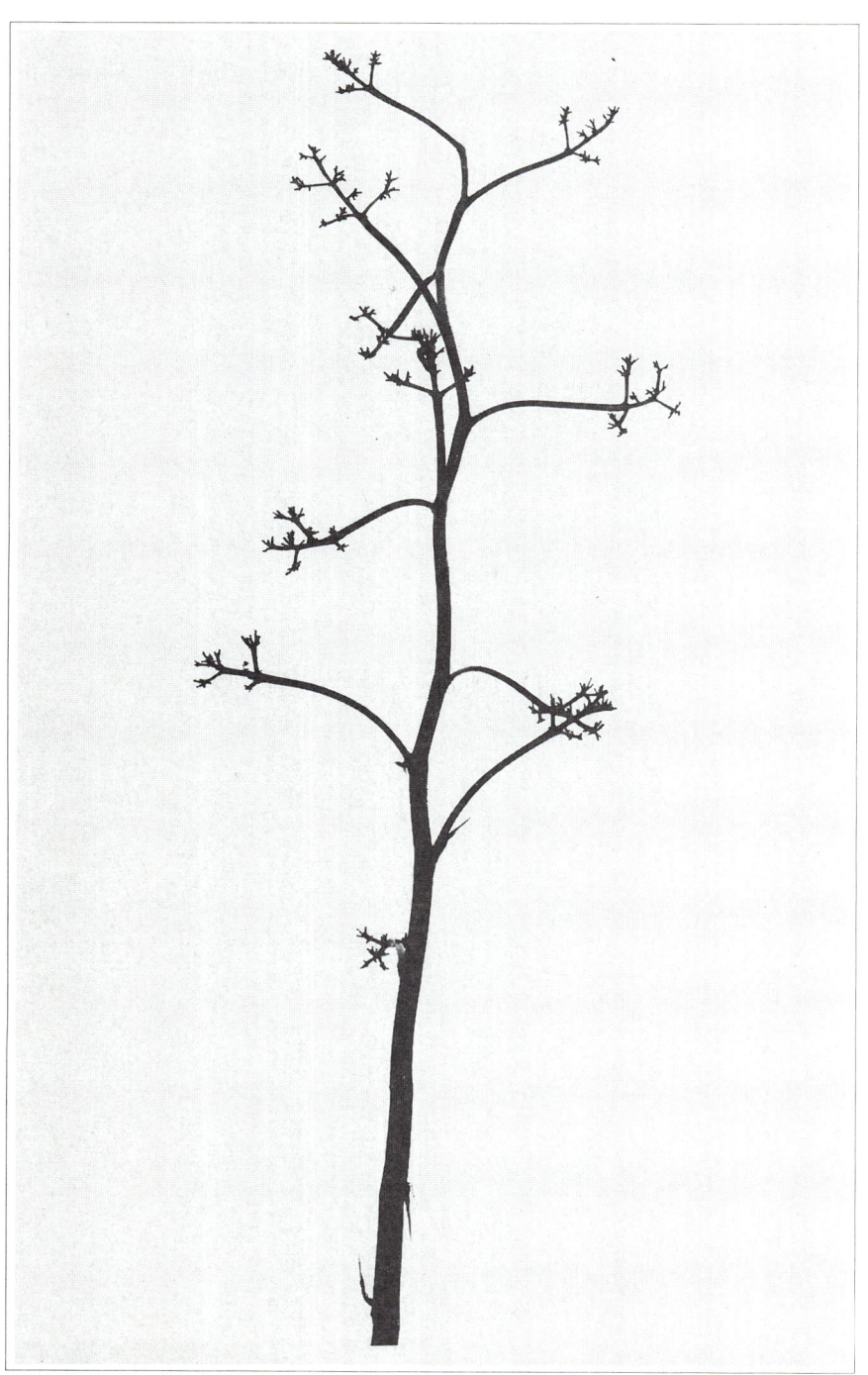

In this chapter we use the polynomial concepts we have studied to form quotients called rational expressions. We learn how to add, subtract, multiply, and divide these expressions. We then solve equations containing rational expressions and end the chapter by applying the ideas studied to number, work, and uniform motion problems.

NAME

CLASS

SECTION

(Answers on pages 256–257)

ANSWERS

1. Write the fraction with the indicated denominator.

   a. $\dfrac{2x^2}{9y^4}$, denominator $27y^6$

   b. $\dfrac{2x+1}{x+1}$, denominator x^2+3x+2

1. a. _____
   b. _____

2. Write in standard form.

   a. $-\dfrac{-3}{y}$

   b. $-\dfrac{x-y}{7}$

2. a. _____
   b. _____

3. Reduce to lowest terms.

   a. $\dfrac{x^4y^5}{xy^2}$

   b. $\dfrac{xy+y^2}{x^2-y^2}$

3. a. _____
   b. _____

4. Reduce $\dfrac{x^2-y^2}{y^3-x^3}$ to lowest terms.

4. _____

5. Divide $\dfrac{28x^5-14x^3+7x^2}{7x^4}$.

5. _____

6. Divide $2x^3-4-4x$ by $2+2x$.

6. _____

7. Factor $2x^3+3x^2-23x-12$ if $x+4$ is one of its factors.

7. _____

8. Multiply $\dfrac{x-2}{3x-2y} \cdot \dfrac{9x^2-4y^2}{2x^2-3x-2}$.

8. _____

9. Divide $\dfrac{x+2}{x-2} \div (x^2+4x+4)$.

9. _____

10. Perform the indicated operations.

$$\dfrac{2-x}{x+3} \div \dfrac{x^3-8}{x+6} \cdot \dfrac{x^3+27}{x+6}$$

10. _____

11. Perform the indicated operations.

a. $\dfrac{x}{x^2-9} + \dfrac{3}{x^2-9}$

b. $\dfrac{x}{x^2-9} - \dfrac{3}{x^2-9}$

11. a. _____
    b. _____

12. Perform the indicated operations.

$$\dfrac{x+2}{x^2+2x-3} + \dfrac{x+4}{x^2-9}$$

12. _____

13. Perform the indicated operations.

$$\dfrac{x-4}{x^2-x-6} - \dfrac{x+1}{x^2-9}$$

13. _____

14. Simplify $\dfrac{x+\dfrac{1}{x^2}}{x-\dfrac{1}{x^3}}$.

14. _____

15. Simplify $3 + \dfrac{a}{3+\dfrac{3}{3+a}}$.

15. _____

16. Solve $\dfrac{x}{x+2} - \dfrac{x}{x-2} = \dfrac{x^2+4}{x^2-4}$.

16. _____

17. Solve $1 + \dfrac{2}{x-3} = \dfrac{12}{x^2-9}$.

17. _____

18. Find two consecutive even integers such that the sum of their reciprocals is $\frac{9}{40}$.

19. Jack can mow the lawn in 3 hr and Jill can mow it in 2 hr. How long would it take to mow the lawn if both work together?

20. A plane traveled 840 m with a 30-mi/hr tail wind in the same time it took to travel 660 mi against the wind. What is the plane's speed in still air?

18. _____

19. _____

20. _____

IF YOU MISSED QUESTION	SECTION	EXAMPLES	PAGE	ANSWERS
1a	4.1	1	261	**1. a.** $\dfrac{6x^2y^2}{27y^6}$
1b	4.1	1	261	**b.** $\dfrac{2x^2 + 5x + 2}{x^2 + 3x + 2}$
2a	4.1	2	262	**2. a.** $\dfrac{3}{y}$
2b	4.1	2	262	**b.** $\dfrac{y - x}{7}$
3a	4.1	3	263	**3. a.** x^3y^3
3b	4.1	3	263	**b.** $\dfrac{y}{x - y}$
4	4.1	4	264	**4.** $\dfrac{-(x + y)}{y^2 + xy + x^2}$
5	4.2	1	270	**5.** $4x - \dfrac{2}{x} + \dfrac{1}{x^2}$
6	4.2	2, 3	271	**6.** $x^2 - x - 1 \, \text{R} - 2$
7	4.2	4	272	**7.** $(x + 4)(x - 3)(2x + 1)$
8	4.3	1, 2	278–279	**8.** $\dfrac{3x + 2y}{2x + 1}$
9	4.3	3	280	**9.** $\dfrac{1}{x^2 - 4}$
10	4.3	4, 5	280–281	**10.** $\dfrac{-(x^2 - 3x + 9)}{x^2 + 2x + 4}$
11a	4.4	1	288	**11. a.** $\dfrac{1}{x - 3}$
11b	4.4	1	288	**b.** $\dfrac{1}{x + 3}$
12	4.4	2, 3	289–290	**12.** $\dfrac{2x^2 + 2x - 10}{(x + 3)(x - 1)(x - 3)}$
13	4.4	2, 3	289–291	**13.** $\dfrac{-4x - 14}{(x + 2)(x + 3)(x - 3)}$
14	4.5	1, 2	298–299	**14.** $\dfrac{x(x^2 - x + 1)}{(x^2 + 1)(x - 1)}$
15	4.5	3, 4	299–301	**15.** $\dfrac{a^2 + 12a + 36}{3a + 12}$
16	4.6	1, 2	308–310	**16.** No solution
17	4.6	3, 4	311–312	**17.** -5

IF YOU MISSED QUESTION	SECTION	EXAMPLES	PAGE	ANSWERS
18	4.7	1	318	**18.** 8 and 10
19	4.7	2	319	**19.** $1\frac{1}{5}$ hr
20	4.7	4	320–321	**20.** 250 mi/hr

4.1 RATIONAL EXPRESSIONS

OBJECTIVES

REVIEW

Before starting this section, you should know:

1. The laws of exponents.
2. How to factor polynomials.

OBJECTIVES

You should be able to:

A. Write a given fraction with the indicated denominator.
B. Write a fraction in standard form.
C. Reduce a fraction to lowest terms.

If you write the value of the coins using fractions, the photo tells us that

$$\frac{1}{2} = \frac{2}{4}$$

$\frac{1}{2}$ and $\frac{2}{4}$ are fractions. The word *fraction* is derived from the Latin word *fractio*, which means "to break" or "to divide." Thus a fraction is an expression denoting a division. We know that any fraction of the form $\frac{a}{b}$ (with a and b as integers and $b \neq 0$) is a *rational number*. We can extend this idea to polynomial expressions: If the numerator and denominator of a fraction are polynomials, then the fraction is called a *rational expression*; that is, a rational expression is an expression of the form $\frac{P}{Q}$, where P and Q are polynomials and $Q \neq 0$. Thus

$$\frac{x + y}{z}, \quad \frac{x^2 - 2x + 1}{x + 2}, \quad \frac{1}{y^2 - 1}, \quad \text{and} \quad \frac{x}{x - 2}$$

are rational expressions. Of course, since any polynomial can be considered as the quotient of itself and 1, all polynomials are rational expressions.

The rational expressions $\frac{x + y}{z}, \frac{x^2 - 2x + 1}{x + 2}, \frac{1}{y^2 - 1}$, and $\frac{x}{x - 2}$ represent real numbers for those replacements of the variable for which the denominator is not zero. If the denominator *is* 0, the expression is said to be *undefined*.

For $\frac{8}{x}$, x cannot be 0 ($x \neq 0$) because then we would have $\frac{8}{0}$, which is not defined.

For

$$\frac{x^2 + 3x + 8}{x + 6}, \qquad x \neq -6$$

For

$$\frac{y^2}{y - 1}, \qquad y \neq 1$$

and for

$$\frac{x^2 + 3x + 8}{x^2 - 4x + 4} = \frac{x^2 + 3x + 8}{(x - 2)^2}, \qquad x \neq 2$$

When $x = 0$, $\dfrac{8}{x} = \dfrac{8}{0}$; not defined.

When $x = -6$, $\dfrac{x^2 + 3x + 8}{x + 6} = \dfrac{36 - 18 + 8}{0}$
$$= \dfrac{26}{0}.$$

When $y = 1$, $\dfrac{y^2}{y - 1} = \dfrac{1}{0}$.

When $x = 2$, $\dfrac{x^2 + 3x + 8}{x^2 - 4x + 4} = \dfrac{4 + 6 + 8}{4 - 8 + 4}$
$$= \dfrac{18}{0}.$$

265 (200), 273, 283

To avoid mentioning over and over that the denominators of algebraic fractions must not be 0, we make the following assumption.

> The variables in a rational expression may not be replaced by values that will make the denominator 0.

We have already mentioned that $\frac{1}{2} = \frac{2}{4}$. In fact,

$$\frac{1}{2} = \frac{2}{4} = \frac{3}{6} = \frac{x}{2x} = \frac{2x}{4x} = \frac{3x}{6x}$$

Do you see why? We multiply both the numerator and denominator of $\frac{1}{2}$ by 2, 3, x, $2x$, and $3x$ and get

$$\frac{1}{2} = \frac{1 \cdot 2}{2 \cdot 2} = \frac{1 \cdot 3}{2 \cdot 3} = \frac{1 \cdot x}{2 \cdot x} = \frac{1 \cdot 2x}{2 \cdot 2x} = \frac{1 \cdot 3x}{2 \cdot 3x}$$

This fact is an application of a basic property of fractions stated next.

> **FUNDAMENTAL PROPERTY OF FRACTIONS**
>
> If P, Q, and K are polynomials,
>
> $$\frac{P}{Q} = \frac{P \cdot K}{Q \cdot K}$$
>
> for all values for which the denominator is not 0.

A. WRITING FRACTIONS WITH AN INDICATED DENOMINATOR

The fraction $\frac{4}{5}$ can be written as an equivalent fraction with a denominator of 10 by multiplying both numerator and denominator by 2, obtaining

$$\frac{4}{5} = \frac{4 \cdot 2}{5 \cdot 2} = \frac{8}{10}$$

Similarly, to write

$$\frac{5x^3}{3y^2}$$

with a denominator of $6y^7$, first write the new equivalent fraction with the old denominator factored out, as shown:

$$\frac{5x^3}{3y^2} = \frac{?}{3y^2(2y^5)}$$

⌞ Multiply by $2y^5$ ⌟

Since the multiplier is $2y^5$, we have

Multiply by $2y^5$

$$\frac{5x^3}{3y^2} = \frac{5x^3(2y^5)}{3y^2(2y^5)} = \frac{10x^3y^5}{6y^7}$$

We are multiplying the denominator by $2y^5$, so we have to multiply the numerator by $2y^5$.

Thus

$$\frac{5x^3}{3y^2} = \frac{10x^3y^5}{6y^7}$$

EXAMPLE 1 Write:

a. $\dfrac{5}{8}$ with a denominator of 16

b. $\dfrac{2x^2}{9y^3}$ with a denominator of $18y^8$

c. $\dfrac{3x+1}{x-1}$ with a denominator of $x^2 + 2x - 3$

Solution

a. $\dfrac{5}{8} \quad = \quad \dfrac{?}{16}$

  ⌞ Multiply by 2 ⌟

  ⌜ Multiply by 2 ⌝

  $\dfrac{5}{6} \quad = \quad \dfrac{10}{16}$

Thus

$$\frac{5}{8} = \frac{10}{16}$$

b. Since $18y^8 = 9y^3(2y^5)$,

  $\dfrac{2x^2}{9y^3} \quad = \quad \dfrac{?}{9y^3(2y^5)}$

  ⌞ Multiply by $2y^5$ ⌟

  ⌜ Multiply by $2y^5$ ⌝

  $\dfrac{2x^2}{9y^3} \quad = \quad \dfrac{2x^2(2y^5)}{9y^3(2y^5)} = \dfrac{4x^2y^5}{18y^8}$

c. We first note that $x^2 + 2x - 3 = (x-1)(x+3)$. Thus

  $\dfrac{3x+1}{x-1} = \dfrac{?}{(x-1)(x+3)}$

  ⌞ Multiply by $(x+3)$ ⌟

  ⌜ Multiply by $(x+3)$ ⌝

  $\dfrac{3x+1}{x-1} = \dfrac{(3x+1)(x+3)}{(x-1)(x+3)} = \dfrac{3x^2 + 10x + 3}{x^2 + 2x - 3}$

Problem 1 Write:

a. $\dfrac{7}{8}$ with a denominator of 16

b. $\dfrac{3x^2}{8y^3}$ with a denominator of $24y^6$

c. $\dfrac{4x+1}{x+2}$ with a denominator of $x^2 - x - 6$

B. THE STANDARD FORM OF A FRACTION

It is important to know that there are three signs associated with a fraction:

1. The sign before the fraction
2. The sign of the numerator
3. The sign of the denominator

Using our definition of a quotient and the fundamental property of fractions, we can conclude that:

$$\frac{-a}{b} = \frac{a}{-b} = -\frac{a}{b} = -\frac{-a}{-b} \text{ and}$$

$$\frac{a}{b} = \frac{-a}{-b} = -\frac{a}{-b} = -\frac{-a}{b}$$

The forms $\frac{-a}{b}$ and $\frac{a}{b}$, in which the sign of the fraction and that of the denominator are positive, are called the *standard forms* of the fractions. Thus, $\frac{-2}{9}$ and $\frac{4}{7}$ are in standard form, but $\frac{2}{-9}$ and $\frac{-4}{-7}$ are not. Of course, in expressions with more than one term in the numerator or denominator, there are alternative standard forms. For example,

$$\underbrace{\frac{-1}{x-y}} = \underbrace{\frac{-1}{-(y-x)}} = \frac{1}{y-x} \qquad \text{Recall that } x - y = -(y - x), \text{ since}$$
$$-(y - x) = -y + x = x - y.$$

| same ↑

Thus either $\frac{-1}{x-y}$ or $\frac{1}{y-x}$ can be used as the standard form. We prefer $\frac{1}{y-x}$ because it has only one negative sign, whereas $\frac{-1}{x-y}$ has two.

EXAMPLE 2 Write the following fractions in standard form.

a. $\dfrac{x}{-2}$ **b.** $-\dfrac{-3}{y}$ **c.** $-\dfrac{x-y}{5}$

Solution

a. $\dfrac{x}{-2} = \dfrac{-x}{2}$ **b.** $-\dfrac{-3}{y} = \dfrac{3}{y}$ **c.** $-\dfrac{x-y}{5} = \dfrac{-(x-y)}{5}$, or $\dfrac{y-x}{5}$

C. REDUCING FRACTIONS

The fundamental principle of fractions can also be used to **reduce** fractions—that is, to write fractions as equivalent ones in which no integers other than 1 can be divided into both the numerator and denominator. For example, the fraction $\frac{14}{21}$ can be reduced by writing the numerator and denominator in factored form and using the fundamental principle of fractions. Thus

$$\frac{14}{21} = \frac{2 \cdot \overset{1}{\cancel{7}}}{3 \cdot \underset{1}{\cancel{7}}} = \frac{2}{3} \qquad \text{Here we are dividing the numerator and denominator}$$
$$\text{by 7. We usually write } \frac{\overset{2}{\cancel{14}}}{\underset{3}{\cancel{21}}}.$$

The fraction

$$\frac{(x+3)(x^2-4)}{3(x+2)(x^2+x-6)}$$

Problem 2 Write in standard form:

a. $\dfrac{y}{-7}$ **b.** $-\dfrac{-4}{x}$ **c.** $-\dfrac{a-b}{8}$

can also be reduced by using the fundamental property of fractions. We do it by steps.

PROCEDURE FOR REDUCING FRACTIONS

1. Write the numerator and denominator of the fraction in factored form.

2. Find the factors that are common to the numerator and denominator.

3. Replace the quotient of the common factors by the number 1, since $\frac{a}{a} = 1$.

4. Rewrite the fraction in simplified form.

We are now ready to reduce

$$\frac{(x + 3)(x^2 - 4)}{3(x + 2)(x^2 + x - 6)}$$

Here are the steps:

1. Write the numerator and denominator in factored form.

$$\frac{(x + 3)(x + 2)(x - 2)}{3(x + 2)(x + 3)(x - 2)}$$

2. Find the factors that are common to the numerator and denominator (we rearranged them so the common factors are in columns).

$$\frac{(x + 2)(x + 3)(x - 2)}{3(x + 2)(x + 3)(x - 2)}$$

3. Replace the quotient of the common factors by the number 1.

$$\frac{\overset{1}{(x + 2)(x + 3)(x - 2)}}{3(x + 2)(x + 3)(x - 2)}$$

4. Rewrite the fraction in simplified form.

$$\frac{1}{3}$$

The whole procedure can be written as

$$\frac{(x + 3)(x^2 - 4)}{3(x + 2)(x^2 + x - 6)} = \frac{\overset{1}{(x + 3)}\overset{1}{(x + 2)}\overset{1}{(x - 2)}}{3(x + 2)(x + 3)(x - 2)} = \frac{1}{3}$$

EXAMPLE 3 Reduce each fraction to lowest terms.

a. $\dfrac{x^3y^4}{xy^6}$ **b.** $\dfrac{xy - y^2}{x^2 - y^2}$ **c.** $\dfrac{2x + xy}{x}$

Solution

a. $\dfrac{x^3y^4}{xy^6} = \dfrac{x^2 \cdot \overset{1}{\cancel{xy^4}}}{y^2 \cdot \cancel{xy^4}}$ Factor numerator and denominator.

$\qquad = \dfrac{x^2}{y^2}$ Divide out the common factor xy^4.

b. $\dfrac{xy - y^2}{x^2 - y^2} = \dfrac{y \cdot \cancel{(x - y)}}{(x + y)\cancel{(x - y)}}$ Factor numerator and denominator.

$\qquad = \dfrac{y}{x + y}$ Divide out the common factor $x - y$.

Problem 3 Reduce to lowest terms.

a. $\dfrac{x^5y^7}{xy^3}$ **b.** $\dfrac{x^2 - xy}{x^2 - y^2}$

c. $\dfrac{3y + xy}{y}$

c. $\dfrac{2x + xy}{x} = \dfrac{\overset{1}{\cancel{x}}(2 + y)}{\cancel{x}}$ Factor numerator and denominator.

$= 2 + y$ Divide out the common factor x.   ▲

In Example 3b, the answer is $\dfrac{y}{x + y}$. The answer cannot be reduced further. A common mistake is to try to divide out the y. This is not correct! You can divide out only *factors* and y is not a factor in the numerator and denominator. Sometimes, however, it seems that there are no common factors in the numerator and denominator. For example, look at the expression $\dfrac{a - b}{b - a}$. Are there any common factors in the numerator and denominator? Since $-(b - a) = -b + a = a - b$, we write:

$$\dfrac{a - b}{b - a} = \dfrac{-(b - a)}{b - a} = -1$$

Thus

$$\dfrac{a - b}{b - a} = -1$$

We use this idea in the next example.

EXAMPLE 4   Reduce.

a. $\dfrac{x^3 - y^3}{y - x}$ **b.** $\dfrac{x^2 - y^2}{y^3 - x^3}$

Solution

a. $\dfrac{x^3 - y^3}{y - x} = \dfrac{(x - y)(x^2 + xy + y^2)}{y - x}$ Factor numerator and denominator.

$= -1(x^2 + xy + y^2)$ $\dfrac{x - y}{y - x} = -1$

$= -(x^2 + xy + y^2)$

b. $\dfrac{x^2 - y^2}{y^3 - x^3} = \dfrac{(x + y)(x - y)}{(y - x)(y^2 + xy + x^2)}$ Factor numerator and denominator.

$= \dfrac{\overset{-1}{\overline{(x - y)}}(x + y)}{(y - x)(y^2 + xy + x^2)}$ $\dfrac{x - y}{y - x} = -1$

$= \dfrac{-(x + y)}{y^2 + xy + x^2}$

Problem 4   Reduce.

a. $\dfrac{y^3 - x^3}{x - y}$ **b.** $\dfrac{y^2 - x^2}{x^3 - y^3}$

ANSWERS   (to problems on pages 262–263)

2. a. $\dfrac{-y}{7}$ **b.** $\dfrac{4}{x}$ **c.** $\dfrac{-(a - b)}{8}$, or $\dfrac{b - a}{8}$

3. a. $x^4 y^4$ **b.** $\dfrac{x}{x + y}$ **c.** $3 + x$

NAME

CLASS

SECTION

ANSWERS

A. In Problems 1–20 write the given rational expression with the indicated denominator.

1. $\dfrac{2x}{3y}$, denominator $6y^3$

2. $\dfrac{-3y}{2x}$; denominator $8x^2$

3. $\dfrac{a}{(x-y)}$; denominator $(x-y)^2$

4. $\dfrac{-b}{y-x}$; denominator $(y-x)^2$

5. $\dfrac{x}{x+y}$; denominator x^2-y^2

6. $\dfrac{-y}{x-y}$; denominator x^2-y^2

7. $\dfrac{-x}{y-x}$; denominator y^2-x^2

8. $\dfrac{-4x}{y-x}$; denominator y^2-x^2

9. $\dfrac{-x}{2x-3y}$; denominator $4x^2-9y^2$

10. $\dfrac{-x}{2x-y}$; denominator $4x^2-y^2$

11. $\dfrac{4x}{x+1}$; denominator x^2-x-2

12. $\dfrac{5y}{y-1}$; denominator y^2+2y-3

13. $\dfrac{-5x}{x+3}$; denominator x^2+x-6

14. $\dfrac{-3y}{y-4}$; denominator y^2-2y-8

15. $\dfrac{3}{x+y}$; denominator x^3+y^3

16. $\dfrac{-4}{x+y}$; denominator x^3+y^3

1. _____

2. _____

3. _____

4. _____

5. _____

6. _____

7. _____

8. _____

9. _____

10. _____

11. _____

12. _____

13. _____

14. _____

15. _____

16. _____

17. $\dfrac{x}{x-y}$; denominator $x^3 - y^3$

17. _____

18. $\dfrac{-y}{x-y}$; denominator $x^3 - y^3$

18. _____

19. $\dfrac{x}{x^2 - xy + y^2}$; denominator $x^3 + y^3$

19. _____

20. $\dfrac{x}{x^2 + xy + y^2}$; denominator $x^3 - y^3$

20. _____

B. In Problems 21–30, write each fraction in standard form.

21. $-\dfrac{y}{-2}$

22. $-\dfrac{x-3}{y}$

21. _____

22. _____

23. $-\dfrac{x}{x-5}$

24. $\dfrac{2x-y}{-x}$

23. _____

24. _____

25. $-\dfrac{-2x}{-5y}$

26. $\dfrac{-x}{-y}$

25. _____

26. _____

27. $\dfrac{-(x+y)}{-(x-y)}$

28. $-\dfrac{-(3x+y)}{-(x-5y)}$

27. _____

28. _____

29. $\dfrac{-1}{-(x-2)}$

30. $\dfrac{-y}{-(x+1)}$

29. _____

30. _____

C. In Problems 31–60, reduce each fraction to lowest terms.

31. $\dfrac{x^4 y^2}{xy^5}$

32. $\dfrac{x^5 y^3 c^2}{x^2 y^6 c^4}$

31. _____

32. _____

33. $\dfrac{3x - 3y}{x - y}$

34. $\dfrac{4x^2}{4x - 4y}$

33. _____

34. _____

35. $\dfrac{3x - 2y}{9x^2 - 4y^2}$

36. $\dfrac{4x^2 - 9y^2}{2x + 3y}$

35. _____

36. _____

37. $\dfrac{(x-y)^3}{x^2 - y^2}$

38. $\dfrac{x^2 - y^2}{(x+y)^3}$

37. _____

38. _____

39. $\dfrac{ay^2 - ay}{ay}$

40. $\dfrac{a^3 + 2a^2 + a}{a}$

39. _____

40. _____

41. $\dfrac{x^2 + 2xy + y^2}{x^2 - y^2}$

42. $\dfrac{x^2 + 3x + 2}{x^2 + 2x + 1}$

41. _____

42. _____

43. $\dfrac{y^2 - 8y + 15}{y^2 + 3y - 18}$

44. $\dfrac{y^2 + 7y - 18}{y^2 - 3y + 2}$

43. _____

44. _____

45. $\dfrac{2-y}{y-2}$

46. $\dfrac{3(x-y)}{4(y-x)}$

47. $\dfrac{9-x^2}{x-3}$

48. $\dfrac{25-9x^2}{3x-5}$

49. $\dfrac{y^3-8}{2-y}$

50. $\dfrac{2+x}{x^3+8}$

51. $\dfrac{3x-2y}{2y-3x}$

52. $\dfrac{5y-2x}{2x-5y}$

53. $\dfrac{x^2+4x-5}{1-x}$

54. $\dfrac{x^2-2x-15}{5-x}$

55. $\dfrac{x^2-6x+8}{4-x}$

56. $\dfrac{x^2-8x+15}{3-x}$

57. $\dfrac{2-x}{x^2+4x-12}$

58. $\dfrac{3-x}{x^2+3x-18}$

59. $-\dfrac{3-x}{x^2-5x+6}$

60. $-\dfrac{4-x}{x^2-3x-4}$

45. _____

46. _____

47. _____

48. _____

49. _____

50. _____

51. _____

52. _____

53. _____

54. _____

55. _____

56. _____

57. _____

58. _____

59. _____

60. _____

✓ **SKILL CHECKER**

Simplify.

61. $\dfrac{4x^3}{2x}$

62. $\dfrac{8x^4}{2x}$

63. $\dfrac{-16x^4}{8x^2}$

64. $\dfrac{-48x^5}{12x^3}$

61. _____

62. _____

63. _____

64. _____

4.1 USING YOUR KNOWLEDGE

We have already mentioned that when reducing rational expressions you can divide out only factors. Use this knowledge to determine what is wrong with the simplifications in Problems 1 and 2.

1. $\dfrac{x+y}{x}=1+y$

2. $\dfrac{y}{x+y}=\dfrac{1}{x+1}$

3. The transmission ratio in your automobile is given by the fraction

$$T=\dfrac{\text{engine speed}}{\text{drive shaft speed}}$$

If the engine is running at 2000 revolutions per minute (rev/min) and the drive shaft speed is 600 rev/min, what is the reduced fraction giving the transmission ratio?

1. _____

2. _____

3. _____

ANSWER (to problem on page 264)

4. **a.** $-(x^2+xy+y^2)$

   b. $\dfrac{-(x+y)}{x^2+xy+y^2}$

4. In Problem 3, if the transmission ratio is $T = \dfrac{10}{3}$ and the engine speed is 3000 rev/min, what is the drive shaft speed?

4. _____

5. In business, many ratios are given as fractions that have to be reduced. For example, the current ratio CR is given by:

$$CR = \frac{\text{current assets}}{\text{current liabilities}}$$

If the current assets of a business amount to \$480,000 and current liabilities are \$220,000, find CR in reduced form.

5. _____

4.2 DIVISION OF POLYNOMIALS

OBJECTIVES

REVIEW

Before starting this section, you should know:

1. How to simplify quotients using the laws of exponents.
2. How to multiply polynomials.

OBJECTIVES

You should be able to:

A. Divide a polynomial by a monomial.
B. Use long division to divide one polynomial by another.
C. Completely factor a polynomial when one of the factors is given.

How efficient is your car engine? The efficiency E of an engine is given by

$$E = \frac{Q_1 - Q_2}{Q_1}$$

where Q_1 is the horsepower rating of the engine and Q_2 is the horsepower delivered to the transmission. Can you do the indicated division in the rational expression $\frac{Q_1 - Q_2}{Q_1}$?

Follow the steps in the procedure.

$$\frac{Q_1 - Q_2}{Q_1} = (Q_1 - Q_2)\left(\frac{1}{Q_1}\right) \qquad$$ Dividing by Q_1 is the same as multiplying by the reciprocal of Q_1, that is, $\frac{1}{Q_1}$.

$$= Q_1\left(\frac{1}{Q_1}\right) - Q_2\left(\frac{1}{Q_1}\right) \qquad$$ Use the distributive law.

$$= \frac{Q_1}{Q_1} - \frac{Q_2}{Q_1} \qquad$$ Multiply.

$$= 1 - \frac{Q_2}{Q_1} \qquad \frac{Q_1}{Q_1} = 1.$$

A. DIVIDING A POLYNOMIAL BY A MONOMIAL

To divide the trinomial $4x^4 - 8x^3 + 12x^2$ by $2x^2$, we proceed similarly. Thus,

$$\frac{4x^4 - 8x^3 + 12x^2}{2x^2} = (4x^4 - 8x^3 + 12x^2)\left(\frac{1}{2x^2}\right)$$

$$= 4x^4\left(\frac{1}{2x^2}\right) - 8x^3\left(\frac{1}{2x^2}\right) + 12x^2\left(\frac{1}{2x^2}\right)$$

$$= \frac{4x^4}{2x^2} - \frac{8x^3}{2x^2} + \frac{12x^2}{2x^2}$$

$$= 2x^2 - 4x + 6$$

Note that

$$\frac{4x^4 - 8x^3 + 12x^2}{2x^2} = \frac{4x^4}{2x^2} - \frac{8x^3}{2x^2} + \frac{12x^2}{2x^2} = 2x^2 - 4x + 6$$

Thus we can deduce the following rule.

RULE FOR DIVIDING A POLYNOMIAL BY A MONOMIAL

To divide a polynomial by a monomial, divide each term in the polynomial by the monomial.

EXAMPLE 1   Divide.

a. $\dfrac{28x^5 - 14x^4 + 7x^3}{7x^2}$ **b.** $\dfrac{20x^4 - 15x^3 + 10x^2 - 30x + 50}{10x^3}$

Solution

a. $\dfrac{28x^5 - 14x^4 + 7x^3}{7x^2} = \dfrac{28x^5}{7x^2} - \dfrac{14x^4}{7x^2} + \dfrac{7x^3}{7x^2}$

$\qquad = 4x^3 - 2x^2 + x$

b. $\dfrac{20x^4 - 15x^3 + 10x^2 - 30x + 50}{10x^3} = \dfrac{20x^4}{10x^3} - \dfrac{15x^3}{10x^3} + \dfrac{10x^2}{10x^3} - \dfrac{30x}{10x^3} + \dfrac{50}{10x^3}$

$\qquad = 2x - \dfrac{3}{2} + \dfrac{1}{x} - \dfrac{3}{x^2} + \dfrac{5}{x^3}$

Note that in this case, the answer is *not* a polynomial.

Problem 1   Divide.

a. $\dfrac{24x^5 - 18x^4 + 12x^3}{6x^2}$

b. $\dfrac{16x^4 - 4x^3 + 8x^2 - 16x + 40}{8x^3}$

B.   DIVIDING ONE POLYNOMIAL BY ANOTHER POLYNOMIAL

If we wish to divide a polynomial called the **dividend** by another polynomial called the **divisor,** we proceed very much as we did in long division in arithmetic. To show you that this is so, we are performing the division of 337 by 16 and $(x^3 + 3x^2 + 3x + 1)$ by $x^2 + x + 1$ side by side.

1. $16 \overline{)\,337\,}^{\,2}$ Divide 33 by 16. It goes twice. Write 2 over the 33.

$x^2 + x + 1 \overline{)\,x^3 + 3x^2 + 3x + 1\,}^{\,x}$ Divide x^3 by x^2. It goes x times. Write x over the $3x$.

2. $16 \overline{)\,337\,}^{\,2}$
$\quad\ -32$
$\quad\ \overline{1}$ Multiply 16 by 2 and subtract the product 32 from 33, obtaining 1.

$x^2 + x + 1 \overline{)\,x^3 + 3x^2 + 3x + 1\,}^{\,x}$
$(-)\ \underline{x^3 + \ x^2 + \ x}$
$\quad\ 0\ + 2x^2 + 2x + 1$ Multiply $x^2 + x + 1$ by x and subtract the product $x^3 + x^2 + x$ from $x^3 + 3x^2 + 3x + 1$, obtaining $0 + 2x^2 + 2x$.

3. $16 \overline{)\,337\,}^{\,21}$
$\quad\ -32$
$\quad\ \overline{17}$ "Bring down" the 7. Now, divide 17 by 16. It goes once. Write 1 after the 2.

$x^2 + x + 1 \overline{)\,x^3 + 3x^2 + 3x + 1\,}^{\,x + 2}$
$(-)\ \underline{x^3 + \ x \ + \ x}$
$\quad\ 0\ + 2x^2 + 2x + 1$ "Bring down" the 1. Now, divide $2x^2 + 2x + 1$ by $x^2 + x + 1$. It goes 2 times. Write $+2$ after the x.

4.
$$16\overline{)\,337}$$
$$\underline{-32}$$
$$17$$
$$\underline{-16}$$
$$1$$

Multiply 16 by 1 and subtract the result from 17. The remainder is 1.

$$\begin{array}{r} x + 2 \\ x^2 + x + 1 \overline{)\, x^3 + 3x^2 + 3x + 1\,} \\ (-)\, x^3 + x^2 + x \\ \hline 0 \;+\; 2x^2 + 2x + 1 \\ (-)\;\; 2x^2 + 2x + 2 \\ \hline -1 \end{array}$$

Multiply $x^2 + x + 1$ by 2, obtaining $2x^2 + 2x + 2$. Subtract this result from $2x^2 + 2x + 1^2$. The remainder is -1.

5. The answer (**quotient**) can be written as 21 R 1 (read "21 remainder 1") or as $21 + \frac{1}{16}$, which is $21\frac{1}{16}$.

The answer (**quotient**) can be written as $x + 2$ R -1 (read "$x + 2$ remainder -1") or as

$$x + 2 - \frac{1}{x^2 + x + 1}$$

6. You can check this answer by multiplying 21 by 16 (336) and adding the remainder 1 to obtain 337, which is the dividend.

You can check the answer by multiplying $(x + 2)(x^2 + x + 1) = x^3 + 3x^2 + 3x + 2$ and adding the remainder -1 to get $x^3 + 3x^2 + 3x + 1$, which is the dividend.

EXAMPLE 2 Divide $x^3 + 2x^2 - 17x$ by $x^2 + x - 3$.

Solution

x^3 divided by x^2 is x

$$\begin{array}{r} x + 1 \\ x^2 + x - 3\overline{)\,x^3 + 2x^2 - 17x} \\ (-)\,\underline{x^3 + x^2 - 3x} \qquad \leftarrow x(x^2 + x - 3) = x^3 + x^2 - 3x \\ x^2 - 14x \\ (-)\,\underline{x^2 + x - 3} \qquad \leftarrow 1(x^2 + x - 3) = x^2 + x - 3 \\ -15x + 3 \qquad \leftarrow \text{Remainder} \end{array}$$ ▲

Problem 2 Divide $x^3 + 4x^2 - 15x$ by $x^2 - 2x + 3$.

If there are missing terms in the polynomial being divided, we insert zero coefficients, as shown in the next example.

EXAMPLE 3 Divide $4x^3 - 4 - 8x$ by $4 + 4x$.

Solution We write the polynomials in **descending** order, inserting $0x^2$ in the dividend. We then have:

$$\begin{array}{r} x^2 - x - 1 \\ 4x + 4\overline{)\,4x^3 + 0x^2 - 8x - 4\,} \\ (-)\,\underline{4x^3 + 4x^2} \\ 0 - 4x^2 - 8x \\ (-)\,\underline{-4x^2 - 4x} \\ -4x - 4 \\ (-)\,\underline{-4x - 4} \\ 0 \end{array}$$

$\begin{cases} 4x^3 \text{ divided by } 4x \text{ is } x^2. \\ \\ x^2(4x + 4) = 4x^2 + 4x^2 \end{cases}$

$\begin{cases} -4x^2 \text{ divided by } 4x \text{ is } -x. \\ -x(4x + 4) = -4x^2 - 4x \end{cases}$

$\begin{cases} -4x \text{ divided by } 4x \text{ is } -1. \\ -1(4x + 4) = -4x - 4 \end{cases}$

The remainder is 0.

Problem 3 Divide $6x^3 + 3x - 9$ by $3x - 3$.

Thus, $\dfrac{4x^3 - 4 - 8x}{4 + 4x} = x^2 - x - 1$. You can check this by multiplying $(4 + 4x)(x^2 - x - 1)$, obtaining $4x^3 - 4 - 8x$.

C. FACTORING WHEN ONE OF THE FACTORS IS KNOWN

Suppose we wish to factor the polynomial

$$6x^3 + 23x^2 + 9x - 18$$

None of the methods we have studied work, so we need some more information. If we find out that $x + 3$ is one of the factors, we can write $6x^3 + 23x^2 + 9x - 18 = (x + 3)P$, where P is a polynomial. Dividing both sides by $x + 3$ gives

$$\frac{6x^3 + 23x^2 + 9x - 18}{x + 3} = P$$

Let us do the division:

$$
\begin{array}{r}
6x^2 + 5x - 6 \\
x + 3 \overline{\smash{)}\ 6x^3 + 23x^2 + 9x - 18} \\
(-)\ \underline{6x^3 + 18x^2} \\
0 + 5x^2 + 9x \\
(-)\ +\ \underline{5x^2 + 15x} \\
0\ - 6x - 18 \\
(-)\ -\ \underline{6x - 18} \\
0
\end{array}
$$

$\begin{cases} 6x^3 \text{ divided by } x \text{ is } 6x^2. \\ \\ 6x^2(x + 3) = 6x^3 + 18x^2 \end{cases}$
$\begin{cases} 5x^2 \text{ divided by } x \text{ is } 5x. \\ 5x(x + 3) = 5x^2 + 15x \end{cases}$
$\begin{cases} -6x \text{ divided by } x \text{ is } -6. \\ -6(x + 3) = -6x - 18 \end{cases}$

Thus the polynomial P is $6x^2 + 5x - 6$. We have:

$$\frac{6x^3 + 23x^2 + 9x - 18}{x + 3} = 6x^2 + 5x - 6$$

Now, $6x^2 + 5x - 6 = (3x - 2)(2x + 3)$, which gives

$$\frac{6x^3 + 23x^2 + 9x - 18}{x + 3} = (3x - 2)(2x + 3)$$

Multiplying both sides by $x + 3$ yields

$$6x^3 + 23x^2 + 9x - 18 = (x + 3)(3x - 2)(2x + 3)$$

Thus, if we wish to factor a polynomial and one of the factors is given, we can divide by this factor. The product of the quotient obtained (factored, if possible) and the given factor gives the complete factorization of the polynomial.

EXAMPLE 4   Factor $6x^3 + 25x^2 + 2x - 8$ if $x + 4$ is one of its factors.

Solution   We start by dividing $6x^3 + 25x^2 + 2x - 8$ by $x + 4$.

$$
\begin{array}{r}
6x^2 + x - 2 \\
x + 4 \overline{\smash{)}\ 6x^3 + 25x^2 + 2x - 8} \\
(-)\ \underline{6x^3 + 24x^2} \\
0 + x^2 + 2x \\
(-)\ \underline{x^2 + 4x} \\
0 - 2x - 8 \\
(-)\ \underline{- 2x - 8} \\
0
\end{array}
$$

$\begin{cases} 6x^3 \text{ divided by } x \text{ is } 6x^2. \\ \\ 6x^2(x + 4) = 6x^3 + 24x^2 \end{cases}$
$\begin{cases} x^2 \text{ divided by } x \text{ is } x. \\ x(x + 4) = x^2 + 4x \end{cases}$
$\begin{cases} -2x \text{ divided by } x \text{ is } -2. \\ -2(x + 4) = -2x - 8 \end{cases}$

Now, we know that the polynomial we are trying to factor is the product of $(x + 4)$ and the quotient we obtained, $6x^2 + x - 2$. To factor completely, we have to factor $6x^2 + x - 2$ as $(3x + 2)(2x - 1)$. Thus

$$6x^3 + 25x^2 + 2x - 8 = (x + 4)(6x^2 + x - 2)$$
$$= (x + 4)(3x + 2)(2x - 1)$$

Problem 4   Factor $2x^3 - x^2 - 18x + 9$ if $x + 3$ is one of its factors.

A. Divide.

1. $\dfrac{3x^3 + 9x^2 - 6x}{3x}$

2. $\dfrac{6x^3 + 8x^2 - 4x}{2x}$

3. $\dfrac{10x^3 - 5x^2 + 15x}{-5x}$

4. $\dfrac{24x^3 - 12x^2 + 6x}{-6x}$

5. $\dfrac{8y^4 - 32y^3 + 12y^2}{-4y^2}$

6. $\dfrac{9y^4 - 45y^3 + 18y^2}{-3y^2}$

7. $\dfrac{10x^5 + 8x^4 - 16x^3 + 6x^2}{2x^3}$

8. $\dfrac{12x^4 + 18x^3 + 16x^2}{4x^3}$

9. $\dfrac{15x^3y^2 - 10x^2y + 15x}{5x^2y}$

10. $\dfrac{18x^4y^4 - 24x^2y^3 + 6xy^2}{3x^2y^2}$

B. Divide.

11. $x^2 + 5x + 6$ by $x + 2$

12. $x^2 + 9x + 20$ by $x + 5$

13. $y^2 + 3y - 10$ by $y - 2$

14. $y^2 + 2y - 15$ by $y - 3$

15. $2x^3 - 4x - 2$ by $2x + 2$

16. $2x^3 + 5x^2 - x - 2$ by $2x - 1$

17. $3x^3 + 14x^2 + 13x - 6$ by $3x - 1$

18. $2x^3 - 5x^2 - 14x + 3$ by $2x - 1$

19. $2x^3 - 10x - 7x^2 + 24$ by $2x - 3$

20. $3x^3 + 8x + 13x^2 - 12$ by $3x - 2$

21. $2x^3 + 2x + 7x^2 - 2$ by $-1 + 2x$

22. $3x^3 + 3x + 8x^2 - 2$ by $-1 + 3x$

23. $y^4 - y^2 - 2y - 1$ by $y^2 + y + 1$

24. $y^4 - y^2 - 4y - 4$ by $y^2 + y + 2$

25. $8x^3 - 6x^2 + 5x - 9$ by $2x - 3$

26. $2x^4 - x^3 + 7x - 2$ by $2x + 3$

27. $x^3 + 8$ by $x + 2$

28. $x^3 + 64$ by $x + 4$

ANSWERS
1. ___
2. ___
3. ___
4. ___
5. ___
6. ___
7. ___
8. ___
9. ___
10. ___
11. ___
12. ___
13. ___
14. ___
15. ___
16. ___
17. ___
18. ___
19. ___
20. ___
21. ___
22. ___
23. ___
24. ___
25. ___
26. ___
27. ___
28. ___

29. $8y^3 - 64$ by $2y - 4$ **30.** $27x^3 - 8$ by $3x - 2$

29. _____

30. _____

31. $a^4 - a^2 - 2a - 1$ by $a^2 + a + 1$

31. _____

32. $b^4 - b^2 - 2b - 1$ by $b^2 - b - 1$

32. _____

33. $x^5 - 5x + 12x^2$ by $x^2 + 5 - 2x$

33. _____

34. $y^5 - y^4 + 10 - 27y + 7y^2$ by $y^2 + 5 - y$

34. _____

35. $4x^4 - 13x^2 + 4x^3 - 3x - 21$ by $2x + 5$

35. _____

36. $8y^4 - 75y^2 - 18y^3 + 46y + 121$ by $4y + 5$

36. _____

C. Factor completely.

37. $x^3 - 4x^2 + x + 6$ if $x + 1$ is one of the factors

37. _____

38. $x^3 - 4x^2 + x + 6$ if $x - 3$ is one of the factors

38. _____

39. $x^4 - 4x^3 + 3x^2 + 4x - 4$ if $x^2 - 4x + 4$ is one of the factors

39. _____

40. $x^4 - 2x^3 - 13x^2 + 14x + 24$ if $x^2 - 6x + 8$ is one of the factors

40. _____

41. $x^4 + 6x^3 + 3x + 140$ if $x^2 - 3x + 7$ is one of the factors

41. _____

42. $x^4 - 22x^2 - 75$ if $x^2 + 3$ is one of the factors

42. _____

D. Applications

In business, the average cost $\bar{C}$ per units is given by

$$\bar{C} = \frac{C}{x}$$

where C is the total cost and x is the number of units.

43. Find the average cost when $C = 500 + 4x$.

43. _____

44. Find the average cost when $C = 200 + 2x^2$.

44. _____

✓ **SKILL CHECKER**

Factor.

45. $x^2 - 9$ **46.** $4x^2 - 9y^2$

45. _____

46. _____

47. $x^2 + 6x + 9$ **48.** $x^2 - 8x + 16$

47. _____

48. _____

49. $x^2 + 3x + 2$ **50.** $x^2 + 5x + 4$

49. _____

50. _____

4.2 USING YOUR KNOWLEDGE

In this section, we factored some polynomials by using a given first-degree factor. How did we find that one factor? Here is one way to do it.

If a polynomial of the form $c_0 x^n + c_1 x^{n-1} + \cdots + c_n$ with integer coefficients has a factor $ax + b$, where a and b are integers, then a

ANSWER (to problem on page 272)

4. $(x + 3)(x - 3)(2x - 1)$

must divide c_0 and b must divide c_n. For example, to find a and b for the polynomial $2x^3 + 3x^2 - 8x + 3$, we need consider only positive divisors of 2. Thus the possibilities for a are 1 and 2 and for b are 1, -1, 3, and -3. This means that only the binomials $x + 1$, $x - 1$, $x + 3$, $x - 3$, $2x + 1$, $2x - 1$, $2x + 3$, and $2x - 3$ have to be checked. It turns out that $x - 1$ is a factor and that division gives

$$2x^3 + 3x^2 - 8x + 3 = (x - 1)(2x^2 + 5x - 3)$$

Since $2x^2 + 5x - 3 = (2x - 1)(x + 3)$, we have

$$2x^3 + 3x^2 - 8x + 3 = (x - 1)(2x - 1)(x + 3)$$

1. What binomials should be checked as possible factors for the polynomial $x^3 + 3x^2 + 5x + 6$ if a and b are positive?

2. What binomials should be checked as possible factors for the polynomial $3x^3 + 5x^2 - 3x - 2$ if a and b are positive?

1. _____

2. _____

4.3 MULTIPLICATION AND DIVISION OF RATIONAL EXPRESSIONS

OBJECTIVES

REVIEW

Before starting this section, you should know:

1. How to multiply and divide rational numbers (Section 1.4).
2. How to factor polynomials.

OBJECTIVES

You should be able to:

A. Multiply rational expressions.
B. Divide rational expressions.
C. Multiply and divide rational expressions.

A regular deck of 52 cards has 4 kings. The probability that you pick 2 kings when drawing two cards is $\frac{4}{52} \cdot \frac{3}{51}$. From Section 1.4,

---

If a, b, c, and d are real numbers ($b \neq 0$, $d \neq 0$),

$$\frac{a}{b} \cdot \frac{c}{d} = \frac{a \cdot c}{b \cdot d}$$

Thus the probability is

$$\frac{4}{52} \cdot \frac{3}{51} = \frac{4 \cdot 3}{52 \cdot 51}$$

$$= \frac{4 \cdot 3}{4 \cdot 13 \cdot 3 \cdot 17} \quad \text{Factor the denominator.}$$

$$= \frac{\overset{1}{4} \cdot \overset{}{3}}{13 \cdot 17 \cdot 4 \cdot 3} \quad \text{Rearrange the denominator.}$$

$$= \frac{1}{221} \quad \text{Use the fundamental property of fractions.}$$

We can save time if we do the simplification in the first step by writing

$$\frac{4}{52} \cdot \frac{3}{51} = \frac{\overset{1}{\cancel{4}} \cdot \overset{1}{\cancel{3}}}{\underset{13}{\cancel{52}} \cdot \underset{17}{\cancel{51}}}$$

$$= \frac{1}{221}$$

A. MULTIPLICATION OF RATIONAL EXPRESSIONS

This same procedure can be used to multiply rational expressions. Thus

$$\frac{6x^3}{2y} \cdot \frac{4y^5}{9x^2} = \frac{6x^3 \cdot 4y^5}{2y \cdot 9x^2}$$ Multiply numerators and denominators.

$$= \frac{\overset{2x}{\cancel{6x^3}} \cdot \overset{2y^4}{\cancel{4y^5}}}{\underset{3}{\cancel{2y} \cdot \cancel{9x^2}}}$$ Reduce.

$$= \frac{4xy^4}{3}$$

If the rational expressions involve binomials or trinomials, they have to be factored so they can be reduced. Thus

$$\frac{x^2 + 2x - 8}{x^2 + 5x + 6} \cdot \frac{x + 2}{x + 4} = \frac{(x^2 + 2x - 8)(x + 2)}{(x^2 + 5x + 6)(x + 4)}$$

$$= \frac{(x - 2)(x + 4)(x + 2)}{(x + 2)(x + 3)(x + 4)}$$ Factor.

$$= \frac{(x - 2)\cancel{(x + 4)}\cancel{(x + 2)}}{\cancel{(x + 2)}(x + 3)\cancel{(x + 4)}}$$ Reduce.

$$= \frac{x - 2}{x + 3}$$

You can save some time by factoring and reducing as shown in Example 1.

EXAMPLE 1 Multiply.

a. $\dfrac{x - 1}{2x - 3y} \cdot \dfrac{4x^2 - 9y^2}{2x^2 - x - 1}$ **b.** $\dfrac{x^2 - 4}{4x^2 - 9y^2} \cdot \dfrac{2x^2 - 3xy}{2x + 4}$

Solution

a. $\dfrac{x - 1}{2x - 3y} \cdot \dfrac{4x^2 - 9y^2}{2x^2 - x - 1} = \dfrac{(x - 1)(4x^2 - 9y^2)}{(2x - 3y)(2x^2 - x - 1)}$ Multiply.

$$= \frac{\cancel{(x - 1)}(2x + 3y)\cancel{(2x - 3y)}}{\cancel{(2x - 3y)}\cancel{(x - 1)}(2x + 1)}$$ Factor and reduce.

$$= \frac{2x + 3y}{2x + 1}$$

b. $\dfrac{x^2 - 4}{4x^2 - 9y^2} \cdot \dfrac{2x^2 - 3xy}{2x + 4} = \dfrac{(x^2 - 4)(2x^2 - 3xy)}{(4x^2 - 9y^2)(2x + 4)}$

$$= \frac{(x + 2)(x - 2)x\cancel{(2x - 3y)}}{\cancel{(2x - 3y)}(2x + 3y)2\cancel{(x + 2)}}$$ Factor and reduce.

$$= \frac{x(x - 2)}{2(2x + 3y)}$$

$$= \frac{x^2 - 2x}{4x + 6y}$$

Problem 1 Multiply.

a. $\dfrac{x - 2}{3x - 2y} \cdot \dfrac{9x^2 - 4y^2}{2x^2 - 3x - 2}$

b. $\dfrac{x^2 - 9}{9x^2 - 4y^2} \cdot \dfrac{3x^2 - 2xy}{3x + 9}$

EXAMPLE 2 Multiply.

a. $\dfrac{2 - x}{x + 1} \cdot \dfrac{x^2 + 3x + 2}{x^2 - 4}$ **b.** $\dfrac{x^2 + 3x + 2}{x^2 + 5x + 4} \cdot \dfrac{x^2 + 2x - 3}{x^2 + x - 2}$

Solution

a. $\dfrac{2 - x}{x + 1} \cdot \dfrac{x^2 + 3x + 2}{x^2 - 4} = \dfrac{(2 - x)(x^2 + 3x + 2)}{(x + 1)(x^2 - 4)}$ Multiply.

$$= \dfrac{\overset{-1}{(2 - x)}(x + 1)(x + 2)}{(x + 1)(x + 2)(x - 2)}$$ Factor and reduce. $\left(\text{Recall that } \dfrac{2 - x}{x - 2} = -1.\right)$

$$= -1$$

b. $\dfrac{x^2 + 3x + 2}{x^2 + 5x + 4} \cdot \dfrac{x^2 + 2x - 3}{x^2 + x - 2} = \dfrac{(x^2 + 3x + 2)(x^2 + 2x - 3)}{(x^2 + 5x + 4)(x^2 + x - 2)}$ Multiply.

$$= \dfrac{(x + 2)(x + 1)(x - 1)(x + 3)}{(x + 4)(x + 1)(x - 1)(x + 2)}$$ Factor and reduce.

$$= \dfrac{x + 3}{x + 4}$$

Problem 2 Multiply.

a. $\dfrac{3 - x}{x + 2} \cdot \dfrac{x^2 + 5x + 6}{x^2 - 9}$

b. $\dfrac{x^2 + 4x + 3}{x^2 + 5x + 6} \cdot \dfrac{x^2 - 2x - 8}{x^2 - 2x - 3}$

B.   DIVISION OF RATIONAL EXPRESSIONS

To divide one rational expression by another, we use the definition of division from Section 1.4.

If a, b, c, and d are real numbers (b, d, and c not 0)

$$\dfrac{a}{b} \div \dfrac{c}{d} = \dfrac{a}{b} \cdot \dfrac{d}{c}$$

Thus the quotient of two rational expressions is the product of the first and the reciprocal of the second. For example,

$$\dfrac{2}{5} \div \dfrac{3}{8} = \dfrac{2}{5} \cdot \dfrac{8}{3} = \dfrac{16}{15}$$ The reciprocal of $\dfrac{3}{8}$ is $\dfrac{8}{3}$.

$\underbrace{\qquad}_{\text{reciprocal}}$

and

$$\dfrac{3x^2}{2y} \div \dfrac{6x}{4y^2} = \dfrac{3x^2}{2y} \cdot \dfrac{4y^2}{6x}.$$

$\underbrace{\qquad}_{\text{reciprocal}}$

$$= \dfrac{\overset{1xy}{12x^2y^2}}{12xy}$$ Note that answers should be given in reduced form.

$$= xy$$

EXAMPLE 3 Divide.

a. $\dfrac{x^3 y}{z} \div \dfrac{x^2 y}{z^4}$ **b.** $\dfrac{5x^2 - 5}{3x + 6} \div \dfrac{x + 1}{3}$ **c.** $\dfrac{x + 3}{x - 3} \div (x^2 + 6x + 9)$

Solution

a. $\dfrac{x^3 y}{z} \div \dfrac{x^2 y}{z^4} = \dfrac{x^3 y}{z} \cdot \dfrac{z^4}{x^2 y}$

   $\quad\quad\quad\quad\quad\quad$ reciprocal

$= \dfrac{\overset{xz^3}{\cancel{x^3 y z^4}}}{\cancel{x^2 y z}}$

$= xz^3$

b. $\dfrac{5x^2 - 5}{3x + 6} \div \dfrac{x + 1}{3} = \dfrac{5(x + 1)(x - 1)}{3(x + 2)} \cdot \dfrac{3}{x + 1}$

   $\quad\quad\quad\quad\quad\quad\quad$ reciprocal

$= \dfrac{5(x + 1)(x - 1)3}{3(x + 2)(x + 1)}$

$= \dfrac{5x - 5}{x + 2}$

c. We first write $x^2 + 6x + 9$ as $\dfrac{x^2 + 6x + 9}{1}$.

$\dfrac{x + 3}{x - 3} \div (x^2 + 6x + 9) = \dfrac{x + 3}{x - 3} \div \dfrac{x^2 + 6x + 9}{1}$

$= \dfrac{x + 3}{x - 3} \cdot \dfrac{1}{x^2 + 6x + 9}$

$= \dfrac{(x + 3) \cdot 1}{(x - 3)(x + 3)(x + 3)}$

$= \dfrac{1}{(x - 3)(x + 3)}$

$= \dfrac{1}{x^2 - 9}$

C. MULTIPLICATION AND DIVISION

Sometimes both multiplications and divisions are involved, as shown in the next example.

EXAMPLE 4 Perform the indicated operations.

$\dfrac{3 - x}{x + 2} \div \dfrac{x^3 - 27}{x + 4} \cdot \dfrac{x^3 + 8}{x + 4}$

Solution We first rewrite the division as a multiplication by the reciprocal, as shown.

Problem 3 Divide.

a. $\dfrac{x^5 y}{z} \div \dfrac{x^3 y}{z^6}$ **b.** $\dfrac{3x^2 - 3}{5x + 10} \div \dfrac{x + 1}{5}$

c. $\dfrac{x + 2}{x - 2} \div (x^2 + 4x + 4)$

Problem 4 Perform the indicated operations.

$\dfrac{2 - x}{x + 3} \div \dfrac{x^3 - 8}{x - 5} \cdot \dfrac{x^3 + 27}{x - 5}$

$$\frac{3-x}{x+2} \div \frac{x^3-27}{x+4} \cdot \frac{x^3+8}{x+4}$$

$$= \frac{3-x}{x+2} \cdot \frac{x+4}{x^3-27} \cdot \frac{x^3+8}{x+4}$$ Change the division to multiplication by the reciprocal.

$$= \frac{(3-x)(x+4)(x^3+8)}{(x+2)(x^3-27)(x+4)}$$ Multiply.

$$= \frac{\overset{-1}{(3-x)}(x+4)(x+2)(x^2-2x+4)}{(x+2)(x-3)(x^2+3x+9)(x+4)}$$ Factor and reduce. (Recall that $\frac{3-x}{x-3} = -1$.)

$$= \frac{-(x^2-2x+4)}{x^2+3x+9}$$

EXAMPLE 5   Perform the indicated operations.

$$\frac{x^3+2x^2-x-2}{x+3} \div \frac{x^3-8}{x^2-9} \cdot \frac{1}{x^2-1}$$

Solution   This time, we have a polynomial with four terms in one numerator. We factor it by grouping.

$$x^3 + 2x^2 - x - 2 = x^2(x+2) - 1 \cdot (x+2)$$
$$= (x+2)(x^2-1)$$
$$= (x+2)(x+1)(x-1)$$

We then rewrite the division as a multiplication by the reciprocal, obtaining

$$\frac{x^3+2x^2-x-2}{x+3} \div \frac{x^3-8}{x^2-9} \cdot \frac{1}{x^2-1}$$

$$= \frac{x^3+2x^2-x-2}{x+3} \cdot \frac{x^2-9}{x^3+8} \cdot \frac{1}{x^2-1}$$

$$= \frac{(x+2)(x+1)(x-1)(x+3)(x-3)}{(x+3)(x+2)(x^2-2x+4)(x+1)(x-1)}$$

$$= \frac{x-3}{x^2-2x+4}$$

Problem 5   Perform the indicated operations.

$$\frac{x^3+x^2-x-1}{x+4} \div \frac{x^3+1}{x^2-16} \cdot \frac{1}{x^2-1}$$

NAME

CLASS

SECTION

ANSWERS

A. Multiply.

1. $\dfrac{3}{4} \cdot \dfrac{2}{5}$

2. $\dfrac{-9}{10} \cdot \dfrac{2}{3}$

3. $\dfrac{14x^2}{15} \cdot \dfrac{5}{7x}$

4. $\dfrac{-5x^3}{7y} \cdot \dfrac{4y^3}{9x^6}$

5. $\dfrac{-2xy^4}{9z^5} \cdot \dfrac{-3z}{7x^3y^3}$

6. $\dfrac{-35x^5z}{24x^3y^9} \cdot \dfrac{84x^3y^8}{15x^4y^7z}$

7. $\dfrac{10x + 50}{6x + 6} \cdot \dfrac{12}{5x + 25}$

8. $\dfrac{x + y}{xy - y^2} \cdot \dfrac{y^2}{x^2 - y^2}$

9. $\dfrac{6y + 3}{2y^2 - 3y - 2} \cdot \dfrac{y^2 - 4}{3y + 6}$

10. $\dfrac{y^2 + 9y + 18}{y - 2} \cdot \dfrac{2y - 1}{5y + 15}$

11. $\dfrac{y - x}{x^2 + 2xy} \cdot \dfrac{5x + 10y}{x^2 - y^2}$

12. $\dfrac{2 - 2x}{9x^2 - 25} \cdot \dfrac{6x - 10}{x^2 - 1}$

13. $\dfrac{3y^2 - 17y + 10}{y^2 - 4y - 5} \cdot \dfrac{y^2 + 3y + 2}{y^2 + y - 2}$

14. $\dfrac{y^2 + 2y - 3}{y^2 - 4y - 5} \cdot \dfrac{y^2 - 3y - 10}{y^2 + 5y - 6}$

15. $\dfrac{y^2 + 2y - 8}{y^2 + 7y + 12} \cdot \dfrac{y^2 + 2y - 3}{y^2 - 3y + 2}$

16. $\dfrac{y^2 + 2y - 15}{y^2 - 7y + 10} \cdot \dfrac{y^2 - 6y + 8}{y^2 - y - 12}$

17. $\dfrac{x^3 - 8}{4 - x^2} \cdot \dfrac{x^2 + x - 2}{x^2 + 2x + 4}$

18. $\dfrac{x^3 + y^3}{y^2 - x^2} \cdot \dfrac{x - y}{x^2 - xy + y^2}$

19. $\dfrac{a^3 + b^3}{a^3 - b^3} \cdot \dfrac{a^2 + ab + b^2}{a^2 - ab + b^2}$

20. $\dfrac{a^3 - 8}{a^2 + 2a + 4} \cdot \dfrac{a^2 + 3a + 9}{a^3 - 27}$

B. Divide.

21. $\dfrac{3}{5} \div \dfrac{10}{9}$

22. $\dfrac{3}{7} \div \dfrac{-9}{14}$

23. $\dfrac{4}{5x^2} \div \dfrac{12}{25x^3}$

24. $\dfrac{6x^2}{7} \div \dfrac{30x}{28}$

25. $\dfrac{24a^2b}{7c^2d} \div \dfrac{8ab}{21cd^2}$

26. $\dfrac{16a^3b}{15a} \div \dfrac{12ab^2}{20b^4}$

27. $\dfrac{3x - 3}{x} \div \dfrac{x^2 - 1}{x^2}$

28. $\dfrac{5x^2 - 45}{x^3} \div \dfrac{x + 3}{x}$

1. _____
2. _____
3. _____
4. _____
5. _____
6. _____
7. _____
8. _____
9. _____
10. _____
11. _____
12. _____
13. _____
14. _____
15. _____
16. _____
17. _____
18. _____
19. _____
20. _____

21. _____
22. _____
23. _____
24. _____
25. _____
26. _____
27. _____
28. _____

29. $\dfrac{y^2 - 25}{y^2 - 4} \div \dfrac{3y - 15}{4y - 8}$

30. $\dfrac{y^2 + y - 12}{y^2 - 1} \div \dfrac{3y + 12}{4y^2 + 4y}$

31. $\dfrac{a^3 + b^3}{a^3 - b^3} \div \dfrac{a^2 - ab + b^2}{a^2 + ab + b^2}$

32. $\dfrac{a^2 + ab + b^2}{a^3 + b^3} \div \dfrac{a^2 + ab + b^2}{a^2 - ab + b^2}$

33. $\dfrac{8a^3 - 1}{6u^4w^3} \div \dfrac{1 - 2a}{3u^2w}$

34. $\dfrac{-b^2c}{27a^3 - 1} \div \dfrac{b^3c^2}{3a - 1}$

35. $\dfrac{x - x^3}{2x^2 + 6x} \div \dfrac{5x^2 - 5x}{2x + 6}$

36. $\dfrac{121y - y^3}{y^2 - 49} \div \dfrac{y^2 - 11y}{y + 7}$

37. $\dfrac{y^2 + y - 12}{y^2 - 8y + 15} \div \dfrac{3y^2 + 7y - 20}{2y^2 - 7y - 15}$

38. $\dfrac{3y^2 + 11y + 6}{4y^2 + 16y + 7} \div \dfrac{3y^2 - y - 2}{2y^2 - y - 28}$

39. $\dfrac{4x^2 - 12x + 9}{25 - 4x^2} \div \dfrac{6x^2 - 5x - 6}{6x^2 + 19x + 10}$

40. $\dfrac{4x^2 + 12x + 9}{9 - 4x^2} \div \dfrac{10x^2 + 27x + 18}{8x^2 - 2x - 15}$

C. Perform the indicated operations.

41. $\dfrac{x^2 + 2x - 3}{x - 5} \div \dfrac{x^2 + 6x + 9}{x^2 - 2x - 15} \cdot \dfrac{1}{x^2 - 1}$

42. $\dfrac{x^2 - 3x + 2}{x^2 - 5x + 6} \div \dfrac{x^2 - 5x + 4}{x^2 - 7x + 12} \cdot \dfrac{x^3 + 1}{x^2 - 1}$

43. $\dfrac{x^2 - 1}{x^2 + 3x - 10} \div \dfrac{x^2 - 3x - 4}{x^2 - 25} \cdot \dfrac{x - 2}{x - 5}$

44. $\dfrac{x^2 - 25}{x^2 - 49} \cdot \dfrac{x^2 - 4x - 21}{x^2 - 10x + 25} \div \dfrac{x^2 + 2x - 3}{x^2 - 6x + 5}$

45. $\dfrac{x - 3}{3 - x} \cdot \dfrac{x^2 + 3x - 4}{x^2 + 7x + 12} \div \dfrac{x^2 + x - 2}{x^2 + 5x + 6}$

46. $\dfrac{x^3 - 125}{x^3 - 8} \cdot \dfrac{x^2 + x - 2}{x^2 + 6x - 7} \div \dfrac{x^2 - 3x - 10}{x^2 + 5x - 14}$

47. $\dfrac{x^2 - y^2}{x^2 - 2xy} \div \dfrac{x^2 + xy - 2y^2}{x^2 - 4y^2} \cdot \dfrac{x^2}{(x + y)^2}$

48. $\dfrac{x^2 + xy - 2y^2}{x^2 - 4y^2} \div \dfrac{x^2 - y^2}{x^2 - 2xy} \cdot \dfrac{(x + y)^2}{x^2}$

49. $\dfrac{x^2 + 2xy - 3y^2}{y^2 - 7y + 10} \div \dfrac{x^2 - 3xy - 2y^2}{y^2 - 3y - 10} \cdot \dfrac{x^2 - 4y^2}{x^2 - 9y^2}$

50. $\dfrac{x^2 + 2xy - 8y^2}{x^2 + 7xy + 12y^2} \div \dfrac{x^2 - 3xy + 2y^2}{x^2 + 2xy - 3y^2} \cdot \dfrac{x^2 - 9}{9 - x^2}$

51. $\dfrac{x^3 + x^2 - x - 1}{x + 2} \div \dfrac{x^3 + 1}{x + 3} \cdot \dfrac{x + 2}{x - 1}$

29. _____

30. _____

31. _____

32. _____

33. _____

34. _____

35. _____

36. _____

37. _____

38. _____

39. _____

40. _____

41. _____

42. _____

43. _____

44. _____

45. _____

46. _____

47. _____

48. _____

49. _____

50. _____

51. _____

52. $\dfrac{x^3 + 3x^2 - 9x - 27}{x + 4} \div \dfrac{x^3 + 27}{x + 4} \cdot \dfrac{x + 1}{x - 3}$

52. _____

53. $\dfrac{x - 2}{x^2 - 9} \div \dfrac{x^3 - 8}{x + 3} \cdot \dfrac{x - 3}{x}$

53. _____

54. $\dfrac{x - 1}{x^2 - 25} \div \dfrac{x - 3}{x^3 + 125} \cdot \dfrac{(x - 5)^2}{x - 1}$

54. _____

D. Applications

55. In business, the product of the price and the demand for a product is constant. If the price for x units of a product is $\dfrac{3x + 9}{4}$ and the demand for these is $\dfrac{600}{x^2 + 3x}$, find the product of the price and the demand.

55. _____

56. If the price for x units of a product is given by $\dfrac{4x + 8}{5}$ and the demand is $\dfrac{200}{x^2 + 2x}$, what is the product of the price and the demand?

56. _____

57. In a simple electric circuit, the current is the quotient of the voltage and the resistance. If the current changes with the time t according to $R = \dfrac{t^2 + 9}{t^2 + 6t + 9}$ and the voltage changes according to the formula $V = \dfrac{4t}{t + 3}$, find the current.

57. _____

58. If in Problem 57 the resistance is $R = \dfrac{t^2 + 5}{t^2 + 4t + 4}$ and the voltage is $V = \dfrac{5t}{t + 2}$, find the current.

58. _____

✓ **SKILL CHECKER**

Write the rational expression with the indicated denominator.

59. $\dfrac{x - 1}{x - 3};$ denominator $x^2 - x - 6$

59. _____

60. $\dfrac{x + 4}{x - 3};$ denominator $x^2 - 9$

60. _____

61. $\dfrac{1}{x^2 + 2x + 4};$ denominator $x^3 - 8$

61. _____

62. $\dfrac{1}{x^2 - 3x + 9};$ denominator $x^3 + 27$

62. _____

4.3 USING YOUR KNOWLEDGE

1. When studying parallel resistors, the expression

$$R \cdot \dfrac{R_T}{R - R_T}$$

1. _____

occurs, where R is a known resistance and R_T is a required one. Multiply this expression.

2. The molecular model predicts that the pressure of a gas is given by

$$\frac{2}{3} \cdot \frac{mv^2}{2} \cdot \frac{N}{v}$$

Multiply this expression.

3. Suppose a store orders 3000 items each year. If it orders x units at a time, the number N of reorders is

$$N = \frac{3000}{x}$$

If there is a fixed $20 reorder fee and a $3 charge per item, the cost of each order is

$$C = 20 + 3x$$

The yearly reorder cost RC is then given by

$$RC = N \cdot C$$

Find RC.

2. _____

3. _____

4.4 ADDITION AND SUBTRACTION OF RATIONAL EXPRESSIONS

OBJECTIVES

REVIEW

Before starting this section, you should know:

1. How to add and subtract real numbers.
2. How to write a fraction with a given denominator.
3. How to factor polynomials.

OBJECTIVES

You should be able to:

A. Add or subtract rational expressions with the same denominator.
B. Add or subtract rational expressions with different denominators.

The photo illustrates the fact that

$$\frac{1}{4} + \frac{2}{4} = \frac{3}{4}$$

A. ADDITION AND SUBTRACTION OF RATIONAL EXPRESSIONS WITH THE SAME DENOMINATOR

In general, if a, b, and c are real numbers, $b \neq 0$,

$$\frac{a}{b} + \frac{c}{b} = \frac{a+c}{b}$$

$$\frac{a}{b} - \frac{c}{b} = \frac{a-c}{b}$$

The important thing to remember is that to add (or subtract) rational expressions with the *same* denominators, we add (or subtract) the numerators and *keep* the denominator. Thus

$$\frac{2}{x} + \frac{6}{x} = \frac{2+6}{x} = \frac{8}{x} \quad \leftarrow \text{Add numerators.}$$
$$\leftarrow \text{Keep the denominator.}$$

Similarly,

$$\frac{5x}{x^2+1} + \frac{2x}{x^2+1} = \frac{5x+2x}{x^2+1} = \frac{7x}{x^2+1}$$

$$\frac{3x}{7(x-1)^2} + \frac{4x}{7(x-1)^2} = \frac{3x+4x}{7(x-1)^2} = \frac{7x}{7(x-1)^2} = \frac{x}{(x-1)^2}$$

In these two problems we add the numerators and keep the same denominator.

and

$$\frac{8x}{x^2+5} - \frac{2x}{x^2+5} = \frac{8x-2x}{x^2+5} = \frac{6x}{x^2+5}$$

$$\frac{10x}{9(x-3)^2} - \frac{x}{9(x-3)^2} = \frac{10x-x}{9(x-3)^2} = \frac{9x}{9(x-3)^2} = \frac{x}{(x-3)^2}$$

In these two problems we subtract the numerators and keep the same denominator.

Note that we write the final answer in reduced form.

EXAMPLE 1 Find.

a. $\dfrac{8x}{3(x-2)} + \dfrac{x}{3(x-2)}$

b. $\dfrac{7x}{5(x+4)^2} + \dfrac{3x}{5(x+4)^2}$

c. $\dfrac{x}{x^2-1} - \dfrac{1}{x^2-1}$

d. $\dfrac{4x}{x+2} - \dfrac{3x-2}{x+2}$

Solution

a. $\dfrac{8x}{3(x-2)} + \dfrac{x}{3(x-2)} = \dfrac{8x+x}{3(x-2)} = \dfrac{\overset{3}{9x}}{\underset{}{3}(x-2)} = \dfrac{3x}{x-2}$

b. $\dfrac{7x}{5(x+4)^2} + \dfrac{3x}{5(x+4)^2} = \dfrac{7x+3x}{5(x+4)^2} = \dfrac{\overset{2}{10x}}{\underset{}{5}(x+4)^2} = \dfrac{2x}{(x+4)^2}$

c. Since both expressions have the same denominator, we simply subtract numerators and use the same denominator. However, the answer can be reduced. You can see that only if the denominator is factored. Make sure your denominators are in factored form so you can spot possible ways of reducing the answer.

$\dfrac{x}{x^2-1} - \dfrac{1}{x^2-1} = \dfrac{x-1}{x^2-1}$ ← Subtract numerators.

    ← Keep denominators.

$\qquad = \dfrac{x-1}{(x+1)(x-1)}$ Divide out the common factor $x-1$.

$\qquad = \dfrac{1}{x+1}$

d. We indicate the subtraction of the numerators by using parentheses. Be careful with the signs when removing the parentheses.

$\dfrac{4x}{x+2} - \dfrac{3x-2}{x+2} = \dfrac{4x-(3x-2)}{x+2}$

$\qquad = \dfrac{4x-3x+2}{x+2}$ Recall that $-(3x-2) = -3x+2$.

$\qquad = \dfrac{x+2}{x+2} = 1$ Combine like terms and reduce.

Problem 1 Find.

a. $\dfrac{4x}{5(x-2)} + \dfrac{6x}{5(x-2)}$

b. $\dfrac{11x}{3(x+2)^2} + \dfrac{4x}{3(x+2)^2}$

c. $\dfrac{x}{x^2-9} - \dfrac{3}{x^2-9}$

d. $\dfrac{5x}{x+3} - \dfrac{4x-3}{x+3}$

B. ADDITION AND SUBTRACTION OF RATIONAL EXPRESSIONS WITH DIFFERENT DENOMINATORS

To add or subtract fractions with different denominators, we must first find a common denominator. It is most convenient to use the smallest one available, called the **lowest common denominator (LCD)**—that is, the smallest multiple of 12 and 18. Thus to add

$$\frac{5}{12} + \frac{7}{18}$$

we start by writing 12 and 18 as a product of *primes*.

A **prime number** is a natural number greater than 1 with exactly two distinct divisors, itself and 1.

The first few primes are 2, 3, 5, 7, 11, and 13. Notice that 1 is not prime (it has only one divisor) and that all the other primes have exactly two divisors. To write 12 as a product of primes, we note that 2 is a factor of 12 and write

$$12 = 2 \cdot 6$$

Since 2 is a factor of 6,

$$12 = 2 \cdot 2 \cdot 3 = 2^2 \cdot 3$$

All the factors are prime, so we are finished.

Do the same for 18. Start by using 2 as a factor and writing

$$18 = 2 \cdot 9$$

Use 3 as a factor for 9:

$$18 = 2 \cdot 3 \cdot 3 = 2 \cdot 3^2$$

Thus we have

$$12 = 2 \cdot 2 \cdot 3 \quad = 2^2 \cdot 3$$
$$18 = \quad 2 \cdot 3 \cdot 3 = 2 \cdot 3^2$$

Note that all the 2s and all the 3s are written in the *same* column.

Since we need the *smallest* number that is a multiple of 12 and 18, we select the factors raised to the *greatest* power in each column. The product of these factors is the LCD. Thus the LCD of 12 and 18 is $2^2 \cdot 3^2 = 4 \cdot 9 = 36$. We then write each fraction with a denominator of 36 and add:

$$\frac{5}{12} = \frac{5 \cdot 3}{12 \cdot 3} = \frac{15}{36}$$

We multiply the denominator of $\frac{5}{12}$ by 3 (to get 36), so we do the same to the numerator.

$$\frac{7}{18} = \frac{7 \cdot 2}{18 \cdot 2} = \frac{14}{36}$$

$$\frac{5}{12} + \frac{7}{18} = \frac{15}{36} + \frac{14}{36} = \frac{29}{36}$$

If the denominators involved have no common factors, the LCD is the product of the denominators. For example, the denominators in $\frac{3}{5}$ and $\frac{1}{7}$ have no common factors. The LCD is $5 \cdot 7$. To subtract $\frac{1}{7}$ from $\frac{3}{5}$, we first write each fraction with a denominator of 35 and then subtract. Here are the steps.

1. The LCD is $5 \cdot 7 = 35$.
2. Write each fraction with 35 as denominator.

$$\frac{3}{5} = \frac{3 \cdot 7}{5 \cdot 7} = \frac{21}{35} \quad \text{and} \quad \frac{1}{7} = \frac{1 \cdot 5}{7 \cdot 5} = \frac{5}{35}$$

3. Subtract: $\dfrac{3}{5} - \dfrac{1}{7} = \dfrac{21}{35} - \dfrac{5}{35} = \dfrac{16}{35}$.

EXAMPLE 2 Subtract $\dfrac{2x}{x+1} - \dfrac{x}{x+2}$.

Solution $(x+1)$ and $(x+2)$ do not have any common factors. Thus the LCD in

$$\frac{2x}{x+1} - \frac{x}{x+2} \quad \text{is } (x+1)(x+2)$$

Problem 2 Subtract $\dfrac{5x}{x+1} - \dfrac{3x}{x+3}$.

Therefore,

$$\frac{2x}{x+1} = \frac{2x(x+2)}{(x+1)(x+2)}$$

$$\frac{x}{x+2} = \frac{x(x+1)}{(x+2)(x+1)}$$

$$\frac{2x}{x+1} - \frac{x}{x+2} = \frac{2x(x+2)}{(x+1)(x+2)} - \frac{x(x+1)}{(x+1)(x+2)}$$

$$= \frac{2x(x+2) - x(x+1)}{(x+1)(x+2)}$$

$$= \frac{2x^2 + 4x - x^2 - x}{(x+1)(x+2)} \qquad \text{Use the distributive law.}$$

$$= \frac{x^2 + 3x}{(x+1)(x+2)} \qquad \text{Combine like terms in the numerator.} \qquad \blacktriangle$$

In general, to add or subtract fractions with different denominators, we use the following procedure.

TO ADD (OR SUBTRACT) FRACTIONS WITH DIFFERENT DENOMINATORS

1. Find the LCD.

2. Write all fractions as equivalent ones with the LCD as the denominator.

3. Add (or subtract) numerators; keep the LCD as denominator.

4. Reduce if possible.

EXAMPLE 3 Perform the indicated operations.

a. $\dfrac{x+1}{x^2+x-2} + \dfrac{x+3}{x^2-1}$ **b.** $\dfrac{x-1}{x^2-x-6} - \dfrac{x+4}{x^2-9}$

Solution

a. We use the four-step procedure to add fractions.

1. We first find the LCD of the denominators. Write the denominators in factored form with the same factors in a column.

$$x^2 + x - 2 = (x+2)(x-1)$$
$$x^2 - 1 = \qquad\quad (x-1)(x+1)$$

Select the factors with the greatest exponents in each column. Thus the LCD is

$$(x+2)(x-1)(x+1)$$

2. We then write each fraction as an equivalent one with the LCD as denominator.

Problem 3 Perform the indicated operations.

a. $\dfrac{x+1}{(x+3)(x-1)} + \dfrac{x+4}{x^2-1}$

b. $\dfrac{x-3}{x^2-x-2} - \dfrac{x+3}{x^2-4}$

ANSWER (to the problem on page 288)

1. a. $\dfrac{2x}{x-2}$ **b.** $\dfrac{5x}{(x+2)^2}$

 c. $\dfrac{1}{x+3}$ **d.** 1

$$\frac{x+1}{x^2+x-2} = \frac{x+1}{(x+2)(x-1)} = \frac{(x+1)(x+1)}{(x+2)(x-1)(x+1)}$$

$$\frac{x+3}{x^2-1} = \frac{x+3}{(x+1)(x-1)} = \frac{(x+3)(x+2)}{(x+1)(x-1)(x+2)}$$

$$= \frac{(x+3)(x+2)}{(x+2)(x-1)(x+1)}$$

3. $\dfrac{x+1}{x^2+x-2} + \dfrac{x+3}{x^2-1} = \dfrac{(x+1)(x+1)}{(x+2)(x-1)(x+1)} + \dfrac{(x+3)(x+2)}{(x+2)(x-1)(x+1)}$

$$= \frac{(x^2+2x+1)+(x^2+5x+6)}{(x+2)(x-1)(x+1)} \quad \text{Add the numerators; keep the denominator.}$$

$$= \frac{2x^2+7x+7}{(x+2)(x-1)(x+1)} \quad \text{Combine like terms in the numerator.}$$

4. The answer is not reducible.

b. To subtract fractions we use the four-step procedure given.

1. To find the LCD, we factor the denominators keeping the same factors in a column:

$$x^2 - x - 6 = \qquad (x-3)(x+2)$$
$$x^2 - 9 = (x+3)(x-3)$$

Thus the LCD is $(x+3)(x-3)(x+2)$.

2. Write each fraction as an equivalent one with the LCD as the denominator:

$$\frac{x-1}{x^2-x-6} = \frac{(x-1)(x+3)}{(x-3)(x+2)(x+3)}$$

$$= \frac{(x-1)(x+3)}{(x+3)(x-3)(x+2)}$$

$$\frac{x+4}{x^2-9} = \frac{(x+4)(x+2)}{(x+3)(x-3)(x+2)}$$

3. $\dfrac{x-1}{x^2-x-6} - \dfrac{x+4}{x^2-9} = \dfrac{(x-1)(x+3)}{(x+3)(x-3)(x+2)} - \dfrac{(x+4)(x+2)}{(x+3)(x-3)(x+2)}$

$$= \frac{(x^2+2x-3)-(x^2+6x+8)}{(x+3)(x-3)(x+2)}$$

$$= \frac{x^2+2x-3-x^2-6x-8}{(x+3)(x-3)(x+2)} \quad \text{Subtract.}$$

$$= \frac{-4x-11}{(x+3)(x-3)(x+2)} \quad \text{Combine terms in the numerator.}$$

4. The answer is not reducible. ▲

How would you add $\frac{6}{12} + \frac{1}{8}$? You can start by finding the LCD, 24. However, it is easier to reduce $\frac{6}{12}$ to $\frac{1}{2}$ first. Then we simply need to add $\frac{1}{2} + \frac{1}{8} = \frac{4}{8} + \frac{1}{8} = \frac{5}{8}$. We illustrate a similar problem in the next example.

EXAMPLE 4 Perform the indicated operations.

a. $\dfrac{x+y}{x^2+2xy+y^2}+\dfrac{x-y}{x^2-2xy+y^2}$ **b.** $\dfrac{x}{(x+2)(x-2)}-\dfrac{2}{(2-x)(x+2)}$

Solution

a. $\dfrac{x+y}{x^2+2xy+y^2}=\dfrac{x+y}{(x+y)^2}=\dfrac{1}{x+y}$ Factor and reduce the first fraction.

$\dfrac{x-y}{x^2-2xy+y^2}=\dfrac{x-y}{(x-y)^2}=\dfrac{1}{x-y}$ Factor and reduce the second fraction.

Since $x+y$ and $x-y$ have no common factors, the LCD is $(x+y)(x-y)$. We have

$$\dfrac{x+y}{x^2-2xy+y^2}+\dfrac{x-y}{x^2-2xy+y^2}=\dfrac{1}{x+y}+\dfrac{1}{x-y}$$

$$=\dfrac{x-y}{(x+y)(x-y)}+\dfrac{x+y}{(x+y)(x-y)}$$ Write each fraction with $(x+y)(x-y)$ as the denominator.

$$=\dfrac{x-y+x+y}{(x+y)(x-y)}$$

$$=\dfrac{2x}{(x+y)(x-y)}$$

b. We note that $x-2=-(2-x)$, so we start by multiplying the numerator and denominator of the second fraction by -1.

$$\dfrac{x}{(x+2)(x-2)}-\dfrac{2}{(2-x)(x+2)}$$

$$=\dfrac{x}{(x+2)(x-2)}-\dfrac{-1\cdot(2)}{-1\cdot(2-x)(x+2)}$$ Multiply numerator and denominator by -1.

$$=\dfrac{x}{(x+2)(x-2)}-\dfrac{-2}{(x-2)(x+2)}$$ $-1\cdot2=-2$

$$=\dfrac{x-(-2)}{(x+2)(x-2)}$$

$$=\dfrac{\cancel{x+2}}{(\cancel{x+2})(x-2)}$$ Divide out $x+2$.

$$=\dfrac{1}{x-2}$$

Problem 4 Perform the indicated operations.

a. $\dfrac{x-y}{x^2-2xy+y^2}+\dfrac{x-y}{x^2-y^2}$

b. $\dfrac{x}{(x+3)(x-3)}+\dfrac{3}{(3-x)(x+3)}$

A. In Problems 1–10 perform the indicated operations.

1. $\dfrac{x}{5} + \dfrac{2x}{5}$

2. $\dfrac{x+1}{3x} + \dfrac{2x+7}{3x}$

3. $\dfrac{7x}{3} - \dfrac{2x}{3}$

4. $\dfrac{2x-1}{5x} - \dfrac{x+1}{5x}$

5. $\dfrac{3}{5x+10} + \dfrac{2x}{5(x+2)}$

6. $\dfrac{2x+1}{3(x+2)} + \dfrac{3x+1}{3x+6}$

7. $\dfrac{2x+1}{2(x+1)} - \dfrac{x-1}{2x+2}$

8. $\dfrac{3x-1}{4(x-1)} - \dfrac{4x-1}{4x-4}$

9. $\dfrac{2x+1}{3(x-1)} + \dfrac{x+3}{3x-3} - \dfrac{x-1}{3(x-1)}$

10. $\dfrac{3x-1}{5(x+1)} - \dfrac{x+1}{5x+5} + \dfrac{2x-5}{5(x+1)}$

B. In Problems 11–50 perform the indicated operations.

11. $\dfrac{x}{x^2+3x-4} + \dfrac{x}{x^2-16}$

12. $\dfrac{x-2}{x^2-9} + \dfrac{x+1}{x^2-x-12}$

13. $\dfrac{3x}{x^2+3x-10} + \dfrac{2x}{x^2+x-6}$

14. $\dfrac{x+3}{x^2-x-2} + \dfrac{x-1}{x^2+2x+1}$

15. $\dfrac{1}{x^2-y^2} - \dfrac{5}{(x+y)^2}$

16. $\dfrac{3}{(x+y)^2} + \dfrac{5}{(x-y)}$

17. $\dfrac{2}{x-5} - \dfrac{3x}{x^2-25}$

18. $\dfrac{x+3}{x^2-x-2} - \dfrac{x-1}{x^2+2x+1}$

19. $\dfrac{x-1}{x^2+3x+2} - \dfrac{x+7}{x^2+5x+6}$

20. $\dfrac{2}{x^2+3xy+2y^2} - \dfrac{1}{x^2-xy-2y^2}$

Hint for Problems 21–28: First reduce the fractions.

21. $\dfrac{x+2}{x^2-4} + \dfrac{x+3}{x^2-9}$

22. $\dfrac{x-3}{x^2-9} + \dfrac{x+3}{x^2-9}$

23. $\dfrac{x-3}{x^2-9} + \dfrac{x+3}{x^2+6x+9}$

24. $\dfrac{a-4}{a^2-16} + \dfrac{a+3}{a^2+5a+6}$

ANSWERS	
1.	
2.	
3.	
4.	
5.	
6.	
7.	
8.	
9.	
10.	
11.	
12.	
13.	
14.	
15.	
16.	
17.	
18.	
19.	
20.	
21.	
22.	
23.	
24.	

25. $\dfrac{a+3}{a^2+5a+6} + \dfrac{a+2}{a^2+6a+8}$

26. $\dfrac{a+3}{a^2+5a+6} - \dfrac{a-4}{a^2-16}$

27. $\dfrac{3a+3}{a^2+5a+4} - \dfrac{a-3}{a^2+a-12}$

28. $\dfrac{2a}{5a-7b} - \dfrac{5a+7b}{25a^2-49b^2}$

29. $\dfrac{5a-15}{a^2+2a-15} - \dfrac{a^2+5a}{a^2+8a+15}$

30. $\dfrac{3}{y^2-9} + \dfrac{2y}{y-3}$

31. $\dfrac{y}{y^2-1} + \dfrac{y}{y-1}$

32. $\dfrac{3y}{y^2-4} - \dfrac{y}{y+2}$

33. $\dfrac{3y+1}{y^2-16} - \dfrac{2y-1}{y-4}$

34. $\dfrac{3x-5y}{2x-3y} + \dfrac{2x-3y}{2x+3y}$

35. $\dfrac{5x+2y}{5x-2y} + \dfrac{5x-2y}{5x+2y}$

36. $\dfrac{x+3y}{x-5y} - \dfrac{x+5y}{x-3y}$

37. $\dfrac{3x-y}{2x-y} - \dfrac{2x+y}{3x+y}$

38. $\dfrac{a+3}{a^2+a-6} + \dfrac{a-2}{a^2+3a-10}$

39. $\dfrac{x+3}{x^2-x-2} + \dfrac{x-1}{x^2+2x+1}$

40. $\dfrac{8x}{x^2-4y^2} - \dfrac{2x}{x^2-5xy+6y^2}$

41. $\dfrac{x+1}{x^2-x-2} - \dfrac{x}{x^2-5x+4}$

42. $\dfrac{3}{x^2-4} + \dfrac{1}{2-x} - \dfrac{1}{2+x}$

43. $\dfrac{2}{5+x} + \dfrac{5x}{x^2-25} + \dfrac{7}{5-x}$

44. $\dfrac{1}{x^2+x-12} + \dfrac{2}{x^2+2x-15} + \dfrac{3}{x^2+9x+20}$

45. $\dfrac{x}{(x-y)(2-x)} - \dfrac{y}{(y-x)(2-x)} + \dfrac{y}{(x-y)(x-2)}$

46. $\dfrac{a}{(b-a)(c-a)} - \dfrac{b}{(b-c)(a-b)} + \dfrac{c}{(a-c)(b-c)}$

47. $\dfrac{4a^2-9b^2}{4a^2-12ab+9b^2} + \dfrac{12a+18b}{4a^2+12ab+9b^2} - \dfrac{2a+3b}{2a+3b}$

48. $\dfrac{x+2y}{x^3+8y^3} + \dfrac{5}{x+2y} + \dfrac{2x-3y}{x^2-2xy+4y^2}$

49. $\dfrac{x+5}{x^3+125} + \dfrac{x-5}{x^2-25} - \dfrac{1}{x+5}$

50. $\dfrac{a}{a+3} + \dfrac{a-2}{a^2-3a+9} + \dfrac{5a-7a}{a^3+27}$

25. _____
26. _____
27. _____
28. _____
29. _____
30. _____
31. _____
32. _____
33. _____
34. _____
35. _____
36. _____
37. _____
38. _____
39. _____
40. _____
41. _____
42. _____
43. _____
44. _____
45. _____
46. _____
47. _____
48. _____
49. _____
50. _____

C. Applications

51. The moment M of a cantilever beam of length L x units from the end is given by

$$-\frac{w_0 x^3}{6L} + \frac{w_0 Lx}{2} - \frac{w_0 L^2}{3}$$

Write this expression as a single rational expression in reduced form.

52. The deflection d of the beam of Problem 51 involves the expression

$$-\frac{x^4}{24L} + \frac{Lx^2}{4} - \frac{L^2 x}{3}$$

Write this expression as a single rational expression in reduced form.

53. In astronomy, when calculating planetary motions, we find the expression

$$\frac{p^2}{2mr^2} - \frac{gmM}{r}$$

Write this expression as a single rational expression in reduced form.

54. When studying the motion of a pendulum, we find the expression

$$\frac{P_1^2 + P_2^2}{2(h_1 + h_2)} + \frac{P_1^2 - P_2^2}{2(h_1 - h_2)}$$

Write this expression as a single rational expression in reduced form.

✓ SKILL CHECKER

Simplify.

55. $9\left(2 + \dfrac{2}{9}\right)$

56. $4\left(60 - \dfrac{15}{2}\right)$

57. $12xy\left(\dfrac{2}{y} + \dfrac{3}{2x}\right)$

58. $6ab\left(\dfrac{3}{a} - \dfrac{4}{b}\right)$

59. $x^2\left(1 - \dfrac{1}{x^2}\right)$

60. $x^3\left(1 - \dfrac{1}{x^3}\right)$

4.4 USING YOUR KNOWLEDGE

In calculus, the derivative of a polynomial is defined as the limiting value of

$$\frac{P(x + h) - P(x)}{h}$$

as h approaches zero.

51. _____

52. _____

53. _____

54. _____

55. _____

56. _____

57. _____

58. _____

59. _____

60. _____

Let $P(x) = x^2$.

1. Find $P(x + h)$ and write it in expanded form.

2. Find $P(x + h) - P(x)$ and simplify it.

3. Find $\dfrac{P(x + h) - P(x)}{h}$ in simplified form.

4. Find $\dfrac{P(x + h) - P(x)}{h}$ for $P(x) = x^2 + x$.

1. _____

2. _____

3. _____

4. _____

CALCULATOR CORNER

You can check the addition of fractions using a calculator. For instance, in Example 3b, we found that

$$\frac{x - 1}{x^2 - x - 6} - \frac{x + 4}{x^2 - 9} = \frac{-4x - 11}{(x + 3)(x - 3)(x + 2)}$$

To check this answer, you can substitute any convenient number for x and see if both sides are equal. Of course, you cannot choose numbers that will give you a 0 denominator. For Example 3b, a simple number to use is $x = 2$. On the left you get

$$\frac{2 - 1}{4 - 2 - 6} - \frac{2 + 4}{4 - 9} = -\frac{1}{4} + \frac{6}{5}$$

Now key in

$\boxed{1}\ \boxed{\div}\ \boxed{4}\ \boxed{+/-}\ \boxed{+}\ \boxed{6}\ \boxed{\div}\ \boxed{5}\ \boxed{=}$

The display shows 0.95. On the right, we have

$$\frac{-8 - 11}{(2 + 3)(2 - 3)(2 + 2)}$$

Now key in

$\boxed{8}\ \boxed{+/-}\ \boxed{-}\ \boxed{1}\ \boxed{1}\ \boxed{=}\ \boxed{\div}\ \boxed{(}\ \boxed{2}\ \boxed{+}\ \boxed{3}\ \boxed{)}\ \boxed{=}$

$\boxed{\div}\ \boxed{(}\ \boxed{2}\ \boxed{-}\ \boxed{3}\ \boxed{)}\ \boxed{=}\ \boxed{\div}\ \boxed{(}\ \boxed{2}\ \boxed{+}\ \boxed{2}\ \boxed{)}\ \boxed{=}$

and the display shows 0.95 again. This will complete the check. Verify the answers in this section using this method. You can also verify the answers for the preceding section.

4.5 COMPLEX FRACTIONS

Photo of Dr. Demento from Gordon/Casady Inc.

OBJECTIVES

REVIEW

Before starting this section, you should know:

1. How to find the LCD of two or more rational expressions.
2. How to remove parentheses by using the distributive law.
3. How to add, subtract, multiply, and divide rational expressions.

OBJECTIVES

You should be able to write a complex fraction as a simple fraction in reduced form.

The picture at the beginning of this section shows the disc jockey of station KMET in Los Angeles. Each hour he devotes $7\frac{1}{2}$ min to commercials, leaving $60 - 7\frac{1}{2}$ min for music. If the records he plays last an average of $3\frac{1}{4}$ min and it takes him about $\frac{1}{2}$ min to get a record going, how many records can he play each hour? The answer is

$$\dfrac{60 - 7\frac{1}{2}}{3\frac{1}{4} + \frac{1}{2}}$$ ← Time allowed for music

             ← Time it takes for playing each record

The fraction

$$\dfrac{60 - 7\frac{1}{2}}{3\frac{1}{4} + \frac{1}{2}}$$

contains other fractions in its numerator and denominator. A fraction whose numerator or denominator (or both) contains other fractions is called a **complex fraction.**

A fraction that is not complex is called a **simple fraction.** Thus,

$$\dfrac{\frac{1}{2}}{\frac{3}{4} + \frac{1}{5}}, \quad \dfrac{\frac{3x}{5} - \frac{1}{8}}{\frac{1x}{7}}, \quad \dfrac{-\frac{1}{3}}{\frac{1}{9}}, \quad \text{and} \quad \dfrac{x}{\frac{7}{8}}$$

are all complex fractions, but $\frac{1}{7}$, $\frac{3}{5}$, and $\frac{x}{9}$ are simple fractions.

To simplify a complex fraction, it is necessary to recall that the main fraction bar indicates that the numerator of the fraction is to be divided by the denominator of the fraction. Thus

$$\dfrac{60 - 7\frac{1}{2}}{3\frac{1}{4} + \frac{1}{2}} \quad \text{means} \quad (60 - 7\tfrac{1}{2}) \div (3\tfrac{1}{4} + \tfrac{1}{2})$$

                                      ↑

                          Use the ÷ sign
                          instead of the bar.

Here is the procedure we use to simplify complex fractions.

We now simplify

$$\frac{60 - 7\frac{1}{2}}{3\frac{1}{4} + \frac{1}{2}} = \frac{60 - \frac{15}{2}}{\frac{13}{4} + \frac{1}{2}}$$

using each of these methods.

METHOD 1. The LCD of $\frac{15}{2}$, $\frac{13}{4}$, and $\frac{1}{2}$ is 4, so we have:

$$\frac{60 - \frac{15}{2}}{\frac{13}{4} + \frac{1}{2}} = \frac{4 \cdot \left(60 - \frac{15}{2}\right)}{4 \cdot \left(\frac{13}{4} + \frac{1}{2}\right)}$$ Multiply numerator and denominator by 4, the LCD of $\frac{15}{2}, \frac{13}{4},$ and $\frac{1}{2}$.

$$= \frac{240 - 30}{13 + 2}$$ $\leftarrow 4(60 - \frac{15}{2}) = 240 - \frac{60}{2}$
$\leftarrow 4(\frac{13}{4} + \frac{1}{2}) = \cancel{4} \cdot \frac{13}{\cancel{4}} + \frac{4}{2}$
$= 13 + 2$

$$= \frac{210}{15}$$ $\leftarrow 240 - 30 = 210$
$\leftarrow 13 + 2 = 15$

$$= 14$$ Divide.

METHOD 2. $$\frac{60 - \frac{15}{2}}{\frac{13}{4} + \frac{1}{2}} = \frac{\frac{120}{2} - \frac{15}{2}}{\frac{13}{4} + \frac{2}{4}}$$ $\leftarrow$ Write 60 as $\frac{120}{2}$.
$\leftarrow$ Write $\frac{1}{2}$ as $\frac{2}{4}$.

$$= \frac{\frac{105}{2}}{\frac{15}{4}}$$ $\leftarrow \frac{120}{2} - \frac{15}{2} = \frac{105}{2}$
$\leftarrow \frac{13}{4} + \frac{2}{4} = \frac{15}{4}$

$$= \frac{105}{2} \div \frac{15}{4}$$ Replace the bar by the division sign, $\div$.

$$= \frac{\overset{7}{\cancel{105}}}{\cancel{2}} \cdot \frac{\overset{2}{\cancel{4}}}{\cancel{15}}$$ Multiply by the reciprocal of $\frac{15}{4}$, which is $\frac{4}{15}$, and reduce.

$$= 14$$

EXAMPLE 1 Write

$$\frac{\frac{3}{a} - \frac{4}{b}}{\frac{1}{2a} + \frac{2}{3b}}$$

as a simple fraction in reduced form.

Problem 1 Write

$$\frac{\frac{2}{b} - \frac{3}{a}}{\frac{1}{2b} + \frac{3}{4a}}$$

as a simple fraction in lowest terms.

Solution The LCD of $\frac{3}{a}$, $\frac{4}{b}$, $\frac{1}{2a}$, and $\frac{2}{3b}$ is $6ab$. Therefore, we multiply the numerator and denominator of the given fraction by $6ab$, obtaining

$$\frac{6ab \cdot \left(\dfrac{3}{a} - \dfrac{4}{b}\right)}{6ab \cdot \left(\dfrac{1}{2a} + \dfrac{2}{3b}\right)} = \frac{6ab \cdot \dfrac{3}{a} - 6ab \cdot \dfrac{4}{b}}{6ab \cdot \dfrac{1}{2a} + 6ab \cdot \dfrac{2}{3b}}$$ Use the distributive law and reduce.

$$= \frac{18b - 24a}{3b + 4a}$$ ← $6b \cdot 3 = 18b$ and $6a \cdot 4 = 24a$.
← $3b \cdot 1 = 3b$ and $2a \cdot 2 = 4a$.

We used Method 1.

EXAMPLE 2 Simplify.

$$\frac{x - \dfrac{1}{x^3}}{x + \dfrac{1}{x^2}}$$

Solution Here the LCD of the fractions involved is x^3. Thus

$$\frac{x - \dfrac{1}{x^3}}{x + \dfrac{1}{x^2}} = \frac{x^3 \cdot \left(x - \dfrac{1}{x^3}\right)}{x^3 \cdot \left(x + \dfrac{1}{x^2}\right)}$$

$$= \frac{x^3 \cdot x - x^3 \cdot \dfrac{1}{x^3}}{x^3 \cdot x + x^3 \cdot \dfrac{1}{x^2}}$$

$$= \frac{x^4 - 1}{x^4 + x}$$ Factor the numerator and denominator.

$$= \frac{(x^2 + 1)(x + 1)(x - 1)}{x(x + 1)(x^2 - x + 1)}$$

$$= \frac{(x^2 + 1)(x - 1)}{x(x^2 - x + 1)}$$ Divide out $x + 1$.

EXAMPLE 3 Simplify.

$$\frac{\dfrac{x}{x - 2} + x}{2 + \dfrac{1}{x^2 - 4}}$$

a. Use the first method.
b. Use the second method.

Solution

a. We first write $x^2 - 4$ as $(x + 2)(x - 2)$, x as $\frac{x}{1}$, and 2 as $\frac{2}{1}$, obtaining

$$\frac{\dfrac{x}{x - 2} + x}{2 + \dfrac{1}{x^2 - 4}} = \frac{\dfrac{x}{x - 2} + \dfrac{x}{1}}{\dfrac{2}{1} + \dfrac{1}{(x + 2)(x - 2)}}$$

Problem 2 Simplify.

$$\frac{x - \dfrac{1}{x^3}}{x - \dfrac{1}{x^2}}$$

Problem 3 Simplify.

$$\frac{\dfrac{x}{x + 2} + x}{1 + \dfrac{1}{x^2 - 4}}$$

a. Use the first method.
b. Use the second method.

The LCD of the denominators is $\dfrac{(x + 2)(x - 2)}{1}$. Multiply numerator and denominator by this LCD.

$$\dfrac{\dfrac{(x + 2)(x - 2)}{1} \cdot \left[\dfrac{x}{x - 2} + x\right]}{\dfrac{(x + 2)(x - 2)}{1} \cdot \left[2 + \dfrac{1}{(x + 2)(x - 2)}\right]}$$

$= \dfrac{x(x + 2) + x(x + 2)(x - 2)}{2(x + 2)(x - 2) + 1}$ Simplify.

$= \dfrac{x^2 + 2x + x^3 - 4x}{2x^2 - 8 + 1}$ Remove parentheses.

$= \dfrac{x(x^2 + x - 2)}{2x^2 - 7}$ Factor x and collect like terms.

$= \dfrac{x(x - 1)(x + 2)}{2x^2 - 7}$ Factor $x^2 + x - 2 = (x - 1)(x + 2)$.

b. $\dfrac{\dfrac{x}{x - 2} + x}{2 + \dfrac{1}{x^2 - 4}} = \dfrac{\dfrac{x}{x - 2} + \dfrac{x(x - 2)}{x - 2}}{\dfrac{2(x^2 - 4)}{x^2 - 4} + \dfrac{1}{x^2 - 4}}$

$\leftarrow$ Rewrite x as $\dfrac{x(x - 2)}{x - 2}$.

$\leftarrow$ Rewrite 2 as $\dfrac{2(x^2 - 4)}{x^2 - 4}$.

$= \dfrac{\dfrac{x + x(x - 2)}{x - 2}}{\dfrac{2(x^2 - 4) + 1}{x^2 - 4}}$ Add in the numerator and denominator.

$= \dfrac{\dfrac{x + x^2 - 2x}{x - 2}}{\dfrac{2x^2 - 8 + 1}{x^2 - 4}}$ Remove parentheses in the numerator and denominator.

$= \dfrac{x^2 - x}{x - 2} \div \dfrac{2x^2 - 7}{x^2 - 4}$ Use the division sign, $\div$, instead of the bar.

$= \dfrac{x(x - 1)}{x - 2} \cdot \dfrac{(x + 2)(x - 2)}{2x^2 - 7}$ Multiply by the reciprocal of $\dfrac{2x^2 - 7}{x^2 - 4}$.

$= \dfrac{x(x - 1)(x + 2) \cdot \overset{1}{\cancel{(x - 2)}}}{\cancel{(x - 2)}(2x^2 - 7)}$ Multiply.

$= \dfrac{x(x - 1)(x + 2)}{2x^2 - 7}$ Divide out $x - 2$.

EXAMPLE 4 Simplify.

$$1 + \dfrac{a}{1 + \dfrac{1}{1 + a}}$$

Problem 4 Simplify.

$$2 + \dfrac{a}{2 + \dfrac{2}{2 + a}}$$

Solution We start by working on the denominator of the fraction to the right of 1. Since we want to add $1 + \dfrac{1}{1 + a}$, we rewrite 1 as $\dfrac{1 + a}{1 + a}$.

$$1 + \frac{a}{1 + \dfrac{1}{1 + a}} = 1 + \frac{a}{\dfrac{1 + a}{1 + a} + \dfrac{1}{1 + a}}$$

$$= 1 + \frac{a}{\dfrac{2 + a}{1 + a}} \qquad \frac{1 + a}{1 + a} + \frac{1}{1 + a} = \frac{1 + a + 1}{1 + a} = \frac{2 + a}{1 + a}$$

$$= 1 + a \div \frac{2 + a}{1 + a}$$

$$= 1 + a \cdot \frac{1 + a}{2 + a} \qquad \frac{a}{\dfrac{2 + a}{1 + a}} = a \cdot \frac{1 + a}{2 + a}$$

$$= 1 + \frac{a(1 + a)}{2 + a}$$

$$= \frac{2 + a}{2 + a} + \frac{a(1 + a)}{2 + a} \qquad \text{Write 1 as } \frac{2 + a}{2 + a}.$$

$$= \frac{2 + 2a + a^2}{2 + a} \qquad \text{Add numerators. keeping the denominator.}$$

ANSWERS (to problems on pages 298–299)

1. $\dfrac{8a - 12b}{2a + 3b}$

2. $\dfrac{(x^2 + 1)(x - 1)}{x(x^2 + x + 1)}$

3. $\dfrac{x(x + 3)(x - 2)}{x^2 - 3}$

ANSWER (to problem on page 300)

4. $\dfrac{12 + 6a + a^2}{6 + 2a}$

NAME

CLASS

SECTION

ANSWERS

Simplify, giving the answer as a simple fraction in reduced form.

1. $\dfrac{\dfrac{3}{5}}{\dfrac{4}{5}}$

2. $\dfrac{\dfrac{-1}{7}}{\dfrac{3}{7}}$

3. $\dfrac{\dfrac{a}{b}}{\dfrac{c}{b}}$

4. $\dfrac{\dfrac{-a^2}{c}}{\dfrac{-b^2}{c}}$

5. $\dfrac{\dfrac{x}{y}}{\dfrac{x^2}{z}}$

6. $\dfrac{\dfrac{x^2}{y^2}}{\dfrac{x}{z}}$

7. $\dfrac{\dfrac{3x}{5y}}{\dfrac{3x}{2z}}$

8. $\dfrac{\dfrac{7x}{3y}}{\dfrac{14x}{5y}}$

9. $\dfrac{\dfrac{1}{2}}{2-\dfrac{1}{2}}$

10. $\dfrac{\dfrac{1}{4}}{3-\dfrac{1}{4}}$

11. $\dfrac{a-\dfrac{a}{b}}{1+\dfrac{a}{b}}$

12. $\dfrac{1-\dfrac{1}{a}}{1+\dfrac{1}{a}}$

13. $\dfrac{y+\dfrac{2}{x}}{y^2-\dfrac{4}{x^2}}$

14. $\dfrac{y-\dfrac{3}{x}}{y^2-\dfrac{9}{x^2}}$

15. $\dfrac{\dfrac{x}{y^2}-\dfrac{y}{x^2}}{x^2+xy+y^2}$

16. $\dfrac{\dfrac{x}{y^2}+\dfrac{y}{x^2}}{x^2-xy+y^2}$

17. $3-\dfrac{3}{3-\dfrac{1}{2}}$

18. $2-\dfrac{2}{2-\dfrac{1}{2}}$

1. _____

2. _____

3. _____

4. _____

5. _____

6. _____

7. _____

8. _____

9. _____

10. _____

11. _____

12. _____

13. _____

14. _____

15. _____

16. _____

17. _____

18. _____

19. $a - \dfrac{a}{a + \dfrac{1}{2}}$

20. $a + \dfrac{a}{a + \dfrac{1}{2}}$

21. $x - \dfrac{x}{1 - \dfrac{x}{1 - x}}$

22. $2x - \dfrac{x}{2 - \dfrac{x}{2 - x}}$

23. $\dfrac{1}{1 + \dfrac{1}{2 + \dfrac{1}{3 + \dfrac{1}{4}}}}$

24. $\dfrac{1}{1 - \dfrac{1}{2 - \dfrac{1}{3 - \dfrac{1}{4}}}}$

25. $\dfrac{\dfrac{x-1}{x+1} + \dfrac{x+1}{x-1}}{\dfrac{x-1}{x+1} - \dfrac{x+1}{x-1}}$

26. $\dfrac{\dfrac{x-1}{x+1} - \dfrac{x+1}{x-1}}{\dfrac{x-1}{x+1} + \dfrac{x+1}{x-1}}$

19. _____

20. _____

21. _____

22. _____

23. _____

24. _____

25. _____

26. _____

Applications

27. When connected in parallel, the combined resistance R of two resistors R_1 and R_2 is given by

$$R = \dfrac{1}{\dfrac{1}{R_1} + \dfrac{1}{R_2}}$$

Simplify this expression.

27. _____

28. When connected in parallel, the combined resistance R of three resistors R_1, R_2, and R_3 is given by

$$R = \dfrac{1}{\dfrac{1}{R_1} + \dfrac{1}{R_2} + \dfrac{1}{R_3}}$$

Simplify this expression.

28. _____

29. The formula for the Doppler effect in light is

$$f = f_s \sqrt{\dfrac{1 + \dfrac{v}{c}}{1 - \dfrac{v}{c}}}$$

Simplify the expression under the radical.

29. _____

30. Balmer's formula for the wavelength λ (lambda) of the hydrogen spectrum light is given by

$$\lambda = \dfrac{1}{\dfrac{1}{m^2} - \dfrac{1}{n^2}}$$

Simplify this expression.

30. _____

✓ SKILL CHECKER

Solve.

31. $4x + 8 = 6x$

32. $5x + 10 = 7x$

33. $x(x + 2) - (x - 3)(x - 4) = 4x + 3$

34. $x(x + 1) - (x - 1)(x - 2) = 2x + 2$

31. _____

32. _____

33. _____

34. _____

4.5 USING YOUR KNOWLEDGE

In the seventeenth century, the Dutch mathematician and astronomer Christian Huygens made a model of the solar system and found out that Saturn takes

$$29 + \cfrac{1}{2 + \cfrac{2}{9}}$$

years to go around the sun. Now,

$$\cfrac{1}{2 + \cfrac{2}{9}} = \frac{9 \cdot 1}{9 \cdot \left(2 + \cfrac{2}{9}\right)}$$

$$= \frac{9}{18 + 2}$$

$$= \frac{9}{20}$$

Thus it takes Saturn $29 + \frac{9}{20} = 29\frac{9}{20}$ yr to go around the sun.

Use your knowledge to simplify the numbers of years it takes the following planets to go around the sun.

1. Mercury: $\cfrac{1}{4 + \frac{1}{6}}$ yr.

2. Venus: $\cfrac{1}{1 + \frac{2}{3}}$ yr.

3. Jupiter: $11 + \cfrac{1}{1 + \frac{7}{43}}$ yr.   (Write your answer as a mixed number.)

4. Mars: $1 + \cfrac{1}{1 + \frac{3}{22}}$ yr.

1. _____

2. _____

3. _____

4. _____

4.6 EQUATIONS INVOLVING RATIONAL EXPRESSIONS

4 out of 5 people say Big John's Beans taste better. We bet 20¢ you'll agree.

Big John's is a registered trademark of Hunt-Wesson Foods, Inc.

OBJECTIVES

REVIEW

Before starting this section, you should know:

1. How to find the LCD for two or more fractions.
2. How to factor polynomials.
3. How to solve linear and quadratic equations.

OBJECTIVE

You should be able to solve equations involving rational expressions.

You have heard surveys claiming that doctors recommend one medicine over another one. Here is a survey that will pay you to agree. If 300 persons said that Big John's Beans did taste better, how many people were surveyed? If we assume that x people were surveyed, since 4 out of 5 agreed with the claim, we have

$$\frac{4}{5}x = 300$$

The first step in solving equations containing rational expressions is to multiply both sides of the equation by the LCD to clear the rational expressions. (See the procedure on page 100). Thus to solve $\frac{4}{5}x = 300$, use the following steps.

1. Multiply both sides by the LCD, 5. $5 \cdot \frac{4}{5}x = 5 \cdot 300$

2. Simplify. $4x = 1500$

3. Divide both sides by 4. $\frac{4x}{4} = \frac{1500}{4}$

$$x = 375$$

Thus 375 people were surveyed.

To solve $\frac{x}{2} + \frac{x}{3} = 10$ we multiply both sides of the equation by 6, the LCD of $\frac{x}{2}$ and $\frac{x}{3}$. Here are the steps.

$$6 \cdot \left(\frac{x}{2} + \frac{x}{3}\right) = 6 \cdot 10 \quad \text{Multiply both sides by the LCD, 6.}$$

$$6 \cdot \frac{x}{2} + 6 \cdot \frac{x}{3} = 6 \cdot 10 \quad \text{Use the distributive law.}$$

$$3x + 2x = 60 \quad \text{Simplify.}$$

$$5x = 60 \quad \text{Collect like terms.}$$

Dividing by 5, we have $x = 12$. The solution set is $\{12\}$. You can check this by substituting 12 for x in $\frac{x}{2} + \frac{x}{3} = 10$, obtaining

$$\frac{12}{2} + \frac{12}{3} = 10$$

$$6 + 4 = 10$$

which is a true statement. Example 1 illustrates the case when the variables are in the denominator.

EXAMPLE 1 Solve.

a. $\dfrac{4}{x} = \dfrac{6}{x+2}$ **b.** $\dfrac{1}{x+1} = \dfrac{2}{x+2}$

Solution

a. The LCD of $\dfrac{4}{x}$ and $\dfrac{6}{x+2}$ is $x(x+2)$. Multiplying both sides of the

equation by the LCD, $x(x+2) = \dfrac{x(x+2)}{1}$, gives

$$\frac{x(x+2)}{1} \cdot \frac{4}{x} = \frac{x(x+2)}{1} \cdot \frac{6}{x+2}$$

$\quad (x+2) \cdot 4 = x \cdot 6$ Divide out x and $x+2$.

$\qquad 4x + 8 = 6x$ Remove parentheses.

$\qquad\quad 8 = 2x$ Subtract $4x$ from both sides.

$\qquad\quad 4 = x$ Divide by 2.

The solution is 4. To check the answer, we substitute 4 for x in the

original equation, obtaining $\dfrac{4}{4} = \dfrac{6}{4+2}$, or $1 = 1$. Therefore, the so-

lution 4 is correct.

b. The LCD of $\dfrac{1}{x+1}$ and $\dfrac{2}{x+2}$ is $(x+1)(x+2)$. Multiplying both sides

of the equation by $\dfrac{(x+1)(x+2)}{1}$, we obtain

$$\frac{(x+1)(x+2)}{1} \cdot \frac{1}{x+1} = \frac{(x+1)(x+2)}{1} \cdot \frac{2}{x+2}$$

$\qquad\quad x + 2 = (x+1) \cdot 2$ Simplify.

$\qquad\quad x + 2 = 2x + 2$ Use the distributive law.

$\qquad\qquad x = 2x$ Subtract 2.

$\qquad\qquad 0 = x$ Subtract x.

The solution is 0. The verification that 0 is the correct solution is
left to the student. ▲

When variables occur in the denominator, it is possible to multiply
both sides of the equation by the LCD of the fractions involved and
obtain a solution of the resulting equation that does *not* satisfy the
original equation. For example, if we assume that there is a solution
for the equation

$$3 + \frac{1}{x-3} = \frac{1}{x-3}$$

we first multiply both sides of the equation by $\dfrac{x-3}{1}$, obtaining

$$\frac{x-3}{1} \cdot \left[3 + \frac{1}{x-3} \right] = \frac{x-3}{1} \cdot \frac{1}{x-3}$$

$$\frac{(x-3)}{1} \cdot 3 + \frac{(x-3)}{1} \cdot \frac{1}{x-3} = \frac{(x-3)}{1} \cdot \frac{1}{x-3} \quad \text{Use the distributive law.}$$

$\qquad\quad 3x - 9 + 1 = 1$ Simplify.

$\qquad\qquad 3x = 9$ Add 8.

$\qquad\qquad x = 3$ Divide by 3.

Problem 1 Solve.

a. $\dfrac{3}{x} = \dfrac{5}{x+2}$ **b.** $\dfrac{2}{x+1} = \dfrac{3}{x+2}$

The solution is 3. However, if we replace x by 3 in the equation $3 + \dfrac{1}{x-3} = \dfrac{1}{x-3}$, we obtain

$$3 + \frac{1}{3-3} = \frac{1}{3-3}$$

or

$$3 + \frac{1}{0} = \frac{1}{0}$$

Since division by 0 is not defined, the equation $3 + \dfrac{1}{x-3} = \dfrac{1}{x-3}$ has no solution. (The solution set is $\varnothing$, the empty set). This example points out the necessity of *checking,* by direct substitution in the original equation, any prospective solutions obtained after multiplying both sides of an equation by factors containing the unknown. If the prospective solution does not satisfy the equation, it is called an **extraneous** solution. In this example, 3 is an extraneous solution.

EXAMPLE 2 Solve.

a. $\dfrac{1}{x-4} - \dfrac{1}{x-2} = \dfrac{2x}{x^2 - 6x + 8}$ **b.** $\dfrac{x}{x-3} - \dfrac{x-4}{x+2} = \dfrac{4x+3}{x^2 - x - 6}$

Solution

a. We first factor the denominator of the right side of the equation, obtaining

$$\frac{1}{x-4} - \frac{1}{x-2} = \frac{2x}{(x-4)(x-2)}$$

Since the LCD of the fractions involved is $\dfrac{(x-4)(x-2)}{1}$, we multiply each side of the equation by this LCD and get

$$\frac{(x-4)(x-2)}{1} \cdot \left[\frac{1}{x-4} - \frac{1}{x-2} \right] = \frac{(x-4)(x-2)}{1} \cdot \left[\frac{2x}{(x-4)(x-2)} \right]$$

$$\frac{(x-4)(x-2)}{1} \cdot \frac{1}{x-4} - \frac{(x-4)(x-2)}{1} \cdot \frac{1}{x-2}$$

$$= \frac{(x-4)(x-2)}{1} \cdot \frac{2x}{(x-4)(x-2)}$$

$$(x-2) - (x-4) = 2x \quad \text{Simplify.}$$
$$x - 2 - x + 4 = 2x \quad \text{Remove parentheses.}$$
$$2 = 2x \quad \text{Collect like terms.}$$
$$x = 1 \quad \text{Divide by 2.}$$

The solution is 1. Substituting 1 for x in the original equation gives

$$\frac{1}{1-4} - \frac{1}{1-2} \overset{?}{=} \frac{2 \cdot 1}{1^2 - 6 \cdot 1 + 8}$$

$$\frac{1}{-3} - \frac{1}{-1} \overset{?}{=} \frac{2}{3}$$

$$-\frac{1}{3} + 1 = \frac{2}{3}$$

Problem 2 Solve.

a. $\dfrac{1}{x-6} - \dfrac{1}{x-4} = \dfrac{6}{(x-6)(x-2)}$

b. $\dfrac{x}{x-4} - \dfrac{x-5}{x+3} = \dfrac{3x+16}{x^2 - x - 12}$

a true statement. Thus our result is correct. The solution is 1 and the solution set is {1}.

b. We first write the right side of the equation with the denominator factored.

$$\frac{x}{x-3} - \frac{x-4}{x+2} = \frac{4x+3}{(x-3)(x+2)}$$

The LCD is $\dfrac{(x-3)(x+2)}{1}$. We multiply both sides by the LCD.

$$\frac{(x-3)(x+2)}{1}\left[\frac{x}{x-3} - \frac{x-4}{x+2}\right] = \frac{(x-3)(x+2)}{1}\left[\frac{4x+3}{(x-3)(x+2)}\right]$$

$$\frac{(x-3)(x+2)}{1} \cdot \frac{x}{x-3} - \frac{(x-3)(x+2)}{1} \cdot \frac{x-4}{x+2}$$

$$= \frac{(x-3)(x+2)}{1} \cdot \frac{4x+3}{(x-3)(x+2)}$$

$$(x+2)\cdot x - (x-3)(x-4) = 4x+3 \qquad \text{Simplify.}$$

$$x^2 + 2x - (x^2 - 7x + 12) = 4x + 3 \qquad \begin{array}{l}\text{Remove parentheses and}\\ \text{combine like terms.}\end{array}$$

$$9x - 12 = 4x + 3$$

$$5x = 15 \qquad \text{Add 12 and subtract } 4x.$$

$$x = 3 \qquad \text{Divide by 5.}$$

The trial solution is 3. However, if x is replaced by 3 in the original equation, the term $\dfrac{x}{x-3}$ yields $\frac{3}{0}$, which is meaningless. Consequently, the equation $\dfrac{x}{x-3} - \dfrac{x-4}{x+2} = \dfrac{4x+3}{x^2-x-6}$ has no solution. Its solution set is $\varnothing$. Thus 3 is an extraneous solution. ▲

The check is important not only to catch errors but also to rule out extraneous solutions.

Finally, we must point out that the equations resulting when clearing denominators are not *always* linear equations. For example, to solve the equation

$$\frac{x^2}{x+3} = \frac{9}{x+3}$$

we first multiply by the LCD $(x+3)$, obtaining

$$(x+3)\frac{x^2}{x+3} = (x+3)\frac{9}{x+3}$$

or

$$x^2 = 9$$

A linear equation is an equation in which the highest exponent of the variable is 1.

In this equation, the variable x has 2 as an exponent; thus it is a quadratic equation and can be solved when written in standard form—that is, by writing the equation as

$$x^2 - 9 = 0$$

$$(x+3)(x-3) = 0 \qquad \text{Factor.}$$

$$x + 3 = 0 \quad \text{or} \quad x - 3 = 0 \qquad \text{Use the zero-factor property.}$$

$$x = -3 \quad \text{or} \qquad x = 3 \qquad \text{Solve each equation.}$$

Recall that a quadratic equation is an equation in which the highest exponent of the variable is 2. A quadratic equation in standard form is written as

$$ax^2 + bx + c = 0$$

In $x^2 - 4 = 0$, $a = 1$, $b = 0$, and $c = -4$.

Obviously, 3 is a solution, since

$$\frac{3^2}{3+3} = \frac{9}{3+3}$$

However, for -3, the denominator $x + 3$ becomes 0. Thus -3 is an extraneous solution. The only solution is 3.

EXAMPLE 3 Solve.

$$1 + \frac{3}{x-2} = \frac{12}{x^2-4}$$

Solution Since $x^2 - 4 = (x+2)(x-2)$, the LCD is $(x+2)(x-2)$. We then write the equation with the denominator $x^2 - 4$ in factored form and multiply each term by the LCD, as before. Here are the steps.

1. Multiply each term by the LCD.

$$(x+2)(x-2) \cdot 1 + (x+2)(x-2) \cdot \frac{3}{x-2} = (x+2)(x-2) \cdot \frac{12}{x^2-4}$$

2. Simplify.
$$(x^2 - 4) + 3(x+2) = 12$$
$$x^2 - 4 + 3x + 6 = 12$$
$$x^2 + 3x + 2 = 12$$

3. Subtract 12 to write in standard form.
$$x^2 + 3x - 10 = 0$$

4. Factor.
$$(x+5)(x-2) = 0$$

5. Use the zero-factor property.
$$x + 5 = 0 \quad \text{or} \quad x - 2 = 0$$

6. Solve each equation.
$$x = -5 \quad \text{or} \quad x = 2$$

7. Since 2 makes the denominator $x - 2$ equal to 0, it is an extraneous solution. The only solution is -5. This solution can be checked in the original equation.

EXAMPLE 4 Solve.

$$\frac{x-3}{x^2-4x} = \frac{2}{x^2-16}$$

Solution As usual, we write all expressions in factored form and then multiply by the LCD, $\dfrac{x(x-4)(x+4)}{1}$.

$$\frac{x(x-4)(x+4)}{1} \cdot \frac{x-3}{x(x-4)} = \frac{x(x-4)(x+4)}{1} \cdot \frac{2}{(x+4)(x-4)}$$

$$(x+4)(x-3) = 2x \qquad \text{Simplify.}$$
$$x^2 + x - 12 = 2x \qquad \text{Remove parentheses.}$$
$$x^2 - x - 12 = 0 \qquad \text{Subtract } 2x.$$
$$(x+3)(x-4) = 0 \qquad \text{Factor.}$$
$$x + 3 = 0 \quad \text{or} \quad x - 4 = 0 \qquad \text{Use the zero-factor property.}$$
$$x = -3 \quad \text{or} \quad x = 4 \qquad \text{Solve.}$$

For $x = -3$, $\dfrac{x^2}{x+3} = \dfrac{(-3)^2}{(-3)+3} = \dfrac{9}{0}$.

Problem 3 Solve.

$$1 - \frac{4}{x^2-1} = \frac{-2}{x-1}$$

Problem 4 Solve.

$$\frac{x-7}{x^2-8x} = \frac{2}{x^2-64}$$

The trial roots are -3 or 4. If we substitute -3 for x in the original equation, we get a true statement. On the other hand, if we substitute 4 for x in the original equation, the denominators on both sides are 0, so the fractions are not defined. Thus, the only solution is -3, and 4 is an extraneous solution.

NAME

CLASS

SECTION

ANSWERS

In Problems 1–36, solve.

1. $\dfrac{x}{3} + \dfrac{x}{6} = 3$

2. $\dfrac{x}{2} + \dfrac{x}{4} = \dfrac{3}{8}$

3. $\dfrac{x}{5} - \dfrac{3x}{10} = \dfrac{1}{2}$

4. $\dfrac{x}{6} - \dfrac{x}{5} = \dfrac{1}{15}$

5. $\dfrac{1}{y} + \dfrac{4}{3y} = 7$

6. $\dfrac{10}{3y} - \dfrac{9}{2y} = \dfrac{7}{30}$

7. $\dfrac{2}{y-8} = \dfrac{1}{y-2}$

8. $\dfrac{2}{y-4} = \dfrac{3}{y-2}$

9. $\dfrac{3}{3z+4} = \dfrac{2}{5z-6}$

10. $\dfrac{2}{4z-1} = \dfrac{3}{2z+1}$

11. $\dfrac{-2}{2x-1} = \dfrac{3}{3x-1}$

12. $\dfrac{-5}{2x+3} = \dfrac{2}{3x-1}$

13. $\dfrac{-1}{x+1} = \dfrac{-2}{2x-1}$

14. $\dfrac{-5}{5x-2} = \dfrac{-3}{3x+1}$

15. $\dfrac{2}{3x+1} = \dfrac{4}{6x+2}$

16. $\dfrac{3}{2x-1} = \dfrac{6}{4x-5}$

17. $\dfrac{2}{x^2-4} + \dfrac{5}{x+2} = \dfrac{7}{x-2}$

18. $\dfrac{3}{x^2-9} + \dfrac{5}{x+3} = \dfrac{8}{x-3}$

19. $\dfrac{t+2}{t^2-3t+2} = \dfrac{3}{t-1} - \dfrac{1}{t-2}$

20. $\dfrac{t+3}{t^2+4t+3} = \dfrac{4}{t+3} - \dfrac{1}{t+1}$

21. $\dfrac{x^2}{x^2-1} = 1 + \dfrac{1}{x+1}$

22. $\dfrac{x^2}{x^2-9} = 1 + \dfrac{1}{x-3}$

23. $\dfrac{1}{x^2-4x+3} + \dfrac{1}{x^2-2x-3} = \dfrac{1}{x^2-1}$

24. $\dfrac{1}{x^2+3x+2} + \dfrac{1}{x^2+x-2} = \dfrac{1}{x^2-1}$

25. $\dfrac{x+2}{3x^2+4x+1} = \dfrac{x+1}{3x^2+7x+2}$

26. $\dfrac{x+2}{2x^2+x-1} = \dfrac{x-2}{2x^2+x-1}$

27. $\dfrac{2z+13}{2z^2+5z-3} + \dfrac{3}{z+3} = \dfrac{4}{2z-1}$

1. _____
2. _____
3. _____
4. _____
5. _____
6. _____
7. _____
8. _____
9. _____
10. _____
11. _____
12. _____
13. _____
14. _____
15. _____
16. _____
17. _____
18. _____
19. _____
20. _____
21. _____
22. _____
23. _____
24. _____
25. _____
26. _____
27. _____

28. $\dfrac{z-14}{2z^2-3z-2} + \dfrac{3}{z-2} = \dfrac{4}{2z+1}$

28. _____

29. $\dfrac{3-x}{5x^2-4x-1} + \dfrac{2}{5x+1} = \dfrac{1}{x-1}$

29. _____

30. $\dfrac{16-x}{4x^2-11x-3} + \dfrac{5}{4x+1} = \dfrac{2}{x-3}$

30. _____

31. $4x^{-1} + 2 = 7$ $\left(\text{Recall that } x^{-1} = \dfrac{1}{x}.\right)$

31. _____

32. $3 + 6x^{-1} = 5$

32. _____

33. $4x^{-1} + 6x^{-1} = 15(x+1)^{-1}$ **34.** $6x^{-1} + 9x^{-1} = 25(x+2)^{-1}$

33. _____

34. _____

35. $2(x-8)^{-1} = (x-2)^{-1}$ **36.** $3(3y+4)^{-1} = 2(5y-6)^{-1}$

35. _____

36. _____

✓ SKILL CHECKER

Use the RSTUV procedure to solve the following problems.

37. The sum of three consecutive odd integers is 69. What are the integers?

37. _____

38. An investor bought some municipal bonds yielding 5% annually and some certificates of deposit yielding 8%. If the total investment amounts to $10,000 and the annual interest is $680, how much is invested in bonds and how much in certificates of deposit?

38. _____

39. How many gallons of a 20% salt solution must be mixed with 40 gal of a 15% solution to obtain an 18% solution?

39. _____

40. A car leaves a town traveling at 50 mi/hr. Two hours later, another car traveling at 60 mi/hr leaves the same town in the same direction. How far from the town does the second car overtake the first?

40. _____

4.6 USING YOUR KNOWLEDGE

There are many instances in which a given formula must be changed to an equivalent form. For example, the formula

$$\frac{P}{R} = \frac{T}{V}$$

is frequently discussed in chemistry. Suppose you know P, R, and T. Can you find V? The idea is to solve for V. As before, we proceed by steps.

1. Since the LCD is RV, we multiply each term by RV, obtaining

$$RV \cdot \frac{P}{R} = \frac{T}{V} \cdot RV$$

2. Simplify. $\qquad VP = TR$

3. Divide by P $\qquad V = \dfrac{TR}{P}$

Thus

$$V = \dfrac{TR}{P}$$

Use your knowledge of fractional equations to solve the given problem for the indicated variable.

1. The area A of a trapezoid is

$$A = \dfrac{h(b_1 + b_2)}{2}$$

Solve for h.

1. _____

2. In an electric circuit, we have

$$\dfrac{1}{R} = \dfrac{1}{R_1} + \dfrac{1}{R_2}$$

Solve for R.

2. _____

3. In refrigeration we find the formula

$$\dfrac{Q_1}{Q_2 - Q_1} = P$$

Solve for Q_1.

3. _____

4. When studying the expansion of metals,

$$\dfrac{L}{1 + at} = L_0$$

Solve for t.

4. _____

5. In photography,

$$\dfrac{1}{f} = \dfrac{1}{a} + \dfrac{1}{b}$$

Solve for f.

5. _____

Reproduction credit: Eric Shaal for Time, Inc., Courtesy of the Vatican.

OBJECTIVES

REVIEW

Before starting this section, you should know:

1. How to solve equations involving rational expressions.
2. How to use the RSTUV procedure (Section 2.6) to solve word problems.

OBJECTIVES

You should be able to solve word problems involving:

A. Integers.
B. Rates of work.
C. Distance, rate, and time

The unfinished canvas pictured here was painted by Leonardo da Vinci and is entitled *St. Jerome*. A Golden Rectangle (black overlay) fits so neatly around St. Jerome that experts conjecture that da Vinci painted the figure to conform to those proportions. For many years it has been said that the Golden Rectangle is one of the most visually satisfying of all geometric forms. Do you know how to construct a Golden Rectangle? Such a rectangle has a special ratio of length to width, about 8 to 5. The situation can be described by writing

$$\frac{\text{length of rectangle}}{\text{width of rectangle}} = \frac{8}{5}$$

Now, suppose you want to make a Golden Rectangle of your own but you want the length to be 6 in. longer than the width. What are the dimensions of your rectangle?

To solve this problem, remember the **RSTUV** procedure given in Section 2.6.

1. **R**ead the problem.
2. **S**elect the variable w to be the width.
3. **T**ranslate the problem. Since you want a Golden Rectangle,

$$\frac{\text{length}}{\text{width}} = \frac{8}{5}$$

the length is 6 in. more $\longrightarrow$ $\dfrac{w + 6}{w} = \dfrac{8}{5}$
the width is w $\longrightarrow$

4. **U**se algebra to solve the equation. The LCD is $5w$, so we multiply both sides by $5w$.

$$5w\left(\frac{w + 6}{w}\right) = 5w \cdot \frac{8}{5}$$

$5w + 30 = 8w$ Simplify.

$30 = 3w$ Subtract $5w$.

$10 = w$ Divide by 3.

Thus, the width is 10 in. and the length is 6 in. more than that, or 16 in.

5. Verify your answer. The rectangle is 10 in. by 16 in., so the ratio of length (16) to width (10) is $\frac{16}{10}$, or $\frac{8}{5}$, as desired.

A. SOLVING NUMBER PROBLEMS

We now discuss problems involving numbers. You can solve them by using the RSTUV procedure.

EXAMPLE 1   There are two consecutive even integers such that the reciprocal of the first added to the reciprocal of the second is $\frac{3}{4}$. What are the integers?

Problem 1   The sum of the reciprocals of two consecutive even integers is $\frac{5}{12}$. What are the integers?

Solution

1. Read the problem.
2. Select n to be the first integer. The next even integer is $n + 2$.
3. Translate the problem:

$$\left[\begin{array}{c}\text{The reciprocal}\\\text{of the first}\end{array}\right]\left[\begin{array}{c}\text{added}\\\text{to}\end{array}\right]\left[\begin{array}{c}\text{the reciprocal}\\\text{of the second}\end{array}\right]\text{ is }\frac{3}{4}$$

$$\frac{1}{n} \quad + \quad \frac{1}{n+2} \quad = \frac{3}{4}$$

4. Use algebra to solve the equation. The LCD is $4n(n + 2)$. Multiply both sides by this LCD.

$$4n(n+2)\left(\frac{1}{n} + \frac{1}{n+2}\right) = 4n(n+2) \cdot \frac{3}{4}$$

$$4n(n+2) \cdot \frac{1}{n} + 4n(n+2) \cdot \frac{1}{n+2} = 4n(n+2) \cdot \frac{3}{4}$$

$$4n + 8 + 4n = 3n^2 + 6n \qquad \text{Remove parentheses.}$$

$$0 = 3n^2 - 2n - 8 \qquad \text{Subtract } 8n \text{ and } 8 \text{ from both sides.}$$

$$0 = (3n + 4)(n - 2) \qquad \text{Factor.}$$

By the zero-factor property,

$$3n + 4 = 0 \quad \text{or} \quad n - 2 = 0$$

$$n = -\frac{4}{3} \quad \text{or} \qquad n = 2 \quad \text{Solve } 3n + 4 = 0, n - 2 = 0.$$

Since n was assumed to be an integer, we discard the answer $-\frac{4}{3}$. Thus, the first even integer is 2 and the next one is 4.

5. Verify that the sum of the reciprocals is $\frac{3}{4}$. Since $\frac{1}{2} + \frac{1}{4} = \frac{3}{4}$, our result is correct.

B. WORK PROBLEMS

Have you ever wished that somebody would help you with your taxes? The Internal Revenue Service estimates that it takes about 7 hours to file your 1040A form. (This time includes record-keeping, familiarizing yourself with the form, and preparing and sending it.)

EXAMPLE 2 A couple is about to file their form 1040A. One of them can complete it in 8 hr, and the other can do it in 6 hr. How long would it take if they work on it together?

Solution

1. Read the problem.
2. Select t to be the time it takes for the couple to complete the form working together.
3. Translate the problem. Here, we shall concentrate on what happens each hour. Since one person can fill the form in 8 hr and the second can do it in 6 hr, the first person will complete $\frac{1}{8}$ of the form and the second will complete $\frac{1}{6}$ of the form each hour. Since they are working together and it takes t hours to do the whole thing, they complete $\frac{1}{t}$ of the form each hour. Here is what happens:

$$\begin{bmatrix} \text{work done by} \\ \text{first person} \end{bmatrix} + \begin{bmatrix} \text{work done by} \\ \text{second person} \end{bmatrix} = \begin{bmatrix} \text{work done} \\ \text{together} \end{bmatrix}$$

$$\frac{1}{8} \qquad + \qquad \frac{1}{6} \qquad = \qquad \frac{1}{t}$$

4. Use algebra to solve this equation. The LCD of the fractions is $24t$.

$$24t\left[\frac{1}{8} + \frac{1}{6}\right] = 24t \cdot \frac{1}{t} \qquad \text{Multiply by } 24t.$$

$$24t \cdot \frac{1}{8} + 24t \cdot \frac{1}{6} = 24t \cdot \frac{1}{t}$$

$$3t + 4t = 24 \qquad \text{Simplify.}$$

$$7t = 24$$

$$t = \frac{24}{7} = 3\frac{3}{7}$$

Thus, it takes $3\frac{3}{7}$ hr (about 3 hr 26 min) to complete the job. ▲

There is another type of problem that can be thought of as a work problem: tank or pool problems. The idea is that the pipes filling or emptying a tank or pool are doing the work to fill or empty the pool. Here is the way we do it.

EXAMPLE 3 A pool can be filled by an intake pipe in 4 hr and can be emptied by a drain pipe in 5 hr. How long would it take to fill the tank with both pipes open?

Solution

1. We are asked for the time it takes to fill the pool.
2. Let this time be T hours.
3. In 1 hr, the intake pipe fills $\frac{1}{4}$ of the pool, the drain pipe empties $\frac{1}{5}$, and together they fill $\frac{1}{T}$ of the pool. Thus, in 1 hr

$$\begin{bmatrix} \text{amount filled} \\ \text{by intake pipe} \end{bmatrix} - \begin{bmatrix} \text{amount emptied} \\ \text{by drain pipe} \end{bmatrix} = \begin{bmatrix} \text{amount} \\ \text{filled by both} \end{bmatrix}$$

$$\frac{1}{4} \qquad - \qquad \frac{1}{5} \qquad = \qquad \frac{1}{T}$$

Problem 2 An accountant can finish form 1040A in 5 hr. An assistant can do it in 10. How long would it take if they work together?

Problem 3 Repeat Example 3 if the intake pipe can fill the pool in 5 hr and the drain pipe can empty it in 6 hr.

4. The LCD is $20T$.

$$20T \cdot \frac{1}{4} - 20T \cdot \frac{1}{5} = 20T \cdot \frac{1}{T}$$
$$5T - 4T = 20$$
$$T = 20$$

Thus, it takes 20 hr to fill the pool if both the intake and drain pipes are open.

5. The intake pipe can fill the pool in 4 hr. It can then fill the pool five times in 20 hr. The drain pipe can empty the pool in 5 hr; thus it can empty the pool four times in 20 hr. Since the intake can fill the pool five times and the drain can empty it four times in 20 hr, the pool would be filled at the end of the 20 hr.

C. DISTANCE PROBLEMS

The ideas we have studied can be used to solve uniform motion problems like the ones discussed in Section 2.6. As you recall, when traveling at a constant rate R, the distance traveled in time T is given by $D = RT$. We use this information in Example 4.

EXAMPLE 4 The world's strongest current is the Saltstraumen in Norway, reaching 18 mi/hr. A speed boat can travel 48 mi downstream in the same time it takes to go 12 mi upstream. What is the speed of the boat in still water?

Solution

1. Read the problem. We want the speed (rate) of the boat in still water.
2. Let R be the rate in still water.
3. Make a chart with D, R, and T as headings.

	D	**R**	**T**
Downstream			
Upstream			

Speed downstream	$R + 18$	Current helps
Speed upstream	$R - 18$	Current hinders

The time T is given by $T = \frac{D}{R}$ (solving for T in $D = RT$). We then have

Time downstream $\dfrac{48}{R + 18}$

Time upstream $\dfrac{12}{R - 18}$

Enter this information in the chart.

Problem 4 Repeat Example 3 if the boat travels 36 mi downstream in the same time it takes it to go 12 mi upstream.

	D	**R**	**T**
Downstream	48	$R + 18$	$\dfrac{48}{R + 18}$
Upstream	12	$R - 18$	$\dfrac{12}{R - 18}$

Since it takes the same time to go upstream as downstream, we have

$$T(\text{up}) = T(\text{down})$$

$$\frac{48}{R + 18} = \frac{12}{R - 18}$$

4. The LCD is $\dfrac{(R + 18)(R - 18)}{1}$.

$$\frac{(R + 18)(R - 18)}{1} \cdot \frac{48}{R + 18} = \frac{(R + 18)(R - 18)}{1} \cdot \frac{12}{R - 18}$$

$$48(R - 18) = 12(R + 18)$$

$$4(R - 18) = R + 18 \qquad \text{Divide by 12.}$$

$$4R - 72 = R + 18 \qquad \text{Simplify.}$$

$$4R = R + 90 \qquad \text{Add 72.}$$

$$3R = 90 \qquad \text{Subtract } R.$$

$$R = 30$$

Thus, the speed of the boat in still water is 30 mi/hr.

5. The verification is left to the student.

NAME

CLASS

SECTION

A. Solve.

ANSWERS

1. The sum of a number and its reciprocal is $\frac{65}{8}$. Find the number.

1. _____

2. The sum of a number and its reciprocal is $\frac{50}{7}$. What is the number?

2. _____

3. One number is twice another. The sum of their reciprocals is $\frac{3}{10}$. Find the numbers.

3. _____

4. One number is three times another. The sum of their reciprocals is $\frac{1}{3}$. Find the numbers.

4. _____

5. Find two consecutive even integers the sum of whose reciprocals is $\frac{7}{24}$.

5. _____

6. Find two consecutive odd integers so that the sum of their reciprocals is $\frac{16}{63}$.

6. _____

7. The denominator of a fraction is 5 more than the numerator. If 3 is added to both numerator and denominator, the resulting fraction is $\frac{1}{2}$. Find the fraction.

7. _____

8. The numerator of a certain fraction is 4 less than the denominator. If the numerator is increased by 8 and the denominator by 35, the resulting fraction is $\frac{1}{2}$. Find the fraction.

8. _____

9. The current ratio of a business is defined by

$$\text{current ratio} = \frac{\text{current assets}}{\text{current liabilities}}$$

By how much can you increase \$40,000 in liabilities if the current assets amount to \$90.000 and you wish to maintain the current ratio at $\frac{3}{2}$?

9. _____

10. Repeat Problem 9 if you want the current ratio to be 2.

10. _____

B. Solve.

11. If one typist can finish a job in 3 hr while another typist can finish in 5 hr, how long will it take both of them working together to finish the job?

11. _____

12. A carpenter can finish a job in 8 hr, and another one can do it in 10 hr. How long will it take them to finish the job working together?

12. _____

13. The world record for riveting is 11,209 rivets in 9 hr, by J. Mair of Ireland. If another person can rivet 11,209 rivets in 10 hr, how long will it take both of them working together to rivet the 11,209 rivets?

13. _____

14. Mr. Gerry Harley of England shaved 130 men in 60 min. If another barber can shave all these men in 5 hr, how long will it take both of them working together to shave the 130 men?

14. _____

15. A printing press can print the evening paper in half the time another press takes to print it. Together, they can print the paper in 2 hr. How long will it take each of them to print the paper?

15. _____

16. A computer can do a job in 4 hr. With the help of a newer computer, the job is completed in 1 hr. How long will it take the newer computer to complete the job alone?

16. _____

17. A tank can be filled by an intake pipe in 9 hr and drained by another pipe in 21 hr. If both pipes are open, how long would it take to fill the tank?

17. _____

18. A faucet can fill a tank in 12 hr, and the drain pipe can empty it in 18 hr. If the faucet and the drain pipe are both open, how long would it take to fill the tank?

18. _____

19. A pipe can fill a pool in 7 hr, and another one can fill it in 21 hr. How long would it take to fill the pool using both pipes?

19. _____

20. One pipe can fill a tank in 6 hr, and another can fill it in 4 hr. How long will it take both pipes together to fill the tank?

20. _____

21. The main engine of a rocket can burn for 60 sec on the fuel in the rocket's tank, while the auxiliary engine can burn for 90 sec on the same amount of fuel. How long can both engines burn if they are operated together on the rocket's tank of fuel?

21. _____

22. An in-flow pipe can fill a pool in 12 hr, and another pipe can drain it in 4 hr. How long will it take to empty the pool if both pipes are open simultaneously? Assume that the pool is full at the start.

22. _____

23. A pipe can fill a tank in 9 hr, but the drain can empty it in 6 hr. How long will it take to empty the tank if both pipes are open simultaneously? Assume that the tank is full at the start.

23. _____

C. Solve.

24. A water skier travels 30 mi downstream in the same time it takes him to go 20 mi upstream. If the river current flows at 5 mi/hr, what is the skier's speed in still water?

24. _____

25. A small plane goes 240 mi against the wind in the same time it takes it to go 360 mi with a tail wind. If the wind velocity is 30 mi/hr, find the plane's speed in still air.

25. _____

26. A jet plane goes 700 mi against the wind in the same time it takes it to go 900 mi with a tail wind. If the wind velocity is 50 mi/hr, what is the plane's speed in still air?

26. _____

27. A small plane can cruise at 120 mi/hr in still air. It takes this plane the same time to move 270 mi against the wind as it does to go 450 mi with a tail wind. What is the wind velocity?

27. _____

28. A small plane can travel 200 mi against the wind in the same time it takes it to travel 260 mi with a tail wind. If the plane's speed in still air is 115 mi/hr, find the wind velocity.

29. An automobile travels 200 mi in the same time in which a small plane travels 1000 mi. Find their rates of speed if the airplane is 100 mi/hr faster than the automobile.

30. A runner ran 1000 m in the same time that another runner ran 950 m. If the speed of the slower runner was $\frac{1}{4}$ m per second less than that of the faster one, what was the speed of the faster runner?

28. _____

29. _____

30. _____

✓ **SKILL CHECKER**

Simplify.

31. $\dfrac{x^{-5}}{x^3}$

32. $\dfrac{x^5}{x^{-3}}$

33. $(x^{-4})^{-5}$

34. $(x^{-4})^5$

35. $(-2xy^2)^3$

36. $(-2x^2y)^{-3}$

37. $x^{-9} \cdot x^7$

38. $x^9 \cdot x^{-11}$

39. $\left(\dfrac{a^{-4}}{b^3}\right)^2$

40. $\left(\dfrac{a^4}{b^{-3}}\right)^{-2}$

31. _____

32. _____

33. _____

34. _____

35. _____

36. _____

37. _____

38. _____

39. _____

40. _____

4.7 USING YOUR KNOWLEDGE

In Section 2.5 we learned how to solve a formula for a specified variable. Many formulas involve rational expressions. You can use the knowledge gained in solving equations involving rational expressions to solve for specified variables in these equations.

1. The formula $\dfrac{1}{F} = \dfrac{1}{f_1} + \dfrac{1}{f_2}$ is used when finding the focal length of lenses. Solve for F.

2. To find the radius of curvature R of a sphere, we use the formula

$$R = \frac{2AS}{L - 2S}$$

Solve for A. Write the answer as a simple fraction.

3. The electric current i in a simple series circuit is given by

$$i = \frac{2E}{R + 2r}$$

Solve for R. Write the answer as a simple fraction.

1. _____

2. _____

3. _____

SUMMARY

SECTION	ITEM	MEANING	EXAMPLE
4.1	Fraction	An expression denoting a division	$\dfrac{3}{4}, \dfrac{-8}{7}$, and $\dfrac{1}{2}$ are fractions.
4.1	Rational expression	An expression of the form $\dfrac{P}{Q}$, where P and Q are polynomials, $Q \neq 0$	$\dfrac{1}{x}, \dfrac{x}{x+y}, \dfrac{x+y}{z}$, and $\dfrac{x^2 + 21x - 1}{x^3 + 3}$ are rational expressions.
4.1	Fundamental property of fractions	If P, Q, and K are polynomials, $\dfrac{P}{Q} = \dfrac{P \cdot K}{Q \cdot K}$ for all values for which the denominator is not 0.	$\dfrac{3}{4} = \dfrac{3 \cdot 8}{4 \cdot 8}, \dfrac{x}{2} = \dfrac{x \cdot 5}{2 \cdot 5}$, and $\dfrac{3}{x+2} = \dfrac{3(x+5)}{(x+2)(x+5)}$
4.1B	Standard form of a fraction	The forms $\dfrac{-a}{b}$ and $\dfrac{a}{b}$ are the standard form of a fraction.	$-\dfrac{5}{4}$ is written as $\dfrac{5}{4}$. $-\dfrac{-8}{-x}$ is written as $\dfrac{-8}{x}$.
4.1C	Reduced fraction	A fraction is reduced if the numerator and denominator have no common factor.	$\dfrac{3}{4}, \dfrac{9}{8}$, and $\dfrac{x}{7}$ are reduced but $\dfrac{3}{6}$ and $\dfrac{x+y}{x^2 - y^2}$ are not.
4.3A	Multiplication of rational expressions	If a, b, c, and d are rational expressions ($b \neq 0$, $d \neq 0$) $\dfrac{a}{b} \cdot \dfrac{c}{d} = \dfrac{a \cdot c}{b \cdot d}$	$\dfrac{3}{4} \cdot \dfrac{5}{7} = \dfrac{3 \cdot 5}{4 \cdot 7} = \dfrac{15}{28}$ $\dfrac{x}{x+y} \cdot \dfrac{3}{x-y} = \dfrac{3x}{(x+y)(x-y)} = \dfrac{3x}{x^2 - y^2}$
4.3B	Division of rational expressions	If a, b, c, and d are rational expressions ($b \neq 0$, $c \neq 0$, $d \neq 0$) $\dfrac{a}{b} \div \dfrac{c}{d} = \dfrac{a}{b} \cdot \dfrac{d}{c}$	$\dfrac{3}{4} \div \dfrac{7}{5} = \dfrac{3}{4} \cdot \dfrac{5}{7}$ $\dfrac{x}{x+y} \div \dfrac{x-y}{3} = \dfrac{x}{x+y} \cdot \dfrac{3}{x-y}$
4.4A	Addition and subtraction of rational expressions	If a, b, and c are rational expressions and $b \neq 0$, $\dfrac{a}{b} + \dfrac{c}{b} = \dfrac{a+c}{b}$ and $\dfrac{a}{b} - \dfrac{c}{b} = \dfrac{a-c}{b}$.	$\dfrac{3}{5} + \dfrac{1}{5} = \dfrac{4}{5}$ and $\dfrac{2}{x} + \dfrac{1}{x} = \dfrac{3}{x}$ $\dfrac{3}{5} - \dfrac{1}{5} = \dfrac{2}{5}$ and $\dfrac{2}{x} + \dfrac{1}{x} = \dfrac{2}{x}$

(Continued)

SECTION	ITEM	MEANING	EXAMPLE
4.4B	Prime number	A natural number greater than 1 with exactly two distinct divisors, itself and 1	2, 17, and 41 are prime.
4.5	Complex fraction	A fraction whose numerator or denominator (or both) contains other fractions	$\dfrac{\frac{1}{2}}{x+1}$, $\dfrac{3}{\frac{1}{5}+x}$, and $\dfrac{\frac{x}{2}}{x+\frac{1}{2}}$ are complex fractions.
4.5	Simple fractions	A fraction that is not complex	$\dfrac{1}{2}$, $\dfrac{2x}{4}$, and $\dfrac{x+y}{x-y}$ are simple fractions.
4.6	Extraneous solution	A trial solution that does not satisfy the equation	3 is an extraneous solution of $3+\dfrac{1}{x-3}=\dfrac{1}{x-3}$.
4.7	**RSTUV** method for solving word problems	**R**ead the problem. **S**elect a variable for the unknown. **T**ranslate. **U**se algebra to solve. **V**erify the answer.	

NAME

CLASS

SECTION

ANSWERS

(If you need help with these exercises, look in the section indicated in brackets.)

1. [4.1A] Write the fraction with the indicated denominator.

   a. $\dfrac{2x^2}{9y^4}$, denominator $36y^7$

   b. $\dfrac{2x^2}{9y^4}$, denominator $45y^8$

   c. $\dfrac{2x^2}{9y^4}$, denominator $54y^9$

   1. a. _____
      b. _____
      c. _____

2. [4.1A] Write the fraction with the indicated denominator.

   a. $\dfrac{2x+1}{x+1}$, denominator $x^2 + 5x + 4$

   b. $\dfrac{2x+1}{x+1}$, denominator $x^2 + 6x + 5$

   c. $\dfrac{2x+1}{x+1}$, denominator $x^2 + 7x + 6$

   2. a. _____
      b. _____
      c. _____

3. [4.1B] Write in standard form.

   a. $-\dfrac{-6}{y}$

   b. $-\dfrac{-7}{y}$

   c. $-\dfrac{-8}{y}$

   3. a. _____
      b. _____
      c. _____

4. [4.1B] Write in standard form.

   a. $-\dfrac{x-y}{6}$

   b. $-\dfrac{x-y}{7}$

   c. $-\dfrac{x-y}{8}$

   4. a. _____
      b. _____
      c. _____

5. [4.1C] Reduce to lowest terms.

a. $\dfrac{x^4 y^7}{xy^2}$

b. $\dfrac{x^4 y^8}{xy^2}$

c. $\dfrac{x^4 y^9}{xy^2}$

5. a. _____
   b. _____
   c. _____

6. [4.1C] Reduce to lowest terms.

a. $\dfrac{xy^2 + y^3}{x^2 - y^2}$

b. $\dfrac{xy^3 + y^4}{x^2 - y^2}$

c. $\dfrac{xy^4 + y^5}{x^2 - y^2}$

6. a. _____
   b. _____
   c. _____

7. [4.1C] Reduce to lowest terms.

a. $\dfrac{4y^2 - x^2}{x^3 + 8y^3}$

b. $\dfrac{4y^2 - x^2}{x^3 - 8y^3}$

c. $\dfrac{9y^2 - x^2}{x^3 - 27y^3}$

7. a. _____
   b. _____
   c. _____

8. [4.2A] Divide.

a. $\dfrac{18x^5 - 12x^3 + 6x^2}{6x^2}$

b. $\dfrac{18x^5 - 12x^3 + 6x^2}{6x^3}$

c. $\dfrac{18x^5 - 12x^3 + 6x^2}{6x^4}$

8. a. _____
   b. _____
   c. _____

9. [4.2B] Divide.

a. $2x^3 - 8 - 4x$ by $2 + 2x$
b. $2x^3 - 9 - 4x$ by $2 + 2x$
c. $2x^3 - 10 - 4x$ by $2 + 2x$

9. a. _____
   b. _____
   c. _____

10. [4.2C] Factor $x^3 - 6x^2 + 11x - 6$ if

a. $x - 1$ is one of its factors
b. $x - 2$ is one of its factors
c. $x - 3$ is one of its factors

10. a. _____
    b. _____
    c. _____

11. [4.3A] Multiply.

a. $\dfrac{x-2}{3x-2y} \cdot \dfrac{9x^2-4y^2}{3x^2-5x-2}$

b. $\dfrac{x-2}{3x-2y} \cdot \dfrac{9x^2-4y^2}{4x^2-7x-2}$

c. $\dfrac{x-2}{3x-2y} \cdot \dfrac{9x^2-4y^2}{5x^2-9x-2}$

11. a. _____
   b. _____
   c. _____

12. [4.3B] Divide.

a. $\dfrac{x+4}{x-2} \div (x^2+8x+16)$

b. $\dfrac{x+5}{x-2} \div (x^2+10x+25)$

c. $\dfrac{x+6}{x-2} \div (x^2+12x+36)$

12. a. _____
   b. _____
   c. _____

13. [4.3C] Perform the indicated operations.

a. $\dfrac{2-x}{x+3} \div \dfrac{x^3-8}{x+6} \cdot \dfrac{x^3+27}{x+6}$

b. $\dfrac{2-x}{x+3} \div \dfrac{x^3-8}{x+7} \cdot \dfrac{x^3+27}{x+7}$

c. $\dfrac{2-x}{x+3} \div \dfrac{x^3-8}{x+8} \cdot \dfrac{x^3+27}{x+8}$

13. a. _____
   b. _____
   c. _____

14. [4.4A] Perform the indicated operations.

a. $\dfrac{x}{x^2-4} + \dfrac{2}{x^2-4}$

b. $\dfrac{x}{x^2-9} + \dfrac{3}{x^2-9}$

c. $\dfrac{x}{x^2-16} + \dfrac{4}{x^2-16}$

14. a. _____
   b. _____
   c. _____

15. [4.4A] Perform the indicated operations.

a. $\dfrac{x}{x^2-9} - \dfrac{3}{x^2-9}$

b. $\dfrac{x}{x^2-16} - \dfrac{4}{x^2-16}$

c. $\dfrac{x}{x^2-25} - \dfrac{5}{x^2-25}$

15. a. _____
   b. _____
   c. _____

16. [4.4B] Perform the indicated operations.

a. $\dfrac{x+1}{x^2+x-2}+\dfrac{x+5}{x^2-1}$

b. $\dfrac{x+1}{x^2+x-2}+\dfrac{x+6}{x^2-1}$

c. $\dfrac{x+1}{x^2+x-2}+\dfrac{x+7}{x^2-1}$

16. a. _____
    b. _____
    c. _____

17. [4.4B] Perform the indicated operations.

a. $\dfrac{x-4}{x^2-x-6}-\dfrac{x+1}{x^2-9}$

b. $\dfrac{x-3}{x^2-x-6}-\dfrac{x+1}{x^2-9}$

c. $\dfrac{x-1}{x^2-x-6}-\dfrac{x+1}{x^2-9}$

17. a. _____
    b. _____
    c. _____

18. [4.5] Simplify.

a. $\dfrac{\dfrac{1}{x}+\dfrac{1}{x^4}}{\dfrac{1}{x}-\dfrac{1}{x^5}}$

b. $\dfrac{\dfrac{1}{x^2}+\dfrac{1}{x^5}}{\dfrac{1}{x^2}-\dfrac{1}{x^6}}$

c. $\dfrac{\dfrac{1}{x^3}+\dfrac{1}{x^6}}{\dfrac{1}{x^3}-\dfrac{1}{x^7}}$

18. a. _____
    b. _____
    c. _____

19. [4.5] Simplify.

a. $4+\dfrac{a}{4+\dfrac{4}{4+a}}$

b. $5+\dfrac{a}{5+\dfrac{5}{5+a}}$

c. $6+\dfrac{a}{6+\dfrac{6}{6+a}}$

19. a. _____
    b. _____
    c. _____

20. [4.6] Solve.

   a. $\dfrac{x}{x+4} - \dfrac{x}{x-4} = \dfrac{x^2+16}{x^2-16}$

   b. $\dfrac{x}{x+5} - \dfrac{x}{x-5} = \dfrac{x^2+25}{x^2-25}$

   c. $\dfrac{x}{x+6} - \dfrac{x}{x-6} = \dfrac{x^2+36}{x^2-36}$

20. a. _____
    b. _____
    c. _____

21. [4.6] Solve.

   a. $1 + \dfrac{4}{x-5} = \dfrac{40}{x^2-25}$

   b. $1 + \dfrac{5}{x-6} = \dfrac{60}{x^2-36}$

   c. $1 + \dfrac{6}{x-7} = \dfrac{84}{x^2-49}$

21. a. _____
    b. _____
    c. _____

22. [4.7A] Find two consecutive even integers such that the sum of their reciprocals is

   a. $\dfrac{11}{60}$

   b. $\dfrac{13}{84}$

   c. $\dfrac{15}{112}$

22. a. _____
    b. _____
    c. _____

23. [4.7B] Jack can paint a room in 4 hr. Find how long it would take to paint the room if he is helped by Jill, who can paint the same room in
   a. 5 hr
   b. 6 hr
   c. 7 hr

23. a. _____
    b. _____
    c. _____

24. [4.7C] A plane traveled 1200 mi with a 25-mi/hr tail wind. What is the plane's speed in still air if it took the plane the same time to travel the given mileage against the wind?
   a. 960 mi
   b. 1000 mi
   c. 1040 mi

24. a. _____
    b. _____
    c. _____

NAME _____

CLASS _____

SECTION _____

(Answers on pages 338–339)

ANSWERS

1. Write the fraction with the indicated denominator.

   a. $\dfrac{2x^2}{9y^4}$, denominator $36y^7$

   b. $\dfrac{2x+1}{x+1}$, denominator $x^2 + 4x + 3$

1. a. _____
   b. _____

2. Write in standard form.

   a. $-\dfrac{-5}{y}$

   b. $-\dfrac{x-y}{5}$

2. a. _____
   b. _____

3. Reduce to lowest terms.

   a. $\dfrac{x^4 y^6}{xy^2}$

   b. $\dfrac{xy + y^2}{x^2 - y^2}$

3. a. _____
   b. _____

4. Reduce $\dfrac{y^2 - x^2}{x^3 - y^3}$ to lowest terms.

4. _____

5. Divide $\dfrac{28x^5 - 14x^3 + 7x^2}{7x^3}$.

5. _____

6. Divide $2x^3 - 6 - 4x$ by $2 + 2x$.

6. _____

7. Factor $2x^3 + 3x^2 - 23x - 12$ if $x - 3$ is one cf its factors.

7. _____

8. Multiply $\dfrac{x-2}{3x-2y} \cdot \dfrac{9x^2 - 4y^2}{2x^2 - x - 6}$.

8. _____

9. Divide $\dfrac{x+3}{x-2} \div (x^2 + 6x + 9)$.

9. _____

10. Perform the indicated operations.

$\dfrac{2-x}{x+3} \div \dfrac{x^3 - 8}{x+5} \cdot \dfrac{x^3 + 27}{x+5}$

10. _____

11. Perform the indicated operations.

   a. $\dfrac{x}{x^2 - 1} + \dfrac{1}{x^2 - 1}$

   b. $\dfrac{x}{x^2 - 4} - \dfrac{2}{x^2 - 4}$

11. a. _____
    b. _____

12. Perform the indicated operations.

$\dfrac{x+1}{x^2 + x - 2} + \dfrac{x+4}{x^2 - 1}$

12. _____

13. Perform the indicated operations.

$\dfrac{x-5}{x^2 - x - 6} - \dfrac{x+1}{x^2 - 9}$

13. _____

14. Simplify $\dfrac{x + \dfrac{1}{x^2}}{x - \dfrac{1}{x^3}}$.

14. _____

15. Simplify $2 + \dfrac{a}{2 + \dfrac{2}{2+a}}$.

15. _____

16. Solve $\dfrac{x}{x+3} - \dfrac{x}{x-3} = \dfrac{x^2 + 9}{x^2 - 9}$.

16. _____

17. Solve $1 + \dfrac{3}{x-4} = \dfrac{24}{x^2 - 16}$.

17. _____

18. Find two consecutive even integers such that the sum of their reciprocals is $\frac{7}{24}$.

19. Jack can mow the lawn in 4 hr and Jill can mow it in 3. How long would it take to mow the lawn if both work together?

20. A plane traveled 990 mi with a 30-mi/hr tail wind in the same time it took to travel 810 mi against the wind. What is the plane's speed in still air?

18. _____

19. _____

20. _____

IF YOU MISSED QUESTION	SECTION	EXAMPLES	PAGE	ANSWERS
1a	4.1	1	261	1. a. $\dfrac{8x^2y^3}{36y^7}$
1b	4.1	1	261	b. $\dfrac{2x^2 + 7x + 3}{x^2 + 4x + 3}$
2a	4.1	2	262	2. a. $\dfrac{5}{y}$
2b	4.1	2	262	b. $\dfrac{y - x}{5}$
3a	4.1	3	263	3. a. x^3y^4
3b	4.1	3	263	b. $\dfrac{y}{x - y}$
4	4.1	4	264	4. $\dfrac{-(y + x)}{x^2 + xy + y^2}$
5	4.2	1	270	5. $4x^2 - 2 + \dfrac{1}{x}$
6	4.2	2, 3	271	6. $x^2 - x - 1 \text{ R } -4$
7	4.2	4	272	7. $(x - 3)(x + 4)(2x + 1)$
8	4.3	1, 2	278–279	8. $\dfrac{3x + 2y}{2x + 3}$
9	4.3	3	280	9. $\dfrac{1}{x^2 + x - 6}$
10	4.3	4, 5	280–281	10. $\dfrac{-(x^2 - 3x + 9)}{x^2 + 2x + 4}$
11a	4.4	1	288	11. a. $\dfrac{1}{x - 1}$
11b	4.4	1	288	b. $\dfrac{1}{x + 2}$
12	4.4	2, 3	289–291	12. $\dfrac{2x^2 + 8x + 9}{(x + 2)(x + 1)(x - 1)}$
13	4.4	2, 3	289–291	13. $\dfrac{-5x - 17}{(x + 2)(x + 3)(x - 3)}$
14	4.5	1, 2	298–299	14. $\dfrac{x(x^2 - x + 1)}{(x^2 + 1)(x - 1)}$
15	4.5	3, 4	299–301	15. $\dfrac{a^2 + 6a + 12}{2a + 6}$
16	4.6	1, 2	308–310	16. No solution
17	4.6	3, 4	311–312	17. -7
18	4.7	1	318	18. 6 and 8

IF YOU MISSED QUESTION	SECTION	EXAMPLES	PAGE	ANSWERS
19	4.7	2	319	**19.** $1\dfrac{5}{7}$ hr
20	4.7	3	320–321	**20.** 300 mi/hr

CHAPTER 5

RATIONAL EXPONENTS AND RADICALS

In this chapter we study square, cube, fourth, and higher roots and the operations of addition, subtraction, multiplication, and division involving such roots. We introduce rational exponents and radicals to write these expressions and use these ideas to solve equations involving radicals. We end the chapter by introducing a new class of numbers, complex numbers.

(*Answers on pages 344–345*)

1. Find, if possible.
   a. $\sqrt[3]{-27}$
   b. $\sqrt{-16}$

2. Find, if possible.
   a. $(-8)^{1/3}$
   b. $\left(\dfrac{1}{81}\right)^{1/4}$

3. Evaluate.
   a. $27^{2/3}$
   b. $(-16)^{3/2}$

4. Evaluate.
   a. $(-8)^{-2/3}$
   b. $27^{-2/3}$

5. If x and y are positive, simplify.
   a. $x^{1/2} \cdot x^{1/3}$
   b. $\dfrac{x^{-1/3}}{x^{1/5}}$

6. If x and y are positive, simplify.
   a. $(x^{1/4}y^{2/5})^{-20}$
   b. $x^{3/4}(x^{-1/4} + y^{2/5})$

7. Simplify.
   a. $\sqrt[4]{(-3)^4}$
   b. $\sqrt[6]{(-x)^6}$

8. Simplify.
   a. $\sqrt[3]{128}$
   b. $\sqrt[3]{54a^5b^9}$

9. Simplify.
   a. $\sqrt{\dfrac{5}{32}}$
   b. $\sqrt[3]{\dfrac{6}{x^3}}$

ANSWERS

1. a. _____
   b. _____

2. a. _____
   b. _____

3. a. _____
   b. _____

4. a. _____
   b. _____

5. a. _____
   b. _____

6. a. _____
   b. _____

7. a. _____
   b. _____

8. a. _____
   b. _____

9. a. _____
   b. _____

10. Rationalize the denominator.

   a. $\dfrac{\sqrt{5}}{\sqrt{7}}$

   b. $\dfrac{\sqrt{2}}{\sqrt{3x}}, x > 0$

11. Rationalize the denominator.

   a. $\dfrac{1}{\sqrt[3]{2x}}$

   b. $\dfrac{\sqrt[5]{3}}{\sqrt[5]{16x^3}}$

12. Reduce the order (index).

   a. $\sqrt[4]{\dfrac{81}{16}}$

   b. $\sqrt[6]{16c^4d^4}$

13. Simplify $\sqrt[6]{\dfrac{2a^2}{32c^{16}}}$.

14. Perform the indicated operations.
   a. $\sqrt{98} + \sqrt{32}$
   b. $\sqrt{175} - \sqrt{28}$

15. Perform the indicated operations.

   a. $2\sqrt{\dfrac{1}{2}} - 3\sqrt{\dfrac{1}{8}}$

   b. $4\sqrt[3]{\dfrac{3}{4x}} - \sqrt[3]{\dfrac{3}{32x}}$

16. Perform the indicated operations.
   a. $\sqrt{2}(\sqrt{8} + \sqrt{3})$
   b. $\sqrt[3]{2x}(\sqrt[3]{4x^2} - \sqrt[3]{16x})$

17. Find the product.
   a. $(\sqrt{27} + \sqrt{50})(\sqrt{12} + \sqrt{8})$
   b. $(\sqrt{3} + 2)(\sqrt{3} + 2)$

18. Find the product.
   a. $(2 - \sqrt{3})^2$
   b. $(\sqrt{5} + \sqrt{3})(\sqrt{5} - \sqrt{3})$

19. Reduce $\dfrac{10 - \sqrt{50}}{5}$.

20. Rationalize the denominator in $\dfrac{\sqrt{y}}{\sqrt{x} + \sqrt{y}}$.

21. Solve $\sqrt{x + 2} = -1$.

22. Solve $\sqrt{x + 2} = x - 4$.

10. a. _____
    b. _____

11. a. _____
    b. _____

12. a. _____
    b. _____

13. _____

14. a. _____
    b. _____

15. a. _____
    b. _____

16. a. _____
    b. _____

17. a. _____
    b. _____

18. a. _____
    b. _____

19. _____

20. _____

21. _____

22. _____

23. Solve $\sqrt{x-2} - x = -2$.

23. _____

24. Solve $\sqrt{x-7} - \sqrt{x} = -1$.

24. _____

25. Solve $\sqrt[3]{x-3} = 2$.

25. _____

26. Write the given expression in terms of i.
    a. $\sqrt{-49}$
    b. $\sqrt{-162}$

26. **a.** _____
    b. _____

27. Find.
    a. $(3 + 5i) + (6 - 14i)$
    b. $(4 + 2i) - (7 - 8i)$

27. **a.** _____
    b. _____

28. Multiply.
    a. $(3 + 2i)(4 - 3i)$
    b. $\sqrt{-9}(4 - \sqrt{-8})$

28. **a.** _____
    b. _____

29. Find.

    a. $\dfrac{4 + 5i}{2 + 3i}$

    b. $\dfrac{4 - 2i}{3 - 5i}$

29. **a.** _____
    b. _____

30. Write the answer as 1, -1, i, or $-i$.
    a. i^{55}
    b. i^{-7}

30. **a.** _____
    b. _____

IF YOU MISSED QUESTION	SECTION	EXAMPLES	PAGE	ANSWERS
1a	5.1	1	348	**1. a.** -3
1b	5.1	1	348	**b.** Not a real number
2a	5.1	2	349	**2. a.** -2
2b	5.1	2	349	**b.** $\dfrac{1}{3}$
3a	5.1	3	350	**3. a.** 9
3b	5.1	3	350	**b.** Not a real number
4a	5.1	4	351	**4. a.** $\dfrac{1}{4}$
4b	5.1	4	351	**b.** $\dfrac{1}{9}$
5a	5.1	5	351	**5. a.** $x^{5/6}$
5b	5.1	5	351	**b.** $\dfrac{1}{x^{8/15}}$
6a	5.1	6	352	**6. a.** $\dfrac{1}{x^5 y^8}$
6b	5.1	6	352	**b.** $x^{1/2} + x^{3/4} y^{2/5}$
7a	5.2	1	360	**7. a.** 3
7b	5.2	1	360	**b.** $|-x|$
8a	5.2	2	360–361	**8. a.** $4\sqrt[3]{2}$
8b	5.2	2	360–361	**b.** $3ab^3 \sqrt[3]{2a^2}$
9a	5.2	3	361	**9. a.** $\dfrac{\sqrt{10}}{8}$
9b	5.2	3	361	**b.** $\dfrac{\sqrt[3]{6}}{x}$
10a	5.2	4	362	**10. a.** $\dfrac{\sqrt{35}}{7}$
10b	5.2	4	362	**b.** $\dfrac{\sqrt{6x}}{3x}$
11a	5.2	5	363	**11. a.** $\dfrac{\sqrt[3]{4x^2}}{2x}$
11b	5.2	5	363	**b.** $\dfrac{\sqrt[5]{6x^2}}{2x}$
12a	5.2	6	364	**12. a.** $\dfrac{3}{2}$
12b	5.2	6	364	**b.** $\sqrt[3]{4c^2 d^2}$

IF YOU MISSED QUESTION	SECTION	EXAMPLES	PAGE	ANSWERS
13	5.2	7	364	**13.** $\dfrac{\sqrt[3]{2ac}}{2c^3}$
14a	5.3	1	370	**14. a.** $11\sqrt{2}$
14b	5.3	1	370	**b.** $3\sqrt{7}$
15a	5.3	2	370–371	**15. a.** $\dfrac{\sqrt{2}}{4}$
15b	5.3	2	370–371	**b.** $\dfrac{7\sqrt[3]{6x^2}}{4x}$
16a	5.3	3	372	**16. a.** $4 + \sqrt{6}$
16b	5.3	3	372	**b.** $2x - 2\sqrt[3]{4x^2}$
17a	5.3	4, 5	372–373	**17. a.** $38 + 16\sqrt{6}$
17b	5.3	4, 5	372–373	**b.** $7 + 4\sqrt{3}$
18a	5.3	5	373	**18. a.** $7 - 4\sqrt{3}$
18b	5.3	5	373	**b.** 2
19	5.3	6	373	**19.** $2 - \sqrt{2}$
20	5.3	7	374–375	**20.** $\dfrac{\sqrt{xy} - y}{x - y}$
21	5.4	1	382	**21.** No real-number solution
22	5.4	1	382	**22.** 7
23	5.4	1	382	**23.** 2 or 3
24	5.4	2	383	**24.** 16
25	5.4	3	384	**25.** 11
26a	5.5	1	389	**26. a.** $7i$
26b	5.5	1	389	**b.** $9\sqrt{2}i$
27a	5.5	2	390	**27. a.** $9 - 9i$
27b	5.5	2	390	**b.** $-3 + 10i$
28a	5.5	3	391	**28. a.** $18 - i$
28b	5.5	4	391–392	**b.** $6\sqrt{2} + 12i$
29a	5.5	5	393	**29. a.** $\dfrac{23}{13} - \dfrac{2}{13}i$
29b	5.5	5	393	**b.** $\dfrac{11}{17} + \dfrac{7}{17}i$
30a	5.5	6	394	**30. a.** $-i$
30b	5.5	6	394	**b.** i

5.1 RATIONAL EXPONENTS

Have you noticed that the speed limit on a curve is lower than on a straight road? The velocity v (miles per hour) that a car can travel on a curved concrete highway of radius r without skidding is $v = \sqrt{9r}$. If the radius r of the curve is 100 ft, the velocity is $\sqrt{900}$ (read "the square root of 900"). The square root of 900 is the number whose square is 900. Since $(30)^2 = 900$, the square root of 900 is 30. Similarly, the square root of 25 is 5 (since $5^2 = 25$) and the square root of 36 is 6 (since $6^2 = 36$). In general, we have the following definition:

> If a and x are real numbers and n is a positive integer, x is an nth root of a if $x^n = a$.

For example,

1. A square (second) root of 4 is 2 because $2^2 = 4$.
2. Another square root of 4 is -2 because $(-2)^2 = 4$.
3. A cube (third) root of 27 is 3 because $(3)^3 = 27$.
4. A cube (third) root of -64 is -4 because $(-4)^3 = -64$.
5. A fourth root of $\frac{16}{81}$ is $\frac{2}{3}$ because $(\frac{2}{3})^4 = \frac{16}{81}$.

Note that 4 has two square roots, 2 and -2, and $\frac{16}{81}$ has two fourth real roots, $\frac{-2}{3}$ and $\frac{2}{3}$. To avoid this situation, we introduce the idea of the principal nth root.

A. RADICALS

> If n is a positive integer, then $\sqrt[n]{a}$ denotes the principal nth root of a, and:
>
> 1. If $a > 0$, $\sqrt[n]{a}$ is the *positive* nth root of a.
>
> 2. If $a < 0$ and n is odd, $\sqrt[n]{a}$ is the *negative* nth root of a.
>
> 3. $\sqrt[n]{0} = 0$.

In this definition, $\sqrt[n]{a}$ is called a **radical expression,** $\sqrt{}$ is called the **radical sign,** a is the **radicand,** and n is the **index** that tells you what root is being considered. By convention, the index 2 for square root is understood but not written. Thus $\sqrt{9}$ means the principal square root of 9 and $\sqrt{25}$ means the principal square root of 25.

Now, let us look at part 1 of the definition. It tells us that whenever the number under the radical sign is positive, the resulting root is also positive.

$$\sqrt{64} = 8 \quad \text{because} \quad 8^2 = 64$$
$$\sqrt[3]{8} = 2 \quad \text{because} \quad 2^3 = 8$$
$$\sqrt[4]{81} = 3 \quad \text{because} \quad 3^4 = 81$$

A common mistake is to assume that $\sqrt{64}$ has two values. This is not correct. By our definition of *principal* root, $\sqrt{64} = 8$. If we wish to refer to the negative nth root, we write $-\sqrt[n]{a}$. Thus $-\sqrt{64} = -8$, $-\sqrt{16} = -4$, and $-\sqrt{25} = -5$. Part 2 of the definition tells us that if a is *negative* and n is *odd*, $\sqrt[n]{a}$ is negative.

$$\sqrt[3]{-27} = -3 \quad \text{because} \quad (-3)^3 = -27$$
$$\sqrt[5]{-32} = -2 \quad \text{because} \quad (-2)^5 = -32$$

What about $\sqrt{-16}$? There is no real number whose square is -16 because the square of a nonzero number is positive. Thus $\sqrt{-16}$ is not a real number. Similarly, $\sqrt[4]{-81}$ is not a real number, since there is no real number whose fourth power is -81. Note that in both cases n (the index) is even and the radicand is negative. This situation is not covered in the definition.

EXAMPLE 1 Find if possible.

a. $\sqrt[3]{-64}$ **b.** $\sqrt{-64}$ **c.** $\sqrt[3]{\left(\dfrac{-1}{8}\right)}$

Problem 1 Find, if possible.

a. $\sqrt{-25}$ **b.** $\sqrt[3]{-125}$ **c.** $\sqrt[3]{\dfrac{-1}{27}}$

Solution

a. $\sqrt[3]{-64} = -4$, since $(-4)^3 = -64$.

b. $\sqrt{-64}$ is not a real number. Note that $\sqrt{-64} \neq -8$, since $(-8)(-8) = 64$ and not -64.

c. $\sqrt[3]{\left(\dfrac{-1}{8}\right)} = \dfrac{-1}{2}$, since $\left(\dfrac{-1}{2}\right)^3 = \dfrac{-1}{8}$. ▲

Here is a summary of our work so far.

If a is a real number and n is a positive integer		
	n even	n odd
For $a > 0$	$\sqrt[n]{a}$ is positive.	$\sqrt[n]{a}$ is positive.
For $a < 0$	$\sqrt[n]{a}$ is not a real number.	$\sqrt[n]{a}$ is negative.
For $a = 0$	$\sqrt[n]{a} = 0$	$\sqrt[n]{a} = 0$

B. FROM RADICALS TO RATIONAL EXPONENTS

We have used radicals to define the nth root of a number. We can also define nth roots using rational exponents. For example, what do you think $a^{1/3}$ means? To find out, let

$$x = a^{1/3}$$
$$x^3 = (a^{1/3})^3 \qquad \text{Cube both sides.}$$
$$\quad = a^{(1/3) \cdot (3)} \qquad \text{Assume } (a^{1/n})^n = a^{(1/n) \cdot (n)}.$$
$$\quad = a^1$$
$$\quad = a$$

Since $x^3 = a$, x must be the cube root of a, that is, $x = \sqrt[3]{a}$. But $x = a^{1/3}$, so

$$a^{1/3} = \sqrt[3]{a}$$

Similarly, if $\sqrt[4]{a}$ is defined,

$$a^{1/4} = \sqrt[4]{a}$$

In general, we have the following definition.

If n is a positive integer and $\sqrt[n]{a}$ is a real number, then
$$a^{1/n} = \sqrt[n]{a}$$

Note that when we write $a^{1/n} = \sqrt[n]{a}$, the denominator of the rational exponent is the index of the radical:

$$16^{1/2} = \sqrt[2]{16} \qquad \text{2 is understood.}$$
$$(-8)^{1/3} = \sqrt[3]{-8} = -2$$
$$\left(\frac{1}{81}\right)^{1/4} = \sqrt[4]{\frac{1}{81}} = \frac{1}{3}$$

EXAMPLE 2 Find.

a. $9^{1/2}$ **b.** $(-125)^{1/3}$ **c.** $\left(\frac{1}{16}\right)^{1/4}$

Solution

a. $9^{1/2} = \sqrt{9} = 3$

b. $(-125)^{1/3} = \sqrt[3]{-125} = -5$

c. $\left(\frac{1}{16}\right)^{1/4} = \sqrt[4]{\left(\frac{1}{16}\right)} = \frac{1}{2}$ ▲

So far, we have defined rational exponents of the form $1/n$. How shall we define $a^{m/n}$, where m and n are positive integers with $n > 1$ and $\sqrt[n]{a}$ a real number? If we assume that $(a^m)^n = a^{m \cdot n}$ then

$$a^{m/n} = (a^{1/n})^m = (a^m)^{1/n} = (\sqrt[n]{a})^m = \sqrt[n]{a^m}$$

From this we arrive at the following definition.

Problem 2 Find.

a. $49^{1/2}$ **b.** $(-216)^{1/3}$ **c.** $\left(\frac{1}{81}\right)^{1/4}$

$a^{m/n} = (\sqrt[n]{a})^m = \sqrt[n]{a^m}$, provided m and n are positive integers with no common factors and $\sqrt[n]{a}$ is a real number.

Note that the numerator of the exponent m/n is the exponent of the radical expression, and the denominator is the index of the radical—that is,

$$\overset{\text{exponent}}{\underset{\text{index}}{a^{m/n} = (\sqrt[n]{a})^m}}$$

For example, $a^{1/5} = \sqrt[5]{a}$ and $a^{2/5} = (\sqrt[5]{a})^2$.

EXAMPLE 3 Evaluate.

a. $8^{2/3}$ **b.** $(-27)^{2/3}$ **c.** $(-25)^{3/2}$

Solution

a. $8^{2/3}$ can be evaluated in two ways.

1. Since $\sqrt[3]{8} = 2$, $8^{2/3} = (\sqrt[3]{8})^2 = (2)^2 = 4$.
2. $8^{2/3} = \sqrt[3]{8^2} = \sqrt[3]{64} = 4$.

b. We can evaluate $(-27)^{2/3}$ in a similar manner.

1. Since $\sqrt[3]{-27} = -3$,

$$(-27)^{2/3} = (\sqrt[3]{-27})^2$$
$$= (-3)^2$$
$$= 9$$

2. $(-27)^{2/3} = \sqrt[3]{(-27)^2} = \sqrt[3]{729} = 9$

c. $(-25)^{3/2} = \sqrt[2]{(-25)^3} = \sqrt{-125}$, which is not a real number. ▲

Finally, to define negative rational exponents, we first note that if m and n are positive integers with no common factors,

$$-\frac{m}{n} = \frac{-m}{n}$$

Thus if $(a^m)^n = a^{m \cdot n}$,

$$a^{-m/n} = (a^{1/n})^{-m}$$
$$= \frac{1}{(a^{1/n})^m} \quad \text{Since } m \text{ is negative}$$

Thus we make the following definition.

$a^{-m/n} = \dfrac{1}{(a^{m/n})}$ m and n positive integers, $a^{1/n}$ a real number, $a \neq 0$.

Using this definition, we have the following.

1. $a^{-1/2} = \dfrac{1}{\sqrt{a}}$

Problem 3 Evaluate.
a. $27^{2/3}$ **b.** $(-64)^{2/3}$ **c.** $(-36)^{3/2}$

ANSWERS (to problems on pages 348–349)

1. **a.** Not a real number **b.** -5 **c.** $\dfrac{-1}{3}$

2. **a.** 7 **b.** -6 **c.** $\dfrac{1}{3}$

2. $32^{-3/5} = \dfrac{1}{32^{3/5}} = \dfrac{1}{(\sqrt[5]{32})^3} = \dfrac{1}{2^3} = \dfrac{1}{8}$

3. $1000^{-2/3} = \dfrac{1}{1000^{2/3}} = \dfrac{1}{(\sqrt[3]{1000})^2} = \dfrac{1}{10^2} = \dfrac{1}{100}$

EXAMPLE 4   Evaluate.

a. $16^{-3/4}$ **b.** $(-8)^{-4/3}$ **c.** $125^{-2/3}$

Solution

a. $16^{-3/4} = \dfrac{1}{16^{3/4}} = \dfrac{1}{(\sqrt[4]{16})^3} = \dfrac{1}{2^3} = \dfrac{1}{8}$

b. $(-8)^{-4/3} = \dfrac{1}{(-8)^{4/3}} = \dfrac{1}{(\sqrt[3]{-8})^4} = \dfrac{1}{(-2)^4} = \dfrac{1}{16}$

c. $125^{-2/3} = \dfrac{1}{125^{2/3}} = \dfrac{1}{(\sqrt[3]{125})^2} = \dfrac{1}{5^2} = \dfrac{1}{25}$

Problem 4   Evaluate.
a. $81^{-3/4}$ **b.** $(-27)^{-4/3}$
c. $216^{-2/3}$

C.   OPERATIONS WITH RATIONAL EXPONENTS

The properties of exponents that we studied in Chapter 1 can be extended to rational exponents. If this is done, we have the following results.

> Let r, s, and t be rational numbers. If a and b are real numbers for which the indicated expressions exist:
>
> I. $a^r \cdot a^s = a^{r+s}$ II. $\dfrac{a^r}{a^s} = a^{r-s}$
>
> III. $(a^r)^s = a^{r \cdot s}$ IV. $(a^r a^s)^t = a^{rt} a^{st}$
>
> V. $\left(\dfrac{a^r}{a^s}\right)^t = \dfrac{a^{rt}}{a^{st}}$

EXAMPLE 5   If x and y are positive, simplify.

a. $x^{1/3} \cdot x^{1/4}$ **b.** $\dfrac{x^{-2/3}}{x^{1/5}}$ **c.** $(y^{3/5})^{-1/6}$

Solution

a. $x^{1/3} \cdot x^{1/4} = x^{1/3+1/4}$ Law I

$\qquad\qquad\quad = x^{4/12+3/12}$ The LCD is 12.

$\qquad\qquad\quad = x^{7/12}$ Add exponents.

b. $\dfrac{x^{-2/3}}{x^{1/5}} = x^{-2/3-1/5}$ Law II

$\qquad\quad = x^{-10/15-3/15}$ The LCD is 15.

$\qquad\quad = x^{-13/15}$

$\qquad\quad = \dfrac{1}{x^{13/15}}$ $a^{-n} = \dfrac{1}{a^n}$

Problem 5   If x and y are positive, simplify.

a. $x^{1/3} \cdot x^{1/5}$ **b.** $\dfrac{x^{-2/3}}{x^{1/4}}$

c. $(y^{3/4})^{-1/5}$

c. $(y^{3/5})^{-1/6} = y^{(3/5) \cdot (-1/6)}$ Law III

$$= y^{-1/10} \qquad \frac{3}{5} \cdot \left(-\frac{1}{6}\right) = -\frac{1}{10}$$

$$= \frac{1}{y^{1/10}}$$

EXAMPLE 6 If x and y are positive, simplify.

a. $(x^{1/5}y^{3/4})^{-20}$ **b.** $\dfrac{x^{1/3}y^{3/4}}{x^{2/3}y^{1/4}}$ **c.** $x^{2/3}(x^{-1/3} + y^{1/5})$

Solution

a. $(x^{1/5}y^{3/4})^{-20} = (x^{1/5})^{-20} \cdot (y^{3/4})^{-20}$ Law IV

$$= x^{(1/5)(-20)} \cdot (y^{3/4})^{-20}$$ Law III

$$= x^{-4} \cdot y^{-15}$$

$$= \frac{1}{x^4 y^{15}}$$ Definition of negative exponent

b. $\dfrac{x^{1/3}y^{3/4}}{x^{2/3}y^{1/4}} = x^{1/3 - 2/3} \cdot y^{3/4 - 1/4}$ Law II

$$= x^{-1/3} \cdot y^{1/2}$$

$$= \frac{y^{1/2}}{x^{1/3}}$$

c. We first use the distributive law and then simplify.

$$x^{2/3}(x^{-1/3} + y^{1/5}) = x^{2/3} \cdot x^{-1/3} + x^{2/3} \cdot y^{1/5}$$

$$= x^{2/3 + (-1/3)} + x^{2/3}y^{1/5}$$

$$= x^{1/3} + x^{2/3}y^{1/5} \qquad \blacktriangle$$

The laws of exponents for rational numbers provide a good way of simplifying some of the expressions we have studied. In Example 3 we evaluated $(8^{2/3})$ and $(-27)^{2/3}$. If we write 8 as 2^3, we can write

$$8^{2/3} = (2^3)^{2/3}$$

$$= 2^{(3) \cdot (2/3)}$$ Law III

$$= 2^2$$

$$= 4$$

Similarly, since $-27 = (-3)^3$,

$$(-27)^{2/3} = [(-3)^3]^{2/3}$$ Law III

$$= (-3)^2$$

$$= 9$$

One word of caution, however. We cannot evaluate $[(-2)^2]^{1/2}$ using law III. If we do, we obtain

$$[(-2)^2]^{1/2} = (-2)^{(2)(1/2)}$$

$$= (-2)^1$$

$$= -2$$

If we use the order of operations we have studied and square -2 first, we have

$$[(-2)^2]^{1/2} = [4]^{1/2}$$

$$= 2$$

Problem 6 If x and y are positive, simplify.

a. $(x^{1/4}y^{3/5})^{-20}$ **b.** $\dfrac{x^{1/4}y^{2/3}}{x^{3/4}y^{1/3}}$

c. $x^{2/5}(x^{-1/5} + y^{1/4})$

The problem is that $a^{m/n}$ was defined provided m and n have no common factors. In this case, $m = 2$ and $n = 2$, which have 2 as a common factor. To remedy this situation, we make the following definition.

If m and n are positive even integers, then
$$(a^m)^{1/n} = |a|^{m/n}$$

Thus, $[(-2)^2]^{1/2} = |2|^{2/2} = |2|^1 = 2$. Note that $[(-2)^{1/2}]^2$ is not defined, since $(-2)^{1/2} = \sqrt{-2}$, which is not a real number.

ANSWER (to problem on page 352)

6. a. $\dfrac{1}{x^5 y^{12}}$ b. $\dfrac{y^{1/3}}{x^{1/2}}$ c. $x^{1/5} + x^{2/5} y^{1/4}$

EXERCISE 5.1

A. In Problems 1–12 evaluate if possible.

1. $\sqrt{4}$

2. $\sqrt{25}$

3. $\sqrt[3]{8}$

4. $\sqrt[3]{125}$

5. $\sqrt[3]{-8}$

6. $\sqrt[3]{-125}$

7. $\sqrt[3]{\dfrac{-1}{64}}$

8. $\sqrt[3]{\dfrac{-1}{27}}$

9. $\sqrt[4]{16}$

10. $\sqrt[4]{625}$

11. $\sqrt[5]{32}$

12. $\sqrt[5]{\dfrac{-1}{243}}$

B. In Problems 13–40 evaluate if possible.

13. $9^{1/2}$

14. $16^{1/2}$

15. $(-4)^{1/2}$

16. $-4^{1/2}$

17. $27^{1/3}$

18. $125^{1/3}$

19. $81^{1/4}$

20. $16^{1/4}$

21. $\left(\dfrac{-1}{8}\right)^{1/3}$

22. $\left(\dfrac{-1}{27}\right)^{1/3}$

23. $\left(\dfrac{-1}{256}\right)^{1/4}$

24. $\left(\dfrac{1}{256}\right)^{1/4}$

25. $27^{2/3}$

26. $(-27)^{2/3}$

27. $125^{2/3}$

28. $216^{2/3}$

29. $\left(\dfrac{1}{8}\right)^{2/3}$

30. $\left(\dfrac{1}{81}\right)^{3/4}$

1. _____
2. _____
3. _____
4. _____
5. _____
6. _____
7. _____
8. _____
9. _____
10. _____
11. _____
12. _____
13. _____
14. _____
15. _____
16. _____
17. _____
18. _____
19. _____
20. _____
21. _____
22. _____
23. _____
24. _____
25. _____
26. _____
27. _____
28. _____
29. _____
30. _____

31. $(-8)^{4/3}$

32. $(-27)^{4/3}$

33. $(32)^{4/5}$

34. $(-32)^{4/5}$

35. $-32^{4/5}$

36. $(-64)^{5/3}$

37. $64^{-2/3}$

38. $27^{-2/3}$

39. $[(-7)^4]^{1/4}$

40. $[(-11)^6]^{1/6}$

C. In Problems 41–70 simplify and write the expression as a product or quotient with positive exponents. All letters represent positive numbers.

41. $x^{1/7} \cdot x^{2/7}$

42. $y^{1/6} \cdot y^{1/6}$

43. $x^{-1/9} \cdot x^{-4/9}$

44. $y^{-5/2} \cdot y^{-3/2}$

45. $\dfrac{x^{4/5}}{x^{2/5}}$

46. $\dfrac{y^{5/7}}{y^{2/7}}$

47. $\dfrac{z^{2/3}}{z^{-1/3}}$

48. $\dfrac{a^{4/5}}{a^{-3/5}}$

49. $(x^{1/5})^{10}$

50. $(y^{1/3})^{12}$

51. $(z^{1/3})^{-6}$

52. $(a^{1/4})^{-8}$

53. $(b^{2/3})^{-6/5}$

54. $(c^{2/7})^{-7/8}$

55. $(a^{2/3}b^{3/4})^{-12}$

56. $(x^{1/8}y^{2/3})^{-24}$

57. $\left(\dfrac{a^{2/3}}{b^{3/5}}\right)^{-15}$

58. $\left(\dfrac{x^{1/2}}{y^{3/5}}\right)^{-20}$

59. $\left(\dfrac{x^{-2/5}}{y^{3/4}}\right)^{-40}$

60. $\left(\dfrac{x^{-1/3}}{y^{3/8}}\right)^{-48}$

61. $x^{1/3}(x^{2/3} + y^{1/2})$

62. $x^{-4/5}(y^{1/3} + x^{-1/5})$

63. $y^{3/4}(x^{1/2} - y^{1/2})$

64. $y^{2/3}(y^{1/2} - x^{2/3})$

31. _____

32. _____

33. _____

34. _____

35. _____

36. _____

37. _____

38. _____

39. _____

40. _____

41. _____

42. _____

43. _____

44. _____

45. _____

46. _____

47. _____

48. _____

49. _____

50. _____

51. _____

52. _____

53. _____

54. _____

55. _____

56. _____

57. _____

58. _____

59. _____

60. _____

61. _____

62. _____

63. _____

64. _____

65. $\dfrac{x^{1/6} \cdot x^{-5/6}}{x^{1/3}}$

66. $\dfrac{(x^{1/3} \cdot x^{1/2})^2}{x^{1/2}}$

67. $\dfrac{(x^{1/3} \cdot y^{-1/2})^6}{(y^{1/2})^{-4}}$

68. $\left(\dfrac{x^{4/3} \cdot y^{1/2}}{x^{1/3}}\right)^{-1/2}$

69. $\dfrac{(x^{1/4} \cdot y^2)^4}{(x^{2/3} \cdot y)^{-3}}$

70. $\left(\dfrac{-8a^{-3}b^{12}}{c^{15}}\right)^{-1/3}$

65. _____

66. _____

67. _____

68. _____

69. _____

70. _____

D. Applications

If air resistance is neglected, the terminal velocity v of a falling body is given by

$v = (20h + v_0)^{1/2}$ meters per second (m/sec)

71. Find v if $h = 10$ and $v_0 = 25$ m/sec.

72. Find v if a body is dropped ($v_0 = 0$) from a height of 45 m.

73. If the velocity as measured in feet per second (ft/sec) is

$v = (64h + v_0)^{1/2}$

Find v if $h = 12$ ft and $v_0 = 16$ ft/sec.

74. Find v if a body is dropped ($v_0 = 0$) from a height of 25 ft.

In Problems 75–80 evaluate $\sqrt{b^2 - 4ac}$.

75. $a = 1, b = 5, c = 4$

76. $a = 1, b = 3, c = 2$

77. $a = 2, b = -3, c = -20$

78. $a = \dfrac{1}{2}, b = -\dfrac{1}{12}, c = -1$

79. $a = \dfrac{1}{12}, b = \dfrac{1}{3}, c = -1$

80. $a = \dfrac{1}{12}, b = \dfrac{1}{2}, c = \dfrac{2}{3}$

71. _____

72. _____

73. _____

74. _____

75. _____

76. _____

77. _____

78. _____

79. _____

80. _____

✓ **SKILL CHECKER**

81. Multiply $(a + b)(a - b)$.

82. Use the result of Problem 81 to multiply $(\sqrt{x} + \sqrt{y})(\sqrt{x} - \sqrt{y})$.

83. Use the result of Problem 81 to multiply $(x^{3/2} + y^{3/2})(x^{3/2} - y^{3/2})$.

84. Write $\dfrac{2x}{xy^2}$ with a denominator of $8x^3y^3$.

85. Write $\dfrac{3xy}{x^2y^3}$ with a denominator of $16x^4y^4$.

81. _____

82. _____

83. _____

84. _____

85. _____

5.1   USING YOUR KNOWLEDGE

1. You already know that $\sqrt{x} = x^{1/2}$. Use your knowledge to write $\sqrt{\sqrt{x}}$ using exponents.

1. _____

2. If you have a calculator with a $\sqrt{}$ button you can find $\sqrt{9}$ by simply pressing 9 $\sqrt{}$. You can also find $\sqrt[4]{16}$ using the $\sqrt{}$ button and the results of Problem 1.
   a. Find $\sqrt[4]{16}$.
   b. Find $\sqrt[4]{4096}$.

3. a. Write $\sqrt[3]{x}$ using exponents.
   b. Write $\sqrt[3]{\sqrt{x}}$ using a single exponent.
   c. If you have a calculator with a $\sqrt{}$ and a $\sqrt[3]{}$ button, find $\sqrt[6]{729}$.

2. a. ___
   b. ___

3. a. ___
   b. ___
   c. ___

CALCULATOR CORNER

Many of the numerical evaluations in this section can be done with the y^x key in your calculator. This key will raise the number y to the x power. Thus to find 2^3, enter 2 y^x 3 $=$. The answer is 8. To find $\sqrt[3]{-64}$, first recall that $\sqrt[3]{-64} = (-64)^{1/3}$. Thus we enter

64 $+/-$ y^x $($ 1 $\div$ 3 $)$

Note that $\frac{1}{3}$ has to be written in parentheses.

If your instructor permits, use the y^x key in your calculator to find each value.

1. $\sqrt[3]{-\dfrac{1}{8}}$

2. $(-125)^{1/3}$

3. $\sqrt[4]{\dfrac{1}{16}}$

1. ___
2. ___
3. ___

5.2 PROPERTIES OF RADICALS

OBJECTIVES

REVIEW

Before starting this sect.
should know:

1. The laws of exponents.
2. How to factor perfect trinomial squares.

OBJECTIVES

You should be able to:

A. Simplify expressions involving radicals by using the laws given in the text.
B. Rationalize the denominator of a rational expression.
C. Reduce the order of an expression containing radicals.

The illustrations at the beginning of this section show a square root sign, derived from the word *radix* (Latin for "root") and first used by Fibonacci, and a cube root sign created in 1525 by Christoff Rudolff, a German mathematician.

These two symbols, which are now written as $\sqrt{}$ and $\sqrt[3]{}$, respectively, are alternative symbols for the square root and the cube root of a number, discussed in the preceding section. In general, the nth root of a number a is defined so that

$$a^{1/n} = \sqrt[n]{a} \qquad (a \geq 0)$$

From this definition, we can derive three important relationships involving radicals that can be proved using the properties of exponents discussed previously. *In the discussion that follows, we shall assume that when the index of a radical is even, the radicand is nonnegative.*

A. PROPERTIES OF RADICALS

$$
\begin{aligned}
\text{I. } & \sqrt[n]{a^n} = a \qquad (a \geq 0) \\
\text{II. } & \sqrt[n]{ab} = \sqrt[n]{a}\,\sqrt[n]{b} \\
\text{III. } & \sqrt[n]{\dfrac{a}{b}} = \dfrac{\sqrt[n]{a}}{\sqrt[n]{b}}
\end{aligned}
$$

The first of these laws is equivalent to the definition of the principal nth root of a. Thus $\sqrt[n]{a^n} = [a^n]^{1/n} = a^{n \cdot 1/n} = a$. The other two laws are obtained as follows.

II. $\sqrt[n]{ab} = (ab)^{1/n} = a^{1/n} \cdot b^{1/n} = \sqrt[n]{a} \cdot \sqrt[n]{b}$

III. $\sqrt[n]{\dfrac{a}{b}} = \left(\dfrac{a}{b}\right)^{1/n} = \dfrac{a^{1/n}}{b^{1/n}} = \dfrac{\sqrt[n]{a}}{\sqrt[n]{b}}$

We have already mentioned that when m and n are even, $(a^m)^n = |a|^{m/n}$. For $m = n$,

$$\sqrt[n]{a^n} = |a|$$

Thus, for even indices, $\sqrt{3^2} = |3| = 3$, $\sqrt{(-3)^2} = |-3| = 3$, and $\sqrt[4]{(-x)^4} = |-x|$. When the index is odd, absolute value is not necessary. Thus, $\sqrt[3]{-8} = -2$ and $\sqrt[5]{-x^5} = -x$.

EXAMPLE 1 Simplify.

a. $\sqrt[4]{(-2)^4}$ **b.** $\sqrt[8]{(-x)^8}$
c. $\sqrt[9]{(-x)^9}$ **d.** $\sqrt{x^2 + 8x + 4}$

Solution

a. Since the index 4 is even, $\sqrt[4]{(-2)^4} = |-2| = 2$.

b. The index is even. Thus $\sqrt[8]{(-x)^8} = |-x|$. No further simplification is possible, since we do not know if x is positive or negative.

c. $\sqrt[9]{(-x)^9} = -x$.   (Absolute value is not necessary, since the index 9 is not even.)

d. We start by factoring: $x^2 + 8x + 16 = (x + 4)^2$. Thus

$$\sqrt{x^2 + 8x + 16} = \sqrt{(x + 4)^2}$$
$$= |x + 4|$$

(Absolute value is needed, since the index 2 is even.)   ▲

If an expression does not have a perfect root, we can factor out factors with perfect roots out of the radicand. To help you, here are the first few squares, cubes, and fourth roots.

SQUARE ROOTS	CUBE ROOTS	FOURTH ROOTS
$\sqrt{0} = 0$	$\sqrt[3]{0} = 0$	$\sqrt[4]{0} = 0$
$\sqrt{1} = 1$	$\sqrt[3]{1} = 1$	$\sqrt[4]{1} = 1$
$\sqrt{4} = 2$	$\sqrt[3]{8} = 2$	$\sqrt[4]{16} = 2$
$\sqrt{9} = 3$	$\sqrt[3]{27} = 3$	$\sqrt[4]{81} = 3$
$\sqrt{16} = 4$	$\sqrt[3]{64} = 4$	$\sqrt[4]{256} = 4$
$\sqrt{25} = 5$	$\sqrt[3]{125} = 5$	$\sqrt[4]{625} = 5$

To simplify $\sqrt[3]{40}$ we have to find a factor of 40 that has a perfect cube root. This factor is $8 = 2^3$. Thus $40 = 2^3 \cdot 5$ and

$$\sqrt[3]{40} = \sqrt[3]{2^3 \cdot 5}$$
$$= \sqrt[3]{2^3} \cdot \sqrt[3]{5} \quad \text{Law II}$$
$$= 2 \cdot \sqrt[3]{5} \quad \text{Law I}$$

EXAMPLE 2 Simplify.

a. $\sqrt[3]{54}$ **b.** $\sqrt[3]{128a^4b^6}$

Problem 1 Simplify.

a. $\sqrt[4]{(-5)^4}$ **b.** $\sqrt[6]{(-x)^6}$
c. $\sqrt[7]{(-x)^7}$ **d.** $\sqrt{x^2 + 10x + 25}$

Problem 2 Simplify.

a. $\sqrt[3]{32}$ **b.** $\sqrt[3]{81a^6b^4}$

Solution

a. Since $54 = 27 \cdot 2 = 3^3 \cdot 2$,

$$\sqrt[3]{54} = \sqrt[3]{3^3 \cdot 2}$$
$$= \sqrt[3]{3^3} \cdot \sqrt[3]{2}$$
$$= 3 \cdot \sqrt[3]{2}$$

b. We have to factor 128 into factors that have perfect cube roots. Since 64 is a perfect cube and $128 = 64 \cdot 2$,

$$\sqrt[3]{128a^4b^6} = \sqrt[3]{4^3 \cdot 2 \cdot a^3 \cdot a \cdot (b^2)^3} \qquad \text{Factor.}$$
$$= \sqrt[3]{4^3 \cdot 2} \cdot \sqrt[3]{a^3} \cdot \sqrt[3]{a} \cdot \sqrt[3]{(b^2)^3} \qquad \text{Law II}$$
$$= 4 \cdot \sqrt[3]{2} \cdot a \cdot \sqrt[3]{a} \cdot b^2 = 4ab^2\sqrt[3]{2a} \qquad \text{Law I}$$

You can save time by writing all perfect cube factors together like this

$$\sqrt[3]{128a^4b^6} = \sqrt[3]{[4^3a^3(b^2)^3] \cdot 2a} = 4ab^2\sqrt[3]{2a} \qquad \blacktriangle$$

The third law mentioned in this section can be used to change a radical into a form in which the radicand contains no fractions. For example,

$$\sqrt{\frac{3}{16}} = \frac{\sqrt{3}}{\sqrt{16}} = \frac{\sqrt{3}}{4}$$

$$\sqrt[3]{\frac{7}{8}} = \frac{\sqrt[3]{7}}{\sqrt[3]{8}} = \frac{\sqrt[3]{7}}{2}$$

If the denominator does not have a perfect root, multiply numerator and denominator by a factor that will yield a perfect root. Thus to simplify $\sqrt{\frac{3}{8}}$, multiply the denominator by 2 to obtain 16, which has a perfect root. Of course, you must multiply the numerator by 2. Thus,

$$\sqrt{\frac{3}{8}} = \sqrt{\frac{3 \cdot 2}{8 \cdot 2}} = \sqrt{\frac{6}{16}} = \frac{\sqrt{6}}{4}$$

EXAMPLE 3 Simplify.

a. $\sqrt{\dfrac{7}{32}}$ **b.** $\sqrt[3]{\dfrac{9}{x^3}}$ **c.** $\sqrt[4]{\dfrac{2}{27x^5}}$

Solution

a. We multiply the denominator and numerator by 2 ($32 \cdot 2 = 64$, a perfect square):

$$\sqrt{\frac{7}{32}} = \sqrt{\frac{7 \cdot 2}{32 \cdot 2}} = \sqrt{\frac{14}{64}} = \frac{\sqrt{14}}{\sqrt{64}} = \frac{\sqrt{14}}{8}$$

b. $\sqrt[3]{\dfrac{9}{x^3}} = \dfrac{\sqrt[3]{9}}{\sqrt[3]{x^3}} = \dfrac{\sqrt[3]{9}}{x}$

c. To obtain a perfect fourth power multiply 27 by 3 and multiply x^5 by x^3:

$$\sqrt[4]{\frac{2}{27x^5}} = \sqrt[4]{\frac{2 \cdot 3x^3}{27x^5 \cdot 3x^3}} = \sqrt[4]{\frac{6x^3}{81x^8}} = \frac{\sqrt[4]{6x^3}}{\sqrt[4]{81x^8}} = \frac{\sqrt[4]{6x^3}}{3x^2}$$

Problem 3 Simplify.

a. $\sqrt{\dfrac{11}{12}}$ **b.** $\sqrt[3]{\dfrac{6}{x^3}}$

c. $\sqrt[4]{\dfrac{3}{8x^6}}$

B. RATIONALIZING THE DENOMINATOR

In some cases, the denominator of a fraction contains expressions involving radicals. For example, $\dfrac{\sqrt{3}}{\sqrt{5}}$ has $\sqrt{5}$ in the denominator. To simplify $\dfrac{\sqrt{3}}{\sqrt{5}}$, we use the fundamental property of fractions to multiply numerator and denominator by $\sqrt{5}$.

$$\frac{\sqrt{3}}{\sqrt{5}} = \frac{\sqrt{3} \cdot \sqrt{5}}{\sqrt{5} \cdot \sqrt{5}}$$

$$= \frac{\sqrt{15}}{\sqrt{5^2}} \qquad \text{Law II}$$

$$= \frac{\sqrt{15}}{5} \qquad \text{Law I}$$

This process is called **rationalizing the denominator.** To rationalize the denominator in the expression $\dfrac{\sqrt{7}}{\sqrt{3x}}$ we have to free the denominator of radicals. We do that by multiplying the denominator and numerator by $\sqrt{3}$. Thus

$$\frac{\sqrt{7}}{\sqrt{3x}} = \frac{\sqrt{7} \cdot \sqrt{3}}{\sqrt{3x} \cdot \sqrt{3}} \qquad \text{Fundamental property of fractions}$$

$$= \frac{\sqrt{21}}{\sqrt{3^2 x}} \qquad \text{Law II}$$

$$= \frac{\sqrt{21}}{3x} \qquad \text{Law I}$$

Note that x remains in the denominator.

EXAMPLE 4   Rationalize the denominators of the given expressions.

a. $\dfrac{\sqrt{11}}{\sqrt{6}}$ **b.** $\dfrac{\sqrt{3}}{\sqrt{5x}}, x > 0$ **c.** $\dfrac{\sqrt{5}}{\sqrt{18x^2}}$

Solution

a. $\dfrac{\sqrt{11}}{\sqrt{6}} = \dfrac{\sqrt{11} \cdot \sqrt{6}}{\sqrt{6} \cdot \sqrt{6}} = \dfrac{\sqrt{66}}{6}$

b. To obtain a perfect square in the denominator multiply by $\sqrt{5x}$.

$$\frac{\sqrt{3}}{\sqrt{5x}} = \frac{\sqrt{3} \cdot \sqrt{5x}}{\sqrt{5x} \cdot \sqrt{5x}} = \frac{\sqrt{15x}}{5x}$$

c. To convert $\sqrt{18}$ to a perfect square root, multiply by $\sqrt{2}$.

$$\frac{\sqrt{5}}{\sqrt{18x^2}} = \frac{\sqrt{5} \cdot \sqrt{2}}{\sqrt{18x^2} \cdot \sqrt{2}} = \frac{\sqrt{10}}{\sqrt{36x^2}} = \frac{\sqrt{10}}{6x^2}. \qquad \blacktriangle$$

When the radical in the denominator is of order n, we must make the radicand an exact nth power. For example, to rationalize $\dfrac{\sqrt[3]{5}}{\sqrt[3]{3x}}$, we

Problem 4   Rationalize the denominator.

a. $\dfrac{\sqrt{7}}{\sqrt{5}}$ **b.** $\dfrac{\sqrt{5}}{\sqrt{6x}}, \quad x > 0$

c. $\dfrac{\sqrt{11}}{\sqrt{32x^3}}$

ANSWERS   (to problems on pages 360–361)

1. a. $|-5| = 5$ **b.** $|-x|$ **c.** $-x$ **d.** $|x + 5|$

2. a. $2\sqrt[3]{4}$ **b.** $3a^2 b \sqrt[3]{3b}$

3. a. $\dfrac{\sqrt{33}}{6}$ **b.** $\dfrac{\sqrt[3]{6}}{x}$ **c.** $\dfrac{\sqrt[4]{6x^2}}{2x^2}$

convert $\sqrt[3]{3}$ to a perfect cube root by multiplying by $\sqrt[3]{3^2}$, and to convert $\sqrt[3]{x}$ to a perfect cube root, we multiply by $\sqrt[3]{x^2}$. We can combine these two steps and multiply by $\sqrt[3]{3^2 x^2}$, obtaining

$$\frac{\sqrt[3]{5}}{\sqrt[3]{3x}} = \frac{\sqrt[3]{5} \cdot \sqrt[3]{3^2 \cdot x^2}}{\sqrt[3]{3x} \cdot \sqrt[3]{3^2 \cdot x^2}} \qquad \text{Multiply the denominator by } \sqrt[3]{3^2 \cdot x^2} \text{ to make the denominator a perfect cube root.}$$

$$= \frac{\sqrt[3]{5 \cdot 9 \cdot x^2}}{\sqrt[3]{3^3 x^3}} \qquad \text{Law II}$$

$$= \frac{\sqrt[3]{45x^2}}{3x} \qquad \text{Law I}$$

EXAMPLE 5 Rationalize the denominator in the following expressions.

a. $\dfrac{1}{\sqrt[3]{5x}}$ **b.** $\dfrac{\sqrt[5]{5}}{\sqrt[5]{8x^4}}$

Solution

a. To convert $\sqrt[3]{5x}$ to a perfect cube root, multiply by $\sqrt[3]{5^2 x^2}$.

$$\frac{1}{\sqrt[3]{5x}} = \frac{1 \cdot \sqrt[3]{5^2 x^2}}{\sqrt[3]{5x} \cdot \sqrt[3]{5^2 x^2}}$$

$$= \frac{\sqrt[3]{25x^2}}{\sqrt[3]{5^3 x^3}}$$

$$= \frac{\sqrt[3]{25x^2}}{5x}$$

b. $\dfrac{\sqrt[5]{5}}{\sqrt[5]{8x^4}} = \dfrac{\sqrt[5]{5}}{\sqrt[5]{2^3 x^4}}$ Write $8x^4$ as $2^3 x^4$.

$$= \frac{\sqrt[5]{5} \cdot \sqrt[5]{2^2 \cdot x}}{\sqrt[5]{2^3 \cdot x^4} \cdot \sqrt[5]{2^2 \cdot x}} \qquad \text{Use the fundamental principle of fractions.}$$

$$= \frac{\sqrt[5]{20x}}{\sqrt[5]{2^5 \cdot x^5}} \qquad \text{Law II}$$

$$= \frac{\sqrt[5]{20x}}{2x} \qquad \text{Law I}$$

C. REDUCING THE INDEX (ORDER) OF A RADICAL

The index of a radical can sometimes be reduced by writing the radical as a power with a rational exponent and then reducing the exponent. For example, if $x \geq 0$,

$$\sqrt[6]{x^3} = (x^3)^{1/6}$$

$$= x^{3 \cdot 1/6}$$

$$= x^{1/2}$$

$$= \sqrt{x}$$

Problem 5 Rationalize the denominator.

a. $\dfrac{1}{\sqrt[3]{54x}}$ **b.** $\dfrac{\sqrt[5]{5}}{\sqrt[5]{27x^4}}$

Similarly, for $x \geq 0$ and $y \geq 0$,

$$
\begin{aligned}
\sqrt[4]{64x^2y^2} &= \sqrt[4]{(8xy)^2} \\
&= [(8xy)^2]^{1/4} \\
&= [8xy]^{2 \cdot 1/4} \\
&= (8xy)^{1/2} = \sqrt{8xy}
\end{aligned}
$$

EXAMPLE 6 Reduce the index (order) of the given expression.

a. $\sqrt[4]{\dfrac{16}{81}}$ **b.** $\sqrt[6]{27c^3d^3}$, $c \geq 0$, $d \geq 0$

Solution

a. $\sqrt[4]{\dfrac{16}{81}} = \left[\left(\dfrac{2}{3}\right)^4\right]^{1/4} = \dfrac{2}{3}$

b. $\sqrt[6]{27c^3d^3} = \sqrt[6]{(3cd)^3}$

$\qquad\qquad = [(3cd)^3]^{1/6}$

$\qquad\qquad = [3cd]^{1/2}$

$\qquad\qquad = \sqrt{3cd}$ ▲

We have used different techniques to *simplify* expressions containing radicals. To make sure that the resulting radicals are simplified, use these steps.

A radical expression is in **simplified** form if

1. The radicand (the expression under the radical) has no factors with exponents greater than or equal to the index.

2. There are no fractions under the radical sign.

3. There are no radicals in the denominator.

4. The index is as low as possible.

EXAMPLE 7 Simplify $\sqrt[6]{\dfrac{a^2}{16c^{10}}}$.

Solution To make the denominator a perfect sixth root, note that $16 = 2^4$ and then multiply numerator and denominator under the radical by 2^2c^2. We have

$$
\begin{aligned}
\sqrt[6]{\dfrac{a^2}{16c^{10}} \cdot \dfrac{2^2c^2}{2^2c^2}} &= \sqrt[6]{\dfrac{2^2a^2c^2}{2^6c^{12}}} \\
&= \dfrac{\sqrt[6]{2^2a^2c^2}}{2c^2} \\
&= \dfrac{(2^2a^2c^2)^{1/6}}{2c^2} \\
&= \dfrac{[(2ac)^2]^{1/6}}{2c^2} \\
&= \dfrac{(2ac)^{1/3}}{2c^2} \\
&= \dfrac{\sqrt[3]{2ac}}{2c^2}
\end{aligned}
$$

Problem 6 Reduce the index (order).

a. $\sqrt[4]{\dfrac{81}{256}}$ **b.** $\sqrt[6]{4c^2d^2}$

Problem 7 Simplify $\sqrt[6]{\dfrac{a^3}{8x^3}}$.

A. In Problems 1–24 simplify. (Hint: Some answers require absolute values.)

1. $\sqrt{(-5)^2}$

2. $\sqrt{(5)^2}$

3. $\sqrt[3]{-64}$

4. $\sqrt[3]{-125}$

5. $\sqrt[6]{(-x)^6}$

6. $\sqrt[5]{(-x)^5}$

7. $\sqrt{x^2 + 12x + 36}$

8. $\sqrt{4x^2 + 12x + 9}$

9. $\sqrt{9x^2 - 12x + 4}$

10. $\sqrt{16x^2 + 8x + 1}$

11. $\sqrt{16x^3 y^3}$

12. $\sqrt{81x^3 y^4}$

13. $\sqrt[3]{40x^4 y}$

14. $\sqrt[3]{81x^3 y^6}$

15. $\sqrt[4]{x^5 y^7}$

16. $\sqrt[4]{162x^4 y^7}$

17. $\sqrt[5]{-243a^{10}b^{17}}$

18. $\sqrt[5]{-32a^{15}b^{20}}$

19. $\sqrt{\dfrac{13}{49}}$

20. $\sqrt{\dfrac{17}{64}}$

21. $\sqrt{\dfrac{17}{4x^2}}$

22. $\sqrt{\dfrac{19}{64x^4}}$

23. $\sqrt[3]{\dfrac{3}{64x^3}}$

24. $\sqrt[3]{\dfrac{-7}{27x^6}}$

B. In Problems 25–40 rationalize the denominator. (Assume all variables represent positive real numbers.)

25. $\sqrt{\dfrac{2}{3}}$

26. $\sqrt{\dfrac{4}{5}}$

27. $\dfrac{-\sqrt{2}}{\sqrt{7}}$

28. $\dfrac{-\sqrt{3}}{\sqrt{11}}$

29. $\sqrt{\dfrac{5}{2a}}$

30. $\sqrt{\dfrac{7}{36}}$

31. $\sqrt{\dfrac{5}{32ab}}$

32. $\sqrt{\dfrac{5}{8ab}}$

33. $-\sqrt{\dfrac{3}{2a^3b^3}}$

34. $-\sqrt{\dfrac{3}{8ab^3}}$

35. $\dfrac{\sqrt{x}\,\sqrt{xy^3}}{\sqrt{y}}$

36. $\dfrac{\sqrt{xy}\,\sqrt{xy^4}}{\sqrt{y}}$

37. $-\sqrt[3]{\dfrac{7}{9}}$

38. $-\sqrt[3]{\dfrac{3}{32}}$

39. $\sqrt[3]{\dfrac{3}{16x^2}}$

40. $\sqrt[3]{\dfrac{5}{16x}}$

29. _____

30. _____

31. _____

32. _____

33. _____

34. _____

35. _____

36. _____

37. _____

38. _____

39. _____

40. _____

C. In Problems 41–50, reduce the order (index) of the given radical and simplify. (Assume the variables represent positive real numbers.)

41. $\sqrt[6]{9}$

42. $\sqrt[6]{4}$

43. $\sqrt[4]{4a^2}$

44. $\sqrt[4]{9a^2}$

45. $\sqrt[4]{25x^6y^2}$

46. $\sqrt[4]{36x^2y^6}$

47. $\sqrt[4]{49x^{10}y^6}$

48. $\sqrt[4]{100x^{10}y^{10}}$

49. $\sqrt[6]{8a^3b^3}$

50. $\sqrt[6]{27a^3b^9}$

41. _____

42. _____

43. _____

44. _____

45. _____

46. _____

47. _____

48. _____

49. _____

50. _____

In Problems 51–55, simplify. (Assume all variables represent positive real numbers.)

51. $\sqrt[6]{\dfrac{a^4}{b^8}}$

52. $\sqrt[4]{\dfrac{c^6}{4b^2}}$

53. $\sqrt[4]{\dfrac{64a^2}{9b^6}}$

54. $\sqrt[4]{\dfrac{4c^2y^6}{9b^4}}$

55. $\sqrt[6]{\dfrac{b^3a^3}{8x^3}}$

51. _____

52. _____

53. _____

54. _____

55. _____

D. Applications

56. A body starting at rest takes t seconds to fall a distance of d feet, where

$$t = \sqrt{\frac{d}{16}}$$

    a. Simplify this expression.

    b. How long would it take an object starting at rest to fall 100 ft?

56. _____

57. The radius of a sphere is given by $r = \sqrt[3]{\dfrac{3V}{4\pi}}$, where V is the volume of the sphere and π is about $\frac{22}{7}$.

    a. Simplify $\sqrt[3]{\dfrac{3V}{4\pi}}$.

    b. If the volume of a sphere is 36π ft^3, what is its radius?

57. _____

58. The root-mean-square velocity, $\bar{v}$, of a gas particle is given by the formula

$$\bar{v} = \frac{\sqrt{3kT}}{\sqrt{m}}$$

where k is a constant, T is the temperature (in degrees Kelvin), and m is the mass of the particle. Rationalize the denominator of the expression on the right side.

58. _____

59. The mass m of an object depends on its speed, v, and the speed of light, c. The relationship is given by the formula

$$m = \frac{m_0}{\sqrt{1 - v^2/c^2}}$$

where m_0 is the *rest mass,* the mass when $v = 0$. Simplify the expression on the right side and rationalize the denominator.

59. _____

60. Have you heard of supersonic airplanes with a speed of Mach 2? Mach 2 means that the speed of the plane is *twice* the speed of sound. The Mach number can be found from the formula

$$M = \sqrt{\frac{2}{\gamma}} \sqrt{\frac{P_2 - P_1}{P_1}}$$

Simplify this expression and write the answer with a rationalized denominator.

60. _____

✓ SKILL CHECKER

Simplify.

61. $5x - (3x + 1) + (5x + 2)$ **62.** $3x - (7x + 3) - (5x + 4)$

61. _____

62. _____

63. $(2x + y)(2x - y)$ **64.** $(3x - 4y)(3x + 4y)$

63. _____

64. _____

EXERCISE 5.2

Divide.

65. $\dfrac{3xy + 6x^2y}{3xy}$

66. $\dfrac{12x^2y^3 + 18xy^3}{6xy}$

Evaluate and simplify $\sqrt{b^2 - 4ac}$ in each case.

67. $a = 2$, $b = -1$, and $c = -6$ **68.** $a = 2$, $b = -5$, and $c = -12$

69. $a = 6$, $b = -4$, and $c = -2$ **70.** $a = -1$, $b = 1$, and $c = 12$

5.2   USING YOUR KNOWLEDGE

We have studied rational exponents and radicals. We are going to use this knowledge to translate one notation to the other.

1. In Problem 59,

$$m = \frac{m_0}{\sqrt{1 - \dfrac{v^2}{c^2}}}$$

Rationalize the denominator and write the result using rational exponents.

2. In Problem 60, simplify the expression defining M and write the result using rational exponents.

3. The period T of a pendulum is $T = \sqrt{\dfrac{2\pi L}{g}}$, where L is the length and g is the gravity constant. Simplify this expression and write the result using rational exponents.

4. The pressure P of a gas is related to its volume V by the formula $P = kV^{-7/5}$. Write this formula using radical notation.

5. The average speed v of oxygen molecules is given by $v = (3kT)^{1/2}m^{-1/2}$. Write this formula in simplified form using radicals.

CALCULATOR CORNER

Some of the simplifications we have made can be checked using a calculator with a $\boxed{\sqrt[x]{y}}$ key. To access this key, you usually have to press the $\boxed{2nd}$ or $\boxed{2ndF}$ button first and then the $\boxed{\sqrt[x]{y}}$ button. The calculator will then find the xth root of y.

In Example 2, we learned that $\sqrt[3]{54} = 3\sqrt[3]{2}$. To check this, enter $\boxed{54}\ \boxed{2nd}\ \boxed{\sqrt[x]{y}}\ \boxed{3}\ \boxed{=}$. The display shows 3.77976315. Now, key in $\boxed{3}\ \boxed{\times}\ \boxed{2}\ \boxed{\sqrt[x]{y}}\ \boxed{3}\ \boxed{=}$. The same result appears, so our answer is correct. If your instructor agrees, check the numerical exercises in this section (Problems 25–28, for example) with a calculator.

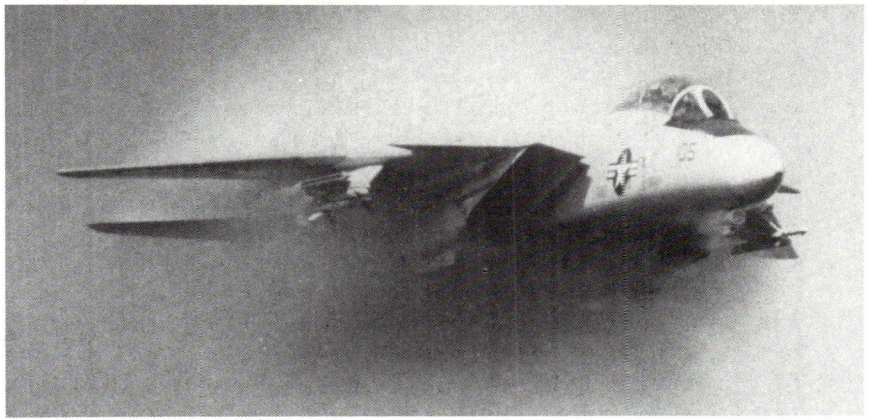

Courtesy of A. Paloumpis

REVIEW

Before starting this section, you should know:

1. How to combine like terms (Section 1.5).
2. How to remove parentheses using the distributive law (Section 1.5).
3. How to write a fraction with a specified denominator (Section 4.1).
4. How to reduce fractions (Section 4.1).

OBJECTIVES

You should be able to:

A. Add and subtract radical expressions that are similar.
B. Multiply and divide radical expressions.
C. Rationalize the denominator of radical expressions involving sums or differences.

How fast can this plane travel? The answer is classified information but it exceeds twice the speed of sound (747 mi/hr). It is then said that the plane's speed is more than Mach 2. The formula for calculating the Mach number is

$$M = \sqrt{\frac{2}{\gamma}} \sqrt{\frac{P_2 - P_1}{P_1}}$$

where P_1 and P_2 are air pressures. This expression can be simplified by multiplying both radical expressions and rationalizing the denominator. In this section we will add, subtract, multiply, and divide radical expressions—that is, expressions containing radicals.

A. ADDITION AND SUBTRACTION OF RADICAL EXPRESSIONS

In Section 1.5 we combined like terms using the distributive law. Thus

$$3x + 5x = (3 + 5)x = 8x$$

Similarly,

$$3\sqrt{2} + 5\sqrt{2} = (3 + 5)\sqrt{2} = 8\sqrt{2}$$

Also,

$$7x - 4x = (7 - 4)x = 3x$$

Similarly,

$$7\sqrt[3]{7} - 4\sqrt[3]{7} = (7 - 4)\sqrt[3]{7} = 3\sqrt[3]{7}$$

Thus, we can combine *like* (*similar*) radical expressions. Here is the definition.

Radical expressions with the same index and the same radicand are **like (similar)** expressions.

If the expressions are not similar, we must simplify them first and see if they are similar before we can add. Thus to add $\sqrt{75} + \sqrt{27}$, we proceed as follows.

$$\begin{aligned}
\sqrt{75} + \sqrt{27} &= \sqrt{25 \cdot 3} + \sqrt{9 \cdot 3} \\
&= \sqrt{25} \cdot \sqrt{3} + \sqrt{9} \cdot \sqrt{3} \quad \sqrt{ab} = \sqrt{a}\sqrt{b} \\
&= 5\sqrt{3} + 3\sqrt{3} \\
&= (5 + 3)\sqrt{3} = 8\sqrt{3} \quad \text{Add like radicals.}
\end{aligned}$$

The subtraction of similar radicals is done in the same way. Thus

$$\begin{aligned}
\sqrt{80} - \sqrt{20} &= \sqrt{16 \cdot 5} - \sqrt{4 \cdot 5} \\
&= \sqrt{16} \cdot \sqrt{5} - \sqrt{4} \cdot \sqrt{5} \\
&= 4\sqrt{5} - 2\sqrt{5} \\
&= (4 - 2) \cdot \sqrt{5} = 2\sqrt{5} \quad \text{Subtract like radicals.}
\end{aligned}$$

EXAMPLE 1 Perform the indicated operations.

a. $\sqrt{175} + \sqrt{28}$ **b.** $\sqrt{98} - \sqrt{32}$
c. $3\sqrt{18x} - 5\sqrt{8x}$ **d.** $5\sqrt[3]{80x} - 3\sqrt[3]{270x}$

Solution

a.
$$\begin{aligned}
\sqrt{175} + \sqrt{28} &= \sqrt{25 \cdot 7} + \sqrt{4 \cdot 7} \\
&= \sqrt{25} \cdot \sqrt{7} + \sqrt{4} \cdot \sqrt{7} \\
&= 5\sqrt{7} + 2\sqrt{7} = 7\sqrt{7}
\end{aligned}$$

b.
$$\begin{aligned}
\sqrt{98} - \sqrt{32} &= \sqrt{49 \cdot 2} - \sqrt{16 \cdot 2} \\
&= \sqrt{49} \cdot \sqrt{2} - \sqrt{16} \cdot \sqrt{2} \\
&= 7\sqrt{2} - 4\sqrt{2} = 3\sqrt{2}
\end{aligned}$$

c.
$$\begin{aligned}
3\sqrt{18x} - 5\sqrt{8x} &= 3\sqrt{9 \cdot 2x} - 5\sqrt{4 \cdot 2x} \\
&= 3\sqrt{9} \cdot \sqrt{2x} - 5\sqrt{4} \cdot \sqrt{2x} \\
&= 3 \cdot 3 \cdot \sqrt{2x} - 5 \cdot 2 \cdot \sqrt{2x} \\
&= 9\sqrt{2x} - 10\sqrt{2x} \\
&= -\sqrt{2x}
\end{aligned}$$

d. This time we must find factors of 80 and 270 that are perfect cubes: $80 = 8 \cdot 10 = 2^3 \cdot 10$ and $270 = 27 \cdot 10 = 3^3 \cdot 10$. Thus

$$\begin{aligned}
5\sqrt[3]{80x} - 3\sqrt[3]{270x} &= 5\sqrt[3]{2^3 \cdot 10} - 3\sqrt[3]{3^3 \cdot 10} \\
&= 5\sqrt[3]{2^3} \cdot \sqrt[3]{10} - 3\sqrt[3]{3^3} \cdot \sqrt[3]{10} \\
&= 5 \cdot 2 \cdot \sqrt[3]{10} - 3 \cdot 3\sqrt[3]{10} \\
&= 10\sqrt[3]{10} - 9\sqrt[3]{10} \\
&= \sqrt[3]{10} \quad \blacktriangle
\end{aligned}$$

We now learn how to combine like radicals involving fractions.

EXAMPLE 2 Perform the indicated operations.

a. $3\sqrt{\dfrac{1}{2}} - 5\sqrt{\dfrac{1}{8}}$ **b.** $3\sqrt[3]{\dfrac{3x}{4x^2}} - \sqrt[3]{\dfrac{3}{32x}}$

Problem 1 Perform the indicated operations.
a. $\sqrt{44} + \sqrt{99}$
b. $\sqrt{98} - \sqrt{50}$
c. $3\sqrt{20x} - 5\sqrt{45x}$
d. $2\sqrt[3]{250x} - 4\sqrt[3]{16x}$

Problem 2 Perform the indicated operations.
a. $5\sqrt{\dfrac{1}{2}} - 7\sqrt{\dfrac{1}{8}}$ **b.** $4\sqrt[3]{\dfrac{5x}{4x^2}} - \sqrt[3]{\dfrac{5}{32x}}$

Solution

a. We first simplify each of the radicals by making the denominator a perfect square.

$$\sqrt{\frac{1}{2}} = \sqrt{\frac{1 \cdot 2}{2 \cdot 2}} = \frac{\sqrt{2}}{2} \quad \text{and} \quad \sqrt{\frac{1}{8}} = \sqrt{\frac{1 \cdot 2}{8 \cdot 2}} = \frac{\sqrt{2}}{4}$$

Thus,

$$
\begin{aligned}
3\sqrt{\frac{1}{2}} - 5\sqrt{\frac{1}{8}} &= 3 \cdot \frac{\sqrt{2}}{2} - 5 \cdot \frac{\sqrt{2}}{4} \\
&= \frac{3\sqrt{2}}{2} - \frac{5\sqrt{2}}{4} \\
&= \frac{2 \cdot 3\sqrt{2}}{2 \cdot 2} - \frac{5\sqrt{2}}{4} \\
&= \frac{6\sqrt{2} - 5\sqrt{2}}{4} \\
&= \frac{\sqrt{2}}{4}
\end{aligned}
$$

b. We first make the denominators perfect cubes, find the LCD of the resulting fractions, and then subtract.

$$\sqrt[3]{\frac{3x}{4x^2}} = \sqrt[3]{\frac{3x \cdot 2x}{4x^2 \cdot 2x}} = \frac{\sqrt[3]{6x^2}}{2x}$$

$$\sqrt[3]{\frac{3}{32x}} = \sqrt[3]{\frac{3 \cdot 2x^2}{32x \cdot 2x^2}} = \frac{\sqrt[3]{6x^2}}{4x}$$

Thus

$$
\begin{aligned}
3\sqrt[3]{\frac{3x}{4x^2}} - \sqrt[3]{\frac{3}{32x}} &= 3 \cdot \frac{\sqrt[3]{6x^2}}{2x} - \frac{\sqrt[3]{6x^2}}{4x} \\
&= \frac{3\sqrt[3]{6x^2}}{2x} - \frac{\sqrt[3]{6x^2}}{4x} \\
&= \frac{2 \cdot 3\sqrt[3]{6x^2}}{2 \cdot 2x} - \frac{\sqrt[3]{6x^2}}{4x} \qquad \text{Since the LCD is } 4x, \text{ multiply} \\
& \qquad\qquad\qquad\qquad\qquad\qquad \text{numerator and denominator of} \\
&= \frac{6\sqrt[3]{6x^2} - \sqrt{6x^2}}{4x} \qquad\qquad \text{the first fraction by 2.} \\
&= \frac{5\sqrt[3]{6x^2}}{4x} \qquad\qquad\qquad x \neq 0
\end{aligned}
$$

B. MULTIPLICATION AND DIVISION OF RADICAL EXPRESSIONS

The distributive law, used in conjunction with the fact that $\sqrt{a} \cdot \sqrt{b} = \sqrt{ab}$, can be used to simplify expressions containing parentheses. For example,

$$
\begin{aligned}
\sqrt{2} \cdot (\sqrt{3} + \sqrt{5}) &= \sqrt{2} \cdot \sqrt{3} + \sqrt{2} \cdot \sqrt{5} \qquad \text{Distributive law} \\
&= \sqrt{6} + \sqrt{10}
\end{aligned}
$$

Similarly, if $x \geq 0$, then

$$\sqrt{2x} \cdot (\sqrt{x} + \sqrt{3}) = \sqrt{2x} \cdot \sqrt{x} + \sqrt{2x} \cdot \sqrt{3}$$
$$= \sqrt{2x^2} + \sqrt{6x} \qquad \text{Since } \sqrt{ab} = \sqrt{a} \cdot \sqrt{b},$$
$$\qquad\qquad\qquad\qquad \sqrt{2x^2} = \sqrt{2} \cdot \sqrt{x^2}.$$
$$= \sqrt{2}\sqrt{x^2} + \sqrt{6x} \qquad \sqrt{x^2} = x \text{ if } x \geq 0.$$
$$= x\sqrt{2} + \sqrt{6x}$$

EXAMPLE 3 Perform the indicated operations.

a. $\sqrt{3}(\sqrt{5} + \sqrt{12})$ **b.** $\sqrt{3x}(\sqrt{x} - \sqrt{5})$, $x \geq 0$
c. $\sqrt[3]{3x}(\sqrt[3]{9x^2} - \sqrt[3]{18x})$

Solution

a. $\sqrt{3}(\sqrt{5} + \sqrt{12}) = \sqrt{3} \cdot \sqrt{5} + \sqrt{3} \cdot \sqrt{12}$
$$= \sqrt{15} + \sqrt{36}$$
$$= \sqrt{15} + 6$$
b. $\sqrt{3x}(\sqrt{x} - \sqrt{5}) = \sqrt{3x}\sqrt{x} - \sqrt{3x}\sqrt{5}$
$$= \sqrt{3x^2} - \sqrt{15x}$$
$$= x\sqrt{3} - \sqrt{15x}$$
c. $\sqrt[3]{3x}(\sqrt[3]{9x^2} - \sqrt[3]{18x}) = \sqrt[3]{3x}\sqrt[3]{9x^2} - \sqrt[3]{3x}\sqrt[3]{18x}$
$$= \sqrt[3]{27x^3} - \sqrt[3]{54x^2}$$
$$= \sqrt[3]{3^3 \cdot x^3} - \sqrt[3]{3^3 \cdot 2 \cdot x^2}$$
$$= 3x - 3\sqrt[3]{2x^2} \qquad \blacktriangle$$

If we wish to obtain the product of two binomials containing radicals, we first simplify the radicals involved (if possible) and then use FOIL. For example, to find the product $(\sqrt{98} + \sqrt{27})(\sqrt{72} + \sqrt{75})$, we proceed as follows.

$$(\sqrt{98} + \sqrt{27})(\sqrt{72} + \sqrt{75})$$
$$= (\sqrt{49 \cdot 2} + \sqrt{9 \cdot 3})(\sqrt{36 \cdot 2} + \sqrt{25 \cdot 3})$$
$$= (7\sqrt{2} + 3\sqrt{3})(6\sqrt{2} + 5\sqrt{3})$$
$$= 7 \cdot 6 \cdot \sqrt{2} \cdot \sqrt{2} + 7 \cdot 5 \cdot \sqrt{2} \cdot \sqrt{3} + 3 \cdot 6 \cdot \sqrt{3} \cdot \sqrt{2} + 3 \cdot 5 \cdot \sqrt{3} \cdot \sqrt{3}$$
$$= 42\sqrt{2^2} + 35\sqrt{2} \cdot \sqrt{3} + 18\sqrt{3} \cdot \sqrt{2} + 15\sqrt{3^2}$$
$$= 42 \cdot 2 + 35\sqrt{6} + 18\sqrt{6} + 15 \cdot 3$$
$$= 84 + 53\sqrt{6} + 45$$
$$= 129 + 53\sqrt{6}$$

$\sqrt{98} = \sqrt{49 \cdot 2}$
$\sqrt{27} = \sqrt{9 \cdot 3}$
$\sqrt{72} = \sqrt{36 \cdot 2}$
$\sqrt{75} = \sqrt{25 \cdot 3}$

EXAMPLE 4 Find the product $(\sqrt{63} + \sqrt{75})(\sqrt{28} - \sqrt{27})$.

Solution We first simplify the radicals and then use FOIL.

$$(\sqrt{63} + \sqrt{75})(\sqrt{28} - \sqrt{27}) = (\sqrt{9 \cdot 7} + \sqrt{25 \cdot 3})(\sqrt{4 \cdot 7} - \sqrt{9 \cdot 3})$$
$$= (3\sqrt{7} + 5\sqrt{3})(2\sqrt{7} - 3\sqrt{3})$$
$$\qquad\qquad \text{F} \qquad\quad \text{O} \qquad\quad \text{I} \qquad\quad \text{L}$$
$$= 6\sqrt{7^2} - 9\sqrt{21} + 10\sqrt{21} - 15\sqrt{3^2}$$
$$= 6 \cdot 7 + \sqrt{21} - 15 \cdot 3$$
$$= 42 + \sqrt{21} - 45$$
$$= -3 + \sqrt{21}$$

Problem 3 Perform the indicated operations.
a. $\sqrt{2}(\sqrt{3} + \sqrt{10})$
b. $\sqrt{5x}(\sqrt{x} - \sqrt{3})$, $x \geq 0$
c. $\sqrt[3]{2x}(\sqrt[3]{4x^2} - \sqrt[3]{16x})$

Problem 4 Find
$(\sqrt{27} + \sqrt{28})(\sqrt{75} - \sqrt{112})$.

ANSWERS (to problems on page 370)
1. **a.** $5\sqrt{11}$ **b.** $2\sqrt{2}$ **c.** $-9\sqrt{5x}$
   d. $2\sqrt[3]{2x}$
2. **a.** $\dfrac{3\sqrt{2}}{4}$ **b.** $\dfrac{7\sqrt[3]{10x^2}}{4x}$

EXAMPLE 5 Multiply.

a. $(\sqrt{3} + 2)^2$ **b.** $(3 - \sqrt{2})^2$

c. $(\sqrt{3} + \sqrt{2})(\sqrt{3} - \sqrt{2})$

Solution

a. Since $(x + a)^2 = x^2 + 2ax + a^2$,

$$\begin{aligned}
(\sqrt{3} + 2)^2 &= (\sqrt{3})^2 + 2 \cdot 2 \cdot \sqrt{3} + 2^2 \\
&= 3 + 4\sqrt{3} + 4 \\
&= 7 + 4\sqrt{3}
\end{aligned}$$

b. Since $(x - a)^2 = x^2 - 2ax + a^2$,

$$\begin{aligned}
(3 - \sqrt{2})^2 &= 3^2 - 2 \cdot \sqrt{2} \cdot 3 + (\sqrt{2})^2 \\
&= 9 - 6\sqrt{2} + 2 \\
&= 11 - 6\sqrt{2}
\end{aligned}$$

c. Since $(x + a)(x - a) = x^2 - a^2$,

$$\begin{aligned}
(\sqrt{3} + \sqrt{2})(\sqrt{3} - \sqrt{2}) &= (\sqrt{3})^2 - (\sqrt{2})^2 \\
&= 3 - 2 \\
&= 1
\end{aligned}$$ ▲

In the next chapter, some of the answers will be of the form $\dfrac{6 + \sqrt{8}}{2}$.

We can write $\dfrac{6 + \sqrt{8}}{2}$ in lowest terms by writing $\sqrt{8}$ as $\sqrt{4 \cdot 2} = 2\sqrt{2}$, factoring, and then reducing. The procedure looks like this:

$$\frac{6 + \sqrt{8}}{2} = \frac{6 + 2\sqrt{2}}{2} = \frac{2(3 + \sqrt{2})}{2} = 3 + \sqrt{2}$$

If we view $\dfrac{6 + \sqrt{8}}{2}$ as a division of a binomial by a monomial, we can solve the problem by writing

$$\frac{6 + \sqrt{8}}{2} = \frac{6 + 2\sqrt{2}}{2} = \frac{6}{2} + \frac{2\sqrt{2}}{2} = 3 + \sqrt{2}$$

as before.

EXAMPLE 6 Reduce $\dfrac{6 + \sqrt{18}}{3}$ to lowest terms.

Solution Since $\sqrt{18} = \sqrt{9 \cdot 2} = 3\sqrt{2}$, we have

$$\frac{6 + \sqrt{18}}{3} = \frac{6 + 3\sqrt{2}}{3} = \frac{3(2 + \sqrt{2})}{3} = 2 + \sqrt{2}$$

We can also do this problem by dividing. Thus

$$\frac{6 + \sqrt{18}}{3} = \frac{6 + 3\sqrt{2}}{3} = \frac{6}{3} + \frac{3\sqrt{2}}{3} = 2 + \sqrt{2}.$$

Problem 5 Multiply.

a. $(\sqrt{2} + 3)^2$ **b.** $(2 - \sqrt{3})^2$

c. $(\sqrt{5} + \sqrt{3})(\sqrt{5} - \sqrt{3})$

Problem 6 Reduce $\dfrac{10 + \sqrt{50}}{5}$ to lowest terms.

C. RATIONALIZING DENOMINATORS

We know how to rationalize the denominator in expressions of the form $\dfrac{a}{\sqrt{b}}$. We now learn how to rationalize radical expressions that contain sums or differences involving radicals in the denominator. The procedure involves the concept of *conjugate expressions*.

> The numbers $a + b$ and $a - b$ are **conjugates** of each other.

Here are some numbers and their conjugates.

NUMBER	CONJUGATE
$3 + \sqrt{2}$	$3 - \sqrt{2}$
$-4 + \sqrt{5}$	$-4 - \sqrt{5}$
$7 - \sqrt{3}$	$7 + \sqrt{3}$
$-8 - \sqrt{6}$	$-8 + \sqrt{6}$

Since $(a + b)(a - b) = a^2 - b^2$, the product of a number and its conjugate is $a^2 - b^2$. Now, suppose we want to rationalize the denominator in $\dfrac{3}{3 + \sqrt{3}}$. We use the fundamental property of fractions and multiply the numerator and denominator of $\dfrac{3}{3 + \sqrt{3}}$ by the conjugate of $3 + \sqrt{3}$—that is, by $3 - \sqrt{3}$—obtaining

$$\frac{3}{3 + \sqrt{3}} = \frac{3 \cdot (3 - \sqrt{3})}{(3 + \sqrt{3})(3 - \sqrt{3})}$$

$$= \frac{3 \cdot (3 - \sqrt{3})}{3^2 - (\sqrt{3})^2} \qquad \text{Since } (a + b)(a - b) = a^2 - b^2,$$
$$\qquad\qquad\qquad (3 + \sqrt{3})(3 - \sqrt{3}) = 3^2 - (\sqrt{3})^2.$$

$$= \frac{3 \cdot (3 - \sqrt{3})}{9 - 3} \qquad \text{Note that } (\sqrt{3})^2 = \sqrt{3} \cdot \sqrt{3}$$
$$\qquad\qquad\qquad\qquad = \sqrt{9}$$
$$\qquad\qquad\qquad\qquad = 3$$

$$= \frac{\overset{1}{\cancel{3}} \cdot (3 - \sqrt{3})}{\underset{2}{\cancel{6}}}$$

$$= \frac{3 - \sqrt{3}}{2} \qquad\qquad \frac{\overset{1}{\cancel{3}} \cdot (3 - \sqrt{3})}{\underset{2}{\cancel{6}}} = \frac{3 - \sqrt{3}}{2}$$

EXAMPLE 7 Rationalize the denominator in the expression $\dfrac{\sqrt{x}}{\sqrt{x} - \sqrt{y}}$, where x and y represent positive numbers.

Solution

We first multiply numerator and denominator by $\sqrt{x} + \sqrt{y}$, the conjugate of $\sqrt{x} - \sqrt{y}$.

Problem 7 Rationalize the denominator.

$$\frac{\sqrt{y}}{\sqrt{y} - \sqrt{x}}$$

$$\frac{\sqrt{x}}{\sqrt{x} - \sqrt{y}} = \frac{\sqrt{x}\,(\sqrt{x} + \sqrt{y})}{(\sqrt{x} - \sqrt{y})(\sqrt{x} + \sqrt{y})}$$ Fundamental property of fractions

$$= \frac{\sqrt{x}(\sqrt{x} + \sqrt{y})}{(\sqrt{x})^2 - (\sqrt{y})^2}$$ $(\sqrt{x} + \sqrt{y})(\sqrt{x} - \sqrt{y}) = (\sqrt{x})^2 - (\sqrt{y})^2$

$$= \frac{\sqrt{x}(\sqrt{x} + \sqrt{y})}{x - y}$$ $(\sqrt{x})^2 = x$ and $(\sqrt{y})^2 = y$

$$= \frac{\sqrt{x^2} + \sqrt{xy}}{x - y}$$ Use the distributive law.

$$= \frac{x + \sqrt{xy}}{x - y}$$ $\sqrt{x^2} = x$ for $x > 0$

NAME

CLASS

SECTION

ANSWERS

A. In Problems 1–30 perform the indicated operations. (Where the index is even, assume all variables are positive.)

1. $12\sqrt{2} + 3\sqrt{2}$

2. $15\sqrt{3} + 2\sqrt{3}$

3. $\sqrt{80a} + \sqrt{125a}$

4. $\sqrt{98a} + \sqrt{32a}$

5. $\sqrt{50} - \sqrt{32}$

6. $\sqrt{75} + \sqrt{12}$

7. $\sqrt{50a^2} - \sqrt{200a^2}$

8. $\sqrt{48a^2} - \sqrt{363a^2}$

9. $2\sqrt{300} - 9\sqrt{12} - 7\sqrt{48}$

10. $\sqrt{175} + \sqrt{567} - \sqrt{63}$

11. $\sqrt[3]{40} + \sqrt[3]{625}$

12. $\sqrt[3]{54} + \sqrt[3]{16}$

13. $\sqrt[3]{81} - 3\sqrt[3]{375}$

14. $\sqrt[3]{24} - \sqrt[3]{81}$

15. $2\sqrt[3]{-24} - 4\sqrt[3]{-81} - \sqrt[3]{375}$

16. $10\sqrt[3]{-40} - 2\sqrt[3]{-135} + 4\sqrt[3]{-320}$

17. $\sqrt[3]{3a} - \sqrt[3]{24a} + \sqrt[3]{375a}$

18. $\sqrt[3]{r^5} - \sqrt[3]{8r^5} - r\sqrt[3]{64r^2}$

19. $\dfrac{3\sqrt[3]{3}}{2} - \dfrac{\sqrt[3]{3}}{3}$

20. $\dfrac{4}{5} - \dfrac{\sqrt[3]{2}}{2}$

21. $\sqrt{\dfrac{1}{2}} + \sqrt{\dfrac{1}{3}} + \sqrt{\dfrac{1}{6}}$

22. $\sqrt{\dfrac{25}{3}} - 2\sqrt{\dfrac{16}{3}} + 2\sqrt{\dfrac{4}{3}}$

23. $3\sqrt{\dfrac{1}{12}} - \sqrt{\dfrac{1}{15}} + 5\sqrt{\dfrac{3}{5}}$

24. $\sqrt{\dfrac{2}{3}} - \sqrt{\dfrac{1}{6}} + \sqrt{\dfrac{1}{2}}$

25. $\sqrt{\dfrac{x}{y}} + \sqrt{\dfrac{y}{x}} - \sqrt{\dfrac{1}{xy}}$

26. $2\sqrt[3]{\dfrac{1}{5}} + 6\sqrt[3]{\dfrac{1}{40}}$

27. $6\sqrt[3]{\dfrac{3}{5}} + 6\sqrt[3]{\dfrac{81}{40}}$

28. $2a\sqrt[3]{\dfrac{a}{5}} + 6\sqrt[3]{\dfrac{a^4}{40}}$

29. $\sqrt[3]{\dfrac{2}{3x}} - \sqrt[3]{\dfrac{3}{32x}} - \sqrt[3]{\dfrac{-2}{9x}}$

30. $3\sqrt[3]{\dfrac{x^5}{4}} - 3x\sqrt[3]{\dfrac{x^2}{108}}$

1. _____
2. _____
3. _____
4. _____
5. _____
6. _____
7. _____
8. _____
9. _____
10. _____
11. _____
12. _____
13. _____
14. _____
15. _____
16. _____
17. _____
18. _____
19. _____
20. _____
21. _____
22. _____
23. _____
24. _____
25. _____
26. _____
27. _____
28. _____
29. _____
30. _____

B. In Problems 31–64 perform the indicated operations. (Where the index is even, assume all variables are positive.)

31. $3(5 - \sqrt{2})$

32. $-2(\sqrt{2} - 3)$

33. $\sqrt[3]{2}(\sqrt[3]{4} + 3)$

34. $\sqrt[3]{3}(\sqrt[3]{9} + 2)$

35. $2\sqrt{3}(7\sqrt{5} + 5\sqrt{3})$

36. $2\sqrt{5}(5\sqrt{2} + 3\sqrt{5})$

37. $3\sqrt[3]{5}(2\sqrt[3]{3} - \sqrt[3]{25})$

38. $4\sqrt[3]{2}(3\sqrt[3]{4} - 3\sqrt[3]{2})$

39. $-4\sqrt{7}(2\sqrt{3} - 5\sqrt{2})$

40. $-3\sqrt{2}(5\sqrt{7} - 2\sqrt{3})$

41. $(5\sqrt{3} + \sqrt{5})(3\sqrt{3} + 2\sqrt{5})$

42. $(2\sqrt{2} + 5\sqrt{3})(3\sqrt{2} + \sqrt{3})$

43. $(3\sqrt{6} - 2\sqrt{3})(4\sqrt{6} + 5\sqrt{3})$

44. $(3\sqrt{5} - 2\sqrt{3})(2\sqrt{5} + 3\sqrt{3})$

45. $(7\sqrt{5} - 11\sqrt{7})(5\sqrt{5} + 8\sqrt{7})$

46. $(2\sqrt{3} - 5\sqrt{2})(3\sqrt{3} + 2\sqrt{2})$

47. $(1 + \sqrt{2})(1 - \sqrt{2})$

48. $(2 + \sqrt{3})(2 - \sqrt{3})$

49. $(2 + 3\sqrt{3})(2 - 3\sqrt{3})$

50. $(5 + 5\sqrt{2})(5 - 5\sqrt{2})$

51. $(\sqrt{3} + \sqrt{2})^2$

52. $(\sqrt{2} + \sqrt{3})^2$

53. $(a + \sqrt{b})^2$

54. $(\sqrt{a} + b)^2$

55. $(\sqrt{3} - \sqrt{2})^2$

56. $(\sqrt{2} - \sqrt{3})^2$

57. $(a - \sqrt{b})^2$

58. $(\sqrt{b} - a)^2$

59. $(\sqrt{a} - \sqrt{b})^2$

60. $(\sqrt{b} - \sqrt{a})^2$

61. $\dfrac{3 + \sqrt{18}}{3}$

62. $\dfrac{5 + \sqrt{50}}{5}$

63. $\dfrac{6 - \sqrt{27}}{12}$

64. $\dfrac{8 - \sqrt{32}}{4}$

31. _____
32. _____
33. _____
34. _____
35. _____
36. _____
37. _____
38. _____
39. _____
40. _____
41. _____
42. _____
43. _____
44. _____
45. _____
46. _____
47. _____
48. _____
49. _____
50. _____
51. _____
52. _____
53. _____
54. _____
55. _____
56. _____
57. _____
58. _____
59. _____
60. _____
61. _____
62. _____
63. _____
64. _____

C. In Problems 65–75 rationalize the denominator. (Assume all variables represent positive numbers.)

65. $\dfrac{3 + \sqrt{3}}{\sqrt{2}}$

66. $\dfrac{2 + \sqrt{5}}{\sqrt{3}}$

67. $\dfrac{2}{3 - \sqrt{2}}$

68. $\dfrac{6}{2 - \sqrt{2}}$

69. $\dfrac{4a}{3 - \sqrt{5}}$

70. $\dfrac{3a}{4 - \sqrt{3}}$

71. $\dfrac{3a + 2b}{3 + \sqrt{2}}$

72. $\dfrac{5a + b}{2 + \sqrt{3}}$

73. $\dfrac{\sqrt{a} + b}{\sqrt{a} - b}$

74. $\dfrac{a + \sqrt{b}}{a - \sqrt{b}}$

75. $\dfrac{\sqrt{a} + \sqrt{2b}}{\sqrt{a} - \sqrt{2b}}$

65. _____

66. _____

67. _____

68. _____

69. _____

70. _____

71. _____

72. _____

73. _____

74. _____

75. _____

✓ **SKILL CHECKER**

In Problems 76–80 solve the equation.

76. $x + 5 = 9$

77. $2x + 3 = 25$

78. $x^2 - 15x + 50 = 0$

79. $x^2 - 3x + 2 = 0$

80. $x^2 + 6x + 5 = 0$

76. _____

77. _____

78. _____

79. _____

80. _____

5.3 USING YOUR KNOWLEDGE

In this section we learned how to rationalize the denominator of a fraction involving the sum or difference of radical expressions. In calculus, sometimes we have to rationalize the *numerator* of a fraction involving the sum or difference of radical expressions. The idea is the same: multiply numerator and denominator by the *conjugate* of the numerator. To rationalize the numerator in

$$\frac{\sqrt{3} + \sqrt{2}}{5}$$

we proceed as follows.

$$\frac{\sqrt{3} + \sqrt{2}}{5} = \frac{(\sqrt{3} + \sqrt{2})(\sqrt{3} - \sqrt{2})}{5(\sqrt{3} - \sqrt{2})}$$ Multiply numerator and denominator by the conjugate.

$$= \frac{3 - 2}{5(\sqrt{3} - \sqrt{2})}$$ $(\sqrt{3} + \sqrt{2})(\sqrt{3} - \sqrt{2}) = (\sqrt{3})^2 - (\sqrt{2})^2$

$$= \frac{1}{5(\sqrt{3} - \sqrt{2})}$$

Rationalize the numerator.

1. $\dfrac{\sqrt{5} + \sqrt{2}}{3}$

2. $\dfrac{\sqrt{5} + \sqrt{3}}{4}$

3. $\dfrac{\sqrt{x} - \sqrt{2}}{5}$

4. $\dfrac{\sqrt{5} - \sqrt{x}}{5}$

5. $\dfrac{\sqrt{x} + \sqrt{y}}{x}$

6. $\dfrac{\sqrt{x} - \sqrt{y}}{x}$

7. $\dfrac{\sqrt{x} + \sqrt{y}}{\sqrt{x}}$

8. $\dfrac{\sqrt{x} + \sqrt{y}}{\sqrt{y}}$

9. $\dfrac{\sqrt{x} - \sqrt{y}}{\sqrt{x}}$

10. $\dfrac{\sqrt{x} - \sqrt{y}}{\sqrt{y}}$

1. _____

2. _____

3. _____

4. _____

5. _____

6. _____

7. _____

8. _____

9. _____

10. _____

5.4 EQUATIONS INVOLVING RADICALS

OBJECTIVES

REVIEW

Before starting this section, you should know:

1. How to solve linear and quadratic equations.
2. How to square an expression involving radical expressions.

OBJECTIVE

You should be able to solve equations involving radicals.

If a traffic engineer wants the speed limit v on this curve to be 45 mi/hr, what radius should the curve have? The speed v (in mi/hr) a car can travel on a concrete highway curve without skidding is $v = \sqrt{9r}$, where r is the radius of the curve in feet. Since $v = 45$, to find the answer we must find r in the equation

$$45 = \sqrt{9r}$$
$$(45)^2 = (\sqrt{9r})^2 \qquad \text{Square both sides.}$$
$$2025 = 9r \qquad (\sqrt{9r})^2 = 9r$$
$$\frac{2025}{9} = r \qquad \text{Divide by 9.}$$
$$225 = r$$

We can check this by substituting 225 for r in the equation

$$45 = \sqrt{9r}, \text{ obtaining}$$
$$45 = \sqrt{9 \cdot 225}$$
$$= \sqrt{9} \cdot \sqrt{225}$$
$$= 3 \cdot 15 \qquad \text{A true statement}$$

Thus, the curve must have a 225-ft radius. In algebra, the equation $45 = \sqrt{9r}$ is called a **radical** equation and can be solved by squaring both sides of the equation. Sometimes, however, squaring both sides of an equation introduces **extraneous solutions**, that is, solutions that do not satisfy the original equation. For example, the equation $x = 2$ has one solution, 2. Squaring both sides gives $x^2 = 4$. The equation $x^2 = 4$ has two solutions, 2 and -2.

We introduced the extraneous solution -2 when we squared both sides. However, all solutions of the equation $x = 2$ are solutions of $x^2 = 4$.

> All solutions of the equation $P = Q$ are solutions of the equation $P^n = Q^n$, n a natural number.

This rule tells us that when we raise both sides of an equation to a power, the solutions of the *original* equation are *always* solutions the new equation. However, the new equation may have extraneous solutions that have to be discarded. Because of this, *the solutions of the new equation must be checked in the original equation.*

EXAMPLE 1 Solve.

a. $\sqrt{x+5} = 3$ **b.** $\sqrt{x+1} = x-1$ **c.** $\sqrt{x-1} - x = -1$

Solution

a. We square both sides to eliminate the radical.

$$\sqrt{x+5} = 3 \qquad \text{Given.}$$
$$(\sqrt{x+5})^2 = 3^2 \qquad \text{Square both sides.}$$
$$x + 5 = 9 \qquad (\sqrt{x+5})^2 = x+5$$
$$x = 4 \qquad \text{Subtract 5 on both sides.}$$

Substituting 4 for x in the original equation gives

$$\sqrt{4+5} \stackrel{?}{=} 3$$
$$\sqrt{9} = 3$$

a true statement. Thus the solution of $\sqrt{x+5} = 3$ is 4.

b. We square both sides to eliminate the radical.

$$\sqrt{x+1} = x-1 \qquad \text{Given.}$$
$$(\sqrt{x+1})^2 = (x-1)^2 \qquad \text{Square both sides.}$$
$$x + 1 = x^2 - 2x + 1 \qquad \text{Expand } (x-1)^2.$$
$$0 = x^2 - 3x \qquad \text{Subtract } x \text{ and 1 to write the equation in standard form.}$$
$$0 = x(x-3) \qquad \text{Factor.}$$
$$x = 0 \quad \text{or} \quad x - 3 = 0 \qquad \text{Use the zero-factor property.}$$
$$x = 0 \quad \text{or} \qquad x = 3 \qquad \text{Solve } x - 3 = 0.$$

Substituting 0 for x in the original equation, we have

$$\sqrt{0+1} \stackrel{?}{=} 0 - 1$$
$$1 \stackrel{?}{=} -1 \qquad \text{A false statement}$$

Thus 0 is not a solution. Substituting 3 for x in $\sqrt{x+1} = x-1$, we have

$$\sqrt{3+1} \stackrel{?}{=} 3 - 1$$
$$\sqrt{4} = 2 \qquad \text{A true statement}$$

The solution of $\sqrt{x+1} = x-1$ is 3.

c. We first have to isolate the radical on one side, so we start by adding x on both sides. Then we proceed as before.

$$\sqrt{x-1} - x = -1 \qquad \text{Given.}$$
$$\sqrt{x-1} = x-1 \qquad \text{Add } x \text{ on both sides.}$$
$$(\sqrt{x-1})^2 = (x-1)^2 \qquad \text{Square both sides.}$$
$$x - 1 = x^2 - 2x + 1 \qquad \text{Expand on the right.}$$
$$0 = x^2 - 3x + 2 \qquad \text{Subtract } x \text{ and add 1.}$$
$$0 = (x-2)(x-1) \qquad \text{Factor.}$$
$$x - 2 = 0 \quad \text{or} \quad x - 1 = 0 \qquad \text{Use the zero-factor property.}$$
$$x = 2 \quad \text{or} \qquad x = 1 \qquad \text{Solve.}$$

Problem 1 Solve.

a. $\sqrt{x+2} = 3$

b. $\sqrt{x+1} = x-5$

c. $\sqrt{x+1} - x = 1$

The trial solutions are 2 and 1. Substituting 2 for x, we have

$$\sqrt{x-1} - x = -1$$
$$\sqrt{2-1} - 2 \overset{?}{=} -1$$
$$\sqrt{1} - 2 = -1 \quad \text{A true statement}$$

Thus 2 is a solution. Now, we check the trial solution 1 by substitution.

$$\sqrt{x-1} - x = -1$$
$$\sqrt{1-1} - 1 \overset{?}{=} -1$$
$$\sqrt{0} - 1 = -1 \quad \text{A true statement}$$

Thus 1 is also a solution. The solutions are 1 and 2. ▲

Sometimes we may have radicals on both sides of the equation. In such cases, we must isolate one of the radicals and square. Since we still have radicals on one side of the equation, we isolate them and square again. Thus to solve $\sqrt{x-11} = \sqrt{x} - 1$, we first square both sides of the equation. Note that the right-hand side contains an expression of the form $(x-a)^2 = x^2 - 2ax + a^2$. Thus we have

$$(\sqrt{x-11})^2 = (\sqrt{x} - 1)^2$$
$$x - 11 = x - 2 \cdot 1 \cdot \sqrt{x} + 1$$
$$x - 11 = x - 2\sqrt{x} + 1 \qquad \text{Subtract } x.$$
$$-11 = -2\sqrt{x} + 1 \qquad \text{Subtract 1.}$$
$$-12 = -2\sqrt{x}$$
$$6 = \sqrt{x} \qquad \text{Divide by } -2.$$
$$36 = x \qquad \text{Square both sides.}$$

The solution is 36. You can verify this by substituting in the original equation.

EXAMPLE 2 Solve $\sqrt{x-5} - \sqrt{x} = -1$.

Solution We first add $\sqrt{x}$ to both sides of the equation so that $\sqrt{x-5}$ is isolated.

$$\sqrt{x-5} = \sqrt{x} - 1$$
$$x - 5 = (\sqrt{x} - 1)^2 \qquad \text{Square both sides.}$$
$$x - 5 = x - 2\sqrt{x} + 1 \qquad (\sqrt{x} - 1)^2 = x - 2\sqrt{x} + 1$$
$$-5 = -2\sqrt{x} + 1 \qquad \text{Subtract } x.$$
$$-6 = -2\sqrt{x} \qquad \text{Subtract 1.}$$
$$3 = \sqrt{x} \qquad \text{Divide both sides by } -2.$$
$$9 = x \qquad \text{Square both sides.}$$

The trial solution is 9. Since $\sqrt{9-5} - \sqrt{9} = \sqrt{4} - \sqrt{9} = 2 - 3 = -1$, 9 is the correct solution. ▲

The sides of an equation can be raised to powers greater than 2. For example, to solve the equation $\sqrt[4]{x} = 2$, we raise both sides of the equation to the fourth power, obtaining

$$(\sqrt[4]{x})^4 = 2^4$$
$$x = 16$$

Here is another example.

Problem 2 Solve $\sqrt{x-3} - \sqrt{x} = -1$.

EXAMPLE 3 Solve $\sqrt[3]{x-2} = 3$.

Solution We cube both sides, obtaining

$$(\sqrt[3]{x-2})^3 = 3^3$$
$$x - 2 = 27$$
$$x = 29 \quad \text{Add 2.}$$

Substituting 29 for x in the original equation gives

$$\sqrt[3]{29 - 2} \stackrel{?}{=} 3$$
$$\sqrt[3]{27} = 3$$

Since this statement is true, 3 is the solution for $\sqrt[3]{x-2} = 3$.

Problem 3 Solve $\sqrt[3]{x-5} = 2$.

NAME

CLASS

SECTION

ANSWERS

Solve each equation.

1. $\sqrt{x} = 4$

2. $\sqrt{3x} = 6$

3. $\sqrt{x + 6} = 7$

4. $\sqrt{x - 3} = 10$

5. $\sqrt{\dfrac{x}{2}} = 3$

6. $\sqrt{\dfrac{3x}{2}} = 3$

7. $\sqrt[3]{x + 1} = 2$

8. $\sqrt[3]{x + 3} = 5$

9. $\sqrt[3]{3x - 1} = \sqrt[3]{5x - 7}$

10. $\sqrt[3]{5x - 3} = \sqrt[3]{7x - 5}$

11. $\sqrt{x + 4} = x + 2$

12. $\sqrt{x + 3} = x + 1$

13. $\sqrt{x + 3} = x - 3$

14. $\sqrt{x + 9} = x - 3$

15. $\sqrt[3]{y + 8} = -2$

16. $\sqrt[3]{y + 4} = -1$

17. $\sqrt{x + 5} - x = -7$

18. $\sqrt{x + 5} - x = -1$

19. $\sqrt{x - 5} - x = -7$

20. $\sqrt{x - 1} - x = -3$

In Problems 21–30, you are required to square twice to eliminate all radicals.

21. $\sqrt{y + 1} = \sqrt{y} + 1$

22. $\sqrt{y - 4} = 2 + \sqrt{y}$

23. $\sqrt{y + 8} - \sqrt{y} = 2$

24. $\sqrt{y + 5} - \sqrt{y} = 1$

25. $\sqrt{x + 3} = \sqrt{x} + \sqrt{3}$

26. $\sqrt{x + 5} = \sqrt{x} + \sqrt{5}$

27. $\sqrt{5x - 1} + \sqrt{x + 3} = 4$

28. $\sqrt{2x - 1} + \sqrt{x + 3} = 3$

ANSWERS

1. _____
2. _____
3. _____
4. _____
5. _____
6. _____
7. _____
8. _____
9. _____
10. _____
11. _____
12. _____
13. _____
14. _____
15. _____
16. _____
17. _____
18. _____
19. _____
20. _____

21. _____
22. _____
23. _____
24. _____
25. _____
26. _____
27. _____
28. _____

29. $\sqrt{x-3} + \sqrt{2x+1} = 2\sqrt{x}$

29. _____

30. $\sqrt{x+4} + \sqrt{3x+9} = \sqrt{x+25}$

30. _____

In Problems 31–40 solve for x or y.

31. $\sqrt{x-a} = b$

32. $\sqrt{x+a} = b$

31. _____

32. _____

33. $\sqrt[3]{a-by} = c$

34. $\sqrt[3]{b^3-y} = a$

33. _____

34. _____

35. $\sqrt{\dfrac{x}{a}} = b$

36. $\sqrt{\dfrac{x}{b}} = \dfrac{a}{b}$

35. _____

36. _____

37. $\sqrt{\dfrac{a}{b-x}} = \sqrt{b}$

38. $\sqrt{\dfrac{b}{a-x}} = \sqrt{a}$

37. _____

38. _____

39. $\sqrt[3]{3x-a} = \sqrt[3]{b-a}$

40. $\sqrt[3]{2x-b} = \sqrt[3]{b-2a}$

39. _____

40. _____

Applications

41. The radius r of a sphere is given by $r = \sqrt{\dfrac{S}{4\pi}}$, where S is the surface area. If the surface area of a sphere is 942 ft², find its radius. (Use $\pi = 3.14$.)

41. _____

42. The radius r of a cone is given by $r = \sqrt{\dfrac{3V}{\pi h}}$, where V is the volume and h is the height. If a 10-cm-high cone contains 94.26 cm³ of ice cream, what is its radius? (Use $\pi = 3.142$.)

42. _____

43. The time t (in seconds) it takes a body to fall d feet is given by

$$t = \sqrt{\dfrac{2d}{g}}$$

a. Solve for d.
b. How far would a body fall in 3 sec? Use $g = 32.2$.

43. _____

44. After traveling d feet, the velocity v (in feet per second) of a falling body starting from rest is given by $v = \sqrt{2gd}$.
a. Solve for d.
b. If a body that started from rest is traveling at 44 ft/sec, how far has it fallen? (Use $g = 32$.)

44. _____

45. A pendulum of length L (feet) takes $t = 2\pi\sqrt{\dfrac{L}{g}}$ seconds to go through a complete cycle.
a. Solve for L.
b. If a pendulum takes 2 sec to go through one complete cycle, how long is the pendulum? (Let $g = 32$ and $\pi = \frac{22}{7}$.)

45. _____

ANSWERS   (to problems on pages 383–384)
2. 4   **3.** 13

EXERCISE 5.4

✓ **SKILL CHECKER**

Simplify by removing parentheses and collecting like terms.

46. $(5 + 4x) + (7 - 2x)$

47. $(3 + 4x) + (8 + 2x)$

48. $(9 + 2x) - (2 + 4x)$

49. $(6 + 5x) - (7 - 3x)$

50. $(8 + 3x) - (5 - 4x)$

In Problems 51–55, rationalize the denominator.

51. $\dfrac{2 + 3\sqrt{2}}{4 + \sqrt{2}}$

52. $\dfrac{4 + 4\sqrt{3}}{3 + 2\sqrt{3}}$

53. $\dfrac{2 - \sqrt{2}}{5 - 3\sqrt{2}}$

54. $\dfrac{3 - \sqrt{3}}{5 - \sqrt{3}}$

55. $\dfrac{\sqrt{x} - \sqrt{y}}{\sqrt{x} + \sqrt{y}}$

46. _____

47. _____

48. _____

49. _____

50. _____

51. _____

52. _____

53. _____

54. _____

55. _____

5.4 USING YOUR KNOWLEDGE

Suppose you are the engineer designing several roads leading to different communities. We mentioned at the beginning of this section that the velocity v (miles per hour) that a car can travel on a concrete highway curve without skidding is $v = \sqrt{9r}$, where r (feet) is the radius of the curve.

Use your knowledge to determine the radius of the curve on a highway exit in which you want the speed to be as follows.

1. 25 mi/hr

2. 30 mi/hr

3. 35 mi/hr

4. 40 mi/hr

1. _____

2. _____

3. _____

4. _____

5.5 COMPLEX NUMBERS

Courtesy of N.Y. Public Library

OBJECTIVES

REVIEW

Before starting this section, you should know:

1. How to remove parentheses and collect like terms in an expression.
2. How to rationalize the denominator of an expression.

OBJECTIVES

You should be able to:

A. Write the square root of a negative integer in terms of i.
B. Add and subtract complex numbers.
C. Multiply and divide complex numbers.
D. Find powers of i.

In Section 5.1 we mentioned that $\sqrt{-16}$, $\sqrt{-2}$, and $\sqrt{-12}$ are not real numbers because the square of a real number is always a nonnegative real number. In mathematics, we say that the set of real numbers is not closed with respect to the operation of taking square roots. To avoid this difficulty, the man in the photo, Carl Friedrich Gauss (1777–1855), developed a new set of numbers containing elements that are square roots of negative numbers. The first of these numbers is i; it is defined as follows.

---

i is a number such that $i^2 = -1$; that is, $i = \sqrt{-1}$.

A. WRITING SQUARE ROOTS OF NEGATIVE NUMBERS IN TERMS OF i

With this definition, the square root of any negative real number can be written as the product of a real number and i. Thus

$$\sqrt{-4} = \sqrt{-1} \cdot \sqrt{4} = i2, \quad \text{or } 2i$$
$$\sqrt{-3} = \sqrt{-1 \cdot 3} = \sqrt{-1} \cdot \sqrt{3} = i\sqrt{3}$$

Since it is easy to confuse $\sqrt{3}i$ and $\sqrt{3i}$, we write products involving radicals and i as factors with the i in front; that is, we write $i\sqrt{5}$ instead of $\sqrt{5}i$.

EXAMPLE 1 Write the given expression in terms of i.

a. $\sqrt{-9}$ **b.** $\sqrt{-18}$

Solution

a. $\sqrt{-9} = \sqrt{-1 \cdot 9} = \sqrt{-1}\sqrt{9} = 3i$
b. $\sqrt{-18} = \sqrt{-1 \cdot 18} = \sqrt{-1}\sqrt{18} = i\sqrt{18} = i\sqrt{9 \cdot 3} = 3i\sqrt{3}$ ▲

The numbers $3i$ and $3i\sqrt{2}$ are called **pure imaginary** numbers. We can form a new set of numbers by adding these imaginary numbers to real numbers as follows.

Problem 1 Write in terms of i.
a. $\sqrt{-25}$ **b.** $\sqrt{-50}$

> If a and b are real numbers, then any number of the form $a + bi$ is called a **complex number.**

In the complex number $a + bi$, a is called the *real* part and bi the *imaginary* part. Thus the number $-3 + 4i$ is a complex number whose real part is -3 and whose imaginary part is $4i$. Similarly, $2 - 3i$ is a complex number with 2 as its real part and $-3i$ as its imaginary part. The sum and the difference of complex numbers are defined next.

$$a \quad + \quad bi$$
$$\uparrow \qquad \quad \uparrow$$
real   imaginary
part   part

B. ADDITION AND SUBTRACTION OF COMPLEX NUMBERS

> **RULE FOR ADDING AND SUBTRACTING**
>
> For a, b, c, and d real numbers,
> $$(a + bi) + (c + di) = (a + c) + (b + d)i$$
> $$(a + bi) - (c + di) = (a - c) + (b - d)i$$

In effect, to add (or subtract) complex numbers, we add (or subtract) the real parts and the imaginary parts separately. For example, $(3 + 4i) + (8 + 2i) = (3 + 8) + (4 + 2)i = 11 + 6i$, and $(9 + 2i) - (2 + 4i) = (9 - 2) + (2 - 4)i = 7 - 2i$.

EXAMPLE 2   Find.

a. $(5 + 4i) + (7 - 2i)$ **b.** $(6 + 5i) - (7 - 3i)$

Solution

a. $(5 + 4i) + (7 - 2i) = (5 + 7) + [4 + (-2)]i$
$$= 12 + 2i$$
b. $(6 + 5i) - (7 - 3i) = (6 - 7) + [5 - (-3)]i$
$$= -1 + 8i$$

Problem 2   Find.
a. $(7 + 3i) + (2 + 4i)$
b. $(2 + 3i) - (4 + 5i)$

C. MULTIPLICATION AND DIVISION OF COMPLEX NUMBERS

To define the product of two complex numbers, we recall that, using FOIL,

$$\begin{array}{cccc} \text{F} & \text{O} & \text{I} & \text{L} \end{array}$$
$$(a + b)(c + d) = ac + ad + bc + bd$$

Thus a reasonable way in which to form the product of two complex numbers is as follows:

$$\begin{array}{cccc} \text{F} & \text{O} & \text{I} & \text{L} \end{array}$$
$$(a + bi)(c + di) = ac + adi + bci + bdi^2$$
$$\qquad\qquad\qquad = ac + adi + bci - bd \qquad \text{Recall that } i^2 = -1, \text{ so } bdi^2 = -bd.$$
$$\qquad\qquad\qquad = (ac - bd) + (ad + bc)i \qquad \text{Group the real and imaginary parts.}$$

We summarize this discussion formally in the next definition.

RULE FOR MULTIPLYING COMPLEX NUMBERS

If a, b, c, and d are real numbers,

$$(a + bi)(c + di) = (ac - bd) + (ad + bc)i$$

In practice, we do not need to memorize this definition to multiply complex numbers—we can merely follow the rule used to multiply binomials (FOIL) and replace i^2 by -1. Thus

$$(3 + 4i)(2 + 3i) = 6 + 9i + 8i + 12i^2$$
$$= 6 + 9i + 8i - 12 \quad \text{Since } i^2 = -1, \ 12i^2 = -12.$$
$$= -6 + 17i$$

Note that the answer is written in the form $a + bi$.

EXAMPLE 3 Find the product.

a. $(2 - 5i)(3 + 7i)$ **b.** $-3i(4 - 7i)$

Solution

a. $(2 - 5i)(3 + 7i) = 6 + 14i - 15i - 35i^2$
$$= 6 - i + 35$$
$$= 41 - i$$

b. $-3i(4 - 7i) = -12i + 21i^2$
$$= -12i - 21$$
$$= -21 - 12i$$ ▲

Expressions such as $\sqrt{-9}$ and $\sqrt{-18}$ should be written in the form bi *before* any other operations are carried out. For example, to multiply $\sqrt{-9} \cdot \sqrt{-4}$, we must first write $\sqrt{-9} \cdot \sqrt{-4} = 3i \cdot 2i$. Then, $3i \cdot 2i = 6i^2 = 6(-1) = -6$. If we were to use the product rule for radicals, we would get $\sqrt{-9} \cdot \sqrt{-4} = \sqrt{(-9) \cdot (-4)} = \sqrt{36} = 6$. This is *not correct*!

EXAMPLE 4 Find.

a. $\sqrt{-16}(3 + \sqrt{-8})$ **b.** $\sqrt{-36}(\sqrt{-3} - \sqrt{-18})$

Solution

a. We first write the square roots of negative numbers in terms of i and then proceed as usual. Since $\sqrt{-16} = 4i$, $\sqrt{-8} = 2i\sqrt{2}$, $\sqrt{-36} = 6i$, and $\sqrt{-18} = 3i\sqrt{2}$, we write

$$\sqrt{-16}(3 + \sqrt{-8}) = 4i(3 + 2i\sqrt{2}) \quad \sqrt{-16} = 4i \text{ and } \sqrt{-8} = 4i\sqrt{2}$$
$$= 12i + 8i^2\sqrt{2} \quad \text{Use the distributive law.}$$
$$= 12i - 8\sqrt{2} \quad \text{Since } i^2 = -1, 8\sqrt{2}i^2 = -8\sqrt{2}.$$
$$= -8\sqrt{2} + 12i \quad \text{Write in the form } a + bi.$$

Problem 3 Find the product.
a. $(2 - 4i)(2 + 6i)$ **b.** $-4i(5 - 8i)$

Problem 4 Find.
a. $\sqrt{-25}(6 + \sqrt{-8})$
b. $\sqrt{-49}(\sqrt{-3} - \sqrt{-27})$

b. $\sqrt{-36}(\sqrt{-3} - \sqrt{-18}) = 6i(i\sqrt{3} - 3i\sqrt{2})$

$$= 6i^2\sqrt{3} - 18i^2\sqrt{2}$$

$$= -6\sqrt{3} + 18\sqrt{2} \qquad \blacktriangle$$

To define the *quotient* of two complex numbers, we use the rationalizing process developed in Section 5.2 and the assumption that $\dfrac{a + bi}{c} = \dfrac{a}{c} + \dfrac{bi}{c}$. For example, to find $\dfrac{2 + 3i}{4 - i}$, we proceed as follows.

$$\dfrac{2 + 3i}{4 - i} = \dfrac{(2 + 3i)(4 + i)}{(4 - i)(4 + i)} \qquad \text{Multiply the numerator and denominator by the conjugate of } 4 - i, \text{ that is, } 4 + i.$$

$$= \dfrac{8 + 2i + 12i + 3i^2}{16 + 4i - 4i - i^2}$$

$$= \dfrac{8 + 2i + 12i - 3}{16 + 4i - 4i + 1}$$

$$= \dfrac{5 + 14i}{17}$$

$$= \dfrac{5}{17} + \dfrac{14i}{17}$$

In general,

$$\dfrac{a + bi}{c + di} = \dfrac{(a + bi)(c - di)}{(c + di)(c - di)}$$

$$= \dfrac{ac - adi + bci - bdi^2}{c^2 - cdi + cdi - d^2i^2}$$

$$= \dfrac{ac - adi + bci + bd}{c^2 + d^2}$$

$$= \dfrac{(ac + bd) + (bc - ad)i}{c^2 + d^2}$$

$$= \dfrac{ac + bd}{c^2 + d^2} + \dfrac{(bc - ad)i}{c^2 + d^2}$$

This fact is stated in the next definition.

RULE FOR DIVIDING COMPLEX NUMBERS

For a, b, c, and d real numbers (c and d not both 0),

$$\dfrac{a + bi}{c + di} = \dfrac{ac + bd}{c^2 + d^2} + \dfrac{(bc - ad)i}{c^2 + d^2}$$

This definition gives the quotient of two complex numbers as another complex number. However, it is not necessary to memorize the formula in this definition; it is better to use the procedure of multiplying numerator and denominator by the conjugate of the denominator, as we did in leading up to the definition. The important fact to know is that $(c + di)(c - di) = c^2 + d^2$; that is, the product of a complex number and its conjugate is a real number, the sum of the squares of the two real-number components. We use this fact in the next example.

Recall that the conjugate of $a + bi$ is $a - bi$.

EXAMPLE 5 Find.

a. $\dfrac{5 + 4i}{3 + 2i}$ **b.** $\dfrac{2 - 4i}{5 - 3i}$ **c.** $\dfrac{3 - 2i}{i}$

Problem 5 Find.

a. $\dfrac{3 + 5i}{2 + 3i}$ **b.** $\dfrac{3 - 5i}{4 - 3i}$ **c.** $\dfrac{2 - 3i}{i}$

Solution

a. $\dfrac{5 + 4i}{3 + 2i} = \dfrac{(5 - 4i)(3 - 2i)}{(3 + 2i)(3 - 2i)}$ Multiply by the conjugate of $3 + 2i$.

$= \dfrac{15 - 10i + 12i - 8i^2}{3^2 + 2^2}$ By definition, the denominator is $c^2 + d^2 = 3^2 + 2^2$.

$= \dfrac{15 - 10i + 12i + 8}{13}$

$= \dfrac{23 + 2i}{13}$

$= \dfrac{23}{13} + \dfrac{2}{13}i$

b. $\dfrac{2 - 4i}{5 - 3i} = \dfrac{(2 - 4i)(5 + 3i)}{(5 - 3i)(5 + 3i)}$

$= \dfrac{10 + 6i - 20i - 12i^2}{5^2 + 3^2}$

$= \dfrac{10 + 6i - 20i + 12}{34}$

$= \dfrac{22 - 14i}{34}$

$= \dfrac{22}{34} - \dfrac{14}{34}i$

$= \dfrac{11}{17} - \dfrac{7}{17}i$

c. Since the conjugate of $a + bi$ is $a - bi$, the conjugate of $0 + 1i$ is $0 - 1i = -i$. Multiplying numerator and denominator of the fraction by $-i$, we have

$\dfrac{3 - 2i}{i} = \dfrac{(3 - 2i)(-i)}{i \cdot (-i)}$

$= \dfrac{-3i + 2i^2}{-i^2}$

$= \dfrac{-3i - 2}{1}$

$= -2 - 3i$

D. POWERS OF i

We already know that, by definition, $i^2 = -1$. If we assume that the laws of exponents hold, we can write any power of i as one of 1, -1, i, or $-i$. Thus

$i = i$ $i^5 = i \cdot i^4 = i \cdot (1) = i$

$i^2 = -1$ $i^6 = i \cdot i^5 = i \cdot i = -1$

$i^3 = i \cdot i^2 = i(-1) = -i$ $i^7 = i \cdot i^6 = i(-1) = -i$

$i^4 = i^2 \cdot i^2 = (-1)(-1) = 1$ $i^8 = i \cdot i^7 = i \cdot (-i) = 1$

Since $i^4 = 1$, the easiest way to simplify higher powers of i is to write them in terms of i^4. Thus to find i^{20}, we write:

$$i^{20} = (i^4)^5 = (1)^5 = 1$$

Similarly,

$$i^{21} = (i^4)^5 \cdot i = 1 \cdot i = i$$
$$i^{22} = (i^4)^5 \cdot i^2 = 1 \cdot (-1) = -1$$
$$i^{23} = (i^4)^5 \cdot i^3 = 1 \cdot i^3 = -i$$

Note that dividing the exponent, 20 in this case, by 4 will give you the answer!

If the remainder is 0, the answer is 1 (as in i^{20}).

If the remainder is 1, the answer is i (as in i^{21}).

If the remainder is 2, the answer is -1 (as in i^{22}).

If the remainder is 3, the answer is $-i$ (as in i^{23}).

After that, the answers repeat.

EXAMPLE 6 Find.

a. i^{53} **b.** i^{47} **c.** i^{-3} **d.** i^{-1}

Solution

a. Dividing 53 by 4, we obtain 13 with a remainder of 1. Thus the answer is i. To show this, write

$$i^{53} = (i^4)^{13} \cdot i$$
$$= 1 \cdot i = i$$

b. If we divide 47 by 4, the remainder is 3. Thus the answer is $-i$. Note that

$$i^{47} = (i^4)^{11} \cdot i^3$$
$$= 1 \cdot i^3$$
$$= -i$$

c. By definition of negative exponents, $i^{-3} = 1/i^3$.

$$i^{-3} = \frac{1 \cdot i}{i^3 \cdot i} \quad \text{Writing with a denominator of } i^4 = 1$$

$$= \frac{i}{1} = i$$

d. $i^{-1} = \dfrac{1}{i} = \dfrac{1 \cdot i^3}{i \cdot i^3} = \dfrac{i^3}{i^4} = i^3 = -i$ ▲

The introduction of the complex numbers completes the development of our number system. We started with the natural numbers (N) and whole numbers (W), studied the integers (I) and the rationals (Q), and then concentrated on the real numbers (R). Now we have developed the complex numbers (C), which include all the other numbers we have discussed. In set language, the relationship is

$$N \subset W \subset I \subset Q \subset R \subset C$$

as shown in the diagram on page 395.

Problem 6 Find.

a. i^{42} **b.** i^{23} **c.** i^{-8} **d.** i^{-2}

ANSWER (to problem on page 393)

5. a. $\dfrac{21}{13} + \dfrac{1}{13}i$ **b.** $\dfrac{27}{25} - \dfrac{11}{25}i$ **c.** $-3 - 2i$

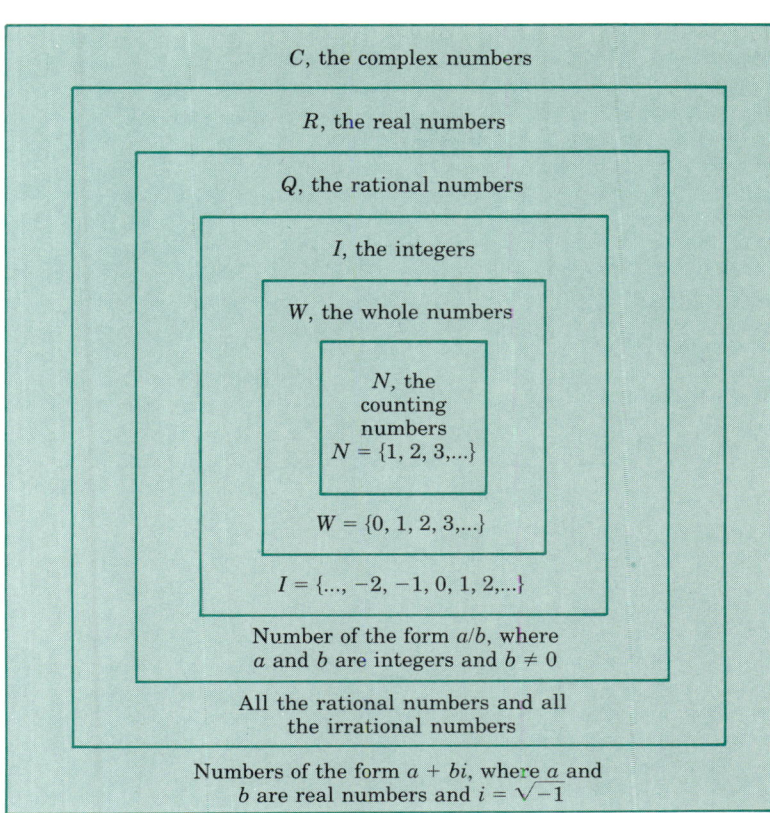

C, the complex numbers

R, the real numbers

Q, the rational numbers

I, the integers

W, the whole numbers

N, the counting numbers
$N = \{1, 2, 3,...\}$

$W = \{0, 1, 2, 3,...\}$

$I = \{..., -2, -1, 0, 1, 2,...\}$

Number of the form a/b, where a and b are integers and $b \neq 0$

All the rational numbers and all the irrational numbers

Numbers of the form $a + bi$, where a and b are real numbers and $i = \sqrt{-1}$

A. In Problems 1–10 write the given expression in terms of i.

1. $\sqrt{-25}$

2. $\sqrt{-81}$

3. $\sqrt{-50}$

4. $\sqrt{-98}$

5. $4\sqrt{-72}$

6. $3\sqrt{-200}$

7. $-3\sqrt{-32}$

8. $-5\sqrt{-34}$

9. $4\sqrt{-28} + 3$

10. $7\sqrt{-18} + 5$

B. In Problems 11–30 perform the indicated operations. (Write the answer in the form $a + bi$.)

11. $(4 + i) + (2 + 3i)$

12. $(7 + 3i) + (2 + i)$

13. $(3 - 2i) - (5 + 4i)$

14. $(4 - 5i) - (2 + 3i)$

15. $(-3 - 5i) + (-2 - i)$

16. $(-7 - 3i) + (-2 - i)$

17. $(3 + \sqrt{-4}) - (5 - \sqrt{-9})$

18. $(-2 - \sqrt{-16}) - (3 - \sqrt{-25})$

19. $(-5 + \sqrt{-1}) + (-2 + 3\sqrt{-1})$

20. $(-3 + 2\sqrt{-1}) + (-4 + 5\sqrt{-1})$

21. $(3 - 4i) + (5 + 3i)$

22. $(3 - 7i) + (3 + 4i)$

23. $(4 + \sqrt{-9}) + (6 + \sqrt{-4})$

24. $(-3 - \sqrt{-25}) + (5 - \sqrt{-16})$

25. $(2 - \sqrt{-2}) - (5 + \sqrt{-2})$

26. $(3 + \sqrt{-50}) - (7 + \sqrt{-2})$

27. $(-5 - \sqrt{-2}) - (-4 - \sqrt{-18})$

28. $(-8 - \sqrt{-125}) - (-2 - \sqrt{-5})$

ANSWERS

1. _____
2. _____
3. _____
4. _____
5. _____
6. _____
7. _____
8. _____
9. _____
10. _____

11. _____
12. _____
13. _____
14. _____
15. _____
16. _____
17. _____
18. _____
19. _____
20. _____
21. _____
22. _____
23. _____
24. _____
25. _____
26. _____
27. _____
28. _____

29. $(-4 + \sqrt{-20}) + (-3 + \sqrt{-5})$

30. $(-7 + \sqrt{-24}) + (-3 + \sqrt{-6})$

C. In Problems 31–70 perform the indicated operations. (Write the answer in the form $a + bi$.)

31. $3(4 + 2i)$

32. $5(4 + 3i)$

33. $-4(3 - 5i)$

34. $-3(7 - 4i)$

35. $\sqrt{-4}(3 + 2i)$

36. $\sqrt{-9}(2 + 5i)$

37. $\sqrt{-3}(3 + \sqrt{-3})$

38. $\sqrt{-5}(2 - \sqrt{-5})$

39. $3i(3 + 2i)$

40. $7i(4 + 3i)$

41. $4i(3 - 7i)$

42. $-5i(2 - 3i)$

43. $-\sqrt{-16}(-5 - \sqrt{-25})$

44. $-\sqrt{-25}(-3 - \sqrt{-9})$

45. $(3 + i)(2 + 3i)$

46. $(2 + 3i)(4 + 5i)$

47. $(3 - 2i)(3 + 2i)$

48. $(4 - 3i)(4 + 3i)$

49. $(3 + 2\sqrt{-4})(4 - \sqrt{-9})$

50. $(-3 + 3\sqrt{-9})(-2 + 5\sqrt{-4})$

51. $(2 + 3\sqrt{-3})(2 - 3\sqrt{-3})$

52. $(4 + 2\sqrt{-5})(4 - 2\sqrt{-5})$

53. $\dfrac{3}{i}$

54. $\dfrac{5}{i}$

55. $\dfrac{6}{-i}$

56. $\dfrac{3}{-2i}$

57. $\dfrac{i}{1 + 2i}$

58. $\dfrac{2i}{1 + 3i}$

59. $\dfrac{3i}{1 - 2i}$

60. $\dfrac{4i}{2 - 3i}$

61. $\dfrac{3 + 4i}{1 - 2i}$

62. $\dfrac{3 + 5i}{1 - 3i}$

29. _____

30. _____

31. _____

32. _____

33. _____

34. _____

35. _____

36. _____

37. _____

38. _____

39. _____

40. _____

41. _____

42. _____

43. _____

44. _____

45. _____

46. _____

47. _____

48. _____

49. _____

50. _____

51. _____

52. _____

53. _____

54. _____

55. _____

56. _____

57. _____

58. _____

59. _____

60. _____

61. _____

62. _____

63. $\dfrac{4 + 3i}{2 + 3i}$

64. $\dfrac{5 + 4i}{3 + 2i}$

65. $\dfrac{3}{\sqrt{-4}}$

66. $\dfrac{-4}{\sqrt{-9}}$

67. $\dfrac{3 + \sqrt{-5}}{4 + \sqrt{-2}}$

68. $\dfrac{2 + \sqrt{-2}}{1 + \sqrt{-3}}$

69. $\dfrac{-1 - \sqrt{-2}}{-3 - \sqrt{-3}}$

70. $\dfrac{-1 - \sqrt{-3}}{-2 - \sqrt{-2}}$

D. In Problems 71–80 write the answer as 1, -1, i, or $-i$.

71. i^{40}

72. i^{28}

73. i^{19}

74. i^{38}

75. i^{21}

76. i^{44}

77. i^{32}

78. i^{53}

79. i^{65}

80. i^{16}

E. Applications

81. The impedance in a circuit is the measure of how much a circuit impedes (hinders) the flow of current through it. If the impedance of a resistor is $Z_1 = 5 + 3i$ ohms and the impedance of another resistor is $Z_2 = 3 - 2i$ ohms, what is the total impedance (sum) of the two resistors when placed in series?

82. Repeat Problem 81 if the impedance of the first resistor is $3 + 7i$ and the impedance of the second is $4 - 5i$.

83. If two resistors Z_1 and Z_2 are connected in parallel, their total impedance is given by $Z_t = \dfrac{Z_1 \cdot Z_2}{Z_1 + Z_2}$. Find the total impedance of the resistors of Problem 81.

84. Use the formula in Problem 83 and find the total impedance of the resistors of Problem 82.

✓ **SKILL CHECKER**

Solve.

85. $x^2 - 4 = 0$

86. $x^2 - 16 = 0$

63. _____

64. _____

65. _____

66. _____

67. _____

68. _____

69. _____

70. _____

71. _____

72. _____

73. _____

74. _____

75. _____

76. _____

77. _____

78. _____

79. _____

80. _____

81. _____

82. _____

83. _____

84. _____

85. _____

86. _____

87. $x^2 = 25$ **88.** $x^2 = 36$

87. _____

88. _____

5.5 USING YOUR KNOWLEDGE

If x is a real number, the absolute value of x is defined as follows.

$$|x| = \begin{cases} x, & \text{if } x > 0 \\ 0, & \text{if } x = 0 \\ -x, & \text{if } x < 0 \end{cases}$$

Thus

$$\begin{aligned} |5| &= 5 \qquad \text{because } 5 > 0 \\ |0| &= 0 \\ |-8| &= -(-8) = 8 \qquad \text{because } -8 < 0. \end{aligned}$$

How can we define the absolute value of a complex number? The definition is

$$|a + bi| = \sqrt{a^2 + b^2}$$

Use this definition to find each value.

1. $|3 + 4i|$ **2.** $|12 + 5i|$

3. $|2 - 3i|$ **4.** $|5 - 7i|$

1. _____

2. _____

3. _____

4. _____

SUMMARY

SECTION	ITEM	MEANING	EXAMPLE
5.1	nth root	If a and x are real numbers and n is a positive integer, x is an nth root of a if $x^n = a$.	The square root of 25 is 5, the cube root of -8 is -2, and the fourth root of -16 is not a real number.
5.1A	$\sqrt{}$	A radical sign	
5.1A	Radicand	In $\sqrt[n]{a}$, a is the radicand.	In $\sqrt[3]{22}$, 22 is the radicand.
5.1A	Index (order)	In $\sqrt[n]{a}$, the index is n.	In $\sqrt[3]{22}$, the index is 3.
5.1A	$\sqrt[n]{a}$, (a radical expression)	The nth root of a	$\sqrt[3]{64} = 4$, $\sqrt[5]{-32} = -2$, and $\sqrt{-25}$ is not a real number.
5.1B	$a^{1/n}$	$a^{1/n} = \sqrt[n]{a}$, if it exists.	$8^{1/3} = \sqrt[3]{8}$ and $32^{1/5} = \sqrt[5]{32}$
5.1B	$a^{m/n}$	$(\sqrt[n]{a})^m = \sqrt[n]{a^m}$	$16^{3/2} = (\sqrt{16})^3 = 4^3 = 64$
5.1B	$a^{-m/n}$	$\dfrac{1}{(a^{mn})}$	$4^{-3/2} = \dfrac{1}{4^{3/2}} = \dfrac{1}{8}$
5.1C	$(a^m)^n$ (m and n positive, even integers)	$|a|^{m/n}$	$[(-4)^2]^{1/2} = |-4| = 4$
5.2	$\sqrt[n]{a^n}$	a, for $a \geq 0$	$\sqrt[4]{2^4} = 2$
5.2	$\sqrt[n]{ab}$	$\sqrt[n]{a} \cdot \sqrt[n]{b}$	$\sqrt{18} = \sqrt{9} \cdot \sqrt{2} = 3\sqrt{2}$
5.2	$\sqrt[n]{\dfrac{a}{b}}$	$\dfrac{\sqrt[n]{a}}{\sqrt[n]{b}}$	$\sqrt[3]{\dfrac{8}{27}} = \dfrac{\sqrt[3]{8}}{\sqrt[3]{27}} = \dfrac{2}{3}$

(Continued)

SECTION	ITEM	MEANING	EXAMPLE
5.2C	Simplified form of a radical	A radical is in simplified form if the radicand has no factors with exponents greater than or equal to the index, there are no fractions under the radical sign or radicals in the denominator, and the index is as low as possible.	$\dfrac{\sqrt[3]{2ac}}{2c^2}$ is in simplified form.
5.3	Like radicals	Radicals with the same index and the same radicand	$\sqrt[3]{3x^2}$ and $-7\sqrt[3]{3x^2}$ are like radicals.
5.3C	Conjugates	$a + b$ and $a - b$ are conjugates.	$3 + \sqrt{2}$ and $3 - \sqrt{2}$ are conjugates.
5.4	Extraneous solution	A trial solution that does not satisfy the original equation	0 is an extraneous solution of $\sqrt{x + 1} = x - 1$.
5.5	i	$\sqrt{-1}$	
5.5A	Complex number	A number that can be written in the form $a + bi$, where a and b are real	$3 + 7i$ and $-4 - 8i$ are complex numbers.
5.5B	Addition and subtraction of complex numbers	$(a + bi) \pm (c + di) =$ $(a \pm c) + (b \pm d)i$	$(3 + 2i) + (5 + 3i) = 8 + 5i$ $(4 - 2i) - (5 + 3i) = -1 - 5i$
5.5C	Multiplication of complex numbers	$(a + bi)(c + di) =$ $(ac - bd) + (ad + bc)i$	$(2 + 3i)(4 - 5i)$ $= [2 \cdot 4 - 3 \cdot (-5)]$ $\quad + [2 \cdot (-5) + 3 \cdot 4]i$ $= 23 + 2i$

ANSWERS

(*If you need help with these exercises, look in the section indicated in brackets.*)

In Problems 1–4, find the real roots.

1. [5.1A]
    a. $\sqrt{-8}$
    b. $\sqrt[3]{-64}$

2. [5.1A]
    a. $\sqrt{-9}$
    b. $\sqrt[3]{-125}$

3. [5.1B]
    a. $(-27)^{1/3}$
    b. $(-64)^{1/3}$

4. [5.1B]
    a. $\left(\dfrac{1}{16}\right)^{1/4}$
    b. $\left(\dfrac{1}{256}\right)^{1/4}$

In Problems 5–8, evaluate if possible.

5. [5.1B]
    a. $125^{2/3}$
    b. $64^{2/3}$

6. [5.1B]
    a. $(-25)^{3/2}$
    b. $(-36)^{3/2}$

7. [5.1B]
    a. $(-8)^{-2/3}$
    b. $(-64)^{-2/3}$

8. [5.1B]
    a. $8^{-2/3}$
    b. $64^{-2/3}$

In Problems 9–18 simplify (*x* and *y* positive).

9. [5.1C]
    a. $x^{1/5} \cdot x^{1/3}$
    b. $x^{1/5} \cdot x^{1/4}$

1. a. _____
   b. _____

2. a. _____
   b. _____

3. a. _____
   b. _____

4. a. _____
   b. _____

5. a. _____
   b. _____

6. a. _____
   b. _____

7. a. _____
   b. _____

8. a. _____
   b. _____

9. a. _____
   b. _____

10. [5.1C]

    a. $\dfrac{x^{-1/4}}{x^{1/5}}$

    b. $\dfrac{x^{-1/3}}{x^{1/5}}$

10. a. _____
    b. _____

11. [5.1C]
    a. $(x^{1/3}y^{2/5})^{-15}$
    b. $(x^{1/6}y^{2/5})^{-30}$

11. a. _____
    b. _____

12. [5.1C]
    a. $x^{3/5}(x^{-1/5} + y^{3/5})$
    b. $x^{4/5}(x^{-1/5} + y^{3/5})$

12. a. _____
    b. _____

13. [5.2A]
    a. $\sqrt[4]{(-7)^4}$
    b. $\sqrt[4]{(-6)^4}$

13. a. _____
    b. _____

14. [5.2A]
    a. $\sqrt[8]{(-x)^8}$
    b. $\sqrt[4]{(-x)^4}$

14. a. _____
    b. _____

15. [5.2A]
    a. $\sqrt[3]{48}$
    b. $\sqrt[3]{56}$

15. a. _____
    b. _____

16. [5.2A]
    a. $\sqrt[3]{16x^4y^6}$
    b. $\sqrt[3]{16x^8y^{15}}$

16. a. _____
    b. _____

17. [5.2A]

    a. $\sqrt{\dfrac{5}{243}}$

    b. $\sqrt{\dfrac{5}{1024}}$

17. a. _____
    b. _____

18. [5.2A]

    a. $\sqrt[3]{\dfrac{1}{x^3}}$

    b. $\sqrt[3]{\dfrac{5}{x^3}}$

18. a. _____
    b. _____

In Problems 19–22 rationalize the denominator.

19. [5.2B]

    a. $\dfrac{\sqrt{5}}{\sqrt{11}}$

    b. $\dfrac{\sqrt{5}}{\sqrt{13}}$

19. a. _____
    b. _____

20. [5.2B]

a. $\dfrac{\sqrt{2}}{\sqrt{5x}}, \quad x > 0$

b. $\dfrac{\sqrt{3}}{\sqrt{5x}}, \quad x > 0$

20. a. _____
    b. _____

21. [5.2B]

a. $\dfrac{1}{\sqrt[3]{5x}}$

b. $\dfrac{1}{\sqrt[3]{7x}}$

21. a. _____
    b. _____

22. [5.2B]

a. $\dfrac{\sqrt[5]{1}}{\sqrt[5]{16x^3}}$

b. $\dfrac{\sqrt[5]{5}}{\sqrt[5]{16x^3}}$

22. a. _____
    b. _____

In Problems 23–24 reduce the index (order) of the expression.

23. [5.2C]

a. $\sqrt[4]{\dfrac{256}{81}}$

b. $\sqrt[4]{\dfrac{625}{81}}$

23. a. _____
    b. _____

24. [5.2C]
a. $\sqrt[6]{81c^4d^4}$
b. $\sqrt[6]{625c^4d^4}$

24. a. _____
    b. _____

25. [5.2C] Simplify.

a. $\sqrt[6]{\dfrac{3a^2}{243c^{16}}}$

b. $\sqrt[6]{\dfrac{9a^2}{81c^{16}}}$

25. a. _____
    b. _____

In Problems 26–31, perform the indicated operations.

26. [5.3A]
a. $\sqrt{8} + \sqrt{32}$
b. $\sqrt{18} + \sqrt{32}$

26. a. _____
    b. _____

27. [5.3A]
a. $\sqrt{63} - \sqrt{28}$
b. $\sqrt{112} - \sqrt{63}$

27. a. _____
    b. _____

28. [5.3A]

   a. $3\sqrt{\dfrac{1}{2}} - 3\sqrt{\dfrac{1}{8}}$

   b. $4\sqrt{\dfrac{1}{2}} - 3\sqrt{\dfrac{1}{8}}$

28. a. _____
   b. _____

29. [5.3A]

   a. $5\sqrt[3]{\dfrac{3}{4x}} - \sqrt[3]{\dfrac{3}{32x}}$

   b. $6\sqrt[3]{\dfrac{3}{4x}} - \sqrt[3]{\dfrac{3}{32x}}$

29. a. _____
   b. _____

30. [5.3B]
   a. $\sqrt{2}(\sqrt{18} + \sqrt{3})$
   b. $\sqrt{2}(\sqrt{32} + \sqrt{3})$

30. a. _____
   b. _____

31. [5.3B]
   a. $\sqrt[3]{2x}(\sqrt[3]{24x^2} - \sqrt[3]{81x})$
   b. $\sqrt[3]{3x}(\sqrt[3]{16x^2} - \sqrt[3]{54x})$

31. a. _____
   b. _____

In Problems 32–35, find the product.

32. [5.3B]
   a. $(\sqrt{27} + \sqrt{18})(\sqrt{12} + \sqrt{8})$
   b. $(\sqrt{12} + \sqrt{18})(\sqrt{12} + \sqrt{8})$

32. a. _____
   b. _____

33. [5.3B]
   a. $(\sqrt{3} + 4)(\sqrt{3} + 4)$
   b. $(\sqrt{3} + 5)(\sqrt{3} + 5)$

33. a. _____
   b. _____

34. [5.3B]
   a. $(3 - \sqrt{3})^2$
   b. $(4 - \sqrt{3})^2$

34. a. _____
   b. _____

35. [5.3B]
   a. $(\sqrt{6} + \sqrt{3})(\sqrt{6} - \sqrt{3})$
   b. $(\sqrt{7} + \sqrt{3})(\sqrt{7} - \sqrt{3})$

35. a. _____
   b. _____

36. [5.3B] Reduce.

   a. $\dfrac{20 - \sqrt{50}}{5}$

   b. $\dfrac{30 - \sqrt{50}}{5}$

36. a. _____
   b. _____

37. [5.3C] Rationalize the denominator.

   a. $\dfrac{\sqrt{y}}{\sqrt{x} - \sqrt{y}}$

   b. $\dfrac{\sqrt{x}}{\sqrt{x} - \sqrt{y}}$

37. a. _____
   b. _____

38. [5.4] Solve.
   a. $\sqrt{x + 2} = -2$
   b. $\sqrt{x + 2} = -3$

39. [5.4] Solve.
   a. $\sqrt{x + 5} = x - 1$
   b. $\sqrt{x + 6} = x - 6$

40. [5.4] Solve.
   a. $\sqrt{x - 3} - x = -3$
   b. $\sqrt{x - 4} - x = -4$

41. [5.4] Solve.
   a. $\sqrt{x - 3} - \sqrt{x} = -1$
   b. $\sqrt{x - 5} - \sqrt{x} = -1$

42. [5.4] Solve.
   a. $\sqrt[3]{x - 3} = 3$
   b. $\sqrt[3]{x - 3} = 4$

43. [5.5A] Write in terms of i.
   a. $\sqrt{-100}$
   b. $\sqrt{-121}$

44. [5.5A] Write in terms of i.
   a. $\sqrt{-72}$
   b. $\sqrt{-50}$

45. [5.5B] Find.
   a. $(3 + 5i) + (7 - 2i)$
   b. $(4 + 7i) + (2 - 4i)$

46. [5.5B] Find.
   a. $(3 + 5i) - (7 - 2i)$
   b. $(4 + 7i) - (2 - 4i)$

47. [5.5C] Multiply.
   a. $(3 + 2i)(5 - 3i)$
   b. $(4 + 5i)(2 - 3i)$

48. [5.5C] Multiply.
   a. $\sqrt{-16}(4 - \sqrt{-72})$
   b. $\sqrt{-36}(4 - \sqrt{-72})$

49. [5.5C] Divide.

   a. $\dfrac{2 + 3i}{4 + 3i}$

   b. $\dfrac{3 - 5i}{4 - 3i}$

50. [5.5D] Write the answer as 1, -1, i, or $-i$.
   a. i^{38}
   b. i^{75}

51. [5.5D] Write the answer as 1, -1, i, or $-i$.
   a. i^{-14}
   b. i^{-27}

38. a. _____
  b. _____

39. a. _____
  b. _____

40. a. _____
  b. _____

41. a. _____
  b. _____

42. a. _____
  b. _____

43. a. _____
  b. _____

44. a. _____
  b. _____

45. a. _____
  b. _____

46. a. _____
  b. _____

47. a. _____
  b. _____

48. a. _____
  b. _____

49. a. _____
  b. _____

50. a. _____
  b. _____

51. a. _____
  b. _____

NAME

CLASS

SECTION

(Answers on pages 412–413)

1. Find, if possible.
   a. $\sqrt[3]{-64}$
   b. $\sqrt{-36}$

2. Find, if possible.
   a. $(-27)^{1/3}$
   b. $\left(\dfrac{1}{16}\right)^{1/4}$

3. Evaluate.
   a. $8^{2/3}$
   b. $(-25)^{3/2}$

4. Evaluate.
   a. $(-27)^{-2/3}$
   b. $8^{-2/3}$

5. If x and y are positive, simplify.
   a. $x^{1/2} \cdot x^{2/3}$
   b. $\dfrac{x^{-1/3}}{x^{2/5}}$

6. If x and y are positive, simplify.
   a. $(x^{1/5}y^{3/4})^{-20}$
   b. $x^{3/5}(x^{-1/5} + y^{3/5})$

7. Simplify.
   a. $\sqrt[4]{(-5)^4}$
   b. $\sqrt[4]{(-x)^4}$

8. Simplify.
   a. $\sqrt[3]{40}$
   b. $\sqrt[3]{54a^4b^{12}}$

9. Simplify.
   a. $\sqrt{\dfrac{5}{243}}$
   b. $\sqrt[3]{\dfrac{5}{x^6}}$

10. Rationalize the denominator.
    a. $\dfrac{\sqrt{5}}{\sqrt{6}}$
    b. $\dfrac{\sqrt{2}}{\sqrt{5x}},\ x > 0$

ANSWERS

1. a. _____
   b. _____

2. a. _____
   b. _____

3. a. _____
   b. _____

4. a. _____
   b. _____

5. a. _____
   b. _____

6. a. _____
   b. _____

7. a. _____
   b. _____

8. a. _____
   b. _____

9. a. _____
   b. _____

10. a. _____
    b. _____

11. Rationalize the denominator.

   a. $\dfrac{1}{\sqrt[3]{3x}}$

   b. $\dfrac{\sqrt[5]{7}}{\sqrt[5]{16x^3}}$

11. a. _____

   b. _____

12. Reduce the order (index).

   a. $\sqrt[4]{\dfrac{16}{81}}$

   b. $\sqrt[6]{81c^4d^4}$

12. a. _____

   b. _____

13. Simplify $\sqrt[6]{\dfrac{a^2}{16c^{16}}}$.

13. _____

14. Perform the indicated operations.
   a. $\sqrt{32} + \sqrt{98}$
   b. $\sqrt{112} - \sqrt{28}$

14. a. _____

   b. _____

15. Perform the indicated operations.

   a. $4\sqrt{\dfrac{1}{2}} - 3\sqrt{\dfrac{1}{8}}$

   b. $5\sqrt[3]{\dfrac{3}{4x}} - \sqrt[3]{\dfrac{3}{32x}}$

15. a. _____

   b. _____

16. Perform the indicated operations.
   a. $\sqrt{2}(\sqrt{8} + \sqrt{5})$
   b. $\sqrt[3]{3x}(\sqrt[3]{9x^2} - \sqrt[3]{16x})$

16. a. _____

   b. _____

17. Find the product.
   a. $(\sqrt{27} + \sqrt{50})(\sqrt{12} + \sqrt{18})$
   b. $(\sqrt{3} + 3)(\sqrt{3} + 3)$

17. a. _____

   b. _____

18. Find the product.
   a. $(3 - \sqrt{3})^2$
   b. $(\sqrt{6} + \sqrt{3})(\sqrt{6} - \sqrt{3})$

18. a. _____

   b. _____

19. Reduce $\dfrac{10 - \sqrt{75}}{5}$.

19. _____

20. Rationalize the denominator in $\dfrac{\sqrt{x}}{\sqrt{x} + \sqrt{y}}$.

20. _____

21. Solve $\sqrt{x + 2} = -2$.

21. _____

22. Solve $\sqrt{x + 2} = x - 10$.

22. _____

23. Solve $\sqrt{x - 3} - x = -3$.

23. _____

24. Solve $\sqrt{x - 9} - \sqrt{x} = -1$.

24. _____

25. Solve $\sqrt[3]{x - 3} = 3$.

25. _____

26. Write the given expression in terms of i.
   a. $\sqrt{-64}$
   b. $\sqrt{-98}$

26. a. _____

   b. _____

27. Find.
   a. $(4 + 5i) + (6 - 14i)$
   b. $(5 + 2i) - (7 - 8i)$

28. Multiply.
   a. $(2 + 2i)(4 - 3i)$
   b. $\sqrt{-9}(3 - \sqrt{-8})$

29. Find.

   a. $\dfrac{4 + 3i}{2 + 3i}$

   b. $\dfrac{4 - 3i}{3 - 5i}$

30. Write the answer as 1, -1, i, or $-i$
   a. i^{56}
   b. i^{-9}

27. a. _____
    b. _____

28. a. _____
    b. _____

29. a. _____
    b. _____

30. a. _____
    b. _____

IF YOU MISSED QUESTION	SECTION	EXAMPLES	PAGE	ANSWERS
1a	5.1	1	348	**1. a.** -4
1b	5.1	1	348	**b.** Not a real number
2a	5.1	2	349	**2. a.** -3
2b	5.1	2	349	**b.** $\dfrac{1}{2}$
3a	5.1	3	350	**3. a.** 4
3b	5.1	3	350	**b.** Not a real number
4a	5.1	4	351	**4. a.** $\dfrac{1}{9}$
4b	5.1	4	351	**b.** $\dfrac{1}{4}$
5a	5.1	5	351	**5. a.** $x^{7/6}$
5b	5.1	5	351	**b.** $\dfrac{1}{x^{11/15}}$
6a	5.1	6	352	**6. a.** $\dfrac{1}{x^4 y^{15}}$
6b	5.1	6	352	**b.** $x^{2/5} + x^{3/5} y^{3/5}$
7a	5.2	1	360	**7. a.** 5
7b	5.2	1	360	**b.** $|-x| = |x|$
8a	5.2	2	360–361	**8. a.** $2\sqrt[3]{5}$
8b	5.2	2	360–361	**b.** $3ab^4 \sqrt[3]{2a}$
9a	5.2	3	361	**9. a.** $\dfrac{\sqrt{15}}{27}$
9b	5.2	3	361	**b.** $\dfrac{\sqrt[3]{5}}{x^2}$
10a	5.2	4	362	**10. a.** $\dfrac{\sqrt{30}}{6}$
10b	5.2	4	362	**b.** $\dfrac{\sqrt{10x}}{5x}$
11a	5.2	5	363	**11. a.** $\dfrac{\sqrt[3]{9x^2}}{3x}$
11b	5.2	5	363	**b.** $\dfrac{\sqrt[5]{14x^2}}{2x}$
12a	5.2	6	364	**12. a.** $\dfrac{2}{3}$
12b	5.2	6	364	**b.** $\sqrt[3]{9c^2 d^2}$

IF YOU MISSED QUESTION	SECTION	EXAMPLES	PAGE	ANSWERS
13	5.2	7	364	**13.** $\dfrac{\sqrt[3]{2ac}}{2c^3}$
14a	5.3	1	370	**14. a.** $11\sqrt{2}$
14b	5.3	1	370	**b.** $2\sqrt{7}$
15a	5.3	2	370–371	**15. a.** $\dfrac{5\sqrt{2}}{4}$
15b	5.3	2	370–371	**b.** $\dfrac{9\sqrt[3]{6x^2}}{4x}$
16a	5.3	3	372	**16. a.** $4 + \sqrt{10}$
16b	5.3	3	372	**b.** $3x - 2\sqrt[3]{6x^2}$
17a	5.3	4, 5	372–373	**17. a.** $48 + 19\sqrt{6}$
17b	5.3	4, 5	372–373	**b.** $12 + 6\sqrt{3}$
18a	5.3	5	373	**18. a.** $12 - 6\sqrt{3}$
18b	5.3	5	373	**b.** 3
19	5.3	6	373	**19.** $2 - \sqrt{3}$
20	5.3	7	374–375	**20.** $\dfrac{x - \sqrt{xy}}{x - y}$
21	5.4	1	382	**21.** No real-number solution
22	5.4	1	382	**22.** 14
23	5.4	1	382	**23.** 3 or 4
24	5.4	2	383	**24.** 25
25	5.4	3	384	**25.** 30
26a	5.5	1	389	**26. a.** $8i$
26b	5.5	1	389	**b.** $7\sqrt{2}i$
27a	5.5	2	390	**27. a.** $10 - 9i$
27b	5.5	2	390	**b.** $-2 + 10i$
28a	5.5	3	391	**28. a.** $14 + 2i$
28b	5.5	4	391–392	**b.** $6\sqrt{2} + 9i$
29a	5.5	5	393	**29. a.** $\dfrac{17}{13} - \dfrac{6}{13}i$
29b	5.5	5	393	**b.** $\dfrac{27}{34} + \dfrac{11}{34}i$
30a	5.5	6	394	**30. a.** 1
30b	5.5	6	394	**b.** $-i$

CHAPTER 6

QUADRATIC EQUATIONS AND INEQUALITIES

In this chapter we solve quadratic equations by completing the square and by the quadratic formula. We then use the discriminant to examine the different types of answers for these equations. We use our knowledge of quadratics to study higher-degree equations that can be written in quadratic form. We end the chapter by covering nonlinear inequalities.

NAME

CLASS

SECTION

(Answers on pages 418–419)

ANSWERS

1. Solve.
   a. $4x^2 - 25 = 0$
   b. $3x^2 + 18 = 0$

1. a. _____
   b. _____

2. Solve.
   a. $(x - 2)^2 = 45$
   b. $2(x - 1)^2 = 49$

2. a. _____
   b. _____

3. Solve $2x^2 + 8x + 8 = 5$.

3. _____

4. Solve by completing the square: $x^2 + 6x - 5 = 0$.

4. _____

5. Solve by completing the square: $4x^2 - 4x - 1 = 0$.

5. _____

6. Solve by the quadratic formula: $5x^2 + 4x - 1 = 0$.

6. _____

7. Solve by the quadratic formula: $2x^2 = 2x + 3$.

7. _____

8. Solve by the quadratic formula: $8x = x^2$.

8. _____

9. Solve by the quadratic formula: $\dfrac{x^2}{9} + \dfrac{5}{3}x = \dfrac{16}{9}$.

9. _____

10. Solve by the quadratic formula: $2x^2 + 3x = -2$.

10. _____

11. Solve $27x^3 - 64 = 0$.

11. _____

12. If the demand (d) is given by $d = \dfrac{100}{p}$ and the supply is given by $s = 200p - 100$, find the equilibrium point (at which $s = d$).

12. _____

13. For the equation $9x^2 - kx = -4$,
   a. Find the discriminant.
   b. Find k so that the equation has exactly one rational root.

13. a. _____
   b. _____

14. Determine whether $15x^2 - 11x - 14$ is factorable into factors with integer coefficients.

14. _____

15. Factor $12x^2 - 32x - 35$ into factors with integer coefficients, if possible.

15. _____

16. **a.** Find the sum and the product of the roots of the equation $6x^2 - 13x - 5 = 0$.
   b. Use the sum and product properties to check if the solutions are $-\frac{1}{3}$ and $\frac{5}{2}$.

16. a. _____
   b. _____

17. Solve.

   a. $\dfrac{6}{x^2 - 9} + \dfrac{1}{x + 3} = 1$

   b. $\dfrac{8}{x^2 - 4} - \dfrac{6}{x - 2} = 1$

17. a. _____
   b. _____

18. Solve $x^4 - 6x^2 + 5 = 0$.

18. _____

19. Solve $(x^2 - 2x)^2 + (x^2 - 2x) - 12 = 0$.

19. _____

20. Solve $x^{1/2} - 9x^{1/4} + 18 = 0$.

21. Solve $x - 3\sqrt{x} + 2 = 0$.

22. Solve $(x - 1)(x + 2) < 0$.

23. Solve $x^2 + 2x \geq 8$.

24. Solve $(x - 1)(x - 2)(x + 3) \leq 0$.

25. Solve $\dfrac{x + 1}{x - 1} \geq 2$.

20. _____

21. _____

22. _____

23. _____

24. _____

25. _____

IF YOU MISSED QUESTION	SECTION	EXAMPLES	PAGE	ANSWERS
1a	6.1A	1a	422	1. a. $\dfrac{5}{2}, -\dfrac{5}{2}$
1b	6.1A	1b	422	b. $\sqrt{6}i, -\sqrt{6}i$
2a	6.1B	2a	423	2. a. $2 \pm 3\sqrt{5}$
2b	6.1B	2b	423	b. $\dfrac{1 \pm 7\sqrt{2}}{2}$
3	6.1B	3	424	3. $\dfrac{-2 \pm \sqrt{10}}{2}$
4	6.1C	4	425	4. $-3 \pm \sqrt{14}$
5	6.1C	5	426	5. $\dfrac{1 \pm \sqrt{2}}{2}$
6	6.2A	1	434	6. $\dfrac{1}{5}, -1$
7	6.2A	2	435	7. $\dfrac{1 \pm \sqrt{7}}{2}$
8	6.2A	3	436	8. $0, 8$
9	6.2A	4	436	9. $1, -16$
10	6.2A	5	437	10. $\dfrac{-3 \pm \sqrt{7}i}{4}$
11	6.2A	6	438	11. $\dfrac{4}{3}, \dfrac{-2 \pm 2\sqrt{3}i}{3}$
12	6.2B	7	439	12. $p = 1$
13a	6.3A	1a	446	13. a. $k^2 - 144$
13b	6.3A	1b	446	b. $k = \pm 12$
14	6.3B	2	447	14. Yes. $D = 31^2$.
15	6.3B	3	447	15. $(2x - 7)(6x + 5)$
16a	6.3C	4a	449	16. a. Sum: $\dfrac{13}{6}$, product: $-\dfrac{5}{6}$
16b	6.3C	4b	449	b. Yes.
17a	6.4A	1a	454	17. a. 4
17b	6.4A	1b	454	b. $0, -6$
18	6.4B	2	454	18. $\pm 1, \pm \sqrt{5}$
19	6.4B	3	455	19. $3, -1, 1 \pm \sqrt{3}i$
20	6.4B	4	455	20. $81, 1296$
21	6.4B	5	456	21. $1, 4$

IF YOU MISSED QUESTION	SECTION	EXAMPLES	PAGE	ANSWERS
22	6.5A	1	460	**22.** $-2 < x < 1$
23	6.5A	2	461	**23.** $x \leq -4$ or $x \geq 2$
24	6.5B	3	462	**24.** $x \leq -3$ or $1 \leq x \leq 2$
25	6.5B	4	462	**25.** $1 < x \leq 3$

6.1 SOLVING QUADRATICS BY COMPLETING THE SQUARE

Courtesy of NASA.

OBJECTIVES

REVIEW

Before starting this section, you should know:

1. How to take the square root of a number.
2. How to rationalize the denominator of a fraction.
3. How to add, subtract, multiply, and divide complex numbers.
4. How to expand $(x \pm a)^2$.

OBJECTIVES

You should be able to:

A. Solve equations of the form $ax^2 + b = 0$.
B. Solve equations of the form $a(x + b)^2 = c$.
C. Solve quadratic equations by completing the square.

In Chapter 2, we studied methods of solving *linear equations*—that is, equations in which the variable involved has an exponent of 1 (the first power). We are now ready to discuss equations containing the *second* (but no higher) power of the unknown.

Such equations are called *second-degree,* or *quadratic,* equations; these can be written in *standard form*. Here is the definition.

$ax^2 + bx + c = 0$ (a, b, c real numbers, $a \neq 0$)

is a **quadratic equation** in **standard form.**

The procedure used to solve these equations is similar to that employed in Chapter 2 and consists of applying certain transformations to obtain equivalent equations whose solution set is evident. For example, look at the square "grids" on the antenna. The area of each one is 36 ft². Can we find the dimensions of each grid? If the length of one side is x feet, the area of the square in the diagram is x^2 square feet. Since the area is also 36 ft², we have $x^2 = 36$.

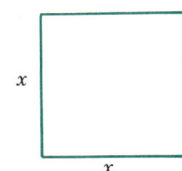

We solve this equation by taking (extracting) the square roots of both sides. Because a nonzero number has two square roots, we have

$$x = \sqrt{36} = 6 \quad \text{or} \quad x = -\sqrt{36} = -6$$

Thus the solutions are 6 and -6. This procedure is usually shortened to

$$x^2 = 36$$
$$x = \pm\sqrt{36} = \pm 6$$

where the notation ± 6 (read "plus or minus 6") means that $x = 6$ or $x = -6$. If x represents the length of a side, the answer -6 must be discarded.

Similarly, the solutions of $x^2 = 25$ are

$$x = \pm\sqrt{25} = \pm 5$$

In general, we have the following.

The solutions of $x^2 = a$, where a is a real number, are $\sqrt{a}$ and $-\sqrt{a}$, abbreviated as $\pm\sqrt{a}$.

A. SOLVING EQUATIONS OF THE FORM $ax^2 + b = 0$

To solve the equation $9x^2 - 16 = 0$, we need to isolate the x^2 and take the square roots of both sides. Here is the way we do it:

$$9x^2 - 16 = 0$$
$$9x^2 = 16 \qquad \text{Add 16 to both sides.}$$
$$x^2 = \frac{16}{9} \qquad \text{Divide both sides by 9 (now } x^2 \text{ is by itself).}$$
$$x = \pm\sqrt{\frac{16}{9}} = \pm\frac{4}{3} \quad \text{Take square roots of both sides.}$$

The solutions are $\frac{4}{3}$ and $-\frac{4}{3}$.

EXAMPLE 1 Solve.

a. $4x^2 - 9 = 0$ **b.** $3x^2 + 24 = 0$ **c.** $5x^2 - 4 = 0$

Solution

a. The idea is to isolate x^2 and then extract square roots.

$$4x^2 - 9 = 0$$
$$4x^2 = 9 \qquad \text{Add 9 to both sides.}$$
$$x^2 = \frac{9}{4} \qquad \text{Divide both sides by 4 (now } x^2 \text{ is isolated).}$$
$$x = \pm\sqrt{\frac{9}{4}} = \pm\frac{3}{2} \quad \text{Extract square roots.}$$

The solutions are $\frac{3}{2}$ and $-\frac{3}{2}$.

Problem 1 Solve.
a. $9x^2 - 4 = 0$ **b.** $3x^2 + 54 = 0$
c. $3x^2 - 16 = 0$

b. $3x^2 + 24 = 0$

$$3x^2 = -24 \qquad \text{Add } -24 \text{ to both sides.}$$
$$x^2 = -8 \qquad \text{Divide both sides by 3.}$$
$$x = \pm\sqrt{-8} \qquad \text{Take square roots.}$$
$$x = \pm 2\sqrt{2}\,i \qquad \sqrt{-8} = \sqrt{4}\sqrt{-2} = 2\sqrt{2}\,i \text{ (simplify).}$$

The solutions are $2\sqrt{2}\,i$ and $-2\sqrt{2}\,i$. Note that the answers are **complex numbers** given in the *simplified* form $\pm 2\sqrt{2}\,i$ instead of $\pm\sqrt{8}\,i$.

c. $5x^2 - 4 = 0$

$$5x^2 = 4 \qquad\qquad\qquad \text{Add 4 to both sides.}$$
$$x^2 = \frac{4}{5} \qquad\qquad\qquad \text{Divide both sides by 5.}$$
$$x = \pm\sqrt{\frac{4}{5}} = \pm\sqrt{\frac{4 \cdot 5}{5 \cdot 5}} \qquad \text{Take square roots.}$$
$$x = \pm\frac{2\sqrt{5}}{5} \qquad\qquad \sqrt{\frac{4 \cdot 5}{5 \cdot 5}} = \frac{\sqrt{20}}{5} = \frac{2\sqrt{5}}{5}$$

The solutions are $\dfrac{2\sqrt{5}}{5}$ and $-\dfrac{2\sqrt{5}}{5}$. Note that the answer is given with a *rationalized* denominator.

B. SOLVING EQUATIONS OF THE FORM $a(x + b)^2 = c$

The method used to solve equations of the form $ax^2 + c = 0$ can also be used in solving equations of the form $(x + a)^2 = b$. Thus, to solve the equation $(x + 3)^2 = 16$, we proceed as follows:

$$(x + 3)^2 = 16$$
$$x + 3 = \pm\sqrt{16} = \pm 4 \qquad \text{Extract square roots.}$$
$$x + 3 = 4 \quad \text{or} \quad x + 3 = -4$$
$$x = 1 \quad \text{or} \qquad x = -7 \qquad \text{Solve } x + 3 = 4 \text{ and } x + 3 = -4.$$

The solutions are 1 and -7.

To solve the equation $(x - 2)^2 - 9 = 0$, we add 9 to both sides and then extract square roots, obtaining

$$(x - 2)^2 - 9 = 0$$
$$(x - 2)^2 = 9$$
$$x - 2 = \pm\sqrt{9} = \pm 3$$
$$x - 2 = 3 \quad \text{or} \quad x - 2 = -3$$
$$x = 5 \quad \text{or} \qquad x = -1$$

Thus the solutions for $(x - 2)^2 - 9 = 0$ are 5 and -1.

EXAMPLE 2 Solve.

a. $(x - 1)^2 = 27$ **b.** $3(x - 2)^2 + 25 = 0$

Solution

a. $(x - 1)^2 = 27$
$$x - 1 = \pm\sqrt{27} = \pm 3\sqrt{3}$$
$$x - 1 = 3\sqrt{3} \qquad \text{or} \quad x - 1 = -3\sqrt{3}$$
$$x = 1 + 3\sqrt{3} \quad \text{or} \qquad x = 1 - 3\sqrt{3}$$

Problem 2 Solve.
a. $(x - 2)^2 = 24$
b. $5(x - 3)^2 + 36 = 0$

The solutions are $1 + 3\sqrt{3}$ and $1 - 3\sqrt{3}$. This can be verified by substituting $1 + 3\sqrt{3}$ and $1 - 3\sqrt{3}$ in the equation $(x - 1)^2 = 27$. Since $[(1 + 3\sqrt{3}) - 1]^2 = (3\sqrt{3})^2 = 3^2 \cdot 3 = 27$ and $[(1 - 3\sqrt{3}) - 1]^2 = (-3\sqrt{3})^2 = 3^2 \cdot 3 = 27$, the result is correct.

b. $3(x - 2)^2 + 25 = 0$

$$3(x - 2)^2 = -25 \qquad \text{Subtract 25 from both sides.}$$

$$(x - 2)^2 = \frac{-25}{3} \qquad \begin{array}{l}\text{Divide both sides by 3 (now the } (x - 2)^2 \\ \text{is isolated).}\end{array}$$

$$x - 2 = \pm\sqrt{\frac{-25}{3}} \qquad \text{Extract square roots.}$$

$$= \pm\sqrt{\frac{-25 \cdot 3}{3 \cdot 3}} \qquad \text{Rationalize the denominator.}$$

$$= \pm\frac{5\sqrt{3}}{3}\,i \qquad \sqrt{\frac{-25 \cdot 3}{3 \cdot 3}} = \frac{\sqrt{-25 \cdot 3}}{3} = \frac{5\sqrt{3}}{3}\,i$$

Thus $x - 2 = \dfrac{5\sqrt{3}}{3}\,i$ or $x - 2 = -\dfrac{5\sqrt{3}}{3}\,i$—that is,

$$x = 2 + \frac{5\sqrt{3}}{3}\,i \qquad \text{Solve the equations for } x.$$

or

$$x = 2 - \frac{5\sqrt{3}}{3}\,i$$

The solutions are $2 + \dfrac{5\sqrt{3}}{3}\,i$ and $2 - \dfrac{5\sqrt{3}}{3}\,i$. The verification is left to the student.

EXAMPLE 3 Solve $3x^2 + 12x + 12 = 20$.

Solution This time, we have to factor the left side first.

$$3x^2 + 12x + 12 = 20 \qquad \text{Given.}$$
$$3(x^2 + 4x + 4) = 20 \qquad \text{Factor out the 3.}$$
$$3(x + 2)^2 = 20 \qquad \text{Factor } x^2 + 4x + 4.$$
$$(x + 2)^2 = \frac{20}{3} \qquad \text{Divide by 3.}$$
$$x + 2 = \pm\sqrt{\frac{20}{3}} \qquad \text{Extract square roots.}$$
$$= \pm\sqrt{\frac{20 \cdot 3}{3 \cdot 3}} \qquad \text{Rationalize the denominator.}$$
$$= \pm\frac{\sqrt{60}}{3}$$
$$= \pm\frac{2\sqrt{15}}{3} \qquad \text{Simplify.}$$
$$x = -2 \pm\frac{2\sqrt{15}}{3} \qquad \text{Subtract 2.}$$

The solutions are $-2 + \dfrac{2\sqrt{15}}{3}$ and $-2 - \dfrac{2\sqrt{15}}{3}$. The verification is left to the student.

Problem 3 Solve $2x^2 + 12x + 18 = 27$.

C. COMPLETING THE SQUARE

The solutions of Examples 2 and 3 were obtained by writing the equation in the form $a(x + b)^2 = c$. Suppose we have an equation that is not of this form. We can make it of this form if we learn a technique called **completing the square.** As you recall

$$(x + a)^2 = x^2 + 2ax + a^2$$

and

$$(x - a)^2 = x^2 - 2ax + a^2$$

In both cases, the last term is the square of one-half the coefficient of x. How can we make $x^2 + 10x$ a perfect square trinomial? Since the coefficient of x is $2a$, a must be 5. Thus we make $x^2 + 10x$ a perfect square trinomial by adding 5^2 to it. We then have

$$x^2 + 10x + 5^2 = (x + 5)^2$$

Now, consider the equation $x^2 + 8x = 12$. To make $x^2 + 8x$ a perfect square trinomial, we note that the coefficient of x is 8, so the number to be added is the square of one-half of 8—that is, 4^2. Adding 4^2 to both sides, we have

$$x^2 + 8x + 4^2 = 12 + 4^2$$

The left side is $(x + 4)^2$, and the right side is $12 + 16 = 28$. Thus $(x + 4)^2 = 28$, the form we want! Take the square root of both sides:

$$x + 4 = \pm\sqrt{28} = \pm 2\sqrt{7}$$
$$x = -4 \pm 2\sqrt{7}$$

The solutions are $-4 + 2\sqrt{7}$ and $-4 - 2\sqrt{7}$.

EXAMPLE 4 Solve $x^2 + 10x - 2 = 0$ by completing the square.

Solution

$$x^2 + 10x - 2 = 0$$

$$x^2 + 10x = 2 \qquad \text{Add 2 to both sides.}$$

$$x^2 + 10x + 5^2 = 2 + 5^2 \qquad \text{Add } 5^2, \text{ the square of one-half of the coefficient of } x\text{— that is, } [\tfrac{1}{2} \cdot 10]^2\text{—to both sides.}$$

$$(x + 5)^2 = 27 \qquad \text{Factor on the left.}$$

$$x + 5 = \pm\sqrt{27} = \pm 3\sqrt{3} \qquad \text{Extract square roots and simplify.}$$

$$x + 5 = 3\sqrt{3} \quad \text{or} \quad x + 5 = -3\sqrt{3}$$

Thus $x = -5 + 3\sqrt{3}$ or $x = -5 - 3\sqrt{3}$, and the solutions are $-5 + 3\sqrt{3}$ and $-5 - 3\sqrt{3}$. ▲

When the coefficient of x^2 is not 1 and the left side does not factor with integer coefficients, we must first divide each term of the equation by the coefficient of x^2. For example, to solve $3x^2 + 2x - 16 = 0$ we add 16 to both sides (so all the variables are on one side) and then divide by 3. Here are the steps:

STEP 1. $3x^2 + 2x - 16 = 0 \qquad$ Given.

$ 3x^2 + 2x = 16 \qquad$ Add 16 to both sides.

Problem 4 Solve $x^2 + 12x - 8 = 0$ by completing the square.

STEP 2. $x^2 + \dfrac{2x}{3} = \dfrac{16}{3}$ Divide each term by 3.

STEP 3. $x^2 + \dfrac{2x}{3} + \left(\dfrac{1}{3}\right)^2 = \dfrac{16}{3} + \left(\dfrac{1}{3}\right)^2$ Add the square of one-half of $\dfrac{2}{3}$—that is, $\left(\dfrac{1}{2} \cdot \dfrac{2}{3}\right)^2 = \left(\dfrac{1}{3}\right)^2$— to each side.

STEP 4. $\left(x + \dfrac{1}{3}\right)^2 = \dfrac{49}{9}$ Factor the left, simplify the right.

STEP 5. $x + \dfrac{1}{3} = \pm\sqrt{\dfrac{49}{9}} = \pm\dfrac{7}{3}$ Extract square roots.

$$x + \dfrac{1}{3} = \dfrac{7}{3} \quad \text{or} \quad x + \dfrac{1}{3} = \dfrac{-7}{3}$$

$$x = \dfrac{6}{3} = 2 \quad \text{or} \quad x = \dfrac{-8}{3} \qquad \text{Subtract } \dfrac{1}{3}.$$

The solutions are 2 and $-\dfrac{8}{3}$.

Here is a summary of this procedure.

TO SOLVE A QUADRATIC EQUATION BY COMPLETING THE SQUARE

1. Write the equation with the variables in descending order on the left and the numbers on the right.

2. If the coefficient of the square term is not 1, divide each term by this coefficient.

3. Add the square of one-half of the coefficient of the first-degree term to both sides.

4. Rewrite the left-hand side as a perfect square.

5. Solve the resulting equation.

EXAMPLE 5   Solve $3x^2 - 3x - 1 = 0$ by completing the square.

Solution

We use the five-step procedure.

$$3x^2 - 3x - 1 = 0 \qquad \text{Given.}$$

STEP 1. $3x^2 - 3x = 1$ Add 1.

STEP 2. $x^2 - x = \dfrac{1}{3}$ Divide each term by 3.

STEP 3. $x^2 - x + \left(\dfrac{1}{2}\right)^2 = \dfrac{1}{3} + \left(\dfrac{1}{2}\right)^2$ Add the square of one-half of -1—that is, $\left[\dfrac{1}{2}(-1)\right]^2 = \left(-\dfrac{1}{2}\right)^2 = \left(\dfrac{1}{2}\right)^2$—to each side.

Problem 5   Solve $5x^2 - 5x - 1 = 0$ by completing the square.

STEP 4. $\left(x - \dfrac{1}{2}\right)^2 = \dfrac{7}{12}$ Factor on the left.

STEP 5. $x - \dfrac{1}{2} = \pm\sqrt{\dfrac{7}{12}} = \pm\sqrt{\dfrac{7 \cdot 3}{12 \cdot 3}}$ Extract square roots and rationalize the denominator.

$$= \pm\dfrac{\sqrt{21}}{6}$$ Simplify.

$$x - \dfrac{1}{2} = \dfrac{\sqrt{21}}{6} \qquad \text{or} \qquad x - \dfrac{1}{2} = -\dfrac{\sqrt{21}}{6}$$

$$x = \dfrac{1}{2} + \dfrac{\sqrt{21}}{6} \qquad \text{or} \qquad x = \dfrac{1}{2} - \dfrac{\sqrt{21}}{6} \qquad \text{Check this.}$$

The solutions are $\dfrac{1}{2} + \dfrac{\sqrt{21}}{6}$ and $\dfrac{1}{2} - \dfrac{\sqrt{21}}{6}$, or $\dfrac{3 + \sqrt{21}}{6}$ and $\dfrac{3 - \sqrt{21}}{6}$.

NAME

CLASS

SECTION

ANSWERS

A. In Problems 1–20 solve the equation.

1. $x^2 = 64$

2. $x^2 = 81$

3. $x^2 = -121$

4. $x^2 = -144$

5. $x^2 - 169 = 0$

6. $x^2 - 100 = 0$

7. $x^2 + 4 = 0$

8. $x^2 + 25 = 0$

9. $36x^2 - 49 = 0$

10. $36x^2 - 81 = 0$

11. $4x^2 + 81 = 0$

12. $9x^2 + 64 = 0$

13. $3x^2 - 25 = 0$

14. $5x^2 - 16 = 0$

15. $5x^2 + 36 = 0$

16. $11x^2 + 49 = 0$

17. $3x^2 - 100 = 0$

18. $4x^2 - 13 = 0$

19. $13x^2 + 81 = 0$

20. $11x^2 + 4 = 0$

B. In Problems 21–40 solve the equation.

21. $(x + 5)^2 = 4$

22. $(x + 3)^2 = 9$

23. $x^2 + 4x + 4 = -25$

24. $x^2 + 2x + 1 = -16$

25. $(x - 6)^2 = 18$

26. $(x - 2)^2 = 50$

27. $x^2 - 2x + 1 = -28$

28. $x^2 - 6x + 9 = -32$

29. $(x - 1)^2 - 50 = 0$

30. $(x - 2)^2 - 18 = 0$

1. _____
2. _____
3. _____
4. _____
5. _____
6. _____
7. _____
8. _____
9. _____
10. _____
11. _____
12. _____
13. _____
14. _____
15. _____
16. _____
17. _____
18. _____
19. _____
20. _____
21. _____
22. _____
23. _____
24. _____
25. _____
26. _____
27. _____
28. _____
29. _____
30. _____

31. $(x - 5)^2 - 32 = 0$

32. $(x - 2)^2 - 4 = 0$

33. $(x - 9)^2 + 64 = 0$

34. $(x - 2)^2 + 25 = 0$

35. $3x^2 + 6x + 3 = 96$

36. $3x^2 + 30x + 75 = 72$

37. $7(x - 2)^2 - 350 = 0$

38. $3(x - 1)^2 - 54 = 0$

39. $7(x - 5)^2 + 189 = 0$

40. $5(x - 3)^2 + 250 = 0$

C. In Problems 41–70 solve by completing the square.

41. $x^2 + 6x + 5 = 0$

42. $x^2 + 4x + 3 = 0$

43. $x^2 + 8x + 15 = 0$

44. $x^2 + 8x + 7 = 0$

45. $x^2 + 6x + 10 = 0$

46. $x^2 + 12x + 37 = 0$

47. $x^2 - 10x + 24 = 0$

48. $x^2 + 12x - 28 = 0$

49. $x^2 - 10x + 21 = 0$

50. $x^2 - 2x - 143 = 0$

51. $x^2 - 8x + 17 = 0$

52. $x^2 - 14x + 58 = 0$

53. $2x^2 + 4x + 3 = 0$

54. $2x^2 + 7x + 6 = 0$

55. $3x^2 + 6x + 78 = 0$

56. $9x^2 + 6x + 2 = 0$

57. $25y^2 - 25y + 6 = 0$

58. $4y^2 - 16y + 15 = 0$

59. $4y^2 - 4y + 5 = 0$

60. $9x^2 - 12x + 13 = 0$

61. $4x^2 - 7 = 4x$

62. $2x^2 - 18 = -9x$

63. $2x^2 + 1 = 4x$

64. $2x^2 + 3 = 6x$

31. _____

32. _____

33. _____

34. _____

35. _____

36. _____

37. _____

38. _____

39. _____

40. _____

41. _____

42. _____

43. _____

44. _____

45. _____

46. _____

47. _____

48. _____

49. _____

50. _____

51. _____

52. _____

53. _____

54. _____

55. _____

56. _____

57. _____

58. _____

59. _____

60. _____

61. _____

62. _____

63. _____

64. _____

65. $(x + 3)(x - 2) = -4$ **66.** $(x + 4)(x - 1) = -6$

67. $2x(x + 5) - 1 = 0$ **68.** $2x(x - 4) = 2(9 - 8x) - x$

69. $2x(x + 3) - 10 = 0$ **70.** $4x(x + 1) - 5 = 0$

65. _____

66. _____

67. _____

68. _____

69. _____

70. _____

D. Applications

71. The distance traveled in t seconds by an object dropped from a height h is given by $h = 16t^2$. How long would it take an object dropped from a height of 64 ft to hit the ground?

72. Use the formula of Problem 71 to find the time it takes an object dropped from a height of 32 ft to hit the ground.

73. The amount of money A received at the end of 2 yr when P dollars are invested at a compound rate r is $A = P(1 + r)^2$. Find the rate of interest r (written as a percent) if a person invested $100 and received $121 at the end of 2 yr.

74. Use the formula of Problem 73 to find the rate of interest r (written as a percent) if a person invested $100 and received $144 at the end of 2 yr.

71. _____

72. _____

73. _____

74. _____

✔ **SKILL CHECKER**

In Problems 75–84 evaluate $\dfrac{-b \pm \sqrt{b^2 - 4ac}}{2a}$ for the given values of a, b, and c.

75. $a = 1, b = -9, c = 0$ **76.** $a = 1, b = -6, c = 0$

77. $a = 1, b = -2, c = -2$ **78.** $a = 1, b = -4, c = -4$

79. $a = 8, b = 7, c = -1$ **80.** $a = 3, b = -2, c = -5$

81. $a = 3, b = -8, c = 7$ **82.** $a = 4, b = -3, c = 5$

83. $a = 1, b = 2, c = 6$ **84.** $a = 1, b = 2, c = 5$

75. _____

76. _____

77. _____

78. _____

79. _____

80. _____

81. _____

82. _____

83. _____

84. _____

6.1 USING YOUR KNOWLEDGE

Many applications of mathematics require finding the maximum or the minimum of certain algebra expressions. Thus a certain business may wish to find the price at which a product will bring *maximum* profits, while engineers may be interested in *minimizing* the amount of carbon monoxide produced by automobiles. Now, suppose you are the manufacturer of a certain product whose average manufacturing cost $\bar{C}$ (in dollars), based on producing x (thousand) units, is given by the expression

$$\bar{C} = x^2 - 8x + 18$$

How many units should be produced to minimize the cost per unit? If we consider the right-hand side of the equation, we can complete the square and leave the equation unchanged by adding and subtracting the appropriate number. Thus

$$\bar{C} = x^2 - 8x + 18$$
$$= (x^2 - 8x + \quad) + 18$$
$$= (x^2 - 8x + 4^2) + 18 - 4^2$$

Then

$$\bar{C} = (x - 4)^2 + 2$$

Now, for $\bar{C}$ to be as small as possible (minimizing the cost), we make $(x - 4)^2$ zero by letting $x = 4$; then $\bar{C} = 2$. This tells us that when 4 (thousand) units are produced, the minimum cost is 2. That is, the minimum average cost per unit is $2.

Use your knowledge about completing the square to solve the following problems.

1. A manufacturer's average cost $\bar{C}$ (in dollars), based on manufacturing x (thousand) items, is given by

   $$\bar{C} = x^2 - 4x + 6$$

   a. How many units should be produced to minimize the cost per unit?
   b. What is the minimum average cost per unit?

   1. a. _____
      b. _____

2. The demand D for a certain product depends on the number x (in thousands) of units produced and is given by

   $$D = x^2 - 2x + 3$$

   for what number of units is the demand at its lowest?

   2. _____

3. Have you seen people adding chlorine to their pools? This is done to reduce the number of bacteria present in the water. Suppose that after t days, the number of bacteria per cubic centimeter is given by the expression

   $$B = 20t^2 - 120t + 200$$

   In how many days will the number of bacteria be at its lowest?

   3. _____

6.2 THE QUADRATIC FORMULA

Courtesy of Clark University.

OBJECTIVES

REVIEW

Before starting this section, you should know:

1. How to find the square root of a number.
2. How to simplify the square root of a number by using the laws of radicals.
3. How to write fractions in lowest terms.
4. How to solve a quadratic equation by completing the square.

OBJECTIVES

You should be able to:

A. Solve quadratic equations using the quadratic formula.
B. Solve applications involving quadratic equations.

The man with the first liquid-fueled rocket is Dr. Robert Goddard. His rocket went up 41 ft! Do you know how long it took to go up this high? The height (in feet) of the rocket is given by

$$h = -16t^2 + v_0t$$

where v_0 is the initial velocity (51.225 ft/sec). We can substitute 41 for h, the height, and 51.225 for v_0, the initial velocity, and solve for t in the equation

$$-16t^2 + 51.225t = 41$$

Unfortunately, this equation is *not* factorable, and to complete the square we would have to divide by -16 and add $\left(-\dfrac{51.225}{32}\right)^2$ to both sides. Instead of doing this, let us solve the general quadratic equation $ax^2 + bx + c = 0$, $a \neq 0$, by using the procedure we studied for completing the square and then substitute the values for a, b, and c. Here are the steps.

$$ax^2 + bx + c = 0 \qquad \text{Given.}$$

1. $\quad ax^2 + bx = -c \qquad \text{Add } -c \text{ to both sides.}$

2. $\quad x^2 + \dfrac{b}{a}x = -\dfrac{c}{a} \qquad \text{Divide each term by } a.$

3. $x^2 + \dfrac{b}{a}x + \left(\dfrac{b}{2a}\right)^2 = \left(\dfrac{b}{2a}\right)^2 - \dfrac{c}{a}$

Add the square of one-half of the coefficient of x—that is, $\left[\dfrac{1}{2} \cdot \dfrac{b}{a}\right]^2$—to both sides.

4. $\quad \left(x + \dfrac{b}{2a}\right)^2 = \dfrac{b^2}{4a^2} - \dfrac{c}{a}$

Factor the left side and multiply exponents on the right.

5. $\quad \left(x + \dfrac{b}{2a}\right)^2 = \dfrac{b^2}{4a^2} - \dfrac{4ac}{4a^2}$

The LCD on the right is $4a^2$; write $-\dfrac{c}{a}$ as $-\dfrac{4ac}{4a^2}$.

6. $\quad \left(x + \dfrac{b}{2a}\right)^2 = \dfrac{b^2 - 4ac}{4a^2} \qquad \text{Combine.}$

7. $\quad x + \dfrac{b}{2a} = \dfrac{\pm\sqrt{b^2 - 4ac}}{2a} \qquad \text{Extract square roots.}$

8. $\quad x = -\dfrac{b}{2a} \pm \dfrac{\sqrt{b^2 - 4ac}}{2a} \qquad \text{Add } -\dfrac{b}{2a}.$

9. $\quad x = \dfrac{-b \pm \sqrt{b^2 - 4ac}}{2a} \qquad \text{Combine.}$

A. SOLVING EQUATIONS USING THE QUADRATIC FORMULA

Step 9 shows that any time we have a quadratic equation in the *standard form* $ax^2 + bx + c = 0$, we can find the solutions by simply substituting the values for a, b, and c in the quadratic formula:

$$x = \frac{-b \pm \sqrt{b^2 - 4ac}}{2a}$$

EXAMPLE 1   Solve $8x^2 + 7x - 1 = 0$.

Solution

1. The equation is written in standard form:

$$8x^2 \; + \; 7x \; - \; 1 = 0$$
$$\uparrow \qquad \uparrow \qquad \uparrow$$
$$a = 8 \qquad b = 7 \quad c = -1$$

2. From the diagram, it is clear that $a = 8$, $b = 7$, and $c = -1$.

Problem 1   Solve $3x^2 + 2x - 5 = 0$.

3. Substituting the values of a, b, and c in the formula, we obtain

$$x = \frac{-7 \pm \sqrt{(7)^2 - 4(8)(-1)}}{2(8)}$$

$$= \frac{-7 \pm \sqrt{49 + 32}}{16}$$

$$= \frac{-7 \pm \sqrt{81}}{16}$$

$$= \frac{-7 \pm 9}{16}$$

Thus

$$x = \frac{-7 + 9}{16} = \frac{2}{16} = \frac{1}{8} \quad \text{or} \quad x = \frac{-7 - 9}{16} = \frac{-16}{16} = -1$$

The solutions are $\frac{1}{8}$ and -1.

EXAMPLE 2 Solve $2x^2 = 2x + 1$.

Solution We proceed by steps as before.

1. To write the equation in standard form, subtract $2x$ and then subtract 1 to obtain

$$2x^2 \quad - \quad 2x \quad - \quad 1 = 0$$
$$\uparrow \qquad\quad \uparrow \qquad\quad \uparrow$$
$$a = 2 \quad\; b = -2 \quad c = -1$$

2. From the diagram, $a = 1$, $b = -2$, and $c = -1$.
3. Substituting these values in the quadratic formula, we have

$$x = \frac{-(-2) \pm \sqrt{(-2)^2 - 4(2)(-1)}}{2(2)}$$

$$= \frac{2 \pm \sqrt{4 + 8}}{4}$$

$$= \frac{2 \pm \sqrt{12}}{4}$$

$$= \frac{2 \pm \sqrt{4 \cdot 3}}{4}$$

$$= \frac{2 \pm 2\sqrt{3}}{4}$$

Thus

$$x = \frac{2 + 2\sqrt{3}}{4} = \frac{2}{4} + \frac{2\sqrt{3}}{4} = \frac{1}{2} + \frac{\sqrt{3}}{2} = \frac{1 + \sqrt{3}}{2}$$

or

$$x = \frac{2 - 2\sqrt{3}}{4} = \frac{2}{4} - \frac{2}{4}\sqrt{3} = \frac{1}{2} - \frac{\sqrt{3}}{2} = \frac{1 - \sqrt{3}}{2}$$

Problem 2 Solve $x^2 = 4x + 4$.

EXAMPLE 3 Solve $9x = x^2$.

Solution

1. Subtracting $9x$, we have

$$0 = x^2 - 9x$$

or

$$\underset{\underset{a=1}{\uparrow}}{x^2} \quad \underset{\underset{b=-9}{\uparrow}}{-\quad 9x} \quad + \quad \underset{\underset{c=0}{\uparrow}}{0} = 0$$

2. From the diagram, $a = 1$, $b = -9$, and $c = 0$ (because the c term is missing).

3. Substituting these values in the formula, we obtain

$$x = \frac{-(-9) \pm \sqrt{(-9)^2 - 4(1)(0)}}{2(1)}$$

$$= \frac{9 \pm \sqrt{81 - 0}}{2}$$

$$= \frac{9 \pm \sqrt{81}}{2}$$

$$= \frac{9 \pm 9}{2}$$

Thus

$$x = \frac{9 + 9}{2} = \frac{18}{2} = 9$$

or

$$x = \frac{9 - 9}{2} = \frac{0}{2} = 0$$

The solutions are 9 and 0.

EXAMPLE 4 Solve $\dfrac{x^2}{4} + \dfrac{2}{3}x = -\dfrac{1}{3}$.

Solution

1. We have to write the equation in standard form, but first we clear fractions by multiplying by the LCD of 4 and 3—that is, by 12:

$$12 \cdot \frac{x^2}{4} + 12 \cdot \frac{2}{3}x = -\frac{1}{3} \cdot 12$$

or

$$3x^2 + 8x = -4$$

We then add 4 to obtain

$$\underset{\underset{a=3}{\uparrow}}{3x^2} \quad + \quad \underset{\underset{b=8}{\uparrow}}{8x} \quad + \quad \underset{\underset{c=4}{\uparrow}}{4} = 0$$

2. From the diagram, $a = 3$, $b = 8$, and $c = 4$.

Problem 3 Solve $6x = x^2$.

Problem 4 Solve $\dfrac{x^2}{4} - \dfrac{3}{8}x = \dfrac{1}{4}$.

3. Substituting in the formula

$$x = \frac{-8 \pm \sqrt{64 - 4(3)(4)}}{2(3)}$$

$$= \frac{-8 \pm \sqrt{64 - 48}}{6}$$

$$= \frac{-8 \pm \sqrt{16}}{6}$$

$$= \frac{-8 \pm 4}{6}$$

Thus

$$x = \frac{-8 + 4}{6} = \frac{-4}{6} = -\frac{2}{3}$$

or

$$x = \frac{-8 - 4}{6} = \frac{-12}{6} = -2$$

The solutions are $-\frac{2}{3}$ and -2. ▲

Now, a word of warning. As you recall, some quadratic equations have complex-number solutions. Such solutions can be obtained by using the quadratic formula, as shown next.

EXAMPLE 5   Solve $3x^2 + 3x = -2$.

Solution

1. We add 2 to write the equation in standard form. We then have

$$3x^2 \; + \; 3x \; + \; 2 = 0$$

$$a = 3 \qquad b = 3 \qquad c = 2$$

2. From the diagram, $a = 3$, $b = 3$, and $c = 2$. Now,

$$x = \frac{-3 \pm \sqrt{(3)^2 - 4(3)(2)}}{2(3)}$$

$$= \frac{-3 \pm \sqrt{9 - 24}}{6}$$

$$= \frac{-3 \pm \sqrt{-15}}{6}$$

Thus

$$x = \frac{-3 \pm \sqrt{-15}}{6} = \frac{-3 \pm \sqrt{15}i}{6}$$

The solutions are $\dfrac{-3 + \sqrt{15}i}{6}$ and $\dfrac{-3 - \sqrt{15}i}{6}$. ▲

Let us now consider the equation $x^3 - 27 = 0$. This equation is *not* a quadratic. We will solve it by factoring. Here are the steps:

$(x - 3)(x^2 + 3x + 9) = 0$ Factor.

$x - 3 = 0$ or $x^2 + 3x + 9 = 0$ Zero-factor property.

$\qquad x = 3$ or $x^2 + 3x + 9 = 0$

Since the second equation does not factor, we use the quadratic formula to solve it. For this equation, $a = 1$, $b = 3$, and $c = 9$. Substituting, in

Problem 5   Solve $3x^2 + 2x = -1$.

the quadratic formula, we have

$$x = \frac{-3 \pm \sqrt{3^2 - 4(1)(9)}}{2 \cdot 1}$$

$$= \frac{-3 \pm \sqrt{9 - 36}}{2}$$

$$= \frac{-3 \pm \sqrt{-27}}{2}$$

$$= \frac{-3 \pm \sqrt{9 \cdot 3 \cdot (-1)}}{2}$$

$$= \frac{-3 \pm 3\sqrt{3}i}{2}$$

The solutions of $x^3 - 27$ are 3, $\dfrac{-3 + 3\sqrt{3}i}{2}$, and $\dfrac{-3 - 3\sqrt{3}i}{2}$. Note that $x^3 - 27 = 0$ *cannot* be solved by extracting roots—that is, by writing

$$x^3 - 27 = 0$$
$$x^3 = 27 \quad \text{Add 27.}$$
$$x = \sqrt[3]{27} \quad \text{Take roots.}$$
$$x = 3$$

As you can see, this method yields only one solution (the other two solutions are imaginary numbers).

EXAMPLE 6 Solve $8x^3 - 27 = 0$.

Solution We factor the equation, use the zero-factor property, and then use the quadratic formula.

$$8x^3 - 27 = 0 \quad \text{Given.}$$
$$(2x - 3)(4x^2 + 6x + 9) = 0 \quad \text{Factor.}$$
$$2x - 3 = 0 \quad \text{or} \quad 4x^2 + 6x + 9 = 0$$
$$x = \frac{3}{2} \quad \text{or} \quad 4x^2 + 6x + 9 = 0$$

The second equation is a quadratic with $a = 4$, $b = 6$, and $c = 9$. We solve it with the quadratic formula, obtaining

$$x = \frac{-6 \pm \sqrt{6^2 - 4(4)(9)}}{2 \cdot 4}$$

$$= \frac{-6 \pm \sqrt{36 - 144}}{8}$$

$$= \frac{-6 \pm \sqrt{-108}}{8}$$

$$= \frac{-6 \pm \sqrt{36 \cdot (3) \cdot (-1)}}{8}$$

$$= \frac{-6 \pm 6\sqrt{3}i}{8}$$

$$= \frac{2(-3 \pm 3\sqrt{3}i)}{2 \cdot 4}$$

$$= \frac{-3 \pm 3\sqrt{3}i}{4}$$

The solutions of $8x^3 - 27 = 0$ are $\dfrac{3}{2}$, $\dfrac{-3 + 3\sqrt{3}i}{4}$, and $\dfrac{-3 - 3\sqrt{3}i}{4}$. ▲

Problem 6 Solve $27x^3 - 8 = 0$

By the way, we left Dr. Goddard's rocket up in the air! How long *did* it fly? The equation was

$$-16t^2 + 51.225t = 41$$

or, in standard form

$$-16t^2 + 51.225t - 41 = 0$$

Here $a = -16$, $b = 51.225$, and $c = -41$, so the quadratic formula gives

$$t = \frac{-51.225 \pm \sqrt{51.225^2 - 4(-16)(-41)}}{2 \cdot (-16)}$$

With a calculator, we get $t \approx 1.6$. Thus, it took the rocket about 1.6 sec to reach 41 ft.

B. APPLICATIONS

In business, when the price p (in dollars) of a product increases, the demand (d) decreases and is given by $d = 300/p$. On the other hand, when the price p increases, the supply s producers are willing to sell increases and is given by $s = 100p - 50$. In economic theory, the point at which the supply equals the demand ($s = d$) is called the **equilibrium point.**

EXAMPLE 7 Find the price p at the equilibrium point.

Solution Since $s = d$ at equilibrium, we have

$$100p - 50 = \frac{300}{p}$$

$$p(100p - 50) = p \cdot \frac{300}{p} \quad \text{Multiply by } p.$$

$$100p^2 - 50p = 300$$

$$2p^2 - p = 6 \qquad \text{Divide by 50.}$$

$$2p^2 - p - 6 = 0 \qquad \text{Subtract 6.}$$

$a = 2$, $b = -1$, and $c = -6$. Thus

$$p = \frac{-(-1) \pm \sqrt{(-1)^2 - 4(2)(-6)}}{2 \cdot 2}$$

$$= \frac{1 \pm \sqrt{1 + 48}}{4}$$

$$= \frac{1 \pm \sqrt{49}}{4}$$

$$= \frac{1 \pm 7}{4}$$

Thus

$$p = \frac{1 + 7}{4} = 2 \quad \text{or} \quad p = \frac{1 - 7}{4} = -\frac{3}{2}$$

Since the price must be positive, we use $p = \$2$. Note that in this case, we obtain two rational roots, which means that the original equation was factorable. Before you use the quadratic formula, try to factor. You will save time!

Problem 7 Find the price p at the equilibrium point if the demand is given by $d = \dfrac{50}{p}$.

ANSWERS   (to problems on pages 438–439)

6. $\dfrac{2}{3}, \dfrac{-1+\sqrt{3}i}{3}, \dfrac{-1-\sqrt{3}i}{3}$ **7.** \$1

A. In Problems 1–34, solve the equation.

1. $x^2 + x - 2 = 0$

2. $x^2 + 4x - 1 = 0$

3. $x^2 + 4x = -1$

4. $x^2 + 6x = -5$

5. $x^2 - 3x = 2$

6. $x^2 - 4x = 12$

7. $7y^2 = 12y - 5$

8. $7x^2 = 6x - 1$

9. $5y^2 + 8y = -5$

10. $5y^2 + 6y = -5$

11. $7y + 6 = -2y^2$

12. $7y + 3 = -2y^2$

13. $\dfrac{x^2}{5} - \dfrac{x}{2} = \dfrac{-3}{10}$

14. $\dfrac{x^2}{4} - \dfrac{x}{2} = -\dfrac{1}{8}$

15. $\dfrac{x^2}{7} + \dfrac{x}{2} = \dfrac{-3}{14}$

16. $\dfrac{x^2}{8} + \dfrac{x}{2} = -\dfrac{1}{8}$

17. $\dfrac{x^2}{2} - \dfrac{3x}{4} = \dfrac{-1}{8}$

18. $\dfrac{x^2}{10} - \dfrac{x}{5} = \dfrac{3}{2}$

19. $\dfrac{x^2}{8} = -\dfrac{x}{4} - \dfrac{1}{8}$

20. $\dfrac{x^2}{12} = -\dfrac{x}{4} - \dfrac{1}{3}$

21. $6x = 4x^2 + 1$

22. $6x = 9x^2 - 4$

23. $3x = 1 - 3x^2$

24. $3x = 2x^2 - 5$

25. $x(x + 2) = 2x(x + 1) - 4$

26. $x(4x - 7) - 10 = 6x^2 - 7x$

27. $6x(x + 5) = (x + 15)^2$

28. $6x(x + 1) = (x + 3)^2$

29. $(x - 2)^2 = 4x(x - 1)$

30. $(x - 4)^2 = 4x(x - 2)$

ANSWERS

1. _____
2. _____
3. _____
4. _____
5. _____
6. _____
7. _____
8. _____
9. _____
10. _____
11. _____
12. _____
13. _____
14. _____
15. _____
16. _____
17. _____
18. _____
19. _____
20. _____
21. _____
22. _____
23. _____
24. _____
25. _____
26. _____
27. _____
28. _____
29. _____
30. _____

31. $x^3 - 8 = 0$ **32.** $x^3 - 1 = 0$

33. $8x^3 - 1 = 0$ **34.** $27x^3 - 1 = 0$

31. _____

32. _____

33. _____

34. _____

B. Applications

35. Find the price p (dollars) at the equilibrium point if the supply s is given by $s = 40p - 40$ and the demand by $d = \dfrac{800}{p}$.

35. _____

36. Find the price p (dollars) at the equilibrium point if the supply is $s = 30p - 50$ and the demand is $d = \dfrac{20}{p}$.

36. _____

37. Find the price p (dollars) at the equilibrium point if the supply is $s = 30p - 50$ and the demand is $d = \dfrac{10}{p}$.

37. _____

38. Find the price p (dollars) at the equilibrium point if the supply is $s = 20p - 60$ and the demand is $d = \dfrac{30}{p}$.

38. _____

39. The bending moment M of a simple beam is given by $M = 20x - x^2$. For what value of x is $M = 40$?

39. _____

40. Use the formula given in Problem 39 to find the value of x for which $M = 60$.

40. _____

The maximum safe length L for which a beam will support a load d is given by $aL^2 + bL + c = d$, where a, b, c, and d depend on the materials and structures used.

41. Find L when $a = 400$, $b = 200$, $c = 200$, and $d = 800$.

41. _____

42. Find L when $a = 5$, $b = 0$, $c = 100$, $d = 180$.

42. _____

✓ SKILL CHECKER

Simplify $\sqrt{b^2 - 4ac}$ given the following values.

43. $a = 3$, $b = -2$, $c = -1$ **44.** $a = 2$, $b = 3$, $c = -1$

43. _____

44. _____

45. $a = 3$, $b = -5$, $c = 4$ **46.** $a = 3$, $b = -1$, $c = 1$

45. _____

46. _____

Find the product.

47. $(2x + 1)(3x - 4)$ **48.** $(3x + 1)(2x - 5)$

47. _____

48. _____

49. $(3x - 7)(4x + 3)$ **50.** $(4x - 8)(3x + 5)$

49. _____

50. _____

6.2 USING YOUR KNOWLEDGE

In this section we derived the quadratic formula by completing the square. The procedure depends on making the coefficient of x^2 one. But there is another way of deriving the quadratic formula. See if you can give reasons for each step given that $ax^2 + bx + c = 0$.

1. $4a^2x^2 + 4abx + 4ac = 0$

2. $4a^2x^2 + 4abx = -4ac$

3. $4a^2\mathrm{x}^2 + 4abx + b^2 = b^2 - 4ac$

4. $(2ax + b)^2 = b^2 - 4ac$

5. $2ax + b = \pm\sqrt{b^2 - 4ac}$

6. $2ax = -b \pm \sqrt{b^2 - 4ac}$

7. $x = \dfrac{-b \pm \sqrt{b^2 - 4ac}}{2a}$

1. _____

2. _____

3. _____

4. _____

5. _____

6. _____

7. _____

CALCULATOR CORNER

Your calculator can be extremely helpful in finding the roots of a quadratic equation by using the quadratic formula. Of course, the roots you obtain are being approximated by decimals. It is most convenient to start with the radical part in the solution of the quadratic equation and then store this value so you can evaluate both roots without having to backtrack or copy down any intermediate steps. Let us look at the equation of Example 1,

$$8x^2 + 7x - 1 = 0$$

Using the quadratic formula, the solution will be obtained by following these keystrokes:

$\boxed{7}\ \boxed{x^2}\ \boxed{-}\ \boxed{4}\ \boxed{\times}\ \boxed{8}\ \boxed{\times}\ \boxed{1}\ \boxed{+/-}\ \boxed{=}$

$\boxed{\sqrt{x}}\ \boxed{\text{STO}}\ \boxed{7}\ \boxed{+/-}\ \boxed{+}\ \boxed{\text{RCL}}\ \boxed{=}\ \boxed{\div}\ \boxed{2}\ \boxed{\div}\ \boxed{8}\ \boxed{=}$

The display shows 0.125 (which was given as $\frac{1}{8}$ in the example). To obtain the other root, key in

$\boxed{7}\ \boxed{+/-}\ \boxed{-}\ \boxed{\text{RCL}}\ \boxed{=}\ \boxed{\div}\ \boxed{2}\ \boxed{\div}\ \boxed{8}\ \boxed{=}$

which yields -1. In general, to solve the equation $ax^2 + bx + c = 0$ using your calculator, key in the following:

$\boxed{b}\ \boxed{x^2}\ \boxed{-}\ \boxed{4}\ \boxed{\times}\ \boxed{a}\ \boxed{\times}\ \boxed{c}\ \boxed{=}$

$\boxed{\sqrt{x}}\ \boxed{\text{STO}}\ \boxed{b}\ \boxed{+/-}\ \boxed{+}\ \boxed{\text{RCL}}\ \boxed{=}\ \boxed{\div}\ \boxed{2}\ \boxed{\div}\ \boxed{a}\ \boxed{=}$

$\boxed{b}\ \boxed{+/-}\ \boxed{-}\ \boxed{\text{RCL}}\ \boxed{=}\ \boxed{\div}\ \boxed{2}\ \boxed{\div}\ \boxed{a}\ \boxed{=}$

Note that if $b^2 - 4ac < 0$, the calculator will give you an error message when you press $\sqrt{x}$. In such cases, you will have to change the sign before pressing $\sqrt{x}$ and supply the i in the final answer. (Try it in Example 6.)

6.3 THE DISCRIMINANT

OBJECTIVES

REVIEW

Before starting this section, you should know:

1. How to evaluate and simplify expressions containing radicals.
2. How to multiply two binomials.

OBJECTIVES

You should be able to:

A. Find the discriminant and determine the character of the roots of a quadratic equation.
B. Use the discriminant to determine if a quadratic is factorable and then factor it.
C. Find a quadratic equation with specified roots.
D. Verify that the solutions of an equation are correct using facts about the sum and product of roots.

A merchant in this mall wants to know if he is going to "break even." First, the daily cost C of his merchandise is $C = C.001x^2 + 10x + 100$, where x is the number of items sold, and the corresponding revenue is $R = 20x - 0.01x^2$. The break-even point occurs when the cost C equals the revenue R—that is, when

$$C = R$$
$$0.001x^2 + 10x + 100 = 20x - 0.01x^2$$
$$0.011x^2 - 10x + 100 = 0 \qquad \text{Standard form}$$

If this equation has real-number solutions, he will break even. How do we ascertain that? Since the quadratic formula

$$x = \frac{-b \pm \sqrt{b^2 - 4ac}}{2a}$$

gives the solutions to any quadratic equation in standard form, we find out what kind of solutions the equation has by looking at the expression under the radical, $b^2 - 4ac$.

> The expression $b^2 - 4ac$ under the radical is called the **discriminant** D; that is, $D = b^2 - 4ac$.

A. USING THE DISCRIMINANT TO CLASSIFY ROOTS

To find the discriminant of $0.011x^2 - 10x + 100 = 0$, note that $a = 0.011$, $b = -10$, and $c = 100$. Thus

$$b^2 - 4ac = (-10)^2 - 4(0.011)(100)$$
$$= 100 - 4.4$$
$$= 95.6$$

This means that the solutions of the equation are

$$\frac{-(-10) + \sqrt{95.6}}{2(0.011)} \quad \text{and} \quad \frac{-(-10) - \sqrt{95.6}}{2(0.011)}$$

That is, there are two real positive solutions for the equation. Thus the merchant can break even. Let us look at the possibilities in general.

The solutions of the equation $ax^2 + bx + c = 0$, where a, b, and c are real numbers, are

$$x = \frac{-b \pm \sqrt{D}}{2a}$$

These solutions will be:

Two distinct irrational numbers when $D > 0$ (D not a perfect square).

Two distinct imaginary numbers when $D < 0$.

One rational number when $D = 0$.

Two distinct rational numbers when D is a perfect square, that is, when $D = N^2$.

Here are some examples.

EQUATION	$b^2 - 4ac$	ROOTS (SOLUTIONS)
$4x^2 - 3x - 5 = 0$	$(-3)^2 - 4(4)(-5) = 89$	Two irrational numbers
$4x^2 - 3x + 5 = 0$	$(-3)^2 - 4(4)(5) = -71$	Two imaginary numbers
$4x^2 - 4x + 1 = 0$	$(-4)^2 - 4(4)(1) = 0$	One rational number
$x^2 - 2x - 3 = 0$	$(-2)^2 - 4(1)(-3) = 16$	Two rational numbers

EXAMPLE 1   Consider the equation $4x^2 - kx = -1$.

a. Find the discriminant.
b. Find k so that the equation has exactly one rational solution.

Solution

a. We first write the equation in standard form by adding 1 to both sides, obtaining $4x^2 - kx + 1 = 0$. Thus $a = 4$, $b = -k$, $c = 1$, and

$$b^2 - 4ac = (-k)^2 - 4(4)(1)$$
$$= k^2 - 16$$

b. For this equation to have one rational solution, the discriminant must be 0:

$$0 = k^2 - 16$$
$$16 = k^2$$
$$\pm\sqrt{16} = k$$
$$\pm 4 = k$$

Thus when k is 4 or -4, the equation has one rational number as its solution.

Problem 1   Consider $4x^2 + kx = -4$.
a. Find the discriminant.
b. Find k so that the equation has one rational solution.

B. DETERMINING IF A QUADRATIC IS FACTORABLE

We have now learned that if D is a perfect square, $ax^2 + bx + c = 0$ has two rational roots, say r and s. Thus, $a(x - r)(x - s) = 0$, which means that $ax^2 + bx + c$ is factorable into factors with *integer* coefficients. Here is our result:

> If $b^2 - 4ac$ is a perfect square, then $ax^2 + bx + c$ is factorable.

Thus to find out if $12x^2 + 20x - 25$ is factorable, we find $b^2 - 4ac = (20)^2 - 4(12)(-25) = 1600$. Since 1600 is a perfect square $((40)^2 = 1600)$, $12x^2 + 20x - 25$ is factorable. (Contrast this technique with the ac test we learned in Section 3.5 or with trial and error!)

EXAMPLE 2 Determine if $20x^2 + 10x - 32$ is factorable.

Solution Here $a = 20$, $b = 10$, and $c = 32$.

$$b^2 - 4ac = (10)^2 - 4(20)(-32)$$
$$= 100 + 2560$$
$$= 2660$$

Since 2660 is not a perfect square, $20x^2 + 10x - 32$ is not factorable. ▲

Now that we know how to use the discriminant to determine if a quadratic is factorable, we shall learn a method to factor it. To factor $12x^2 + 20x - 25$, we first solve the corresponding quadratic equation $12x^2 + 20x - 25 = 0$. Recall that $b^2 - 4ac = 1600$; thus

$$x = \frac{-20 \pm \sqrt{1600}}{2(12)} = \frac{-20 \pm 40}{24}$$

The solutions are

$$\frac{-20 + 40}{24} = \frac{5}{6} \quad \text{and} \quad \frac{-20 - 40}{24} = -\frac{5}{2}$$

We now reverse the steps we take to solve a quadratic by factoring. Since

$$x = \frac{5}{6} \quad \text{or} \quad x = -\frac{5}{2}$$
$$6x = 5 \quad \text{or} \quad 2x = -5$$
$$6x - 5 = 0 \quad \text{or} \quad 2x + 5 = 0$$

That is, $(6x - 5)(2x + 5) = 0$. (You can check this by multiplying.) Thus, $12x^2 + 20x - 25 = (6x - 5)(2x + 5)$.

EXAMPLE 3 Factor $12x^2 + x - 35$ if possible.

Solution

$$b^2 - 4ac = (1)^2 - 4(12)(-35) = 1 + 1680 = 1681$$

Problem 2 Determine if $20x^2 + 10x - 30$ is factorable.

Problem 3 Factor $21x^2 + x - 10$ if possible.

Since $1681 = 41^2$ the expression is factorable. We now use the quadratic formula to solve the related equation $12x^2 + x - 35 = 0$.

$$x = \frac{-1 + \sqrt{1681}}{2(12)} = \frac{-1 \pm 41}{24}$$

$$x = \frac{5}{3} \quad \text{and} \quad x = -\frac{7}{4}$$

We now reverse the steps we use in solving a quadratic equation by factoring.

$$x = \frac{5}{3} \quad \text{or} \qquad x = -\frac{7}{4}$$

$$3x = 5 \quad \text{or} \qquad 4x = -7$$

$$3x - 5 = 0 \quad \text{or} \quad 4x + 7 = 0$$

Multiplying, $(3x - 5)(4x + 7) = 0$. Thus $12x^2 + x - 35 = (3x - 5)(4x + 7)$. You can check that the factorization is correct by multiplying the factors to obtain $12x^2 + x - 35$.

C. VERIFYING SOLUTIONS OF QUADRATICS

In the process of solving Example 3, we found out that the solutions of the equation $12x^2 + x - 35 = 0$ are $\frac{5}{3}$ and $-\frac{7}{4}$. To verify this, we can substitute these values in the original equation. But there is another way. If the equation is

$$ax^2 + bx + c = 0$$

then, dividing by a, we can rewrite it as

$$x^2 + \frac{b}{a}x + \frac{c}{a} = 0 \tag{1}$$

If the roots are r_1 and r_2, we can also write

$$(x - r_1)(x - r_2) = 0$$

or

$$x^2 - (r_1 + r_2)x + r_1 r_2 = 0 \tag{2}$$

Comparing Equations (1) and (2), we see that

$$\frac{b}{a} = -(r_1 + r_2) \quad \text{and} \quad \frac{c}{a} = r_1 r_2$$

This discussion can be summarized as follows.

If r_1 and r_2 are the solutions of the equation $ax^2 + bx + c = 0$, then

$$r_1 + r_2 = -\frac{b}{a} \quad \text{and} \quad r_1 r_2 = \frac{c}{a}$$

That is, the sum of the roots of a quadratic is $-\frac{b}{a}$, and the product of the roots is $\frac{c}{a}$.

We can now verify that $\frac{5}{3}$ and $-\frac{7}{4}$ are solutions of $12x^2 + x - 35 = 0$. The sum of the roots is

$$\frac{5}{3} + \left(-\frac{7}{4}\right) = \frac{20}{12} + \left(-\frac{21}{12}\right) = -\frac{1}{12} = -\frac{b}{a}$$

The product is

$$\frac{5}{3} \cdot \left(-\frac{7}{4}\right) = -\frac{35}{12} = -\frac{c}{a}$$

so our results are correct.

EXAMPLE 4   Use the sum and product properties to see if the solutions of $3x^2 + 5x - 2 = 0$ are

a. $-\frac{1}{3}$ and 2 **b.** $\frac{1}{3}$ and -2

Solution

a. In the equation $3x^2 + 5x - 2 = 0$, $a = 3$, $b = 5$, and $c = -2$. The sum of the roots is

$$-\frac{b}{a} = -\frac{5}{3}$$

Since $-\frac{1}{3} + 2 = -\frac{1}{3} + \frac{6}{3} = \frac{5}{3}$, $-\frac{1}{3}$ and 2 cannot be the roots.

b. The sum of the proposed roots is $\frac{1}{3} + \left(-\frac{6}{3}\right) = -\frac{5}{3}$. The product of the roots of $3x^2 + 5x - 2 = 0$ must be

$$\frac{c}{a} = \frac{-2}{3}$$

The product of the proposed roots is $\frac{1}{3} \cdot (-2) = \frac{-2}{3}$. Thus $\frac{1}{3}$ and -2 are the correct roots.

Problem 4   Use the sum and product properties to see if the solutions of $4x^2 - 12x + 5 = 0$ are
a. $\frac{1}{2}$ and $-\frac{5}{2}$ **b.** $\frac{1}{2}$ and $\frac{5}{2}$

NAME

CLASS

SECTION

ANSWERS

A. In Problems 1–10 find the discriminant and determine the character of the roots.

1. $3x^2 + 5x - 2 = 0$

2. $3x^2 - 2x + 5 = 0$

3. $4x^2 = 4x - 1$

4. $2x^2 = 2x + 5$

5. $x^2 - 10x = -25$

6. $x^2 - 5x = 5$

7. $4x^2 - 5x + 3 = 0$

8. $5x^2 - 7x + 8 = 0$

9. $x^2 - 2 = \dfrac{5}{2}x$

10. $x^2 + \dfrac{1}{5} = \dfrac{2}{5}x$

1. _____

2. _____

3. _____

4. _____

5. _____

6. _____

7. _____

8. _____

9. _____

10. _____

In Problems 11–20 determine the value of k that will make the given equation have exactly one rational root.

11. $x^2 - 4kx + 64 = 0$

12. $3x^2 + kx + 3 = 0$

13. $kx^2 - 10x = 5$

14. $2kx^2 - 12x = -9$

15. $2x^2 = kx - 8$

16. $3x^2 = kx - 3$

17. $25x^2 - kx = -4$

18. $4x^2 + 9kx = -1$

19. $x^2 + 8x = k$

20. $2x^2 - 4x = k$

11. _____

12. _____

13. _____

14. _____

15. _____

16. _____

17. _____

18. _____

19. _____

20. _____

B. In Problems 21–30 use the discriminant to determine if the given polynomial is factorable into factors with integer coefficients. If it is, use the technique of Example 3 to factor it.

21. $10x^2 - 7x + 8$

22. $10x^2 - 7x + 1$

23. $12x^2 - 17x + 6$

24. $12x^2 - 17x + 2$

25. $27x^2 + 51x - 56$

26. $15x^2 + 52x - 83$

21. _____

22. _____

23. _____

24. _____

25. _____

26. _____

27. $15x^2 + 52x - 84$

28. $27x^2 - 57x - 40$

29. $12x^2 - 61x + 60$

30. $30x^2 - 19x - 140$

27. _____

28. _____

29. _____

30. _____

C. In Problems 31–35 find (a) The sum of the roots and (b) The product of the roots. (c) Determine if the two given values are the roots of the given equation.

31. $4x^2 - 6x + 5 = 0$ The proposed roots are $\dfrac{1}{2}$ and $\dfrac{5}{2}$.

31. _____

32. $2x^2 + 9x = 35$ The proposed roots are $-\dfrac{7}{2}$ and -1.

32. _____

33. $5x^2 + 13x = 6$ The proposed roots are $\dfrac{2}{5}$ and -3.

33. _____

34. $4 - 3x = 7x^2$ The proposed roots are $\dfrac{7}{4}$ and 1.

34. _____

35. $-2 - 5x = 2x^2$ The proposed roots are $\dfrac{1}{2}$ and 2.

35. _____

36. If d is a constant and 3 is one root of the equation $2x^2 - dx + 5 = 0$, use the product property to find the other root.

36. _____

37. If k is a constant and -5 is one root of the equation $3x^2 + kx = 40$, use the product property to find the other root.

37. _____

38. If the sum of the roots of the equation $2x^2 - kx = 4$ is 3, find the value of k.

38. _____

39. If the sum of the roots of the equation $10x^2 + (k - 2)x = 3$ is $-\frac{13}{10}$, find k.

39. _____

40. If the sum of the roots of the equation $3x^2 + (2k - 5)x + 8 = 0$ is 4, find k.

40. _____

✓ **SKILL CHECKER**

Solve.

41. $x + 2\sqrt{x} - 3 = 0$

42. $x + 4\sqrt{x} - 8 = 0$

43. $x^2 + 6x + 5 = 0$

44. $x^2 - 14x - 15 = 0$

41. _____

42. _____

43. _____

44. _____

6.4 SOLVING EQUATIONS IN QUADRATIC FORM

OBJECTIVES

REVIEW

Before starting this section, you should know:

1. How to find the LCD of two or more rational expressions.
2. How to solve quadratic equations by factoring or by using the quadratic formula.

OBJECTIVES

You should be able to:

A. Solve equations involving rational expressions by converting them to quadratic equations.
B. Solve equations that are quadratic in form by substitution.

The two people in the picture can finish laying the blocks in 2 days. If they work alone, one of them takes 3 days more than the other. How long would it take each of them working alone to finish the job? If we assume that the first worker can finish in d days, that worker will do $\dfrac{1}{d}$ of the work each day. The other one will finish in $d + 3$ days and do $\dfrac{1}{d + 3}$ of the work each day. The work done each day is

$$\underset{\substack{\text{work done} \\ \text{by first}}}{} + \underset{\substack{\text{work done} \\ \text{by second}}}{} = \underset{\substack{\text{work done} \\ \text{together}}}{}$$

$$\frac{1}{d} \quad + \quad \frac{1}{d + 3} \quad = \quad \frac{1}{2}$$

To solve this equation, we multiply each term by the LCD, $2d(d + 3)$, obtaining

$$2d(d + 3) \cdot \frac{1}{d} + 2d(d + 3) \cdot \frac{1}{d + 3} = 2d(d + 3) \cdot \frac{1}{2}$$

$2(d + 3) + 2d = d(d + 3)$	Simplify.
$2d + 6 + 2d = d^2 + 3d$	Remove parentheses.
$0 = d^2 - d - 6$	Write in standard form.
$0 = (d - 3)(d + 2)$	Factor.
$d - 3 = 0 \quad \text{or} \quad d + 2 = 0$	Zero-factor principle.
$d = 3 \quad \text{or} \qquad d = -2$	Solve.

Since d is the number of days, $d = -2$ has to be discarded. Thus it takes one person working alone 3 days to finish and the other one working alone 3 days more—that is, 6 days—to finish.

A. SOLVING EQUATIONS CONTAINING RATIONAL EXPRESSIONS

As we have seen, equations involving rational expressions can lead to quadratic equations that can be solved by factoring or by using the quadratic formula. When solving such equations, make sure that the solutions are checked in the original equations to avoid zero denominators.

EXAMPLE 1 Solve.

a. $\dfrac{4}{x^2 - 4} - \dfrac{1}{x - 2} = 1$ **b.** $\dfrac{-12}{x^2 - 9} + \dfrac{1}{x - 3} = 1$

Problem 1 Solve

a. $\dfrac{6}{x^2 - 9} - \dfrac{1}{x - 3} = 1$

b. $\dfrac{6}{x^2 - 4} + \dfrac{1}{x - 2} = 1$

Solution

a. Multiplying each term by the LCD, $(x + 2)(x - 2)$, we have

$$(x + 2)(x - 2) \cdot \frac{4}{(x^2 - 4)} - (x + 2)(x - 2) \cdot \frac{1}{x - 2} = (x + 2)(x - 2) \cdot 1$$

$$4 - (x + 2) = x^2 - 4 \quad \text{Simplify.}$$
$$x^2 + x - 6 = 0 \quad \text{Write in standard form.}$$
$$(x + 3)(x - 2) = 0 \quad \text{Factor.}$$

$x = -3 \quad \text{or} \quad x = 2$

The possible solutions are -3 and 2. However, $x = 2$ is not a solution, since $\dfrac{1}{x - 2}$ is not defined for $x = 2$. Thus the only solution is -3, which you can check in the original equation.

b. Multiplying each term by the LCD, $(x + 3)(x - 3)$, we have

$$(x + 3)(x - 3) \cdot \frac{-12}{x^2 - 9} + (x + 3)(x - 3) \cdot \frac{1}{x - 3} = (x + 3)(x - 3) \cdot 1$$

$$-12 + x + 3 = x^2 - 9 \quad \text{Simplify.}$$
$$x^2 - x = 0 \quad \text{Write in standard form.}$$
$$x(x - 1) = 0 \quad \text{Factor.}$$

$x = 0 \quad \text{or} \quad x = 1$

This time both solutions satisfy the original equation. Thus the solutions are 0 and 1. (Check this.)

B. SOLVING EQUATIONS BY SUBSTITUTION

Some equations that are not quadratics can be written in quadratic form and solved by an appropriate substitution. We shall illustrate several of these substitutions in the next examples.

EXAMPLE 2 Solve $x^4 - 10x^2 + 9 = 0$.

Problem 2 Solve $x^4 - 5x^2 + 4 = 0$.

Solution This equation can be written as

$$(x^2)^2 - 10(x^2) + 9 = 0$$
$$u^2 - 10u + 9 = 0 \quad \text{Let } u = x^2.$$
$$(u - 9)(u - 1) = 0 \quad \text{Factor.}$$
$$u = 9 \quad \text{or} \quad u = 1 \quad \text{Solve.}$$

$x^2 = 9$　　or　$x^2 = 1$　　　Substitute $x^2 = u$.

　$x = \pm 3$　or　　$x = \pm 1$　Extract roots.

Thus the solutions of $x^4 - 10x^2 + 9 = 0$ are 3, 1, -3, and -1.

EXAMPLE 3　Solve $(x^2 - x)^2 - (x^2 - x) - 30 = 0$.

Solution　This equation is already quadratic in form. If we let $u = x^2 - x$, we can write

$$u^2 - u - 30 = 0$$

$$(u - 6)(u + 5) = 0 \qquad\qquad \text{Factor.}$$

$$u = 6 \quad \text{or} \qquad\quad u = -5 \quad \text{Solve.}$$

$$x^2 - x = 6 \quad \text{or} \qquad x^2 - x = -5 \quad \text{Substitute } u = x^2 - x.$$

$$x^2 - x - 6 = 0 \quad \text{or} \quad x^2 - x + 5 = 0 \quad \text{Standard form.}$$

The first equation can be solved by factoring:

$$x^2 - x - 6 = (x - 3)(x + 2) = 0$$

Thus $x = 3$ or $x = -2$. We use the quadratic formula for $x^2 - x + 5 = 0$. Here $a = 1$, $b = -1$, and $c = 5$. Thus

$$x = \frac{-(-1) \pm \sqrt{(-1)^2 - 4(1)(5)}}{2(1)} = \frac{1 \pm \sqrt{-19}}{2}$$

$$= \frac{1 \pm \sqrt{19}i}{2}$$

Thus the solutions of $(x^2 - x)^2 - (x^2 - x) - 30 = 0$ are 3, -2, $\dfrac{1 + \sqrt{19}i}{2}$, and $\dfrac{1 - \sqrt{19}i}{2}$.

EXAMPLE 4　Solve $x^{1/2} - 8x^{1/4} + 15 = 0$.

Solution　To write the equation in quadratic form, we let $u = x^{1/4}$, which makes $u^2 = x^{1/2}$.

$$x^{1/2} - 8x^{1/4} + 15 = 0$$

$$u^2 - 8u + 15 = 0 \qquad\qquad \text{Substitute.}$$

$$(u - 3)(u - 5) = 0 \qquad\qquad \text{Factor.}$$

$$u = 3 \quad \text{or} \quad\;\; u = 5 \qquad \text{Solve.}$$

$$x^{1/4} = 3 \quad \text{or} \quad x^{1/4} = 5 \qquad \text{Substitute } u = x^{1/4}.$$

$$x = 3^4 \quad \text{or} \qquad x = 5^4 \qquad \text{Raise each side to the fourth power.}$$

$$x = 81 \quad \text{or} \qquad x = 625$$

You can verify that the answers are correct by substituting in the original equation.　▲

Finally, we solve some equations involving radicals. For example, the equation $x + 2\sqrt{x} - 3 = 0$ can be written as a quadratic if we let $u = \sqrt{x}$. This makes $u^2 = x$, so that we can write

$$x + 2\sqrt{x} - 3 = 0$$

$$u^2 + 2u - 3 = 0$$

$$(u + 3)(u - 1) = 0 \qquad\qquad \text{Factor.}$$

$$u = -3 \quad \text{or} \quad\;\; u = 1 \quad \text{Solve.}$$

$$\sqrt{x} = -3 \quad \text{or} \quad \sqrt{x} = 1 \quad \text{Substitute } u = \sqrt{x}.$$

Problem 3　Solve $(x^2 - x)^2 - (x^2 - x) - 2 = 0$.

Problem 4　Solve $x^{1/2} - 5x^{1/4} + 6 = 0$.

But the square root of x is never negative, so the equation $\sqrt{x} = -3$ has no solution. The solution of $\sqrt{x} = 1$ is 1. Thus the equation $x + 2\sqrt{x} - 3 = 0$ has only one solution, 1. You can verify that this solution is correct by direct substitution into the equation.

EXAMPLE 5 Solve $x - 4\sqrt{x} + 3 = 0$.

Solution We let $u = \sqrt{x}$, which makes $u^2 = x$.

$$x - 4\sqrt{x} + 3 = 0$$
$$u^2 - 4u + 3 = 0$$
$$(u - 3)(u - 1) = 0 \qquad \text{Factor.}$$
$$u = 3 \quad \text{or} \quad u = 1 \quad \text{Solve.}$$
$$\sqrt{x} = 3 \quad \text{or} \quad \sqrt{x} = 1 \quad \text{Substitute } u = \sqrt{x}.$$
$$x = 9 \quad \text{or} \quad x = 1 \quad \text{Square both sides.}$$

You can verify that both solutions satisfy the original equation.

Problem 5 Solve $x - 5\sqrt{x} + 4 = 0$.

A. In Problems 1–10 solve the equation.

1. $\dfrac{x}{x+4} + \dfrac{x}{x+1} = 0$

2. $\dfrac{x}{x+2} + \dfrac{x}{x+3} = 0$

3. $\dfrac{x-1}{x+11} - \dfrac{2}{x-1} = 0$

4. $\dfrac{x+1}{x-2} - \dfrac{8}{x-1} = 0$

5. $\dfrac{x}{x-1} - \dfrac{x}{x+1} = 0$

6. $\dfrac{x}{x+4} + \dfrac{x}{x-2} = \dfrac{-1}{2}$

7. $\dfrac{x}{x+2} - \dfrac{x}{x+1} = \dfrac{-1}{6}$

8. $\dfrac{x}{x+1} - \dfrac{x}{x-1} = \dfrac{-3}{4}$

9. $\dfrac{x}{x+4} + \dfrac{x}{x+2} = \dfrac{-4}{3}$

10. $\dfrac{2x}{x-2} + \dfrac{x}{x-1} = \dfrac{7}{6}$

B. In Problems 11–35 solve by substitution.

11. $x^4 - 13x^2 + 36 = 0$

12. $x^4 - 5x^2 + 4 = 0$

13. $4x^4 + 35x^2 = 9$

14. $3x^4 + 2x^2 = 8$

15. $3y^4 = 5y^2 + 2$

16. $6y^4 = 7y^2 - 2$

17. $x^6 + 7x^3 - 8 = 0$

18. $x^6 - 26x^3 - 27 = 0$

19. $(x+1)^2 - 3(x+1) = 40$

20. $(x+2)^2 - 2(x+2) = 8$

21. $(y^2 - y)^2 - 8(y^2 - y) = 9$

22. $(y^2 - y)^2 + 4(y^2 - y) = 12$

23. $x^{1/2} + 3x^{1/4} - 10 = 0$

24. $x^{1/2} + 4x^{1/4} - 12 = 0$

25. $y^{2/3} - 5y^{1/3} = -6$

26. $y^{2/3} + 5y^{1/3} = -6$

27. $x + \sqrt{x} - 6 = 0$

28. $x - \sqrt{x} - 30 = 0$

29. $(x^2 - 4x) - 8\sqrt{x^2 - 4x} + 15 = 0$

ANSWERS

1. _____

2. _____

3. _____

4. _____

5. _____

6. _____

7. _____

8. _____

9. _____

10. _____

11. _____

12. _____

13. _____

14. _____

15. _____

16. _____

17. _____

18. _____

19. _____

20. _____

21. _____

22. _____

23. _____

24. _____

25. _____

26. _____

27. _____

28. _____

29. _____

30. $(x^2 + 3x) + 5\sqrt{x^2 + 3x} - 14 = 0$ 30. _____

31. $z + 3 - \sqrt{z + 3} - 6 = 0$ 31. _____

32. $z + 4 - \sqrt{z + 4} - 12 = 0$ 32. _____

33. $x^{-2} + 2x^{-1} - 3 = 0$ 33. _____

34. $x^{-2} + 2x^{-1} - 8 = 0$ 34. _____

35. $3\sqrt{x} - 5\sqrt[4]{x} + 2 = 0$ 35. _____

C. Applications

36. Working together, two workers can complete a job in 6 hr. If   36. _____
they work alone, one of them takes 9 hr more than the other to
finish. How long does it take for each of them working alone to
finish the job?

37. Working together, Jack and Jill can shovel the snow in the   37. _____
driveway in 6 hr. It takes Jack 5 hr more than Jill to do the job
by himself. How long does it take each of them working alone
to finish the job?

✔ **SKILL CHECKER**

Solve.

38. $\dfrac{x}{5} - \dfrac{x}{3} \leq \dfrac{x - 5}{5}$ **39.** $\dfrac{7x + 2}{6} \geq \dfrac{3x - 2}{4}$

38. _____

39. _____

40. $\dfrac{8x - 23}{6} + \dfrac{1}{3} \geq \dfrac{5}{2}x$ 40. _____

6.4 USING YOUR KNOWLEDGE

In Section 4.2 we studied how to factor a third-degree polynomial
when one of the factors was known. Thus to factor $6x^3 + 23x^2 +
9x - 18$ if it is known that $x + 3$ is one of the factors, we divided
$6x^3 + 23x^2 + 9x - 18$ by $x + 3$ and found that $6x^3 + 23x^2 + 9x - 18 =
(x + 3)(3x - 2)(2x + 3)$. (You can go to Section 4.2 C and see how
this was done.) If we use a similar procedure to solve the third-degree
equation $6x^3 + 23x^2 + 9x - 18 = 0$, the fact that $x + 3$ is a factor of
the original polynomial tells us that -3 is a root of the equation.
Then, by division and the zero-factor property, we obtain $3x - 2 = 0$
or $2x + 3 = 0$; that is, $x = \frac{2}{3}$ and $x = -\frac{3}{2}$ are the other two roots.

Use your knowledge of division to solve the following third-degree
equations when one root is given.

1. $6x^3 + 25x^2 + 2x - 8 = 0$ if -4 is one of the roots   (See Example   1. _____
4, Section 4.2.)

2. $x^3 - 4x^2 + x + 6 = 0$ if -1 is one of the roots   (See Problem 37,   2. _____
Exercise 4.2, if you don't want to do the division again.)

3. $x^3 - 4x^2 + x + 6 = 0$ if 3 is one of the roots   (See Problem 38,   3. _____
Section 4.2.)

ANSWER   (to problems on page 456)
5. 1, 16

6.5 NONLINEAR INEQUALITIES

OBJECTIVES

REVIEW

Before starting this section, you should know:

1. How to solve a quadratic equation.
2. How to solve and graph linear inequalities.

OBJECTIVES

You should be able to:

A. Solve quadratic inequalities.
B. Solve inequalities of degree three or higher.
C. Solve rational inequalities.

Have you seen an officer measuring skid marks at the scene of an accident? The distance d (in feet) in which a car traveling v miles per hour can be stopped is given by

$$d = 0.05v^2 + v$$

The skid marks in an accident were more than 40 ft long. The accident occurred in a 20 mi/hr zone. Was the driver over the speed limit? To answer the question we must solve the quadratic inequality $0.05v^2 + v > 40$.

A. QUADRATIC INEQUALITIES

$0.05v^2 + v > 40$	Given.
$5v^2 + 100v > 4000$	Multiply by 100.
$v^2 + 20v > 800$	Divide by 5.
$v^2 + 20v - 800 > 0$	Standard form.
$(v + 40)(v - 20) > 0$	Factor.

If the product of $v + 40$ and $v - 20$ is *positive* (>0), both factors must be positive or both must be negative. Let us start by examining the factor $v + 40$ more closely:

$v + 40 = 0$ when $v = -40$ (a **critical value**)

$v + 40 > 0$ (positive, $+$) when $v > -40$

$v + 40 < 0$ (negative, $-$) when $v < -40$

On the number line, we have the following.

$v + 40 < 0$ $v + 40 > 0$ in this interval

```
  - - - - - |+ + + + + + + + + + + + + + + + + + + + + + + + +
 -50  -40  -30  -20  -10   0    10   20   30
```

$v + 40 = 0$
when $v = -40$

$v - 20 = 0$ when $v = 20$ (a *critical value*)

$v - 20 > 0$ (positive, $+$) when $v > 20$

$v - 20 < 0$ (negative, $-$) when $v < 20$

On the number line, we have the following.

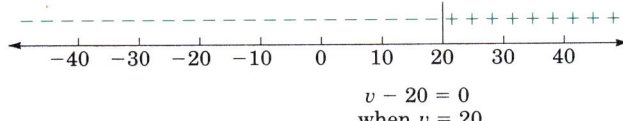

$$v - 20 = 0$$
when $v = 20$

Drawing the two number lines together, we have:

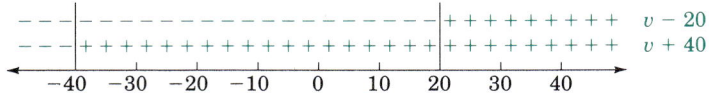

Since $(v + 40)(v - 20) > 0$ when both factors are positive ($v > 20$ in the diagram) or when both factors are negative ($v < -40$ in the diagram), the solutions of $0.5v^2 + v > 40$ are as shown in the next figure.

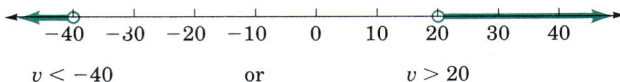

$v < -40$ or $v > 20$

Since the velocity is positive, we discard $v < -40$ and conclude that the car was going more than 20 mi/h.

Here is the procedure we have used.

TO SOLVE A QUADRATIC INEQUALITY

1. Write it in standard form.

2. Find the critical values (where the quadratic is 0) by factoring or by using the quadratic formula.

3. Graph the critical values and the regions in which each of the factors is positive or negative on the number line.

4. Determine the intervals that satisfy the conditions of the original inequality.

EXAMPLE 1   Solve $(x - 1)(x + 3) < 0$.

Solution   Since the inequality is in standard form, we find the critical values: $(x - 1)(x + 3) = 0$ when $x = 1$ or $x = -3$. The product $(x - 1)(x + 3)$ is negative (<0) when the factors have opposite signs. We draw a number line showing the critical values and the intervals where the factors are positive or negative.

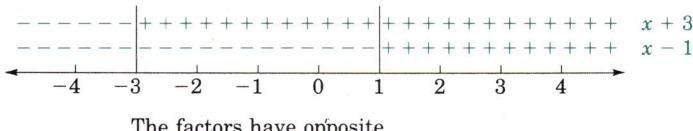

The factors have opposite
signs in this interval
$-3 < x < 1$

Problem 1   Solve $(x + 3)(x - 2) < 0$.

Thus, the graph of the solutions is as follows.

EXAMPLE 2   Solve $x^2 + x \geq 6$.

Solution   We first add -6 to both sides.

$$x^2 + x - 6 \geq 0$$
$$(x + 3)(x - 2) \geq 0 \quad \text{Factor.}$$

The critical values as well as the regions in which each of the factors is positive or negative are shown on the number line.

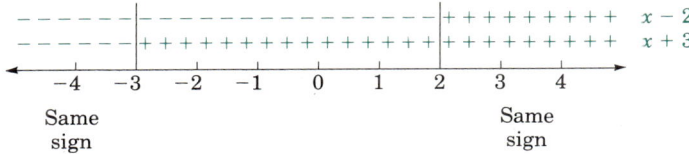

Note that this time the critical values -3 and 2 are part of the solution set, since they make the product 0. To be positive, $(x + 3)(x - 2)$ must have factors with the same signs. Since this occurs to the left of -3 and to the right of 2, the solution is $x \leq -3$ or $x \geq 2$, and the graph of the solutions is as shown.

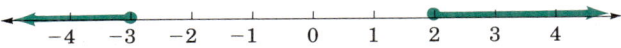

B.   POLYNOMIAL INEQUALITIES

Suppose we wish to solve the inequality

$$(x + 3)(x - 2)(x - 4) \geq 0$$

For this product to be positive, we need all the factors positive or two of the factors to be negative and one to be positive. As before, we graph the critical values -3, 2, and 4, which are part of the solution because they will make the product 0, as well as the regions in which each of the factors is positive or negative on the number line.

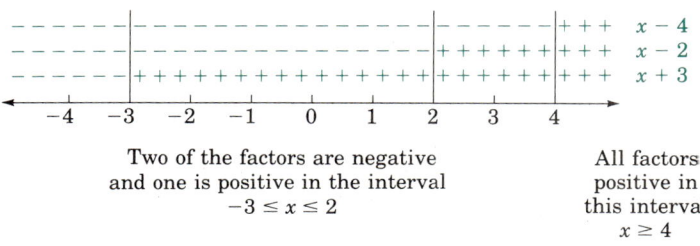

Thus, the solutions are $-3 \leq x \leq 2$ or $x \geq 4$, and the graph of these solutions is as shown.

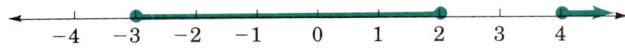

Problem 2   Solve $x^2 + x \geq 2$.

EXAMPLE 3 Solve $(x + 2)(x - 1)(x + 3) \le 0$.

Problem 3 Solve $(x + 3)(x - 2)(x + 1) \le 0$.

Solution The critical values are -2, 1, and -3. They are part of the solution, since they make the product 0. These values and the interval in which the factors are positive or negative are shown.

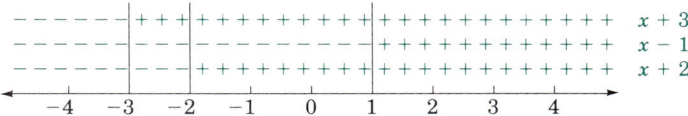

Product negative
$x \le -3$

Product negative
$-2 \le x \le 1$

Thus, the solutions are $x \le -3$ or $-2 \le x \le 1$, as shown.

C. RATIONAL INEQUALITIES

In Chapter 4 we studied rational expressions, quotients of polynomials. We are now ready to consider rational inequalities. For example, to solve the rational inequality $\dfrac{x - 3}{x + 2} \le 0$, we note that the quotient of two factors is negative when the factors have different signs. As before, we find the critical points of $x - 3$ and $x + 2$—that is, 3 and -2—and make sure to exclude -2 from the solution, since substituting -2 in the expression will cause the denominator to be 0. We then find the intervals in which $x + 2$ and $x - 3$ are positive or negative, as shown on the number line.

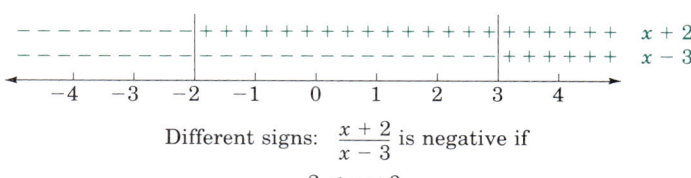

Different signs: $\dfrac{x + 2}{x - 3}$ is negative if

$-2 < x \le 3$

Thus the solution is $-2 < x \le 3$. Note that -2 is *not* included. The graph of the solution set is shown.

Thus the solution is $-2 < x \le 3$. Note that -2 is *not* included. The graph of the solution set is shown.

EXAMPLE 4 Solve $\dfrac{x}{x - 2} \ge 2$.

Problem 4 Solve $\dfrac{x}{x - 1} \le 1$.

Solution This inequality is slightly different. Since we know only when quotients are negative (<0) or positive (>0), we must have a 0 on the right side of the inequality. To do this, we first subtract 2 from both sides and then get a common denominator to simplify:

$$\frac{x}{x-2} \geq 2 \quad \text{Given.}$$

$$\frac{x}{x-2} - 2 \geq 0 \quad \text{Subtract 2.}$$

$$\frac{x}{x-2} - \frac{2(x-2)}{x-2} \geq 0 \quad 2 = \frac{2(x-2)}{x-2}.$$

$$\frac{x - 2x + 4}{x-2} \geq 0 \quad \text{Remove parentheses.}$$

$$\frac{4 - x}{x-2} \geq 0 \quad \text{Simplify.}$$

The critical points are 4 and 2. The 2 will cause the denominator to be 0, so it is not included in the solution set. The four is included, since it will make the rational expression 0. Note that $4 - x > 0$ when $4 > x$ (or $x < 4$), so plus signs appear to the *left* of 4.

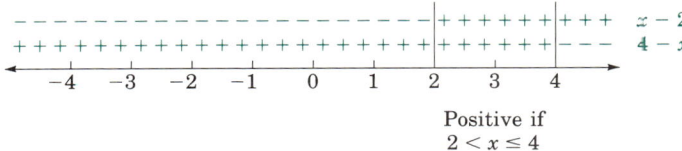

Positive if
$2 < x \leq 4$

The graph of the solution set is shown.

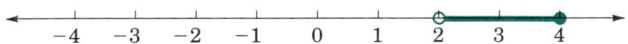

ANSWERS

A. In Problems 1–16 solve the given inequality.

1. $(x + 1)(x - 3) > 0$

2. $(x - 1)(x + 2) < 0$

3. $x(x + 4) \leq 0$

4. $(x - 1)x \geq 0$

5. $x^2 - x - 2 \leq 0$

6. $x^2 - x - 6 \leq 0$

7. $x^2 - 3x \geq 0$

8. $x^2 + 2x \leq 0$

9. $x^2 - 3x + 2 < 0$

10. $x^2 - 2x - 3 > 0$

11. $x^2 + 2x - 3 < 0$

12. $x^2 + x - 2 < 0$

13. $x^2 + 10x \leq -25$

14. $x^2 + 8x \leq -16$

15. $x^2 - 8x \geq -16$

16. $x^2 - 6x \geq -9$

In Problems 17–20 use the quadratic formula to find the critical values.

17. $x^2 - x \geq 1$ $(\sqrt{5} \approx 2.2)$

18. $x^2 - x \geq 3$ $(\sqrt{13} \approx 3.6)$

19. $x^2 - x \leq 4$ $(\sqrt{17} \approx 4.1)$

20. $x^2 - x \leq 5$ $(\sqrt{21} \approx 4.6)$

B. In Problems 21–24, solve the inequality.

21. $(x + 1)(x - 2)(x + 3) \geq 0$

22. $(x - 1)(x + 2)(x - 3) \geq 0$

23. $(x - 1)(x - 2)(x - 3) \leq 0$

24. $(x - 2)(x - 3)(x - 4) \leq 0$

C. In Problems 25–34, solve the inequality.

25. $\dfrac{2}{x - 2} \geq 0$

26. $\dfrac{3}{x - 1} \leq 0$

ANSWERS

1. _____
2. _____
3. _____
4. _____
5. _____
6. _____
7. _____
8. _____
9. _____
10. _____
11. _____
12. _____
13. _____
14. _____
15. _____
16. _____

17. _____
18. _____
19. _____
20. _____

21. _____
22. _____
23. _____
24. _____

25. _____
26. _____

27. $\dfrac{x+5}{x-1} > 2$

28. $\dfrac{2x-3}{x+3} < 1$

29. $\dfrac{3x-4}{2x-1} < 1$

30. $\dfrac{x-1}{x+5} > 1$

31. $\dfrac{1}{x-1} < \dfrac{1}{x+2}$

32. $\dfrac{1}{x+1} > \dfrac{1}{x-2}$

33. $\dfrac{4}{x} + 6 > \dfrac{2}{x} + 2$

34. $\dfrac{3}{x} + 1 < \dfrac{1}{x} - 2$

27. _____

28. _____

29. _____

30. _____

31. _____

32. _____

33. _____

34. _____

In Problems 35–38 find all values of x for which the given expression is a real number. (*Hint:* $\sqrt{a}$ is a real number if $a \geq 0$.)

35. $\sqrt{x^2 - 9}$

36. $\sqrt{x^2 - 4x + 4}$

37. $\sqrt{x^2 - 6x + 5}$

38. $\sqrt{3x - 8}$

35. _____

36. _____

37. _____

38. _____

D. Applications

39. The equivalent resistance of two electric circuits is given by $R^2 - 3R + 1$. When is this resistance more than 5 ohms?

40. The bending moment of a beam is $M = 20 - x^2$, where x is the distance, in feet, from one end of the beam. At what distances will the bending moment be more than 4?

41. The number N of water mites in a water sample depends on the temperature T in degrees Fahrenheit and is given by $N = 110T - T^2$. At what temperatures will the number of mites exceed 1000?

42. The profit P in a business varies in an 8-hr day according to the formula $P = 15t - 5t^2$, where t is the time in hours. During what hours is there a profit—that is, during what hours is $P > 0$?

43. The height h (in feet) of a projectile is $h = 48t - 16t^2$, where t is the time in seconds. During what time interval will the projectile be higher than 32 ft above the ground?

44. The distance d (in feet) in which a car traveling v miles per hour can be stopped is given by

$$d = 0.05v^2 + v$$

At what speed will it take more than 120 ft to stop a car?

39. _____

40. _____

41. _____

42. _____

43. _____

44. _____

✓ SKILL CHECKER

In Problems 45–48 find y for the given value of x.

45. $y = 3x + 6$, $x = 2$

46. $y = 3x + 6$, $x = -2$

45. _____

46. _____

47. $y = -\dfrac{2}{3}x + 4, \quad x = 3$

48. $y = -\dfrac{2}{3}x + 4, \quad x = -3$

49. If $y = 3x + 6$ and $y = 0$, find x.

50. If $y = 2x + 8$ and $y = 0$, find x.

6.5 USING YOUR KNOWLEDGE

We already know that the distance d (in feet) in which a car traveling v miles per hour can be stopped is given by

$$d = 0.05v^2 + v$$

1. For what values of v is $50 \le d \le 60$? (*Hint:* $\sqrt{11} \approx 3.32$ and $\sqrt{13} \approx 3.61$.)

2. For what values of v is $80 \le d \le 90$? (*Hint:* $\sqrt{17} \approx 4.12$ and $\sqrt{19} \approx 4.36$.)

3. According to the *Guinness Book of Records,* the longest skid mark on a public road was made by a Jaguar automobile involved in an accident in England. The skid mark was 950 ft long. If you assume that the car actually stopped after the brakes were applied and it traveled for 950 ft, how fast was the car traveling? (Use a calculator to solve this problem!)

SUMMARY

SECTION	ITEM	MEANING	EXAMPLE
6.1	Quadratic equation	An equation of the form $ax^2 + bx + c = 0$ ($a \ne 0$)	$2x^2 - 3x + 7 = 0$ is a quadratic equation.
6.1	$\pm\sqrt{a}$	The solutions of $x^2 = a$	$\pm\sqrt{3}$ are the solutions of $x^2 = 3$.
6.1C	Completing the square	1. Write the equation with the variables in descending order on the left and numbers on the right.	$5x^2 - 5x - 1 = 0$ Given 1. $5x^2 - 5x = 1$
		2. Divide each term by the coefficient of x^2 ($\ne 1$).	2. $x^2 - x = \dfrac{1}{5}$
		3. Add the square of one-half the coefficient of x to both sides.	3. $x^2 - x + \left(-\dfrac{1}{2}\right)^2$ $= \dfrac{1}{5} + \left(-\dfrac{1}{2}\right)^2$
		4. Factor the perfect square trinomial on the left	4. $\left(x - \dfrac{1}{2}\right)^2 = \dfrac{9}{20}$

(Continued)

SECTION	ITEM	MEANING	EXAMPLE
		5. Solve the resulting equation.	5. $x - \dfrac{1}{2} = \pm\sqrt{\dfrac{9}{20}}$ $= \pm\dfrac{3\sqrt{5}}{10}$ $x = \dfrac{1}{2} \pm \dfrac{3\sqrt{5}}{10}$
6.2	Quadratic formula	The solutions of $ax^2 + bx + c = 0$ are $x = \dfrac{-b \pm \sqrt{b^2 - 4ac}}{2a}.$	The solutions of $3x^2 + 2x - 5 = 0$ are $\dfrac{-2 \pm \sqrt{4 + 60}}{6} = 1$ and $-\dfrac{5}{3}.$
6.3	Discriminant	The discriminant of $ax^2 + bx + c = 0$ is $D = b^2 - 4ac.$	The discriminant of $3x^2 + 2x - 5 = 0$ is 64.
6.3A	Types of solutions	1. If $D > 0$ and D is not a perfect square, two distinct irrational roots 2. If $D < 0$, two distinct imaginary roots 3. If $D = 0$, one rational root 4. If $D = N^2$, two distinct rational roots	1. $4x^2 - 3x - 5 = 0$ ($D = 89$) has two distinct irrational roots. 2. $4x^2 - 3x + 5 = 0$ ($D = -71$) has two distinct imaginary roots. 3. $4x^2 - 4x + 1 = 0$ ($D = 0$) has one rational root. 4. $x^2 - 2x - 3 = 0$ ($D = 16$) has two rational roots.
6.3B	Factorable quadratics	If D is a perfect square, $ax^2 + bx + c$ is factorable.	$20x^2 + 10x - 30$ ($D = 2500$) is factorable.
6.3C	Sum and product of roots	The sum and product of the roots of $ax^2 + bx + c = 0$ are $-\dfrac{b}{a}$ and $\dfrac{c}{a}$, respectively.	The sum and product of the roots of $4x^2 - 12x + 5 = 0$ are 3 and $\frac{5}{4}$, respectively.
6.4	Equations that are quadratic in form	Equations that can be written as quadratics by use of appropriate substitutions.	$x^4 - 5x^2 + 4 = 0$, $(x^2 - x)^2 - (x^2 - x) - 2 = 0$ and $x - 5\sqrt{x} + 4 = 0$
6.5A	Quadratic inequality	An inequality that can be written as $ax^2 + bx + c < 0$ (The symbol $<$ can be replaced by $>$, $\leq$, or $\geq$.)	$x^2 + x - 6 < 0$ and $x^2 + x - 2 \geq 0$

(If you need help with these exercises, look in the section indicated in brackets.)

1. [6.1A] Solve.
   a. $16x^2 - 49 = 0$
   b. $25x^2 - 16 = 0$

2. [6.1A] Solve.
   a. $5x^2 + 30 = 0$
   b. $6x^2 + 42 = 0$

3. [6.1B] Solve.
   a. $(x - 3)^2 = 32$
   b. $(x - 5)^2 = 50$

4. [6.1B] Solve.
   a. $2(x - 2)^2 + 25 = 0$
   b. $3(x - 3)^2 + 64 = 0$

5. [6.1B] Solve.
   a. $5x^2 - 10x + 5 = 12$
   b. $12x^2 + 12x + 3 = 16$

6. [6.1C] Solve by completing the square.
   a. $x^2 - 8x - 9 = 0$
   b. $x^2 + 12x + 32 = 0$

7. [6.1C] Solve by completing the square.
   a. $4x^2 + 4x - 3 = 0$
   b. $16x^2 - 24x + 7 = 0$

8. [6.2A] Solve by the quadratic formula.
   a. $3x^2 + 5x - 2 = 0$
   b. $5x^2 - 9x - 2 = 0$

9. [6.2A] Solve by the quadratic formula.
   a. $3x^2 = 2x + 4$
   b. $4x^2 = 6x + 3$

10. [6.2A] Solve by the quadratic formula.
    a. $16x = x^2$
    b. $12x = x^2$

11. [6.2A] Solve by the quadratic formula.

    a. $x^2 + \dfrac{x}{15} = \dfrac{1}{3}$

    b. $\dfrac{x^2}{2} + \dfrac{9x}{10} = \dfrac{1}{5}$

12. [6.2A] Solve by the quadratic formula.
    a. $3x^2 - 2x = -1$
    b. $5x^2 - 2x = -4$

ANSWERS

1. a. _____
   b. _____

2. a. _____
   b. _____

3. a. _____
   b. _____

4. a. _____
   b. _____

5. a. _____
   b. _____

6. a. _____
   b. _____

7. a. _____
   b. _____

8. a. _____
   b. _____

9. a. _____
   b. _____

10. a. _____
    b. _____

11. a. _____
    b. _____

12. a. _____
    b. _____

13. [6.2A] Solve.
    a. $8x^3 - 125 = 0$
    b. $125x^3 - 8 = 0$

13. a. _____
   b. _____

14. [6.2B]

    a. If the demand (d) is given by $d = \dfrac{450}{p}$ and the supply (s) is given by $s = 100p - 150$, find the equilibrium point (at which $d = s$).

    b. In part a, replace the values of d and s by $d = \dfrac{50}{p}$ and $s = 150p - 100$ and find the equilibrium point.

14. a. _____
   b. _____

15. [6.3A] For the given equation, find the discriminant and determine the value of k so that the equation has exactly one rational root.
    a. $16x^2 - kx = -1$
    b. $8x^2 - kx = -2$

15. a. _____
   b. _____

16. [6.3B] Determine whether the given quadratic is factorable into factors with integer coefficients.
    a. $3x^2 - 11x - 6$
    b. $18x^2 + 13x + 2$

16. a. _____
   b. _____

17. [6.3B] Factor into factors with integer coefficients if possible.
    a. $18x^2 - 9x - 5$
    b. $18x^2 + 13x + 1$

17. a. _____
   b. _____

18. [6.3C] Without solving the equation, find the sum and the product of the roots.
    a. $15x^2 + 4x - 3 = 0$
    b. $9x^2 - 12x - 5 = 0$

18. a. _____
   b. _____

19. [6.3C]
    a. Use the sum and product properties to check whether $\frac{1}{3}$ and $-\frac{3}{5}$ are the roots of the equation in part a of Problem 18.
    b. Do the same for $\frac{1}{3}$ and $-\frac{5}{3}$ for the equation in part b of Problem 18.

19. a. _____
   b. _____

20. [6.4A] Solve.

    a. $\dfrac{8}{x^2 - 16} - \dfrac{1}{x - 4} = 1$

    b. $\dfrac{-24}{x^2 - 36} + \dfrac{2}{x + 6} = 1$

20. a. _____
   b. _____

21. [6.4B] Solve.
    a. $(x^2 + x)^2 + 2(x^2 + x) - 8 = 0$
    b. $(x^2 - 3x)^2 - 4(x^2 - 3x) - 12 = 0$

21. a. _____
   b. _____

22. [6.4B] Solve.
    a. $x^{1/2} - 4x^{1/4} = -3$
    b. $x^{1/2} + x^{1/4} = 6$

22. a. _____
   b. _____

23. [6.4B] Solve.
    a. $x^{2/3} + x^{1/3} = 12$
    b. $x^{2/3} - 5x^{1/3} = 6$

23. a. _____
   b. _____

24. [6.4B] Solve.
    a. $x - 2\sqrt{x} = 3$
    b. $x - 4\sqrt{x} = 5$

25. [6.5A] Solve.
    a. $(x - 2)(x + 3) < 0$
    b. $(x + 2)(x - 3) < 0$

26. [6.5A] Solve.
    a. $x^2 + 4x \geq 0$
    b. $x^2 - 3x \geq 0$

27. [6.5A] Solve.
    a. $x^2 + 4x \geq 8$
    b. $x^2 + 6x \geq 18$

28. [6.5B] Solve.
    a. $(x - 1)(x - 2)(x - 3) \leq 0$
    b. $(x + 1)(x + 2)(x - 3) \leq 0$

29. [6.5B] Solve.
    a. $(x + 1)(x + 2)(x - 3) \geq 0$
    b. $(x - 1)(x + 2)(x + 3) \geq 0$

30. [6.5B] Solve.

    a. $\dfrac{x + 2}{x - 2} \leq 2$

    b. $\dfrac{x - 2}{x + 2} \geq 3$

24. a. _____
    b. _____

25. a. _____
    b. _____

26. a. _____
    b. _____

27. a. _____
    b. _____

28. a. _____
    b. _____

29. a. _____
    b. _____

30. a. _____
    b. _____

_____
_____
_____
_____
_____

(Answers on pages 476–477)

(Answers on pages 476–477)

ANSWERS

1. Solve.
   a. $25x^2 - 4 = 0$
   b. $18x^2 + 3 = 0$

1. a. _____
   b. _____

2. Solve.
   a. $(x - 1)^2 = 45$
   b. $2(x - 2)^2 = 49$

2. a. _____
   b. _____

3. Solve $3x^2 + 6x + 3 = 5$.

3. _____

4. Solve by completing the square: $x^2 - 6x + 5 = 0$.

4. _____

5. Solve by completing the square: $9x^2 - 6x - 1 = 0$.

5. _____

6. Solve by the quadratic formula: $7x^2 + 5x - 2 = 0$

6. _____

7. Solve by the quadratic formula: $3x^2 = 3x + 2$.

7. _____

8. Solve by the quadratic formula: $32x = x^2$.

8. _____

9. Solve by the quadratic formula: $\dfrac{x^2}{16} + \dfrac{5x}{4} = \dfrac{11}{4}$.

9. _____

10. Solve by the quadratic formula: $4x^2 + 3x = -2$.

10. _____

11. Solve $64x^3 - 27 = 0$.

11. _____

12. If the demand (d) is given by $d = \dfrac{500}{p}$ and the supply is given by $s = 200p - 150$, find the equilibrium point (at which $s = d$).

12. _____

13. For the equation $4x^2 - kx = -9$
  a. Find the discriminant.
  b. Find k so that the equation has exactly one rational root.

13. **a.** _____
   b. _____

14. Determine whether $12x^2 - 4x - 21$ is factorable into factors with integer coefficients.

14. _____

15. Factor $35x^2 - 32x - 12$ into factors with integer coefficients if possible.

15. _____

16. **a.** Find the sum and the product of the roots of the equation $6x^2 + 7x - 5 = 0$.
  b. Use the sum and product properties to check if the solutions are $-\frac{5}{3}$ and $\frac{1}{2}$.

16. **a.** _____
   b. _____

17. Solve.

  a. $\dfrac{6}{x^2 - 9} - \dfrac{1}{x - 3} = 1$

  b. $\dfrac{2}{x^2 - 1} - \dfrac{3}{x - 1} = 1$

17. **a.** _____
   b. _____

18. Solve $x^4 - 7x^2 + 6 = 0$.

18. _____

19. Solve $(x^2 - 2x)^2 - (x^2 - 2x) - 6 = 0$.

19. _____

20. Solve $x^{1/2} - 3x^{1/4} + 2 = 0$.

21. Solve $x - 5\sqrt{x} + 6 = 0$.

22. Solve $(x + 1)(x - 2) < 0$.

23. Solve $x^2 + 2x \geq 15$.

24. Solve $(x + 1)(x - 2)(x - 3) \leq 0$.

25. Solve $\dfrac{x + 1}{x - 1} \leq 3$

20. _____

21. _____

22. _____

23. _____

24. _____

25. _____

IF YOU MISSED QUESTION	SECTION	EXAMPLES	PAGE	ANSWERS
1a	6.1A	1a	422	1. a. $\dfrac{2}{5}, -\dfrac{2}{5}$
1b	6.1A	1b	422–423	b. $\dfrac{\sqrt{6}}{6}i, -\dfrac{\sqrt{6}}{6}i$
2a	6.1B	2a	423–424	2. a. $1 \pm 3\sqrt{5}$
2b	6.1B	2b	423–424	b. $\dfrac{2 \pm 7\sqrt{2}}{2}$
3	6.1B	3	424	3. $\dfrac{-1 \pm \sqrt{15}}{3}$
4	6.1C	4	425	4. $1, 5$
5	6.1C	5	426–427	5. $\dfrac{1 \pm \sqrt{2}}{3}$
6	6.2A	1	434–435	6. $\dfrac{2}{7}, -1$
7	6.2A	2	435	7. $\dfrac{3 \pm \sqrt{33}}{6}$
8	6.2A	3	436	8. $0, 32$
9	6.2A	4	436–437	9. $2, -22$
10	6.2A	5	437	10. $\dfrac{-3 \pm \sqrt{23}i}{8}$
11	6.2A	6	438	11. $\dfrac{3}{4}, \dfrac{-3 \pm 3\sqrt{3}i}{8}$
12	6.2B	7	439	12. $p = 2$
13a	6.3A	1a	446	13. a. $k^2 - 144$
13b	6.3A	1b	446	b. $k = \pm 12$
14	6.3B	2	447	14. Yes. $D = 32^2$.
15	6.3B	3	447–448	15. $(5x - 6)(7x + 2)$
16a	6.3C	4a	449	16. a. Sum: $-\dfrac{7}{6}$, product: $-\dfrac{5}{6}$
16b	6.3C	4b	449	b. Yes
17a	6.4A	1a	454	17. a. -4
17b	6.4A	1b	454	b. $0, -3$
18	6.4B	2	454–455	18. $\pm 1, \pm \sqrt{6}$
19	6.4B	3	455	19. $3, -1, 1 \pm i$
20	6.4B	4	455	20. $1, 16$
21	6.4B	5	456	21. $4, 9$

IF YOU MISSED QUESTION	SECTION	EXAMPLES	PAGE	ANSWERS
22	6.5A	1	460–461	**22.** $-1 < x < 2$
23	6.5A	2	461	**23.** $x \le -5$ or $x \ge 3$
24	6.5B	3	462	**24.** $x \le -1$ or $2 \le x \le 3$
25	6.5B	4	462–463	**25.** $x < 1$ or $x \ge 2$

CHAPTER 7

CARTESIAN COORDINATE SYSTEMS

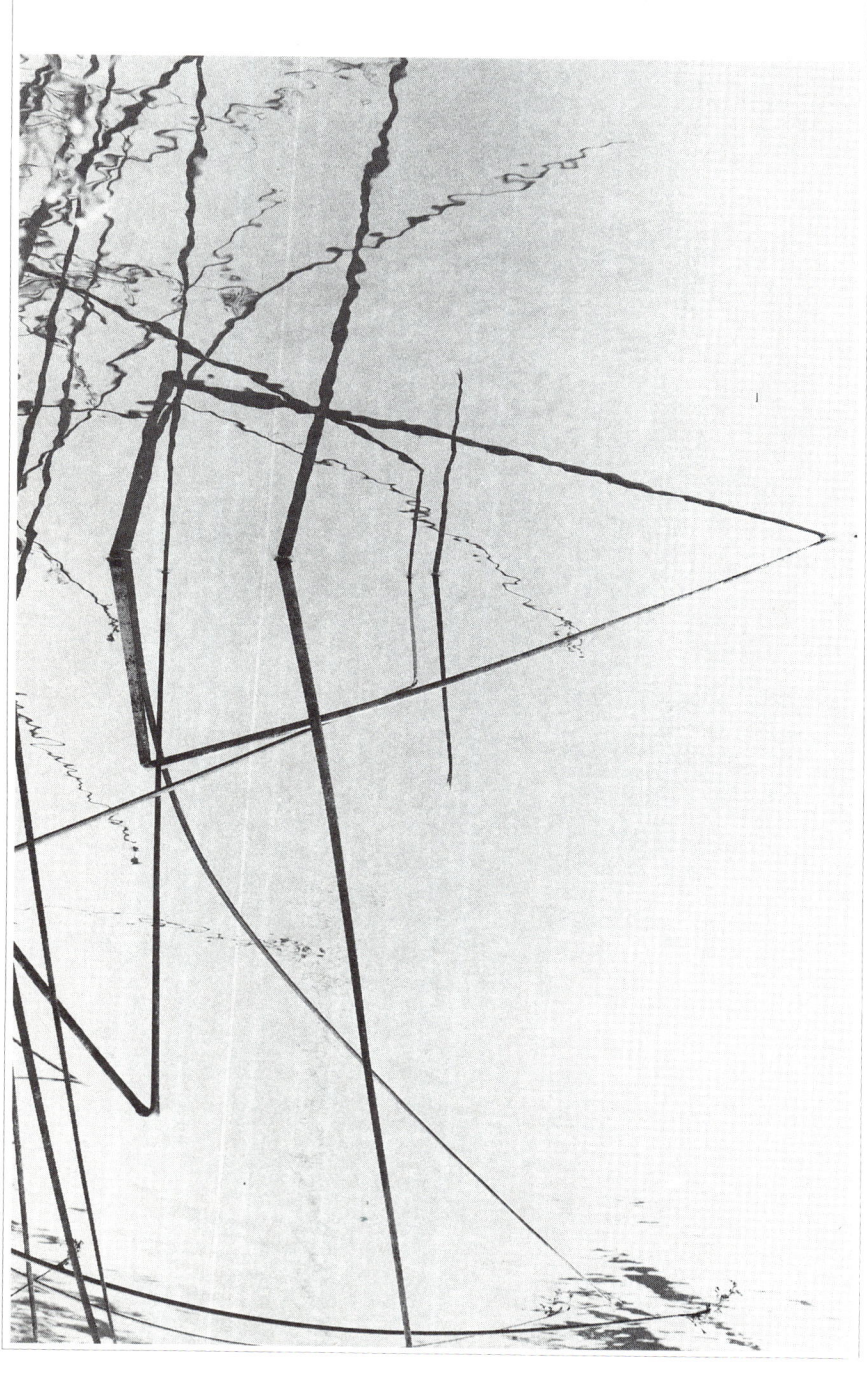

In this chapter we generalize our work with equations and include equations in two variables and their graphs. This is done by first introducing the Cartesian coordinate system and learning how to graph points. We then discuss the distance between two points, the slope of a line, and the different ways in which the equation for a line can be written. As before, after equations, we study inequalities. We end the chapter with applications to three types of variation: direct, inverse, and joint.

NAME

CLASS

SECTION

ANSWERS

(Answers on pages 484–489)

1. Graph $A(2, 3)$, $B(-2, 4)$, $C(-2, -1)$, and $D(1, -3)$.

1.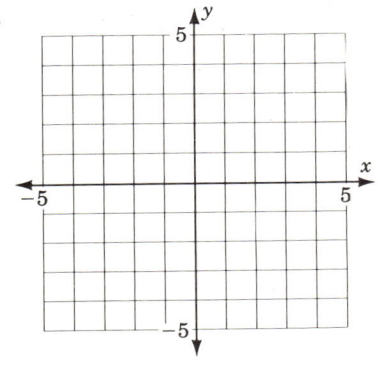

2. Graph $2x - y = 4$.

2.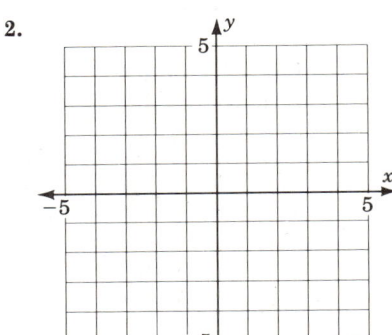

3. Find the x- and y-intercepts of $y = 2x + 4$ and then graph the line.

3.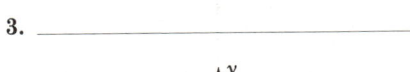

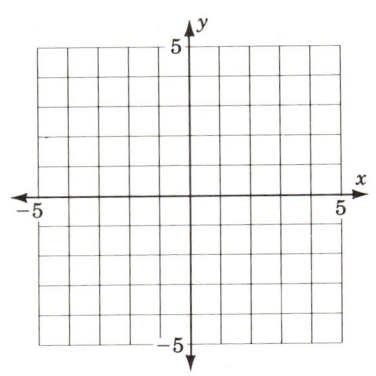

4. Graph.

   a. $2x = -4$

4. a.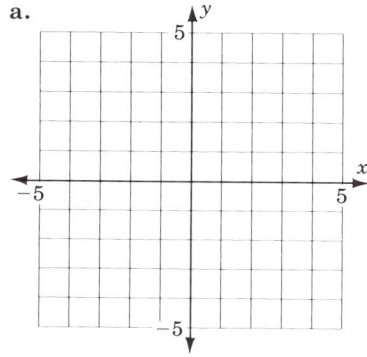

   b. $3y = -9$

b.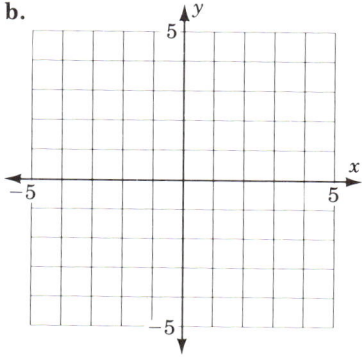

5. Find the distance between the two given points.

   a. $A(3, 3)$ and $B(6, 7)$

   b. $A(3, 2)$ and $B(5, -2)$

   c. $A(-4, 2)$ and $B(-4, 5)$

5. a. _____

   b. _____

   c. _____

6. Find the slope of the line through the two given points.

   a. $A(-2, 2)$ and $B(0, 1)$

   b. $A(-2, 4)$ and $B(-2, 7)$

   c. $A(3, 2)$ and $B(2, 3)$

   d. $A(4, 3)$ and $B(5, 3)$

6. a. _____

   b. _____

   c. _____

7. A line L_1 has slope $\frac{3}{2}$. Find whether the line through the two given points is parallel or perpendicular to L_1.

   a. $A(4, 2)$ and $B(7, 0)$

   b. $A(-1, -4)$ and $B(-3, -6)$

7. a. _____

   b. _____

8. The line through $A(1, -4)$ and $B(4, y)$ is perpendicular to a line with slope $-\frac{2}{3}$. Find y.

8. _____

9. A line goes through the point $(2, -1)$ and has slope $-\frac{1}{2}$. Graph this line.

9.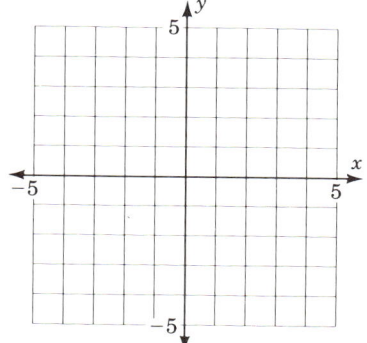

10. Find an equation of the line through (4, 2) and (3, 4). Then write the equation in standard form and graph it.

10. _____

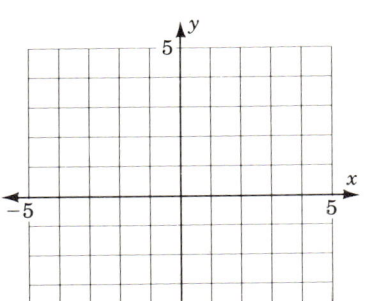

11. Find an equation of the line with slope -1 and passing through the point $(2, -3)$. Then graph this line.

11. _____

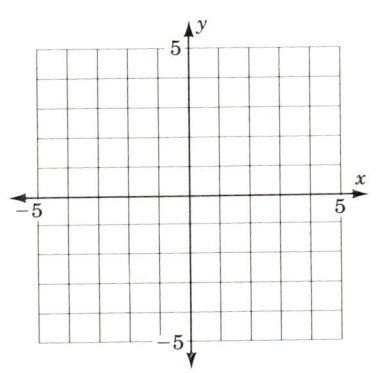

12. A line has slope 2 and y-intercept 3. Find the slope-intercept equation of this line and graph the line.

12. _____

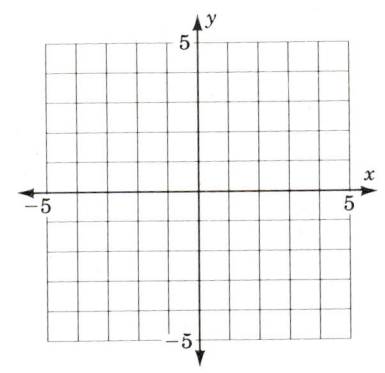

13. Find the slope and the y-intercept of the line $16x + 8y = 40$.

13. _____

14. Find an equation of the line through the point $(1, 2)$ that is
  a. Parallel to the line $3x - 2y = 5$
  b. Perpendicular to the line $3x - 2y = 5$

14. **a.** _____
   b. _____

15. Graph $2x - y < -4$.

15.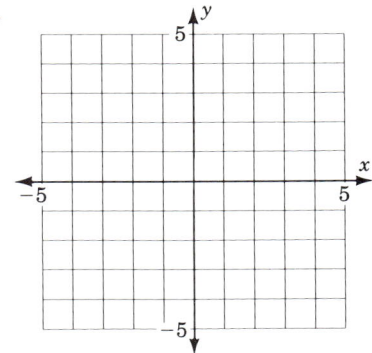

16. Graph $y \leq 2x - 4$.

16.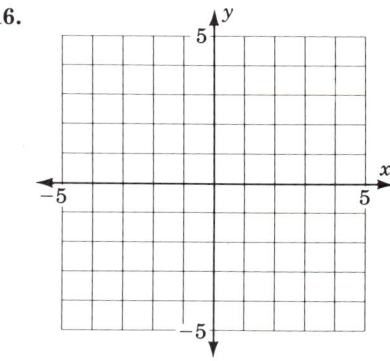

17. Graph $x \geq -2$.

17.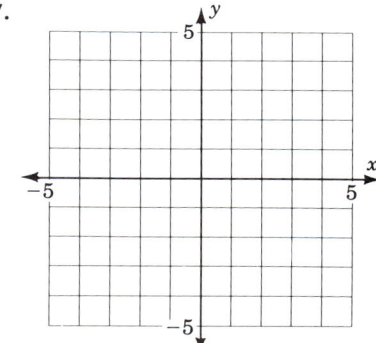

18. Graph $|y| \leq 3$.

18.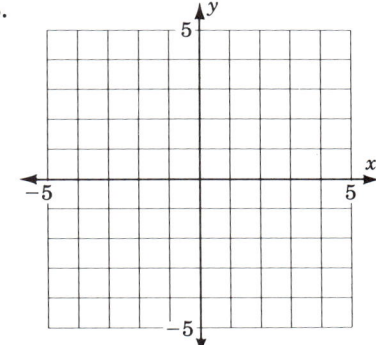

19. Graph $|x - 1| > 2$.

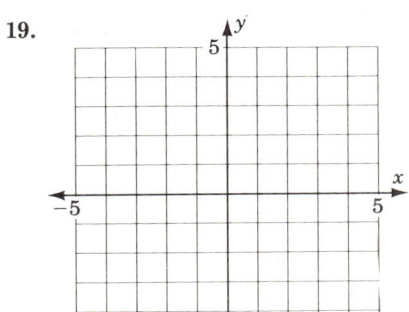

20. An enclosed gas exerts a pressure P on the walls of the container. This pressure is directly proportional to the temperature T of the gas.
   a. Write an equation of variation with k as the constant of variation.
   b. If the pressure is 5 lb/in.² when the temperature is 460°F, find k.

20. a. _____
    b. _____

21. If the temperature of a gas is held constant, the pressure P varies inversely as the volume V.
   a. Write an equation of variation with k as the constant of variation.
   b. A pressure of 1760 lb/in.² is exerted by 3 ft³ of air in a cylinder fitted with a piston. Find k.

21. a. _____
    b. _____

22. The force F with which the earth attracts an object above the earth's surface varies inversely as the square of the distance d from the center of the earth.
   a. Write an equation of variation with k as the constant of variation.
   b. A meteorite weighs 50 lb on the earth's surface. If the radius of the earth is about 4000 mi, find k.

22. a. _____
    b. _____

23. What would be the weight of the meteorite of Problem 22 at a distance of 1000 mi above the earth's surface?

23. _____

24. The horsepower that a rotating shaft can safely transmit varies jointly as the cube of its diameter d and the number n of revolutions it makes per minute.
   a. Write an equation of variation with k as the constant of variation.
   b. If a 2-in. shaft at a speed of 1400 rev/min can safely transmit 500 hp, find k.

24. a. _____
    b. _____

25. Use the value of k from Problem 24 to find the horsepower that a 4-in. shaft can safely transmit at a speed of 2100 rev/min.

25. _____

IF YOU MISSED QUESTION	SECTION	EXAMPLES	PAGE	ANSWERS
1	7.1A	1	491	1.
2	7.1B	2	493–494	2.
3	7.1C	3	495	3. x: $(-2, 0)$; y: $(0, 4)$ x-int. -2, y-int. 4

IF YOU MISSED QUESTION	SECTION	EXAMPLES	PAGE	ANSWERS
4a	7.1D	4a	496	**4. a.** $2x = -4$
4b	7.1D	4b	496	**b.** $3y = -9$
5a	7.2A	1	506	**5. a.** 5
5b	7.2A	1	506	**b.** $2\sqrt{5}$
5c	7.2A	1	506	**c.** 3
6a	7.2B	2	508–509	**6. a.** $-\frac{1}{2}$
6b	7.2B	2	508–509	**b.** Undefined
6c	7.2B	2	508–509	**c.** -1
6d	7.2B	2	508–509	**d.** 0
7a	7.2C	3	512	**7. a.** Perpendicular
7b	7.2C	3	512	**b.** Neither
8	7.2C	4	512–513	**8.** $y = \frac{1}{2}$

IF YOU MISSED QUESTION	SECTION	EXAMPLES	PAGE	ANSWERS
9	7.2D	5	513	9.
10	7.3A	1	521–522	10. $2x + y = 10$
11	7.3B	2	522–523	11. $x + y = -1$

IF YOU MISSED QUESTION	SECTION	EXAMPLES	PAGE	ANSWERS
12	7.3C	3	523	**12.** $y = 2x + 3$
13	7.3C	4	524	**13.** $m = -2, y = 5$
14a	7.3D	5	524–525	**14. a.** $3x - 2y = -1$
14b	7.3D	5	524–525	**b.** $2x + 3y = 8$
15	7.4A	1	534–535	**15.**
16	7.4A	2	535	**16.**

IF YOU MISSED QUESTION	SECTION	EXAMPLES	PAGE	ANSWERS
17	7.4A	3	535	17. $x \geq -2$
18	7.4B	4	536	18. $|y| \leq 3$
19	7.4B	5	536–537	19. $|x - 1| > 2$

IF YOU MISSED QUESTION	SECTION	EXAMPLES	PAGE	ANSWERS
20a	7.5A	1	548	**20. a.** $P = kT$
20b	7.5A	1	548	**b.** $k = \frac{1}{92}$
21a	7.5B	2	549	**21. a.** $P = \dfrac{k}{V}$
21b	7.5B	2	549	**b.** $k = 5280$
22a	7.5B	3	549–550	**22. a.** $F = \dfrac{k}{d^2}$
22b	7.5B	3	549–550	**b.** $k = 800{,}000{,}000$
23	7.5B	3	549–550	**23.** 32 lb
24a	7.5C	4	550–551	**24. a.** $H = kd^3 n$
24b	7.5C	4	550–551	**b.** $k = \frac{5}{112}$
25	7.5C	4	550–551	**25.** 6000 hp

7.1 THE RECTANGULAR COORDINATE SYSTEM

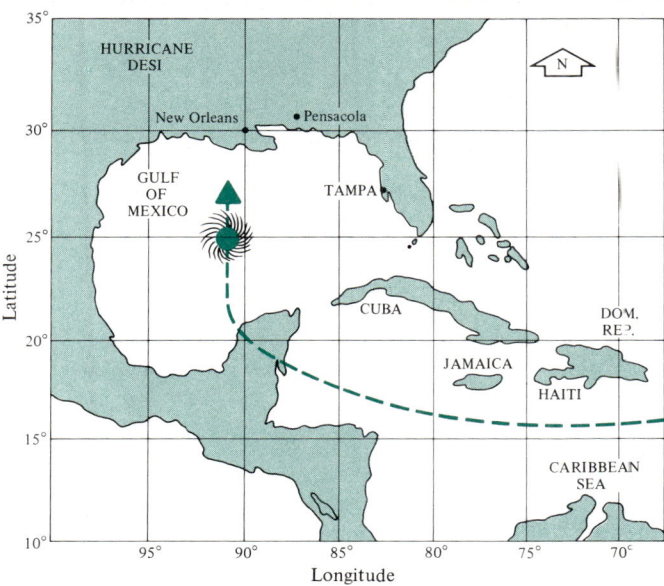

REVIEW

Before starting this section, you should know:

1. How to evaluate an expression.
2. How to solve linear equations.

OBJECTIVES

You should be able to:

A. Graph an ordered pair of numbers in the Cartesian plane.
B. Graph lines by finding two or more points satisfying the equation of the line.
C. Graph lines by finding the x- and y-intercepts.
D. Graph horizontal and vertical lines.

The map shows the **coordinates** of the hurricane to be near the vertical line indicating 90° of longitude and the horizontal line indicating 25° of latitude. If we agree to list the longitude first, we can identify this point by using the ordered pair

$$(90, \quad 25)$$

This is the ⟶ longitude.   ⟵ This is the latitude.

A. GRAPHING ORDERED PAIRS

In algebra we use a similar system to locate points by using **ordered pairs** of numbers. We draw two perpendicular number lines called the x-axis and the y-axis, intersecting at a point O called the **origin.** (See the art in the margin.) On the x-axis, *positive is to the right*, whereas on the y-axis, *positive is up.* The two axes divide the plane into four regions called **quadrants.** These quadrants are numbered counterclockwise using Roman numerals and starting in the upper right-hand region, as shown. The whole arrangement is called a **Cartesian coordinate system,** a **rectangular coordinate system,** or simply a **coordinate plane.**

Every point P in the plane can be associated with an ordered pair (x, y). For example, the point A can be associated with the ordered pair $(1, 3)$, the point B with the ordered pair $(-1, 4)$, and the point C with the ordered pair $(0, -2)$. The **graph** of the points $A(1, 3)$, $B(-1, 4)$, and $C(0, -2)$ is indicated by the heavy dots in the figure.

EXAMPLE 1 Graph the ordered pairs $A(2, 3)$, $B(-1, 2)$, $C(-2, -1)$, and $D(1, -3)$.

Solution To graph the ordered pair $(2, 3)$, we start at the origin and move 2 units to the right and then 3 units up, reaching the point whose

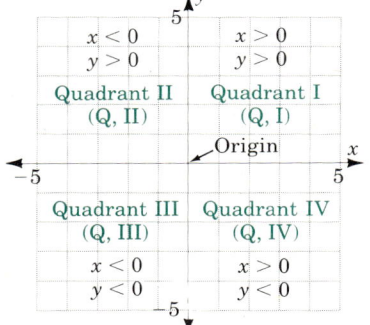

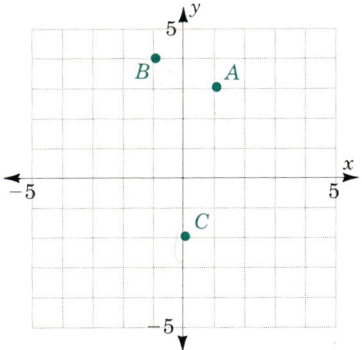

Problem 1 Graph $A(3, 2)$, $B(-1, 4)$, $C(-1, -2)$, and $D(3, -1)$.

coordinates are (2, 3). The other three pairs are graphed in a similar manner and are shown in the figure.

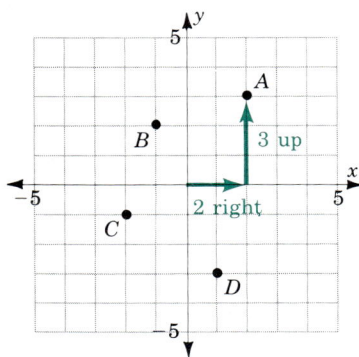

B. GRAPHING LINES

The equation $2x + 3y = 12$ is an equation in two variables, x and y. If in this equation x is replaced by 3 and y by 2, we obtain the *true* statement $2 \cdot 3 + 3 \cdot 2 = 12$. Thus we say that the *ordered pair* $(3, 2)$ is a *solution* of the equation $2x + 3y = 12$, or that it *satisfies* the equation. On the other hand, $(2, 3)$ is *not* a solution of $2x + 3y = 12$, because $2 \cdot 2 + 3 \cdot 3 = 4 + 9 \neq 12$.

The *solution set* of an equation in two variables can be written using set notation. For example, the solution set of $2x + 3y = 12$ can be written as $\{(x, y) \mid 2x + 3y = 12\}$ or, if we decide to solve for y, as $\left\{ (x, y) \mid y = \dfrac{-2}{3}x + 4. \right\}$. The solution set of the equation $2x + 3y = 12$ consists of *infinitely* many points, so it would be impossible to list all these points. However, we can find some of these points by substituting values for one of the variables and then computing the corresponding values for the other variable. For example, if we replace x by -3 in the equation $y = \dfrac{-2}{3}x + 4$, we have

$$y = \frac{-2}{3} \cdot (-3) + 4 = 6$$

If $x = 0$, $\quad y = \dfrac{-2}{3} \cdot (0) + 4 = 4$

If $x = 3$, $\quad y = \dfrac{-2}{3} \cdot (3) + 4 = 2$

If $x = 6$, $\quad y = \dfrac{-2}{3} \cdot (6) + 4 = 0$

If $x = 9$, $\quad y = \dfrac{-2}{3} \cdot (9) + 4 = -2$

These ordered pairs can be entered in a table as follows:

x	y
-3	6
0	4
3	2
6	0
9	-2

ANSWER   (to problem on page 491)

1.
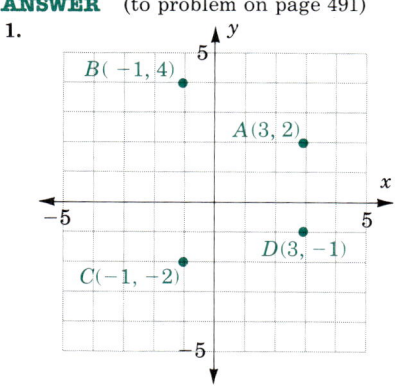

The points $(-3, 6)$, $(0, 4)$, $(3, 2)$, $(6, 0)$, and $(9, -2)$ appearing in the table can also be graphed, as shown in Figure 1.

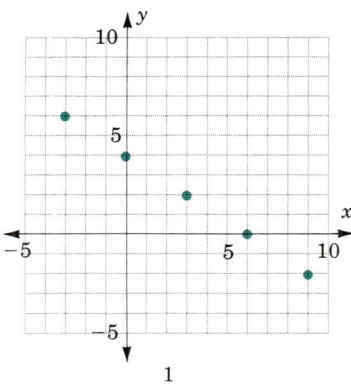

1

FIG. 1

The points appear to be on a straight line. In fact, it can be proved that every *solution* of $y = \dfrac{-2}{3}x + 4$ (or $2x + 3y = 12$) corresponds to a point on the line, and vice versa. Thus the line shown in Figure 2, obtained by joining the points shown in Figure 1, is the *graph* of $y = \dfrac{-2}{3}x + 4$ (or $2x + 3y = 12$). The line in Fig. 2 represents only a *part* of the complete graph, which continues without end in both directions, as indicated by the arrows in the figure. Although it takes only *two* points to determine a straight line, we shall include a third point to make sure that we have drawn the correct line.

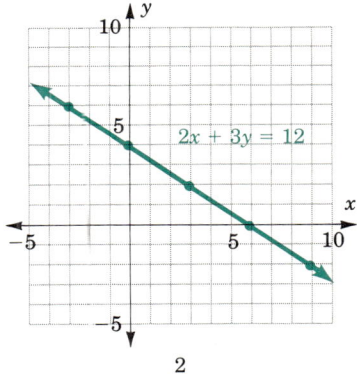

2

FIG. 2

EXAMPLE 2 Graph $x + 2y = 8$.

Solution

For $x = -2$,	For $x = 0$,	For $x = 2$,
$-2 + 2y = 8$	$0 + 2y = 8$	$2 + 2y = 8$
$2y = 10$	$2y = 8$	$2y = 6$
$y = 5$	$y = 4$	$y = 3$
We have $(-2, 5)$.	We have $(0, 4)$.	We have $(2, 3)$.

Thus we graph the points $(-2, 5)$, $(0, 4)$, and $(2, 3)$ and then draw a line, through them, as shown in the figure on page 494.

Problem 2 Graph $3x - y = 6$.

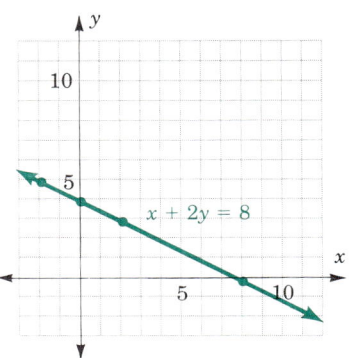

C. GRAPHING LINES USING INTERCEPTS

The equation $x + 2y = 8$ has a *straight* line as its graph. It can be shown that any equation in two variables x and y that is of the form $Ax + By = C$ (A and B not both zero) has a straight line for its graph. (This is why $Ax + By = C$ is called a *linear* equation.) Here is the definition.

> Any equation that can be written in the form $Ax + By = C$ is a **linear equation** in two variables, and the graph is a *straight line*. The form $Ax + By = C$ is called the *standard form* of the equation.

Since two points determine a straight line, it is enough to locate two points in order to graph a linear equation. The easiest points to compute are those involving zeros—that is, points of the form $(x, 0)$ and $(0, y)$. For example, if we let $x = 0$ in the equation $x + 2y = 8$, we obtain $2y = 8$, or $y = 4$. Thus $(0, 4)$ is a point on the graph. Similarly, if we let $y = 0$ in the equation $x + 2y = 8$, we have $x = 8$. Thus $(8, 0)$ is also on the graph. Since the points $(8, 0)$ and $(0, 4)$ are the points at which the line crosses the x- and y-axes, respectively, the numbers 8 and 4 are called the **x- and y-intercepts.** These two intercepts, as well as the line determined by them (the graph of $x + 2y = 8$), are shown in the figure. Here is the procedure we used to find the intercepts.

> To find the x-intercept, let $y = 0$ and find x.
> To find the y-intercept, let $x = 0$ and find y.

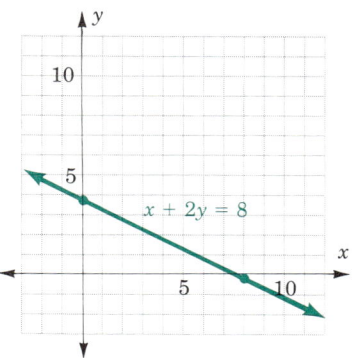

EXAMPLE 3   Find the x- and y-intercepts of $y = 3x + 6$ and then graph the line.

Solution   To find the x-intercept, we let $y = 0$, obtaining $0 = 3x + 6$, or $x = -2$. Hence $(-2, 0)$ is the point at which the line crosses the x-axis. To find the y-intercept, we let $x = 0$, so $y = 6$. Hence $(0, 6)$ is the point at which the line crosses the y-axis. To graph the equation $y = 3x + 6$, we graph the points $(-2, 0)$ and $(0, 6)$ and join them with a line, as shown in the figure.

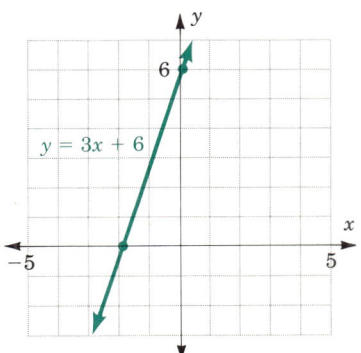

Problem 3   Find the x- and y-intercepts of $2x - 3y = 6$ and then graph the line.

D.   GRAPHING HORIZONTAL AND VERTICAL LINES

It is worth noting that the procedure we have discussed works *only* for equations of the form $Ax + By = C$, where *none* of the quantities A, B, and C is 0. Thus, since the equation $2y = 6$ can be written in the form $0 \cdot x + 2y = 6$, we *cannot* graph $2y = 6$ by finding its intercepts. The equation $2y = 6$ assigns to every value of x a y-value of 3. As a matter of fact, if you solve for $2y = 6$, you get $y = 3$, which has no x-coordinate. Thus for $x = 1$, $y = 3$, for $x = 2$, $y = 3$ (y is always 3). If we graph the points $(1, 3)$ and $(2, 3)$ and connect them with a straight line, we see that the result is a horizontal line, as shown in Figure 3. Similarly, the equation $2x = 6$ assigns an x value of 3 to every y. Thus, for $y = 1$, $x = 3$, and for $y = 5$, $x = 3$. If we graph the points $(3, 1)$ and $(3, 5)$ and draw a straight line through them, we see that the result is a vertical line, as shown in Figure 4. In general, we have the following.

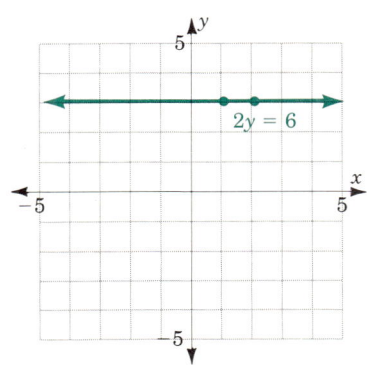

FIG. 3

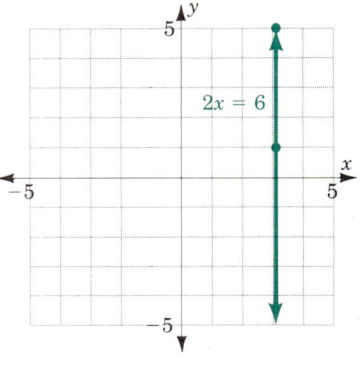

FIG. 4

ANSWER   (to problem on page 493)
2.

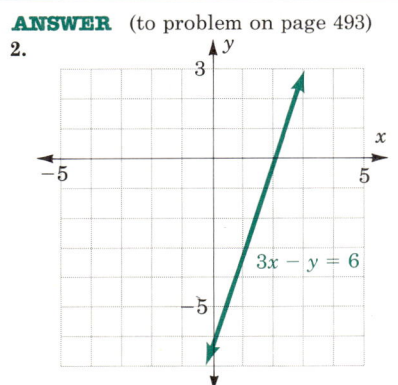

The graph of the equation $y = C$ is a *horizontal* line.
The graph of $x = C$ is a *vertical* line.

EXAMPLE 4 Graph the equations.

a. $2x = 8$ **b.** $5y = -10$

Solution

a. Since $2x = 8$ is equivalent to $x = 4$, the graph of $2x = 8$ is a vertical line for which $x = 4$. If we choose the solutions $(4, 1)$ and $(4, 2)$ and draw a straight line through them, we obtain the graph of $2x = 8$ shown in Figure 5.

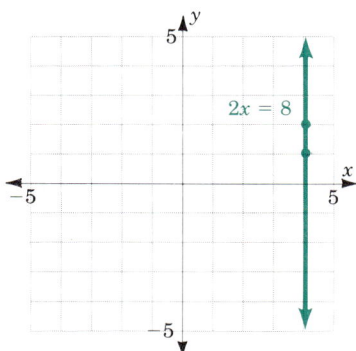

FIG. 5

b. Since $5y = -10$ is equivalent to $y = -2$, $5y = -10$ is a horizontal line for which $y = -2$. If we choose the solutions $(1, -2)$ and $(2, -2)$ and draw a straight line through them, we obtain the graph of the equation $5y = -10$ shown in Figure 6.

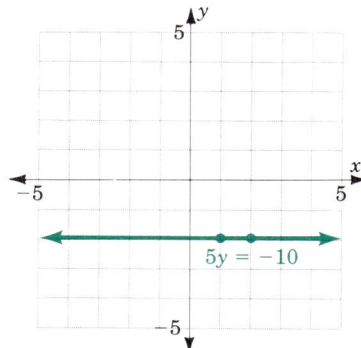

FIG. 6   ▲

Problem 4 Graph.
a. $3x = -9$ **b.** $2y = -6$

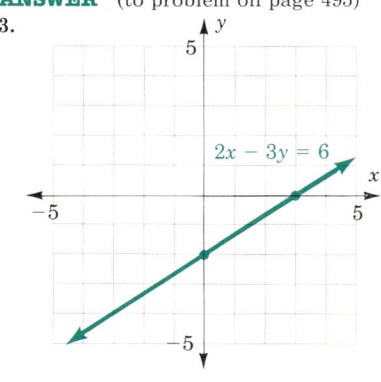

_____
_____
_____
_____
_____

A. In Problems 1–10 graph the ordered pair and state in which quadrant (if any) each lies.

ANSWERS

1. (3, 4)   **2.** (4, 3)

1.–10.

3. (−4, 3)   **4.** (−3, 4)

5. (−3, −2)   **6.** (−2, −3)

7. (0, 2)   **8.** (2, 0)

9. $\left(\dfrac{1}{2}, -3\right)$ **10.** $\left(3, -\dfrac{1}{2}\right)$

B. In Problems 11–16 complete the ordered pairs to satisfy the equation and then graph it.

11. $y = x + 3$,
(−2,   ), (−1,   ), (0,   ),
(1,   ), (2,   )

12. $y = 2x + 1$
(−1,   ), (0,   ), (1,   )

11. _____

12. _____

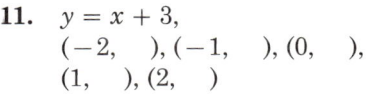

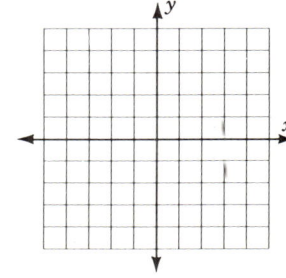

13. $x - y = 4$
$(-1, \quad), (0, \quad), (1, \quad)$

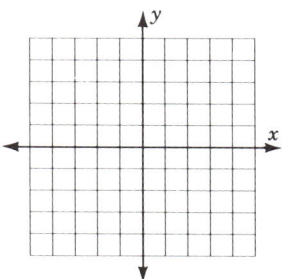

14. $x - 3y = 6$
$(0, \quad), (3, \quad), (-3, \quad)$

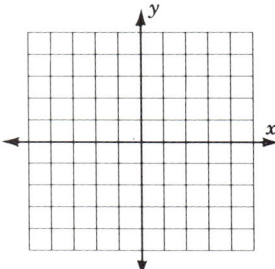

13. _____

14. _____

15. $2x - y - 3 = 0$
$(-1, \quad), (0, \quad), (1, \quad)$

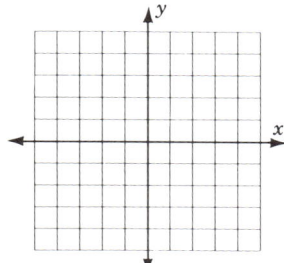

16. $2x + y + 2 = 0$
$(-1, \quad), (0, \quad), (1, \quad)$

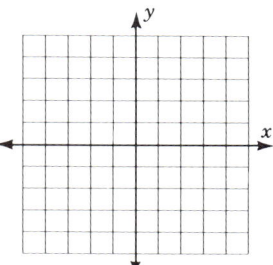

15. _____

16. _____

ANSWER (to problem on page 496)
4. a.

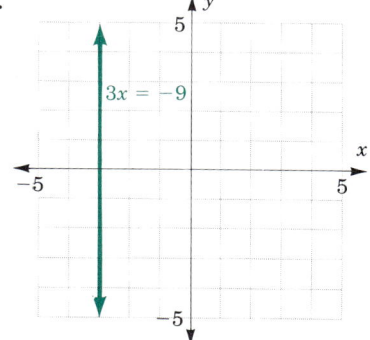

$3x = -9$

b.

$2y = -6$

C. In Problems 17–30, find the x- and y-intercepts and then graph the equation.

17. $y = x - 5$

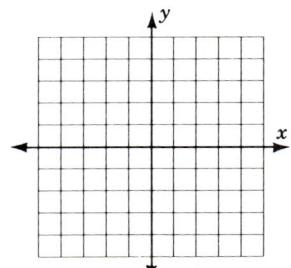

18. $2y = 4x - 2$

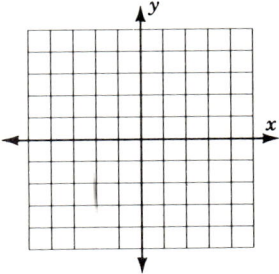

17. _____

18. _____

19. $2x + 3y = 6$

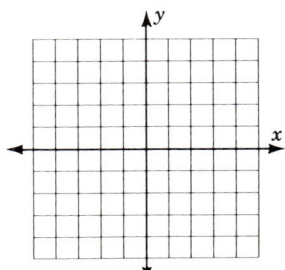

20. $3x + 2y = 6$

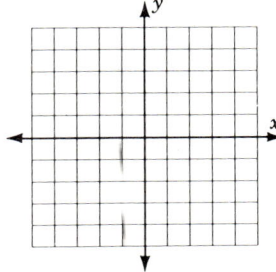

19. _____

20. _____

21. $2x - y = 4$

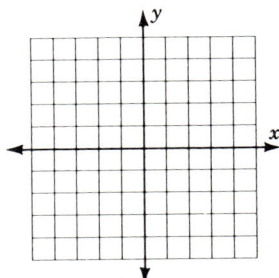

22. $3x - y = 3$

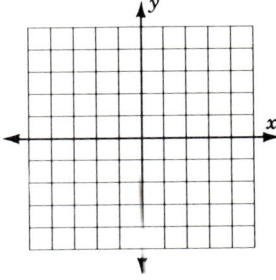

21. _____

22. _____

23. $2x + y - 4 = 0$

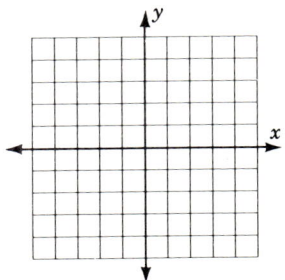

24. $3x + y - 3 = 0$

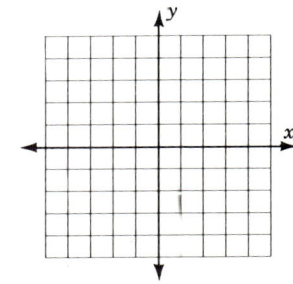

23. _____

24. _____

25. $y + 2x = 4$

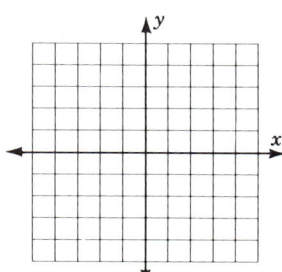

26. $y + 3x = -3$

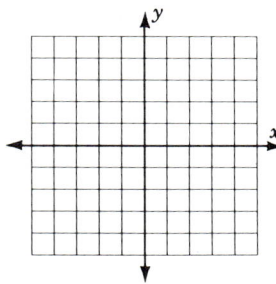

25. _____

26. _____

27. $2x - 5y = -10$

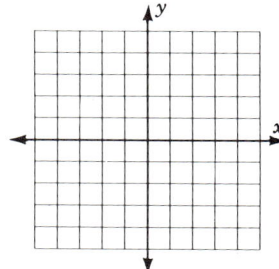

28. $2x - 3y = -6$

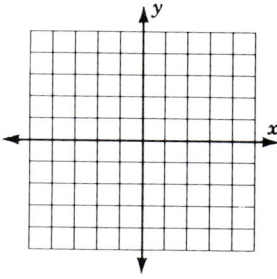

27. _____

28. _____

29. $x - y - 5 = 0$

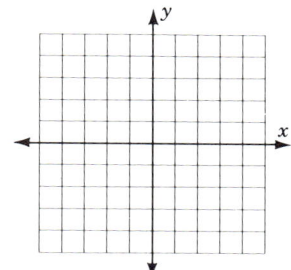

30. $2x - 3y - 12 = 0$

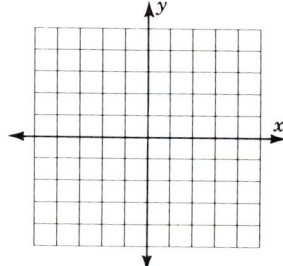

29. _____

30. _____

D. In Problems 31–40 determine if the given line is horizontal or vertical and then graph the line.

31. $-\dfrac{7}{2}x = 14$

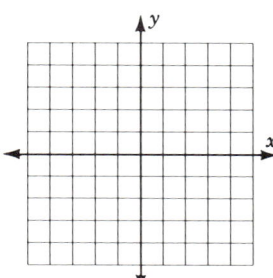

32. $-\dfrac{1}{2}x = 2$

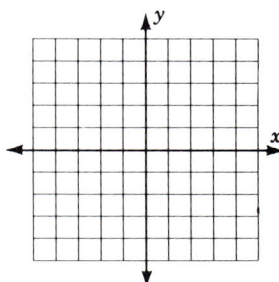

31. _____

32. _____

33. $\dfrac{3}{2}x = 6$

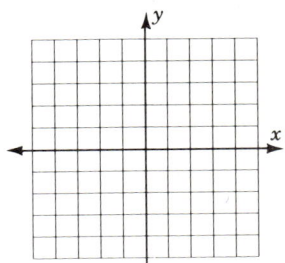

34. $-\dfrac{5}{2}y = 10$

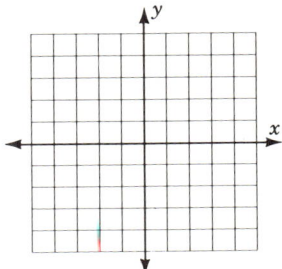

35. $-\dfrac{3}{4}x = 3$

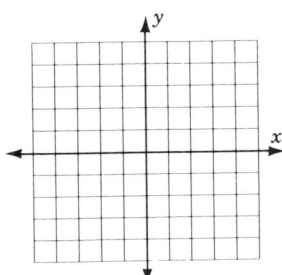

36. $-\dfrac{3}{7}x = -\dfrac{6}{7}$

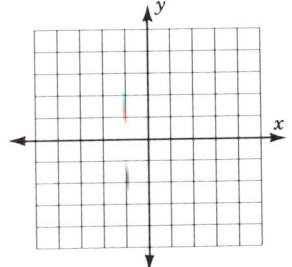

37. $-\dfrac{1}{3} + y = \dfrac{2}{3}$

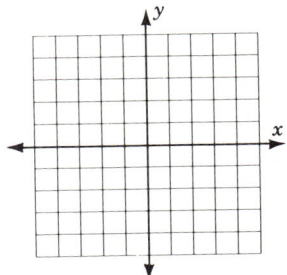

38. $-\dfrac{1}{5} + y = \dfrac{4}{5}$

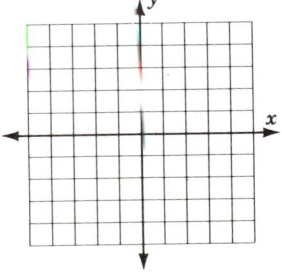

39. $\dfrac{2}{3} = x - \dfrac{4}{3}$

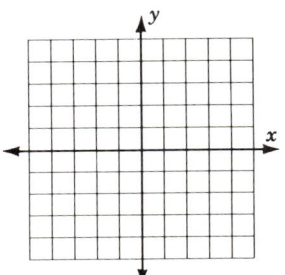

40. $\dfrac{3}{4} = x - \dfrac{5}{4}$

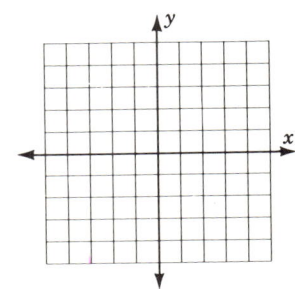

33. _____

34. _____

35. _____

36. _____

37. _____

38. _____

39. _____

40. _____

E. Applications

41. The number N of chirps a cricket makes per minute is given by

$$N = 4(T - 40)$$

where T is the temperature in degrees Fahrenheit. The ordered pairs corresponding to this equation are of the form (T, N).
a. Does the ordered pair (60, 80) satisfy the equation?
b. Based on your answer to (a), how many chirps does a cricket make when the temperature is 60°F?
c. How many chirps does a cricket make when the temperature is 80°F?

41. **a.** _____
      b. _____
      c. _____

42. Based on the formula of Problem 41, crickets stop chirping when $N = 0$. The corresponding ordered pair will be $(T, 0)$.
a. Find T.
b. At what temperature will crickets stop chirping?

42. **a.** _____
      b. _____

Read the margin note and use the following graph in Problems 43–46.

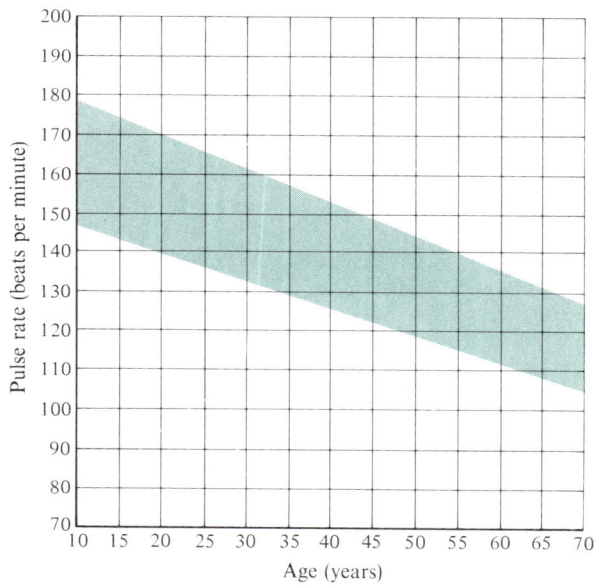

Are you exercising hard? Your target zone can tell you. It works like this. Take your pulse after exercising, find your age on the x-axis and follow a vertical line up to the lower edge of the shaded area, then go across to the number on the y-axis at the left. That pulse rate is the lower limit for your target zone. To find the upper limit, continue vertically on your age line to the top of the shaded area and then go across to the y-axis to locate your pulse rate.

43. What is the lower limit pulse rate for a 20-year-old? Write the answer as an ordered pair (age, limit).

43. _____

44. What is the upper limit pulse rate for a 20-year-old? Write the answer as an ordered pair.

44. _____

45. What is the upper limit pulse rate for a 45-year-old? Write the answer as an ordered pair.

45. _____

46. What is the lower limit pulse rate for a 50-year-old? Write the answer as an ordered pair.

46. _____

✓ SKILL CHECKER

Find the indicated quotients.

47. $\dfrac{3-6}{3-1}$

48. $\dfrac{6-3}{1-3}$

49. $\dfrac{3-(-6)}{1-4}$

50. $\dfrac{4-(-2)}{2-5}$

51. $\dfrac{4-(-2)}{3-(-6)}$

52. $\dfrac{2-(-6)}{4-(-8)}$

47. _____
48. _____
49. _____
50. _____
51. _____
52. _____

7.1 USING YOUR KNOWLEDGE

The ideas presented in this section are vital for understanding graphs. For example, have you been exercising in the summer? To determine the risk of exercising in the heat, you must know how to read the graph. To do this, first find the temperature on the y-axis, then read across from it to the right, stopping at the vertical line representing the relative humidity. Thus, on a 90°F day, if the humidity is less than 30%, the weather is in the safe zone.

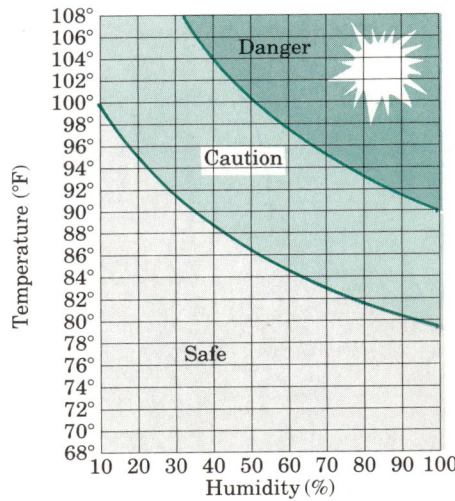

1. If the humidity is 50%, how high can the temperature go and still be in the safe zone for exercising? (Answer to the nearest degree.)

1. _____

2. If the humidity is 70%, at what temperature will the danger zone start?

2. _____

3. If the temperature is 100°F, what does the humidity have to be so that it is safe to exercise?

3. _____

4. Between what temperatures should you use caution when exercising if the humidity is 80%?

4. _____

5. If you start jogging at 1 P.M. when the temperature is 86°F and the humidity is 60%, how many degrees can the temperature rise before you get to the danger zone?

5. _____

7.2 THE DISTANCE FORMULA AND THE SLOPE OF A LINE

In the BMW test, Texaco's new System³ gasoline removed performance-robbing deposits left by ordinary gasoline.

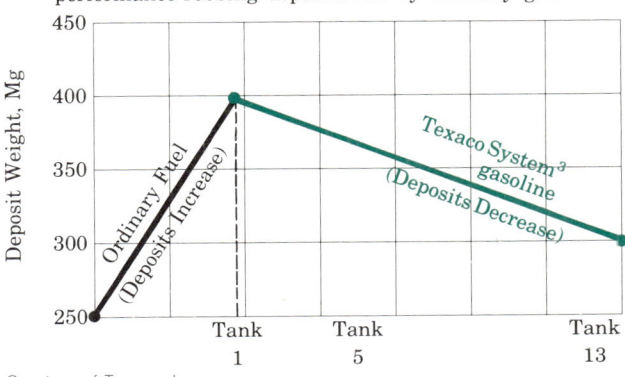

Courtesy of Texaco, Inc.

OBJECTIVES

REVIEW

Before starting this section, you should know:

1. The Pythagorean theorem (Section 3.8).
2. How to add, subtract, multiply, and divide integers.
3. How to recognize the reciprocal of a number.

OBJECTIVES

You should be able to:

A. Find the distance between two points in the Cartesian plane.
B. Find the slope of a line passing through two given points.
C. Use the slope formula to decide whether two lines are perpendicular, parallel, or neither.
D. Graph a line given its slope and a point on the line.

In the ad, the distance along the x-axis shows the number of weeks elapsed, and the distance on the y-axis shows the amount of deposits (in milligrams) in your engine. The black line shows that the deposits *increase* when using ordinary fuels but *decrease* after adding Texaco.

In mathematics we have a formula for finding the distance between any two points in the plane. We also measure how fast lines increase or decrease by measuring *slopes*.

A. THE DISTANCE FORMULA

In the triangle of Figure 7, the distance a between $(-2, -1)$ and $(6, -1)$ is $6 - (-2) = 8$, and the distance b between $(6, 5)$ and $(6, -1)$ is $5 - (-1) = 6$. To find the distance c we need to use the Pythagorean theorem studied in Chapter 3. According to that theorem, if the legs of a right triangle are a and b and the hypotenuse is c, then

$$c^2 = a^2 + b^2$$

Thus

$$c^2 = 8^2 + 6^2 = 64 + 36$$
$$c^2 = 100$$
$$c = \pm\sqrt{100} = \pm 10$$

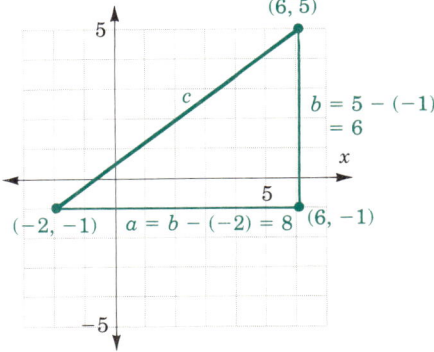

FIG. 7

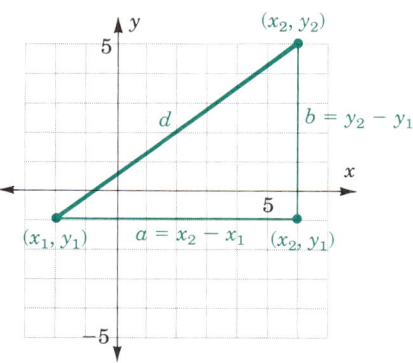

FIG. 8

Since c represents a distance, we discard the negative answer and conclude that $c = 10$. We can repeat a similar argument to find the distance d between any two points (x_1, y_1) and (x_2, y_2) (Figure 8). As before, the distance a is $|x_2 - x_1|$ and the distance b is $|y_2 - y_1|$. By the Pythagorean theorem, we have

$$d^2 = (x_2 - x_1)^2 + (y_2 - y_1)^2$$
$$d = \sqrt{(x_2 - x_1)^2 + (y_2 - y_1)^2}$$

Here is the summary of what we have done.

THE DISTANCE FORMULA

The distance between the points (x_1, y_1) and (x_2, y_2) is
$$d = \sqrt{(x_2 - x_1)^2 + (y_2 - y_1)^2}$$

EXAMPLE 1 Find the distance between the two given points.

a. $A(1, 1)$ and $B(5, 4)$ **b.** $C(2, 3)$ and $D(-2, 5)$
c. $E(-2, 1)$ and $F(-2, 3)$

Solution

a. If we let $x_1 = 1$, $y_1 = 1$, $x_2 = 5$, and $y_2 = 4$, then
$$d = \sqrt{(5 - 1)^2 + (4 - 1)^2} = \sqrt{(4)^2 + (3)^2} = \sqrt{25} = 5$$

b. Here $x_1 = 2$, $y_1 = 3$, $x_2 = -2$, and $y_2 = 5$. Thus
$$d = \sqrt{[2 - (-2)]^2 + (3 - 5)^2} = \sqrt{[4]^2 + (-2)^2} = \sqrt{20} = 2\sqrt{5}$$

c. Now $x_1 = -2$, $y_1 = 1$, $x_2 = -2$, and $y_2 = 3$. Hence,
$$d = \sqrt{[-2 - (-2)]^2 + (1 - 3)^2} = \sqrt{[0]^2 + (-2)^2} = \sqrt{4} = 2$$

Note that EF is a vertical line, so that $d = |3 - 1| = 2$. ▲

Problem 1 Find the distance between the two given points.
a. $A(2, 2)$ and $B(8, 10)$
b. $C(-3, 2)$ and $D(5, 4)$
c. $E(-4, 3)$ and $F(-4, 7)$

B. THE SLOPE OF A LINE

A second property of a line segment joining two points in a plane is its inclination. For example, suppose you want to find the average yearly raise in the teachers' average salary, as shown in Figure 9.

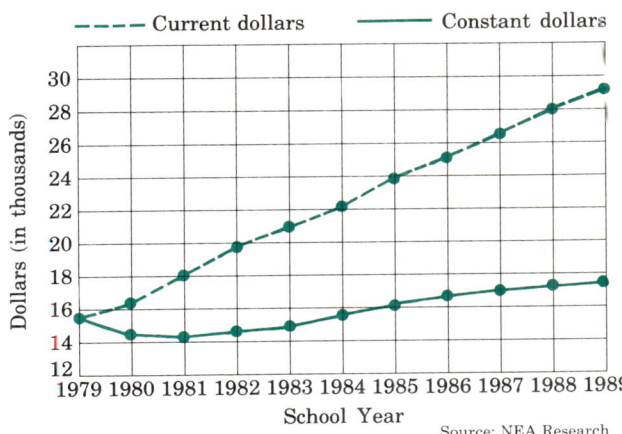

Classroom Teacher Average Salary
School Years 1978-79 to 1988-89

- - - - Current dollars   ———— Constant dollars

FIG. 9

According to the graph, the salary in 1979 was about 16 (thousand); in 1989 it was about 30 (thousand). The change is

$$30 - 16 = 14 \text{ (thousand)}$$

during the 10-yr period. Thus the average yearly raise is given by the ratio

$$\frac{\text{increase in ten years}}{\text{number of years}} = \frac{30 - 16}{10} = \frac{14}{10} = 1.4$$

Thus the teachers' average salary went up about $1400, 1.4 (thousand), each year. In mathematics, the ratio of the change in y to the change in x is called the *slope* of the line segment and is denoted by the letter m. Thus

$$m = \frac{\text{change in } y}{\text{change in } x}$$

We summarize this discussion in the following definition.

DEFINITION OF SLOPE

If $A(x_1, y_1)$ and $B(x_2, y_2)$ are any two distinct points on a line L (which is not parallel to the y-axis), then the **slope** of L, denoted by m, is

$$m = \frac{y_2 - y_1}{x_2 - x_1}$$

It should be noted that when using this definition it does not matter which point is taken for A and which for B. For example, the slope of the line passing through the points $A(0, -6)$ and $B(3, 3)$ is

$$m = \frac{3 - (-6)}{3 - 0} = \frac{9}{3} = 3$$

If we choose A to be $(3, 3)$ and B to be $(0, -6)$, the slope is

$$m = \frac{-6 - 3}{0 - 3} = \frac{-9}{-3} = 3$$

Thus since

This is the second point (x_1, y_1) This is the first point (x_1, y_1)

This is the first point (x_2, y_2) This is the second point (x_2, y_2)

$$\frac{y_2 - y_1}{x_2 - x_1} = \frac{y_1 - y_2}{x_1 - x_2}$$

A and B can be interchanged without changing the resulting slope.

EXAMPLE 2 Find the slope of the line passing through the given points.

a. $A(-3, 1)$, $B(-1, -2)$ **b.** $A(-1, 5)$, $B(-1, 6)$
c. $A(1, 1)$, $B(2, 3)$ **d.** $A(3, 4)$, $B(1, 4)$

Solution

a. The slope is $m = \dfrac{-2 - 1}{-1 - (-3)} = -\dfrac{3}{2}$.

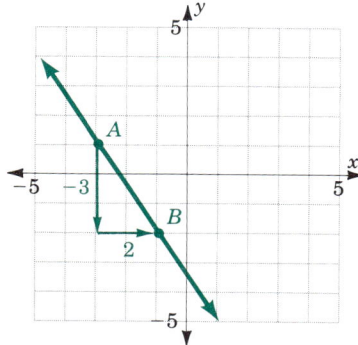

b. The slope is $m = \dfrac{6 - 5}{-1 - (-1)} = \dfrac{1}{0}$, which is undefined. Notice that the line passing through $A(-1, 5)$ and $B(-1, 6)$ has as its equation $x = -1$ and is a line *parallel* to the y-axis. The fact that the change in x is zero in lines parallel to the y-axis (that is, *vertical* lines) requires that they be excluded from the definition; their slope is *undefined*.

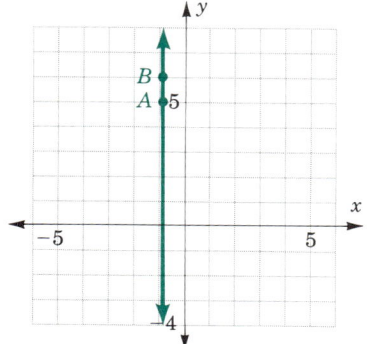

Problem 2 Find the slope of the line passing through the given points.
a. $A(-4, 2)$, $B(-2, -3)$
b. $A(-2, 4)$, $B(-2, 7)$
c. $A(2, 1)$, $B(4, 5)$
d. $A(2, 3)$, $B(1, 3)$

ANSWER (to problem on page 506)
1. a. 10 **b.** $2\sqrt{17}$ **c.** 4

c. The slope is $m = \dfrac{3-1}{2-1} = \dfrac{2}{1} = 2$.

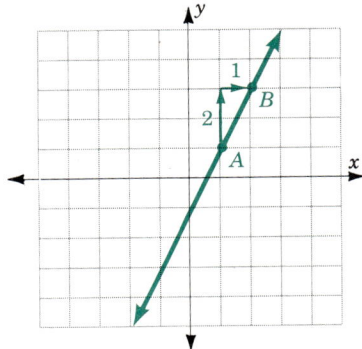

d. The slope is $m = \dfrac{4-4}{1-3} = \dfrac{0}{-2}$. Since $\dfrac{0}{-2} = 0$, the slope is 0. Note that the line is parallel to the x-axis and has equation $y = 4$, a horizontal line. Horizontal lines have 0 slope.

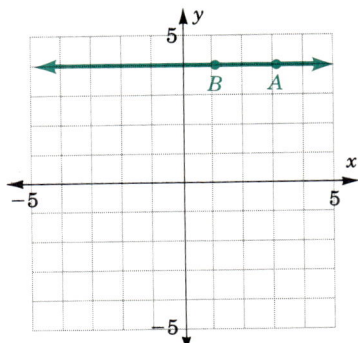

We can see from Example 2 that the following hold true.

1. A line that falls from left to right has a *negative* slope.

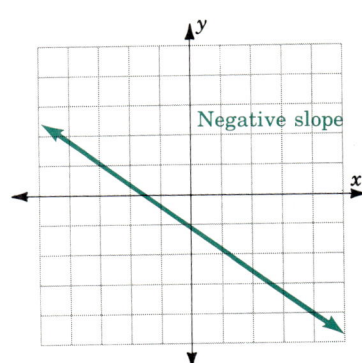

2. The slope of a *vertical* line is undefined. $\left(\text{Since } x_2 - x_1 = 0,\right.$

$m = \dfrac{y_2 - y_1}{x_2 - x_1} = \dfrac{y_2 - y_1}{0}$ so that m is undefined. $\Big)$

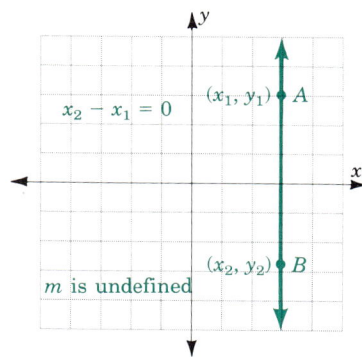

3. A line that rises from left to right has a *positive* slope.

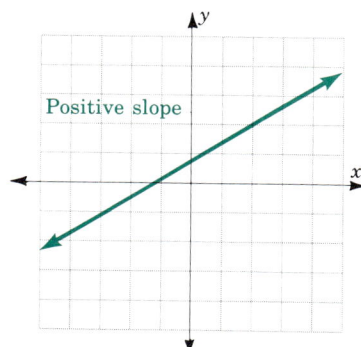

4. As for lines that are parallel to the *x*-axis, a *horizontal* line has a slope of 0. $\left(\text{Since } y_2 - y_1 = 0, \ m = \dfrac{y_2 - y_1}{x_2 - x_1} = \dfrac{0}{x_2 - x_1} = 0. \right)$

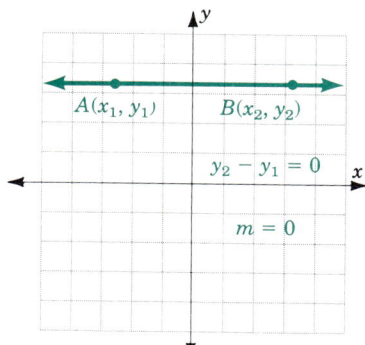

C. PARALLEL AND PERPENDICULAR LINES

The definition of slope can be used to determine when two line segments are parallel. Since two parallel lines must have the same inclination and thus the same slope, we have the following.

The lines L_1 and L_2 with slopes m_1 and m_2 are parallel if and only if $m_1 = m_2$

For example, consider the lines passing through points $A(4, 2)$, $B(5, -1)$ and points $C(-1, 3)$, $D(0, 0)$ shown in Figure 10. The slope of AB is

$$m_1 = \frac{-1 - 2}{5 - 4} = \frac{-3}{1} = -3$$

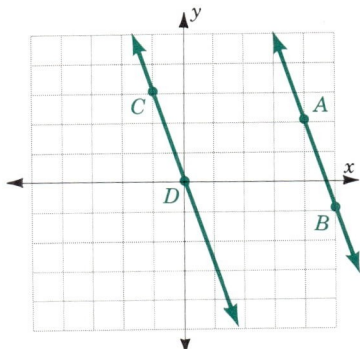

FIG. 10

and the slope of CD is

$$m_2 = \frac{0 - 3}{0 - (-1)} = \frac{-3}{1} = -3$$

Thus, $m_1 = m_2$, so both lines have the same slope and are parallel.

It can also be shown that two lines with slopes m_1 and m_2 are perpendicular if $m_1 = \frac{-1}{m_2}$; that is, if the slopes are negative reciprocals. You can check this by making sure that $m_1 \cdot m_2 = -1$.

SLOPES OF PERPENDICULAR LINES

The lines L_1 and L_2 with slopes m_1 and m_2, respectively, are perpendicular if and only if the slopes are negative reciprocals; that is, $m_1 \cdot m_2 = -1$.

We can show that the line passing through $A(4, 2)$ and $B(5, -1)$ is perpendicular to the line passing through $C(2, 1)$ and $D(5, 2)$. Since the slope of AB is

$$m_1 = \frac{-1 - 2}{5 - 4} = \frac{-3}{1} = -3$$

and that of CD is

$$m_2 = \frac{2 - 1}{5 - 2} = \frac{1}{3}$$

we have $m_1 \cdot m_2 = -3 \cdot \frac{1}{3} = -1$. These perpendicular lines can be graphed as shown in Figure 11.

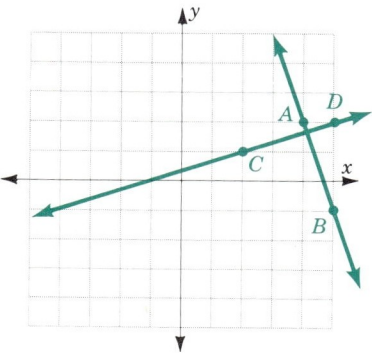

FIG. 11

EXAMPLE 3 A line L_1 has slope $\frac{2}{3}$. Find.

a. Whether the line passing through $A(5, 1)$ and $B(8, 3)$ is parallel or perpendicular to L_1

b. Whether the line passing through $C(-1, -4)$ and $D(-3, -1)$ is parallel or perpendicular to L_1

Solution

a. The slope of AB is $m = \dfrac{3 - 1}{8 - 5} = \dfrac{2}{3}$. Since the slope of L_1 is also $\frac{2}{3}$, line AB is parallel to L_1.

b. The slope of CD is $m = \dfrac{-1 - (-4)}{-3 - (-1)} = \dfrac{3}{-2} = -\dfrac{3}{2}$. Since the slope of L_1 is $\dfrac{2}{3}$ and $\dfrac{2}{3} \cdot \left(-\dfrac{3}{2}\right) = -1$, line CD is perpendicular to L_1. ▲

EXAMPLE 4 The line through $A(4, y)$ and $B(-2, -5)$ is perpendicular to a line whose slope is $-\frac{2}{3}$. Find y.

Solution The slope of the line through points A and B is

$$\frac{y - (-5)}{4 - (-2)} = \frac{y + 5}{6}$$

If the line through A and B is perpendicular to a line whose slope is $-\frac{2}{3}$, the slope of the line through A and B must be $\frac{3}{2}$ (the negative reciprocal of $-\frac{2}{3}$). Thus

$$(6)\frac{(y + 5)}{6} = \frac{3}{2} \cdot (6)$$

$$2(y + 5) = 6 \cdot 3$$

$$2y + 10 = 18$$

$$2y = 8$$

$$y = 4$$

Since the line through $A(4, 4)$ and $B(-2, -5)$ has slope

$$\frac{4 - (-5)}{4 - (-2)} = \frac{9}{6} = \frac{3}{2}$$

Problem 3 A line L_1 has slope $\frac{3}{2}$. Find.

a. Whether the line passing through $A(6, 3)$ and $B(3, 5)$ is parallel or perpendicular to L_1

b. Whether the line passing through $A(-6, -8)$ and $B(-4, -5)$ is parallel or perpendicular to L_1

Problem 4 The line through $A(5, y)$ and $B(-1, -4)$ is perpendicular to a line whose slope is $-\frac{3}{4}$. Find y.

which makes it perpendicular to a line whose slope is $-\frac{2}{3}$, our result is correct. ▲

D. GRAPHING LINES USING THE SLOPE AND A POINT

We can graph a line if we know its slope and a point on the line. For example, suppose a line goes through the point $(1, -2)$ and has slope $\frac{2}{3}$. To graph the line, we recall that the slope of a line is the ratio

$$\frac{\text{change in } y}{\text{change in } x} = \frac{2}{3}$$

We start at the point $(1, -2)$ and go 3 units right (change in x is 3) and 2 units up (change in y is 2), ending at $(4, 0)$. We then draw a line through the points $(1, -2)$ and $(4, 0)$ to obtain the graph. See Figure 12. If the slope of the line had been $-\frac{2}{3}$, we would still move 3 units right (change in x) but then go 2 units *down* (change in y), ending at $(4, -4)$. This second line is shown in color.

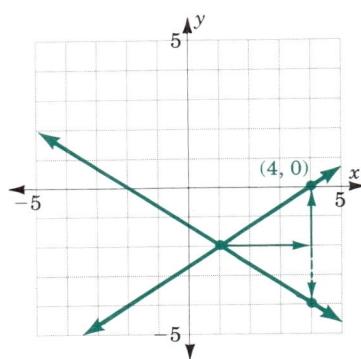

FIG. 12

EXAMPLE 5 A line goes through the point $(1, 2)$ and has slope $-\frac{3}{4}$. Graph this line.

Solution We start at the point $(1, 2)$ and go four units right (the change in x). We then go *down* 3 units (the change in y), ending at $(5, -1)$, as shown in the figure. The graph is obtained by drawing a line through the points $(1, 2)$ and $(5, -1)$.

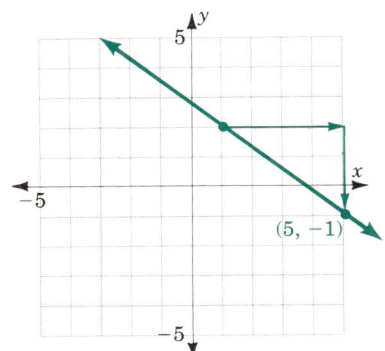

Problem 5 A line goes through the point $(2, 1)$ and has slope $-\frac{1}{3}$. Graph this line.

▲

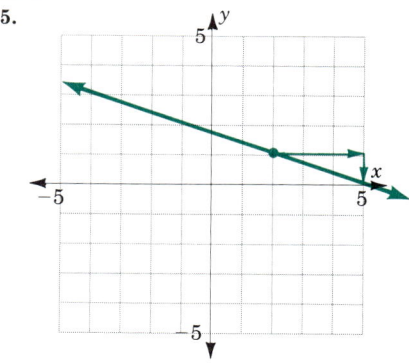

_____
_____
_____
_____
_____

A., B. In Problems 1–10 find the distance between the points, and the slope of the line passing through the points.

1. $A(2, 4)$, $B(-1, 0)$

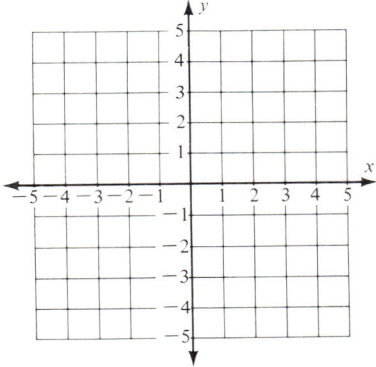

2. $A(3, -2)$, $B(8, 10)$

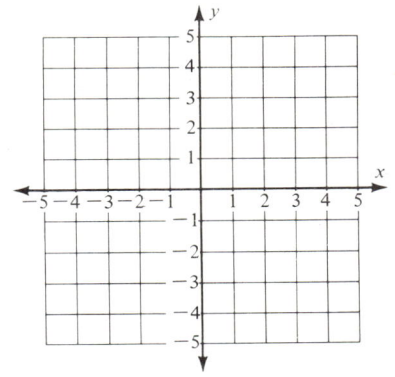

1. _____

2. _____

3. $C(-4, -5)$, $D(-1, 3)$

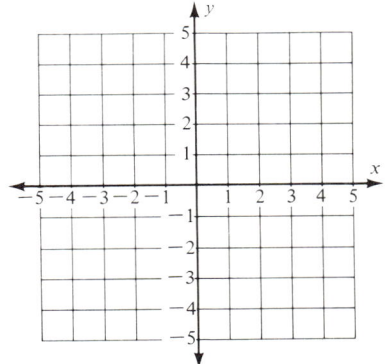

4. $C(5, 7)$, $D(-2, 3)$

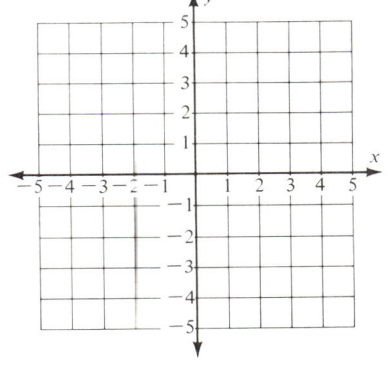

3. _____

4. _____

5. $E(4, 8)$, $G(1, -1)$

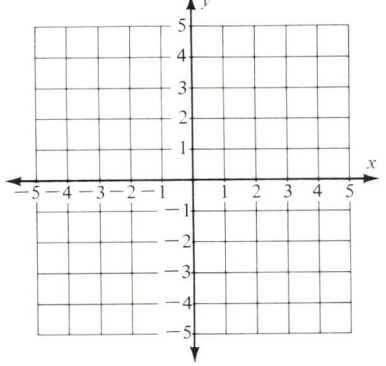

6. $H(-2, -2)$, $I(6, -4)$

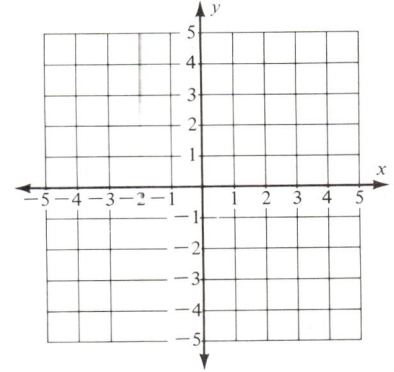

5. _____

6. _____

7. $A(3, -1)$, $B(-2, -1)$

8. $C(-2, 3)$, $D(4, 3)$

7. _____

8. _____

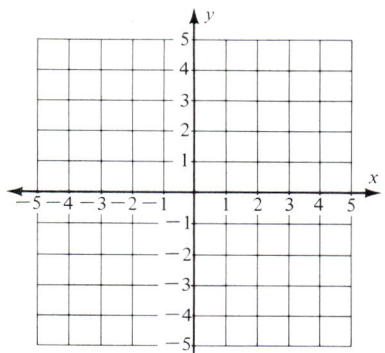

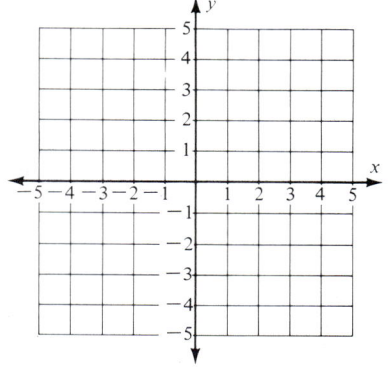

9. $E(-1, 2)$, $F(-1, -4)$

10. $G(-3, 2)$, $H(-3, 5)$

9. _____

10. _____

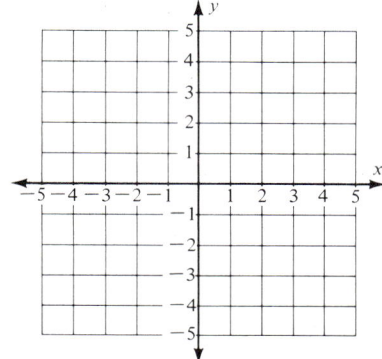

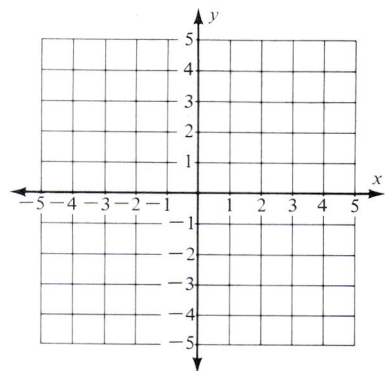

C. In Problems 11–20 determine whether lines AB and CD are parallel, perpendicular, or neither.

11. $A(1, 6)$, $B(-1, 4)$ and $C\left(1, \dfrac{-7}{2}\right)$, $D\left(\dfrac{7}{2}, -1\right)$

11. _____

12. $A(0, 4)$, $B(1, -1)$ and $C(0, 1)$, $D\left(\dfrac{7}{2}, -1\right)$

12. _____

13. $A(2, 0)$, $B(4, 5)$ and $D\left(\dfrac{7}{2}, 0\right)$, $E(1, 1)$

13. _____

14. $A(1, 1)$, $B(-1, 2)$ and $E(1, -1)$, $F(0, -3)$

14. _____

15. $A(-1, 1)$, $B(1, 2)$ and $C(1, -1)$, $D(0, -1)$

15. _____

16. $A(1, 1)$, $B(3, 3)$ and $C(1, -1)$, $D(0, 2)$

16. _____

17. $A(1, -1)$, $B\left(2, \dfrac{-1}{2}\right)$ and $C(2, -2)$, $D(1, 0)$

17. _____

18. $A(1, 1)$, $B\left(\dfrac{1}{5}, 0\right)$ and $C(1, 1)$, $D\left(0, \dfrac{9}{5}\right)$

18. _____

19. $A(0, 1)$, $B(14, -1)$ and $C\left(0, \dfrac{3}{2}\right)$, $D\left(\dfrac{7}{2}, 1\right)$

                                19. _____

20. $A(2, -2)$, $B(1, -7)$ and $C(1, -3)$, $D(0, -8)$

                                20. _____

In Problems 21–30, find the unknown.

21. The line through $A(x, 4)$ and $B(6, 8)$ is parallel to a line whose slope is 1. Find x.

                                21. _____

22. The line through $A(x, 5)$ and $(-2, 3)$ is parallel to a line whose slope is $\frac{2}{5}$. Find x.

                                22. _____

23. The line through $A(x, 2)$ and $B(2, 6)$ is perpendicular to a line whose slope is $\frac{1}{2}$. Find x.

                                23. _____

24. The line through $A(x, 4)$ and $B(-3, -4)$ is perpendicular to a line whose slope is $\frac{2}{5}$. Find x.

                                24. _____

25. The line through $A(x, -6)$ and $B(-2, -1)$ is perpendicular to a line whose slope is $-\frac{2}{3}$. Find x.

                                25. _____

26. The line through $A(2, y)$ and $B(3, 4)$ is parallel to a line whose slope is 2. Find y.

                                26. _____

27. The line through $A(3, y)$ and $B(1, -2)$ is parallel to a line whose slope is -3. Find y.

                                27. _____

28. The line through $A(2, y)$ and $B(1, -4)$ is perpendicular to a line whose slope is $\frac{1}{3}$. Find y.

                                28. _____

29. The line through $A(x, 4)$ and $(3, 5)$ is perpendicular to the horizontal line $y = 5$. Find x.

                                29. _____

30. The line through $A(-4, 2)$ and $(x, 7)$ is perpendicular to the horizontal line $y = -3$. Find x.

                                30. _____

D. In Problems 31–40, graph the line with the indicated slope and passing through the given point.

31. Slope 2 through $(1, 1)$

                                31. _____

32. Slope $\dfrac{2}{3}$ through $(1, 2)$

                                32. _____

33. Slope $-\dfrac{2}{3}$ through $(1, 1)$

                                33. _____

34. Slope $-\dfrac{3}{4}$ through $(-1, -1)$

                                34. _____

35. Slope 0 through $(2, 3)$

                                35. _____

36. Slope 0 through $(3, 2)$

                                36. _____

37. Slope 0 through $(0, 0)$

                                37. _____

38. Slope undefined through $(-1, 2)$

                                38. _____

39. Slope undefined through $(2, -1)$

                                39. _____

40. Slope undefined through $(0, 0)$

                                40. _____

E. Applications

In Problems 41–46, use the Pythagorean theorem to determine if the three points form the vertices (corners) of a right triangle.

41. $A(2, 2), B(0, 5), C(-20, 12)$

42. $A(0, 6), B(-3, 0), C(9, -6)$

43. $A(2, 2), B(0, 5), C(-19, -12)$

44. $A(0, 0), B(6, 0), C(3, 3)$

45. $A(2, 2), B(-4, -14), C(-20, -8)$

46. $A(3, 2), B(0, -4), C(12, -10)$

41. _____

42. _____

43. _____

44. _____

45. _____

46. _____

✓ **SKILL CHECKER**

In Problems 47–50, solve for y.

47. $6x + 3y = 12$

48. $2x + 3y = 6$

49. $3y - 2x = 12$

50. $5y - 2x = 10$

47. _____

48. _____

49. _____

50. _____

7.2 USING YOUR KNOWLEDGE

The following problems require using your knowledge about a topic previously discussed.

1. A line has x-intercept 2 and y-intercept -3. What is the slope of the line?

2. A line has x-intercept -3 and y-intercept 2. What is the slope of the line?

3. Find the slope of a line parallel to the line through the points $(3, \sqrt{3})$ and $(2, \sqrt{27})$.

4. Find the slope of a line parallel to the line through the points $(5, \sqrt{8})$ and $(6, \sqrt{2})$.

5. Find the slope of a line perpendicular to a second line passing through the points $(3, \sqrt{8})$ and $(2, \sqrt{32})$.

6. Find the slope of a line perpendicular to a second line passing through the points $(5, \sqrt{27})$ and $(4, -\sqrt{3})$.

7. The line through the points $(1, c)$ and $(2, c^2)$ is parallel to another line whose slope is 3. What are the possible values for c?

8. The line through the points $(3, 2c)$ and $(4, c^2)$ is parallel to another line whose slope is 3. What are the possible values for c?

9. The line through the points $(4, c)$ and $(5, c^2)$ is perpendicular to another line whose slope is $-\frac{1}{2}$. What are the possible values for c?

1. _____

2. _____

3. _____

4. _____

5. _____

6. _____

7. _____

8. _____

9. _____

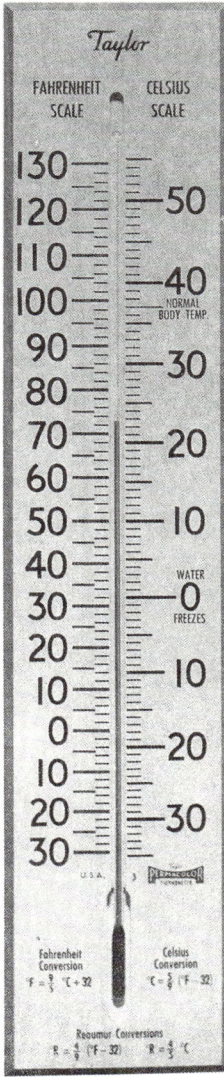

REVIEW

Before starting this section, you should know how to:

1. Graph lines when two points are given.
2. Write an equation in standard form.
3. Solve an equation for a specified variable.

OBJECTIVES

You should be able to find the equation and the graph of a line given:

A. Two points.
B. One point and the slope.
C. The slope and the *y*-intercept.
D. One point and the fact that the line is parallel or perpendicular to a given line.

The photo at the beginning of this section shows a thermometer with both the Celsius (centigrade) and Fahrenheit scales. Do you know how to convert temperatures in the Celsius scale to Fahrenheit? One way of doing this is simply to graph ordered pairs of numbers in which the first coordinate represents the Celsius temperature and the second, the corresponding Fahrenheit temperature. Since at 0° Celsius the corresponding Fahrenheit temperature is 32°, one such point is (0, 32). Another one is (100, 212). Thus, the graph of the equation can be drawn as shown in Figure 13 (page 520). As you know, the graph of a line can be obtained by using *any* two given points on the line. To obtain the equation of the line, that is, to find *F* in terms of *C*, we select a point on the line and assign it coordinates (*C*, *F*). The slope of the line going through (0, 32) and (100, 212) is

$$m = \frac{212 - 32}{100 - 0} = \frac{180}{100} = \frac{9}{5}$$

The points (0, 32) and (100, 212) correspond to the freezing and boiling points of water, respectively.

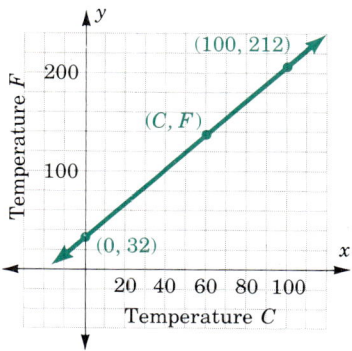

FIG. 13

The slope of the line going through $(0, 32)$ and (C, F) is

$$m = \frac{F - 32}{C - 0} = \frac{F - 32}{C}$$

Since these slopes are the same, we write

$$\frac{F - 32}{C} = \frac{9}{5}$$

$$F - 32 = \frac{9}{5}C \qquad \text{Multiply both sides by } C.$$

$$F = \frac{9}{5}C + 32 \qquad \text{Add 32 to both members.}$$

A. FINDING EQUATIONS GIVEN TWO POINTS

What we did here was to work with the graph of a line to find its equation. (In Section 7.1, we worked from an equation of a line to draw its graph.) In general, if a line goes through two points $P_1(x_1, y_1)$ and $P_2(x_2, y_2)$, as shown in Figure 14, an equation for this line can be found as follows:

1. Select a general point $P(x, y)$ on the line.

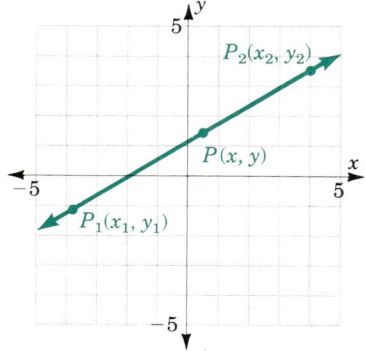

FIG. 14

2. The slope of the line P_1P_2 is

$$m = \frac{y_2 - y_1}{x_2 - x_1}, \qquad x_2 \neq x_1$$

3. The slope of the line P_1P is

$$m = \frac{y - y_1}{x - x_1}, \qquad x \neq x_1$$

4. Since the slopes are equal,

$$\frac{y - y_1}{x - x_1} = \frac{y_2 - y_1}{x_2 - x_1}$$

or

$$y - y_1 = \frac{y_2 - y_1}{x_2 - x_1} \cdot (x - x_1) \qquad \text{Multiply both sides by } (x - x_1).$$

We summarize this discussion as follows.

THE TWO-POINT FORM

An equation of a line going through the points (x_1, y_1) and (x_2, y_2) is given by

$$y - y_1 = \frac{y_2 - y_1}{x_2 - x_1} \cdot (x - x_1), \qquad x_2 \neq x_1 \qquad\qquad (1)$$

Equation 1 is called the **two-point form**.

EXAMPLE 1 Find an equation of the line going through the points (5, 2) and (6, 4). Then write the equation in standard form and graph it.

Solution Letting $(x_1, y_1) = (5, 2)$ and $(x_2, y_2) = (6, 4)$, and substituting in the two-point form equation, we have

$$y - 2 = \frac{4 - 2}{6 - 5} \cdot (x - 5)$$

$$\begin{aligned}
y - 2 &= 2(x - 5) &&\text{Two-point form} \\
y - 2 &= 2x - 10 &&\text{Distributive law} \\
8 &= 2x - y &&\text{Subtract } y \text{ and add 10.}
\end{aligned}$$

In standard form an equation of the line is $2x - y = 8$. The graph is shown in Figure 15 (page 522).

Problem 1 Find an equation of the line going through (3, 1) and (4, 3). Then write the equation in standard form and graph it.

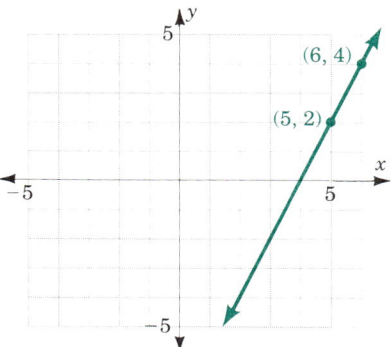

FIG. 15 ▲

B. FINDING EQUATIONS GIVEN A POINT AND THE SLOPE

If we replace $\dfrac{y_2 - y_1}{x_2 - x_1}$ by m in Equation 1, we obtain the following.

THE POINT-SLOPE FORM

$$y - y_1 = m(x - x_1) \tag{2}$$

The point-slope form enables us to find an equation of a line when a point $P(x_1, y_1)$ and the slope m are given. We shall use this equation in Example 2.

EXAMPLE 2 Find the equation of the line with slope $m = -2$ and passing through the point $(3, 5)$. Then graph the line.

Solution Here $m = -2$, $(x_1, y_1) = (3, 5)$. Substituting in Equation 2, we get

$$y - 5 = -2(x - 3)$$
$$y - 5 = -2x + 6 \qquad \text{Distributive law}$$
$$2x + y = 11 \qquad\quad \text{Add } 2x \text{ and } 5 \text{ to both sides.}$$

To graph the equation, start at $(3, 5)$. Since $m = -2 = \dfrac{-2}{1}$, go 1 unit

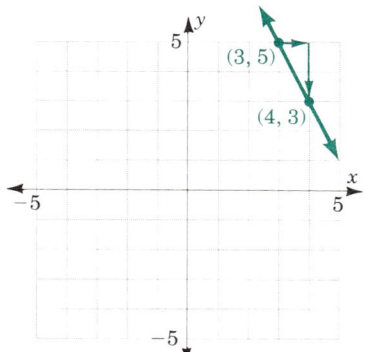

FIG. 16

Problem 2 Find the equation of the line with slope $m = -3$ and passing through $(1, 2)$. Then graph the line.

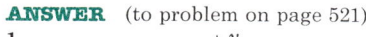
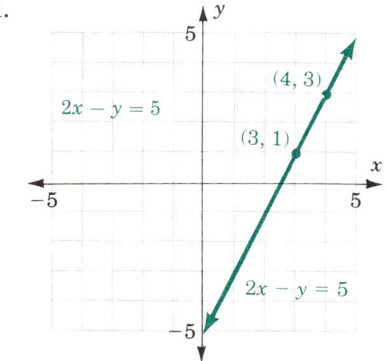

right (the change in x) and 2 units down (the change in y), ending at (4, 3). The graph is the line through the points (3, 5) and (4, 3) (Figure 16). ▲

C. FINDING EQUATIONS GIVEN THE SLOPE AND THE y-INTERCEPT

If, in Equation 2, the point $P(x_1, y_1)$ is on the y-axis, then $x_1 = 0$. If we let $y_1 = b$, then $P(x_1, y_1) = P(0, b)$. Thus the point-slope form of the equation is $y - b = m(x - 0) = mx$. Solving for y, by adding b to both sides, we have the following.

THE SLOPE-INTERCEPT FORM

$$y = mx + b \qquad (3)$$

slope ⎯⎯⎯⎯⎯⎯⎯⎯ y-intercept

Note that in Equation 3, m is the *slope* and b is the *y-intercept* of the line.

EXAMPLE 3 A line has slope 5 and y-intercept 3. Find the slope-intercept form of the equation of this line and then graph it.

Solution Using Equation 3 with $m = 5$ and $b = 3$, we find the required equation to be $y = 5x + 3$. To graph this line, we start at the y-intercept, 3. Since the slope is $5 = \frac{5}{1}$, we go 1 unit right and 5 units up, ending up at (1, 8). The graph is the line drawn through the points (0, 3) and (1, 8) (Figure 17).

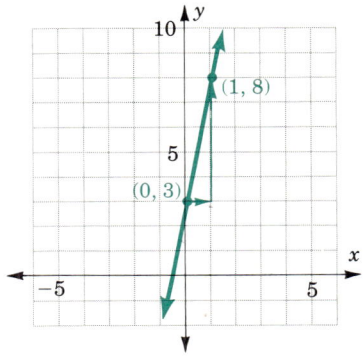

FIG. 17 ▲

Equation 3 is especially convenient because we can find the slope and y-intercept of a given line by just writing the equation of the line in the form $y = mx + b$. For example, to find the slope and y-intercept of the line $2x + 5y = 7$, we solve for y, obtaining

$$y = -\frac{2}{5}x + \frac{7}{5}$$

Thus comparing with Equation 3, we see that m (the slope) is $-\frac{2}{5}$, and b (the y-intercept) is $\frac{7}{5}$.

Problem 3 A line has slope 3 and y-intercept 2. Find the slope-intercept form of the equation of this line and graph it.

EXAMPLE 4   Find the slope and the y-intercept of the line $6x + 3y = 12$.

Solution   We have to solve

$$6x + 3y = 12 \quad \text{for } y.$$

Thus

$$3y = -6x + 12 \quad \text{Subtract } 6x.$$
$$y = -2x + 4 \quad \text{Divide by 3.}$$

slope ⎯⎯⎯⎤   ⎣⎯⎯⎯ intercept

Thus the slope is -2 and the y-intercept is 4.   ▲

D.   FINDING THE EQUATION OF A LINE THROUGH A GIVEN POINT AND PARALLEL OR PERPENDICULAR TO A GIVEN LINE

We are already familiar with particular aspects of the slopes of parallel lines and perpendicular lines. For example, we can use the fact that two parallel lines have identical slopes to find an equation of a line that is parallel to a given line. Likewise, the fact that two perpendicular lines have slopes that are negative reciprocals of each other can be used to find an equation of a line that is perpendicular to a given line. We illustrate these ideas next.

EXAMPLE 5   Find the equation of the line passing through the point $(2, 1)$ and

a. Parallel to the line $y - x = 1$
b. Perpendicular to the line $y - x = 1$

Solution

a. We first write the equation $y - x = 1$ in the slope-intercept form: $y = x + 1$. The slope of this line is 1. If we wish to construct another line parallel to $y = x + 1$ passing through $(2, 1)$, we simply use the point-slope form, with $(x_1, y_1) = (2, 1)$ and $m = 1$, the same slope as that of $y = x + 1$. Thus

$$y - 1 = 1(x - 2)$$
$$y = x - 1$$

b. The line $y - x = 1$ has slope 1. A line perpendicular to this line must have a slope $\dfrac{-1}{1} = -1$. Using the point-slope form again, we find that an equation of the line perpendicular to $y - x = 1$ ($m = -1$) and passing through $(2, 1)$ is

$$y - 1 = -1(x - 2)$$
$$y = -x + 3$$

All the lines involved are shown in the figure.

Problem 4   Find the slope and the y-intercept of the line $8x + 4y = 16$.

Problem 5   Find an equation of the line passing through the point $(1, 1)$ and
a. Parallel to $2y - 6x = 5$
b. Perpendicular to $2y - 6x = 5$

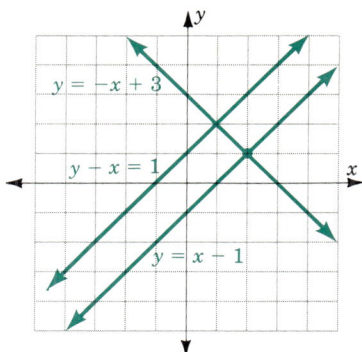

Before you go on to the exercises, here is a reference guide to most of the formulas dealing with linear equations.

IF YOU WANT AN EQUATION	WRITE
In standard form	$Ax + By = C$
For a vertical line	$x = C$ (C a constant)
For a horizontal line	$y = C$ (C a constant)
For a line with slope m and y-intercept b	$y = mx + b$
For the distance d between (x_1, y_1) and (x_2, y_2)	$d = \sqrt{(x_2 - x_1)^2 + (y_2 - y_1)^2}$

IF YOU WANT THE FORMULA	WRITE
For the slope of a line passing through (x_1, y_1) and (x_2, y_2)	$m = \dfrac{y_2 - y_1}{x_2 - x_1}$
For the slope of a vertical line	Undefined
For the slope of a horizontal line	$m = 0$
For the slope of a line parallel to the line $y = mx + b$	m
For the slope of a line perpendicular to the line $y = mx + b$	$-\dfrac{1}{m}$

ANSWERS (to problems on pages 522–523)

2. $3x + y = 5$

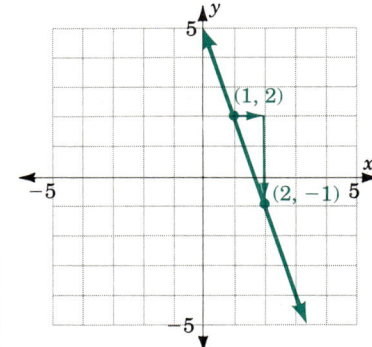

3. $y = 3x + 2$

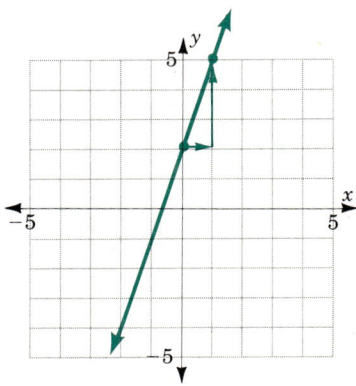

NAME

CLASS

SECTION

ANSWERS

A. In Problems 1–4, find the equation in standard form and graph the line passing through the given points.

1. $(1, -1)$ and $(2, 2)$

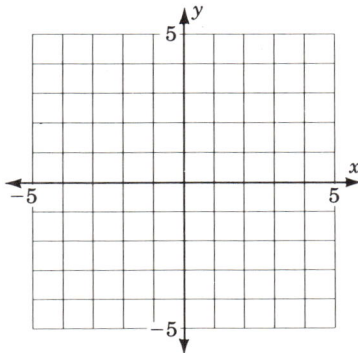

2. $(-3, -4)$ and $(-2, 0)$

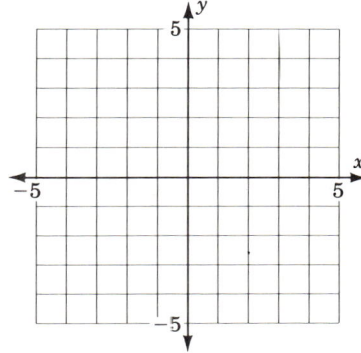

1. _____

2. _____

3. $(3, 2)$ and $(2, 3)$

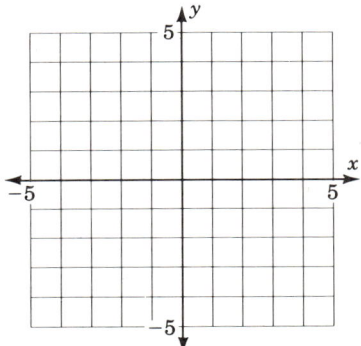

4. $(3, 0)$ and $(0, 5)$

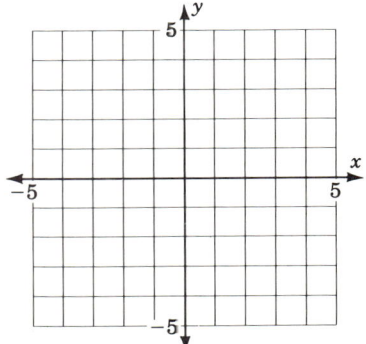

3. _____

4. _____

B. In Problems 5–8, find the point-slope form and graph the line.

5. Slope 2, through $(-3, 5)$

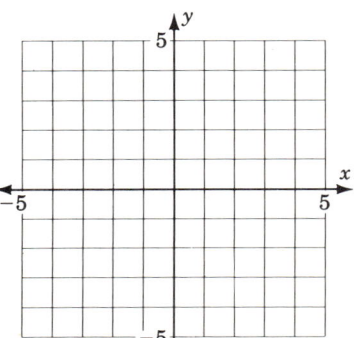

6. Slope $\frac{1}{2}$, through $(2, 3)$

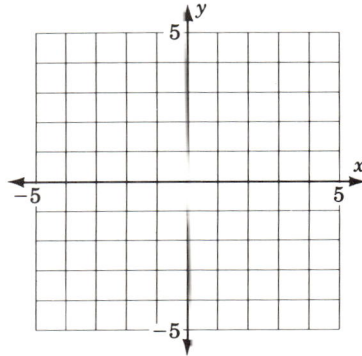

5. _____

6. _____

7. Slope -3, through $(-1, -2)$ **8.** Slope $\dfrac{-1}{3}$, through $(2, -4)$

7. _____

8. _____

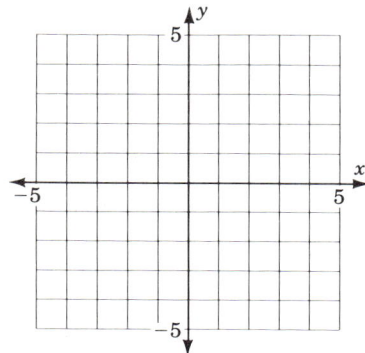   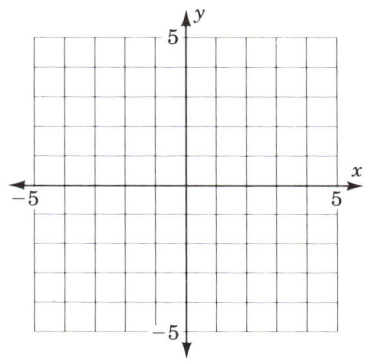

C. In Problems 9–12, find the slope-intercept equation of the line.

9. Slope 5, y-intercept 2

9. _____

10. Slope $\dfrac{1}{4}$, y-intercept 2

10. _____

11. Slope $\dfrac{-1}{5}$, y-intercept $\dfrac{-1}{3}$

11. _____

12. Slope $\dfrac{-1}{7}$, y-intercept $\dfrac{-1}{9}$

12. _____

D. In Problems 13–20, find the slope-intercept form of the equation of the line passing

13. Through $(1, -2)$ and parallel to $y = 2x + 1$

13. _____

14. Through $(-1, -2)$ and parallel to $2y = -4x + 5$

14. _____

15. Through $(-5, 3)$ and parallel to $2y + 6x = 8$

15. _____

16. Through $(-3, -5)$ and parallel to $3y - 6x = 12$

16. _____

17. Through $(1, 1)$ and perpendicular to $2y = x + 6$

17. _____

18. Through $(2, 3)$ and perpendicular to $3y = -x + 5$

18. _____

19. Through $(-2, -4)$ and perpendicular to $y - x = 3$

19. _____

20. Through $(-3, 5)$ and perpendicular to $2y - x = 5$

20. _____

In Problems 21–26, find the slope-intercept equation of the given line.

21. $x - 3y = 5$ **22.** $4x + 5y = 20$

21. _____

22. _____

23. $2y = 6 - 5x$ **24.** $2y = -3x + 6$

23. _____

24. _____

25. $x = 4 - 8y$ **26.** $2x = 6 - 4y$

25. _____

26. _____

27. A line passes through the point $(3, 4)$ and has slope 0. Find an equation of the line.

27. _____

28. A line goes through the point $(-2, -4)$ and has slope 0. Find an equation of the line.

28. _____

29. The slope of a line is undefined and it passes through the point $(-2, 4)$. Find an equation of the line.

29. _____

30. The slope of a line is undefined and it passes through the point $(-4, -5)$. Find an equation of the line.

30. _____

31. A vertical line goes through $(-2, 3)$. Find an equation of the the line.

31. _____

32. A vertical line goes through $(-3, -1)$. Find an equation of the line.

32. _____

33. A horizontal line goes through $(3, 2)$. Find an equation of the line.

33. _____

34. A horizontal line goes through $(-3, -4)$. Find an equation of the line.

34. _____

35. Find an equation of the line with x-intercept 2 and y-intercept 4.

35. _____

36. Find an equation of the line with x-intercept -3 and y-intercept -1.

36. _____

37. Find an equation of the line with x-intercept -3 and y-intercept 2.

37. _____

38. Find an equation of the line with x-intercept -4 and y-intercept 3.

38. _____

E. Applications

39. When sunglasses are sold at the regular price of $6, a student purchases two pairs. If they are sold for $4 at the flea market, the student purchases six pairs. Let the ordered pair (d, p) represent the demand and price for sunglasses.
  a. Find the demand equation.
  b. How many pairs would the student buy if sunglasses are selling for $2?

39. a. _____
    b. _____

40. When skateboards are sold for $80, 5 of them are sold each day. When they are on sale for $40, 13 of them are sold daily. Let (d, p) represent the demand and price for skateboards.
  a. Find the demand equation.
  b. For 10 skateboards to be sold on a given day, what should be the price?

40. a. _____
    b. _____

41. A programmer will supply 50 sets of video games at $35. If the price drops to $20, she will provide only 20. Let (s, p) represent the supply and the price for video games.
  a. Find the supply equation.
  b. At what price would the programmer stop selling the video games; that is, when is the supply 0?
  c. If the price went up to $40 per game, how many games would she be willing to supply?

41. a. _____
    b. _____
    c. _____

42. When T-shirts are selling for $6, a retailer is willing to supply 6 of them each day. If the price increases to $12, the number of available T-shirts is 12.
   a. Find the supply equation.
   b. If the T-shirts were free, how many would the retailer supply?

43. The supply s of a product is given by $s = 3p - 6$, while the demand d is $d = -2p + 14$. What will the price be when the supply s equals the demand d? (This price is called the *equilibrium price*.)

44. The supply and demand for a product are given by $s = 3p - 10$ and $d = -2p + 40$, respectively. At what price will the supply equal the demand?

✓ **SKILL CHECKER**

Find the x- and y-intercepts of the line.

45. $2x - 4y = 8$ **46.** $3x - 2y = 6$

47. $y = 2x + 6$ **48.** $y = -4x + 8$

Solve.

49. $3(x - 1) \leq 6x + 3$ **50.** $4(x - 1) \leq 8x + 4$

42. a. _____
    b. _____

43. _____

44. _____

45. _____
46. _____
47. _____
48. _____
49. _____
50. _____

7.3   USING YOUR KNOWLEDGE

In economics and business, the slope m and the y-intercept of an equation play an important part. Let us see how.

Suppose you wish to go into the business of manufacturing fancy candles. First, you have to buy some ingredients such as wax, paint, and so on. Assume all these ingredients cost you $100. This is the *fixed cost*. Now, suppose it costs $2 to manufacture each candle. This is the *marginal cost*. What would be the total cost y if the marginal cost is $2, x units are produced, and the fixed cost is $100? The answer is

$$y \;=\; \underset{\substack{\uparrow \\ \text{total} \\ \text{cost}}}{} \underset{\substack{\uparrow \\ \text{cost for} \\ x \text{ units}}}{2x} \;+\; \underset{\substack{\uparrow \\ \text{fixed} \\ \text{cost}}}{100}$$

In general, an equation of the form

$$y = mx + b$$

gives the total cost y of producing x units, when m is the cost of producing 1 unit and b is the fixed cost.

1. Find the total cost y of producing x units of a product costing $2 per unit if the fixed cost is $50.

2. Find the total cost y of producing x units of a product whose production cost is $7 per unit if the fixed cost is $300.

3. The total cost y of producing x units of a certain product is given by

   $y = 2x + 75$

   a. What is the production cost for each unit?
   b. What is the fixed cost?

1. _____

2. _____

3. a. _____
   b. _____

OBJECTIVES

REVIEW

Before starting this section, you should know how to:

1. Find the x- and y-intercepts of a line.
2. Solve linear equations.
3. Graph lines.
4. Solve inequalities involving absolute values (Section 2.4).

OBJECTIVES

You should be able to:

A. Graph linear inequalities.
B. Graph inequalities involving absolute values.

We know how to graph linear equations in two variables. Can we graph *linear inequalities* in two variables? The answer is yes.

LINEAR INEQUALITY

A **linear inequality** is a statement that can be written in the form

$$Ax + By \leq C \quad \text{or} \quad Ax + By \geq C$$

where A and B are not both 0.

A. LINEAR INEQUALITIES

Suppose you want to rent a car for a few days. Here are prices for an intermediate car rented in Florida in a recent year:

Rental A: $36 per day, $0.15 per mile
Rental B: $49 per day, $0.33 per mile

The total cost C for the Rental A car is

$$C = \underbrace{36d}_{\substack{\text{cost for } d \\ \text{days}}} + \underbrace{0.15m}_{\substack{\text{cost for} \\ m \text{ miles}}}$$

Now, suppose you want the cost C to be $180. Then $180 = 36d + 0.15m$. We graph this equation by finding the intercepts. When $d = 0$, $180 = 0.15m$, or $m = \dfrac{180}{0.15} = 1200$. When $m = 0$, $180 = 36d$, or $d = \dfrac{180}{36} = 5$. We join (0, 1200) and (5, 0) with a line and then graph the discrete points corresponding to 1, 2, 3, or 4 days.

Now, suppose you want the cost to be less than $180. We then have

$$36d + 0.15m < 180$$

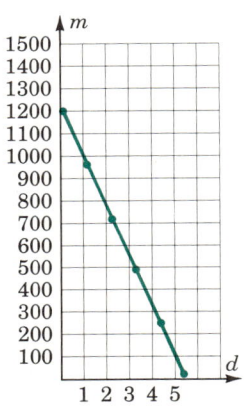

We have graphed the points on the line $36d + 0.15m = 180$. Where are the points for which $36d + 0.15m < 180$? As the graph shows, the line $36d + 0.15m < 180$ divides the plane into three parts:

1. The points *below* the line
2. The points *on* the line
3. The points *above* the line

It can be shown that if any point on one side of the line $Ax + By = C$ satisfies the inequality $Ax + By < C$, then all points on that side satisfy the inequality and no point on the other side of the line does. Let us select $(0, 0)$ as a test point. Since

$$36 \cdot 0 + 0.15 \cdot 0 < 180$$
$$0 < 180$$

is true, all points below the line (shown shaded) satisfy the inequality. (As a check, note that $(5, 500)$, which is above the line, does not satisfy the inequality, since $36 \cdot 5 + 0.15 \cdot 500 < 180$ is a false statement.) The line $36d + 0.15m = 180$ is *not* part of the graph and is shown dashed. To apply this result to our rental-car problem, note that d must be an integer. The solution to our problem consists of the points on the heavy line segments at $d = 1, 2, 3,$ and 4. The graph shows, for instance, that you can rent a car for 2 days and go about 700 mi at a cost less than $180.

EXAMPLE 1   Graph $x - 2y < -4$.

Solution   We first graph the line $x - 2y = -4$ by finding the intercepts.

x	y
0	2
-4	0

When $x = 0$, $-2y = -4$, and $y = 2$
When $y = 0$, $x - 2 \cdot 0 = -4$ and $x = -4$

Note that the line itself is shown dashed to indicate that it is not part of the solution (Figure 18). We select an easy test point and see if it satisfies the inequality. If it does, the solution lies on the same side of the line as the test point; otherwise the solution is on the other side of the line.

An easy point is $(0, 0)$, which is *below* the line. If we substitute $x = \mathbf{0}$ and $y = \mathbf{0}$ in the inequality $x - 2y < -4$, we obtain

$$0 - 2 \cdot 0 < -4$$
$$0 < -4$$

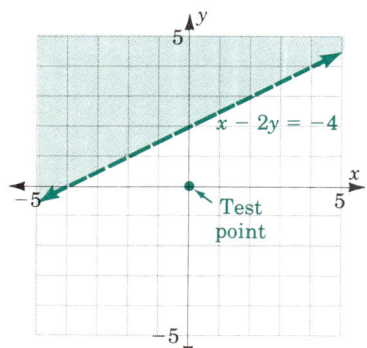

FIG. 18

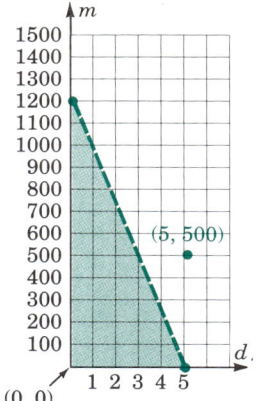

Test point

Shaded region is $36d + 0.15\,m < 180$

Problem 1   Graph $3x - 2y < -6$.

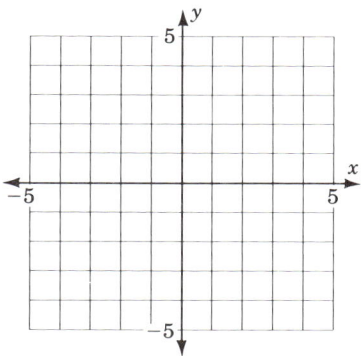

which is false. Thus the point (0, 0) is not part of the solution. Because of this, the solution consists of the points *above* (on the other side of) the line $x - 2y = -4$, as shown shaded. ▲

In the preceding problem, the test point was not part of the solution for the given inequality. Next, we give an example in which the test point is part of the solution for the inequality.

EXAMPLE 2 Graph $y \leq -2x + 4$.

Solution As usual, we first graph the line $y = -2x + 4$.

x	y	
0	4	When $x = 0$, $y = 4$.
2	0	When $y = 0$, $0 = -2x + 4$, or $x = 2$.

The graph of the line is shown in the following figure. Now, we select the point (0, 0) as a test point. When $x = 0$ and $y = 0$, we have

$$y \leq -2x + 4$$
$$0 \leq -2 \cdot 0 + 4$$
$$0 \leq 4$$

which is true. Thus all the points on the same side of the line as (0, 0), that is, the points *below* the line, are solutions of $y \leq -2x + 4$. These solutions are shown shaded in the figure. This time, the line is *solid* because it is part of the solution.

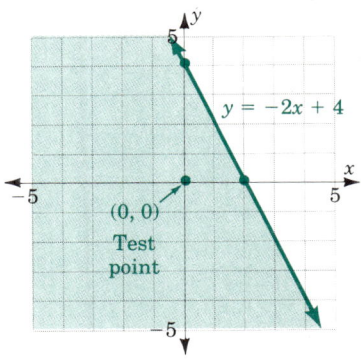

EXAMPLE 3 Graph $x \geq -1$

Solution We first graph the vertical line $x = -1$. All points to the *right* of this line have x-coordinates greater than -1 (points to the left have x-coordinates less than -1). The graph, which includes the line $x = -1$, is shown.

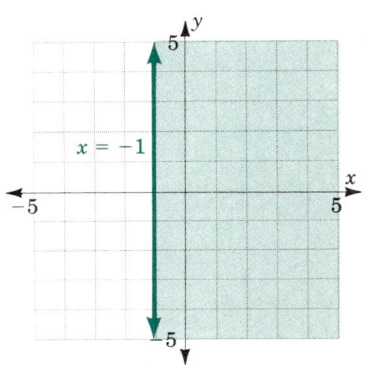

Problem 2 Graph $y \leq -4x + 4$.

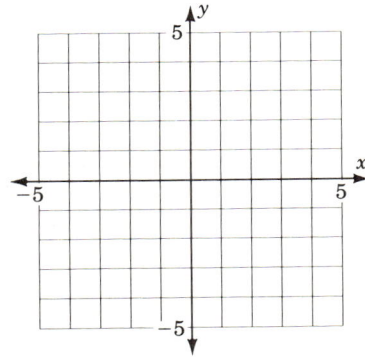

Problem 3 Graph $x \geq -2$.

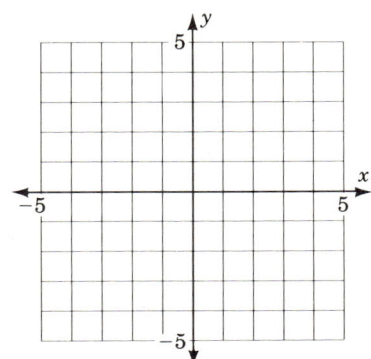

B. ABSOLUTE VALUE INEQUALITIES

As you recall from Section 2.4,

$|x| \le a$ is equivalent to $-a \le x \le a$

Thus if we graph $|x| \le 1$, we must graph all the points satisfying the inequality $-1 \le x \le 1$—that is, the points between -1 and 1—as well as the lines $x = -1$ and $x = 1$. These points are shown in Figure 19.

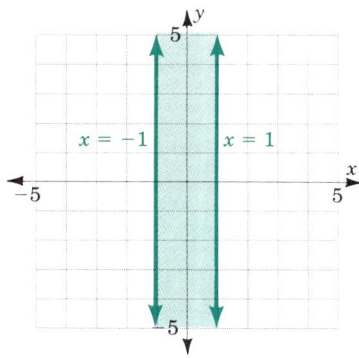

FIG. 19

EXAMPLE 4   Graph $|y| \le 2$.

Solution   Since $|y| \le 2$ is equivalent to $-2 \le y \le 2$, the graph consists of all points bounded by the horizontal lines $y = -2$ and $y = 2$, as well as these two lines, as shown in the graph.

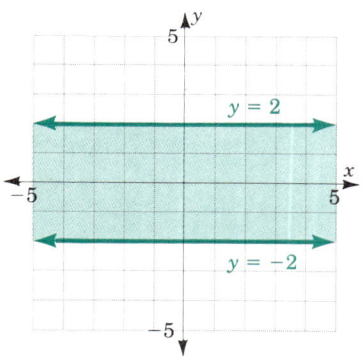

EXAMPLE 5   Graph $|x + 1| > 2$.

Solution   The inequality $|x + 1| > 2$ is equivalent to

$x + 1 > 2$ or $x + 1 < -2$
$x > 1$ or $x < -3$

Thus the graph of $|x + 1| > 2$ consists of all points to the *right* of the vertical line $x = 1$ and all points to the *left* of the line $x = -3$. Note that the boundary lines $x = 1$ and $x = -3$ are *not* part of the graph. They are shown dashed.

Problem 4   Graph $|y| \le 1$.

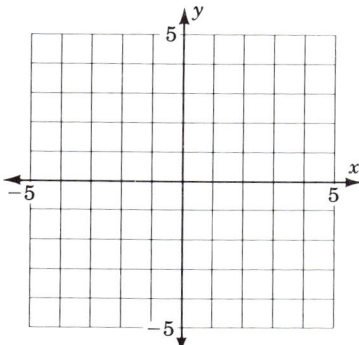

Problem 5   Graph $|x + 2| > 3$.

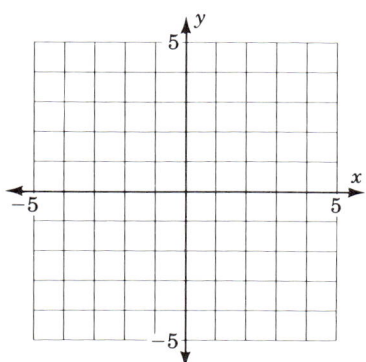

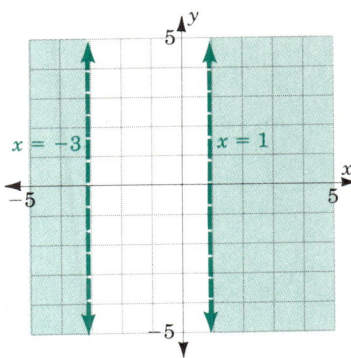

To help you with the exercises, here is a summary of the procedure we have used to graph linear inequalities.

TO GRAPH A LINEAR INEQUALITY

1. Graph the line associated with the inequality. If the inequality involves $\leq$ or $\geq$, include the line. If the inequality involves $<$ or $>$, draw the line dashed.

2. Choose a test point ($(0, 0)$ if possible) not on the line.

3. If the test point satisfies the inequality, shade the region containing the test point; otherwise, shade the region on the other side of the line.

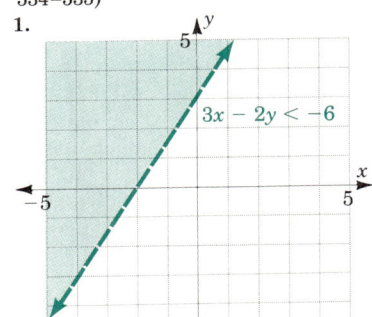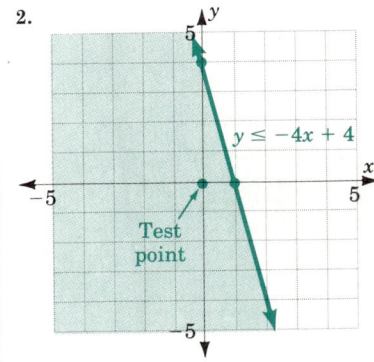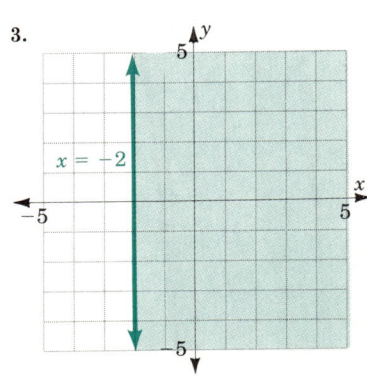

4.

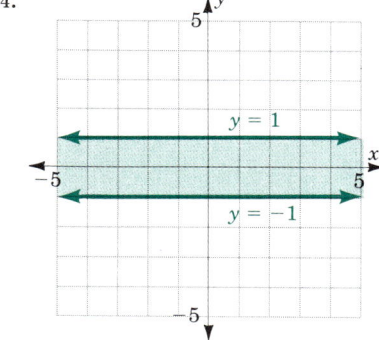

5.

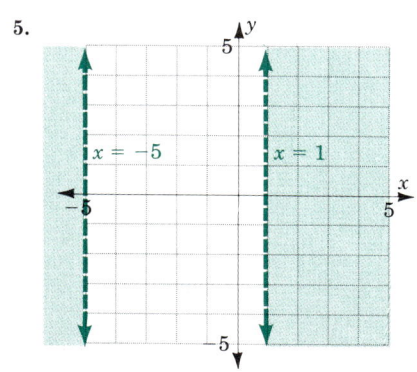

NAME

CLASS

SECTION

A.   Graph.

1. $x + 2y > 4$

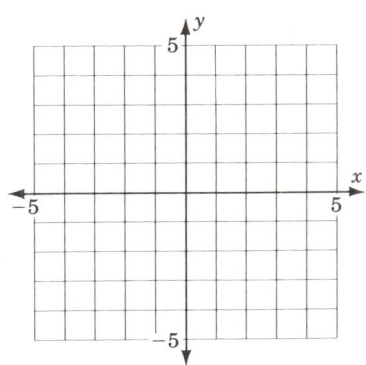

2. $x + 3y > 3$

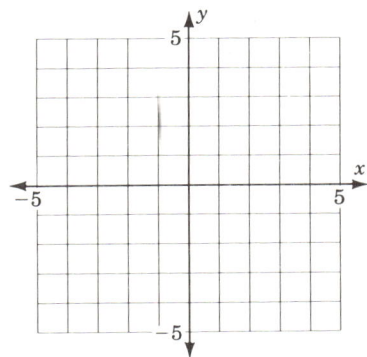

3. $-2x - 5y \leq -10$

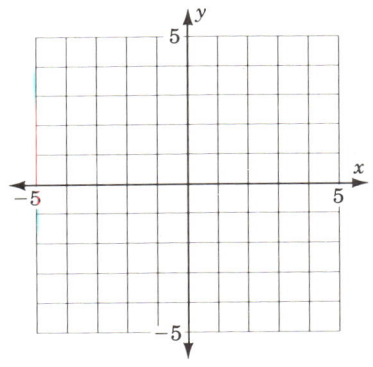

4. $-3x - 2y \leq -6$

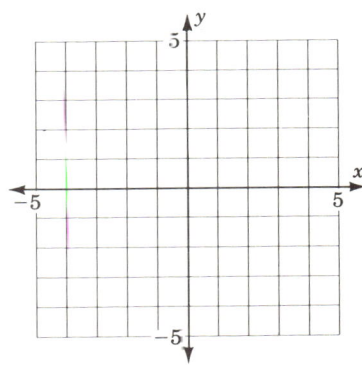

5. $y \geq 2x - 2$

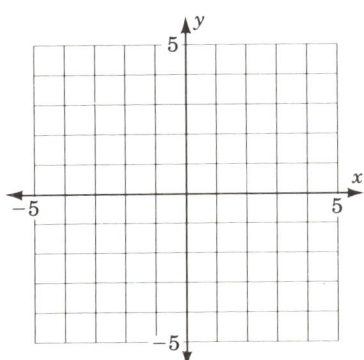

6. $y \geq -2x + 4$

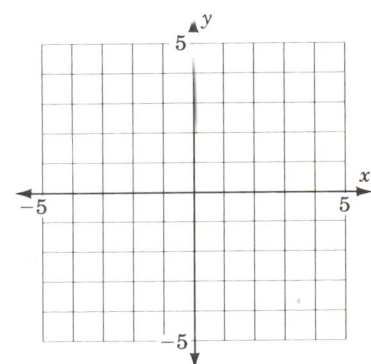

7. $6 < 3x - 2y$

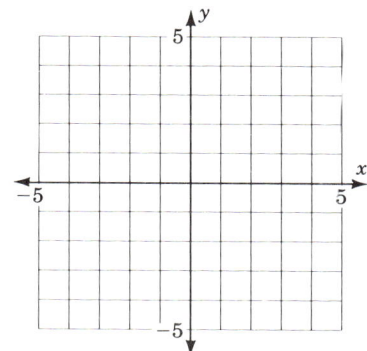

8. $6 < 2x - 3y$

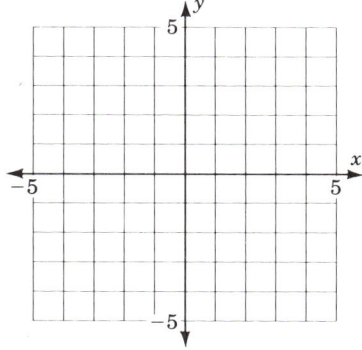

9. $4x + 3y \geq 12$

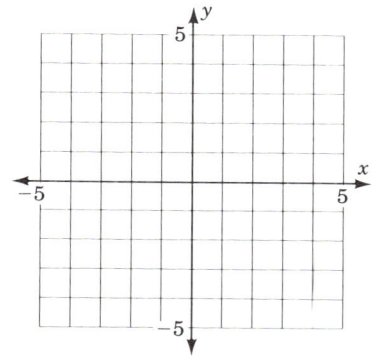

10. $-3y \geq 6x + 6$

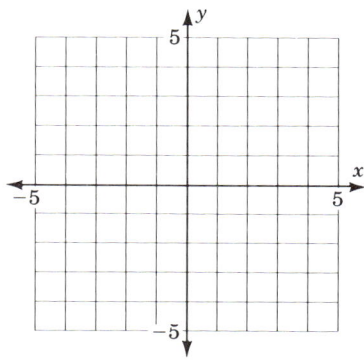

11. $10 < -5x + 2y$

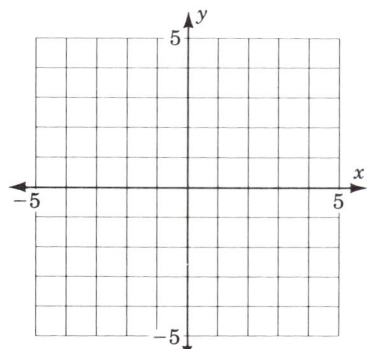

12. $4 < -2x - 4y$

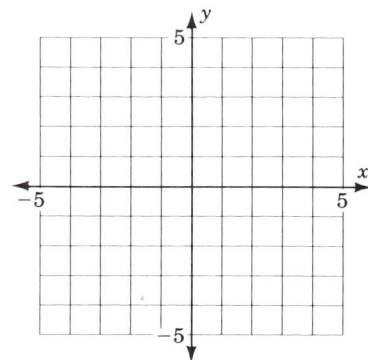

13. $2x \geq 2y - 4$

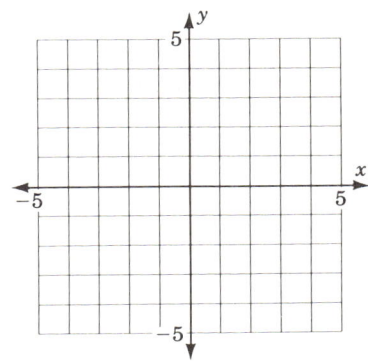

14. $2x \geq 4y + 2$

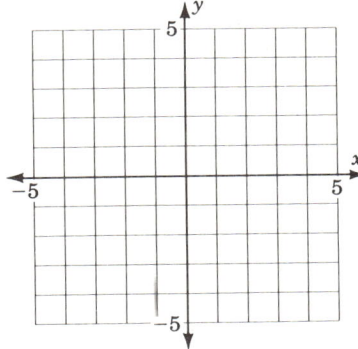

15. $2y < -4x + 8$

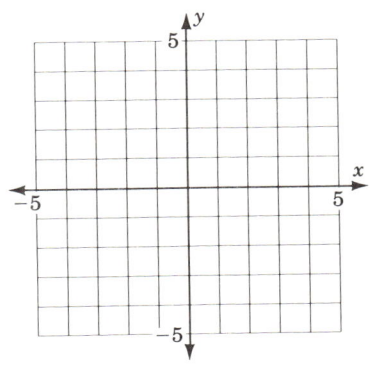

16. $3y < -6x + 9$

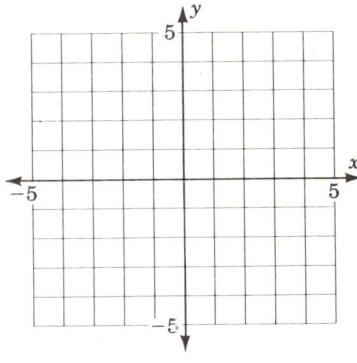

17. $x \geq -3$

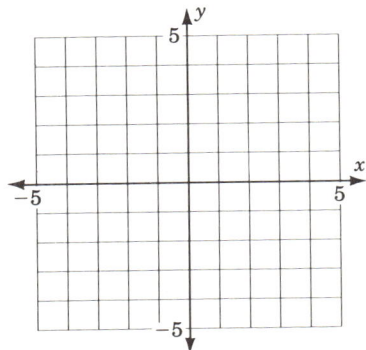

18. $x \geq -4$

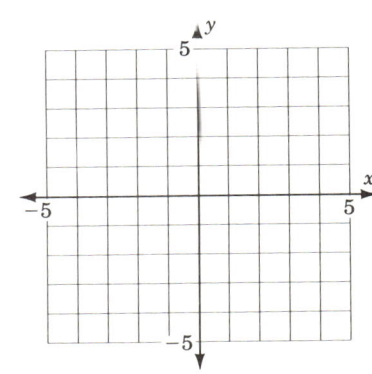

19. $y < 3$

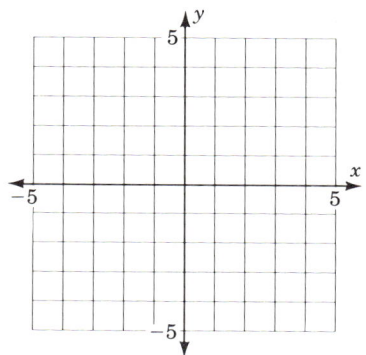

20. $y < -2$

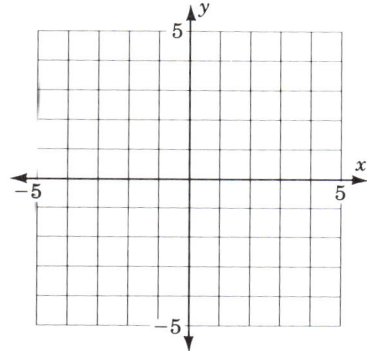

B. Graph.

21. $|x| < 1$

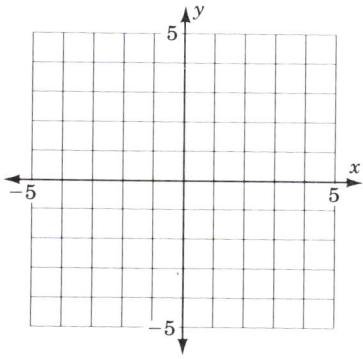

22. $|x| < 3$

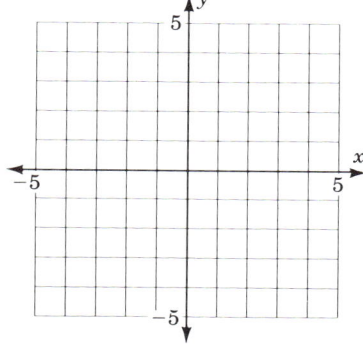

23. $|y| < 2$

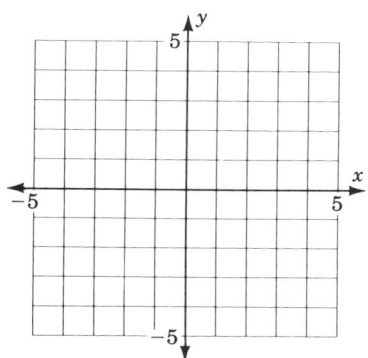

24. $|y| < 3$

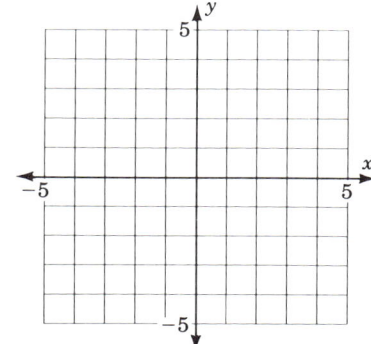

25. $|x| \geq 1$

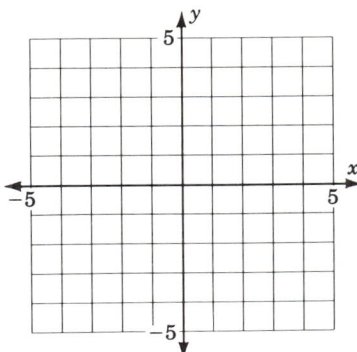

26. $|x| \geq 2$

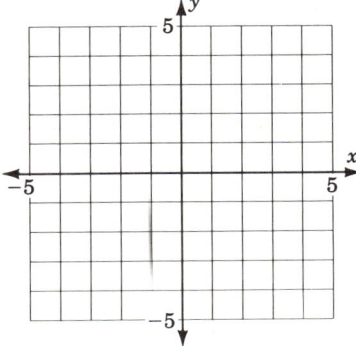

27. $|y| \geq 2$

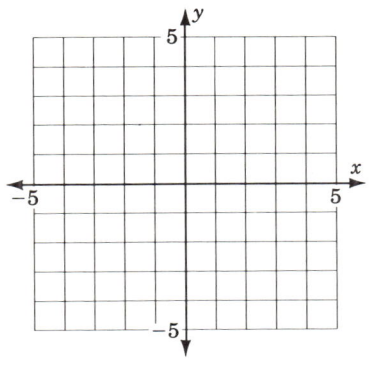

28. $|y| \geq 4$

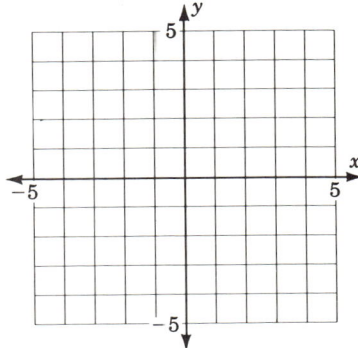

29. $|x + 2| < 1$

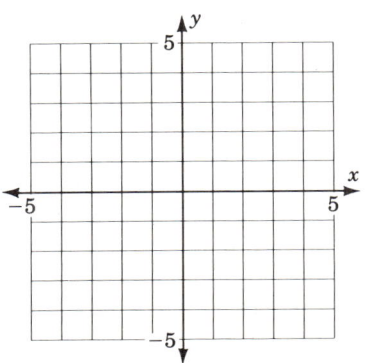

30. $|x + 3| < 1$

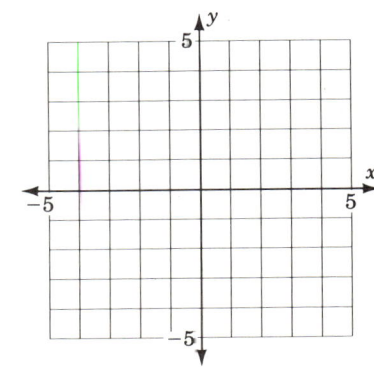

31. $|y + 2| < 1$

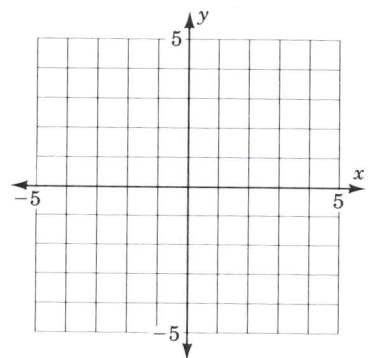

32. $|y + 2| < 2$

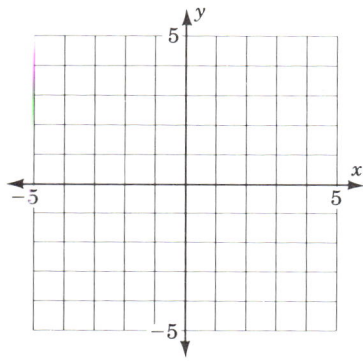

33. $|x + 1| \geq 3$

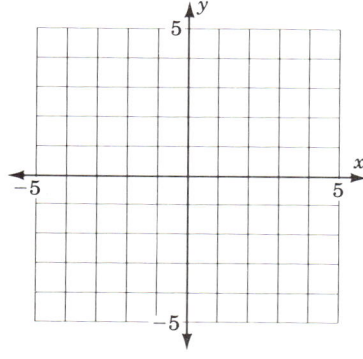

34. $|x + 2| \geq 1$

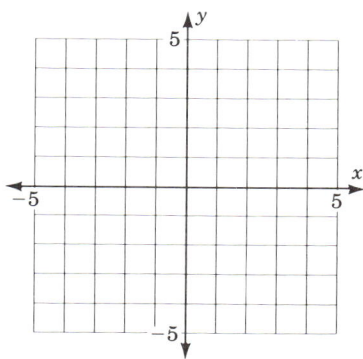

35. $|x - 1| \leq 2$

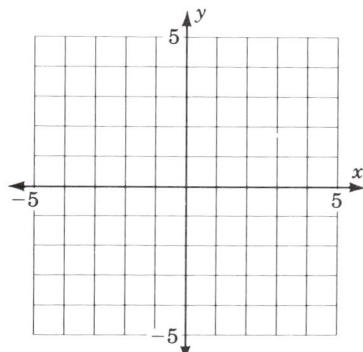

36. $|x - 2| \leq 1$

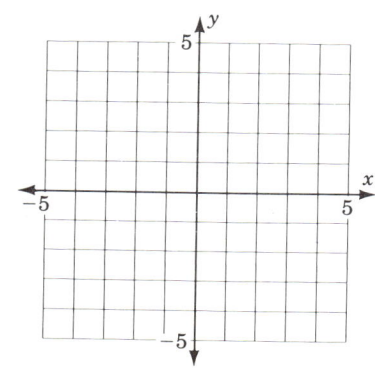

EXERCISE 7.4

37. $|y - 2| < 1$

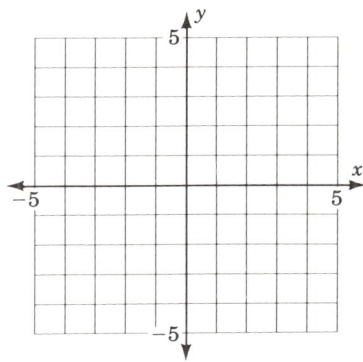

38. $|y - 3| < 1$

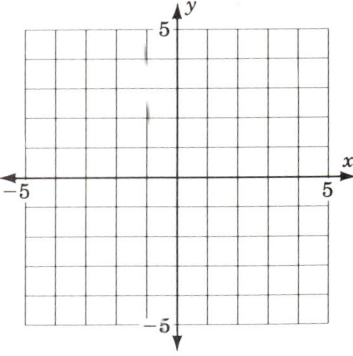

✓ **SKILL CHECKER**

Solve.

39. $56 = k \cdot 14$

40. $39 = k \cdot 13$

41. $60 = \dfrac{k}{3}$

42. $40 = \dfrac{k}{5}$

39. _____
40. _____
41. _____
42. _____

7.4 USING YOUR KNOWLEDGE

The information in this section can save you money when renting a car. Here are the rental prices for an intermediate car obtained from a telephone survey conducted in Tampa, Florida, in a recent year:

Rental A: $41 a day, unlimited mileage
Rental B: $36 a day, $0.15 per mile
Rental C: $42 a day, unlimited mileage

1. If you compare the Rental A price ($41) with the Rental B price, you see that Rental B is initially cheaper.
   a. How far can you drive a Rental B car in one day if you wish to spend exactly $41? (Answer to the nearest mile.)
   b. If you are planning on driving 100 mi, which car would you rent, Rental B or Rental A?

2. How far can you drive a Rental B car in one day if you wish to spend exactly $42? (Answer to the nearest mile.)

3. Based on your answers to Problems 1 and 2, which is the cheapest rental price? (*Hint:* It has to do with the miles you drive.)

1. a. _____
   b. _____

2. _____

3. _____

7.5 VARIATION

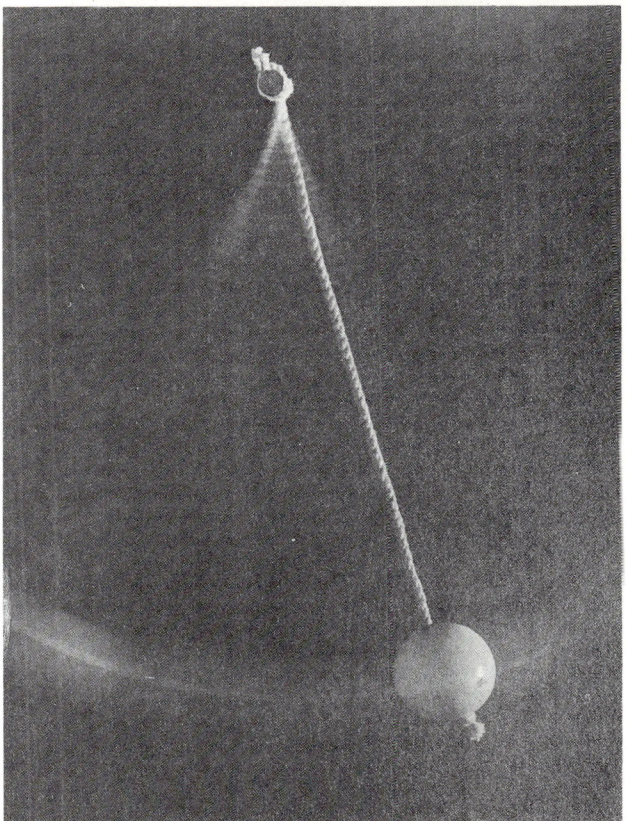

OBJECTIVES

REVIEW

Before starting this section, you should know:

1. How to evaluate an expression.
2. How to solve a linear equation.

OBJECTIVES

You should be able to:

A. Write an equation expressing direct variation.
B. Write an equation expressing inverse variation.
C. Write an expression involving joint variation.
D. Solve applications involving direct, inverse, and joint variation.

As the length L of the string increases, the time T it takes the pendulum to make a full back and forth swing increases. What is the formula relating the length L and the time T? Galileo Galilei discovered that the time T (in seconds) it takes for one swing of the pendulum varies *directly* as the square root of the length L of the pendulum. In the same manner, the number m of miles you drive a car is *proportional to,* or *varies directly as,* the number g of gallons of gas used. This means that the ratio $\dfrac{m}{g}$ is a constant, that is, $\dfrac{m}{g} = k$, or, equivalently, $m = kg$.

A. DIRECT VARIATION

Here is the definition for direct variation.

DIRECT VARIATION

y **varies directly as** x if there is a constant k such that

$$y = kx$$

(k is usually called the constant of variation.)

Here are some other words that indicate direct variation and the corresponding translation.

ENGLISH PHRASE	TRANSLATION
y varies with x	$y = kx$
y varies directly as t	$y = kt$
y is proportional to v	$y = kv$
v varies as the square of t	$v = kt^2$
p varies as the cube of r	$p = kr^3$
T varies as the square root of L	$T = k\sqrt{L}$

EXAMPLE 1 The length L of a moustache varies directly as the time t that it takes to grow.

a. Write an equation of variation.

b. The longest moustache on record was owned by Masuriya Din. His moustache grew 56 in. (on each side) over a 14-year period. Find k and explain what it represents.

Problem 1 Hair length L is proportional to time t.
a. Write an equation of variation.
b. If your hair grew 6 in. in 2 mos, find k.

Solution

a. Since the length L varies directly as the time t,

$$L = kt$$

b. We know that when $L = 56$, $t = 14$. Thus

$$56 = k \cdot 14$$
$$4 = k$$

This means that the moustache grew 4 in. each year. ▲

B. INVERSE VARIATION

Sometimes, as one quantity increases, a related quantity decreases proportionately. For example, the more time you spend practicing a task, the less time it will take you to do the task. In such cases, we say that the quantities *vary inversely as* each other.

y **varies inversely as** x if there is a constant k such that

$$y = \frac{k}{x}$$

Here are some other words that also mean vary inversely.

ENGLISH PHRASE	TRANSLATION
y varies inversely with x	$y = \dfrac{k}{x}$
y is inversely proportional to x	$y = \dfrac{k}{x}$
v varies inversely as the square of t	$v = \dfrac{k}{t^2}$
p varies inversely as the cube of r	$p = \dfrac{k}{r^3}$
T varies inversely as the square root of L	$T = \dfrac{k}{\sqrt{L}}$

EXAMPLE 2 The speed s that a car travels is inversely proportional to the time t it takes to travel a given distance.

a. Write the equation of variation.

b. If a car travels at 60 mi/hr for 3 hr, what is k, and what does it represent?

Solution

a. The equation is

$$s = \frac{k}{t}$$

b. We know that $s = 60$ when $t = 3$. Substituting 60 for s and 3 for t,

$$60 = \frac{k}{3}$$

$$k = 180$$

In this case, k represents the distance traveled, 180 miles. ▲

EXAMPLE 3 Have you ever heard one of those loud boom boxes or a car sound system that makes your stomach tremble? The loudness L of sound is inversely proportional to the square of the distance d that you are from the source.

a. Write an equation of variation.

b. The loudness of rock music coming from a boom box 5 ft away is 100 dB (decibels). Find k.

c. If you move to 10 ft away from the boom box, how loud is the sound?

Solution

a. The equation is

$$L = \frac{k}{d^2}$$

Problem 2 The principal P invested is inversely proportional to the annual rate of interest r.
a. Write an equation of variation.
b. Find k if $r = 10\%$ and $P = \$100$.

Problem 3 The f-number on a camera varies inversely as the diameter a of the aperture when the distance is set at infinity.
a. Write an equation of variation.
b. Find k when the f-number is 8 and $a = \frac{1}{2}$.
c. Find a if the f-number is 16.

b. We know that $L = 100$ for $d = 5$, so that

$$100 = \frac{k}{5^2} = \frac{k}{25}$$

Multiplying both sides by 25, we find that $k = 2500$.

c. Since $k = 2500$,

$$L = \frac{2500}{d^2} \quad \text{Substitute 2500 for } k.$$

When $d = 10$,

$$L = \frac{2500}{10^2} = 25 \text{ dB}$$ ▲

C. JOINT VARIATION

Besides the direct and inverse variations we have discussed so far, there can be variation involving a third variable. A variable z can vary *jointly* with the variables x and y. For example, labor costs c vary jointly with the number of workers w used and the number of hours h that they work. The formal expression of joint variation is given next.

---

z **varies jointly** with x and y if there is a constant k such that

$$z = kxy$$

The statement *z is proportional to x and y* is sometimes used to mean *z varies jointly with the variables x and y*.

Thus the fact that labor costs c vary jointly with the number w of workers used and the number h of hours worked can be expressed as $c = kwh$, k a constant.

EXAMPLE 4 The lifting force P exerted by the atmosphere on the wings of an airplane varies jointly with the wing area A in square feet and the square of the plane's speed V in miles per hour. Suppose the lift is 1200 lb for a wing area of 100 ft^2 and a speed of 75 mi/hr.

a. Find an equation of variation.
b. Find k.
c. Find the lifting force on a wing area of 60 ft^2 when $V = 125$.

Solution

a. Since P varies jointly with the area A and the square of the velocity V, we have $P = kAV^2$.
b. When the lift $P = 1200$, we know that $A = 100$ and $V = 75$. Substituting these values in the equation $P = kAV^2$, we obtain

$$1200 = k \cdot 100 \cdot (75)^2$$

Problem 4 The wind force F on a vertical surface varies jointly with the area A of the surface and the square of the wind velocity V. Suppose the wind force on 1 ft^2 of surface is 1.8 lb when $V = 20$ mi/hr.

a. Find an equation of variation.
b. Find k.
c. Find the force on a 2-ft^2 vertical surface when $V = 60$ mi/hr.

Dividing both sides by $100 \cdot 75^2$,

$$k = \frac{1200}{100 \cdot 75^2} = \frac{12}{75^2} = \frac{4}{1875}$$

c. $P = \dfrac{4}{1875} A V^2$

$$= \frac{4}{1875}(60)(125^2)$$

$$= 2000 \text{ lb}$$

▲

ANSWER (to problem on page 550)
4. **a.** $F = kAV^2$ **b.** 0.0045 **c.** 32.4 lb

NAME

CLASS

SECTION

ANSWERS

A. In Problems 1–5 write an equation of variation using k as the constant.

1. The tension T on a spring varies directly with the distance s it is stretched.

2. The distance s a body falls in t seconds is directly proportional to the square of t.

3. The weight W of a dam varies directly with the cube of its height h.

4. The kinetic energy KE of a moving body is proportional to the square of its velocity v.

5. The weight W of a human brain is directly proportional to the body weight B.

B. In Problems 6–8 write an equation of variation using k as the constant.

6. In a circuit with constant voltage, the current I varies inversely with the resistance R of the circuit.

7. For a wire of fixed length, the resistance R varies inversely with the square of its diameter D.

8. The intensity of illumination I from a source of light varies inversely with the square of the distance d from the source.

C. In Problems 9–20 write an equation of variation using k as the constant.

9. The annual interest I received on a savings account varies jointly with the principal P (the amount in the account) and the interest rate r paid by the bank.

10. The cost C of a building varies jointly as the number w of workers used to build it and the cost of materials m.

11. The amount of oil A used by a ship traveling at a uniform speed varies jointly with the distance s and the square of the speed v.

12. The power P in an electric circuit varies jointly with the resistance R and the square of the current I.

13. The volume V of a rectangular container of fixed length varies jointly with its depth d and width w.

14. The force of attraction F between two spheres of mass m_1 and m_2, respectively, varies directly as the product of the masses and inversely as the square of the distance d between their centers.

ANSWERS

1. _____

2. _____

3. _____

4. _____

5. _____

6. _____

7. _____

8. _____

9. _____

10. _____

11. _____

12. _____

13. _____

14. _____

15. The illumination I in foot-candles upon a wall varies directly with the intensity i in candlepower of the source of light and inversely with the square of the distance d from the light.

15. _____

16. The strength S of a horizontal beam of rectangular cross section and of length L varies jointly as the breadth b and the square of the depth d and inversely as the length L.

16. _____

17. The electrical resistance R of a wire of uniform cross section varies directly as its length L and inversely as its cross-sectional area A.

17. _____

18. The electrical resistance R of a wire varies directly as the length L and inversely as the square of its diameter d.

18. _____

19. The weight W of a body varies inversely as the square of its distance d from the center of the earth.

19. _____

20. z varies directly as the cube of x and inversely as the square of y.

20. _____

D. Applications

21. The amount of annual interest I you receive on a savings account is directly proportional to the amount of money m you have in the account.
   a. Write an equation of variation.
   b. If $480 produces $26.40 in interest, what is k?
   c. How much annual interest would you receive if the account had $750?

21. a. _____
    b. _____
    c. _____

22. The number of revolutions R a record makes as it is being played varies directly as the time t that it is on the turntable.
   a. Write an equation of variation.
   b. A record that lasted $2\frac{1}{2}$ min made 112.5 rev. What is k?
   c. If a record makes 108 rev, how long does it take to play it?

22. a. _____
    b. _____
    c. _____

23. The distance d an automobile travels after the brakes have been applied varies directly as the square of its speed s.
   a. Write an equation of variation.
   b. If the stopping distance for a car going 30 mi/hr is 54 ft, what is k?
   c. What is the stopping distance for a car going 60 mi/h?

23. a. _____
    b. _____
    c. _____

24. The weight of a person varies directly as the cube of the person's height h (in inches). The **threshhold weight** T (in pounds) for a person is defined as "the crucial weight, above which the mortality (risk) for the patient rises astronomically."
   a. Write an equation of variation relating T and h.
   b. If $T = 196$ when $h = 70$, find k written as a fraction.
   c. To the nearest pound, what is the threshhold weight T for a person 75 in. tall?

24. a. _____
    b. _____
    c. _____

25. The number S of new songs a rock band needs each year is inversely proportional to the number y of years the band has been in the business.
   a. Write an equation of variation.
   b. If, after 3 years in the business, the band needs 50 new songs, how many songs will it need after 5 years?

25. a. _____
    b. _____

26. When the distance is set at infinity, the *f*-number on a camera lens varies inversely as the diameter *d* of the aperture (opening).

    a. Write an equation of variation.

    b. If the *f*-number on a camera is 8 when the aperture is $\frac{1}{2}$ in., what is *k*?

    c. Find the *f*-number when the aperture is $\frac{1}{4}$ in.

26. a. _____
    b. _____
    c. _____

27. The weight *W* of an object varies inversely as the square of its distance *d* from the center of the earth.

    a. Write an equation of variation.

    b. An astronaut weighs 121 lb on the surface of the earth. If the radius of the earth is 3960 mi, find the value of *k* for this astronaut. (Do not multiply out your answer.)

    c. What will this astronaut weigh when she is 880 mi above the surface of the earth?

27. a. _____
    b. _____
    c. _____

✓ **SKILL CHECKER**

Graph.

28. $x + y = 3$

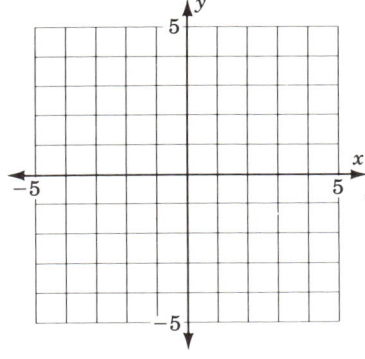

29. $2x - y = 2$

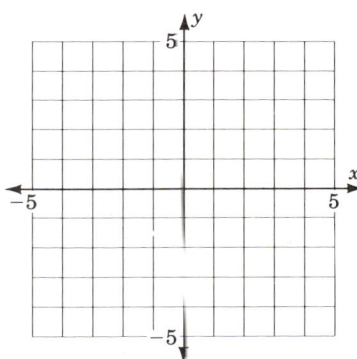

30. $2x + \frac{1}{2}y = 2$

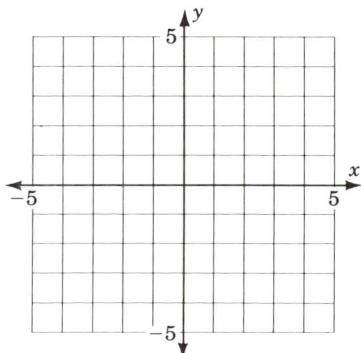

31. $y = -x - 3$

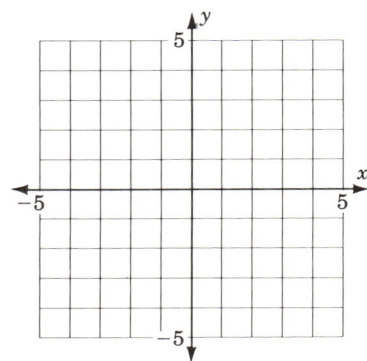

32. $y = -4x + 4$

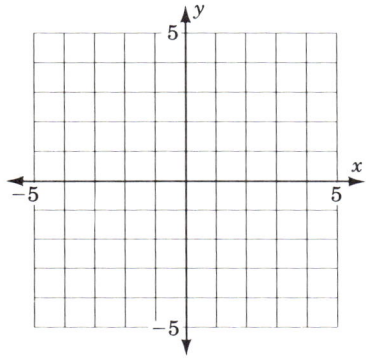

7.5 USING YOUR KNOWLEDGE

The equation for direct variation between x and y ($y = kx$) and the equation of a line of slope m passing through the origin ($y = mx$) are very similar. Let us look at the table giving the water pressure in pounds per square inch (lb/in.^{2}) exerted on a diver.

Depth of diver (ft)	10	25	40	55
Pressure on diver (lb/in.^{2})	4.2	10.5	16.8	23.1

1. Graph these points.

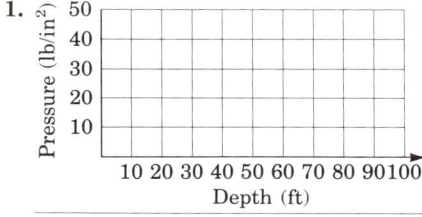

2. What is the slope of the resulting line?

3. As it turns out, the pressure p on the diver is directly proportional to the depth d. Write an equation of variation.

4. Use one of the points in the table to find k.

5. What is the relationship between k and the slope found in 2?

6. Predict the pressure on the diver at a depth of 125 ft.

1.

2. _____

3. _____

4. _____

5. _____

6. _____

SUMMARY

SECTION	ITEM	MEANING	EXAMPLES
7.1C	Linear equation	Any equation that can be written in the form $Ax + By = C$	$7x + 8y = 3$ and $2x = 3y - 2$ are linear equations.
7.1C	x-intercept	The x-coordinate of the point at which the graph crosses the x-axis	The x-intercept of $2x + y = 6$ is $x = 3$.
7.1C	y-intercept	The y-coordinate of the point at which the graph crosses the y-axis	The y-intercept of $2x + y = 6$ is $y = 6$.

(Continued)

SECTION	ITEM	MEANING	EXAMPLES
7.1D	Horizontal line	A line whose equation can be written in the form $y = C$, C a constant	$y = 4$, $y = -3$, $y = 0.25$, and $2y = 6$ are equations of horizontal lines.
7.1D	Vertical line	A line whose equation can be written in the form $x = C$, C a constant	$x = 4$, $x = -3$, $x = 0.25$, and $2x = 6$ are equations of horizontal lines.
7.2A	Distance formula	The distance between the points (x_1, y_1) and (x_2, y_2) is $\sqrt{(x_2 - x_1)^2 + (y_2 - y_1)^2}$.	The distance between $(3, 4)$ and $(6, 8)$ is $\sqrt{3^2 + 4^2} = 5$.
7.2B	Slope	The slope of the line through (x_1, y_1) and (x_2, y_2) is $m = \dfrac{y_2 - y_1}{x_2 - x_1}$.	The slope of the line through $(3, 5)$ and $(9, 7)$ is $m = \dfrac{7 - 5}{9 - 3} = \dfrac{1}{3}$.
7.2C	Slopes of parallel lines	Two lines L_1 and L_2 with slopes m_1 and m_2 are parallel if and only if $m_1 = m_2$.	The lines $y = -2x + 3$ and $y = -2x + 9$ are parallel.
7.2C	Slopes of perpendicular lines	Two lines L_1 and L_2 with slopes m_1 and m_2 are perpendicular if and only if $m_1 = -\dfrac{1}{m_2}$.	A line perpendicular to the line $y = -2x + 3$ will have a slope of $\frac{1}{2}$, since the slope of $y = -2x + 3$ is -2.
7.3A	Two-point form of a line	An equation of the line going through the points (x_1, y_1) and (x_2, y_2) is $y - y_1 = \dfrac{y_2 - y_1}{x_2 - x_1}(x - x_1)$.	An equation of the line through the points $(3, 1)$ and $(4, 3)$ is $y - 1 = \dfrac{3 - 1}{4 - 3}(x - 3)$.
7.3B	Point-slope form of a line	An equation of the line going through the point (x_1, y_1) and with slope m is $y - y_1 = m(x - x_1)$.	An equation of the line going through the point $(2, 5)$ and with slope -3 is $y - 5 = -3(x - 2)$.
7.3C	Slope-intercept form of a line	An equation of the line with slope m and y-intercept b is $y = mx + b$.	An equation of the line with slope 3 and y-intercept -4 is $y = 3x - 4$.
7.4A	Linear inequality	A linear inequality is a statement that can be written in the form $Ax + By \le C$ or $Ax + By \ge C$.	$3x + 5y \le 10$ is a linear inequality.
7.4B	Absolute value inequality	An inequality of the form $|x| \le a$ or $|x| \ge a$	$|x| < 3$ and $|x - 1| \ge 5$ are absolute value inequalities.
7.5A	Direct variation	y varies directly as x if there is a constant k such that $y = kx$.	If you are paid an hourly rate, the salary s you receive varies directly as the number of hours h you work, so $s = kh$.
7.5B	Inverse variation	y varies inversely as x if there is a constant k such that $y = \dfrac{k}{x}$.	The intensity I of the light you get from a reflector varies inversely with the square of the distance d you are from the reflector, so $I = \dfrac{k}{d^2}$.
7.5C	Joint variation	z varies jointly with x and y if there is a constant k such that $z = kxy$.	The interest I received varies jointly with the principal P and the interest rate r, so $I = kPr$.

(If you need help with these exercises, look in the section indicated in brackets.)

1. [7.1A] Graph.
   a. $A(1, 4)$, $B(-3, 1)$, $C(-3, -2)$, and $D(3, -1)$

   1. a.

   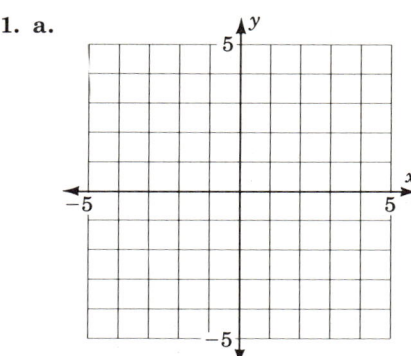

   b. $A(4, 1)$, $B(-1, 3)$, $C(-2, -3)$, and $D(1, -3)$

   b.

   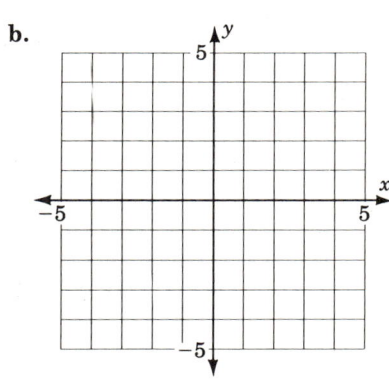

2. [7.1B] Graph.
   a. $x + 2y = 4$

   2. a.

   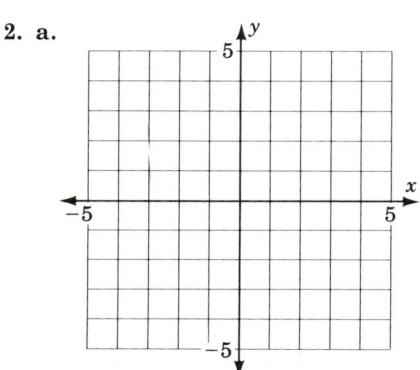

b. $2x - y = 2$

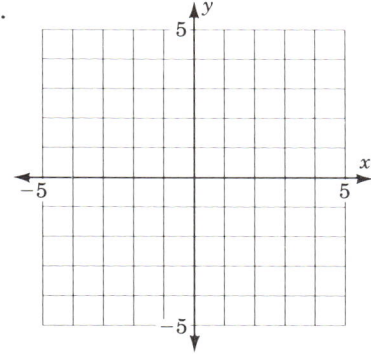

b.

3. [7.1C] Find the x and y-intercepts and graph the line.
   a. $y = 3x + 3$

3. **a.** _____

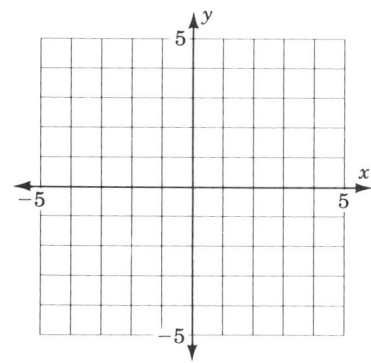

b. $y = 2x - 4$

b. _____

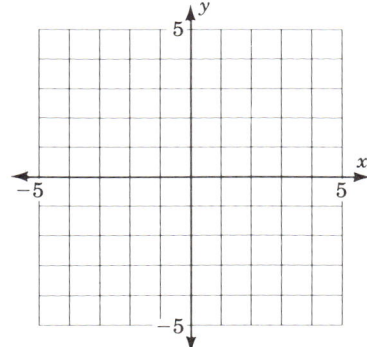

4. [7.1D] Graph on the same coordinate system.
   a. $2x = 6$
   b. $3y = 6$

4.

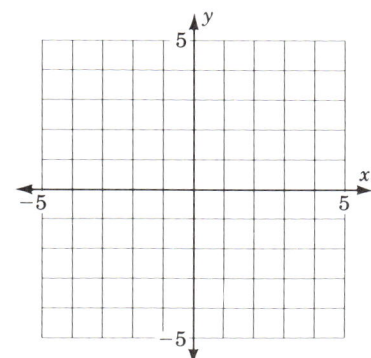

5. [7.1D] Graph on the same coordinate system.
   a. $2x = -6$
   b. $3y = -6$

5.

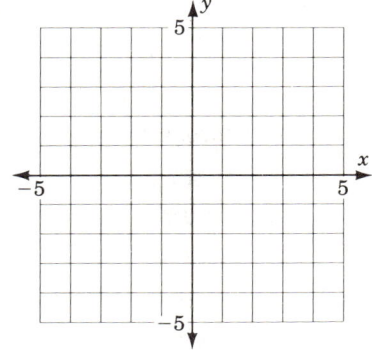

6. [7.2A] Find the distance between the two given points.
   a. $A(1, 3)$ and $B(6, 15)$
   b. $A(2, 3)$ and $B(-2, 5)$
   c. $A(2, -4)$ and $B(6, -4)$

6. a. _____
   b. _____
   c. _____

7. [7.2B] Find the slope of the line through the given points.
   a. $A(-3, 2)$ and $B(1, 0)$
   b. $A(4, -2)$ and $B(4, -7)$

7. a. _____
   b. _____

8. [7.2B] Find the slope of the line through the given points.
   a. $A(3, 4)$ and $B(4, 3)$
   b. $A(1, -1)$ and $B(-3, 5)$

8. a. _____
   b. _____

9. [7.2C] A line L has slope $\frac{3}{4}$. Find whether the line through the two given points is parallel or perpendicular to L.
   a. $A(1, 3)$ and $B(-2, 7)$
   b. $A(1, 3)$ and $B(5, 6)$

9. a. _____
   b. _____

10. [7.2C] A line L has slope -2. Find whether the line through the two given points is parallel or perpendicular to L.
   a. $A(2, -1)$ and $(1, 1)$
   b. $A(3, -1)$ and $(2, -3)$

10. a. _____
   b. _____

11. [7.2C] The line through $(2, 4)$ and $(5, y)$ is perpendicular to a line with the given slope. Find y.

   a. $m = \dfrac{3}{2}$

   b. $m = -2$

11. a. _____
   b. _____

12. [7.2D] A line passes through the point $(-1, 2)$ and has the given slope. Graph the line.

   a. $m = -\dfrac{1}{2}$

12. a.

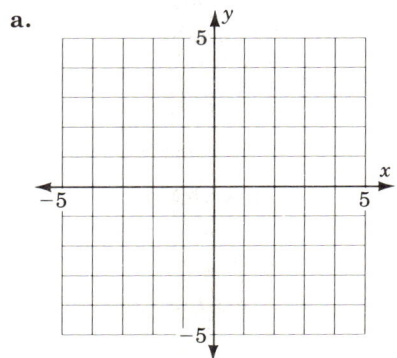

b. $m = \dfrac{2}{3}$

b.

13. [7.3A] Find an equation of the line through the two given points, and write the equation in standard form.
   a. $A(2, 5)$ and $B(-1, 2)$
   b. $A(-4, 3)$ and $B(-2, -2)$

13. a. _____
    b. _____

14. [7.3B] Find an equation of the line with slope 2 and passing through the given point, and write the equation in standard form.
   a. $A(-1, -3)$
   b. $A(5, 0)$

14. a. _____
    b. _____

15. [7.3C]
   a. A line has slope 3 and y-intercept 2. Find the slope-intercept equation of this line.
   b. Repeat part a if the slope is -3 and the y-intercept is 4.

15. a. _____
    b. _____

16. [7.3C] Find the slope and the y-intercept of the line.
   a. $4x - 2y = 8$
   b. $3x + 6y = 12$

16. a. _____
    b. _____

17. [7.3C] Find the slope and the y-intercept of the line.

   a. $\dfrac{x}{2} + \dfrac{y}{4} = 1$

   b. $\dfrac{x}{3} - \dfrac{y}{4} = -1$

17. a. _____
    b. _____

18. [7.3D] Find an equation of the line through the point $(2, 1)$ and that is parallel to the line
   a. $2x + y = 7$
   b. $3x - y = 4$

18. a. _____
    b. _____

19. [7.3D] Find an equation of the line through the point $(2, 1)$ and that is perpendicular to the line
   a. $2x + 3y = 7$
   b. $3x - 2y = 4$

19. a. _____
    b. _____

20. [7.4A] Graph.
  a. $2x + y < -4$

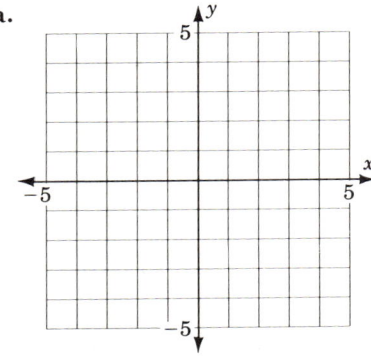

20. a.

  b. $x - 2y < 2$

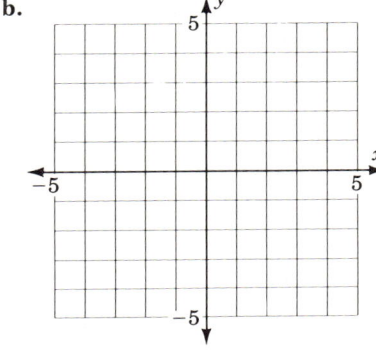

b.

21. [7.4A] Graph.
  a. $y \leq 2x + 2$

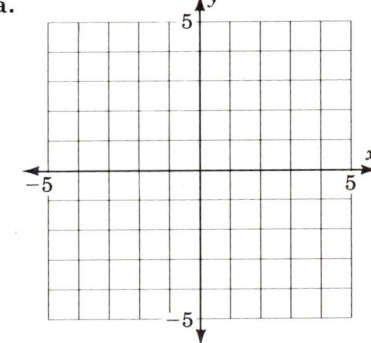

21. a.

  b. $y \leq -x + 3$

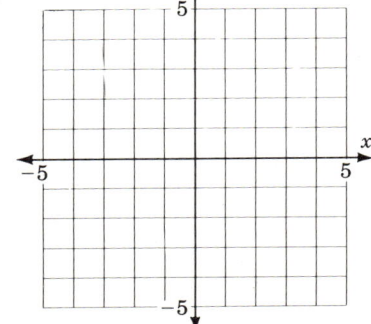

b.

NAME

22. [7.4A] Graph.
   a. $x \geq -3$

22. **a.**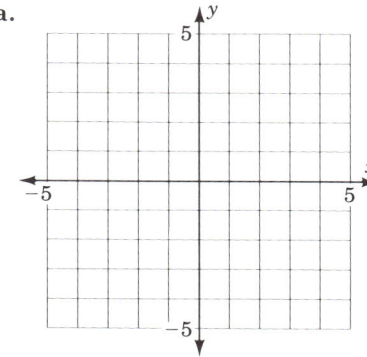

   b. $x \geq 2$

b.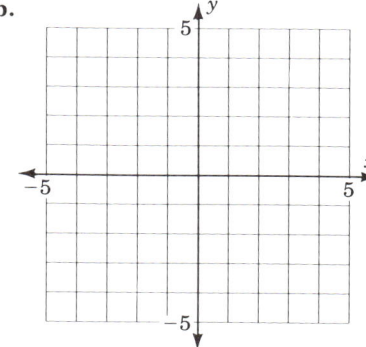

23. [7.4B] Graph.
   a. $|y| \leq 2$

23. **a.**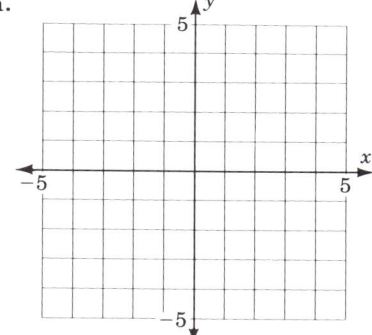

   b. $|y| \geq 2$

b.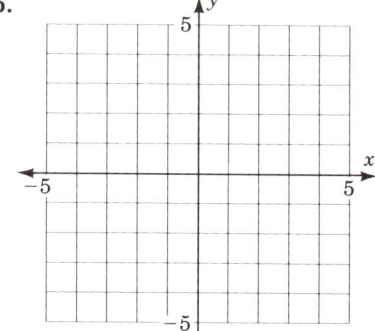

24. [7.4B] Graph.

    a. $|x - 2| > 1$

24. **a.**

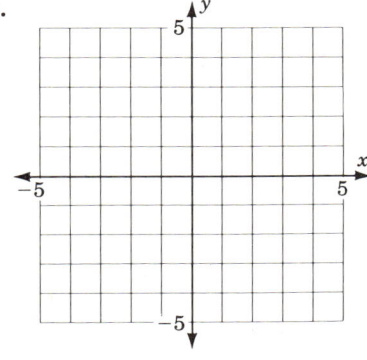

    b. $|x - 1| \le 1$

b.

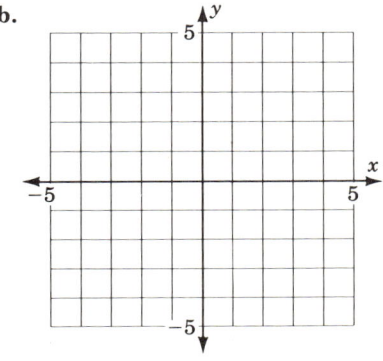

25. [7.5A] The gas in a closed container exerts a pressure P on the walls of the container. This pressure varies directly as the temperature T of the gas.

    a. Write an equation of variation using k for the constant of variation.

    b. If the pressure is 3 lb/in.^{2} when the temperature is 360°F, find k.

25. **a.** _____

    b. _____

26. [7.5B] If the temperature of a gas is held constant, the pressure P varies inversely as the volume V.

    a. Write an equation of variation using k for the constant of variation.

    b. A pressure of 1600 lb/in.^{2} is exerted by 2 ft^3 of air in a cylinder fitted with a piston. Find k.

26. **a.** _____

    b. _____

27. [7.5B] The force F with which the Earth attracts an object above the Earth's surface varies inversely as the square of the distance d from the center of the Earth.

    a. Write an equation of variation using k for the constant of variation.

    b. An astronaut weighs 120 lb on the Earth's surface. If the radius of the Earth is about 4000 mi, find the value of k.

27. **a.** _____

    b. _____

28. [7.5B] What would the astronaut of Problem 27(b) weigh if she is on a space voyage 1000 mi above the Earth's surface?

28. _____

29. [7.5C] The horsepower h that a rotating shaft can safely transmit varies jointly as the cube of its diameter d and the number of revolutions r it makes per minute.
   a. Write an equation of variation using k for the constant of variation.
   b. A 2-in. shaft at a speed of 1000 rev/min can safely transmit 400 hp. Find k for this shaft.

29. a. _____
    b. _____

30. [7.5C] Find the horsepower that the shaft of Problem 29(b) can safely transmit at a speed of 1500 rev/min.

30. _____

NAME

CLASS

SECTION

(Answers on pages 572–577)

ANSWERS

1. Graph $A(3, 2)$, $B(4, -2)$, $C(-1, -2)$, and $D(-1, 3)$.

1.
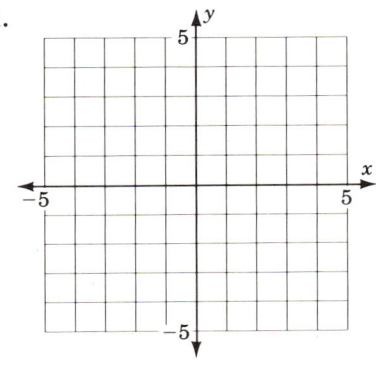

2. Graph $3x - y = 3$.

2.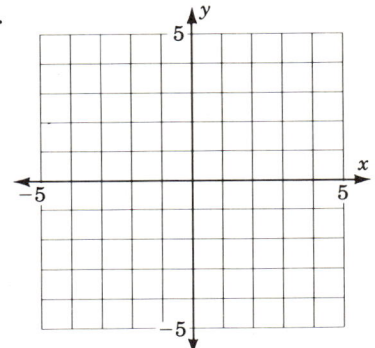

3. Find the x- and y-intercepts of $y = 3x + 2$ and then graph the line.

3.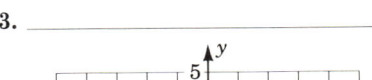

4. Graph.

a. $3x = -9$

4. **a.**

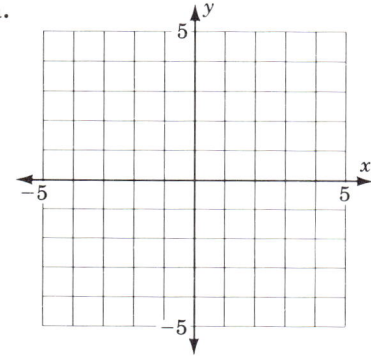

b. $2y = -4$

b.

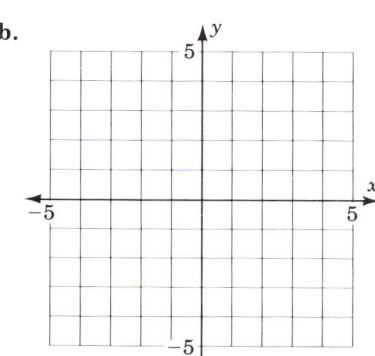

5. Find the distance between the two given points.
a. $A(4, 3)$ and $B(7, 7)$
b. $A(3, 4)$ and $B(5, -2)$
c. $A(-3, 2)$ and $B(-3, 5)$

5. **a.** _____
b. _____
c. _____

6. Find the slope of the line through the two given points.
a. $A(-2, 2)$ and $B(1, 1)$
b. $A(-2, 4)$ and $B(-2, 8)$
c. $A(3, 4)$ and $B(4, 3)$
d. $A(4, 4)$ and $B(5, 4)$

6. **a.** _____
b. _____
c. _____
d. _____

7. A line L_1 has slope $\frac{3}{2}$. Find whether the line through the two given points is parallel or perpendicular to L_1.
a. $A(4, 2)$ and $B(1, 4)$
b. $A(-1, -4)$ and $B(-4, -6)$

7. **a.** _____
b. _____

8. The line through $A(1, -2)$ and $B(-1, y)$ is perpendicular to a line with slope $-\frac{2}{3}$. Find y.

8. _____

9. A line goes through the point $(3, -1)$ and has slope $-\frac{1}{2}$. Graph this line.

9.

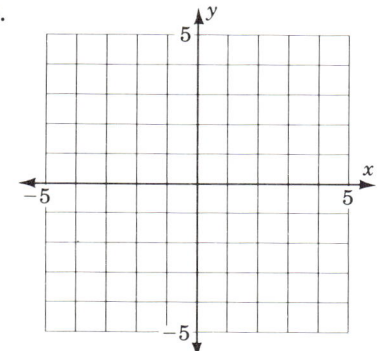

10. Find an equation of the line through (4, 3) and (2, 4). Then write the equation in standard form and graph it.

10. _____

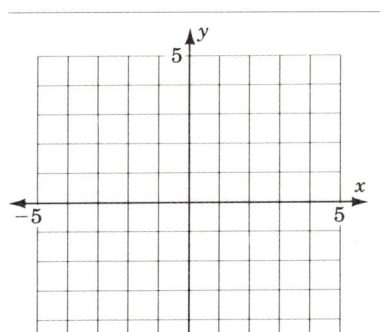

11. Find an equation of the line with slope -2 and passing through the point (2, -3). Then graph this line.

11. _____

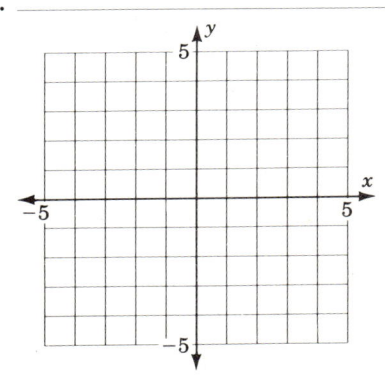

12. A line has slope 3 and y-intercept 2. Find the slope-intercept equation of this line and graph the line.

12. _____

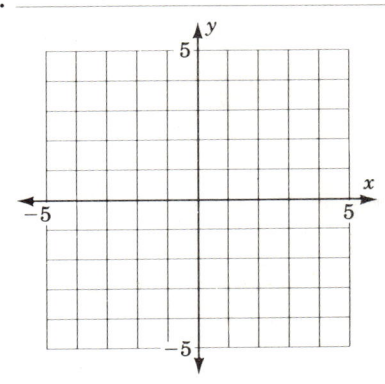

13. Find the slope and the y-intercept of the line $6x + 3y = 12$.

13. _____

14. Find an equation of the line through the point (1, 2) and
   a. Parallel to the line $2x - 3y = 5$
   b. Perpendicular to the line $2x - 3y = 5$

14. **a.** _____
   b. _____

15. Graph $2x - y < -2$.

15.

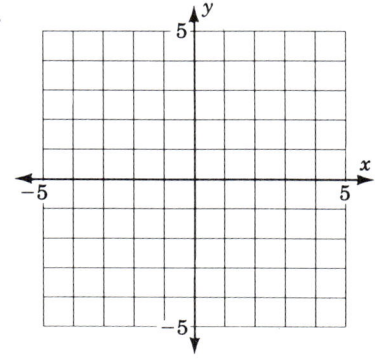

16. Graph $y \leq 2x - 2$.

16.

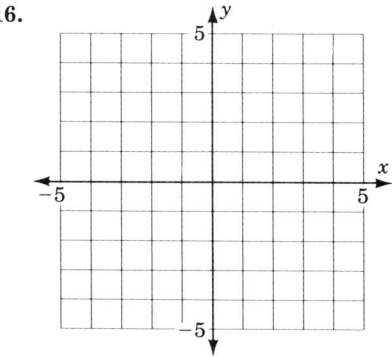

17. Graph $x \geq -3$.

17.

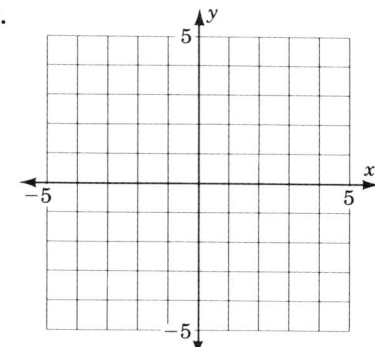

18. Graph $|y| \leq 2$.

18.

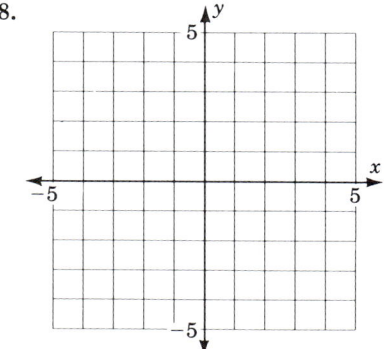

19. Graph $|x - 1| > 2$.

19.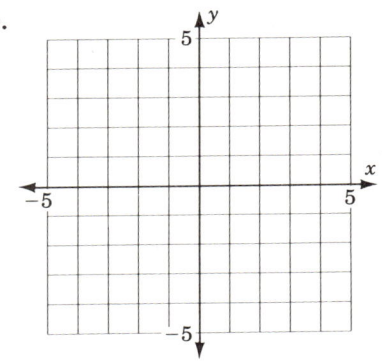

20. An enclosed gas exerts a pressure P on the walls of the container. This pressure is directly proportional to the temperature T of the gas.
   a. Write an equation of variation with k as the constant of variation.
   b. If the pressure is 4 lb/in.^{2} when the temperature is 460°F, find k.

20. a. _____
 b. _____

21. If the temperature of a gas is held constant, the pressure P varies inversely as the volume V.
   a. Write an equation of variation with k as the constant of variation.
   b. A pressure of 1760 lb/in.^{2} is exerted by 4 ft^3 of air in a cylinder fitted with a piston. Find k.

21. a. _____
 b. _____

22. The force F with which the earth attracts an object above the earth's surface varies inversely as the square of the distance d from the center of the earth.
   a. Write an equation of variation with k as the constant of variation.
   b. A meteorite weighs 40 lb on the earth's surface. If the radius of the earth is about 4000 mi, find k.

22. a. _____
 b. _____

23. What would be the weight of the meteorite of Problem 22 at a distance of 1000 mi above the earth's surface?

23. _____

24. The horsepower that a rotating shaft can safely transmit varies jointly as the cube of its diameter d and the number n of revolutions it makes per minute.
   a. Write an equation of variation with k as the constant of variation.
   b. If a 2-in. shaft at a speed of 1400 rev/min can safely transmit 400 hp, find k.

24. a. _____
 b. _____

25. Use the value of k from Problem 24 to find the horsepower that a 4-in. shaft can safely transmit at a speed of 2100 rev/min.

25. _____

IF YOU MISSED QUESTION	SECTION	EXAMPLES	PAGE	ANSWERS
1	7.1A	1	491	**1.**
2	7.1B	2	493–494	**2.**
3	7.1C	3	495	**3.**

1.

2.

$3x - y = 3$

3.

$y = 3x + 2$

x-int. $-\frac{2}{3}$; y-int. 2

IF YOU MISSED QUESTION	SECTION	EXAMPLES	PAGE	ANSWERS
4a	7.1D	4	496	**4. a.**

$3x = -9$

4b	7.1D	4	496	**b.**

$2y = -4$

5a	7.2A	1	506	**5. a.** 5
5b	7.2A	1	506	**b.** $2\sqrt{10}$
5c	7.2A	1	506	**c.** 3
6a	7.2B	2	508–509	**6. a.** $-\frac{1}{3}$
6b	7.2B	2	508–509	**b.** Undefined
6c	7.2B	2	508–509	**c.** -1
6d	7.2B	2	508–509	**d.** 0
7a	7.2C	3	512	**7. a.** Perpendicular
7b	7.2C	3	512	**b.** Neither
8	7.2C	4	512–513	**8.** $y = -5$

IF YOU MISSED QUESTION	SECTION	EXAMPLES	PAGE	ANSWERS
9	7.2D	5	513	**9.**
10	7.3A	1	521–522	**10.** $x + 2y = 10$
11	7.3B	2	522–523	**11.**

IF YOU MISSED QUESTION	SECTION	EXAMPLES	PAGE	ANSWERS
12	7.3C	3	523	**12.** $y = 3x + 2$
13	7.3C	4	524	**13.** $m = -2, y = 4$
14a	7.3D	5	524–525	**14. a.** $2x - 3y = -4$
14b	7.3D	5	524–525	**b.** $3x + 2y = 7$
15	7.4A	1	534–535	**15.**
16	7.4A	2	535	**16.**

IF YOU MISSED QUESTION	SECTION	EXAMPLES	PAGE	ANSWERS
17	7.4A	3	535	17.
18	7.4B	4	536	18.
19	7.4B	5	536–537	19.

IF YOU MISSED QUESTION	SECTION	EXAMPLES	PAGE	ANSWERS
20a	7.5A	1	548	**20. a.** $P = kT$
20b	7.5A	1	548	**b.** $k = \dfrac{1}{115}$
21a	7.5B	2	549	**21. a.** $P = \dfrac{k}{V}$
21b	7.5B	2	549	**b.** $k = 7040$
22a	7.5B	3	549–550	**22. a.** $F = \dfrac{k}{d^2}$
22b	7.5B	3	549–550	**b.** $k = 640{,}000{,}000$
23	7.5B	3	549–550	**23.** 25.6 lb
24a	7.5C	4	550–551	**24. a.** $H = kd^3 n$
24b	7.5C	4	550–551	**b.** $k = \dfrac{1}{28}$
25	7.5C	4	550–551	**25.** 4800 hp

CHAPTER 8

SYSTEMS OF LINEAR EQUATIONS

In this chapter we solve systems of two or three linear equations by using four different methods: graphical, substitution, elimination, and Cramer's rule. In connection with Cramer's rule, we introduce the notion of 2×2 and 3×3 determinants. We end the chapter by studying various applications of these systems.

(Answers on pages 582–583)

ANSWERS

1. Use the graphical method to solve the system.

    $x - 2y = 2$

    $\quad y = x - 3$

1. _____

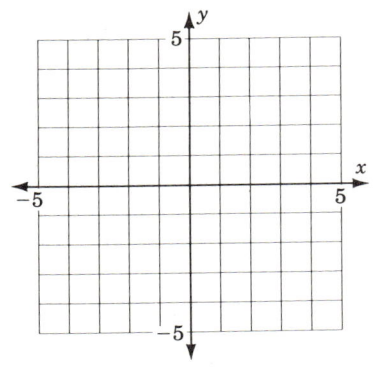

2. Use the graphical method to solve the system.

    $y - 2x = 2$

    $\quad 2y = 4x + 6$

2. _____

3. Use the graphical method to solve the system.

    $2x + 3y = 6$

    $\quad x = 3 - \dfrac{3}{2}y$

3. _____

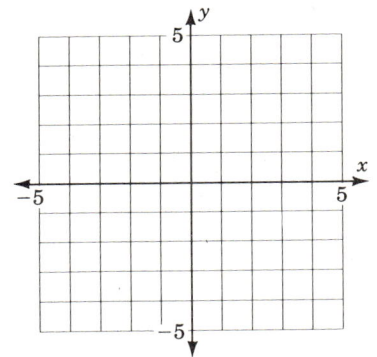

4. Use the substitution method to solve the system.

    $x - 2y = 6$

    $x + \ y = 0$

4. _____

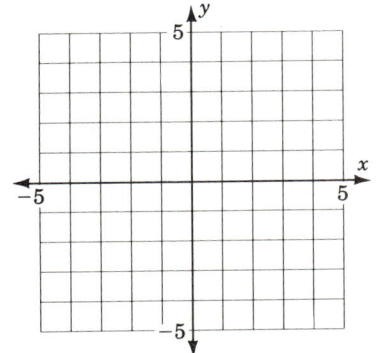

5. Use the substitution method to solve the system.

$x + 2y = 6$

$\quad 4y = -2x$

5. _____

6. Use the substitution method to solve the system.

$2x - y = 5$

$\quad 2y = 4x - 10$

6. _____

7. Solve by the elimination method.

$2x - 3y = 7$

$x + \ y = 1$

7. _____

8. Solve by the elimination method.

$2x - 3y = \ 7$

$4x - 6y = 10$

8. _____

9. Solve the system.

$\dfrac{x}{3} + \dfrac{y}{2} = 2$

$\dfrac{x}{6} + \dfrac{y}{4} = 1$

9. _____

10. Solve the system.

$3x = 10 + 2y$

$3y = 2 - 4x$

10. _____

11. Solve the system.

$x + y + z = \ \ \ 2$

$2x + y - z = -1$

$x + y - z = -2$

11. _____

12. Solve the system.

$x + 2y - 2z = \ \ 4$

$x - \ y + 2z = \ \ 2$

$3x \qquad + 2z = 12$

12. _____

13. Solve the system.

$x + 2y + \ z = \ \ 4$

$2x - \ y + \ z = \ \ 3$

$9x - 2y + 5z = 16$

13. _____

14. Evaluate.

a. $\begin{vmatrix} 2 & 4 \\ -6 & -5 \end{vmatrix}$

b. $\begin{vmatrix} -2 & -4 \\ 3 & 6 \end{vmatrix}$

14. a. _____

  b. _____

15. Solve by Cramer's rule.

$$3x - 2y = 5$$
$$4x - 6y = 7$$

15. $D = \underline{\hspace{1cm}}$, $D_x = \underline{\hspace{1cm}}$,
$D_y = \underline{\hspace{1cm}}$, $x = \underline{\hspace{1cm}}$,
$y = \underline{\hspace{1cm}}$

16. Evaluate.

$$\begin{vmatrix} 2 & -4 & 2 \\ 1 & 0 & 3 \\ 4 & 5 & 2 \end{vmatrix}$$

16. _____

17. Solve by Cramer's rule.

$$x + y + 2z = 6$$
$$x - y + z = 7$$
$$2x + 2y - z = -3$$

17. $D = \underline{\hspace{1cm}}$, $D_x = \underline{\hspace{1cm}}$,
$D_y = \underline{\hspace{1cm}}$, $D_z = \underline{\hspace{1cm}}$,
$x = \underline{\hspace{1cm}}$, $y = \underline{\hspace{1cm}}$,
$z = \underline{\hspace{1cm}}$

18. Solve by Cramer's rule.

$$x + y + 2z = 6$$
$$x - y + z = 7$$
$$3x + y + 5z = 8$$

18. $D = \underline{\hspace{1cm}}$, $D_x = \underline{\hspace{1cm}}$,
$D_y = \underline{\hspace{1cm}}$, $D_z = \underline{\hspace{1cm}}$,
$x = \underline{\hspace{1cm}}$, $y = \underline{\hspace{1cm}}$,
$z = \underline{\hspace{1cm}}$

19. Expand by minors along the first row.

$$\begin{vmatrix} 1 & 1 & -1 \\ 1 & 3 & 2 \\ 3 & 2 & -1 \end{vmatrix}$$

19. _____

20. Expand by minors along the third column.

$$\begin{vmatrix} 0 & 2 & 1 \\ 1 & 1 & -2 \\ 2 & -2 & 1 \end{vmatrix}$$

20. _____

21. Expand by minors along the second row.

$$\begin{vmatrix} 1 & 2 & 0 \\ 0 & 1 & -1 \\ 2 & 1 & 3 \end{vmatrix}$$

21. _____

22. José has $3 in nickels and dimes. He has three more nickels than dimes. How many of each coin does he have?

22. _____

23. The total height of a building and a flagpole on the roof is 180 ft. The building is nine times as high as the flagpole. How high is the building?

23. _____

24. A motorboat can go 8 mi downstream on a river in 20 min. It takes 30 min for this boat to go the same 8 mi upstream. Find the speed of the current.

24. _____

25. Annie has three investments, totaling $50,000. These investments earn interest at 4%, 6%, and 8%, respectively. Annie's total annual income from these investments is $3400. The income from the 8% investment exceeds the total income from the other two investments by $1400. Find how much she has invested at each rate.

25. _____

IF YOU MISSED QUESTION	SECTION	EXAMPLES	PAGE	ANSWERS
1	8.1A, B	1, 4	586–587 589–591	1. $(4, 1)$
2	8.1A, B	2, 4	587–588 589–591	2. Inconsistent; no solution.
3	8.1A, B	3, 4	588–591	3. Dependent; infinitely many solutions.
4	8.1C	5	592–593	4. $(2, -2)$
5	8.1C	6	593	5. Inconsistent; no solution.
6	8.1C	7	593	6. Dependent; infinitely many solutions.
7	8.2A	1	602	7. $(2, -1)$
8	8.2A	2	603	8. Inconsistent; no solution.

IF YOU MISSED QUESTION	SECTION	EXAMPLES	PAGE	ANSWERS
9	8.2A	3	603–604	9. Dependent; infinitely many solutions.
10	8.2B	4	605	10. $(2, -2)$
11	8.3A	1	612–613	11. $(1, -1, 2)$; consistent.
12	8.3A	2	613	12. Inconsistent; no solution.
13	8.3A	3	613–614	13. Dependent; infinitely many solutions.
14a	8.4A	1	622	14. a. 14
14b	8.4A	1	622	b. 0
15	8.4B	2	624	15. $\left(\frac{8}{5}, -\frac{1}{10}\right)$
16	8.4C	3	625	16. -60
17	8.4D	4	626–627	17. $(2, -2, 3)$; consistent.
18	8.4D	5	627	18. Inconsistent; no solution.
19	8.4E	6	628	19. 7
20	8.4E	7	629	20. -14
21	8.4E	7	629	21. 0
22	8.5A	1	636	22. 22 nickels, 19 dimes
23	8.5B	2	637	23. 162 ft
24	8.5C	3	637–638	24. 4 mi/hr
25	8.5D	4	638–639	25. $10,000 at 4%, $10,000 at 6%, $30,000 at 8%

8.1 SYSTEMS OF LINEAR EQUATIONS IN TWO VARIABLES

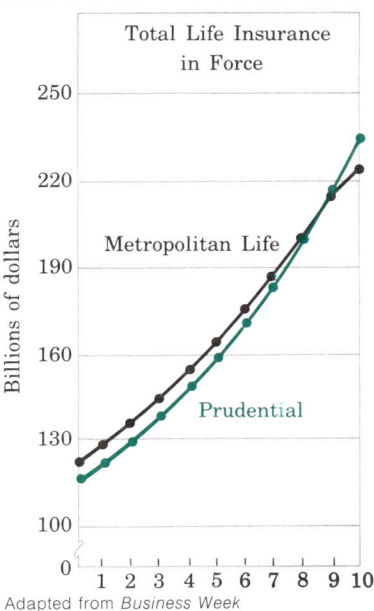

Total Life Insurance in Force

Adapted from *Business Week*

OBJECTIVES

REVIEW

Before starting this section, you should know:

1. How to find the *x*- and *y*-intercepts of a line.
2. How to graph lines.

OBJECTIVES

You should be able to:

A. Find (if possible) the solution of a system of two linear equations using the graphical method.
B. Classify a system of two linear equations as consistent, inconsistent, or dependent.
C. Find the solution of a system of two linear equations using the substitution method.

The graph at the beginning of this section shows the amount (in billions of dollars) of total life insurance in force for a 10-yr period. Can you tell from the graph in what year Metropolitan Life and Prudential had the same amount of insurance in force? This happened during the eighth year. This is because the value representing the total life insurance in force for each of the companies is the same (approximately 200 billion) in the middle of the eighth year. If we agree to denote the year by x and the total life insurance in force by y, the point $(x, y) = (8.5, 200)$ represents the point at which the two curves *intersect*. The coordinates of the point of intersection (if there is one) are the common solution of both equations. Suppose that for company A the total life insurance in force, y, at the end of x years is $y = 10x + 110$, and for company B it is $y = 5x + 120$. We can find the year when both companies had the same amount of life insurance in force by graphing these two equations.

For company A

When $x = 0$, $y = 10(0) + 110 = 110$.
When $x = 5$, $y = 10(5) + 110 = 160$.

We then graph the points (0, 110) and (5, 160) and draw a line through them. See Figure 1 (page 586).

For company B

When $x = 0$, $y = 5(0) + 120 = 120$.
When $x = 5$, $y = 5(5) + 120 = 145$.

We then graph the points (0, 120) and (5, 145) and draw a line through them. The point of intersection seems to be (2, 130). Thus at the end of the second year both companies had $130 billion of life insurance in force.

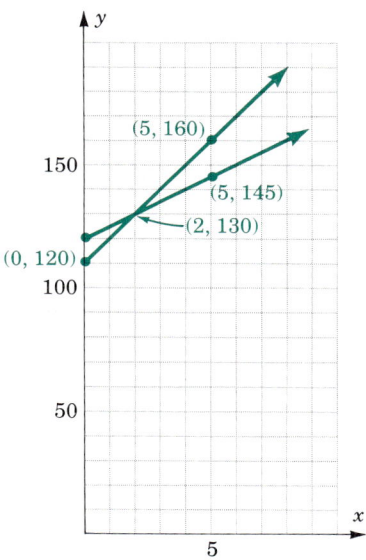

FIG. 1

We can check this by substituting $x = 2$ and $y = 130$ in both equations. $y = 10x + 110$ becomes

$$130 = 10(2) + 110$$
$$= 20 + 110$$
$$= 130$$

a true statement. $y = 5x + 120$ becomes

$$130 = 5(2) + 120$$
$$= 10 + 120$$
$$= 130$$

a true statement.

As you can see from the graph, this is the *only* ordered pair that is a solution of both equations.

A. FINDING SOLUTIONS USING THE GRAPHICAL METHOD

EXAMPLE 1   Use the graphical method to find the solution of the system.

$$2x - y = 2$$
$$y = x - 1$$

Solution   We first graph the equation $2x - y = 2$ using the x- and y-intercepts.

When $x = 0$, $2x - y = 2$ becomes $2(0) - y = 2$, or $y = -2$.

When $y = 0$, $2x - y = 2$ becomes $2x - 0 = 2$, or $x = 1$.

Thus the points $(0, -2)$ and $(1, 0)$ are on the graph. We draw a line through these two points. See Figure 2. The two points and the complete graph are shown in color. We then graph $y = x - 1$.

When $x = 0$, $y = x - 1$ becomes $y = 0 - 1$, or $y = -1$.

When $y = 0$, $y = x - 1$ becomes $0 = x - 1$, or $x = 1$.

Problem 1   Use the graphical method to find the solution of the system.

$$x - y = -1$$
$$y = -x - 1$$

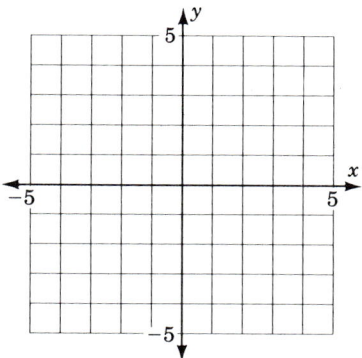

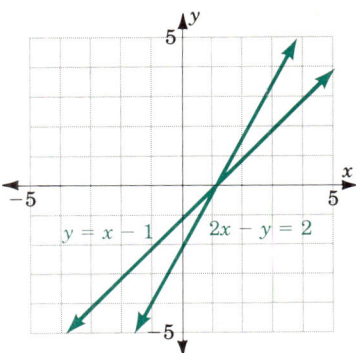

FIG. 2

We then join the points $(0, -1)$ and $(1, 0)$ with a line. The graph for $y = x - 1$ is shown in black in Figure 2. Since the lines intersect at $(1, 0)$, the point $(1, 0)$ is the solution of the system of equations.

Check: For $x = 1$, $y = 0$, $2x - y = 2$ becomes $2(1) - 0 = 2$ (true).
For $x = 1$, $y = 0$, $y = x - 1$ becomes $0 = 1 - 1$ (true).
Thus, $(1, 0)$ is the correct solution for the system.   ▲

EXAMPLE 2   Use the graphical method to find the solution of the system.

$$y - 2x = \ 4$$
$$3y - 6x = 18$$

Solution   We first graph the equation $y - 2x = 4$ by using the x- and y-intercepts shown in the table.

	x	y
$y - 2(0) = 4$, so $y = 4$	0	4
$0 - 2x = 4$, so $x = -2$	-2	0

The two points, as well as the completed graph, are shown in color in the figure. We then graph $3y - 6x = 18$ using the accompanying table.

	x	y
$3y - 6(0) = 18$, so $y = 6$	0	6
$3(0) - 6x = 18$, so $x = -3$	-3	0

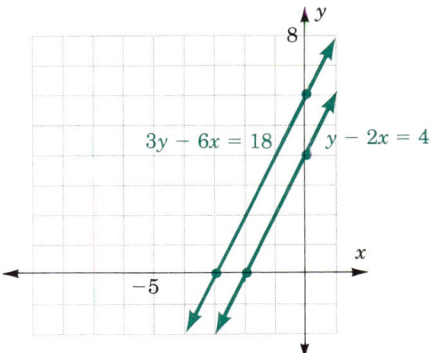

The graph of $3y - 6x = 18$ is shown in black in the figure. The two lines appear to be parallel and not intersecting. If we examine the equations more carefully, we see that by dividing the second equation

Problem 2   Use the graphical method to find the solution of the system.

$$y - 3x = \ 3$$
$$2y - 6x = 12$$

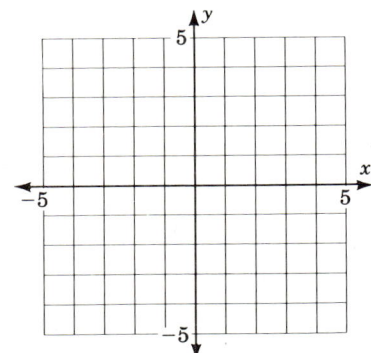

by 3, we get $y - 2x = 6$. Thus one equation says $y - 2x = 4$ and the other says that $y - 2x = 6$. Obviously, both equations cannot be true at the same time and their graphs cannot intersect. Thus there is *no solution* for this system; the system is said to be **inconsistent**. Note that the slope of $y - 2x = 4$ is 2 and the slope of $3y - 6x = 18$ is also 2. The lines are parallel. ▲

EXAMPLE 3 Use the graphical method to solve the system.

$$2x + \frac{1}{2}y = 2$$
$$y = -4x + 4$$

Solution We use the x- and y-intercepts shown in the table to graph $2x + \frac{1}{2}y = 2$.

x	y
0	4
1	0

$2(0) + \frac{1}{2}y = 2$, so $y = 4$
$2x + \frac{1}{2}(0) = 2$, so $x = 1$

The graph of $2x + \frac{1}{2}y = 0$ is shown in color in Figure 3. To graph $y = -4x + 4$, we first let $x = 0$ obtaining $y = 4$. For $y = 0$, $0 = -4x + 4$, or $x = 1$. Thus, the two points in our table will be

x	y
0	4
1	0

But these points are exactly the same as those obtained in the preceding table. What does this mean? It means that the graphs of the lines $2x + \frac{1}{2}y = 2$ and $y = -4x + 4$ **coincide** (are the same). Thus, a solution of one equation is automatically a solution of the other. In fact, there are *infinitely* many solutions. Such a system is called **dependent.**

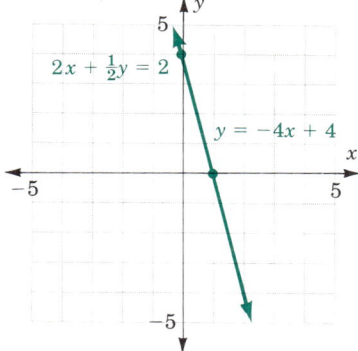

FIG. 3 ▲

B. CONSISTENT, INCONSISTENT, AND DEPENDENT EQUATIONS

As you can see from the examples we have given, a system of equations can have *one* solution (when the lines intersect, as in Example 1), *no* solution (when the lines are parallel, as in Example 2), and *infinitely*

Problem 3 Use the graphical method to solve the system

$$x + \tfrac{1}{2}y = -2$$
$$y = -2x - 4$$

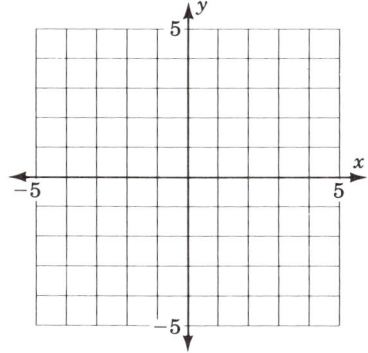

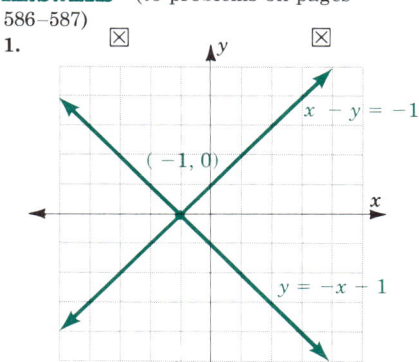
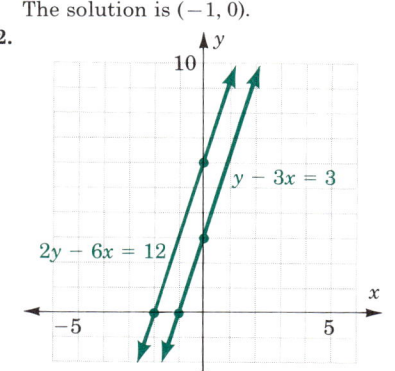

many solutions (when the graphs of the two lines are identical, as in Example 3). These examples illustrate that there are three possibilities when solving a system of simultaneous linear equations in two variables:

1. Consistent and independent equations: The graphs of the equations intersect at one point whose coordinates give the solution of the system.

2. Inconsistent equations: The graphs of the equations are *parallel* lines; there is *no* solution for the system.

3. Dependent equations: The graphs of the equations *coincide* (are the same). There are *infinitely many* solutions for this system.

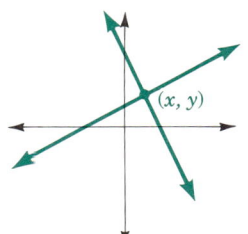

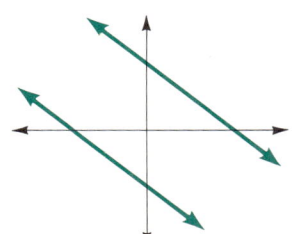

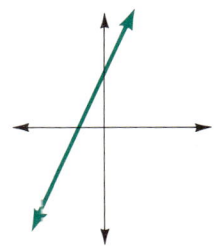

The system is consistent and independent. The solution is (x, y).

The system is inconsistent. There is no solution.

The system is dependent. There are infinitely many solutions.

EXAMPLE 4   Use the graphical method to solve the given system of equations. Classify each system as consistent (one solution), inconsistent (no solution), or dependent (infinitely many solutions).

a. $2x + 2y = 8$ b. $x + 2y = 4$ c. $x + 2y = 4$
    $2y - x = -1$ $2x + 4y = 2$ $4y + 2x = 3$

Solution

a. The table showing the x- and y-intercepts for $2x + 2y = 8$ is

x	y
0	4
4	0

The table showing the x- and y-intercepts for $2y - x = -1$ is

x	y
0	$\dfrac{-1}{2}$
1	0

Problem 4   Use the graphical method to solve the given system. Classify each system as consistent (one solution), inconsistent (no solution), or dependent (infinitely many solutions).

a. $2x + 2y = 8$ b. $2x + y = 4$
    $2y - x = 2$ $2y + 4x = 6$

c. $2x + y = 4$
    $2y + 4x = 8$

The graphs of the two lines are shown. As you can see, the solution is (3, 1). The system is *consistent*.

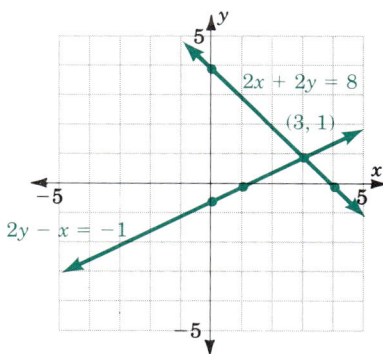

b. The table showing the x- and y-intercepts for $x + 2y = 4$ is

x	y
0	2
4	0

The table showing the x- and y-intercepts for $2x + 4y = 2$ is

x	y
0	$\dfrac{1}{2}$
1	0

The graphs of the two lines are shown. There is no solution because the lines are parallel. (The slope of $x + 2y = 4$ is $-\frac{1}{2}$, the same as that of $2x + 4y = 2$.) The system is *inconsistent*.

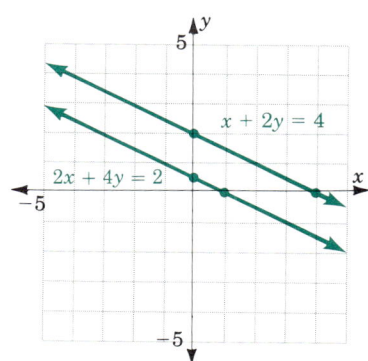

c. The table for $x + 2y = 4$ is

x	y
0	2
4	0

The table for $4y + 2x = 8$ is

x	y
0	2
4	0

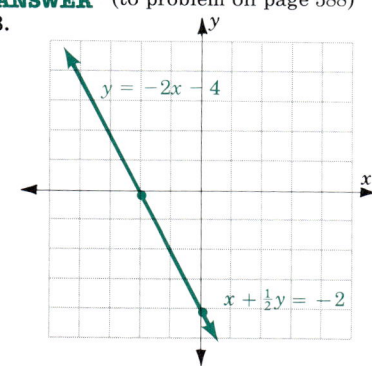

which is identical to the previous table. There are many solutions because the lines coincide (see the graph). The system is dependent.

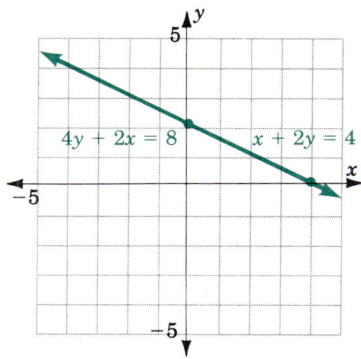

Note that if we multiply both sides of $x + 2y = 4$ by 2, we obtain

$2x + 4y = 8$

or

$4y + 2x = 8$

the second equation. In a *dependent* system, one equation is always a multiple of the other. ▲

C. THE SUBSTITUTION METHOD

In some cases the graphical method is not accurate enough to determine the exact solutions of a system of equations. For example, to solve the system

$y = 2.5x + 72$
$y = 3x + 70$

we may start by letting $x = 0$ in the first equation, obtaining $y = 2.5(0) + 72$, or $y = 72$. To graph this point, we need a piece of graph paper with 72 units, or else we have to make each division on the graph paper 10 units. But there is a way out. Keep in mind that we are looking for a point that satisfies both equations and notice that the second equation tells us that y is $3x + 70$. Thus, we can *substitute* $3x + 70$ for y in the first equation to obtain

$$3x + 70 = 2.5x + 72$$
$$3x = 2.5x + 2 \qquad \text{Subtract 70.}$$
$$0.5x = 2 \qquad \text{Subtract } 2.5x.$$
$$x = \frac{2}{0.5} \qquad \text{Divide by 0.5.}$$
$$x = 4$$

Now, substitute 4 for x in the second equation, obtaining $y = 3(4) + 70 = 82$.

The solution of the system is (4, 82) and the method we used is called the **substitution method.** This method is most useful when both equations are solved for one of the variables in terms of the other (as in the system we just solved) or when at least one of the equations is in

4. a.

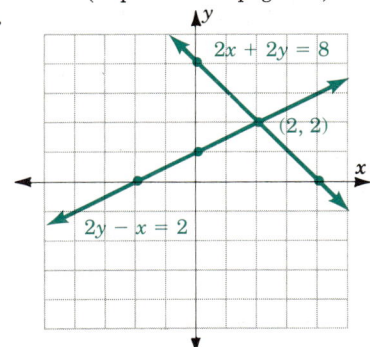

The solution is (2, 2)
The system is consistent.

b.

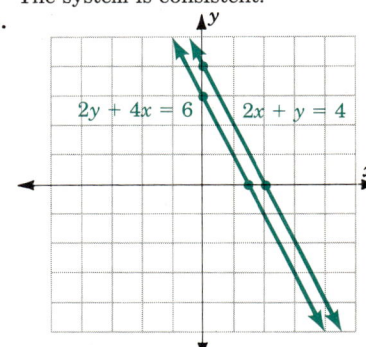

The lines are parallel.
There are no solutions.
The system is inconsistent.

c.

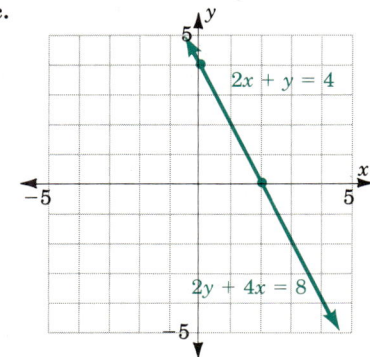

The lines coincide.
There are many solutions.
The system is dependent.

this form, as in Example 1. Thus to solve the system of Example 1 by substitution, write

$$2x - y = 2$$
$$y = x - 1$$

Since $y = x - 1$, we substitute $x - 1$ for y in the first equation, $2x - y = 2$, which becomes

$$2x - (x - 1) = 2$$
$$2x - x + 1 = 2$$
$$x + 1 = 2$$
$$x = 1$$

Writing 1 for x in the second equation, $y = x - 1$, gives

$$y = 1 - 1 = 0$$

Thus the solution is $(1, 0)$, as was shown before.

EXAMPLE 5   Use the substitution method to solve the system.

$$3y + x = 9 \tag{1}$$
$$x - 3y = 0 \tag{2}$$

Solution   We solve Equation 2 for x, obtaining $x = 3y$. Substituting $3y$ for x in Equation 1,

$$x = 3y$$

$$3y + x = 9 \quad \text{becomes} \quad 3y + 3y = 9$$
$$6y = 9$$
$$y = \frac{3}{2}$$

Since $x = 3y$ and $y = \frac{3}{2}$,

$$x = 3\left(\frac{3}{2}\right) = \frac{9}{2}$$

Thus the solution of $3y + x = 9$ and $x - 3y = 0$ is $\left(\frac{9}{2}, \frac{3}{2}\right)$.

Check: $3\left(\frac{3}{2}\right) + \frac{9}{2} = 9$, and $\frac{9}{2} - 3 \cdot \left(\frac{3}{2}\right) = 0$.

The system is consistent.   ▲

How do we recognize that a system such as $x + y = 3$ and $y = -x - 3$ is inconsistent? We can do this with the substitution method as follows. The system is

$$x + y = 3 \tag{1}$$
$$y = -x - 3 \tag{2}$$

We substitute $-x - 3$ for y in Equation 1

$$y = -x - 3$$

$$x + y = 3 \quad \text{becomes} \quad x + (-x - 3) = 3$$
$$-3 = 3$$

Problem 5   Use the substitution method to solve.

$$2y + x = 3$$
$$x - 3y = 0$$

Since this is a contradiction, the system $x + y = 3$ and $y = -x - 3$ is inconsistent; it has *no* solution. In other words, the solution set is $\varnothing$.

EXAMPLE 6 Solve the system

$$x + y = 5 \qquad (1)$$
$$y = -x \qquad (2)$$

Solution Substituting $y = -x$ in Equation 1, we get

$$x + (-x) = 5$$
$$0 = 5$$

Since this is a contradiction, the system $x + y = 5$ and $y = -x$ has *no* solution. The system is inconsistent. ▲

EXAMPLE 7 Solve the system.

$$x + 2y = 4 \qquad (1)$$
$$2x = 8 - 4y \qquad (2)$$

Solution Solving Equation 1 for x, we obtain $x = 4 - 2y$. We now substitute $4 - 2y$ for x in Equation 2.

$$2x = 8 - 4y$$
$$2(4 - 2y) = 8 - 4y$$
$$8 - 4y = 8 - 4y$$

Since this equation is an identity, any value of y will make it true. Thus there are infinitely many solutions. The system is dependent. For example, if $x = 0$ in Equation 1,

$$2y = 4 \quad \text{and} \quad y = 2$$

Thus $(0, 2)$ is a solution of the system. Similarly, if we let $y = 0$ in equation (2),

$$2x = 8 - 4(0) \quad \text{and} \quad x = 4$$

Thus $(4, 0)$ is another solution. You can continue to obtain solutions by assigning numbers to one of the variables in either equation and solving for the other variable. ▲

Problem 6 Solve.

$$x - y = 2$$
$$y = x$$

Problem 7 Solve.

$$x - 3y = 4$$
$$2x = 8 + 6y$$

NAME

CLASS

SECTION

A., B. Solve by graphing. Label each system as consistent (one solution), inconsistent (no solution), or dependent (many solutions).

ANSWERS

1. $x - 2y = 6$

$y = 2x$

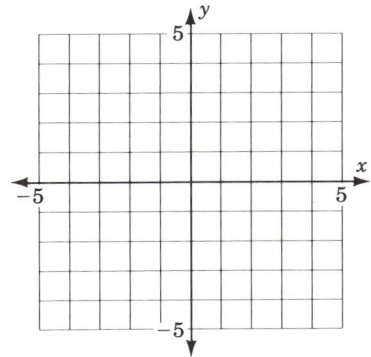

2. $2x + y = 4$

$y = 2x$

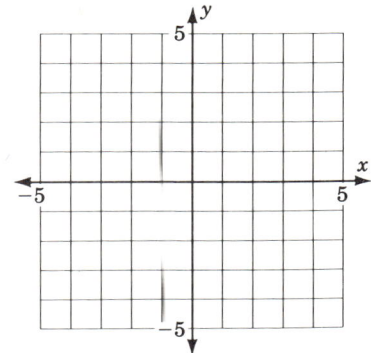

1. _____

2. _____

3. $y = x - 3$

$y = 2x - 4$

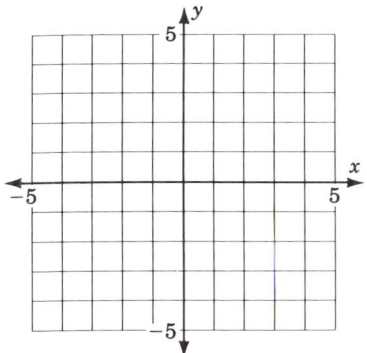

4. $y = x - 1$

$y = 3x - 3$

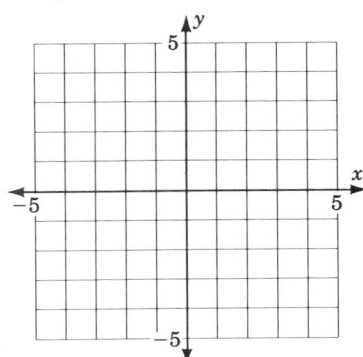

3. _____

4. _____

5. $2y = -x + 4$

$y = -2x + 4$

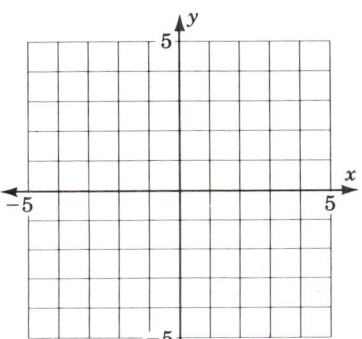

6. $2y = -x + 2$

$y = -2x + 2$

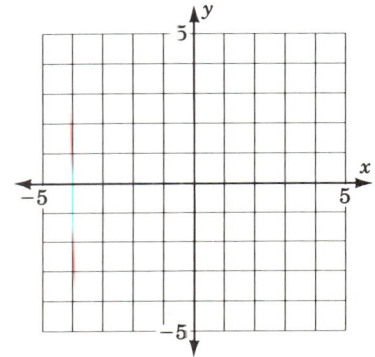

5. _____

6. _____

7. $y = 3x + 3$
$-y = -2x - 2$

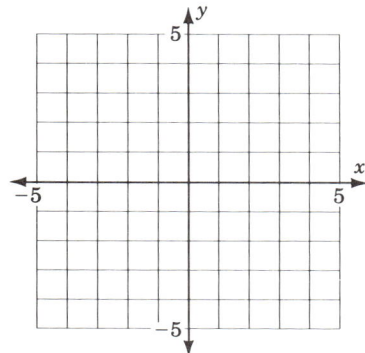

8. $y = x - 1$
$-y = -3x - 1$

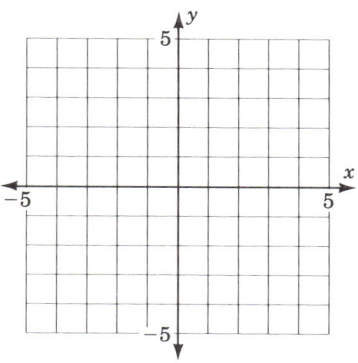

7. _____

8. _____

9. $2x - y = -2$
$y = 2x + 4$

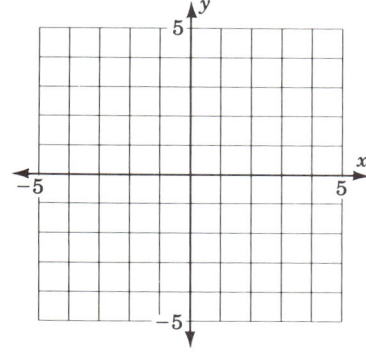

10. $2x + y = -2$
$y = -2x + 4$

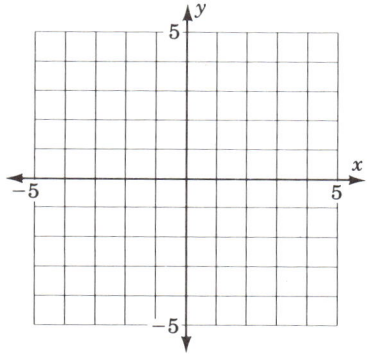

9. _____

10. _____

11. $y = -1$
$2y = x + 2$

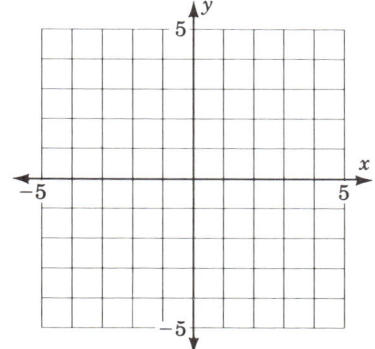

12. $3y = 6 - x$
$y = 3$

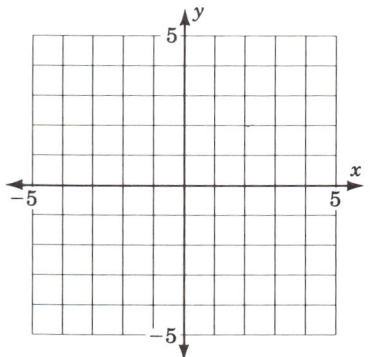

11. _____

12. _____

13. $x = 3$

$\quad\quad y = 2x - 4$

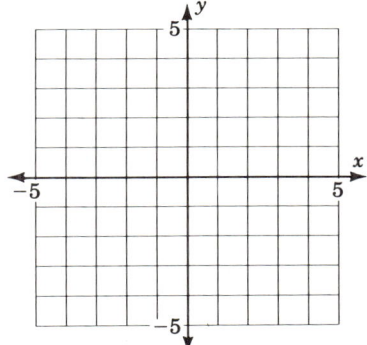

14. $y = -x + 2$

$\quad\quad x = -1$

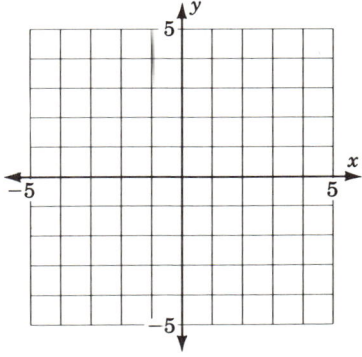

13. _____

14. _____

15. $x + y = 3$

$\quad\quad 2x - y = 0$

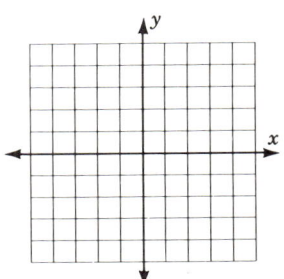

16. $x + y = 5$

$\quad\quad x - 4y = 0$

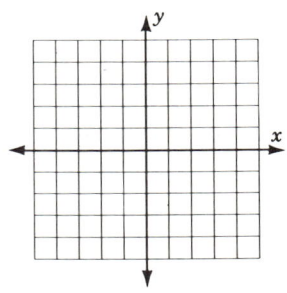

15. _____

16. _____

17. $5x + y = 5$

$\quad\quad 5x = 15 - 3y$

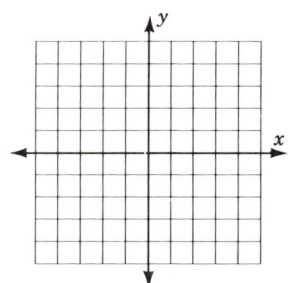

18. $2x - y = -4$

$\quad\quad 4x = 4 + 2y$

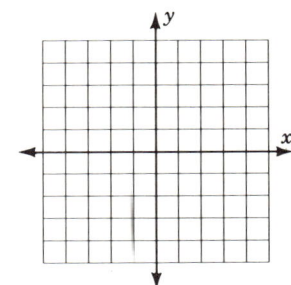

17. _____

18. _____

19. $3x + 4y = 12$

$\quad\quad 8y = 24 - 6x$

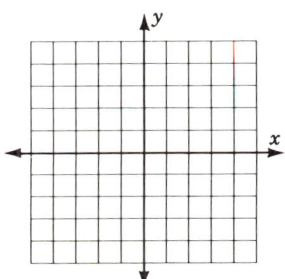

20. $2x - 3y = 6$

$\quad\quad 6x = 18 + 9y$

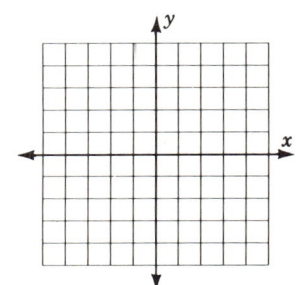

19. _____

20. _____

B., C. Solve by the substitution method. Label each system as consistent (one solution), inconsistent (no solution), or dependent (many solutions).

21. $y = 2x - 4$
    $-2x = y - 4$

22. $y = 2x + 2$
    $-x = y + 1$

23. $x + y = 5$
    $3x + y = 9$

24. $x + y = 5$
    $3x + y = 3$

25. $y - 4 = 2x$
    $y = 2x + 2$

26. $y + 5 = 4x$
    $y = 4x + 7$

27. $x = 8 - 2y$
    $x + 2y = 4$

28. $x = 4 - 2y$
    $x - 2y = 0$

29. $x + 2y = 4$
    $x = -2y + 4$

30. $x + 3y = 6$
    $x = -3y + 6$

31. $x = 2y + 1$
    $y = 2x + 1$

32. $y = 3x + 2$
    $x = 3y + 2$

33. $2x - y = -4$
    $4x = 4 + 2y$

34. $5x + y = 5$
    $5x = 15 - 3y$

35. $x = 5 - y$
    $0 = x - 4y$

36. $x = 3 - y$
    $0 = 2x - y$

D. Applications

37. The supply y of a certain item is given by the equation $y = 2x + 8$, where x is the number of days elapsed. If the demand is given by $y = 4x$, in how many days will the supply equal the demand?

38. The supply of a certain item is $y = 3x + 8$, where x is the number of days elapsed. If the demand is given by $y = 4x$, in how many days will the supply equal the demand?

39. A company has 10 units of a certain item and can manufacture 5 items each day. If the demand for the item is $y = 7x$, in how many days will the demand equal the supply?

40. Clonker Manufacturing has 12 clonkers in stock. They manufacture 3 other clonkers each day. If the clonker demand is 7 each day, in how many days will the supply equal the demand?

✓ **SKILL CHECKER**

Solve.

41. $0.03x = 1.50$

42. $0.05x = 2.50$

21. _____
22. _____
23. _____
24. _____
25. _____
26. _____
27. _____
28. _____
29. _____
30. _____
31. _____
32. _____
33. _____
34. _____
35. _____
36. _____

37. _____

38. _____

39. _____

40. _____

41. _____

42. _____

43. $1 - y = -1$

44. $2 - y = -3$

45. $-8 + 6y = -2$

46. $-6 + 4y = -2$

43. _____

44. _____

45. _____

46. _____

8.1 USING YOUR KNOWLEDGE

The ideas presented in this section are very important in other fields. Use your knowledge to solve the following problems.

1. The total inductance of the inductors L_1 and L_2 in an oscillator must be 400 μH (microhenrys). Thus

$L_1 = -L_2 + 400$

To provide the correct regeneration for the oscillator circuit, the ratio $L_2/L_1 = 4$—that is,

$L_2 = 4L_1$

Solve the system

$L_1 = -L_2 + 400$
$L_2 = 4L_1$

by substitution.

2. The equations for the resistors in a voltage divider must be such that

$\qquad R_1 = 3R_2$
$R_1 + R_2 = 400$

Solve for R_1 and R_2 using the substitution method.

3. The total revenue R for a certain manufacturer is

$R = 5x$

His total cost C is

$C = 4x + 500$

where x represents the number of units produced and sold.
a. Use your knowledge of the substitution method to write the equation that will result when

$R = C$

b. The point at which $R = C$ is called the *break-even* point. Find the number of units the manufacturer must produce and sell in order to break even.

1. _____

2. _____

3. a. _____
   b. _____

CALCULATOR CORNER

If a system of equations has a solution, this solution can be substituted into the original equations and the result checked with a calculator. Thus, in Example 4a, the solution of the system is the ordered pair (3, 1). To check that this is indeed the case, we let $x = 3$ and $y = 1$ in the first equation using the keystrokes

$$\boxed{2}\times\boxed{3}+\boxed{2}\times\boxed{1}=$$

which yields 8, a true statement. The second equation is $2y - x = -1$. When $x = 3$ and $y = 1$, we key in

$$\boxed{2}\times\boxed{1}-\boxed{3}=$$

obtaining -1. (If your calculator does not have algebraic logic, you have to use parentheses around 2×3 and 2×1.) If you can use calculators, check the solutions you obtain in Exercise 8.1.

8.2 SOLVING SYSTEMS OF EQUATIONS BY ELIMINATION

Have you been to a wedding reception or a breakfast buffet lately? How do they figure out how much to charge each person? Olympia Catering has determined that the expenses for a buffet consist of two parts: the fixed costs f (such as labor, utensils, and transportation) and the cost per guest g. For 60 guests, Olympia charges a total of $560. The charge for 40 guests is $440. What is the fixed cost f and the cost per guest g? If we have 60 guests, the total cost is the amount charged for the guests $60g$, plus the fixed costs f. Thus,

$$60g + f = 560 \tag{1}$$

For 40 guests, the total cost is $40g$, plus the fixed costs f. We then have

$$40g + f = 440 \tag{2}$$

To find g and f we must find an ordered pair (g, f) that satisfies the system of equations

$$60g + f = 560 \tag{1}$$
$$40g + f = 440 \tag{2}$$

If we subtract Equation 2 from Equation 1 or, equivalently, multiply Equation 2 by -1 and add, we obtain

$$
\begin{aligned}
60g + f &= 560 \quad &(1)\\
(-)\ 40g + f &= 440 \quad &(2)\\
\hline
20g &= 120
\end{aligned}
$$

The result is an equation in one variable. We have *eliminated* the variable f (thus explaining why this method is called the **elimination method**). Solving for g gives

$$20g = 120$$
$$g = 6$$

Substituting 6 for c in Equation 2,

$$40(6) + f = 440$$
$$240 + f = 440$$
$$f = 200$$

Thus the fixed costs are \$200 and the cost per guest is \$6.

A. SOLVING BY ELIMINATION

EXAMPLE 1   Solve the given system by the elimination method.

$$2x + 3y = 8 \qquad (1)$$
$$x - y = -1 \qquad (2)$$

Solution   This time we multiply the second equation by 3 to eliminate y. We proceed by steps.

STEP 1. Write the system.

$$2x + 3y = 8 \qquad (1)$$
$$x - y = -1 \qquad (2)$$

STEP 2. Multiply Equation 2 by 3:

$$2x + 3y = 8 \quad \leftarrow \text{Same.}$$
$$\underline{3x - 3y = -3} \quad \leftarrow \text{Multiply by 3.}$$

STEP 3.
$$5x = 5 \quad \text{Add.}$$
$$x = 1 \quad \text{Solve.}$$

STEP 4. Substituting $x = 1$ in Equation 2, we get $1 - y = -1$; so $2 = y$.

STEP 5. The solution is $(1, 2)$, as can easily be checked, since

$$2(1) + 3(2) = 8 \text{ and} \qquad (1)$$
$$1 - 2 = -1 \qquad (2)$$

The system is *consistent*.   ▲

In many cases it is not enough to multiply one of the equations by a number to eliminate one of the variables. For example, to find the common solution of

$$2x + 3y = -1 \qquad (1)$$
$$3x + 2y = -4 \qquad (2)$$

we multiply *both* equations by numbers chosen so that the coefficients of one of the variables become opposites in the resulting equations. Thus we have the following steps.

STEP 1. Write the system.

$$2x + 3y = -1 \qquad (1)$$
$$3x + 2y = -4 \qquad (2)$$

STEP 2.
$$4x + 6y = -2 \quad \leftarrow \text{Multiply Equation 1 by 2.}$$
$$\underline{-9x - 6y = 12} \quad \leftarrow \text{Multiply Equation 2 by } -3.$$

STEP 3.
$$-5x = 10 \quad \text{Add.}$$
$$x = -2 \quad \text{Solve.}$$

Problem 1   Solve by elimination.

$$3x + 2y = 8$$
$$x - y = 1$$

STEP 4. Substituting $x = -2$ in (1), we get

$$2(-2) + 3y = -1$$
$$-4 + 3y = -1$$
$$3y = 3$$
$$y = 1$$

STEP 5. The solution of the system is $(-2, 1)$. The system is *consistent*.

EXAMPLE 2 Solve the system.

$$3x + 2y = 6 \tag{1}$$
$$6x + 4y = 10 \tag{2}$$

Solution We proceed by steps as before.

STEP 1. Write the system.

$$3x + 2y = 6 \tag{1}$$
$$6x + 4y = 10 \tag{2}$$

STEP 2. $-6x - 4y = -12 \leftarrow$ Multiply Equation 1 by -2.
$6x + 4y = 10 \leftarrow$ Same.

STEP 3. $ 0 = -2$ Add.

STEP 4. Since we obtained the contradiction $0 = -2$, this system is *inconsistent* and has no solution.

STEP 5. There is no ordered pair (x, y) that is a common solution of both equations. Note that the slope of $3x + 2y = 6$ is $-\frac{3}{2}$, the same as that of the line $6x + 4y = 10$. Thus the lines are parallel. ▲

EXAMPLE 3 Solve the system.

$$\frac{x}{6} + \frac{y}{2} = 1 \tag{1}$$

$$\frac{x}{9} + \frac{y}{3} = \frac{2}{3} \tag{2}$$

Solution Since it is easier to work with integer coefficients, we multiply the first equation by 6, the LCD of $\frac{x}{6}$ and $\frac{y}{2}$, and the second equation by 9, the LCD of $\frac{x}{9}$ and $\frac{y}{3}$.

STEP 1. Write the system.

$$\frac{x}{6} + \frac{y}{2} = 1 \tag{1}$$

$$\frac{x}{9} + \frac{y}{3} = \frac{2}{3} \tag{2}$$

STEP 2. $x + 3y = 6 \leftarrow$ Multiply Equation 1 by 6.
$x + 3y = 6 \leftarrow$ Multiply Equation 2 by 9.

STEP 3. $ 0 = 0$ Subtract.

STEP 4. Since we obtain the true statement $0 = 0$, the equations are

Problem 2 Solve the system.

$$3x + 5y = 2$$
$$6x + 10y = 5$$

Problem 3 Solve.

$$\frac{x}{2} + \frac{y}{6} = 1$$

$$\frac{x}{3} + \frac{y}{9} = \frac{2}{3}$$

dependent. The system has infinitely many solutions. For example, if we substitute 6 for x in Equation 1, we have:

$$\frac{6}{6} + \frac{y}{2} = 1$$

$$1 + \frac{y}{2} = 1$$

$$\frac{y}{2} = 0$$

$$y = 0$$

Thus $(6, 0)$ is a solution. For $x = 12$,

$$\frac{12}{6} + \frac{y}{2} = 1$$

$$2 + \frac{y}{2} = 1$$

$$\frac{y}{2} = -1$$

$$y = -2$$

Thus $(12, -2)$ is another solution.

STEP 5. Note that in step 2, we came up with identical equations. Any time we have two equivalent (or identical) equations, the system is dependent. ▲

B. CONSISTENT, INCONSISTENT, AND DEPENDENT

Before going on, here are two important reminders.

1. There are three possibilities when solving equations by elimination.

> **a. Consistent and independent** equations have *one solution*.
> **b. Inconsistent** equations have *no solution*. You can recognize them when you get a contradiction (a false statement) in your work, as we did in Example 2.
> **c. Dependent** equations have *many solutions*. You can recognize them when you get a true statement, such as $0 = 0$ in Example 3. Remember that any solution of one of these equations is a solution of the other.

2. The second reminder has to do with the position of the variables in the equations. All the equations with which we have worked were written in the form

$$ax + by = c$$
$$dx + ey = f$$

Constant terms

y column

x column

ANSWERS (to problems on pages 602–603)
1. $(2, 1)$; the system is consistent.
2. No solution; the system is inconsistent.
3. Infinitely many solutions; the system is dependent.

If the equations are not in this form, rewrite them using this form. It helps to keep things straight! Practice this in the next example.

EXAMPLE 4 Solve the system.

$$2x = 9 - 5y$$
$$2y = 8 - 3x$$

Solution

STEP 1. We first write the system in standard form—that is, the x's first, then the y's, then the constants. The result is the equivalent system

$$2x + 5y = 9 \tag{1}$$
$$3x + 2y = 8 \tag{2}$$

STEP 2. This time we multiply the first equation by 3 and the second one by -2 so that the x's will cancel upon addition.

$$2x + 5y = 9 \xrightarrow{\text{Multiply by 3}} 6x + 15y = 27$$

$$3x + 2y = 8 \xrightarrow{\text{Multiply by } -2} \underline{-6x - 4y = -16}$$

STEP 3.
$$0 + 11y = 11 \qquad \text{Add.}$$
$$y = 1$$

STEP 4. Substitute $y = 1$ in (1)

$$2x + 5(1) = 9$$
$$2x + 5 = 9$$
$$2x = 4$$
$$x = 2$$

STEP 5. The solution is $(2, 1)$. The system is consistent. You should check this result to make sure it satisfies both equations. ▲

Problem 4 Solve.

$$5x = 6 - 4y$$
$$3y = 4x - 11$$

NAME

CLASS

SECTION

ANSWERS

A., B. Solve each system by elimination. Label the system consistent (one solution), inconsistent (no solution), or dependent (infinitely many solutions).

1. $x + y = 8$
$x - y = 2$

2. $x + y = 3$
$x - y = 1$

3. $x + 4y = 2$
$x - 4y = -2$

4. $x - 5y = 15$
$x + 5y = -5$

5. $-x - 2y = -2$
$x - 2y = -2$

6. $x + 3y = -7$
$-x + 2y = -3$

7. $2x + y = 7$
$3x - 2y = 0$

8. $2x + y = 4$
$3x - 2y = -1$

9. $2x - 2y = 6$
$x + y = 2$

10. $3x - 2y = 0$
$x + y = -5$

11. $3x + 5y = 1$
$-6x - 10y = 2$

12. $5x - 2y = 4$
$-10x + 4y = 1$

13. $2x + y = 8$
$3x - y = 7$

14. $x - 3y = -2$
$x + 3y = 4$

15. $2x + 5y = 9$
$3x + 2y = 8$

16. $3x + 5y = 26$
$5x + 3y = 22$

17. $6x + 5y = 12$
$9x - 4y = -5$

18. $5x + 4y = 6$
$4x - 3y = 11$

19. $2x - 3y = 16$
$x - y = 7$

20. $3x - 2y = 35$
$x - 5y = 42$

21. $18x - 15y = 1$
$10x - 12y = 3$

22. $6x - 9y = -2$
$3x - 5y = -6$

23. $\dfrac{x}{3} + \dfrac{y}{6} = \dfrac{2}{3}$

$\dfrac{2}{5}x + \dfrac{y}{4} = \dfrac{1}{5}$

24. $\dfrac{x}{6} + \dfrac{y}{3} = \dfrac{1}{2}$

$\dfrac{3}{5}x + \dfrac{y}{4} = \dfrac{17}{20}$

25. $\dfrac{5}{6}x + \dfrac{y}{4} = 7$

$\dfrac{2}{3}x - \dfrac{y}{8} = 3$

26. $\dfrac{1}{5}x + \dfrac{2}{5}y = 1$

$\dfrac{1}{4}x - \dfrac{1}{3}y = \dfrac{-5}{12}$

1. _____

2. _____

3. _____

4. _____

5. _____

6. _____

7. _____

8. _____

9. _____

10. _____

11. _____

12. _____

13. _____

14. _____

15. _____

16. _____

17. _____

18. _____

19. _____

20. _____

21. _____

22. _____

23. _____

24. _____

25. _____

26. _____

27. $\dfrac{2}{x} + \dfrac{3}{y} = \dfrac{-1}{2}$

$\dfrac{3}{x} - \dfrac{2}{y} = \dfrac{17}{12}$

(*Hint:* Multiply the first equation by 2, and the second by 3 and add.)

27. _____

28. $\dfrac{4}{x} + \dfrac{2}{y} = \dfrac{26}{21}$

$\dfrac{2}{x} - \dfrac{1}{y} = \dfrac{-1}{21}$

(*Hint:* Multiply the second equation by 2 and add.)

28. _____

29. $\dfrac{2}{x} - \dfrac{1}{y} = 0$

$\dfrac{3}{x} + \dfrac{5}{y} = \dfrac{13}{4}$

(*Hint:* Multiply the first equation by 5 and add.)

29. _____

30. $\dfrac{1}{x} - \dfrac{3}{y} = \dfrac{-13}{10}$

$\dfrac{5}{x} + \dfrac{2}{y} = 2$

(*Hint:* Multiply the first equation by -5 and add.)

30. _____

C. Applications

31. According to the *Guinness Book of World Records,* the sum of the ages of the two oldest cats is 70 yr. The difference of their ages is 2 yr. What are their ages?

31. _____

32. According to the *Guinness Book of World Records,* the sum of the heights of the shortest and tallest persons on record (measured in inches) is 130. If the difference in their heights was 84 in., how tall was each?

32. _____

33. The height of the Empire State Building and its antenna is 1472 ft. The difference in height between the building and the antenna is 1028 ft. How tall is the antenna and how tall is the building?

33. _____

34. The height of the Eiffel Tower and its antenna is 1052 ft. The difference in height between the tower and the antenna is 920 ft. How tall is the antenna and how tall is the tower?

34. _____

35. At one time, the combined weight of the McCreary Brothers was 1300 lb. Their weight difference was 20 lb. What was the weight of each of the McCreary Brothers?

35. _____

✓ **SKILL CHECKER**

Find the x- and y-intercepts of each line.

36. $2x + y = 6$

37. $3x + 2y = 12$

38. $-2x + 3y = 9$

39. $-3x + 2y = 4$

40. $-2x - 3y = 7$

36. _____

37. _____

38. _____

39. _____

40. _____

8.2 USING YOUR KNOWLEDGE

Have you read *Alice in Wonderland*? Do you know who the author of this book is? The answer is Lewis Carroll. Although better known as the author of *Alice in Wonderland,* Lewis Carroll was also a mathematician and logician. He also wrote another book called *Through the Looking Glass.* In this book, one of the characters, Tweedledee, is talking to Tweedledum. Here is the conversation.

Tweedledee: The sum of your weight and twice mine is 361 pounds.

Tweedledum: Contrariwise, the sum of your weight and twice mine is 360 pounds.

1. If Tweedledee weighs x pounds and Tweedledum weighs y pounds, can you use the knowledge gained in this section to find their weights?

1. _____

8.3 SOLVING SYSTEMS OF EQUATIONS IN THREE VARIABLES

OBJECTIVES

REVIEW

Before starting this section, you should know:

1. How to solve linear equations.
2. How to evaluate an expression.
3. How to solve a system of two equations in two unknowns.

OBJECTIVES

You should be able to:

A. Solve a system of three equations and three unknowns by the elimination method.
B. Determine if a system of three equations in three unknowns is consistent, inconsistent, or dependent.

The man in the photograph has coffees worth \$1.30, \$1.40, and \$1.50 per pound. If he calls these coffees A, B, and C and decides to make 50 lb of a mixture containing x pounds of A, y pounds of B, and z pounds of C, then

$$x + y + z = 50 \tag{1}$$

Assume that he decides to have twice as much Brand B as C. Thus

$$y = 2z \tag{2}$$

Finally, if he sells the 50 lb at \$1.42 per pound, the total price will be \$71.00, so that

$$1.30x + 1.40y + 1.50z = 71.00 \tag{3}$$

Now we rewrite Equations 1, 2, and 3 in standard form:

$$x + \quad y + \quad z = \ 50 \tag{4}$$
$$y - \ 2z = \ \ 0 \quad \text{Subtract } 2z \text{ from both members of (2).} \tag{5}$$
$$13x + 14y + 15z = 710 \quad \text{Multiply each member of (3) by 10.} \tag{6}$$

We have obtained a system of linear equation in three unknowns. There are many ways to solve this system, but the idea is to take two *different* pairs of equations and eliminate the same variable from each pair. We first note that Equation 2 *does not* contain x. Because of this fact, we select Equations 4 and 6 and eliminate x by multiplying Equation 4 by -13 and adding Equation 6, as follows:

$$-13x - 13y - 13z = -650 \tag{7}$$
$$13x + 14y + 15z = \ \ \ 710 \tag{6}$$
$$\overline{\qquad\qquad y + \ \ 2z = \ \ \ \ 60} \quad \text{Add.} \tag{8}$$

The new system, consisting of Equations 5 and 8, is a system of two equations in two unknowns. We solve it using the techniques of Section 8.2. To eliminate z from this system, we add Equations 5 and 8, obtaining

$$y - 2z = \ 0 \tag{5}$$
$$y + 2z = 60 \tag{8}$$
$$\overline{2y \qquad = 60} \quad \text{Add.} \tag{9}$$
$$y = 30$$

Substituting $y = 30$ in Equation 5 gives

$$y - 2z = 0$$
$$30 - 2z = 0$$
$$z = 15$$

Putting $y = 30$ and $z = 15$ in Equation 4 gives

$$x + 30 + 15 = 50$$
$$x + 45 = 50$$
$$x = 5$$

Thus, the solution for the system is the ordered triple $(5, 30, 15)$. You can check this by substituting 5 for x, 30 for y, and 15 for z in each of the original equations.

A. SOLVING EQUATIONS BY ELIMINATION

As you can see from this example, to solve a system of three linear equations in three unknowns, we can proceed as follows.

PROCEDURE FOR SOLVING BY ELIMINATION

1. Select a pair of equations and eliminate one variable from this pair.

2. Select a *different* pair of equations and eliminate the same variable as in step 1.

3. Solve the pair of equations resulting from steps 1 and 2. (Use the procedure outlined in Section 8.2.)

4. Substitute the values found in step 3 in the simplest of the original equations, and then solve for the third variable.

5. Check by substituting the values in each of the original equations.

EXAMPLE 1   Solve the system.

$$x + y + z = 12 \qquad (1)$$
$$2x - y + z = 7 \qquad (2)$$
$$x + 2y - z = 6 \qquad (3)$$

If we add these two, z cancels.
If we add these two, z cancels.

Solution   It is easiest to eliminate z from the two pairs of equations, (1) and (3), and (2) and (3).

STEP 1. Adding (1) and (3), we obtain

$$2x + 3y = 18 \qquad (4)$$

STEP 2. Adding (2) and (3) we have

$$3x + y = 13 \qquad (5)$$

STEP 3. We now have the system

$$2x + 3y = 18 \qquad (4)$$
$$3x + y = 13 \qquad (5)$$

Problem 1   Solve the system.

$$x + y + z = 4$$
$$x - y + z = 2$$
$$2x + 2y - z = -4$$

Multiplying Equation 5 by -3, we have

$$\begin{aligned} 2x + 3y &= 18 \quad &(4) \\ -9x - 3y &= -39 \quad &(6) \\ \hline -7x &= -21 \quad &\text{Add.} \\ x &= 3 \end{aligned}$$

Substituting 3 for x in (5), we get $9 + y = 13$, or $y = 4$.

STEP 4. In Equation 1, $x + y + z = 12$. Since we know now that x is 3 and that y is 4 we can substitute these values in Equation 1, to obtain $3 + 4 + z = 12$. Solving, we find $z = 5$.

STEP 5. The solution of the system is $(3, 4, 5)$, as can easily be verified:

$$\begin{aligned} 3 + 4 + 5 &= 12 \quad &(1) \\ 2(3) - 4 + 5 &= 7 \quad &(2) \\ 3 + 2(4) - 5 &= 6 \quad &(3) \end{aligned}$$

The system is consistent. ▲

EXAMPLE 2 Solve the system.

$$\begin{aligned} x + 3y - z &= 1 \quad &(1) \\ x - y + z &= 4 \quad &(2) \\ 3x + y + z &= 3 \quad &(3) \end{aligned}$$

If we add these two, z cancels.
If we add these two, z cancels.

Solution Adding first (1) and (2) and then (1) and (3), we obtain

$$2x + 2y = 5 \quad \text{This is the sum of (1) and (2).} \quad (4)$$
$$4x + 4y = 4 \quad \text{This is the sum of (1) and (3).} \quad (5)$$

Multiplying (4) by -2 to eliminate x, we have

$$\begin{aligned} -4x - 4y &= -10 \quad &\text{This is } -2(2x + 2y) = -2 \cdot 5. \quad &(6) \\ 4x + 4y &= 4 \quad & &(5) \\ \hline 0 &= -6 \quad &\text{Add.} \quad &(7) \end{aligned}$$

Since it is impossible for 0 to be equal to -6, this system has *no solution;* it is an *inconsistent system.* ▲

EXAMPLE 3 Solve the system.

$$\begin{aligned} x - 2y + 3z &= 4 \quad &(1) \\ 2x - y + z &= 1 \quad &(2) \\ x + y - 2z &= -3 \quad &(3) \end{aligned}$$

Solution

STEP 1. Multiplying (2) by -2 and adding to (1) (to eliminate y), we get

$$\begin{aligned} x - 2y + 3z &= 4 \quad &(1) \\ -4x + 2y - 2z &= -2 \quad &(4) \\ \hline -3x + z &= 2 \quad &(5) \end{aligned}$$

STEP 2. Adding (2) and (3) to eliminate y, we obtain

$$\begin{aligned} 2x - y + z &= 1 \quad &(2) \\ x + y - 2z &= -3 \quad &(3) \\ \hline 3x - z &= -2 \quad &(6) \end{aligned}$$

Problem 2 Solve the system.

$$\begin{aligned} x + 8y - z &= 8 \\ -x + 2y - z &= 4 \\ 2x + y + z &= 2 \end{aligned}$$

Problem 3 Solve the system.

$$\begin{aligned} x + 2y + z &= -10 \\ x + y - z &= -3 \\ 5x + 7y - z &= -29 \end{aligned}$$

STEP 3. We now have the system

$$-3x + z = 2 \tag{5}$$
$$\underline{3x - z = -2} \tag{6}$$
$$0 = 0 \quad \text{Add.} \tag{7}$$

Thus the system is *dependent* and has infinitely many solutions. One such solution is obtained if we let $x = 0$ in (5), obtaining $-3(0) + z = 2$ or $z = 2$.

STEP 4. Substituting $x = 0$ and $z = 2$ in (2), we have

$$2 \cdot 0 - y + 2 = 1 \tag{2}$$
$$-y + 2 = 1$$
$$y = 1$$

So $(0, 1, 2)$ is *one* of the solutions for the system, as can be easily checked. ▲

B. CONSISTENT, INCONSISTENT, AND DEPENDENT

As was the case with the solution of two equations in two unknowns, the solution of three equations in three unknowns can produce three different possibilities.

> The system is *consistent* and *independent,* as in Example 1; it has one solution consisting of an ordered triple (x, y, z).
>
> The system is *inconsistent,* as in Example 2. It has no solution.
>
> The system is *dependent,* as in Example 3. It has infinitely many solutions.

In the case of two unknowns, the fact that a system is *consistent* tells us that if we graph the lines associated with the system, the lines *intersect.* If we graph a linear equation in three unknowns, the graph is a plane. Thus if three linear equations in three unknowns have a solution, it means that the three planes corresponding to their equations intersect at a point, as shown in Figure 4.

If the equations are *inconsistent,* the planes do not intersect at a common point, as shown in Figures 5 and 6. Finally, if the equations

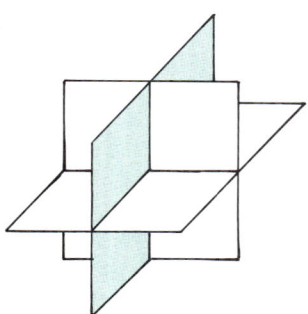

FIG. 4

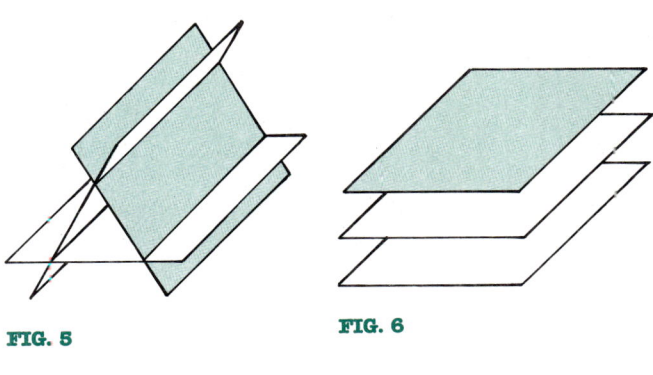

FIG. 5

FIG. 6

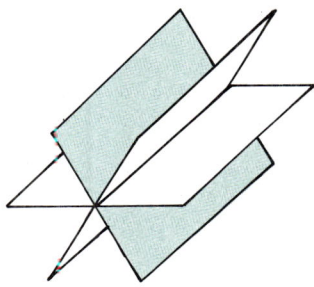

FIG. 7

are *dependent,* the three planes intersect in a line, as in Figure 7. Any point on the line is a solution; consequently, there are infinitely many solutions.

NAME

CLASS

SECTION

A., B. In Problems 1–20 solve the system. Label the system as consistent, inconsistent, or dependent.

ANSWERS

1. $x + y + z = 12$
$x - y + z = 6$
$x + 2y - z = 7$

2. $x + y + z = 13$
$x - 2y + 4z = 10$
$3x + y - 3z = 5$

3. $x + y + z = 4$
$2x + 2y - z = -4$
$x - y + z = 2$

4. $2x - y + z = 3$
$x + 4y - z = 6$
$3x + 2y + 3z = 16$

5. $2x - y + z = 3$
$x + 2y + z = 12$
$4x - 3y + z = 1$

6. $x - 3y - 2z = -12$
$2x + y - 3z = -1$
$3x - 2y - z = -5$

7. $x - 2y - 3z = 2$
$x - 4y - 13z = 14$
$-3x + 5y + 4z = 2$

8. $2x + 2y + z = 3$
$-x + y - z = 5$
$3x + 5y + z = 8$

9. $2x + 4y + 3z = 3$
$10x - 8y - 9z = 0$
$4x + 4y - 3z = 2$

10. $9x + 4y - 10z = 6$
$6x - 8y + 5z = -1$
$12x + 12y - 15z = 10$

11. $x - 2y - z = 3$
$2x - 5y + z = -1$
$x - 2y - z = -3$

12. $x - 3y - 6z = -8$
$3x - 2y - 10z = 11$
$5x - 6y - 2z = 7$

13. $2x + y + z = 5$
$-x + 2y - z = 3$
$3x + 4y + z = 10$

14. $x - 3y + z = 2$
$x + 2y - z = 1$
$-7x + y + z = -10$

15. $x + y = 5$
$y + z = 3$
$x + z = 7$

16. $x + 2y = -1$
$2y + z = 0$
$x + 2z = 11$

17. $x - 2y = 0$
$y - 2z = 5$
$x + y + z = 8$

18. $2y + z = 9$
$z - 2y = 1$
$x + y + z = 1$

19. $5x - 3z = 2$
$2z - y = -5$
$x + 2y - 4z = 8$

20. $5x - 2z = 1$
$3z - y = 6$
$x + 2y - z = -1$

ANSWERS

1. _____

2. _____

3. _____

4. _____

5. _____

6. _____

7. _____

8. _____

9. _____

10. _____

11. _____

12. _____

13. _____

14. _____

15. _____

16. _____

17. _____

18. _____

19. _____

20. _____

21. Find a condition on a, b, and c so that the system

$$-4x + 3y = a$$
$$5x - 4y = b$$
$$-3x + 2y = c$$

has a solution.

21. _____

22. Show that the system

$$2x + 4z = 6 \qquad (1)$$
$$3x + y + z = -1 \qquad (2)$$
$$ 2y - z = -2 \qquad (3)$$
$$x - y + z = -5 \qquad (4)$$

does not have a solution. (*Hint:* Solve the system consisting of Equations 1, 2, and 4 and then show that the solution does not satisfy Equation 3.)

22. _____

23. Find the solution set of

$$2x + 4z = 6$$
$$3x + y + z = -1$$
$$ 2y - z = -2$$
$$x - y - 2z = -5$$

23. _____

24. Find a value of k so that the system

$$5x - y + 2z = 2 \qquad (1)$$
$$3x + y - 3z = 7 \qquad (2)$$
$$x + 5y + z = 5 \qquad (3)$$
$$x + ky - z = 9 \qquad (4)$$

has a solution. (*Hint:* Solve the system consisting of Equations 1, 2, and 3. Then substitute the values of x, y, and z in Equation 4 and solve for k.)

24. _____

25. Repeat Problem 24 for the system

$$2x + 4z = 6 \qquad (1)$$
$$3x + y + z = -1 \qquad (2)$$
$$ 2y - z = -2 \qquad (3)$$
$$x - y + kz = -5 \qquad (4)$$

25. _____

✓ SKILL CHECKER

Evaluate.

26. $(13)(-4) - (21)(3)$

27. $(2)(-4) - (5)(93)$

28. $\dfrac{(5)(-1) - (3)(1)}{(1)(-1) - (3)(1)}$

29. $\dfrac{(1)(3) - (3)(5)}{(1)(-1) - (1)(93)}$

30. $\dfrac{(-3)(1) - (-2)(5)}{(2)(1) - (-1)(-3)}$

26. _____

27. _____

28. _____

29. _____

30. _____

EXERCISE 8.3

8.3 USING YOUR KNOWLEDGE

Can you tell how hot it is by listening to the crickets? You can if you know the formula! The number of chirps a cricket makes per minute is given by

$$F = \frac{n}{4} + 40 \tag{1}$$

where F is the number of degrees in the Fahrenheit temperature. Ants are also affected by temperature changes. As a matter of fact, the crawling speed d (in centimeters per second) of a certain ant is given by

$$C = 6d + 4 \tag{2}$$

where C is the number of degrees in the Celsius temperature. The relationship between C and F is given by

$$C = \frac{5}{9}(F - 32) \tag{3}$$

1. Use the substitution method with Equations 2 and 3 to solve for F in terms of d.

2. Substitute the expression obtained for F in Problem 1 into Equation 1 to find the relationship between d and n.

3. If you got the answer to Problem 2, you can answer this question: If the cricket is chirping 112 times a minute, how fast is the ant crawling?

1. _____

2. _____

3. _____

OBJECTIVES

REVIEW

Before starting this section, you should know how to evaluate an expression involving integers.

OBJECTIVES

You should be able to:

A. Evaluate a 2×2 determinant.
B. Use determinants to solve a system of two equations in two unknowns.
C. Evaluate a 3×3 determinant.
D. Use determinants to solve a system of three equations in three unknowns.
E. Expand determinants by minors.

The illustration at the top of this page shows Gottfried Wilhelm Leibniz, the originator of the theory of determinants. In 1693, Leibniz studied and used determinants to solve systems of simultaneous equations. A **determinant** is a square array of numbers of the form

$$\begin{vmatrix} a_1 & b_1 \\ a_2 & b_2 \end{vmatrix}$$

The numbers a_1, a_2, b_1, and b_2 are called the *elements* of the determinant. As you can see, this determinant has *two* rows and *two* columns. For this reason,

$$\det A = \begin{vmatrix} a_1 & b_1 \\ a_2 & b_2 \end{vmatrix}$$

is called a two-by-two (2×2) determinant.

A.   EVALUATING DETERMINANTS

The value of det A is defined to be

$$a_1 b_2 - a_2 b_1$$

which can be obtained by multiplying along the diagonals, as indicated by the diagram in the following definition.

$$\begin{array}{c} + \\ \begin{vmatrix} a_1 & b_1 \\ a_2 & b_2 \end{vmatrix} \end{array} \,{}^{-} = a_1b_2 - a_2b_1$$

For example,

$$\begin{array}{c} + \\ \begin{vmatrix} 2 & 5 \\ -3 & -9 \end{vmatrix} \end{array} \,{}^{-} = (2)(-9) - (-3)(5) = -18 + 15 = -3$$

EXAMPLE 1 Evaluate.

a. $\begin{vmatrix} -3 & 7 \\ -5 & 4 \end{vmatrix}$ **b.** $\begin{vmatrix} -3 & 6 \\ -5 & 10 \end{vmatrix}$

Solution

a. $\begin{vmatrix} -3 & 7 \\ -5 & 4 \end{vmatrix} = (-3)(4) - (-5)(7) = -12 - (-35) = 23$

b. $\begin{vmatrix} -3 & 6 \\ -5 & 10 \end{vmatrix} = (-3)(10) - (-5)(6) = -30 - (-30) = 0$ ▲

Problem 1 Evaluate.

a. $\begin{vmatrix} -2 & 6 \\ -3 & 5 \end{vmatrix}$ **b.** $\begin{vmatrix} -2 & 8 \\ -3 & 12 \end{vmatrix}$

B. CRAMER'S RULE FOR A SYSTEM OF TWO EQUATIONS

Notice in Example 1(b) that the second-column elements of the determinant are both the same multiple of the corresponding first-column elements:

$$6 = (-2)(-3) \quad \text{and} \quad 10 = (-2)(-5)$$

In general, if the elements of one row are just some constant k times the elements of the other row (or if this is true of the columns), then the value of the determinant is 0. This is very easy to see as follows.

$$\begin{vmatrix} a & b \\ ka & kb \end{vmatrix} = kab - kab = 0$$

This result will be of use to us in Using Your Knowledge 8.5.

One of the important applications of determinants is in the solution of a system of linear equations. Let us look at the simplest case, two equations in two unknowns. Such a system can be written

$$a_1x + b_1y = d_1 \tag{1}$$
$$a_2x + b_2y = d_2 \tag{2}$$

We can eliminate y from this system by multiplying the first equation by b_2 and the second equation by b_1 and then subtracting, as follows.

$$a_1 b_2 x + b_1 b_2 y = d_1 b_2 \qquad \text{This is } b_2 \text{ times equation (1).}$$
$$\underline{(-)\, a_2 b_1 x + b_1 b_2 y = d_2 b_1} \qquad \text{This is } b_1 \text{ times equation (2).}$$
$$a_1 b_2 x - a_2 b_1 x = d_1 b_2 - d_2 b_1 \qquad \text{Subtract the second equation from the first.}$$

$$(a_1 b_2 - a_2 b_1)x = d_1 b_2 - d_2 b_1 \qquad \text{Factor out } x.$$

Now, if the quantity $a_1 b_2 - a_2 b_1 \neq 0$, we can divide by this quantity to get

$$x = \frac{d_1 b_2 - d_2 b_1}{a_1 b_2 - a_2 b_1} \qquad \text{Solve for } x.$$

Notice that the denominator, which we shall denote by D, can be written as the determinant

$$D = \begin{vmatrix} a_1 & b_1 \\ a_2 & b_2 \end{vmatrix}$$

This determinant is naturally called the **determinant of the coefficients.** Notice also that the numerator can be obtained from the denominator by replacing the coefficients of x (the a's) by the corresponding constant terms, the d's. We denote the numerator of x by D_x. Thus

constant terms replace the a's $\qquad D_x = \begin{vmatrix} d_1 & b_1 \\ d_2 & b_2 \end{vmatrix} \qquad$ y coefficients unchanged

We can now write the solution for x in the form

$$x = \frac{D_x}{D}$$

A similar procedure shows that the solution for y is

$$y = \frac{a_1 d_2 - a_2 d_1}{a_1 b_2 - a_2 b_1}$$

Notice that the denominator is again the determinant D. The numerator, which we shall denote by D_y, can be formed from D by replacing the coefficients of y (the b's) by the corresponding d's. Hence,

x coefficients unchanged $\qquad D_y = \begin{vmatrix} a_1 & d_1 \\ a_2 & d_2 \end{vmatrix} \qquad$ constant terms replace the b's

and we can write the solution for y in the form

$$y = \frac{D_y}{D}$$

We summarize these results as follows.

EXAMPLE 2 Use Cramer's rule to solve the system.

$$2x + 3y = 7$$
$$5x + 9y = 11$$

Solution

$$D = \begin{vmatrix} 2 & 3 \\ 5 & 9 \end{vmatrix} = 18 - 15 = 3$$

$$D_x = \begin{vmatrix} 7 & 3 \\ 11 & 9 \end{vmatrix} = 63 - 33 = 30$$

$$D_y = \begin{vmatrix} 2 & 7 \\ 5 & 11 \end{vmatrix} = 22 - 35 = -13$$

Therefore,

$$x = \frac{D_x}{D} = \frac{30}{3} = 10 \qquad y = \frac{D_y}{D} = \frac{-13}{3} = -\frac{13}{3}$$

Check: Substituting $x = 10$, $y = -\frac{13}{3}$ in the left sides of the equations, we obtain

$$(2)(10) + (3)\left(-\frac{13}{3}\right) = 20 - 13 = 7$$

$$(5)(10) + (9)\left(-\frac{13}{3}\right) = 50 - 39 = 11$$

Thus the answers do satisfy the given equations, and the solution is $x = 10$, $y = -\frac{13}{3}$, or $(10, -\frac{13}{3})$. ▲

C. EVALUATING 3 × 3 DETERMINANTS

We can also solve a system of three linear equations in three unknowns by using determinants. First, we have to define the value of a 3×3 determinant—that is, a determinant with three rows and three columns. We write

$$\det A = \begin{vmatrix} a_1 & b_1 & c_1 \\ a_2 & b_2 & c_2 \\ a_3 & b_3 & c_3 \end{vmatrix}$$

Problem 2 Use Cramer's rule to solve the system.

$$2x + 3y = 13$$
$$5x - 4y = 21$$

and we define the value of this determinant in terms of the 2×2 determinants that we can form from the second and third rows. The 2×2 determinant that remains when we cross out the row and column of any element is called the *minor* of that element. For example, deleting the first row and the second column of

$$\begin{vmatrix} a_1 & b_1 & c_1 \\ a_2 & b_2 & c_2 \\ a_3 & b_3 & c_3 \end{vmatrix}$$

leaves the 2×2 determinant

$$\begin{vmatrix} a_2 & c_2 \\ a_3 & c_3 \end{vmatrix}$$

which is the *minor* of b_1. The following definition gives the value of the 3×3 determinant in terms of the minors of the first row.

$$\begin{matrix} + & - & + \\ \end{matrix}$$
$$\begin{vmatrix} a_1 & b_1 & c_1 \\ a_2 & b_2 & c_2 \\ a_3 & b_3 & c_3 \end{vmatrix} = a_1 \begin{vmatrix} b_2 & c_2 \\ b_3 & c_3 \end{vmatrix} - b_1 \begin{vmatrix} a_2 & c_2 \\ a_3 & c_3 \end{vmatrix} + c_1 \begin{vmatrix} a_2 & b_2 \\ a_3 & b_3 \end{vmatrix}$$

Note that we have written the plus and minus signs above the first row to help you remember this definition.

EXAMPLE 3 Evaluate the determinant.

$$\begin{vmatrix} 2 & -4 & 3 \\ 3 & 0 & 5 \\ 1 & 1 & 2 \end{vmatrix}$$

Solution By definition,

$$\begin{matrix} + & - & + \\ \end{matrix}$$
$$\begin{vmatrix} 2 & -4 & 3 \\ 3 & 0 & 5 \\ 1 & -1 & 2 \end{vmatrix} = (2)\begin{vmatrix} 0 & 5 \\ -1 & 2 \end{vmatrix} - (-4)\begin{vmatrix} 3 & 5 \\ 1 & 2 \end{vmatrix} + (3)\begin{vmatrix} 3 & 0 \\ 1 & -1 \end{vmatrix}$$

$$= (2)[0 \cdot 2 - (-1)(5)] - (-4)[3 \cdot 2 - 1 \cdot 5]$$
$$+ (3)[3 \cdot (-1) - 1 \cdot 0]$$
$$= (2)(5) + 4(1) + 3(-3)$$
$$= 10 + 4 - 9 = 5$$

▲

Problem 3 Evaluate the determinant.

$$\begin{vmatrix} 2 & -3 & 4 \\ -1 & 2 & 1 \\ 1 & 1 & -2 \end{vmatrix}$$

D. CRAMER'S RULE FOR A SYSTEM OF THREE EQUATIONS

We can solve the system

$$a_1 x + b_1 y + c_1 z = d_1$$
$$a_2 x + b_2 y + c_2 z = d_2 \tag{1}$$
$$a_3 x + b_3 y + c_3 z = d_3$$

in exactly the same manner as we solved the system of equations with two unknowns. The details are similar to those for a system of two equations so we just state the result.

CRAMER'S RULE FOR A SYSTEM OF THREE EQUATIONS IN THREE UNKNOWNS

The system

$$a_1 x + b_1 y + c_1 z = d_1$$
$$a_2 x + b_2 y + c_2 z = d_2$$
$$a_3 x + b_3 y + c_3 z = d_3$$

1. Has the unique solution

$$x = \frac{D_x}{D}, \qquad y = \frac{D_y}{D}, \qquad z = \frac{D_z}{D}$$

where

$$D = \begin{vmatrix} a_1 & b_1 & c_1 \\ a_2 & b_2 & c_2 \\ a_3 & b_3 & c_3 \end{vmatrix}, \qquad D_x = \begin{vmatrix} d_1 & b_1 & c_1 \\ d_2 & b_2 & c_2 \\ d_3 & b_3 & c_3 \end{vmatrix},$$

$$D_y = \begin{vmatrix} a_1 & d_1 & c_1 \\ a_2 & d_2 & c_2 \\ a_3 & d_3 & c_3 \end{vmatrix}, \quad \text{and} \quad D_z = \begin{vmatrix} a_1 & b_1 & d_1 \\ a_2 & b_2 & d_2 \\ a_3 & b_3 & d_3 \end{vmatrix}, \quad D \neq 0$$

2. Is *inconsistent* and has no solution if $D = 0$ and any one of D_x, D_y, D_z is different from zero.

3. Has *no unique* solution if $D = 0$ and $D_x = D_y = D_z = 0$. (In this case, the system either has no solution or infinitely many solutions. This situation is studied more fully in advanced algebra.)

Note carefully that D is the determinant of the coefficients; D_x is formed from D by replacing the coefficients of x (the a's) by the corresponding d's; D_y is formed from D by replacing the coefficients of y (the b's) by the corresponding d's, and D_z is formed similarly.

EXAMPLE 4   Use Cramer's rule to solve the system.

$$x + y + 2z = 7$$
$$x - y - 3z = -6$$
$$2x + 3y + z = 4$$

Solution   To use Cramer's rule, we first have to evaluate D, the determinant of the coefficients. If D is not 0, then we calculate the other three determinants, D_x, D_y, and D_z. For the given system,

$$D = \begin{vmatrix} 1 & 1 & 2 \\ 1 & -1 & -3 \\ 2 & 3 & 1 \end{vmatrix} = (1)\begin{vmatrix} -1 & -3 \\ 3 & 1 \end{vmatrix} - (1)\begin{vmatrix} 1 & -3 \\ 2 & 1 \end{vmatrix} + (2)\begin{vmatrix} 1 & -1 \\ 2 & 3 \end{vmatrix}$$

$$= (1)(-1 + 9) - (1)(1 + 6) + (2)(3 + 2)$$
$$= 8 - 7 + 10 = 11$$

$$D_x = \begin{vmatrix} 7 & 1 & 2 \\ -6 & -1 & -3 \\ 4 & 3 & 1 \end{vmatrix} = (7)\begin{vmatrix} -1 & -3 \\ 3 & 1 \end{vmatrix} - (1)\begin{vmatrix} -6 & -3 \\ 4 & 1 \end{vmatrix} + (2)\begin{vmatrix} -6 & -1 \\ 4 & 3 \end{vmatrix}$$

$$= (7)(-1+9) - (1)(-6+12) + (2)(-18+4)$$

$$= 56 - 6 - 28 = 22$$

$$D_y = \begin{vmatrix} 1 & 7 & 2 \\ 1 & -6 & -3 \\ 2 & 4 & 1 \end{vmatrix} = (1)\begin{vmatrix} -6 & -3 \\ 4 & 1 \end{vmatrix} - 7\begin{vmatrix} 1 & -3 \\ 2 & 1 \end{vmatrix} + (2)\begin{vmatrix} 1 & -6 \\ 2 & 4 \end{vmatrix}$$

$$= (1)(-6+12) - (7)(1+6) + (2)(4+12)$$

$$= 6 - 49 + 32 = -11$$

$$D_z = \begin{vmatrix} 1 & 1 & 7 \\ 1 & -1 & -6 \\ 2 & 3 & 4 \end{vmatrix} = (1)\begin{vmatrix} -1 & -6 \\ 3 & 4 \end{vmatrix} - (1)\begin{vmatrix} 1 & -6 \\ 2 & 4 \end{vmatrix} + (7)\begin{vmatrix} 1 & -1 \\ 2 & 3 \end{vmatrix}$$

$$= (1)(-4+18) - (1)(4+12) + (7)(3+2)$$

$$= 14 - 16 + 35 = 33$$

Thus by Cramer's rule,

$$x = \frac{D_x}{D} = \frac{22}{11} = 2 \qquad y = \frac{D_y}{D} = \frac{-11}{11} = -1 \qquad z = \frac{D_z}{D} = \frac{33}{11} = 3$$

You can check the solution $(2, -1, 3)$ by substituting into the given equations. ▲

EXAMPLE 5 Use Cramer's rule to solve the system.

$$x + y - z = 2$$
$$2x + y + z = 4$$
$$-x - y + z = 3$$

Solution By Cramer's rule, if there is a unique solution, it is given by

$$x = \frac{D_x}{D}, \qquad y = \frac{D_y}{D}, \quad \text{and} \quad z = \frac{D_z}{D}$$

However,

$$D = \begin{vmatrix} 1 & 1 & -1 \\ 2 & 1 & 1 \\ -1 & -1 & 1 \end{vmatrix} = (1)\begin{vmatrix} 1 & 1 \\ -1 & 1 \end{vmatrix} - (1)\begin{vmatrix} 2 & 1 \\ -1 & 1 \end{vmatrix} + (-1)\begin{vmatrix} 2 & 1 \\ -1 & -1 \end{vmatrix}$$

$$= 2 - 3 + 1 = 0$$

and

$$D_x = \begin{vmatrix} 2 & 1 & -1 \\ 4 & 1 & 1 \\ 3 & -1 & 1 \end{vmatrix}$$

$$= (2)\begin{vmatrix} 1 & 1 \\ -1 & 1 \end{vmatrix} - (1)\begin{vmatrix} 4 & 1 \\ 3 & 1 \end{vmatrix} + (-1)\begin{vmatrix} 4 & 1 \\ 3 & -1 \end{vmatrix}$$

$$= 4 - 1 + 7 = 10 \neq 0$$

Hence, the system is inconsistent and has no solution. ▲

Problem 5 Use Cramer's rule to solve the system.

$$x - y - z = 2$$
$$x + 2y + z = 6$$
$$-x + y + z = 4$$

E. EXPANDING DETERMINANTS BY USING MINORS

Our evaluation of a 3×3 determinant uses the idea of a minor of an element in a determinant. Here is the definition.

In the determinant

$$\begin{vmatrix} a_1 & b_1 & c_1 \\ a_2 & b_2 & c_2 \\ a_3 & b_3 & c_3 \end{vmatrix}$$

the **minor** of an element is the determinant that remains after deleting the row and column in which the element appears.

For example, in the determinant $\begin{vmatrix} a_1 & b_1 & c_1 \\ a_2 & b_2 & c_2 \\ a_3 & b_3 & c_3 \end{vmatrix}$,

the minor of a_1 is $\begin{vmatrix} b_2 & c_2 \\ b_3 & c_3 \end{vmatrix}$

the minor of b_1 is $\begin{vmatrix} a_2 & c_2 \\ a_3 & c_3 \end{vmatrix}$

and the minor of c_1 is $\begin{vmatrix} a_2 & b_2 \\ a_3 & b_3 \end{vmatrix}$

Earlier, we presented the value of a 3×3 determinant as

$$\begin{vmatrix} a_1 & b_1 & c_1 \\ a_2 & b_2 & c_2 \\ a_3 & b_3 & c_3 \end{vmatrix} = a_1 \begin{vmatrix} b_2 & c_2 \\ b_3 & c_3 \end{vmatrix} - b_1 \begin{vmatrix} a_2 & c_2 \\ a_3 & c_3 \end{vmatrix} + c_1 \begin{vmatrix} a_2 & b_2 \\ a_3 & b_3 \end{vmatrix} \qquad (4)$$

This is the expansion of the determinant by minors along the first row.

The 2×2 determinants in this equation are clearly minors of the elements in the first row of the 3×3 determinant on the left; the right side is called the expansion of the 3×3 determinant by minors along the first row. An expansion with numbers is carried out in Example 3.

EXAMPLE 6   Expand

$$\begin{vmatrix} 1 & 1 & 1 \\ 1 & 2 & 1 \\ 1 & 1 & 2 \end{vmatrix}$$

by minors along the first row.

Problem 6   Expand the determinant in Example 6 by minors along the first column.

Solution

$$\begin{vmatrix} 1 & 1 & 1 \\ 1 & 2 & 1 \\ 1 & 1 & 2 \end{vmatrix} = (1)\begin{vmatrix} 2 & 1 \\ 1 & 2 \end{vmatrix} - (1)\begin{vmatrix} 1 & 1 \\ 1 & 2 \end{vmatrix} + (1)\begin{vmatrix} 1 & 2 \\ 1 & 1 \end{vmatrix}$$

$$= 3 - 1 - 1 = 1 \qquad \blacktriangle$$

In our work so far, the given determinant has been evaluated by minors along the first row. Actually, it is possible to expand a determinant by the minors of *any row* or *any column*. To do this, it is neces-

ANSWERS   (to problems on pages 626–627)

4. $(1, 2, 3)$

5. Since $D = 0$ and $D_x \neq 0$, the system is inconsistent. There is no solution.

sary to define the *sign array* of a determinant. For a three by three (3×3) determinant, the sign array is the following arrangement of alternating signs:

$$\begin{array}{ccc} + & - & + \\ - & + & - \\ + & - & + \end{array} \tag{5}$$

To obtain the expansion of

$$\begin{vmatrix} a_1 & b_1 & c_1 \\ a_2 & b_2 & c_2 \\ a_3 & b_3 & c_3 \end{vmatrix}$$

along a particular row or column, we simply write in front of each term in the expansion the corresponding sign from the array. For example, to expand

$$\begin{vmatrix} 1 & 1 & 1 \\ 1 & 2 & 1 \\ 1 & 1 & 2 \end{vmatrix}$$

along the *second row,* we write

$$\begin{vmatrix} 1 & 1 & 1 \\ 1 & 2 & 1 \\ 1 & 1 & 2 \end{vmatrix} = \overset{-}{-}(1)\begin{vmatrix} 1 & 1 \\ 1 & 2 \end{vmatrix} + \overset{+}{(2)}\begin{vmatrix} 1 & 1 \\ 1 & 2 \end{vmatrix} - \overset{-}{(1)}\begin{vmatrix} 1 & 1 \\ 1 & 1 \end{vmatrix}$$

$$= -1(1) + 2(1) - 1(0) = 1 \qquad \text{Note that we used the signs of the } \textit{second row} \text{ of (5), } - + -, \text{ for the first, second, and third terms, respectively.}$$

EXAMPLE 7 Expand the given determinant along the third column.

a. $\begin{vmatrix} 0 & 1 & 1 \\ 1 & 2 & -1 \\ 1 & -1 & 3 \end{vmatrix}$ **b.** $\begin{vmatrix} 1 & 1 & 0 \\ 0 & -1 & 1 \\ 2 & -1 & -3 \end{vmatrix}$

Solution

a. $\begin{vmatrix} 0 & 1 & 1 \\ 1 & 2 & -1 \\ 1 & -1 & 3 \end{vmatrix} = +(1)\begin{vmatrix} 1 & 2 \\ 1 & -1 \end{vmatrix} - (-1)\begin{vmatrix} 0 & 1 \\ 1 & -1 \end{vmatrix} + (3)\begin{vmatrix} 0 & 1 \\ 1 & 2 \end{vmatrix}$

$$= (1)(-3) + (1)(-1) + (3)(-1) = -3 - 1 - 3 = -7$$

b. $\begin{vmatrix} 1 & 1 & 0 \\ 0 & -1 & 1 \\ 2 & -1 & -3 \end{vmatrix} = +(0)\begin{vmatrix} 0 & -1 \\ 2 & -1 \end{vmatrix} - (1)\begin{vmatrix} 1 & 1 \\ 2 & -1 \end{vmatrix} + (-3)\begin{vmatrix} 1 & 1 \\ 0 & -1 \end{vmatrix}$

$$= 0 - (1)(-3) + (-3)(-1) = 6 \qquad \blacktriangle$$

A last word before you do the problems. When expanding a 3×3 determinant, it is easier to expand along the row or column containing the most zeros, since the coefficient of the resulting minors would then be zero. Do it this way; it will save you time.

Problem 7 Expand the determinants of Example 7 along the third row.

ANSWERS   (to problems on pages 628–629)
6. 1
7. a. -7 **b.** 6

ANSWERS

A. In Problems 1–10 evaluate the determinant.

1. $\begin{vmatrix} 1 & 1 \\ 0 & 2 \end{vmatrix}$

2. $\begin{vmatrix} 2 & -1 \\ 4 & 3 \end{vmatrix}$

3. $\begin{vmatrix} -3 & -2 \\ 5 & 1 \end{vmatrix}$

4. $\begin{vmatrix} 2 & -1 \\ -3 & 1 \end{vmatrix}$

5. $\begin{vmatrix} -2 & 0 \\ 5 & -3 \end{vmatrix}$

6. $\begin{vmatrix} 5 & 2 \\ -10 & -4 \end{vmatrix}$

7. $\begin{vmatrix} \dfrac{1}{2} & \dfrac{-1}{4} \\ \dfrac{1}{2} & \dfrac{3}{4} \end{vmatrix}$

8. $\begin{vmatrix} \dfrac{1}{5} & \dfrac{1}{10} \\ \dfrac{1}{2} & \dfrac{1}{4} \end{vmatrix}$

9. $\begin{vmatrix} \dfrac{3}{5} & \dfrac{1}{2} \\ \dfrac{-1}{4} & \dfrac{-1}{2} \end{vmatrix}$

10. $\begin{vmatrix} \dfrac{4}{5} & \dfrac{-1}{3} \\ \dfrac{-1}{2} & \dfrac{1}{2} \end{vmatrix}$

B. In Problems 11–30 solve the system. If the system is dependent or inconsistent, state that fact.

11. $\begin{aligned} x + y &= 5 \\ 3x - y &= 3 \end{aligned}$

12. $\begin{aligned} x + y &= 9 \\ x - y &= 3 \end{aligned}$

13. $\begin{aligned} x + y &= 9 \\ x - y &= -1 \end{aligned}$

14. $\begin{aligned} 2x + y &= -1 \\ x - 2y &= -13 \end{aligned}$

15. $\begin{aligned} 4x + 9y &= 3 \\ 3x + 7y &= 2 \end{aligned}$

16. $\begin{aligned} 5x + 2y &= 32 \\ 3x + y &= 18 \end{aligned}$

17. $\begin{aligned} x - y &= -1 \\ x - 2y &= -6 \end{aligned}$

18. $\begin{aligned} x - 2y &= -13 \\ 3x - 2y &= -19 \end{aligned}$

19. $\begin{aligned} 2x + 3y &= -13 \\ 6x + 9y &= -39 \end{aligned}$

20. $\begin{aligned} 4x + 5y &= -2 \\ 12x + 15y &= -6 \end{aligned}$

21. $\begin{aligned} x - y &= 1 \\ x - 2y &= 4 \end{aligned}$

22. $\begin{aligned} x - 2y &= 4 \\ 4x - 5y &= 7 \end{aligned}$

23. $\begin{aligned} x + 3y &= 6 \\ 2x + 6y &= 5 \end{aligned}$

24. $\begin{aligned} x - 2y &= 3 \\ -x + 2y &= 6 \end{aligned}$

25. $\begin{aligned} x &= 7y + 3 \\ 2x + 3y &= 23 \end{aligned}$

26. $\begin{aligned} x &= 3y + 1 \\ 2x + 3y &= 20 \end{aligned}$

1. _____
2. _____
3. _____
4. _____
5. _____
6. _____
7. _____
8. _____
9. _____
10. _____
11. _____
12. _____
13. _____
14. _____
15. _____
16. _____
17. _____
18. _____
19. _____
20. _____
21. _____
22. _____
23. _____
24. _____
25. _____
26. _____

27. $y = -3x + 17$
 $2x - y = 8$

28. $y = -2x + 14$
 $3x - y = 11$

29. $\dfrac{x}{2} - \dfrac{y}{3} = \dfrac{-1}{6}$

 $\dfrac{x}{3} + \dfrac{y}{4} = \dfrac{-7}{12}$

30. $\dfrac{x}{3} - \dfrac{y}{5} = \dfrac{4}{3}$

 $\dfrac{x}{4} - \dfrac{y}{3} = \dfrac{1}{12}$

27. _____

28. _____

29. _____

30. _____

C., E. In Problems 31–40 evaluate the determinant.

31. $\begin{vmatrix} 1 & 3 & 2 \\ 2 & 4 & 1 \\ 3 & 6 & 5 \end{vmatrix}$

32. $\begin{vmatrix} 1 & 3 & 5 \\ 2 & 0 & 10 \\ -3 & 1 & -15 \end{vmatrix}$

33. $\begin{vmatrix} 1 & 2 & 3 \\ 4 & 5 & 6 \\ 7 & 8 & 9 \end{vmatrix}$

34. $\begin{vmatrix} 1 & 1 & 1 \\ 2 & 3 & 1 \\ 2 & 4 & 1 \end{vmatrix}$

35. $\begin{vmatrix} 2 & 1 & 3 \\ 1 & 2 & -1 \\ 3 & 1 & 5 \end{vmatrix}$

36. $\begin{vmatrix} -1 & 1 & -1 \\ -2 & 2 & -6 \\ 3 & -3 & 4 \end{vmatrix}$

37. $\begin{vmatrix} 1 & 1 & 6 \\ 1 & 1 & 4 \\ 1 & -1 & 2 \end{vmatrix}$

38. $\begin{vmatrix} 1 & 4 & 0 \\ 1 & -3 & 1 \\ 0 & 8 & -1 \end{vmatrix}$

39. $\begin{vmatrix} 0 & -1 & 2 \\ 2 & 1 & -3 \\ 1 & -3 & 1 \end{vmatrix}$

40. $\begin{vmatrix} -3 & 2 & -4 \\ 1 & -1 & 3 \\ 1 & 2 & 10 \end{vmatrix}$

31. _____

32. _____

33. _____

34. _____

35. _____

36. _____

37. _____

38. _____

39. _____

40. _____

D., E. In Problems 41–50 solve the system using Cramer's rule. If the system is inconsistent or has no unique solution, state that fact. You can expand the resulting determinants by using minors.

41. $x + y + z = 6$
 $2x - 3y + 3z = 5$
 $3x - 2y - z = -4$

42. $x + y + z = 13$
 $3x + y - 3z = 5$
 $x - 2y + 4z = 10$

43. $6x + 5y + 4z = 5$
 $5x + 4y + 3z = 5$
 $4x + 3y + z = 7$

44. $3x + 2y + z = 4$
 $4x + 3y + z = 5$
 $5x + y + z = 9$

45. $x - 2y + 3z = 15$
 $5x + 7y - 11z = -29$
 $-13x + 17y + 19z = 37$

46. $2x - y + z = 3$
 $x + 2y + z = 12$
 $4x - 3y + z = 1$

47. $5x + 3y + 5z = 3$
 $3x + 5y + z = -5$
 $2x + 2y + 3z = 7$

48. $x + y = 5$
 $y + z = 3$
 $x + z = 7$

49. $2y + z = 9$
 $-2y + z = 1$
 $x + y + z = 1$

50. $x - y = 3$
 $y - z = 3$
 $x + z = 9$

41. _____

42. _____

43. _____

44. _____

45. _____

46. _____

47. _____

48. _____

49. _____

50. _____

F.   Applications.

Show that the statement is true.

51. $\begin{vmatrix} a & b & 0 \\ c & d & 0 \\ e & f & 0 \end{vmatrix} = 0$

52. $\begin{vmatrix} a & b & c \\ d & e & f \\ 0 & 0 & 0 \end{vmatrix} = 0$

53. $\begin{vmatrix} a & b & c \\ 1 & 2 & 3 \\ a & b & c \end{vmatrix} = 0$

54. $\begin{vmatrix} 1 & a & a \\ 2 & b & b \\ 3 & c & c \end{vmatrix} = 0$

55. $\begin{vmatrix} 1 & 2 & 3 \\ 3 & 1 & 2 \\ k & 2k & 3k \end{vmatrix} = k\begin{vmatrix} 1 & 2 & 3 \\ 3 & 1 & 2 \\ 1 & 2 & 3 \end{vmatrix}$

56. $\begin{vmatrix} 1 & 2 & 3k \\ 3 & 2 & k \\ 0 & 1 & 2k \end{vmatrix} = k\begin{vmatrix} 1 & 2 & 3 \\ 3 & 2 & 1 \\ 0 & 1 & 2 \end{vmatrix}$

57. $\begin{vmatrix} kb_1 & b_1 & 1 \\ kb_2 & b_2 & 2 \\ kb_3 & b_3 & 3 \end{vmatrix} = 0$

58. $\begin{vmatrix} b_1 & b_2 & b_3 \\ kb_1 & kb_2 & kb_3 \\ 1 & 2 & 3 \end{vmatrix} = 0$

59. $\begin{vmatrix} 1 & 1 & 1 \\ 2 & a & a \\ 3 & b & b \end{vmatrix} = 0$

60. $\begin{vmatrix} 0 & 0 & 0 \\ a & b & c \\ d & e & f \end{vmatrix} = 0$

51. _____
52. _____
53. _____
54. _____
55. _____
56. _____
57. _____
58. _____
59. _____
60. _____

✓ **SKILL CHECKER**

Solve the following problems.

61. A person bought some bonds yielding 5% annually and some certificates yielding 7%. If the total investment amounts to $20,000 and the interest received is $1160, how much is invested in bonds and how much in certificates?

62. A car leaves a town going north at 30 mi/hr. Two hours later, another car leaves the same town traveling on the same road in the same direction at 40 mi/hr. How far from the town does the second car overtake the first one?

63. How many gallons of a 30% solution must be mixed with 40 gal of a 12% solution to obtain a 20% solution?

64. A person can do a job in 3 hr. Another person can do it in 2. How long would it take to complete the job if both people work together?

65. A plane traveled 840 mi with a 30-mi/hr tail wind in the same time it took to travel 660 mi against the wind. What is the plane's speed in still air?

61. _____
62. _____
63. _____
64. _____
65. _____

8.4   USING YOUR KNOWLEDGE

Determinants provide a very convenient and neat way of writing the equation of a line through two points. This is one of the problems that we studied in Section 7.3. Suppose the line is to pass through the points (x_1, y_1) and (x_2, y_2). Then, an equation of the line can be written in the form

$$\begin{vmatrix} x & y & 1 \\ x_1 & y_1 & 1 \\ x_2 & y_2 & 1 \end{vmatrix} = 0$$

It is quite easy to verify this fact. First, think of expanding the determinant by minors along the first row. The coefficients of x and y will be constants, so the equation is linear. Next, you can see that (x_1, y_1) and (x_2, y_2) both satisfy the equation, because if you substitute either of these pairs for (x, y) in the first row of the determinant, the result is a determinant with two identical rows and this you know is 0. See Problems 57 and 58. Thus the equation is that of the desired line.

As an illustration, let us find an equation of the line through $(1, 3)$ and $(-5, -2)$. The determinant form of this equation is

$$\begin{vmatrix} x & y & 1 \\ 1 & 3 & 1 \\ -5 & -2 & 1 \end{vmatrix} = 0$$

or

$$[3 - (-2)]x - [1 - (-5)]y + [-2 - (-15)](1) = 0$$

where the quantities in brackets are the expanded minors of the first-row elements. The final equation is

$$5x - 6y + 13 = 0$$

Use the determinant method to find the equation of the line through the two given points.

1. $(2, 7)$ and $(0, 3)$

2. $(10, 12)$ and $(-7, 1)$

3. $(-1, 4)$ and $(8, 2)$

4. $(5, 0)$ and $(0, -3)$

5. $(a, 0)$ and $(0, b)$, $ab \neq 0$

6. (a, b) and $(0, 0)$, $(a, b) \neq (0, 0)$

1. _____

2. _____

3. _____

4. _____

5. _____

6. _____

8.5 APPLICATIONS: COIN, RATE-TIME-DISTANCE, AND INVESTMENT PROBLEMS

OBJECTIVES

REVIEW

Before starting this section, you should know:

1. The RSTUV procedure used to solve word problems (Section 2.6).
2. How to solve a system of equations in two or three unknowns by one of the methods studied.

OBJECTIVES

You should be able to:

A. Solve coin problems with two or more unknowns.
B. Solve general problems with two or more unknowns.
C. Solve rate, time, and distance problems with two or more unknowns.
D. Solve investment problems with two or more unknowns.

A. COIN PROBLEMS

The pile of money contains $3.25 in nickels and dimes. There are five more nickels than dimes. How many nickels and how many dimes are in the pile?

As usual, we use the RSTUV procedure to solve this problem.

1. Read the problem carefully. We are asking for the number of nickels and dimes.
2. Select n to be the number of nickels and d, the number of dimes.
3. Translate the problem. We have to translate two statements, yielding two equations and two unknowns. First note that:

If you have 1 nickel, you have $0.05(1)

If you have 2 nickels, you have $0.05(2)

If you have n nickels, you have $0.05(n)

Similarly,

If you have d dimes, you have $0.10(d)

Now, we can translate the statement.

$$\begin{bmatrix} \text{total} \\ \text{amount} \end{bmatrix} = \begin{bmatrix} \text{amount} \\ \text{in nickels} \end{bmatrix} + \begin{bmatrix} \text{amount} \\ \text{in dimes} \end{bmatrix}$$

$$3.25 = 0.05n + 0.10d$$
$$325 = 5n + 10d \qquad \text{Multiply by 100.}$$
$$65 = n + 2d \qquad \text{Divide by 5.}$$

The statement there are five more nickels than dimes means that

$$\begin{bmatrix} \text{the number} \\ \text{of nickels} \end{bmatrix} \text{[is]} \begin{bmatrix} \text{5 more} \\ \text{than} \end{bmatrix} + \begin{bmatrix} \text{the number} \\ \text{of dimes} \end{bmatrix}$$

$$n = 5 + d$$

Thus the complete problem can be reduced to the system of equations

$$65 = n + 2d$$
$$n = 5 + d$$

4. Use algebra to solve this system. Substituting $5 + d$ for n in the first equation gives

$65 = n + 2d$

$65 = (5 + d) + 2d$

$\qquad = 5 + d + 2d$ Remove parentheses.

$\qquad = 5 + 3d$ Collect like terms.

$65 = 5 + 3d$

$60 = 3d$ Subtract 5.

$20 = d$

Since $n = 5 + d$, we substitute 20 for d:

$n = 5 + 20 = 25$

Hence, we have 25 nickels and 20 dimes.

5. Verify the answer. We do have 5 more nickels (25 in total) than dimes (20 in total) and $3.25 ($1.25 in nickels and $2.00 in dimes).

EXAMPLE 1   Jack has $3 in nickels and dimes. He has twice as many nickels as he has dimes. How many nickels and how many dimes does he have?

Solution   As usual, we use the RSTUV method.

1. Read the problem. We are asking for the number of nickels and dimes.

2. Select n to be the number of nickels and d the number of dimes.

3. Translate the problem. Jack has $3 (300 cents) in nickels and dimes:

$300 = 5n + 10d$

He has twice as many nickels as he has dimes:

$n = 2d$

We then have the system

$5n + 10d = 300$

$\qquad\quad n = 2d$

4. Use algebra to solve this system. This time it is easy to use the substitution method.

$5n + 10d = 300 \rightarrow 5(2d) + 10d = 300$ Let $n = 2d$.

$\qquad\qquad\qquad\qquad\quad 10d + 10d = 300$ Simplify.

$\qquad\qquad\qquad\qquad\qquad\quad 20d = 300$ Combine like terms.

$\qquad\qquad\qquad\qquad\qquad\qquad d = 15$ Divide by 20.

$\qquad\qquad\qquad\quad n = 2(15) = 30$ Substitute $d = 15$ in $n = 2d$.

Thus Jack has 15 dimes ($1.50) and 30 nickels ($1.50).

5. It is easy to verify this answer since Jack has $3 and he does have twice as many nickels as dimes.   ▲

Problem 1   Jill has $2 in nickels and dimes. She has twice as many nickels as she has dimes. How many nickels and how many dimes does she have?

B. GENERAL PROBLEMS

We can also use systems of equations to solve other problems. Here is an interesting one.

EXAMPLE 2 The newest greatest weight differential recorded for a married couple is 1300 lb in the case of Jon Brower Minnoch and his wife Jeannette. Their combined weight is 1520 lb. What is the weight of each of the Minnochs? (He is the heavy one.)

Problem 2 The McGuire twins weighed a total of 1466 lb. If their weight differential was 20 lb, what was the weight of each of the twins?

Solution

1. We are asked for the weight of the Minnochs.
2. Let h be the weight of the husband and w be the weight of the wife.
3. Translate the problem. The weight differential is 1300 lb.

$$h - w = 1300$$

Their combined weight is 1520 lb.

$$h + w = 1520$$

We then have the system

$$h - w = 1300$$
$$h + w = 1520$$

4. Use algebra to solve the system.

$$
\begin{aligned}
h - w &= 1300 \\
h + w &= 1520 \\
\hline
2h &= 2820 \quad \text{Add.} \\
h &= 1410 \quad \text{Divide by 2.}
\end{aligned}
$$

Substitute $h = 1410$ in $h + w = 1520$:

$$1410 + w = 1520$$
$$w = 110$$

Thus Jeannette weighs 110 lb and Jon weighs 1410 lb.

5. You can verify this by going to the *Guinness Book of Records*! ▲

C. DISTANCE PROBLEMS

Remember the distance problems we solved earlier? They can also be done using two variables. The procedure is about the same! We write the given information in a chart labeled $R \times T = D$, and as usual, use the RSTUV method.

EXAMPLE 3 The world's strongest current is the Saltstraumen in Norway. The current is so strong that a boat, which can go 48 mi downstream (with the current) in 1 hr, takes 4 hr to go the same 48 mi upstream (against the current). How fast is the current flowing?

Problem 3 A plane goes 1200 mi with a tail wind in 3 hr. It takes 4 hr to travel the same distance against the wind. Find the velocity of the wind and the velocity of the plane in still air.

Solution

1. Read the problem carefully.
2. Let x be the speed of the boat in still water; y be the speed of the current. Then $(x + y)$ is the speed of the boat downstream; $(x - y)$ is the speed of the boat upstream.
3. We enter this information in a chart:

	R ×	**T** =	**D**	
Downstream:	$x + y$	1	$1 \cdot (x + y)$	$\rightarrow$ $x + y = 48$
Upstream:	$x - y$	4	$4 \cdot (x - y)$	$\rightarrow 4(x - y) = 48$

4. Our system of equations is simplified as follows:

$$x + y = 48 \xrightarrow{\text{Leave as is}} x + y = 48$$

$$4(x - y) = 48 \xrightarrow{\text{Divide by 4}} \underline{x - y = 12}$$

$$2x = 60 \quad \text{Add.}$$

$$x = 30 \quad \text{Divide by 2.}$$

$$30 + y = 48 \quad \text{Substitute } x = 30 \text{ in } x + y = 48.$$

$$y = 18 \quad \text{Subtract 30.}$$

Thus the speed of the boat in still water is $x = 30$ mi/hr and the speed of the current is 18 mi/hr.

5. The verification is left for the student. ▲

D. INVESTMENT PROBLEMS

The investment problems we studied in Section 2.6 are easier when done using three variables. We illustrate their solution next.

EXAMPLE 4 An investor divides $20,000 among three investments at 6%, 8%, and 10%, respectively. If the total income is $1700 annually and the income of the 10% investment exceeds the income from the 6% and 8% investments by $300, how much was invested at each rate?

Problem 4 Repeat the example if the total income is $1740 and the income of the 10% investment exceeds the income from the 6% and 8% investments by $260.

Solution

1. Read the problem. We are asked how much was invested at each rate.

2. Let x be the amount invested at 6%, y the amount invested at 8%, and z the amount invested at 10%.

3. Let us enter the information in a chart:

	Principal	·	Rate	=	Interest
1st investment	x		6%		$0.06x$
2nd investment	y		8%		$0.08y$
3rd investment	z		10%		$0.10z$
TOTAL	20,000				$1700

Looking at the column labeled *principal*, we see that

$$x + y + z = 20{,}000 \tag{1}$$

From the column labeled *interest,* we can see that the total interest earned is $1700; thus

$$0.06x + 0.08y + 0.10z = 1700$$

Multiplying by 100 gives

$$6x + 8y + 10z = 170{,}000 \tag{2}$$

We also know that

$$\begin{bmatrix} \text{the income} \\ \text{from the 10\%} \\ \text{investment} \end{bmatrix} = \begin{bmatrix} \text{exceeds the income} \\ \text{from the 6\% and} \\ \text{8\% investments} \end{bmatrix} \quad [\text{by \$300}]$$

$$0.10z \quad = \quad 0.06x + 0.08y \quad + \quad 300$$

$$10z \quad = \quad 6x + 8y \quad + \quad 30{,}000 \quad \text{Multiply by 100.}$$

$$-6x - 8y + 10z \quad = \quad 30{,}000 \tag{3}$$

ANSWERS (to problems on pages 636–637)
1. 20 nickels; 10 dimes
2. 723 and 743 lb
3. Wind velocity 50 mi/hr, plane velocity 350 mi/hr

We now have the system

$$x + y + z = 20,000 \qquad (1)$$
$$6x + 8y + 10z = 17,000 \qquad (2)$$
$$-6x - 8y + 10z = 30,000 \qquad (3)$$

Adding Equations 2 and 3, we have

$$20z = 200,000 \qquad (4)$$
$$z = 10,000$$

Now, multiply Equation 1 by 6 and add it to Equation 3:

$$
\begin{array}{rl}
6x + 6y + 6z = 120,000 & (5) \\
-6x - 8y + 10z = 30,000 & (3) \\
\hline
-2y + 16z = 150,000 & (6)
\end{array}
$$

Now, substitute 10,000 for z in Equation 6:

$$-2y + 16(10,000) = 150,000 \qquad (7)$$
$$-2y + 160,000 = 150,000$$
$$-2y = -10,000$$
$$y = 5000$$

Finally, substitute 10,000 for z and 5000 for y in Equation 1:

$$x + 5000 + 10,000 = 20,000$$
$$x + 15,000 = 20,000$$
$$x = 5000$$

Thus $x = 5000$, $y = 5000$, and $z = 10,000$.

5. You can verify that if we invest \$5000 at 6%, \$5000 at 8%, and \$10,000 at 10% the conditions of the problem are satisfied. ▲

NAME

CLASS

SECTION

ANSWERS

A. Solve (coin problems).

1. Natasha has $6.25 in nickels and dimes. If she has twice as many dimes as she has nickels, how many dimes and how many nickels does she have?

2. Mida has $2.25 in nickels and dimes. She has four times as many dimes as nickels. How many dimes and how many nickels does she have?

3. Dora has $5.50 in nickels and quarters. She has twice as many quarters as she has nickels. How many of each coin does she have?

4. Mongo has 20 coins consisting of nickels and dimes. If the nickels were dimes and the dimes were nickels, he would have 50¢ more than he now has. How many nickels and how many dimes does he have?

5. Desi has 10 coins consisting of pennies and nickels. Strangely enough, if the nickels were pennies and the pennies were nickels, she would have the same amount of money as she now has. How many pennies and nickels does she have?

6. Don has $26 in his pocket. If he had only one dollar bills and five dollar bills, and he had a total of 10 bills, how many of each of the bills did he have?

7. A person went to the bank to deposit $300. The money was in 10 and 20 dollar bills, 25 bills in all. How many of each did the person have?

8. A woman has $5.95 in nickels and dimes. If she has a total of 75 coins, how many nickels and how many dimes does she have?

9. A man has $7.05 in nickels, dimes, and quarters. The quarters are worth $4.60 more than the dimes and the dimes are worth 25¢ more than the nickels. How many nickels, dimes, and quarters does the man have?

10. Amy has $2.50 consisting of nickels, dimes, and quarters in her piggy bank. She has the same amount in nickels and dimes, and twice as much in nickels as she has in quarters. How many nickels, dimes, and quarters does Amy have?

B. Solve (general problems).

11. The sum of two numbers is 102. Their difference is 16. What are the numbers?

12. The difference between two numbers is 28. Their sum is 82. What are the numbers?

ANSWERS

1. _____

2. _____

3. _____

4. _____

5. _____

6. _____

7. _____

8. _____

9. _____

10. _____

11. _____

12. _____

13. The sum of two integers is 126. If one of the integers is five times the other, what are the integers?

13. _____

14. The difference between two integers is 245. If one of the integers is eight times the other, find the integers.

14. _____

15. The difference between two numbers is 16. One of the numbers is 5 times the other. What are the numbers?

15. _____

16. The sum of two numbers is 116. One of the numbers is 50 less than the other. What are the numbers?

16. _____

17. Longs Peak is 145 ft higher than Pikes Peak. If you were to put these two peaks one on top of the other, you would still be 637 ft short of reaching the elevation of Mt. Everest, 29,002 ft. Find the elevation of Longs Peak and of Pikes Peak.

17. _____

18. The height of the Empire State building and its antenna is 1472 ft. The difference in height between the building and the antenna is 1028 ft. How tall is the antenna and how tall is the building?

18. _____

19. The largest sundae ever made contained about 6700 lb of topping. The topping flavors were chocolate, butterscotch, and caramel. There was the same amount of butterscotch as caramel but 600 lb more of chocolate than butterscotch. How many pounds of each were included in the topping?

19. _____

20. The largest pancake ever made contained buckwheat flour, Puritan mix, and 15 gal of syrup. The wheat and the mix weighed 100 lb more than the 15 gal of syrup. What was the weight of the syrup if the whole pancake weighed 4100 lb? By the way, 68 lb of butter were added before it was consumed.

20. _____

C. Solve (distance problems).

21. A plane flies 540 mi with a tail wind in $2\frac{1}{4}$ hr. The plane makes the return trip against the same wind and takes 3 hr. Find the speed of the plane in still air and the speed of the wind.

21. _____

22. A motor boat runs 45 mi downstream in $2\frac{1}{2}$ hr and 39 mi upstream in $3\frac{1}{4}$ hr. Find the speed of the boat in still water and the speed of the current.

22. _____

23. A motorboat can travel 15 mi/hr downstream and 9 mi/hr upstream in a certain river. Find the rate of the current and the rate at which the boat can travel in still water.

23. _____

24. It takes a motorboat $1\frac{1}{3}$ hr to go 20 mi downstream and $2\frac{2}{9}$ hr to return. Find the rate of the current and the rate at which the boat can travel in still water.

24. _____

25. A plane flying with the wind took 2 hr for a 1000-mi flight and $2\frac{1}{2}$ hr for the return flight. Find the wind velocity and the speed of the plane in still air.

25. _____

D. Solve (investment problems).

26. Two sums of money totaling $20,000 earn 8% and 10% annual interest, respectively. If the interest from both investments amounts to $1900, how much is invested at each rate?

26. _____

EXERCISE 8.5

27. An investor invested $10,000, part at 6% and the rest at 8%. Find the amount invested at each rate if the annual income from the two investments is $720.

28. Andy Cabazos has $20,000 in three investments paying 6%, 8%, and 10%, respectively. The total interest on the 6% and 8% investments is $300 less than that obtained from the 10% investment. If his annual income from these investments is $1700, how much does he have invested at each rate?

29. Marlene McGuire invested $25,000 in municipal bonds. The first investment paid 6%, the second 8%, and the third, 10%. If her annual income from these bonds was $2000 and the interest she received on the combined 6% and 8% investments equaled the interest on the 10% investment, how much money did she have in each category?

30. Marc Goldstein divided $20,000 into three parts. One part yielded 4%, another, 8%, and the third one, 6%. If his total return was $1080 and he made $80 less on his 8% investment than on his 4% investment, what amount did he invest in each category?

✓ **SKILL CHECKER**

Factor.

31. $x^2 + 4x + 3$

32. $x^2 - 4x + 4$

33. $x^2 + 2x - 3$

34. $-x^2 - 4x - 3$

35. $-x^2 + 4x - 3$

27. _____

28. _____

29. _____

30. _____

31. _____

32. _____

33. _____

34. _____

35. _____

8.5 USING YOUR KNOWLEDGE

In this section we did coin, general, distance, and investment problems. What kind of problem is left out? Mixture problems! Use your knowledge to solve these mixture problems.

A dietitian wants to arrange a diet composed of three basic foods A, B, and C. The diet must include 170 units of calcium, 90 units of iron, and 110 units of vitamin B. The table gives the number of units per ounce of each of the needed ingredients contained in each of the basic foods.

	UNITS PER OUNCE		
	Food A	Food B	Food C
Calcium	15	5	20
Iron	5	5	10
Vitamin B	10	15	10

1. If a, b, and c are the number of ounces of basic foods A, B, and C taken by an individual, write an equation indicating the amount of calcium needed.

1. _____

2. Write an equation indicating the amount of iron needed.

3. Write an equation indicating the amount of Vitamin B needed.

4. Use the equations obtained in 1, 2, and 3 to find the number of ounces of each of the basic foods needed to meet the diet requirements.

2. _____

3. _____

4. _____

SUMMARY

SECTION	ITEM	MEANING	EXAMPLE
8.1B	Consistent system	Graphs intersect at one point	$2x - y = 2$ and $y = x - 1$ form a consistent system intersecting at $(1, 0)$.
8.1B	Inconsistent system	Graphs are parallel lines.	$y - 2x = 4$ and $3y - 6x = 18$ form an inconsistent system.
8.1B	Dependent system	Graphs coincide.	$2x + \frac{1}{2}y = 2$ and $y = -4x + 4$ form a dependent system.
8.2A	Elimination method	A method where equations are multiplied by suitable numbers so that addition eliminates one of the variables.	For the system $$x - 2y = 4$$ $$x + y = 6$$ multiplying the second equation by 2 yields $$x - 2y = 4$$ $$2x + 2y = 12$$ so that addition eliminates the y.
8.4	Determinant	$\det A = \begin{vmatrix} a_1 & b_1 \\ a_2 & b_2 \end{vmatrix}$	$\begin{vmatrix} 2 & -4 \\ -3 & 5 \end{vmatrix}$ is a 2×2 determinant.
8.4A	The value of $\det A$	$a_1 b_2 - a_2 b_1$	The value of $\begin{vmatrix} 2 & -4 \\ -3 & 5 \end{vmatrix}$ is $(2)(5) - (-3)(-4) = -2$.
8.4B	Cramer's rule for a 2×2 system	The solution to the system $$a_1 x + b_1 y = d_1$$ $$a_2 x + b_2 y = d_2$$ is given by $$x = \frac{D_x}{D} \quad \text{and} \quad y = \frac{D_y}{D} \quad (D \neq 0)$$ where $$D = \begin{vmatrix} a_1 & b_1 \\ a_2 & b_2 \end{vmatrix}, \quad D_x = \begin{vmatrix} d_1 & b_1 \\ d_2 & b_2 \end{vmatrix},$$ and $D_y = \begin{vmatrix} a_1 & d_1 \\ a_2 & d_2 \end{vmatrix}$	
8.4C	The value of $\begin{vmatrix} a_1 & b_1 & c_1 \\ a_2 & b_2 & c_2 \\ a_3 & b_3 & c_3 \end{vmatrix}$	$a_1 \begin{vmatrix} b_2 & c_2 \\ b_3 & c_3 \end{vmatrix} - b_1 \begin{vmatrix} a_2 & c_2 \\ a_3 & c_3 \end{vmatrix} + c_1 \begin{vmatrix} a_2 & b_2 \\ a_3 & b_3 \end{vmatrix}$	
8.4E	Minor	The minor of an element of a determinant is the determinant that remains after deleting the row and column in which the element appears.	The minor of 6 in $$\begin{vmatrix} 3 & 4 & 6 \\ 1 & 2 & 3 \\ 0 & 1 & 4 \end{vmatrix}$$ is $\begin{vmatrix} 1 & 2 \\ 0 & 1 \end{vmatrix}$.

(*If you need help with these exercises, look in the section indicated in brackets.*)

1. [8.1A, B] Use the graphical method to solve the system.

    a. $2x - y = 2$
    $\quad\quad y = 3x - 4$

    b. $x - 2y = 0$
    $\quad\quad y = x - 2$

2. [8.1A, B] Use the graphical method to solve the system.

    a. $2y - x = 3$
    $\quad\quad 4y = 2x + 7$

1. a. _____

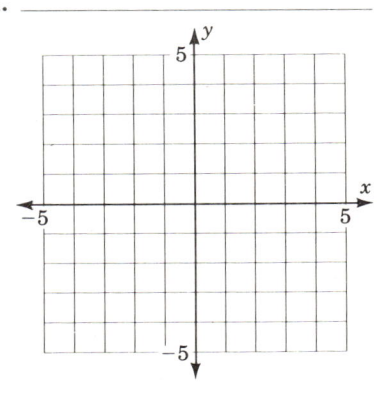

b. _____

2. a. _____

b. $3y + x = 5$

$2x = 8 - 6y$

b. _____

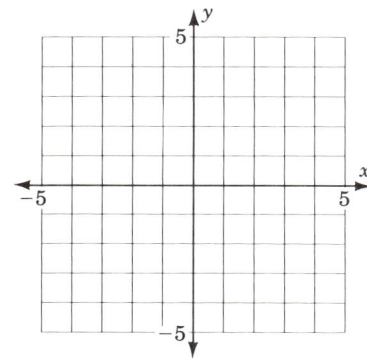

3. [8.1A, B] Use the graphical method to solve the system.

a. $3x + 2y = 6$

$y = 3 - \dfrac{3}{2}x$

3. a. _____

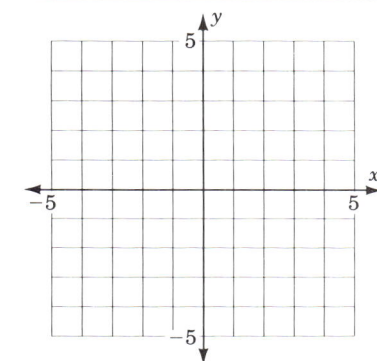

b. $x + 2y = 4$

$2x = 8 - 4y$

b. _____

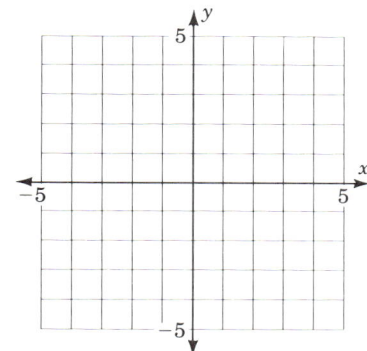

4. [8.1C] Solve by the substitution method.

a. $2x - y = 4$

$x + y = 5$

b. $2x + 3y = 10$

$x - y = -1$

4. a. _____

b. _____

5. [8.1C] Solve by the substitution method.

a. $2x + 4y = 7$

$x = -2y - 1$

b. $2y + x = 5$

$3x = 10 - 6y$

5. a. _____

b. _____

6. [8.1C] Solve by the substitution method.

a. $2y - x = 5$

$2x = 4y - 10$

b. $x + 5y = 5$

$y = 1 - \dfrac{x}{5}$

7. [8.2A] Solve by the elimination method.

a. $x - 3y = 7$

$2x - y = 9$

b. $2x + 3y = 4$

$x + y = 1$

8. [8.2A] Solve by the elimination method.

a. $2x + 3y = 7$

$6x + 9y = 14$

b. $3x - 4y = 5$

$6x - 8y = 15$

9. [8.2A] Solve by the elimination method.

a. $\dfrac{x}{5} + \dfrac{y}{2} = \dfrac{1}{5}$

$2x + 5y = 2$

b. $\dfrac{x}{3} - \dfrac{y}{4} = 2$

$\dfrac{x}{6} - \dfrac{y}{8} = 1$

10. [8.2A] Solve the system.

a. $4y = -5x - 2$

$4x = -3y - 1$

b. $2x = 3y - 1$

$2y = 3x - 1$

11. [8.3A] Solve the system.

a. $x - y + z = 4$

$x + y - z = 0$

$2x - y + z = 6$

b. $2x - 3y + z = 3$

$2x + 3y + z = -3$

$2x - 9y + z = 9$

12. [8.3A] Solve the system.

   a. $\begin{aligned} 2x + y - 2z &= 4 \\ -x + y + 2z &= 2 \\ 3y + 2z &= 12 \end{aligned}$

   b. $\begin{aligned} 2x + 2y + z &= 4 \\ -2x - y + z &= 2 \\ -2x + 3z &= 9 \end{aligned}$

13. [8.3A] Solve the system.

   a. $\begin{aligned} x + 2y + 3z &= 6 \\ x - 2y - z &= -2 \\ x + z &= 2 \end{aligned}$

   b. $\begin{aligned} x + 2y &= 4 \\ y + 2z &= 6 \\ 2x + 2y - 4z &= -4 \end{aligned}$

14. [8.4A] Evaluate.

   a. $\begin{vmatrix} 3 & 5 \\ 2 & -4 \end{vmatrix}$

   b. $\begin{vmatrix} -4 & 5 \\ -6 & 4 \end{vmatrix}$

15. [8.4B] Solve by Cramer's rule.

   a. $\begin{aligned} 2x + 5y &= -8 \\ 3x - 4y &= 11 \end{aligned}$

   b. $\begin{aligned} 4x + 2y &= 1 \\ 2x - 6y &= 4 \end{aligned}$

16. [8.4C] Evaluate.

   a. $\begin{vmatrix} 1 & -2 & -2 \\ 3 & 0 & -1 \\ 4 & 1 & 2 \end{vmatrix}$

   b. $\begin{vmatrix} 0 & 2 & 4 \\ 1 & 2 & 0 \\ 2 & 1 & 3 \end{vmatrix}$

17. [8.4D] Solve by Cramer's rule.

   a. $\begin{aligned} x + 2y + z &= 6 \\ x + y - z &= 7 \\ 2x - y + 2z &= -3 \end{aligned}$

   b. $\begin{aligned} x + y + 2z &= -3 \\ x - y + 2z &= 1 \\ x + 2y - z &= -2 \end{aligned}$

12. a. _____
  b. _____

13. a. _____
  b. _____

14. a. _____
  b. _____

15. a. $D = $ _____, $D_x = $ _____,
    $D_y = $ _____, $x = $ _____,
    $y = $ _____
  b. $D = $ _____, $D_x = $ _____,
    $D_y = $ _____, $x = $ _____,
    $y = $ _____

16. a. _____
  b. _____

17. a. $D = $ _____, $D_x = $ _____,
    $D_y = $ _____, $D_z = $ _____,
    $x = $ _____, $y = $ _____,
    $z = $ _____
  b. $D = $ _____, $D_x = $ _____,
    $D_y = $ _____, $D_z = $ _____,
    $x = $ _____, $y = $ _____,
    $z = $ _____

18. [8.4D] Solve by Cramer's rule.

   a. $2x + y + z = 6$
   $x - y + z = 7$
   $5x + y + 3z = 8$

   b. $x + 2y = 0$
   $y - z = 2$
   $2x + y + 3z = 5$

19. [8.4E] Expand by minors along the first row.

   a. $\begin{vmatrix} 1 & -1 & 1 \\ 2 & 3 & 1 \\ 1 & 3 & 2 \end{vmatrix}$

   b. $\begin{vmatrix} 4 & -2 & -1 \\ 2 & 5 & -2 \\ 1 & -2 & 2 \end{vmatrix}$

20. [8.4E] Expand by minors along the second column.

   a. $\begin{vmatrix} 1 & 0 & 5 \\ 3 & 2 & 1 \\ 5 & 3 & -1 \end{vmatrix}$

   b. $\begin{vmatrix} 1 & 2 & 1 \\ 0 & 4 & -2 \\ 3 & 6 & -2 \end{vmatrix}$

21. [8.4E] Expand by minors along the third column.

   a. $\begin{vmatrix} 1 & 3 & 0 \\ 0 & 1 & -2 \\ 2 & 4 & 3 \end{vmatrix}$

   b. $\begin{vmatrix} 3 & 1 & 5 \\ 1 & 0 & -2 \\ 6 & 1 & 3 \end{vmatrix}$

22. [8.5A]
   a. Joey has $4 in nickels and dimes. If he has five more nickels than dimes, how many of each coin does he have?
   b. Alice has $2 in nickels and dimes. If she has five fewer nickels than dimes, how many of each coin does she have?

23. [8.5B]
   a. The total height of a building and a flagpole on the roof is 200 ft. The building is nine times as high as the flagpole. How high is the building?
   b. The total height of a building and a flagpole on the roof is 180 ft. The building is eight times as high as the flagpole. How high is the building?

18. a. $D = $ _____ , $D_x = $ _____ ,
   $D_y = $ _____ , $D_z = $ _____ ,
   $x = $ _____ , $y = $ _____ ,
   $z = $ _____

   b. $D = $ _____ , $D_x = $ _____ ,
   $D_y = $ _____ , $D_z = $ _____ ,
   $x = $ _____ , $y = $ _____ ,
   $z = $ _____

19. a. _____
   b. _____

20. a. _____
   b. _____

21. a. _____
   b. _____

22. a. _____
   b. _____

23. a. _____
   b. _____

24. [8.5C]

    a. A motorboat can go 12 mi downstream on a river in 20 min. It takes this boat 30 min to go upstream the same 12 mi. Find the speed of the current.

    b. A motorboat can go 6 mi downstream on a river in 15 min. It takes this boat 20 min to go upstream the same 6 mi. Find the speed of the current.

25. [8.5D]

    a. Bill has three investments totaling $40,000. These investments earn interest at 4%, 6%, and 8%, respectively. Bill's annual income from these investments is $2600. The income from the 8% investment exceeds the total income from the other two investments by $600. Find how much Bill has invested at each rate.

    b. Betty has three investments totaling $45,000. These investments earn interest at 4%, 6%, and 8%, respectively. Betty's annual income from these investments is $2900. The income from the 8% investment exceeds the total income from the other two investments by $300. Find how much Betty has invested at each rate.

24. a. _____

    b. _____

25. a. _____

    b. _____

NAME

CLASS

SECTION

(Answers on pages 654–655)

ANSWERS

1. Use the graphical method to solve the system.

    $x - 3y = 3$

    $y = x - 1$

1.
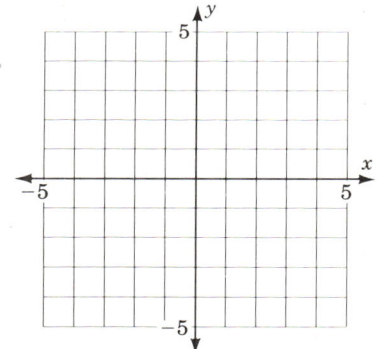

2. Use the graphical method to solve the system.

    $y - 3x = -3$

    $3y = 9x + 9$

2.
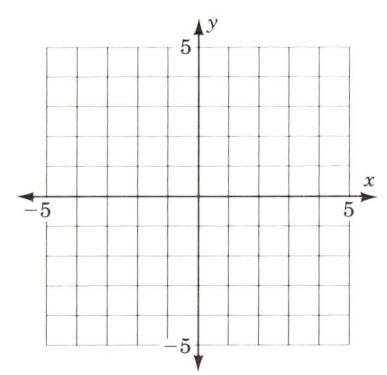

3. Use the graphical method to solve the system.

    $3x + 2y = 6$

    $x = 2 - \dfrac{2}{3}y$

3.

4. Use the substitution method to solve the system.

    $x - 2y = 4$

    $x = 1 + y$

4.

5. Use the substitution method to solve the system.

$$2x - 3y = 6$$
$$4x = 6y + 7$$

5. _____

6. Use the substitution method to solve the system.

$$2x - y = 6$$
$$2y = 4x - 12$$

6. _____

7. Solve by the elimination method.

$$3x + 4y = 5$$
$$x + y = 1$$

7. _____

8. Solve by the elimination method.

$$3x + 4y = 5$$
$$6x + 8y = 9$$

8. _____

9. Solve the system.

$$\frac{x}{2} + \frac{y}{3} = 2$$

$$\frac{x}{4} + \frac{y}{6} = 1$$

9. _____

10. Solve the system.

$$2x = 3y - 10$$
$$2y = 3x + 10$$

10. _____

11. Solve the system.

$$x + y + z = 2$$
$$2x + y - z = 5$$
$$x + y - z = 4$$

11. _____

12. Solve the system.

$$2x + y - 2z = 4$$
$$-x + y + 2z = 2$$
$$3y + 2z = 12$$

12. _____

13. Solve the system.

$$2x + y + z = 4$$
$$-x + 2y + z = 3$$
$$-2x + 9y + 5z = 16$$

13. _____

14. Evaluate.

a. $\begin{vmatrix} 4 & -2 \\ 5 & -6 \end{vmatrix}$

b. $\begin{vmatrix} 4 & -2 \\ 6 & -3 \end{vmatrix}$

14. a. _____

b. _____

15. Solve by Cramer's rule.

$2x - 3y = 5$

$6x - 4y = 7$

16. Evaluate.

$$\begin{vmatrix} -2 & 4 & -2 \\ 1 & 0 & 3 \\ 4 & 5 & 2 \end{vmatrix}$$

17. Solve by Cramer's rule.

$x + 2y + z = 6$

$x + y - z = 7$

$2x - y + 2z = -3$

18. Solve by Cramer's rule.

$x + 2y + z = 6$

$x + y - z = 7$

$3x + 5y + z = 8$

19. Expand by minors along the first row.

$$\begin{vmatrix} 3 & 2 & -1 \\ 1 & 3 & 2 \\ 1 & 1 & -1 \end{vmatrix}$$

20. Expand by minors along the third column.

$$\begin{vmatrix} 1 & 1 & -2 \\ 2 & -2 & 1 \\ 0 & 2 & 1 \end{vmatrix}$$

21. Expand by minors along the second row.

$$\begin{vmatrix} 1 & 0 & 2 \\ 0 & 1 & -2 \\ 2 & 1 & 3 \end{vmatrix}$$

22. José has $3.50 in nickels and dimes. He has 10 more nickels than dimes. How many of each coin does he have?

23. The total height of a building and a flagpole on the roof is 240 ft. The building is nine times as high as the flagpole. How high is the building?

24. A motorboat can go 10 mi downstream on a river in 20 min. It takes 30 min for this boat to go back upstream the same 10 mi. Find the speed of the current.

25. Annie has three investments totaling $60,000. These investments earn interest at 4%, 6%, and 8%, respectively. Annie's total annual income from these investments is $4000. The income from the 8% investment exceeds the total income from the other two investments by $800. Find how much she has invested at each rate.

15. $D =$ _____, $D_x =$ _____,
$D_y =$ _____, $x =$ _____,
$y =$ _____

16. _____

17. $D =$ _____, $D_x =$ _____,
$D_y =$ _____, $D_z =$ _____,
$x =$ _____, $y =$ _____, $z =$ _____

18. $D =$ _____, $D_x =$ _____,
$D_y =$ _____, $D_z =$ _____,
$x =$ _____, $y =$ _____, $z =$ _____

19. _____

20. _____

21. _____

22. _____

23. _____

24. _____

25. _____

IF YOU MISSED QUESTION	SECTION	EXAMPLES	PAGE	ANSWERS
1	8.1A, B	1, 4	586–587 589–591	**1.** The solution is $(0, -1)$.
2	8.1A, B	2, 4	587–588 589–591	**2.** Inconsistent; no solution.
				Dependent; infinitely
3	8.1A, B	3, 4	588–591	**3.** many solutions.
4	8.1C	5	592–593	**4.** $(-2, -3)$
5	8.1C	6	593	**5.** Inconsistent; no solution.
				Dependent; infinitely
6	8.1C	7	593	**6.** many solutions.
7	8.2A	1	602	**7.** $(-1, 2)$
8	8.2A	2	603	**8.** Inconsistent; no solution.

IF YOU MISSED QUESTION	SECTION	EXAMPLES	PAGE	ANSWERS
9	8.2A	3	603–604	**9.** Dependent; infinitely many solutions.
10	8.2B	4	605	**10.** $(-2, 2)$
11	8.3A	1	612–613	**11.** $(1, 2, -1)$; consistent.
12	8.3A	2	613	**12.** Inconsistent; no solution.
13	8.3A	3	613–614	**13.** Dependent; infinitely many solutions.
14a	8.4A	1	622	**14. a.** -14
14b	8.4A	1	622	**b.** 0
15	8.4B	2	624	**15.** $(\frac{1}{10}, -\frac{8}{5})$
16	8.4C	3	625	**16.** 60
17	8.4D	4	626–627	**17.** $(2, 3, -2)$; consistent.
18	8.4D	5	627	**18.** Inconsistent; no solution.
19	8.4E	6	628	**19.** -7
20	8.4E	7	629	**20.** -14
21	8.4E	7	629	**21.** 1
22	8.5A	1	636	**22.** 30 nickels, 20 dimes
23	8.5B	2	637	**23.** 216 ft
24	8.5C	3	637–638	**24.** 5 mi/hr
25	8.5D	4	638–639	**25.** $10,000 at 4%, $20,000 at 6%, $30,000 at 8%

CHAPTER 9

CONIC SECTIONS AND NONLINEAR SYSTEMS OF EQUATIONS

In this chapter we study curves called conic sections because they can be constructed by slicing a cone with a plane. The conic sections are the parabola, circle, ellipse, and hyperbola. We then study second-degree inequalities and nonlinear systems of equations involving the conic sections.

(Answers on pages 664–669)

ANSWERS

1. Graph the parabola $y = -x^2 + 3$.

1.

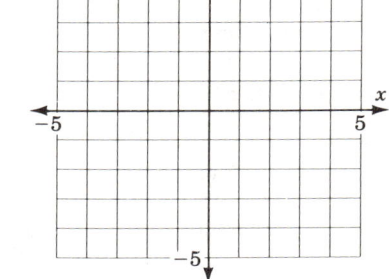

2. Find the vertex and graph the parabola $y = (x - 2)^2 + 1$.

2. _____

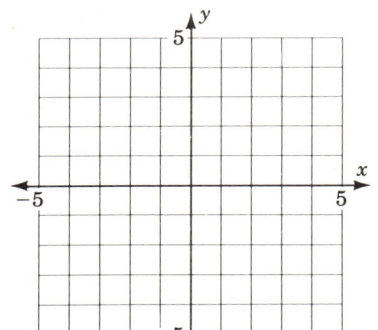

3. Find the vertex and graph the parabola $y = x^2 + 4x - 1$.

3. _____

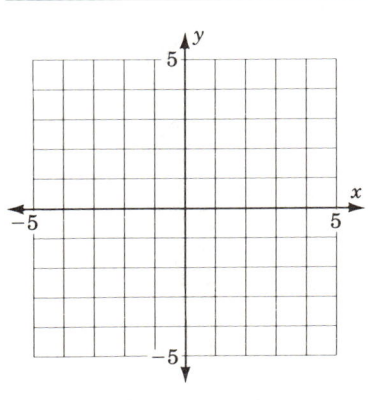

4. Find the vertex and graph the parabola $y = -2x^2 + 4x + 1$.

4. _____

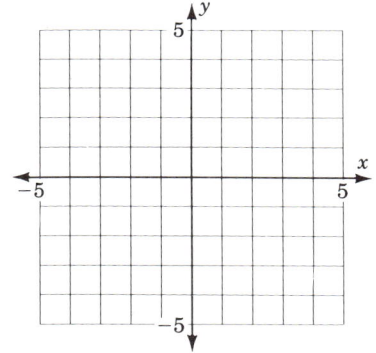

5. Find the vertex and graph the parabola $x = 4(y-1)^2 - 1$.

5. _____

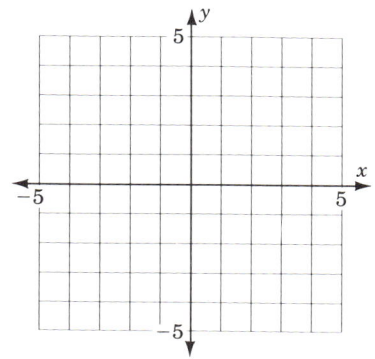

6. Find the vertex and graph the parabola $x = y^2 - 3y + 2$.

6. _____

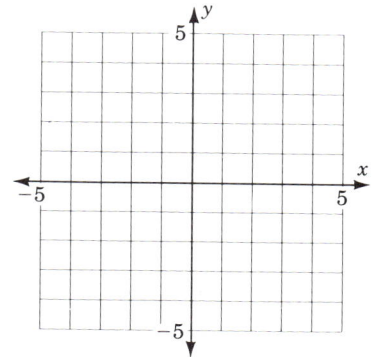

7. If the revenue is given by $R = 40x - 0.02x^2$, find the value of x that yields the maximum revenue.

7. _____

8. Find an equation of the circle of radius 3 and with center at $(-1, 2)$.

8. _____

9. Find an equation of the circle of radius 2 and with center at the origin.

9. _____

10. Find the center and the radius and sketch the graph of $(x - 1)^2 + (y + 2)^2 = 4$.

10. _____

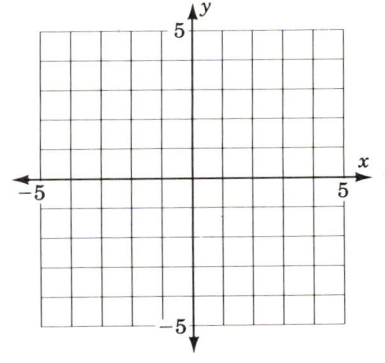

11. Sketch the graph of $x^2 + y^2 = 9$.

11.

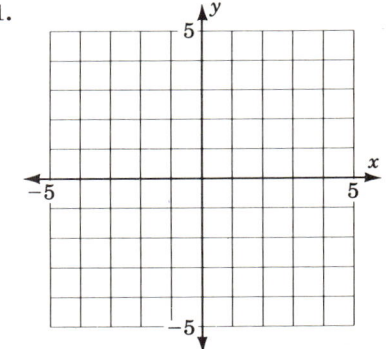

12. Find the center and the radius and sketch the graph of $x^2 + y^2 - 2x + 4y + 1 = 0$.

12. _____

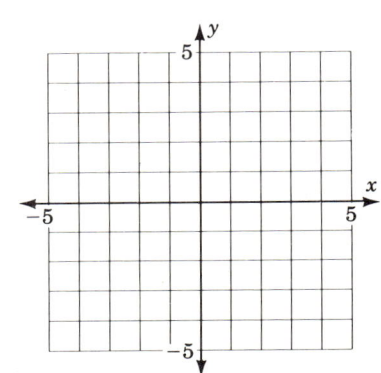

13. Graph $x^2 + 4y^2 = 4$.

13.

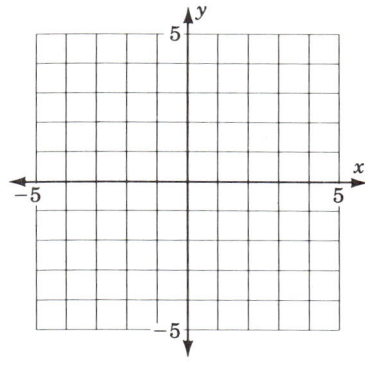

14. Graph $\dfrac{(x - 2)^2}{9} + \dfrac{(y + 1)^2}{4} = 1$.

14.

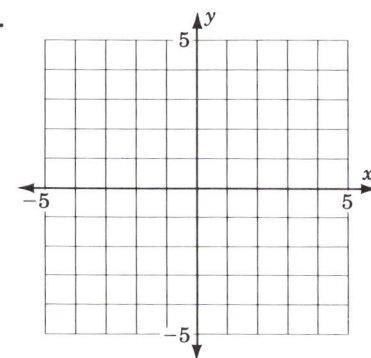

15. Graph $\dfrac{y^2}{9} - \dfrac{x^2}{16} = 1$.

15.

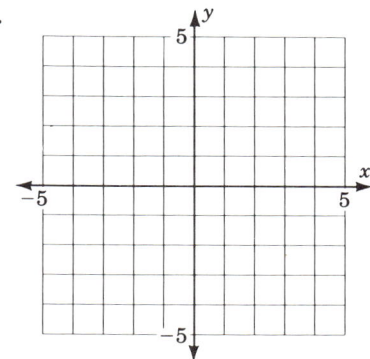

16. Graph $\dfrac{x^2}{16} - \dfrac{y^2}{9} = 1$.

16.

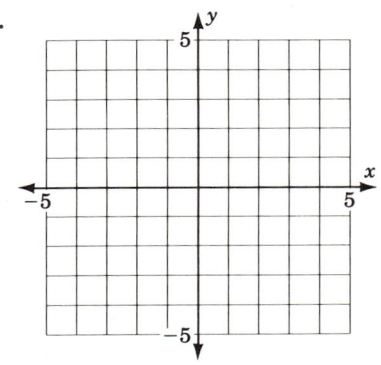

17. Identify each of the following curves.
 a. $x^2 = 4 - y^2$
 b. $y = x^2 - 9$
 c. $9x^2 = 144 - 16y^2$
 d. $16x^2 = 144 + 9y^2$

17. a. _____
 b. _____
 c. _____
 d. _____

18. Graph the inequality $y \leq -x^2 + 4$.

18.

19. Graph the inequality $x^2 + y^2 < 4$.

19.

20. Graph the inequality $4y^2 - x^2 \leq 4$.

20.

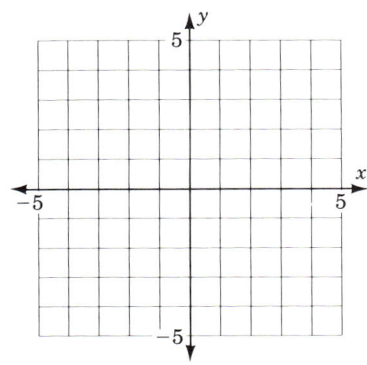

21. Graph the system $x^2 + y^2 \leq 9$ and $y \leq x^2$.

21.

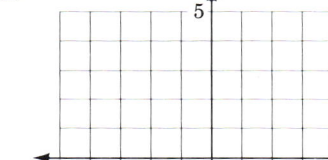

22. Use the substitution method to solve the system.

$$x^2 + y^2 = 9$$
$$x + y = 3$$

22. _____

23. Use the substitution method to solve the system.

$$x^2 + y^2 = 4$$
$$x + y = 4$$

23. _____

24. Solve the system.

$$x^2 - \ y^2 = 2$$
$$x^2 + 2y^2 = 8$$

24. _____

25. The cost C of manufacturing and selling x units of a product is $C = 20x + 75$, and the corresponding revenue R is $R = x^2 - 50$. Find the break-even value of x.

25. _____

IF YOU MISSED QUESTION	SECTION	EXAMPLES	PAGE	ANSWERS
1	9.1A	1, 2	672–674	1.
2	9.1A	3	674	2. $(2, 1)$
3	9.1B	4	676–677	3. $(-2, -5)$

IF YOU MISSED QUESTION	SECTION	EXAMPLES	PAGE	ANSWERS
4	9.1B	5	678	4. $(1, 3)$
5	9.1C	6	679	5. $(-1, 1)$
6	9.1C	7	679	6. $\left(-\dfrac{1}{4}, \dfrac{3}{2}\right)$
7	9.1D	8	680	7. $x = 1000$
8	9.2A	1	689	8. $(x + 1)^2 + (y - 2)^2 = 9$
9	9.2A	2	690	9. $x^2 + y^2 = 4$

IF YOU MISSED QUESTION	SECTION	EXAMPLES	PAGE	ANSWERS
10	9.2B	3	690	**10.** Center, $(1, -2)$; $r = 2$
11	9.2B	4	691	**11.**
12	9.2B	5	691	**12.** Center, $(1, -2)$; $r = 2$

IF YOU MISSED QUESTION	SECTION	EXAMPLES	PAGE	ANSWERS
13	9.2C	6	692	**13.**
14	9.2C	7	693	**14.**
15	9.3A	1a	705	**15.**
16	9.3A	1b	705	**16.**

IF YOU MISSED QUESTION	SECTION	EXAMPLES	PAGE	ANSWERS
17a	9.3B	2	707	**17. a.** Circle
17b	9.3B	2	707	**b.** Parabola
17c	9.3B	2	707	**c.** Ellipse
17d	9.3B	2	707	**d.** Hyperbola
18	9.4A	1	717–718	**18.**

| 19 | 9.4A | 2 | 718 |

19.

| 20 | 9.4A | 3 | 718–719 |

20.

IF YOU MISSED QUESTION	SECTION	EXAMPLES	PAGE	ANSWERS
21	9.4B	4	719	**21.**
22	9.5A	1	728–729	**22.** $(3, 0)$ and $(0, 3)$
23	9.5A	2	729–730	**23.** $(2 + \sqrt{2}i, 2 - \sqrt{2}i)$, $(2 - \sqrt{2}i, 2 + \sqrt{2}i)$
24	9.5B	3	730–731	**24.** $(2, \sqrt{2}), (2, -\sqrt{2})$, $(-2, \sqrt{2}), (-2, -\sqrt{2})$
25	9.5C	4	731–732	**25.** $x = 25$

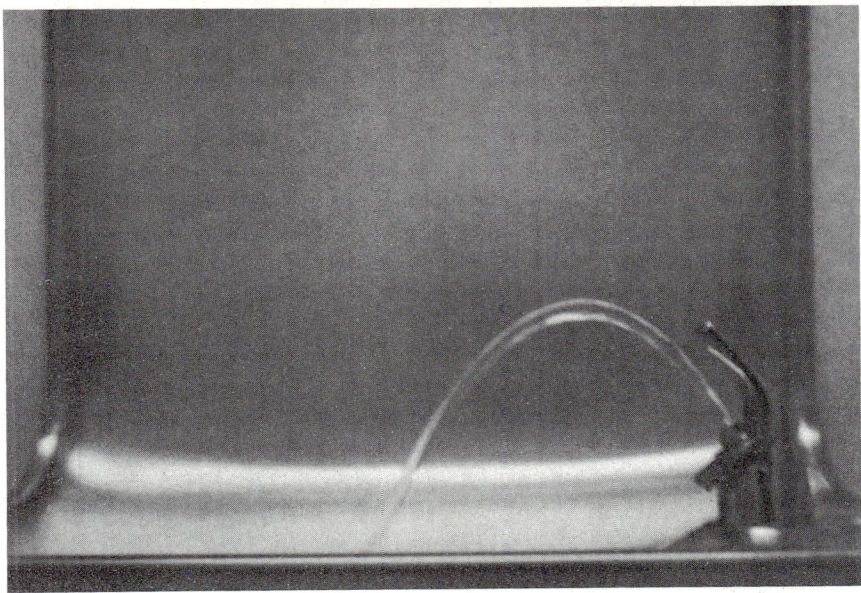

OBJECTIVES

REVIEW

Before starting this section, you should know:

1. How to graph points in the Cartesian coordinate system.
2. How to find *x*- and *y*-intercepts.
3. The quadratic formula.
4. How to complete the square in a quadratic equation.
5. How to find the discriminant of a quadratic equation.

OBJECTIVES

You should be able to:

A. Graph a parabola of the form $y = a(x - h)^2 + k$.
B. Graph a parabola $y = ax^2 + bx + c$ by using the procedure in the text.
C. Graph parabolas of the form $x = a(y - k)^2 + h$ or $x = ay^2 + by + c$.
D. Solve applications using the material studied.

Have you seen any parabolas lately? They are as near as your water fountain. Parabolas, ellipses, circles, and hyperbolas are called **conic sections** because they can be obtained by slicing a cone with a plane, as shown in Figure 1.

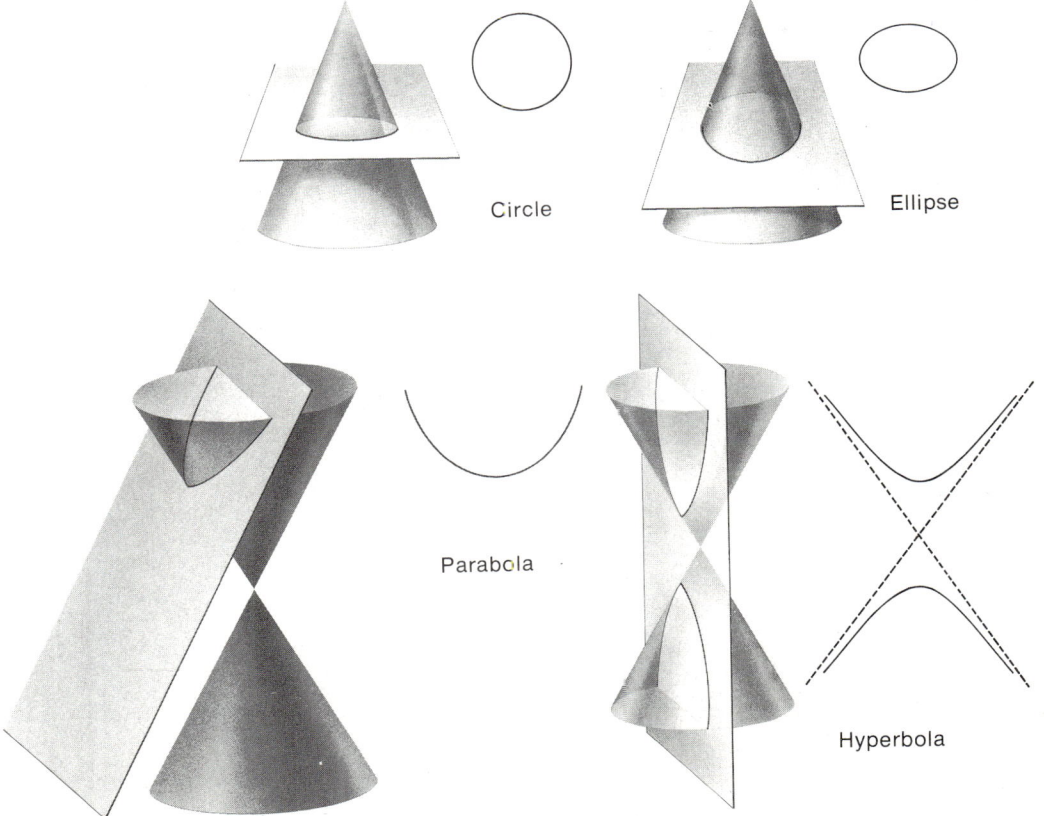

Circle   Ellipse

Parabola

Hyperbola

A. THE PARABOLA $y = a(x - h)^2 + k$

The simplest parabola is the graph of $y = x^2$. To draw this graph, we select values for x, find the corresponding values of y, and make a table, as shown.

x-VALUE	y-VALUE
$x = -2$	$y = x^2 = (-2)^2 = 4$
$x = -1$	$y = x^2 = (-1)^2 = 1$
$x = 0$	$y = x^2 = (0)^2 = 0$
$x = 1$	$y = x^2 = (1)^2 = 1$
$x = 2$	$y = x^2 = (2)^2 = 4$

x	y
-2	4
-1	1
0	0
1	1
2	4

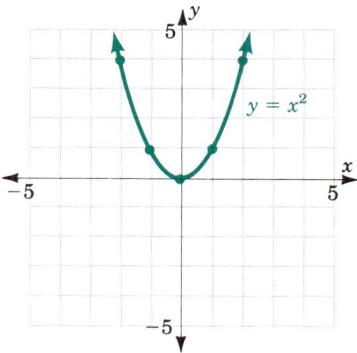

We then plot the ordered pairs on a coordinate system and draw a smooth curve through the plotted points as in the figure. A very important feature of this parabola is its symmetry to the y-axis. This symmetry follows because the same value of y is obtained for any value of x and its opposite. For instance, $x = 2$ and $x = -2$ both give $y = 4$. (See the table.) Because of this symmetry, the y-axis is called the **axis** of the parabola. The point $(0, 0)$, where the parabola crosses its axis, is called the **vertex** of the curve. Note that the arrows on the curve in the figure mean that the parabola goes on without end.

EXAMPLE 1. Graph $y = -x^2$.

Solution We could make a table of x and y values as before. However, note that for any x-value, the y-value will be the *negative* of the y-value on the parabola $y = x^2$. (If you don't believe this, go ahead and make the table and check it.) Thus, the parabola $y = -x^2$ has the same shape as $y = x^2$, but it is turned in the *opposite* direction (opens *downward*). The graph of $y = -x^2$ is shown.

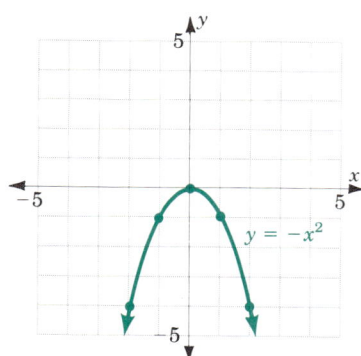

Problem 1 Graph $y = -2x^2$.

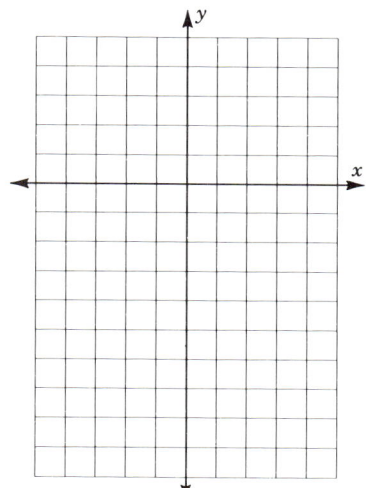

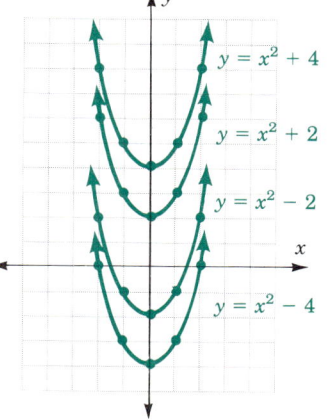

As you can see from the two preceding examples, when the coefficient of x^2 is positive (as in $y = x^2 = 1x^2$), the parabola is **concave up,** but when the coefficient of x^2 is negative (as in $y = -x^2 = -1x^2$) the parabola is **concave down.**

What do you think will happen if we graph the parabola $y = x^2 + 2$? Two things: First, the parabola opens upward, since the coefficient of x^2 is understood to be $+1$. Second, all of the points will be two units higher than those for the same value of x on the parabola $y = x^2$. Thus we can make the graph of $y = x^2 + 2$ by following the pattern of $y = x^2$. The graphs of $y = x^2 + 2$, $y = x^2 + 4$, $y = x^2 - 2$, and $y = x^2 - 4$ are shown in the margin.

For $y = x^2 + 2$		For $y = x^2 + 4$		For $y = x^2 - 2$		For $y = x^2 - 4$	
x	y	x	y	x	y	x	y
0	2	0	4	0	-2	0	-4
± 1	3	± 1	5	± 1	-1	± 1	-3
± 2	6	± 2	8	± 2	2	± 2	0

Note that adding or subtracting a number k on the right side of the equation $y = x^2$ raises or lowers the graph by k units.

EXAMPLE 2 Graph $y = -x^2 - 2$.

Solution Since the coefficient of x^2 (which is understood to be -1) is negative, the parabola opens downward. It is also 2 units lower than the graph of $y = -x^2$. Thus the graph of $y = -x^2 - 2$ is as shown.

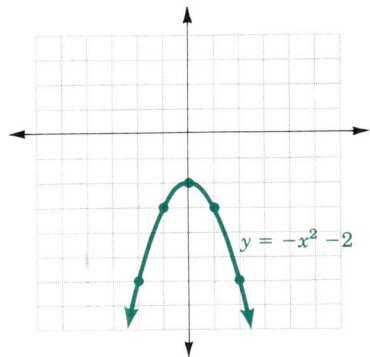

Problem 2 Graph $y = -x^2 - 1$.

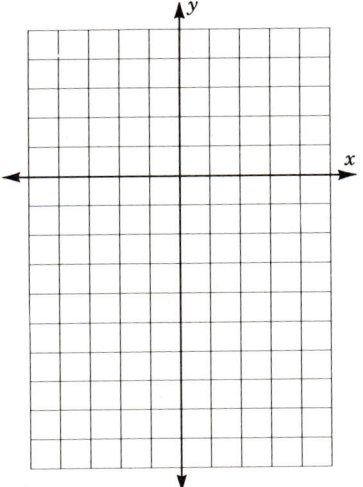

So far, we have graphed only parabolas of the form $y = ax^2 + k$. How do you think the graph of $y = (x - 1)^2$ looks? As before, we make a table of values. For example, for

$$x = -1, \quad y = (-1 - 1)^2 = (-2)^2 = 4$$
$$x = 0, \quad y = (0 - 1)^2 = (-1)^2 = 1$$
$$x = 1, \quad y = (1 - 1)^2 = (0)^2 = 0$$
$$x = 2, \quad y = (2 - 1)^2 = 1^2 = 1$$
$$x = 3, \quad y = (3 - 1)^2 = 2^2 = 4$$

The table and the graph appear below.

x	y
-1	4
0	1
1	0
2	1
3	4

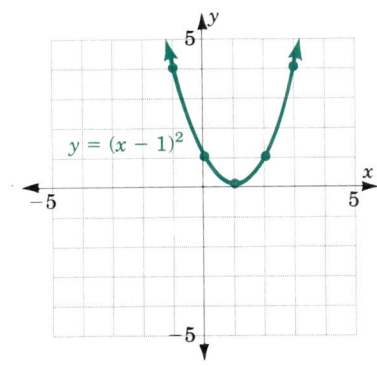

Note that the shape of the graph is identical to that of $y = x^2$, but it is shifted 1 unit to the *right*. Similarly, the graph of $y = -(x + 1)^2$ is identical to that of $y = -x^2$ but shifted 1 unit to the *left,* as shown in the figure.

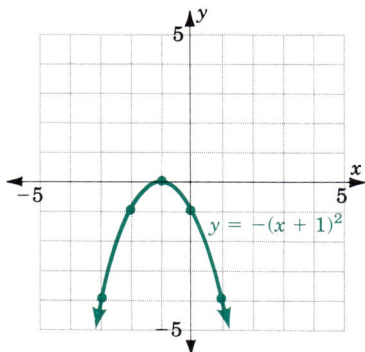

$y = -(x + 1)^2$

EXAMPLE 3 Graph $y = (x - 1)^2 - 2$.

Solution The graph of this equation is identical to the graph of $y = x^2$ except for its position. The new parabola is shifted one unit to the right (because of the -1) and 2 units down (because of the -2). The diagram indicates these two facts and shows the finished graph of $y = (x - 1)^2 - 2$.

$$y = (x - 1)^2 - 2$$

opens upward (positive)   shifted 1 unit right   shifted 2 units down.

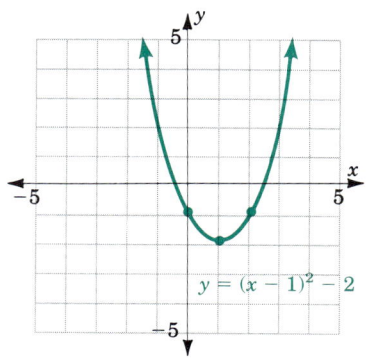

$y = (x - 1)^2 - 2$

The graph of $y = -x^2 - 2$ (Example 2) is exactly the same as the graph of $y = -x^2$ (Example 1) but moved vertically two units *down*. In general, the graph of $y = ax^2 + k$ is the same as the graph of $y = ax^2$ but moved vertically k units. In both cases, the vertex is at $(0, k)$. The graph of $y = (x - 1)^2$ is the same as that of $y = x^2$ but moved horizontally one unit *right*. Moreover, the graph of $y = (x - 1)^2 - 2$ (Example 3) is exactly the same as the graph of $y = (x - 1)^2$ but moved vertically two units *down*.

Here is the summary of this discussion.

> The graph of the parabola $y = a(x - h)^2 + k$ is the same as that of $y = ax^2$ but moved h units horizontally and k units vertically. The *vertex* is at the point (h, k).

Problem 3 Graph $y = (x - 2)^2 - 1$.

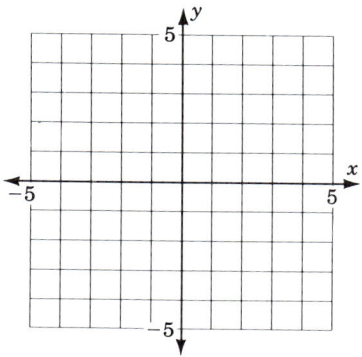

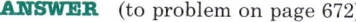

ANSWER (to problem on page 672)

1.

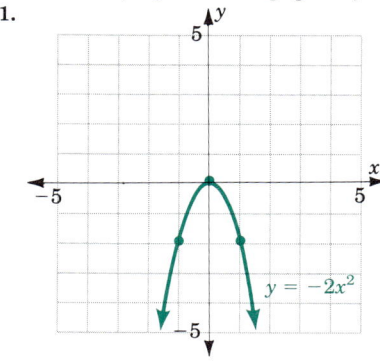

$y = -2x^2$

In conclusion, follow the given directions to graph an equation of the form

$$y = a(x - h)^2 + k$$

opens upward for $a > 0$, downward for $a < 0$ shifts the graph right or left   moves the graph up or down

The graphs of $y = 2(x - 1)^2 + 1$, $y = 2(x - 1)^2 + 3$, $y = -2(x - 1)^2 - 1$, and $y = -2(x - 1)^2 - 3$ are shown.

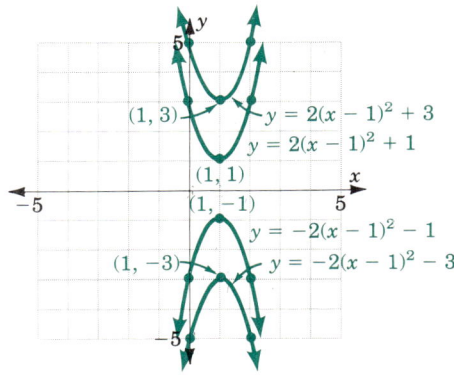

$(1, 3)$ $y = 2(x - 1)^2 + 3$
$y = 2(x - 1)^2 + 1$
$(1, 1)$
$(1, -1)$
$y = -2(x - 1)^2 - 1$
$(1, -3)$ $y = -2(x - 1)^2 - 3$

B. GRAPHING THE PARABOLA $y = ax^2 + bx + c$

We can graph the parabola $y = ax^2 + bx + c$ the same way we graphed $y = ax^2$. The **intercepts** of $y = ax^2 + bx + c$ are found by letting $x = 0$ and finding y and then letting $y = 0$ and finding x. For $x = 0$, the equation $y = ax^2 + bx + c$ gives $y = c$, the y-intercept of the parabola. Similarly, for $y = 0$, we have $ax^2 + bx + c = 0$, and x can be found by using the quadratic formula. These x-values will be real numbers only if $b^2 - 4ac \geq 0$.

We can find the *vertex,* the lowest or highest point on the parabola, by completing the square to write the equation in the form

$$y = a(x - h)^2 + k$$

which we have already studied.

Here are the steps:

$y = ax^2 + bx + c$ Given.

$\displaystyle = a\left[x^2 + \frac{b}{a} + \right] + c$ Factor a.

$\displaystyle = a\left[x^2 + \frac{b}{a} + \left(\frac{b}{2a}\right)^2\right] + c - a\left(\frac{b}{2a}\right)^2$ Add and subtract $a\left(\dfrac{b}{2a}\right)^2$.

$\displaystyle = a\left(x + \frac{b}{2a}\right)^2 + c - a\left(\frac{b}{2a}\right)^2$ Factor.

$\displaystyle = a\left(x + \frac{b}{2a}\right)^2 + c - a \cdot \frac{b^2}{4a^2}$ Square.

$\displaystyle = a\left(x + \frac{b}{2a}\right)^2 + c - \frac{b^2}{4a}$ Multiply.

$\displaystyle = a\left(x + \frac{b}{2a}\right)^2 + \frac{4ac - b^2}{4a}$ Find the LCD.

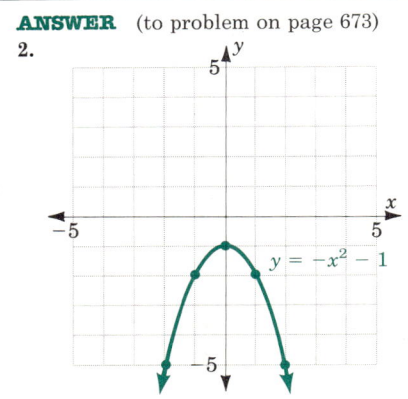

To write

$$y = a\left(x + \frac{b}{2a}\right)^2 + \frac{4ac - b^2}{4a}$$

in the form $y = a(x - h)^2 + k$ we must have $h = -\dfrac{b}{2a}$ and $k = \dfrac{4ac - b^2}{4a}$, the coordinates of the vertex. Note that you *do not* have to memorize the y-coordinate of the vertex. After you find the x-coordinate, substitute in the equation and find y.

Here is a summary of this discussion.

TO GRAPH THE PARABOLA $y = ax^2 + bx + c$

1. To find the vertex let $x = -\dfrac{b}{2a}$ in the equation and solve for y, or complete the square and compare with $y = a(x - h)^2 + k$.

2. Let $x = 0$. The result, c, is the y-intercept.

3. Let $y = 0$. Find x by solving $ax^2 + bx + c = 0$. If the roots are real numbers, they are the x-intercepts. If not, the parabola does not intersect the x-axis.

4. Draw a smooth curve through the points found in steps 1–3. Remember that if $a > 0$, the parabola opens *upward;* if $a < 0$, the parabola opens *downward.*

5. If you need more points, keep in mind that the parabola is symmetric to its axis. Use this symmetry to get additional points.

We use this procedure in the next example.

EXAMPLE 4   Graph the parabola $y = x^2 + 3x + 2$.

Solution

1. We first find the vertex.

METHOD 1.

Use the formula.

Since $a = 1$, $b = 3$, and $c = 2$,

$$x = -\frac{b}{2a} = -\frac{3}{2}$$

Substituting for x
in the equation gives

$$y = x^2 + 3x + 2$$
$$= \left(-\frac{3}{2}\right)^2 + 3\left(-\frac{3}{2}\right) + 2$$
$$= \frac{9}{4} - \frac{9}{2} + 2$$
$$= \frac{9}{4} - \frac{18}{4} + \frac{8}{4} = -\frac{1}{4}$$

The vertex is at $\left(-\frac{3}{2}, -\frac{1}{4}\right)$.

METHOD 2.

Complete the square.

$$y = [x^2 + 3x + \quad] + 2$$
$$= \left[x^2 + 3x + \left(\frac{3}{2}\right)^2\right] + 2 - \left(\frac{3}{2}\right)^2$$
$$= \left(x + \frac{3}{2}\right)^2 + 2 - \frac{9}{4}$$
$$= \left(x + \frac{3}{2}\right)^2 - \frac{1}{4}$$

The vertex is at $x = -\frac{3}{2}$, $y = -\frac{1}{4}$—that is, at the point $\left(-\frac{3}{2}, -\frac{1}{4}\right)$.

Problem 4   Graph $y = x^2 + 2x - 3$

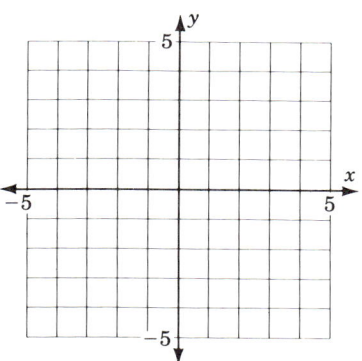

2. Let $x = 0$; then $y = x^2 + 3x + 2$ becomes $y = 2$. The y-intercept is 2.

3. Let $y = 0$. $y = x^2 + 3x + 2$ becomes

$$0 = x^2 + 3x + 2$$
$$= (x + 2)(x + 1)$$

Thus $x = -2$ or $x = -1$. The graph intersects the x-axis at $(-2, 0)$ and $(-1, 0)$.

4. Since the coefficient of x^2 is 1, $a > 0$ and the parabola opens upward.

5. By symmetry, the point $(-3, 2)$ is also on the graph.

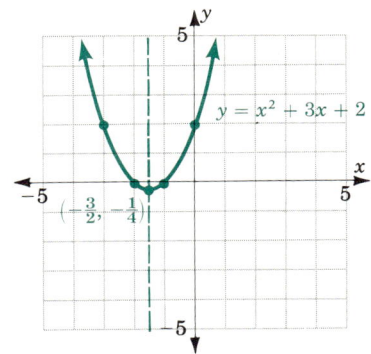

We draw a smooth curve through these points to obtain the graph of the parabola as shown. ▲

In Example 4, $x^2 + 3x + 2$ can be factored, and thus we can find the points at which the parabola crosses the x-axis. If the equation of the parabola cannot be factored, look at the discriminant $D = b^2 - 4ac$ and determine what kind of roots the equation has.

1. If $D < 0$, there are no real roots and the equation will *not* cross the x-axis.

2. If $D \geq 0$, use the quadratic formula to find x and approximate the answers so you can graph them.

Here are the possibilities.

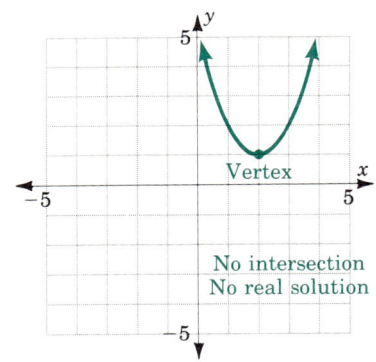

No intersection
No real solution

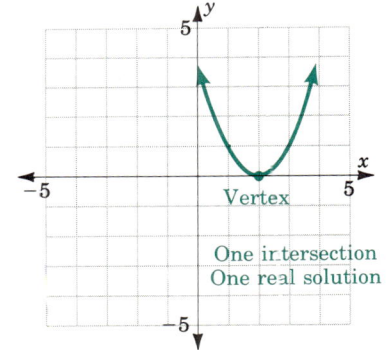

One intersection
One real solution

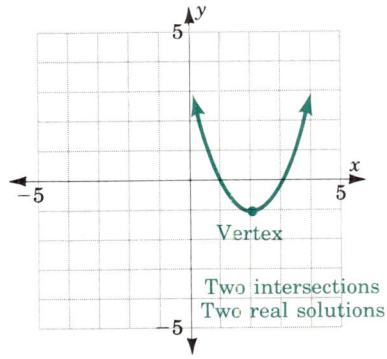

Two intersections
Two real solutions

ANSWER   (to problem on page 674)

3.

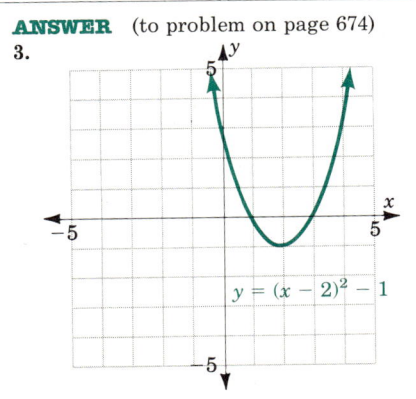

$$y = (x - 2)^2 - 1$$

EXAMPLE 5 Graph $y = -2x^2 + 4x - 3$.

Problem 5 Graph $y = -2x^2 - 4x - 3$.

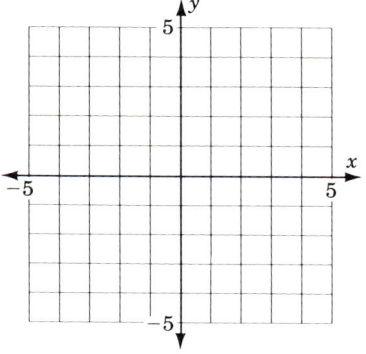

Solution

1. To find the vertex, we can use:

METHOD 1.

Use the vertex formula.

Here $a = -2$ and $b = 4$, so

$$x = -\frac{b}{2a}$$

$$= -\frac{4}{-2 \cdot 2}$$

$$= 1$$

If we substitute $x = 1$
in $y = -2x^2 + 4x - 3$,

$$y = -2(1)^2 + 4(1) - 3$$

$$= -2 + 4 - 3$$

$$= -1$$

Thus the vertex is at $(1, -1)$.

METHOD 2.

Complete the square.

$$y = -2x^2 + 4x - 3$$

$$= -2(x^2 - 2x + \quad) - 3$$

$$= -2(x^2 - 2x + 1) - 3 + 2$$

$$= -2(x - 1)^2 - 1$$

The vertex is at $(1, -1)$.

2. If $x = 0$, $y = -2x^2 + 4x - 3 = -3$, the y-intercept.
3. For $y = 0$, $0 = -2x^2 + 4x - 3$. However, the right-hand side is not factorable. As a matter of fact, the discriminant of the equation is $4^2 - 4(-2)(-3) = 16 - 24 = -8$. This means that this equation has no solution, and there are no x-intercepts. The graph does not cross the x-axis.
4. Since $a = -2 < 0$, the parabola opens downward.
5. We graph the vertex $(1, -1)$ and the y-intercept -3. To make a more accurate graph, we need some more points. Since the parabola is symmetric, we can get a point across from the y-intercept by letting $x = 2$. Then

$$y = -2(2)^2 + 4(2) - 3$$

$$= -8 + 8 - 3 = -3$$

as expected. The completed graph is shown.

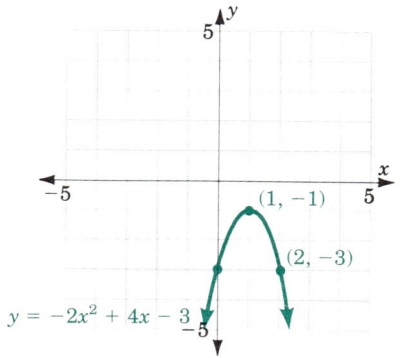

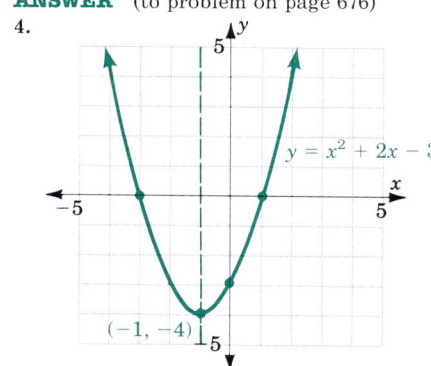

C. GRAPHING $x = a(y - k)^2 + h$ OR $x = ay^2 + by + c$

EXAMPLE 6   Graph $x = 2(y - 1)^2 + 1$.

Solution   In this problem, the roles of x and y are reversed, so the graph will look like that of $y = 2(x - 1)^2 + 1$ but opening horizontally.

The vertex of $x = 2(y - 1)^2 + 1$ is at $(1, 1)$. The curve opens to the right, the positive x-direction. The graph is shown.

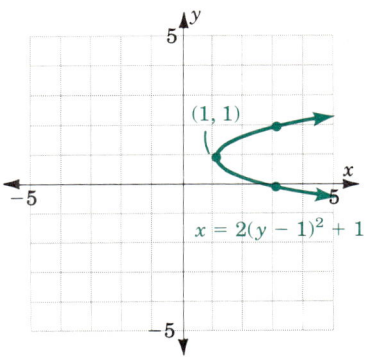

EXAMPLE 7   Graph $x = y^2 + 3y + 2$.

Solution   The graph is similar to that of $y = x^2 + 3x + 2$, but it opens horizontally (see Example 4). The vertex occurs where $y = -\frac{3}{2}$. Substituting for y in the equation gives $x = (-\frac{3}{2})^2 + 3(-\frac{3}{2}) + 2 = -\frac{1}{4}$. Thus the vertex is at $(-\frac{1}{4}, -\frac{3}{2})$. The x-intercept is 2 and the y-intercepts are where

$$0 = y^2 + 3y + 2$$
$$= (y + 2)(y + 1)$$

That is, $y = -2$ or $y = -1$. The parabola opens to the right, and the completed graph is shown.

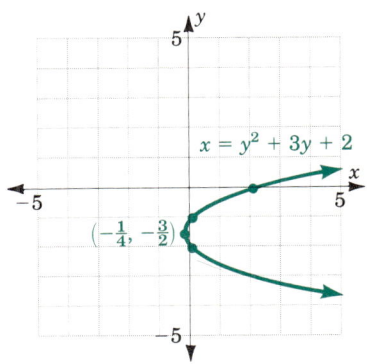

D. APPLICATIONS

Many applied problems involve finding the highest or lowest point of a parabola. For example, suppose that a record company manufactures and sells x records per week. If the revenue is given by $R = 10x - 0.01x^2$, we can use the techniques studied to maximize the revenue. We do that next.

Problem 6   Graph $x = 2(y - 1)^2 + 3$.

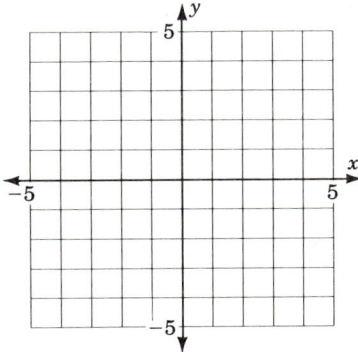

Problem 7   Graph $x = y^2 + 2y - 3$.

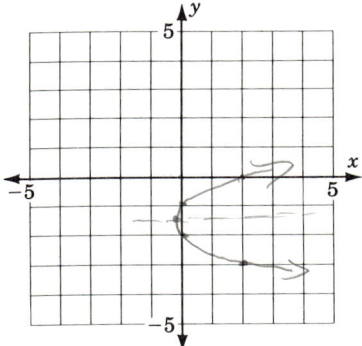

EXAMPLE 8 If $R = 10x - 0.01x^2$, how many records does the company have to sell in order to obtain maximum revenue?

Solution We first write the equation as $R = -0.01x^2 + 10x$. Since the coefficient of x^2 is negative, the parabola is concave down, and the vertex is its highest point. Letting $x = -\dfrac{b}{2a} = -\dfrac{10}{-0.02} = 500$, $R = 10(500) - 0.01(500)^2 = 5000 - 2500 = 2500$. Thus when the company produces $x = 500$ records a week, the revenue is a maximum: \$2500.

▲

Problem 8 Repeat the example if the revenue is $R = 20x - 0.01x^2$.

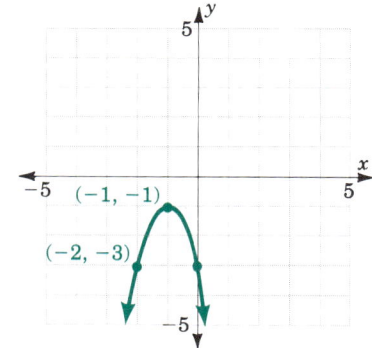
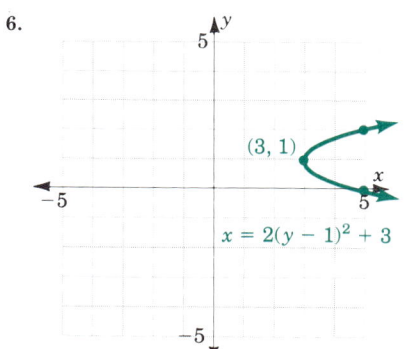
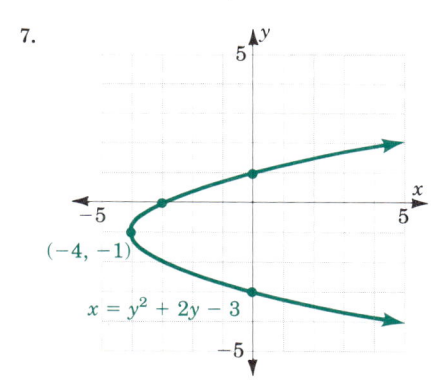

NAME

CLASS

SECTION

A.   Graph on the same coordinate axes.

1. **a.** $y = 2x^2$
   b. $y = 2x^2 + 2$
   c. $y = 2x^2 - 2$

2. **a.** $y = 3x^2 + 1$
   b. $y = 3x^2 + 3$
   c. $y = 3x^2 - 2$

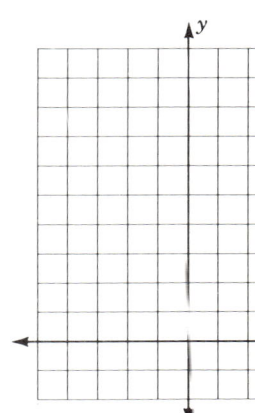

3. **a.** $y = (x + 2)^2 + 3$
   b. $y = (x + 2)^2$
   c. $y = (x + 2)^2 - 2$

4. **a.** $y = (x - 2)^2 + 2$
   b. $y = (x - 2)^2$
   c. $y = (x - 2)^2 - 2$

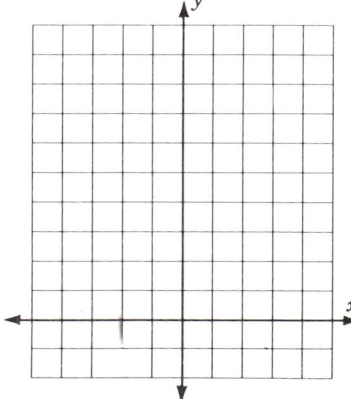

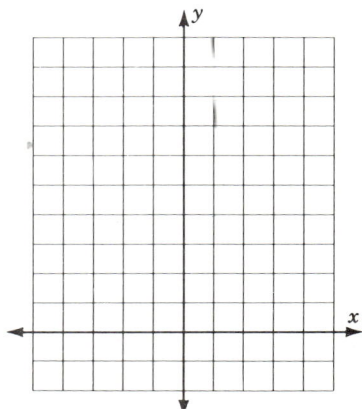

5. **a.** $y = -(x + 2)^2 - 2$
   b. $y = -(x + 2)^2$
   c. $y = -(x + 2)^2 - 4$

6. **a.** $y = -(x - 1)^2 + 1$
   b. $y = -(x - 1)^2$
   c. $y = -(x - 1)^2 + 2$

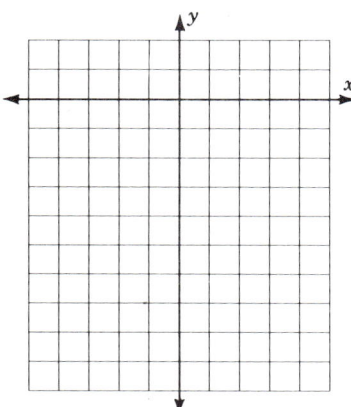 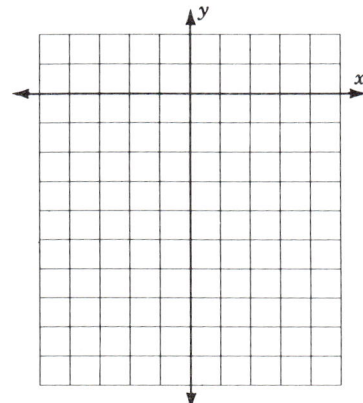

7. **a.** $y = -2(x + 2)^2 - 2$
   b. $y = -2(x + 2)^2$
   c. $y = -2(x + 2)^2 - 4$

8. **a.** $y = -2(x - 1)^2 + 1$
   b. $y = -2(x - 1)^2$
   c. $y = -2(x - 1)^2 + 2$

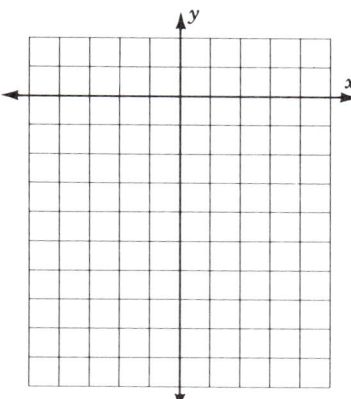 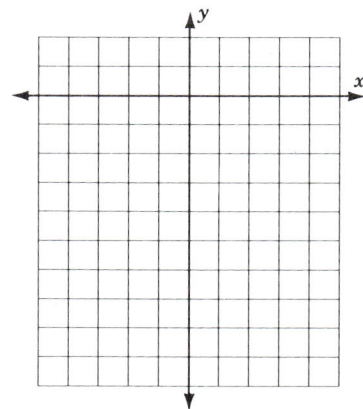

9. **a.** $y = 2(x + 1)^2 + \dfrac{1}{2}$

   b. $y = 2(x + 1)^2$

10. **a.** $y = 2(x + 1)^2 - \dfrac{1}{2}$

   b. $y = 2(x + 1)^2$

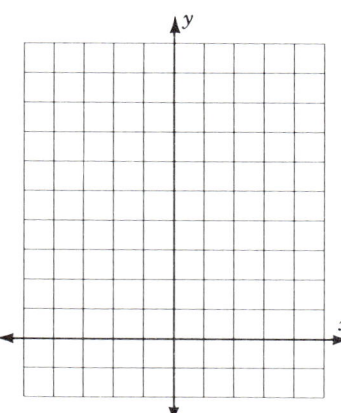 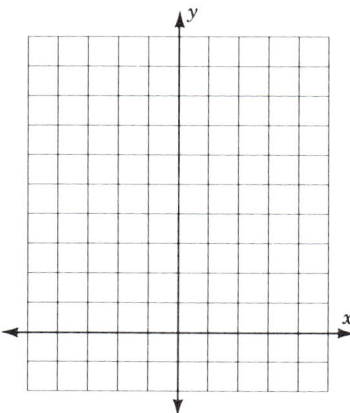

ANSWER (to problem on page 680)
8. The revenue is a maximum ($10,000) when the company produces and sells 1000 records.

B. In Problems 11–22 use the five-step procedure in the text to sketch the graph. Label the vertex and the intercepts.

11. $y = x^2 + 2x + 1$

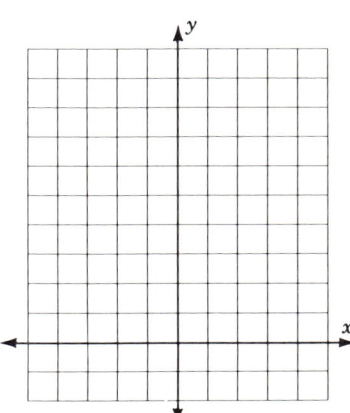

12. $y = x^2 + 4x + 4$

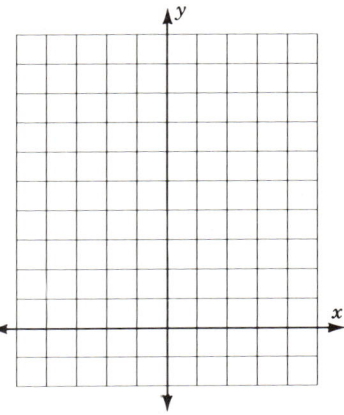

13. $y = -x^2 + 2x + 1$

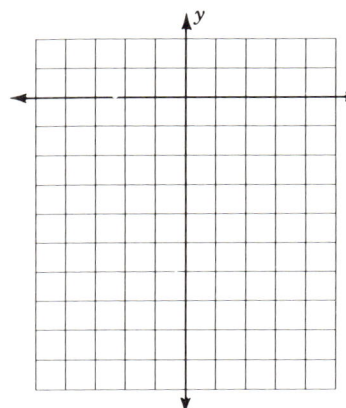

14. $y = -x^2 + 4x - 2$

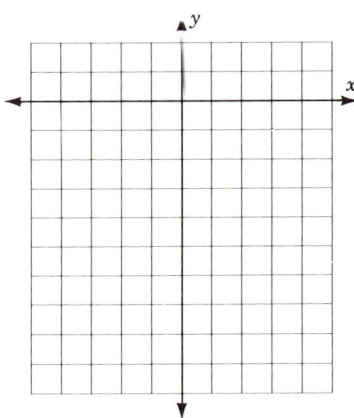

15. $y = -x^2 + 4x - 5$

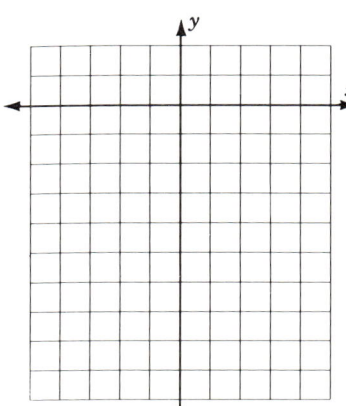

16. $y = -x^2 + 4x - 3$

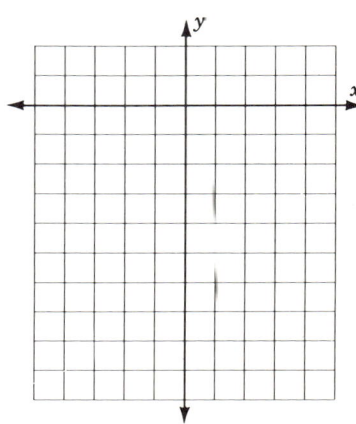

17. $y = -3 - 5x + 2x^2$

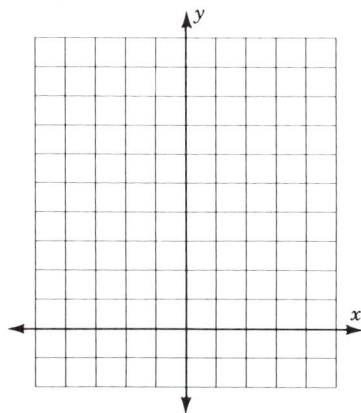

18. $y = -3 + 5x + 2x^2$

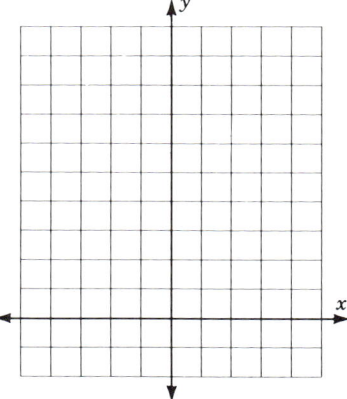

19. $y = 5 - 4x - 2x^2$
Hint: $\sqrt{56} = 7.5$

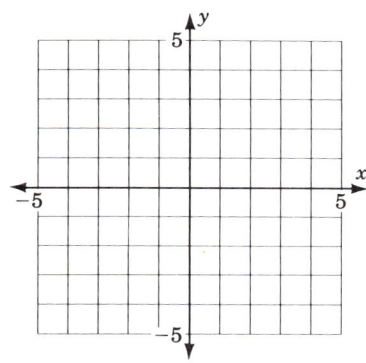

20. $y = 3 - 4x - 2x^2$
Hint: $\sqrt{40} = 6.3$

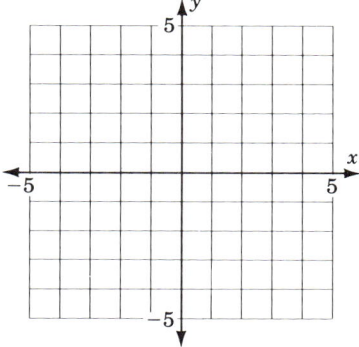

21. $y = -3x^2 + 3x + 2$
Hint: $\sqrt{33} = 5.7$

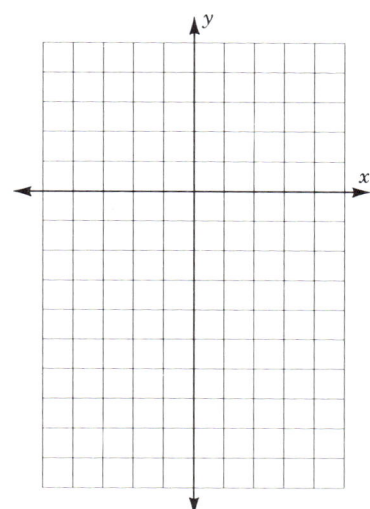

22. $y = -3x^2 + 3x + 1$
Hint: $\sqrt{21} = 4.6$

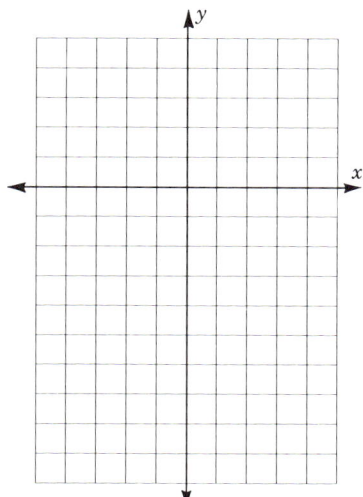

C. Graph.

23. **a.** $x = (y + 2)^2 + 3$
     b. $x = (y + 2)^2$

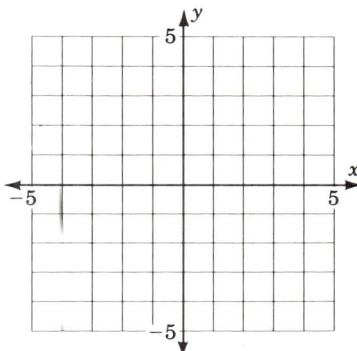

24. **a.** $x = (y - 2)^2 + 2$
     b. $x = (y - 2)^2$

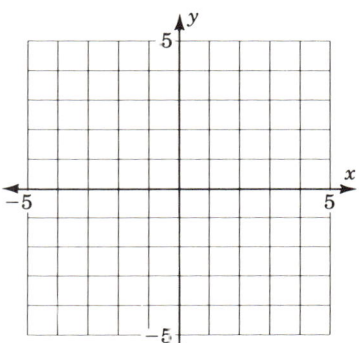

25. **a.** $x = -(y + 2)^2 - 2$
     b. $x = -(y + 2)^2$

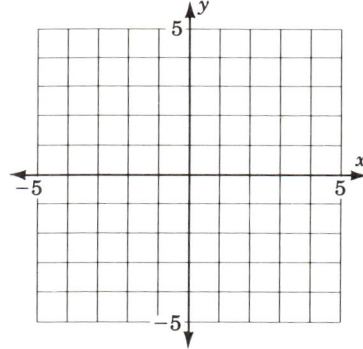

26. **a.** $x = -(y - 1)^2 + 1$
     b. $x = -(y - 1)^2$

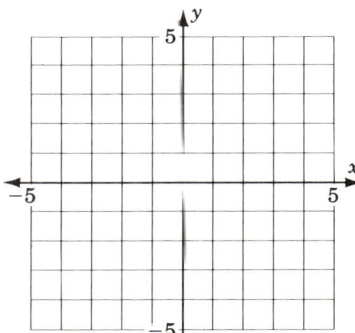

27. $x = y^2 + 2y + 1$

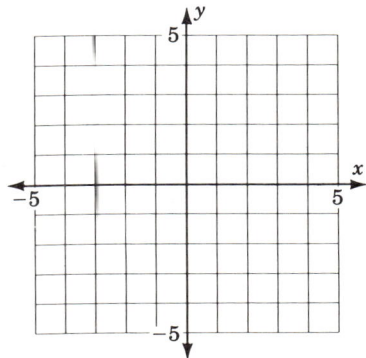

28. $x = y^2 + 4y + 4$

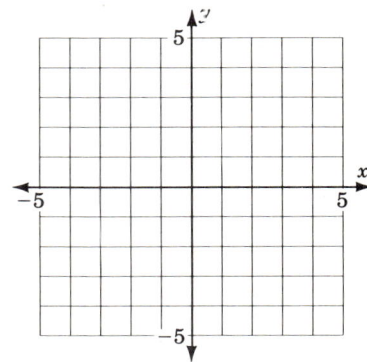

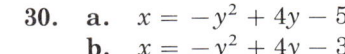

29.   **a.** $x = -y^2 + 2y + 1$
      b. $x = -y^2 + 2y + 4$

30.   **a.** $x = -y^2 + 4y - 5$
        b. $x = -y^2 + 4y - 3$

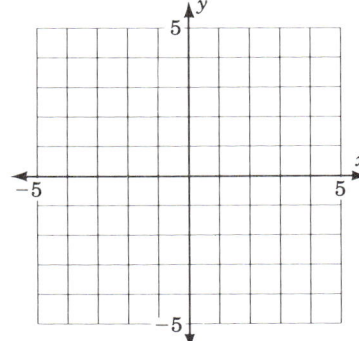 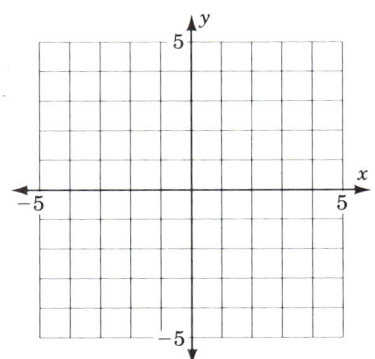

D.   Applications

31.   The profit P (dollars) for a company is $P = -5000 + 8x - 0.001x^2$, where x is the number of items produced each month. How many items does the company have to produce in order to obtain maximum profit? What is the profit?

31. _____

32.   The revenue R for Shady Glasses is given by $R = 1500p - 75p^2$, where p is the price of each pair of sunglasses. What should the price be to maximize revenues? (R and p in dollars.)

32. _____

33.   After spending x thousand dollars in an advertising campaign, the number of units N sold is given by $N = 50x - x^2$. How much should be spent in the campaign to obtain maximum sales?

33. _____

34.   The number N of units of a product sold after a television commercial blitz is $N = 40x - x^2$, where x is the number of thousands of dollars spent. How much should be spent on television commercials to obtain maximum sales?

34. _____

35.   If a ball is batted up at 160 ft/sec, its height h feet after t seconds is given by $h = -16t^2 + 160t$. Find the maximum height reached by the ball.

35. _____

36.   If a ball is thrown upward at 20 ft/sec, its height h feet after t seconds is given by $h = -16t^2 + 20t$. How many seconds does it take for the ball to reach its maximum height, and what is this height?

36. _____

37.   If a farmer digs potatoes today, she will have 600 bushels worth \$1 per bushel. Every week she waits, the crop increases by 100 bushels, but the price decreases 10¢ a bushel. Show that she should dig and sell her potatoes at the end of two weeks.

37. _____

38.   A man has a large piece of property along Washington Street. He wants to fence the sides and back of a rectangular plot. If he has 400 ft of fencing, what dimensions will give him the maximum area?

38. _____

EXERCISE 9.1

✓ SKILL CHECKER

Find the distance between each pair of points.

39. $A(3, 4)$ and $B(6, 8)$

40. $A(2, -3)$ and $B(4, 2)$

39. _____

40. _____

9.1 USING YOUR KNOWLEDGE

Here is another way of defining a parabola.

> A **parabola** is the set of all points equidistant from a fixed point $F(0, p)$ (called the **focus**) and a fixed line $y = -p$ (called the **directrix**).

If $P(x, y)$ is a point on the parabola, the definition says that $FP = DP$ (see the figure).

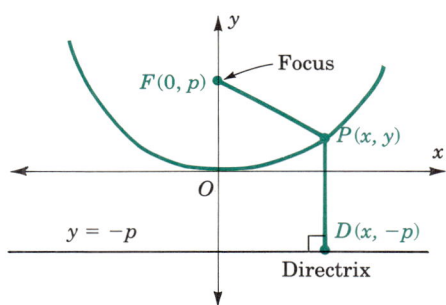

1. Find FP.

2. Find DP.

3. Set $FP = DP$ and solve for x^2.

4. For the parabola $x^2 = 4y$
  a. Locate the focus.
  b. Write the equation of the directrix.

1. _____

2. _____

3. _____

4. a. _____
   b. _____

Many applications of the parabola depend on an important focal property of the curve. If the parabola were a mirror, a ray of light parallel to the axis would be reflected to the focus, and a ray originating at the focus would be reflected parallel to the axis. (This can be proved by methods of calculus.)

If the parabola is revolved about its axis, a surface called a *paraboloid of revolution* is formed. This is the shape used for automobile headlights and for searchlights that throw a parallel beam of light when the light source is placed at the focus; it is also the shape of a radar dish or a reflecting telescope mirror that collects parallel rays of energy (light) and reflects them to the focus.

We can find the equation of the parabola needed to generate a paraboloid of revolution by using the equation $x^2 = 4py$ as follows: Suppose a parabolic mirror is to have a diameter of 6 ft and a depth of 1 ft. Then, we find the value of p that makes the parabola pass through the point (3, 1). This means that we substitute into the equation and solve for p. Thus, we have

$3^2 = 4p(1)$

so that $4p = 9$ and $p = 2.25$. The equation of the parabola is $x^2 = 9y$ and the focus is at (0, 2.25).

5. A radar dish has a diameter of 10 ft and a depth of 2 ft. The dish is in the shape of a paraboloid of revolution. Find an equation for a parabola that would generate this dish and locate the focus.

6. The cables of a suspension bridge hang very nearly in the shape of a parabola. A cable on such a bridge spans a distance of 1000 ft and sags 50 ft in the middle. Find an equation for this parabola.

5. _____

6. _____

9.2 CIRCLES AND ELLIPSES

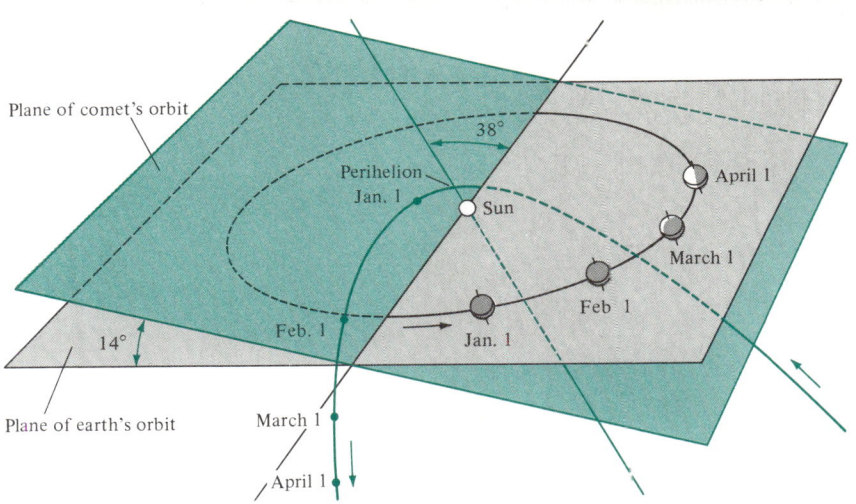

Plane of comet's orbit

38°

Perihelion
Jan. 1

Sun

April 1

March 1

Feb 1

Feb. 1

Jan. 1

14°

Plane of earth's orbit

March 1

April 1

OBJECTIVES

REVIEW

Before starting this section, you should know:

1. The distance formula.
2. How to complete the square on a quadratic equation.

OBJECTIVES

You should be able to:

A. Find the equation of a circle with a given center and radius.
B. Find the center and radius and sketch the graph of a circle when its equation is given.
C. Graph an ellipse when its equation is given.

The diagram at the beginning of this section shows the orbit of the comet Kohoutek with respect to the orbit of the earth. The comet's orbit is an *ellipse,* whereas the earth's orbit is nearly a perfect *circle.*

A. THE EQUATION OF A CIRCLE

Do you know what a circle is? A *circle* is defined as a set of points in a plane that are equidistant from a fixed point. The fixed point is called the *center* and the given distance is the *radius.* To find the equation of a circle of radius r, suppose the center is at a point $C(h, k)$. The distance from C to any point $P(x, y)$ on the circle is found by the distance formula. Since r is the radius, this distance must be r. Thus

$$\sqrt{(x - h)^2 + (y - k)^2} = r$$
$$(x - h)^2 + (y - k)^2 = r^2 \quad \text{Square both sides.}$$

We then have the following.

EQUATION OF A CIRCLE OF RADIUS r AND WITH CENTER AT (h, k)

The equation of a circle with **radius** r and with **center** at $C(h, k)$ is

$$(x - h)^2 + (y - k)^2 = r^2$$

EXAMPLE 1 Find the equation of the circle with center at $(3, -5)$ and radius 2.

Solution Here, the center $(h, k) = (3, -5)$, and $r = 2$. This means $h = 3$, $k = -5$, and $r = 2$. Using the formula, we have

$$(x - h)^2 + (y - k)^2 = r^2$$
$$(x - 3)^2 + [(y - (-5)]^2 = 2^2$$
$$(x - 3)^2 + (y + 5)^2 = 4$$

▲

Problem 1 Find the equation of the circle with center at $(-3, 6)$ and radius 3.

EXAMPLE 2   Find the equation of a circle of radius 3 and with center at the origin.

Solution   The center is at $(h, k) = (0, 0)$. Thus, $h = 0$, $k = 0$, and $r = 3$. Substituting in the formula gives

$$(x - 0)^2 + (y - 0)^2 = 3^2$$
$$x^2 + y^2 = 9 \qquad \blacktriangle$$

In general, we have following.

> The equation of a circle of radius r with center at the origin is
> $$x^2 + y^2 = r^2$$

B.   FINDING THE CENTER AND RADIUS

If we have the equation of a circle, we can write it in the form $(x - h)^2 + (y - k)^2 = r^2$ and find the center and radius. For example, if a circle has equation $(x - 3)^2 + (y - 4)^2 = 5^2$, then $h = 3$, $k = 4$, and $r = 5$. Thus the equation $(x - 3)^2 + (y - 4) = 5^2$ is the equation of a circle of radius 5 with center at $(3, 4)$.

EXAMPLE 3   Find the center and radius and sketch the graph of the circle whose equation is

$$(x + 2)^2 + (y - 1)^2 = 9$$

Solution   We first write the equation in the form

$$(x - h)^2 + (y - k)^2 = r^2$$

Thus

$$[(x - (-2)]^2 + (y - 1)^2 = 3^2$$

We have $h = -2$, $k = 1$, and $r = 3$. The center is at $(h, k) = (-2, 1)$ and the radius is 3. The sketch is shown.

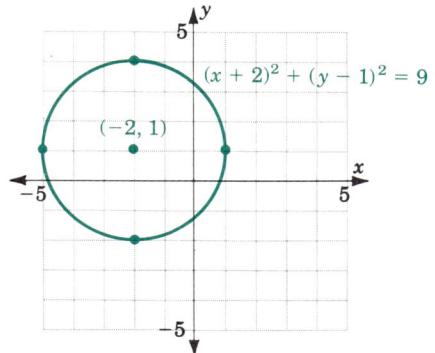

Problem 2   Find the equation of a circle of radius 5 and with center at the origin.

Problem 3   Find the center and radius and sketch the graph of the circle whose equation is $(x - 3)^2 + (y - 1)^2 = 4$.

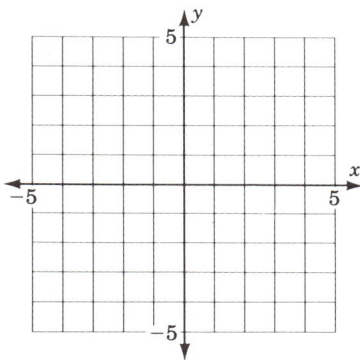

EXAMPLE 4   Sketch the graph of $x^2 + y^2 = 25$.

Solution   We find the center and radius by writing the equation in the form $(x - h)^2 + (y - k)^2 = r^2$.

$$(x - 0)^2 + (y - 0)^2 = 5^2$$

This means that $h = 0$, $k = 0$, and $r = 5$. The circle is centered at the origin and has radius 5. The sketch is shown.

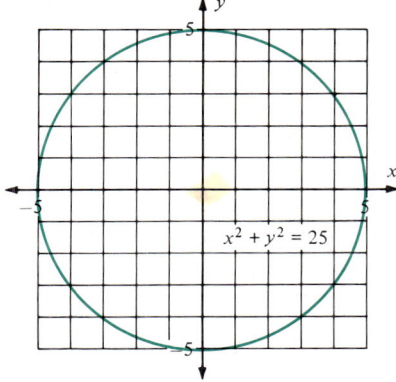

$x^2 + y^2 = 25$

▲

EXAMPLE 5   Sketch the graph of $x^2 - 4x + y^2 + 6y + 9 = 0$.

Solution   As before, we must find the center and the radius by writing the equation in the form $(x - h)^2 + (y - k)^2 = r^2$. We can do this by completing the square on x and y.

$$x^2 - 4x + y^2 + 6y + 9 = 0$$
$$x^2 - 4x + \underline{\quad} + y^2 + 6y + \underline{\quad} = -9$$
$$x^2 - 4x + 4 + y^2 + 6y + 9 = -9 + 4 + 9$$
$$(x - 2)^2 + (y + 3)^2 = 4$$
$$(x - 2)^2 + (y + 3)^2 = 2^2$$

Thus, the center is at $(2, -3)$ and the radius is 2. The graph is shown.

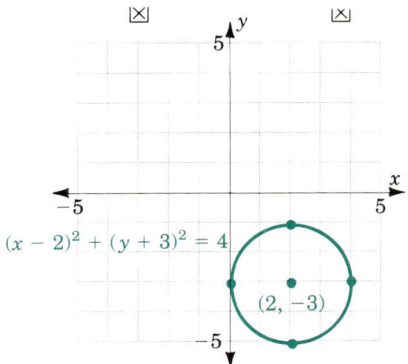

$(x - 2)^2 + (y + 3)^2 = 4$

$(2, -3)$

▲

C.   ELLIPSES

If you look at the diagram at the beginning of this section you will see the drawing of part of an ellipse. An *ellipse* is the set of points in a plane such that the sum of the distances of each point from two fixed points called the **foci** (singular, *focus*) is a constant. If the coordinates of the foci are $(c, 0)$ and $(-c, 0)$, then the center of the ellipse is at the origin, and the x- and y-intercepts are given by $x = \pm a$ and $y = \pm b$.

Problem 4   Sketch the graph of $x^2 + y^2 = 4$.

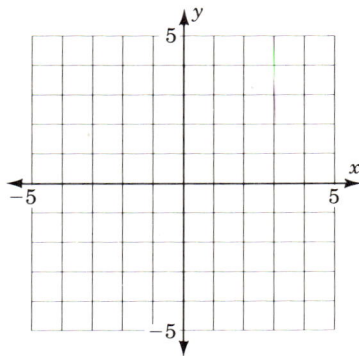

Problem 5   Sketch the graph of $x^2 - 4x + y^2 - 6y + 9 = 0$.

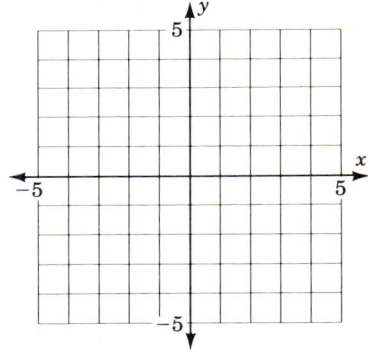

We can use the distance formula to find the equation of an ellipse (see Using Your Knowledge 9.2) but for the time being, we assume the following.

EQUATION OF AN ELLIPSE WITH FOCI ON THE AXIS AND CENTER AT THE ORIGIN

The equation of the ellipse whose x-intercepts are $x = \pm a$ and whose y-intercepts are $y = \pm b$ is

$$\frac{x^2}{a^2} + \frac{y^2}{b^2} = 1$$

If a and b are equal, the ellipse is a circle.

EXAMPLE 6   Graph $4x^2 + 25y^2 = 100$.

Solution   To make sure we have an ellipse, we must write the equation in the form $\frac{x^2}{a^2} + \frac{y^2}{b^2} = 1$. If we divide each term by 100 (to make the right side 1), we have

$$\frac{x^2}{25} + \frac{y^2}{4} = 1 \qquad \text{(elips +)}$$

The x-intercepts are $x = \pm 5$ and the y-intercepts are $y = \pm 2$. We then pass the ellipse through the four intercepts, as shown in the graph.

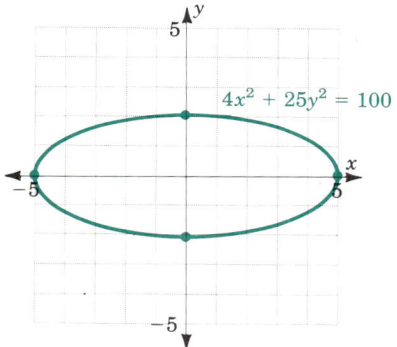

Note that we can actually graph the original equation $4x^2 + 25y^2 = 100$ by letting $x = 0$, obtaining

$$25y^2 = 100$$
$$y^2 = 4$$
$$y = \pm\sqrt{4} = \pm 2$$

Letting $y = 0$ will yield $x = \pm 5$, as before.   ▲

As in the case of circles, ellipses can be centered away from the origin, as in the next example.

Problem 6   Graph $4x^2 + 9y^2 = 36$.

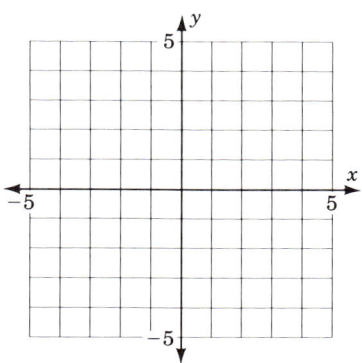

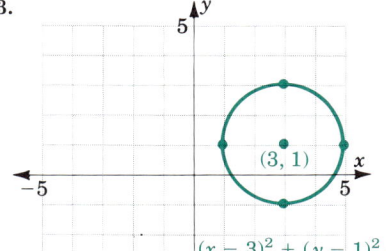

EXAMPLE 7 Graph $\dfrac{(x-3)^2}{4} + \dfrac{(y+1)^2}{9} = 1$.

Solution The center of this ellipse is at $(3, -1)$. We construct a new coordinate system with origin at $(3, -1)$, as shown in the figure. The x-intercepts are at ± 2 units from 3 and the y-intercepts at ± 3 units from -1.

The graph of the ellipse is shown in the second figure.

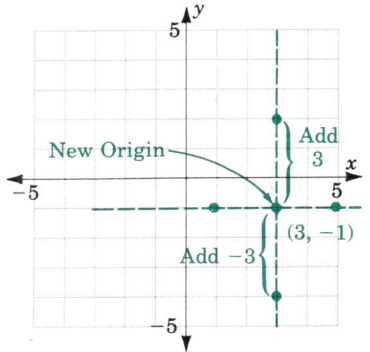

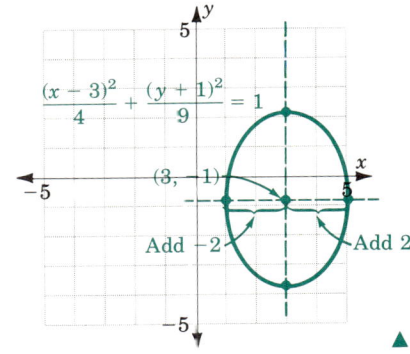

Problem 7 Graph

$\dfrac{(x+3)^2}{4} + \dfrac{(y-1)^2}{9} = 1$.

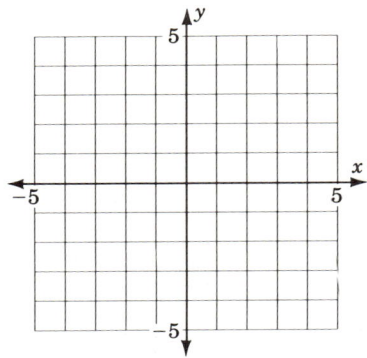

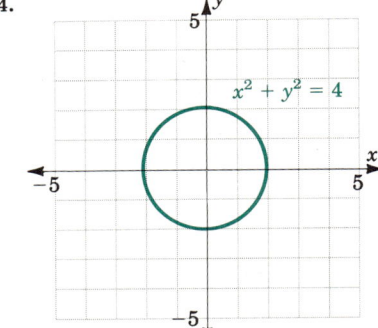
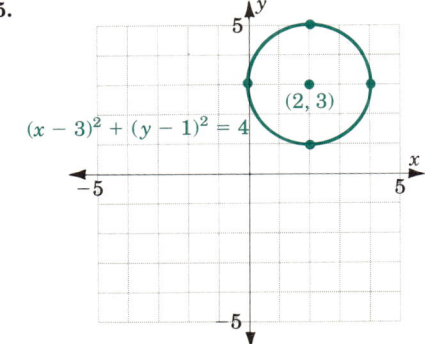

6.

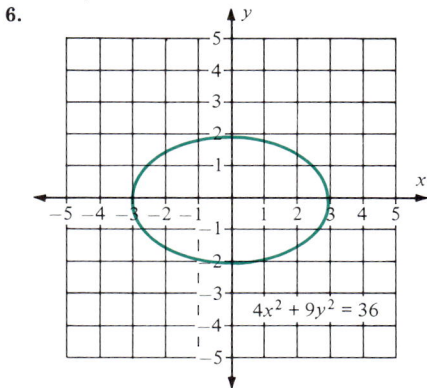

$4x^2 + 9y^2 = 36$

7.

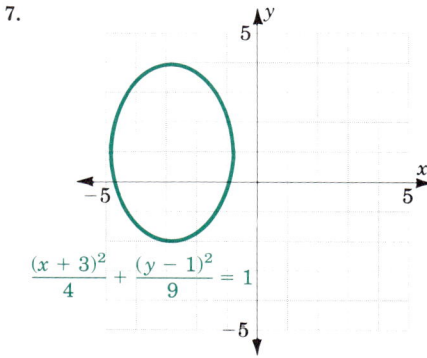

$$\frac{(x + 3)^2}{4} + \frac{(y - 1)^2}{9} = 1$$

A. In Problems 1–10 find the equation of a circle with the given center and radius.

1. Center $(3, 8)$, radius 2

2. Center $(2, 5)$, radius 3

3. Center $(-3, 4)$, radius 5

4. Center $(-5, 2)$, radius 5

5. Center $(-3, -2)$, radius 4

6. Center $(-1, -7)$, radius 9

7. Center $(2, -4)$, radius $\sqrt{5}$

8. Center $(3, -5)$, radius $\sqrt{7}$

9. Radius 3, center at the origin

10. Radius 4, center at the origin

ANSWERS

1. _____
2. _____
3. _____
4. _____
5. _____
6. _____
7. _____
8. _____
9. _____
10. _____

B. In Problems 11–32 find the center and radius of the circle and sketch the graph.

11. $(x - 1)^2 + (y - 2)^2 = 9$

12. $(x - 2)^2 + (y - 1)^2 = 4$

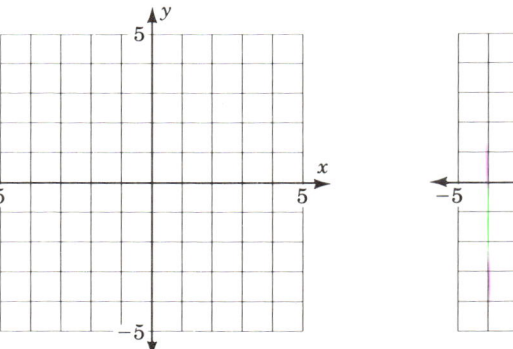
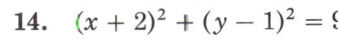

13. $(x + 1)^2 + (y - 2)^2 = 4$

14. $(x + 2)^2 + (y - 1)^2 = 9$

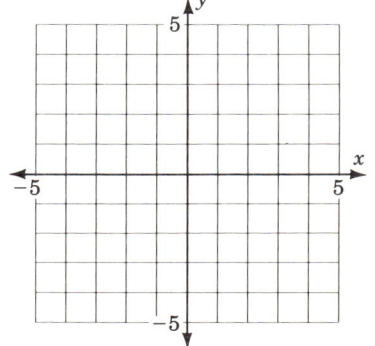
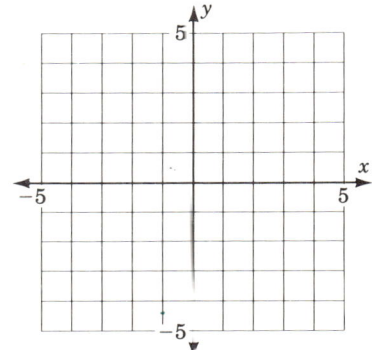

15. $(x - 1)^2 + (y + 2)^2 = 1$

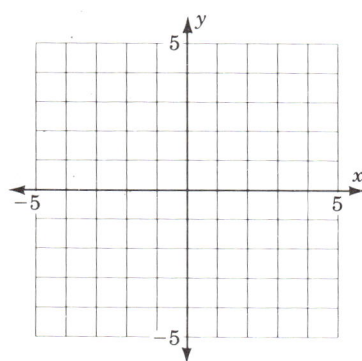

16. $(x - 2)^2 + (y + 1)^2 = 4$

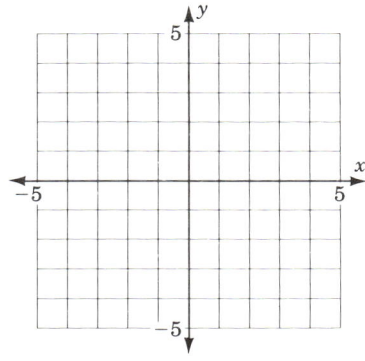

17. $(x + 2)^2 + (y + 1)^2 = 9$

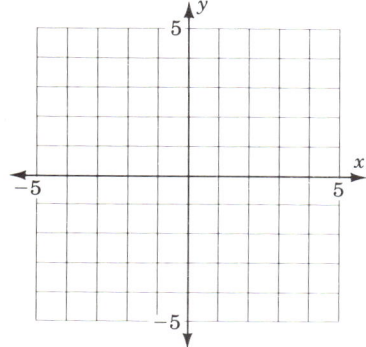

18. $(x + 3)^2 + (y + 1)^2 = 4$

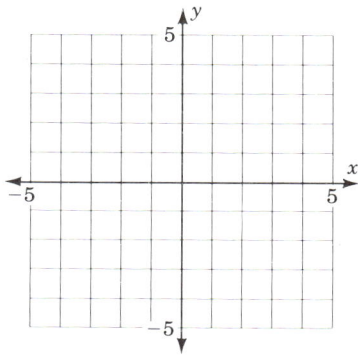

19. $(x - 1)^2 + (y - 1)^2 = 7$

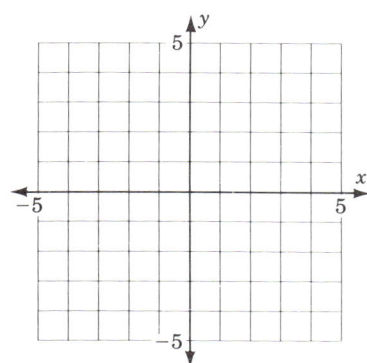

20. $(x - 1)^2 + (y - 1)^2 = 3$

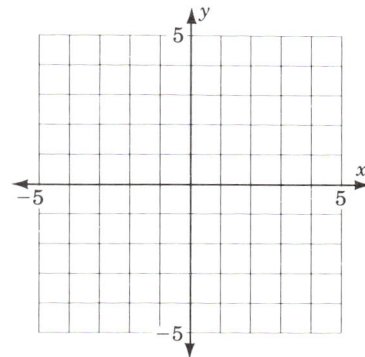

21. $x^2 - 6x + y^2 - 4y + 9 = 0$

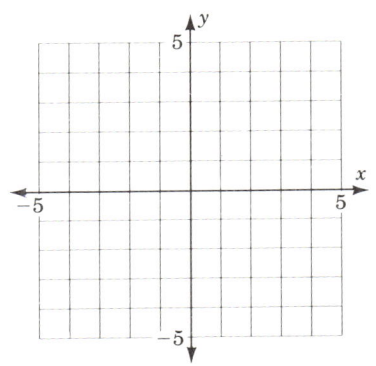

22. $x^2 - 6x + y^2 - 2y + 9 = 0$

23. $x^2 + y^2 - 4x + 2y - 4 = 0$

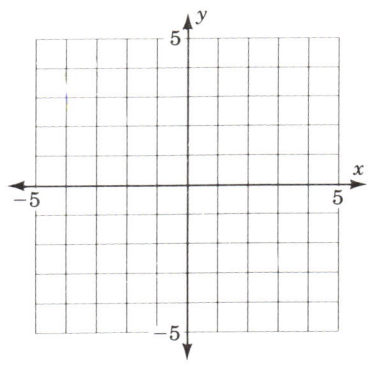

24. $x^2 + y^2 + 2x - 4y - 4 = 0$

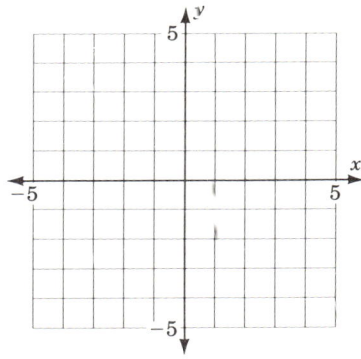

25. $x^2 + y^2 - 25 = 0$

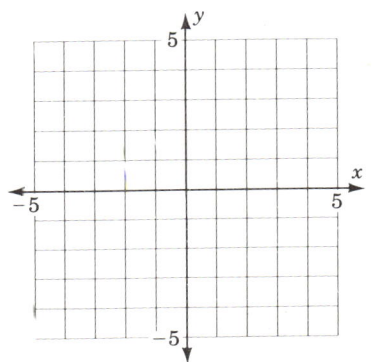

26. $x^2 + y^2 - 9 = 0$

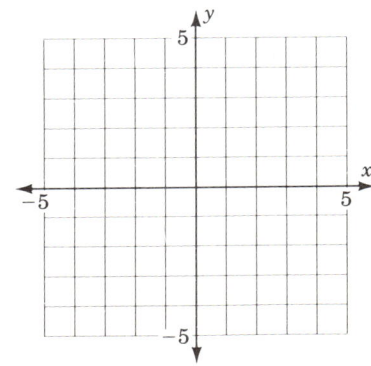

27. $x^2 + y^2 - 7 = 0$

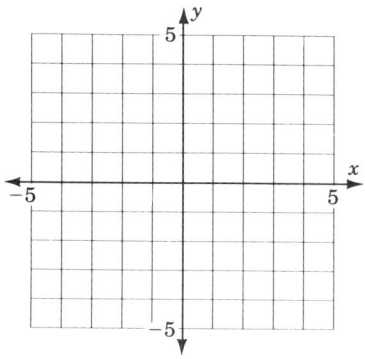

28. $x^2 + y^2 - 3 = 0$

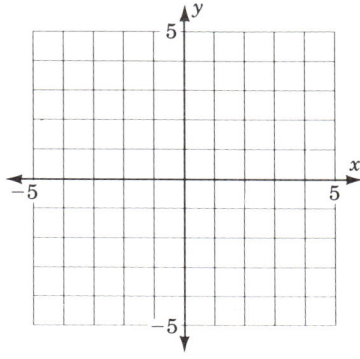

29. $x^2 + y^2 + 6x - 2y = -6$

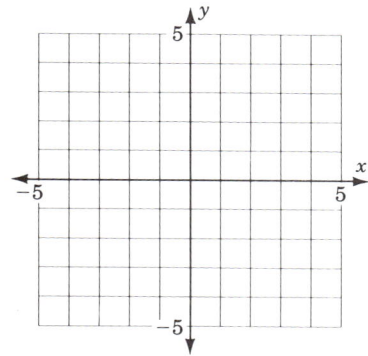

30. $x^2 + y^2 + 4x - 2y = -4$

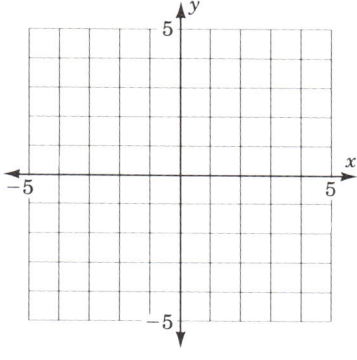

31. $x^2 + y^2 - 6x - 2y + 6 = 0$

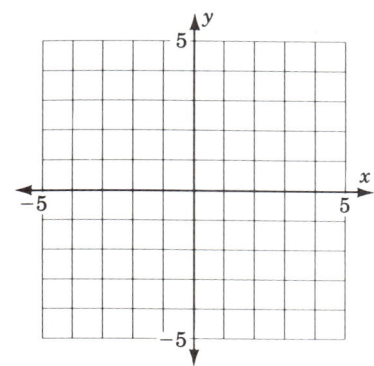

32. $x^2 + y^2 - 4x - 6y + 12 = 0$

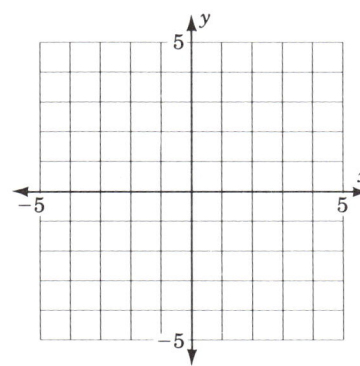

C. In Problems 33–46 graph the ellipse.

33. $25x^2 + 4y^2 = 100$

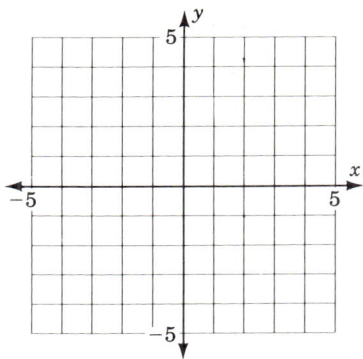

34. $9x^2 + 4y^2 = 36$

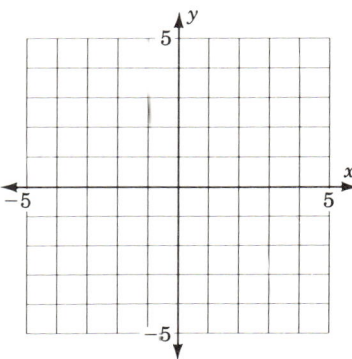

35. $x^2 + 4y^2 = 4$

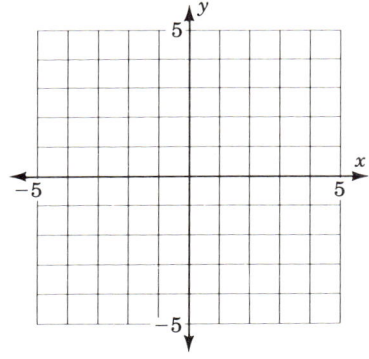

36. $x^2 + 9y^2 = 9$

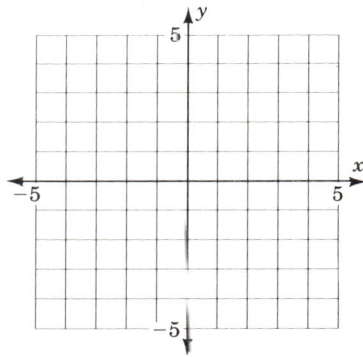

37. $x^2 + 4y^2 = 16$

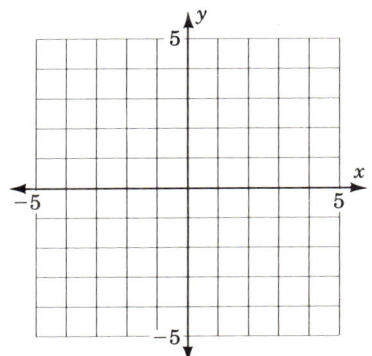

38. $x^2 + 9y^2 = 25$

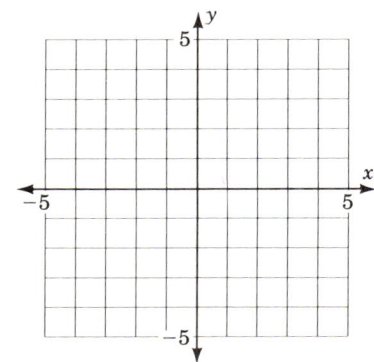

39. $\dfrac{x^2}{9} + \dfrac{y^2}{16} = 1$

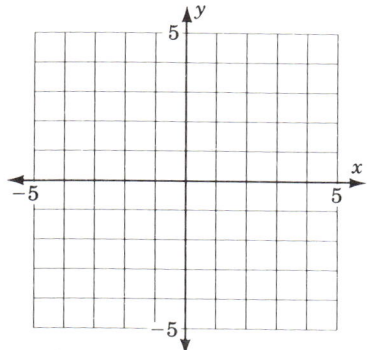

40. $\dfrac{x^2}{4} + \dfrac{y^2}{1} = 1$

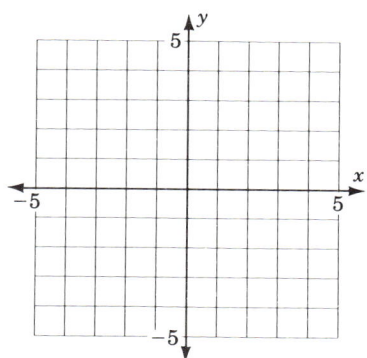

41. $\dfrac{(x-1)^2}{4} + \dfrac{(y-2)^2}{9} = 1$

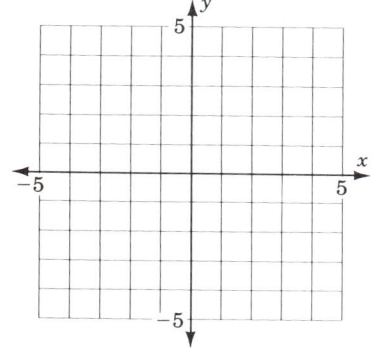

42. $\dfrac{(x-2)^2}{9} + \dfrac{(y-1)^2}{4} = 1$

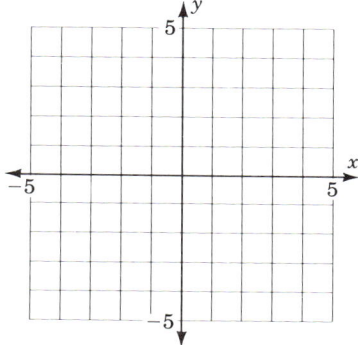

43. $\dfrac{(x-2)^2}{9} + \dfrac{(y+3)^2}{4} = 1$

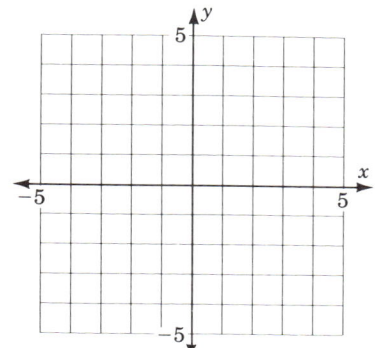

44. $\dfrac{(x-1)^2}{4} + \dfrac{(y+2)^2}{9} = 1$

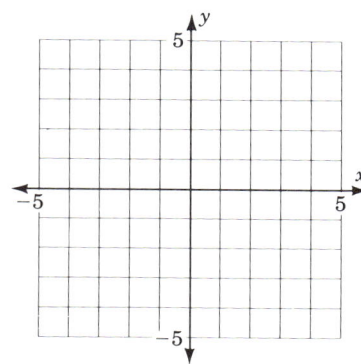

45. $\dfrac{(x-1)^2}{16} + \dfrac{(y-1)^2}{9} = 1$

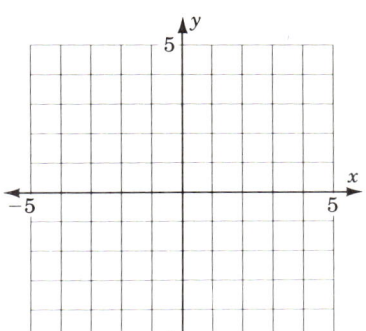

46. $\dfrac{(x-2)^2}{9} + \dfrac{(y-1)^2}{16} = 1$

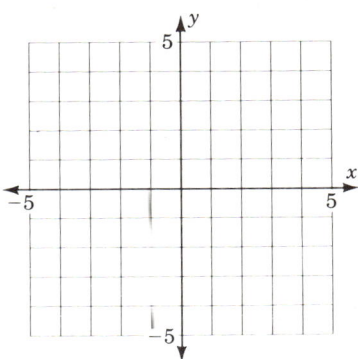

47. Find an equation of the circle with center at the origin and passing through the point (4, 3).

47. _____

48. Find an equation of the circle with center at the origin and passing through the point (3, 4).

48. _____

49. Find an equation of the circle with center at the origin and passing through the point $(-5, -12)$.

49. _____

50. Find an equation of the circle with center at the origin and x-intercepts ± 5.

50. _____

51. Find an equation of the circle with center at the origin and y-intercepts ± 3.

51. _____

✓ **SKILL CHECKER**

52. If you solve the equation $x^2 + y^2 = 25$ for x, you obtain two answers. What are these answers?

52. _____

53. One of the answers in Problem 52 is always non-negative. Look at the graph of $x^2 + y^2 = 25$ in Example 4. To what part of the graph does the positive answer correspond?

53. _____

54. One of the answers in Problem 52 is always nonpositive. Look at the graph of $x^2 + y^2 = 25$ in Example 4. To what part of the graph does the negative answer correspond?

54. _____

9.2 USING YOUR KNOWLEDGE

The definition of an ellipse is as follows.

An **ellipse** is the set of all points the sum of whose distances from two fixed points $(c, 0)$ and $(-c, 0)$ is a constant $a > c$. Each fixed point is called a **focus.** (See the figure)

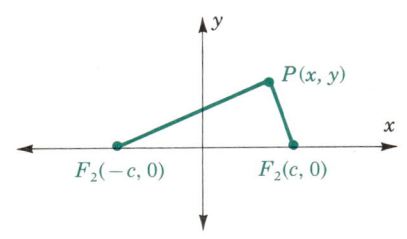

In Problems 1–7 we will prove that this definition leads to the equation we have given as a definition for the ellipse.

1. Suppose $P(x, y)$ is a point on the ellipse. Find the distance from P to F_1.

2. Find the distance from P to F_2.

3. The sum of the distances found in Problems 1 and 2 must be $2a$. Thus

$$\sqrt{(x - c)^2 + y^2} + \sqrt{(x + c)^2 + y^2} = 2a$$

   Rewrite this equation with one radical on each side. Then square both sides and simplify. What is your answer?

4. Rewrite the answer obtained in Problem 3 with the radical on one side and then square both sides again and simplify. What is your answer?

5. In your answer for Problem 4, since $a > c$, let $a^2 - c^2 = b^2$. Isolate all the variables on the left side. What is your answer?

6. Divide all terms of the answer you obtained in Problem 5 by $a^2 b^2$. What is your answer? If everything went well, you should have $\dfrac{x^2}{a^2} + \dfrac{y^2}{b^2} = 1$, the equation of an ellipse.

1. _____

2. _____

3. _____

4. _____

5. _____

6. _____

OBJECTIVES

REVIEW

Before starting this section, you should know:

1. How to graph points on the cartesian plane.
2. How to find the x- and y-intercepts of a given curve.

OBJECTIVES

You should be able to:

A. Graph hyperbolas.
B. Identify conic sections by examining their equations.

Have you seen any hyperbolas lately? It is more likely that you do so at night. The beams of light you see shining on the sides of many buildings at night are *hyperbolas*.

A. HYPERBOLAS

A *hyperbola* is the set of points in a plane such that the difference of the distances of each point from two fixed points (called the **foci**) is a constant. Consider the equation

$$\frac{x^2}{4} - \frac{y^2}{9} = 1 \qquad (hyp. -)$$

When $x = 0$, $y^2 = -9$, so there are no y-intercepts. When $y = 0$, $x^2 = 4$ and $x = \pm 2$ are the x-intercepts. The graph is shown in Figure 1(a).
Similarly, the graph of

$$\frac{y^2}{9} - \frac{x^2}{4} = 1$$

has no x-intercept, since $y = 0$ yields $x^2 = -4$, which has no solution. The y-intercepts are ± 3. The graph is shown in Figure 1(b).

The hyperbola $\dfrac{x^2}{4} - \dfrac{y^2}{9} = 1$ has intercepts $x = \pm 2$. We can use the denominator of y^2 to help us with the graph. If we draw a rectangle with sides parallel to the x- and y-axes and passing through the x-intercept and the points on the y-axis corresponding to the square root of the denominator of y^2, in this case ± 3, and then connect opposite corners of the rectangle with a line, the graph of the hyperbola will

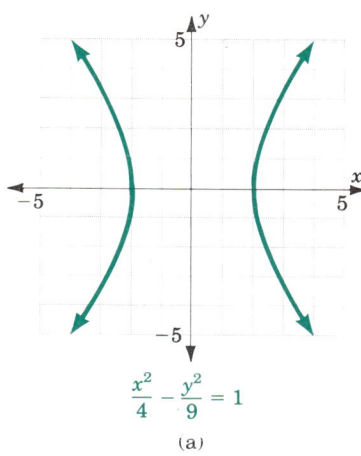

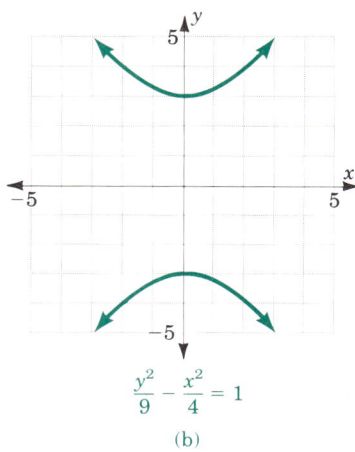

$$\frac{x^2}{4} - \frac{y^2}{9} = 1$$
(a)

$$\frac{y^2}{9} - \frac{x^2}{4} = 1$$
(b)

FIG. 1

approach these lines, called *asymptotes.* The graphs of the hyperbolas

$$\frac{x^2}{4} - \frac{y^2}{9} = 1 \quad \text{and} \quad \frac{y^2}{9} - \frac{x^2}{4} = 1$$

will be as shown in Figure 2.

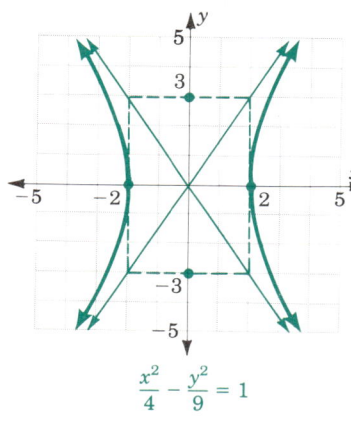

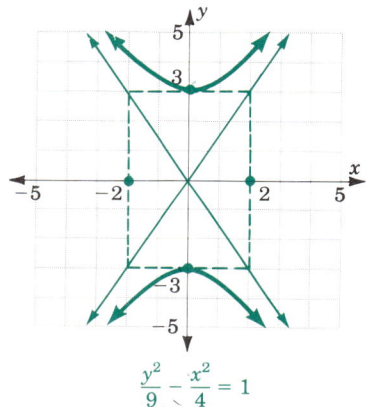

$$\frac{x^2}{4} - \frac{y^2}{9} = 1$$

$$\frac{y^2}{9} - \frac{x^2}{4} = 1$$

FIG. 2

Here is a summary of this discussion.

TO GRAPH HYPERBOLAS

The graph of the equation

(1) $\dfrac{x^2}{a^2} - \dfrac{y^2}{b^2} = 1$

is a **hyperbola** centered at the origin with x-intercepts $\pm a$.

The graph of the equation

(2) $\dfrac{y^2}{a^2} - \dfrac{x^2}{b^2} = 1$

is a **hyperbola** centered at the origin with y-intercepts $\pm a$.

The **asymptotes** of either hyperbola are the lines through opposite corners of the auxiliary rectangle whose sides pass through $(\pm a, 0)$ and $(0, \pm b)$ for (1), and $(0, \pm a)$ and $(\pm b, 0)$ for (2).

EXAMPLE 1 Graph.

a. $\dfrac{y^2}{4} - \dfrac{x^2}{25} = 1$ **b.** $\dfrac{x^2}{4} - \dfrac{y^2}{25} = 1$

Problem 1 Graph.

a. $\dfrac{y^2}{16} - \dfrac{x^2}{9} = 1$ **b.** $\dfrac{x^2}{16} - \dfrac{y^2}{9} = 1$

Solution

a. Since the y^2-term is positive, the hyperbola is centered at the origin and has y-intercepts $y = \pm a = \pm 2$. (There are no x-intercepts). Our auxiliary rectangle will pass through $y = \pm 2$ and through $x = \pm 5$, the square root of the denominator of x^2. We then connect opposite corners to complete our asymptotes, as shown in the figure at left. Since our hyperbola has y-intercepts $y = \pm 2$, we start our graph from $y = 2$ and approach the asymptotes, obtaining the top half of the hyperbola. The bottom half is obtained similarly by starting at $y = -2$.

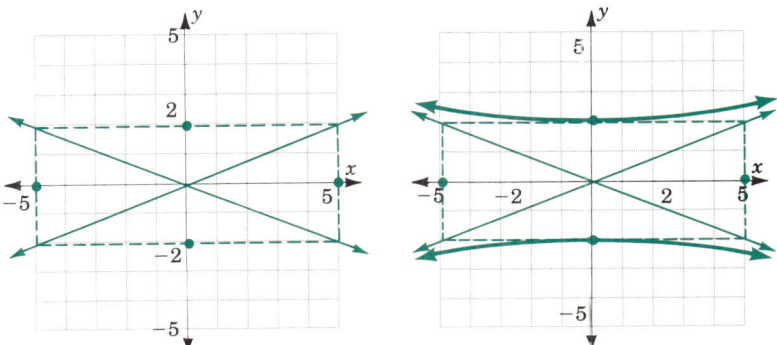

b. This time, we will show our auxiliary rectangle and the hyperbola on the same graph. Since the x^2-term is positive, the hyperbola has x-intercepts ± 2. Our auxiliary rectangle will pass through $x = \pm 2$ and through $y = \pm 5$. We then complete the auxiliary rectangle, the asymptotes, and the graph of the hyperbola with x-intercepts at ± 2 as shown.

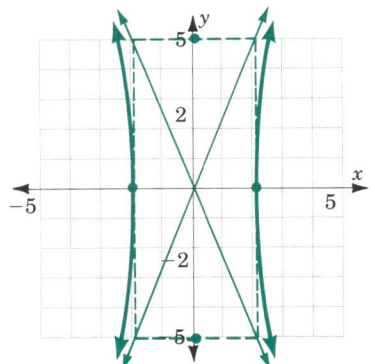

B. IDENTIFYING CONIC SECTIONS BY THEIR EQUATIONS

How would you know the shape of the graph of a conic by studying its equation? The chart on the next page will help you with this.

EQUATION	GRAPH	DESCRIPTION	IDENTIFICATION
$y = a(x - h)^2 + k$		Parabola with vertex at (h, k). Opens upward for $a > 0$, downward for $a < 0$.	y is not squared.
$y = ax^2 + bx + c$		Parabola with vertex at $x = -\dfrac{b}{2a}$. Opens upward for $a > 0$, downward for $a < 0$.	y is not squared.
$x = a(y - k)^2 + h$		Parabola with vertex at (h, k). Opens right if $a > 0$, left if $a < 0$.	x is not squared.
$x = ay^2 + by + c$		Parabola with vertex at $y = -\dfrac{b}{2a}$. Opens right if $a > 0$, left if $a < 0$.	x is not squared.
$(x - h)^2 + (y - k)^2 = r^2$		Circle of radius r, centered at (h, k).	The coefficients of x^2 and y^2 are positive and equal.
$\dfrac{x^2}{a^2} + \dfrac{y^2}{b^2} = 1$		Ellipse with x-intercepts $\pm a$, y-intercepts $\pm b$.	The coefficients of x^2 and y^2 are positive and not equal.
$\dfrac{x^2}{a^2} - \dfrac{y^2}{b^2} = 1$		Hyperbola with x-intercepts $\pm a$. Auxiliary rectangle passing through $(\pm a, 0)$ and $(0, \pm b)$. Asymptotes drawn through the corners of the auxiliary rectangle.	x^2 has positive coefficient, y^2 has negative coefficient.
$\dfrac{y^2}{a^2} - \dfrac{x^2}{b^2} = 1$		Hyperbola with y-intercepts $\pm a$. Auxiliary rectangle passing through $(0, \pm a)$ and $(\pm b, 0)$. Asymptotes drawn through the corners of the auxiliary rectangle.	y^2 has positive coefficient, x^2 has negative coefficient.

EXAMPLE 2 Identify the following.

a. $x^2 = 9 - y^2$ **b.** $y = x^2 - 4$
c. $4x^2 = 36 - 9y^2$ **d.** $9x^2 = 36 + 4y^2$

Solution If both variables appear to the second power, we shall write all variables on the left to make the identification easier.

a. In $x^2 = 9 - y^2$ both variables appear to the second power. Thus $x^2 = 9 - y^2$ is written as $x^2 + y^2 = 9$. The square terms have the same coefficient (1) and are added. The equation $x^2 = 9 - y^2$ represents a circle centered at the origin and with radius 3.

b. In this case, only one variable is squared, the x. Thus the conic is a parabola with the vertex at $(0, -4)$ and opening upward.

c. Both variables are squared. We then write $4x^2 = 36 - 9y^2$ as $4x^2 + 9y^2 = 36$. Here the square terms have different coefficients and are added. The equation corresponds to an ellipse centered at the origin. The x-intercepts are found by letting $y = 0$, obtaining $x = \pm 3$. Similarly, the y-intercepts are $y = \pm 2$.

d. Again, both variables are squared. Thus we write $9x^2 = 36 + 4y^2$ as $9x^2 - 4y^2 = 36$. The minus sign indicates that the conic is a hyperbola with x-intercepts $x = \pm 2$. ▲

Problem 2 Identify.
a. $4x^2 = 36 + 9y^2$ **b.** $y = x^2 + 3$
c. $y^2 = 9 - x^2$ **d.** $9x^2 = 36 - 4y^2$

ANSWER (to problem on page 705)

1. a.

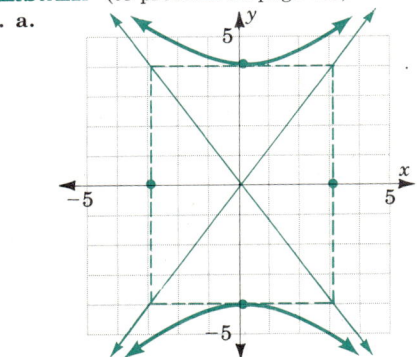

b.

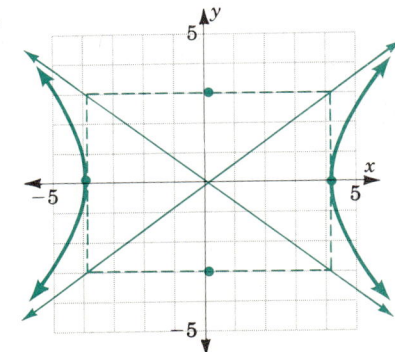

A. Graph.

1. $\dfrac{x^2}{25} - \dfrac{y^2}{9} = 1$

2. $\dfrac{y^2}{9} - \dfrac{x^2}{25} = 1$

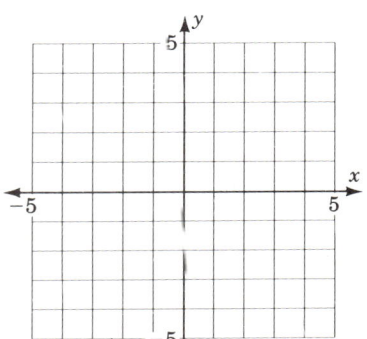

3. $\dfrac{y^2}{9} - \dfrac{x^2}{9} = 1$

4. $\dfrac{x^2}{9} - \dfrac{y^2}{9} = 1$

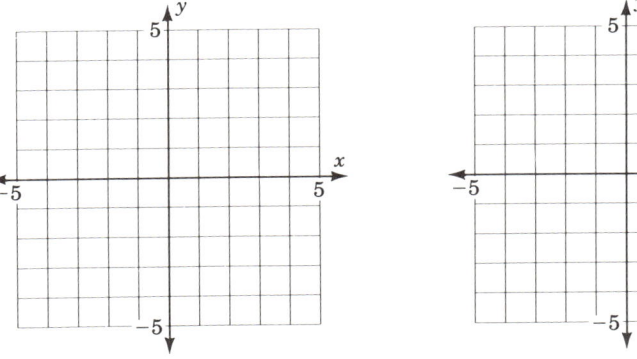

5. $\dfrac{x^2}{9} - \dfrac{y^2}{1} = 1$

6. $\dfrac{y^2}{16} - \dfrac{x^2}{1} = 1$

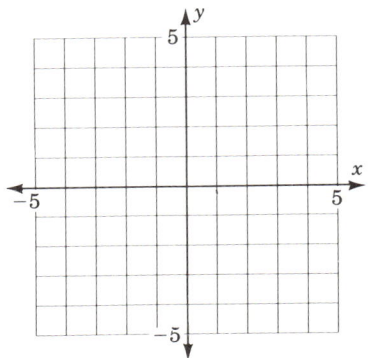
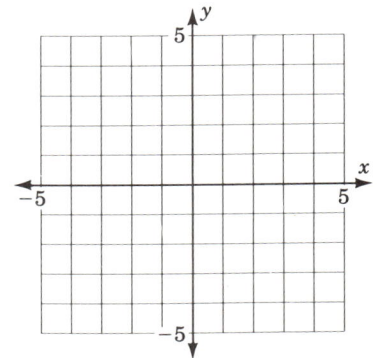

7. $\dfrac{x^2}{64} - \dfrac{y^2}{49} = 1$ **8.** $\dfrac{y^2}{49} - \dfrac{x^2}{64} = 1$

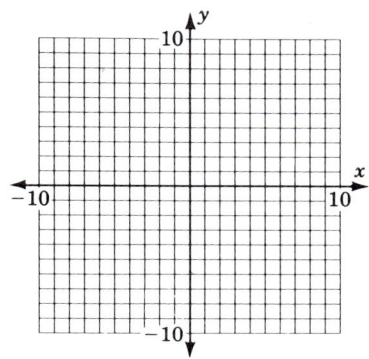   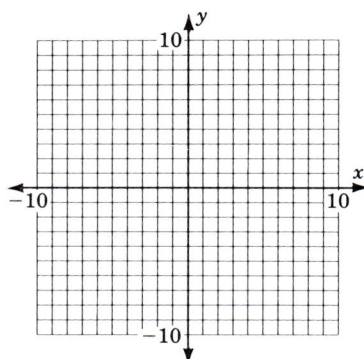

9. $\dfrac{y^2}{\frac{16}{9}} - \dfrac{x^2}{\frac{9}{16}} = 1$ **10.** $\dfrac{x^2}{\frac{9}{4}} - \dfrac{y^2}{\frac{4}{9}} = 1$

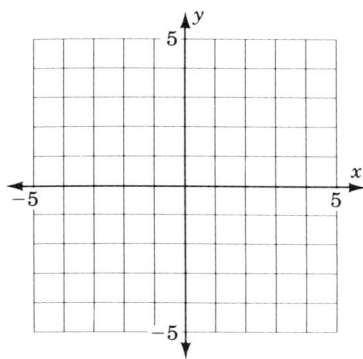   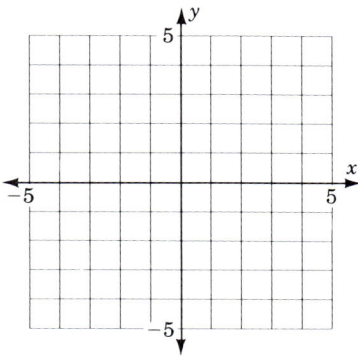

11. $x^2 - 9y^2 = 9$ **12.** $y^2 - 16x^2 = 16$

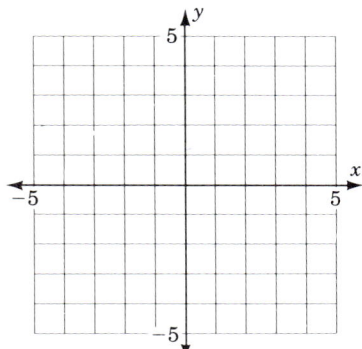   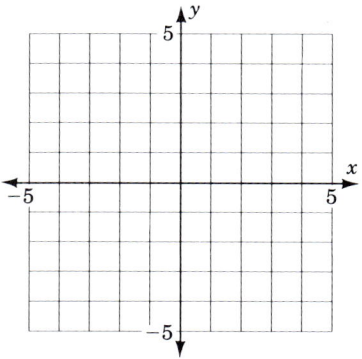

B. In Problems 13–26 identify the conic and give the intercepts. If the conic is a parabola, give the vertex.

13. $x^2 + y^2 = 25$ **14.** $x^2 - y^2 = 25$

13. _____

14. _____

15. $x^2 - y^2 = 36$

16. $x^2 + y^2 = 36$

17. $x^2 - y = 9$

18. $x^2 + y = 9$

19. $y^2 - x = 4$

20. $y^2 + x = 4$

21. $9x^2 = 36 - 9y^2$

22. $4x^2 = 16 - 4y^2$

23. $9x^2 = 36 + 9y^2$

24. $4y^2 = 36 - 9x^2$

25. $x^2 = 9 - 9y^2$

26. $y^2 = 4 - 4x^2$

15. _____

16. _____

17. _____

18. _____

19. _____

20. _____

21. _____

22. _____

23. _____

24. _____

25. _____

26. _____

In Problems 27–30, sketch the hyperbolas. (They are not centered at the origin.)

27. $\dfrac{(x-1)^2}{4} - \dfrac{(y+1)^2}{9} = 1$

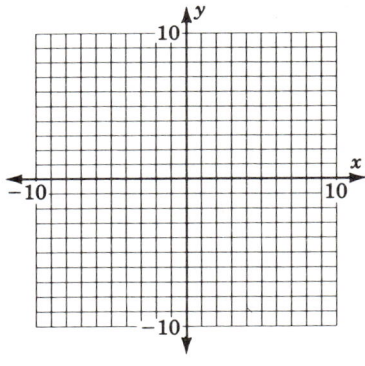

28. $\dfrac{(x-2)^2}{9} - \dfrac{(y+1)^2}{4} = 1$

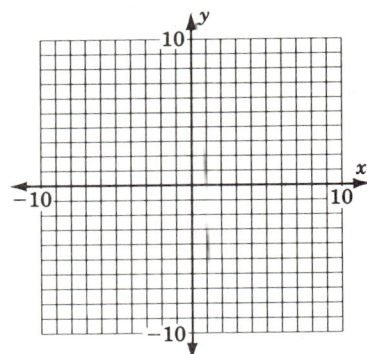

29. $\dfrac{(x-1)^2}{9} - \dfrac{(y-2)^2}{4} = 1$

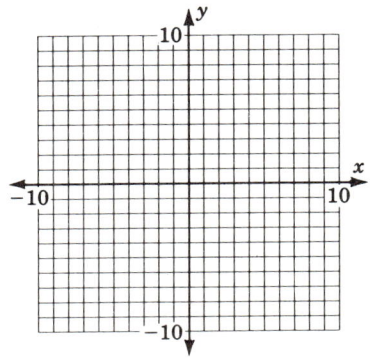

30. $\dfrac{(x-2)^2}{4} - \dfrac{(y-1)^2}{9} = 1$

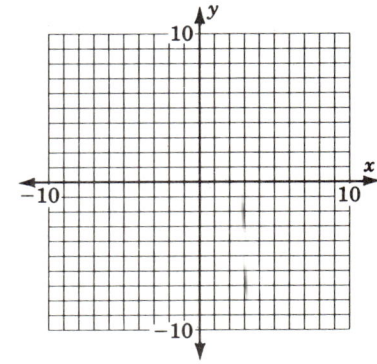

Applications

31. A semicircular plate of diameter D with a circular opening of diameter d is to be constructed. If the area of the plate is π in.^{2}, the relationship between D and d is given by

$$\frac{D^2}{8} - \frac{d^2}{4} = 1$$

a. What type of conic is this?

b. Sketch the graph of $\dfrac{D^2}{8} - \dfrac{d^2}{4} = 1$. (Use $\sqrt{8} \approx 2.8$.)

31. a. _____

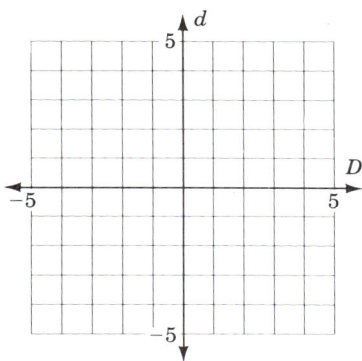

32. If three holes of diameter d were drilled in the plate of Problem 31 and the remaining area was still π in.^{2}, the relationship between D and d would be

$$\frac{D^2}{8} - \frac{3d^2}{4} = 1$$

a. What type of conic is this?
b. Show that you can write the equation of the conic as

$$\frac{D^2}{8} - \frac{d^2}{\frac{4}{3}} = 1$$

c. Sketch the graph of $\dfrac{D^2}{8} - \dfrac{3d^2}{4} = 1$. $\left(\text{Use } \sqrt{\dfrac{4}{3}} \approx 1.15.\right)$

32. a. _____
b. _____

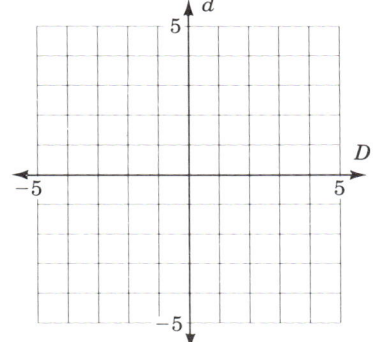

33. The total kinetic energy of a spinning body moving through the air is 144 ft-lb. If the velocity v through the air and the spinning velocity ω are related by the equation $4v^2 + 9\omega^2 = 144$, graph the equation.

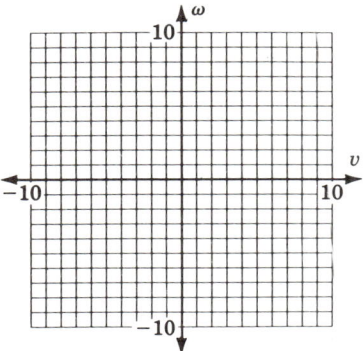

34. If the equation in Problem 33 is $16v^2 + 4\omega^2 = 256$, graph the equation.

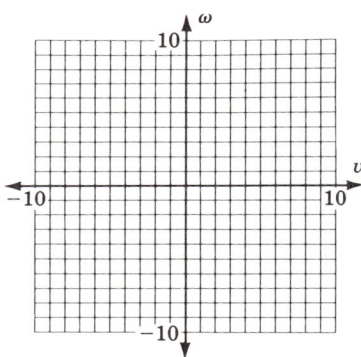

✓ **SKILL CHECKER**

Graph.

35. $x - y < 4$

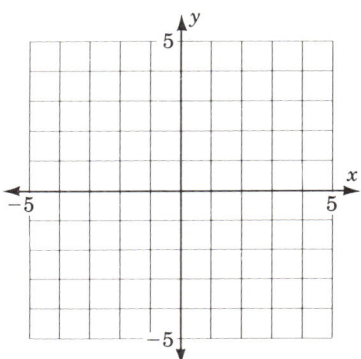

36. $y - x < 4$

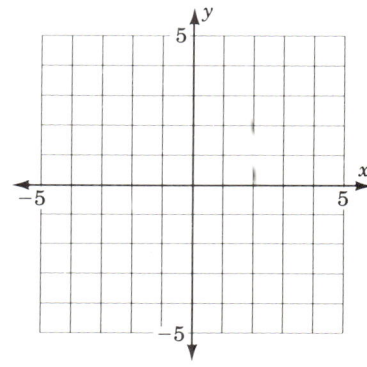

37. $2x - 3y \geq 6$

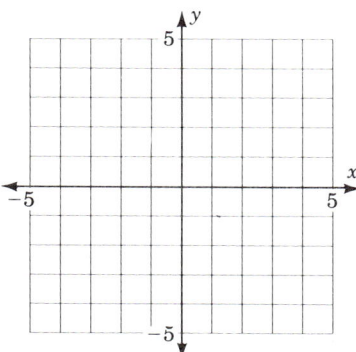

38. $3x - 2y \geq 6$

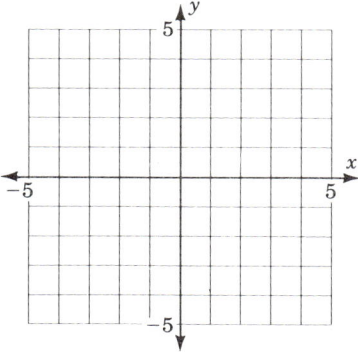

39. $y \geq 2x + 4$

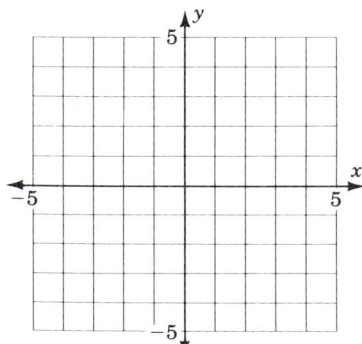

40. $y \geq 3x + 6$

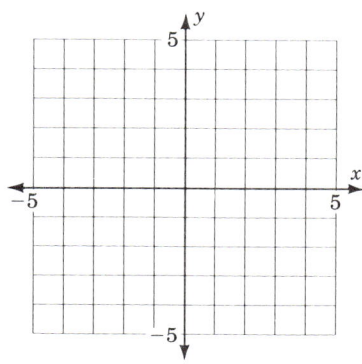

9.3   USING YOUR KNOWLEDGE

The definition for a hyperbola is as follows.

> A **hyperbola** is the set of points in a plane such that the difference of the distances of each point from two fixed points (called the **foci**) is a constant. (See Figure 3.)

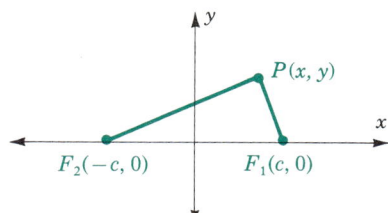

FIG. 3

In Problems 1–6 we will prove that this definition leads to the equation we have given as a definition for the hyperbola.

1. Suppose (x, y) is a point on the hyperbola. Find the distance from P to F_1.

1. _____

2. Find the distance from P to F_2.

2. _____

3. Let the difference of the distances found in Problems 1 and 2 be $2a$. Thus

$$\sqrt{(x - c)^2 + y^2} - \sqrt{(x + c)^2 + y^2} = \pm 2a$$

Rewrite this equation with one radical on each side. Then square both sides and simplify. What is your answer?

3. _____

4. Rewrite the answer obtained in Problem 3 with the radical on one side. Then square both sides again and simplify. What is your answer?

4. _____

5. In your answer for Problem 4, let $c^2 - a^2 = b^2$, where $a < c$. Isolate all the variables on the left side. What is your answer?

5. _____

6. Divide all terms of the answer you obtained in Problem 5 by $a^2 b^2$. What is your answer? If everything went well, you should have $\dfrac{x^2}{a^2} - \dfrac{y^2}{b^2} = 1$, the equation of a hyperbola.

6. _____

We have shown you how to graph the asymptotes of a hyperbola. We are now ready to use this knowledge to get the equation of these asymptotes. Consider the hyperbola

$$\frac{x^2}{a^2} - \frac{y^2}{b^2} = 1$$

7. The expression on the left is the difference of two squares. Factor it.

7. _____

8. Isolate $\dfrac{x}{a} - \dfrac{y}{b}$ on the left. The expression on the right is a complex fraction with 1 as numerator. What is it?

8. _____

9. Look at the denominator of the complex fraction in Problem 8. If x and y are positive and very large, what happens to the denominator? What happens to the complex fraction?

9. _____

10. If you answered that the complex fraction is very small, you are correct. In mathematics, we say that $\dfrac{x}{a} - \dfrac{y}{b} \to 0$ (the expression approaches 0). Thus for very large positive x and y, $\dfrac{x}{a} - \dfrac{y}{b} \approx 0$. This means that $\dfrac{x}{a} - \dfrac{y}{b} = 0$ is an asymptote. Solve for y and find its equation.

10. _____

11. We can show in the same way that $\dfrac{x}{a} + \dfrac{y}{b} = 0$ is an asymptote. Solve for y and find its equation.

11. _____

12. In summary, what are the equations of the asymptotes for the hyperbola $\dfrac{x^2}{a^2} - \dfrac{y^2}{b^2} = 1$? What about the asymptotes for the hyperbola $\dfrac{y^2}{a^2} - \dfrac{x^2}{b^2} = 1$?

12. _____

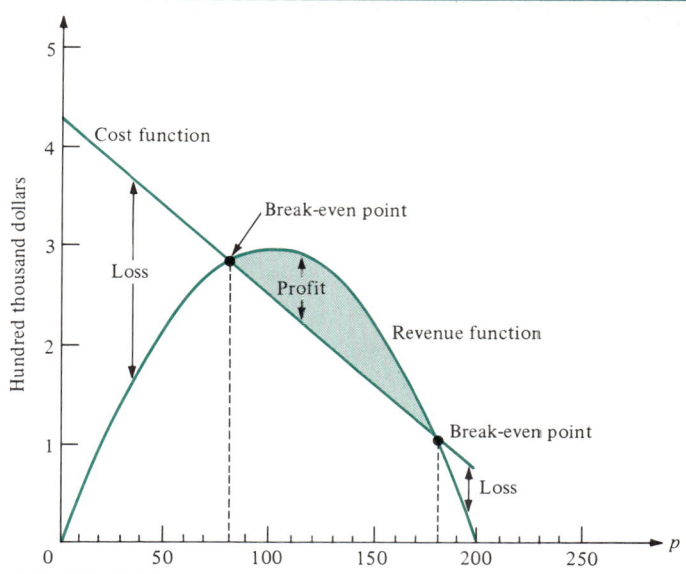

REVIEW

Before starting this section, you should know:

1. How to graph linear inequalities in two variables (Section 7.4).
2. How to graph conic sections.

OBJECTIVES

You should be able to:

A. Graph second-degree inequalities.
B. Graph the solution set of a system of inequalities.

The graph offers a wealth of information if you know how to read it. The line represents the cost of a product, whereas the parabola represents the revenue obtained when the product is sold. Here are some facts about the graph:

1. Where the graph representing the cost lies above the graph of the revenue, there is a loss (region shown in color).
2. The revenue equals the cost at a point called the break-even point.
3. Where the graph representing the revenue lies above the graph of the cost, there is a profit (region shown in black).

A. SECOND-DEGREE INEQUALITIES

In Section 7.4 we graphed linear inequalities by graphing the boundary, selecting a test point not on the boundary, and determining the regions corresponding to the solution set. The region under the parabola in the graph is obtained by graphing a **second-degree inequality** using a similar procedure. Let us do an example.

EXAMPLE 1 Graph $y \leq -x^2 + 3$.

Solution The boundary of the required region is the parabola $y = -x^2 + 3$ with its vertex at $(0, 3)$, as shown. Since the inequality sign is

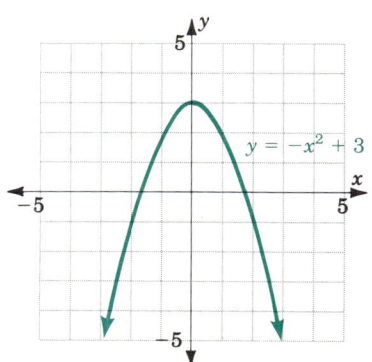

Problem 1 Graph $y \leq -x^2 - 3$.

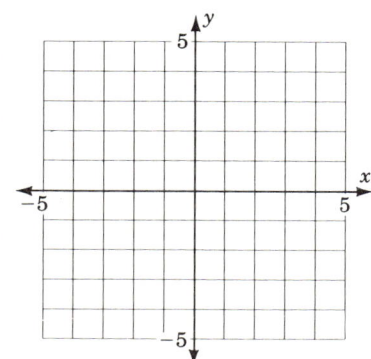

$\le$, the boundary *is* included in the solution set. To determine which region represents the solution set, select a test point not on the boundary and test the original inequality. A convenient point is $(0, 0)$. When $x = 0$ and $y = 0$, $y \le -x^2 + 3$ becomes $0 \le 0 + 3$, a true statement. Thus, all points on the same side as $(0, 0)$ will be in the solution set. The graph of the solution set is shown in color.

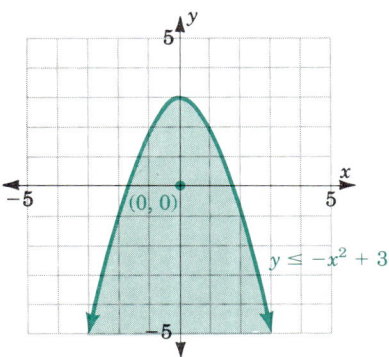

EXAMPLE 2 Graph $x^2 + y^2 > 4$.

Solution This time, the boundary is a circle of radius 2 centered at the origin. The inequality sign is $>$, so that the boundary is *not* included in the solution set. It is shown dashed in the figure. For the point $(0, 0)$, $x = 0$ and $y = 0$, and $x^2 + y^2 > 4$ becomes $0 + 0 > 4$, a false statement. This means that the region containing $(0, 0)$ is *not* in the solution set. We then color the region outside the circle to represent the solution set, as shown in the figure.

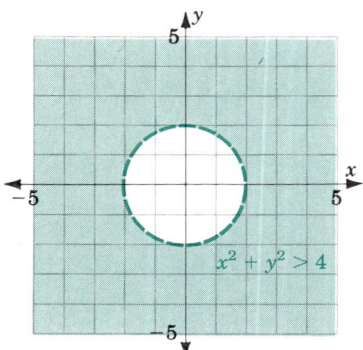

EXAMPLE 3 Graph $9y^2 - 4x^2 \le 36$.

Solution Since all the variables are on the left and there is a $(-)$ sign between the terms, the boundary is a hyperbola with y-intercepts $y = \pm 2$. Since the inequality is $\le$, the boundary *is* part of the solution set. If we use $(0, 0)$ for our test point, $9y^2 - 4x^2 \le 36$ becomes $0 - 0 \le 36$, which is true. Thus the point $(0, 0)$ is part of the complete solution set, which is shown in color.

Problem 2 Graph $x^2 + y^2 > 9$.

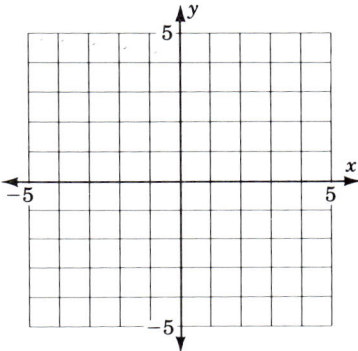

Problem 3 Graph $4y^2 - 9x^2 \le 36$.

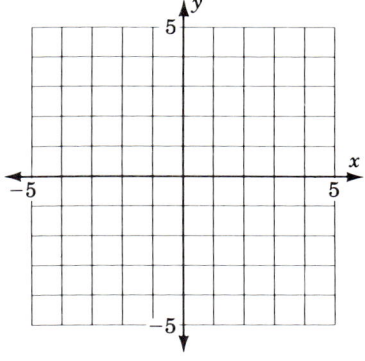

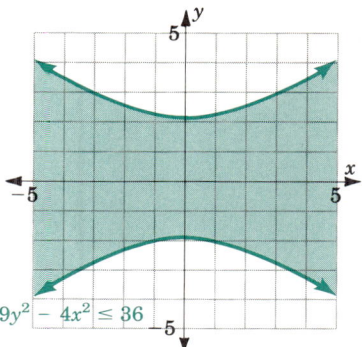

$9y^2 - 4x^2 \leq 36$

▲

B. SYSTEMS OF INEQUALITIES

We have already solved a system of linear equations in two variables using the graphical method. To solve a system of inequalities using this method, we graph each inequality on the same set of axes and find the region common to both graphs—that is, the intersection of both graphs. The result is the solution set of the system. We illustrate the idea next.

EXAMPLE 4 Graph the system $x^2 + y^2 \leq 4$ and $y \geq x^2 - 3$.

Solution The boundary for the first inequality is a circle of radius 2 centered at the origin. The solution set of this inequality includes the boundary and the points inside the circle and is shown in color. The boundary for the second inequality is a parabola with vertex at $(0, -3)$, and the solution set includes the boundary and all the points above the parabola.

The solution set for the system is the intersection of the two solution sets, the region inside *both* the circle and the parabola.

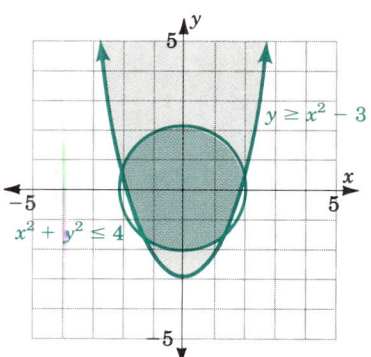

$y \geq x^2 - 3$

$x^2 + y^2 \leq 4$

▲

Problem 4 Graph the system $x^2 + y^2 \leq 9$ and $y \geq x^2 + 2$.

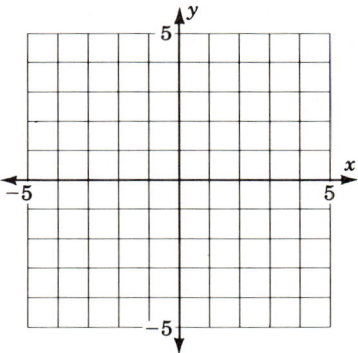

ANSWER (to problem on page 717)
1.

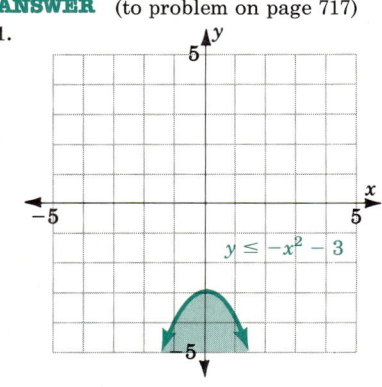

$y \leq -x^2 - 3$

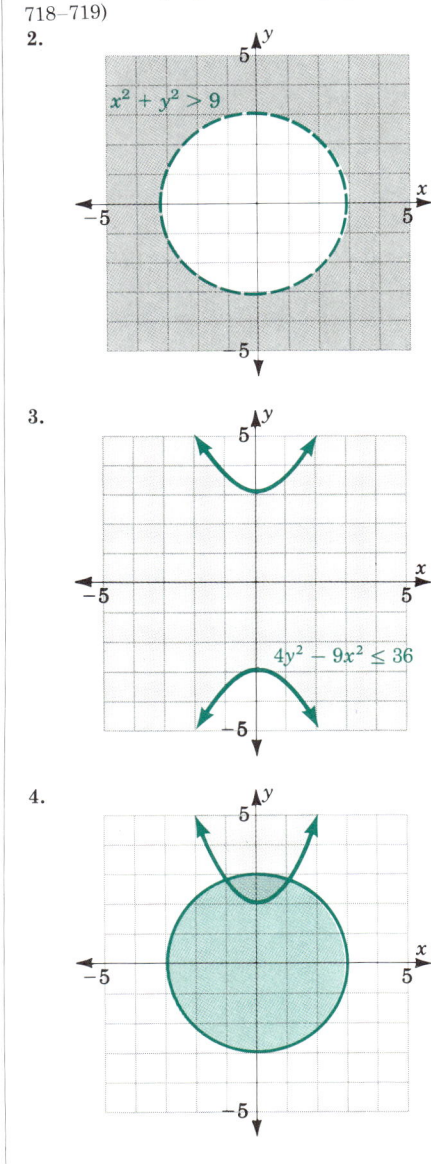

2.

$x^2 + y^2 > 9$

3.

$4y^2 - 9x^2 \leq 36$

4.

A. Graph.

1. $x^2 + y^2 > 16$

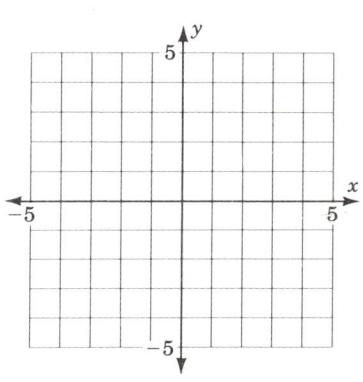

2. $x^2 + y^2 < 16$

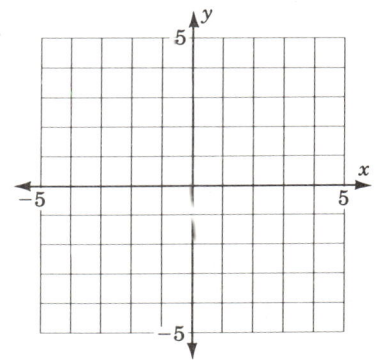

3. $x^2 + y^2 \leq 1$

4. $x^2 + y^2 \geq 1$

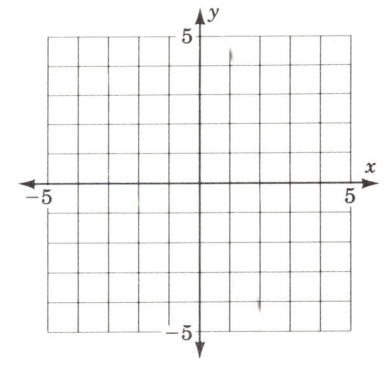

5. $y < x^2 - 2$

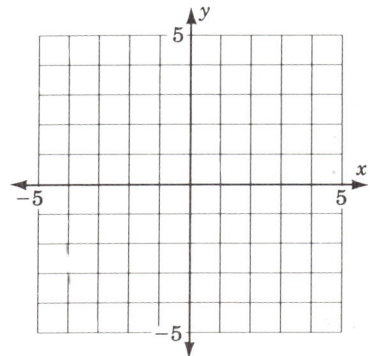

6. $y > x^2 - 2$

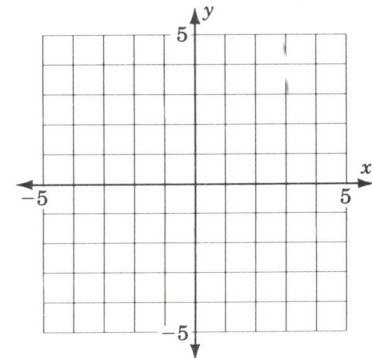

7. $y \leq -x^2 + 3$

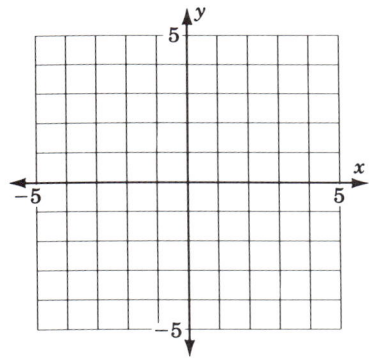

8. $y \geq -x^2 + 3$

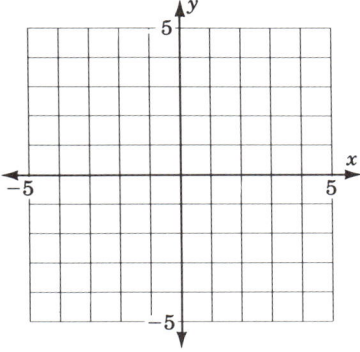

9. $4x^2 - 9y^2 > 36$

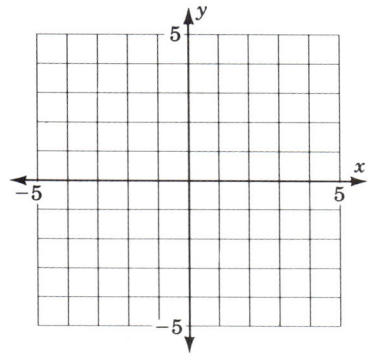

10. $4x^2 - 9y^2 < 36$

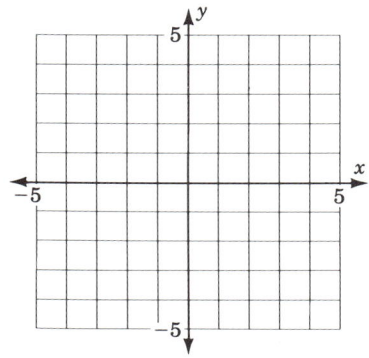

11. $x^2 - y^2 \geq 1$

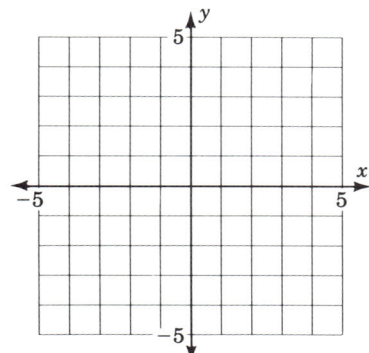

12. $x^2 - y^2 \leq 1$

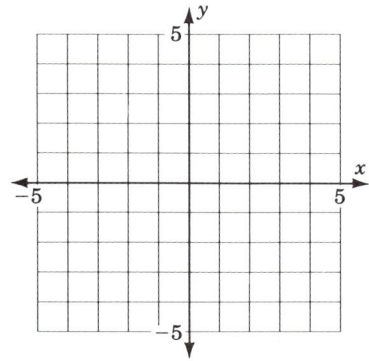

B. In Problems 13–24 graph the solution set of the system.

13. $x^2 + y^2 \leq 25$
$\qquad y \geq x^2$

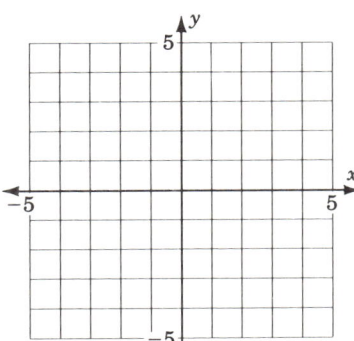

14. $x^2 + y^2 \leq 25$
$\qquad y \leq x^2$

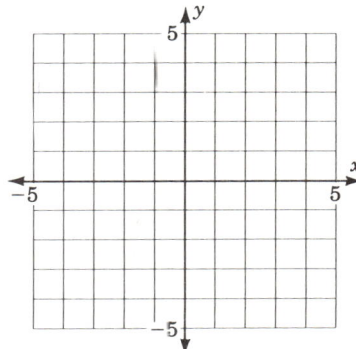

15. $x^2 + y^2 \geq 25$
$\qquad y \leq x^2$

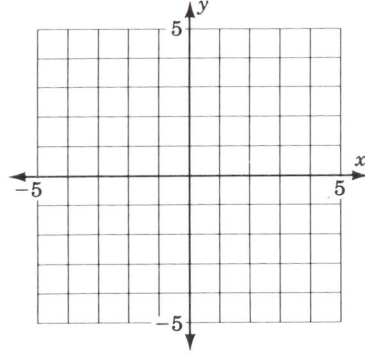

16. $x^2 + y^2 \geq 25$
$\qquad y \geq x^2$

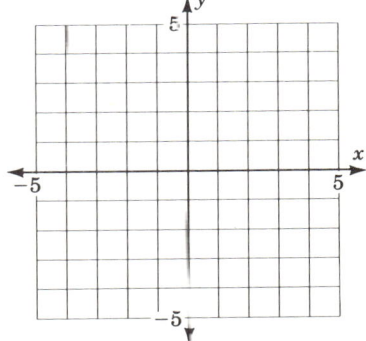

17. $y < x^2 + 2$
$\qquad y > x^2 - 2$

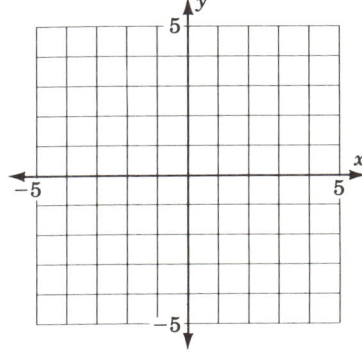

18. $y < x^2 + 2$
$\qquad y < x^2 - 2$

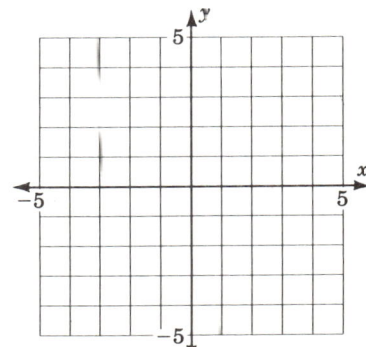

19. $y \geq x^2 + 2$
$\quad\;\; y \geq x^2 - 2$

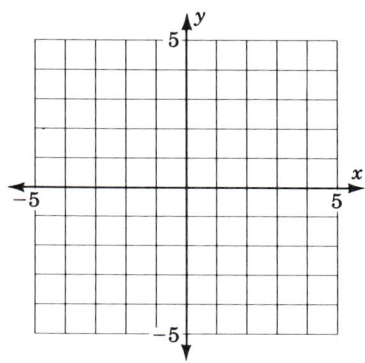

20. $y \geq x^2 + 2$
$\quad\;\; y \leq x^2 - 2$

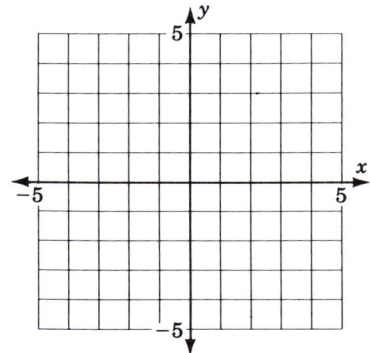

21. $\dfrac{x^2}{4} - \dfrac{y^2}{4} \geq 1$

$\quad\;\; \dfrac{x^2}{25} + \dfrac{y^2}{4} \leq 1$

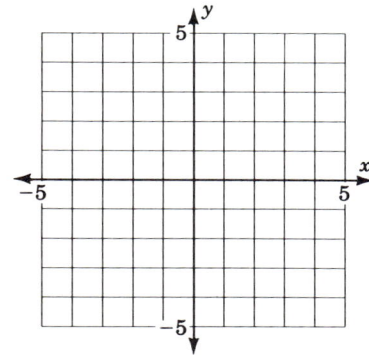

22. $\dfrac{x^2}{4} - \dfrac{y^2}{4} \geq 1$

$\quad\;\; \dfrac{x^2}{25} + \dfrac{y^2}{4} \geq 1$

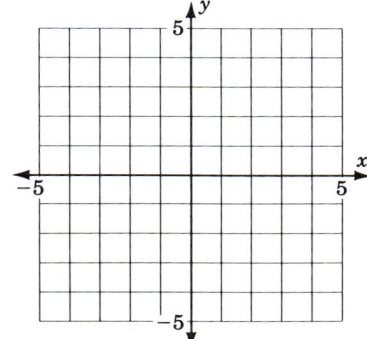

23. $\dfrac{x^2}{36} + \dfrac{y^2}{16} < 1$

$\quad\;\; \dfrac{x^2}{16} + \dfrac{y^2}{36} < 1$

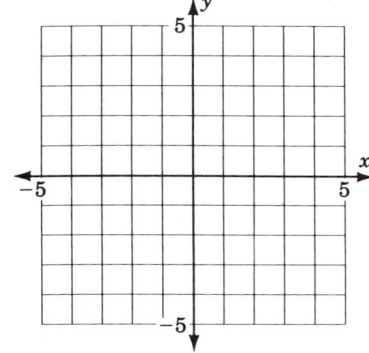

24. $\dfrac{x^2}{36} + \dfrac{y^2}{16} > 1$

$\quad\;\; \dfrac{x^2}{16} + \dfrac{y^2}{36} < 1$

25. Fill in each blank with an inequality symbol so that the system

$$\frac{x^2}{36} + \frac{y^2}{16} \underline{\quad} 1 \qquad \text{(ellipse)}$$

$$\frac{x^2}{16} + \frac{y^2}{16} \underline{\quad} 1 \qquad \text{(circle)}$$

will have
a. No solution
b. Two solutions

26. Can you fill in each blank with an inequality symbol so that the system

$$\frac{x^2}{25} - \frac{y^2}{4} \underline{\quad} 1 \qquad \text{(hyperbola)}$$

$$\frac{x^2}{25} + \frac{y^2}{4} \underline{\quad} 1 \qquad \text{(ellipse)}$$

will have no solution?

✓ **SKILL CHECKER**

Solve the system by substitution.

27. $2x - 3y = 3$
$\quad\quad y = -x + 4$

28. $3x - y = 13$
$\quad\quad y = -x + 3$

29. $2y + x = -1$
$\quad\quad 11y - 3x = 71$

30. $2y + x = -1$
$\quad\quad 10y - 3x = 51$

9.4 USING YOUR KNOWLEDGE

You may have wondered about the equations representing the cost C and the revenue R in the graph at the beginning of this section. Note that both the cost C and the revenue R are given in terms of p, the price per unit. Can we find the break-even point—that is, the point at which the revenue R equals the cost C? Let us try.

1. The demand equation—that is, the number x of units retailers are likely to buy at p dollars per unit—is given by $x = 6000 - 30p$, and the cost C is given by $C = 72{,}000 + 60x$. Substitute $x = 6000 - 30p$ in $C = 72{,}000 + 60x$ and find C in terms of the price p.

2. The revenue is $R = xp$—that is, the revenue is the product of the number of units retailers are likely to buy and the price p per unit. To find R in terms of p, substitute $x = 6000 - 30p$ into $R = xp$. What is R in terms of p? What shape does the graph of R have?

3. The graph at the beginning of this section shows two points at which the revenue R equals the cost—that is, $R = C$. Substitute the expressions for R and C from Problems 1 and 2 in the equation $R = C$ and solve for p.

25. a. _____
    b. _____

26. _____

27. _____
28. _____
29. _____
30. _____

1. _____

2. _____

3. _____

9.5 NONLINEAR SYSTEMS

HOW SUPPLY AND DEMAND DETERMINE
MARKET PRICE AND QUANTITY

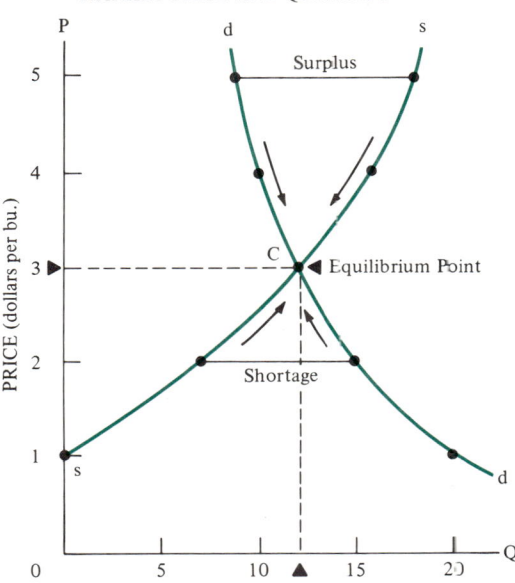

QUANTITY (million bu. per month)

In the graph, the demand curve is labeled *d* and the supply curve *s*. (From *Economics*, 10th ed., by Paul A. Samuelson. Copyright 1976. Used with permission of McGraw-Hill Book Company.)

The graph shows how supply and demand determine the market price and quantity of wheat available for sale. As the price of wheat *decreases,* the quantity demanded by consumers *increases*. If the price *increases,* the demand *decreases*. The point *C* of intersection of the two curves is called the **equilibrium point.** At this point, the price of a bushel of wheat is $3, and the amount demanded by the consumers, 12 million bushels per month, exactly equals the amount supplied by producers. Since the graphical method depends on how accurately we graph the equations, we shall concentrate on the substitution method to solve nonlinear systems.

A. SOLVING NONLINEAR SYSTEMS BY SUBSTITUTION

Suppose the demand curve for a product is given by $y = (x - 5)^2$, where x is the number of units produced and y is the price, and the supply curve is given by $y = x^2 + 2x + 13$, where y is the price and x the number of units available. To find the equilibrium point, we sketch both curves.

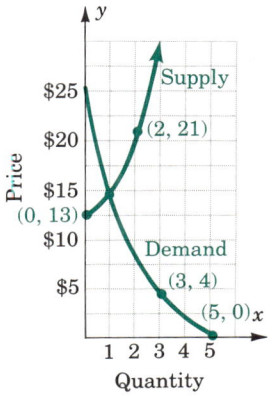

REVIEW

Before starting this section, you should know:

1. How to graph lines and conic sections.
2. How to use the graphical and substitution methods to solve systems of equations.
3. How to solve quadratic equations.

OBJECTIVES

You should be able to:

A. Solve a nonlinear system by substitution.
B. Solve a system with two second-degree equations by using the method of elimination.
C. Solve applications using the concepts involved.

The equilibrium point seems to be the point (1, 16). How can we be sure? We use the substitution method to solve the system.

$$y = (x - 5)^2 \qquad (1)$$
$$y = x^2 + 2x + 13 \qquad (2)$$

We wish to find an ordered pair (x, y) that is a solution of *both* (1) and (2). Using the substitution method, we substitute $(x - 5)^2$ for y on the left side of (2), obtaining

$$(x - 5)^2 = x^2 + 2x + 13$$

$x^2 - 10x + 25 = x^2 + 2x + 13$ Expand.

$-10x + 25 = 2x + 13$ Subtract x^2.

$-12x = -12$ Subtract $2x$ and 25.

$x = 1$ Divide by -12.

If $x = 1$ in Equation 2, then

$$y = (1)^2 + 2(1) + 13 = 16$$

Hence, the solution for the given system is (1, 16), as you can verify by substituting $x = 1$ and $y = 16$ in the two equations.

EXAMPLE 1 Find the solution set of the given system by the substitution method. Check the solution by sketching the graphs of the equations.

$$x^2 + y^2 = 25 \qquad (1)$$
$$x + y = 5 \qquad (2)$$

Solution We first rewrite Equation 2 in the equivalent form $y = 5 - x$, obtaining

$$x^2 + y^2 = 25 \qquad (1)$$
$$y = 5 - x \qquad (3)$$

Replacing y in Equation 1 by $5 - x$, we get

$x^2 + (5 - x)^2 = 25$ Substitute in (1).   (4)

$x^2 + 25 - 10x + x^2 = 25$ Expand.

$2x^2 - 10x = 0$ Simplify; subtract 25.

$x^2 - 5x = 0$ Divide by 2.

$x(x - 5) = 0$ Factor.

$x = 0$ or $x - 5 = 0$

$x = 0$ or $x = 5$ Solve for x.

We now place $x = 0$ and $x = 5$ in Equation 3 to obtain the corresponding y-values:

$$y = 5 - 0 = 5 \quad \text{and} \quad y = 5 - 5 = 0$$

Thus when $x = 0$, $y = 5$, and when $x = 5$, $y = 0$. Therefore, the solutions of the system are (0, 5) and (5, 0).

Note that if we had substituted $x = 0$ and $x = 5$ in Equation 1 rather than in Equation 3, we would have obtained

$$0^2 + y^2 = 25 \quad \text{and} \quad 5^2 + y^2 = 25$$

Problem 1 Find the solution set of

$$x^2 + y^2 = 16$$
$$x + y = 4$$

That is, $y = \pm 5$ and $y = 0$. In this case, the solutions obtained would have been $(0, 5)$, $(0, -5)$, $(5, 0)$. However, $(0, -5)$ is *not* a solution of Equation 3, since $-5 \neq 5 - 0$. Therefore, the only solutions are $(0, 5)$ and $(5, 0)$. As you can see, if the degrees of the equations are different, one component of a solution should be substituted in the *lower-degree* equation to find the ordered pairs satisfying *both* equations. For this reason we double-check our work by graphing the given system (see the figure) to verify that our solutions are correct.

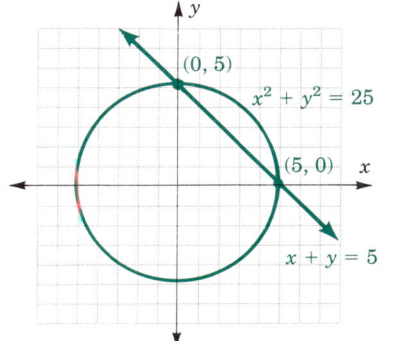

$x^2 + y^2 = 25$ is a circle of radius 5

▲

EXAMPLE 2 Find the solution of the given system by the substitution method. Check the solution by sketching the graphs of the equations.

$$x^2 + y^2 = 9 \tag{1}$$
$$x + y = 5 \tag{2}$$

Solution Rewriting (2) in the form $y = 5 - x$, we obtain the equivalent system

$$x^2 + y^2 = 9 \tag{1}$$
$$y = 5 - x \tag{3}$$

Substituting $y = 5 - x$ in (1), we get

$$x^2 + (5 - x)^2 = 9$$
$$x^2 + 25 - 10x + x^2 = 9 \quad \text{Expand.}$$
$$2x^2 - 10x + 25 = 9 \quad \text{Simplify.}$$
$$2x^2 - 10x + 16 = 0 \quad \text{Subtract 9.}$$
$$x^2 - 5x + 8 = 0 \quad \text{Divide by 2.}$$

Using the quadratic formula with $a = 1$, $b = -5$, $c = 8$, we get

$$x = \frac{5 \pm \sqrt{25 - 4 \cdot 8}}{2} = \frac{5 \pm \sqrt{-7}}{2} = \frac{5 \pm \sqrt{7}i}{2}$$

Substituting these values in (3), we obtain

$$y = 5 - \frac{(5 + \sqrt{7}i)}{2} \quad \text{and} \quad y = 5 - \frac{(5 - \sqrt{7}i)}{2}$$

That is,

$$y = \frac{5}{2} - \frac{\sqrt{7}}{2}i \quad \text{and} \quad y = \frac{5}{2} + \frac{\sqrt{7}}{2}i$$

Problem 2 Find the solution set of

$$x^2 + y^2 = 3$$
$$x + y = 3$$

Hence, the solutions of the system are

$$\left(\frac{5 + \sqrt{7}i}{2}, \frac{5 - \sqrt{7}i}{2}\right) \quad \text{and} \quad \left(\frac{5 - \sqrt{7}i}{2}, \frac{5 + \sqrt{7}i}{2}\right)$$

as can be checked in the original equations. The graphs of the two equations are shown in the figure. As you can see, the graphs *do not intersect*. When the solutions of a system of equations are imaginary numbers, there are no points of intersection for the graphs. This is because the coordinates of points in the real plane are *real* numbers.

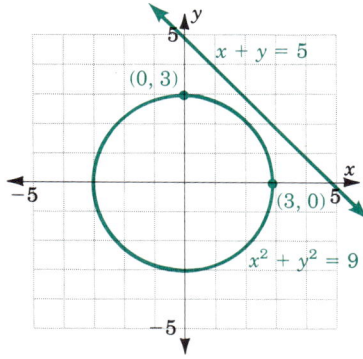

B. SOLVING SYSTEMS WITH TWO SECOND-DEGREE EQUATIONS

When both equations in a system are of second degree, it is easier to use the elimination method, as in the next example.

EXAMPLE 3 Solve the system $x^2 - 2y^2 = 1$ and $x^2 + 4y^2 = 25$ and verify the solution by graphing.

Solution To eliminate y^2, we multiply the first equation by 2 and add the result to the second equation:

$$\begin{array}{r} 2x^2 - 4y^2 = 2 \\ x^2 + 4y^2 = 25 \\ \hline 3x^2 = 27 \\ x^2 = 9 \\ x = \pm 3 \end{array}$$

The x-coordinates of the point of intersection are 3 and -3. Substituting into the second equation,

$$\begin{aligned} (\pm 3)^2 + 4y^2 &= 25 \\ 4y^2 &= 16 \\ y^2 &= 4 \\ y &= \pm 2 \end{aligned}$$

Problem 3 Solve.

$$\begin{aligned} x^2 - 9y^2 &= 9 \\ x^2 + y^2 &= 9 \end{aligned}$$

Check by graphing.

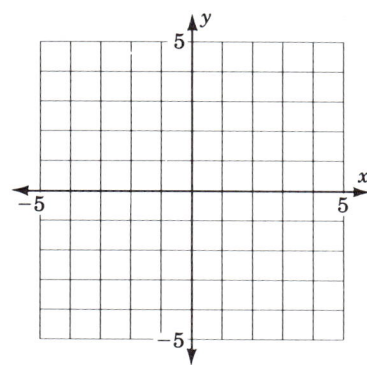

Thus, the four points of intersection are $(3, 2)$, $(3, -2)$, $(-3, 2)$, $(-3, -2)$, as you can check in the original equations. The graph for the two equations is shown.

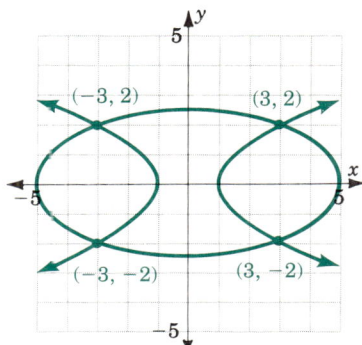

▲

Note that a system with at least one quadratic equation can have four, three, two, one, or no real-number solutions.

C. APPLICATIONS

At the beginning of Section 9.4 we mentioned the break-even point, the point at which the cost C and the revenue R are equal. The next example shows how to use the methods we have studied to find the break-even point.

EXAMPLE 4 The total cost C for manufacturing and selling x items of a product each week is given by $C = 30x + 100$, whereas the revenue R is given by $R = 81x - 0.5x^2$. How many items must be manufactured and sold for the company to break even?

Solution We want the value for which $C = R$, that is,

$$30x + 100 = 81x - 0.5x^2$$

or, in standard form,

$$0.5x^2 - 51x + 100 = 0$$

where $a = 0.5$, $b = -51$, and $c = 100$. Using the quadratic formula, we get

$$x = \frac{51 \pm \sqrt{(-51)^2 - 4(0.5)(100)}}{2 \cdot 0.5}$$

$$= \frac{51 \pm \sqrt{2601 - 200}}{1}$$

$$= \frac{51 \pm \sqrt{2401}}{1}$$

$$= 51 \pm 49$$

Problem 4 Repeat Example 4 if $R = 57x - 0.5x^2$.

Thus x is 2 or 100; that is, the company will break even when 2 or 100 items are sold. Note that the company makes a profit when they sell between 2 and 100 items.

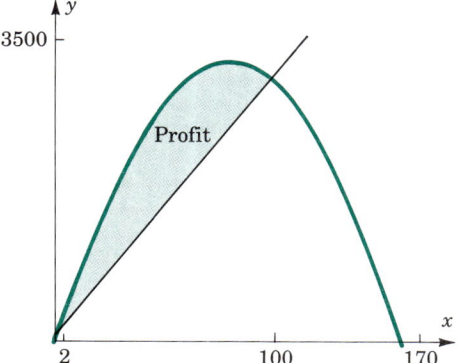

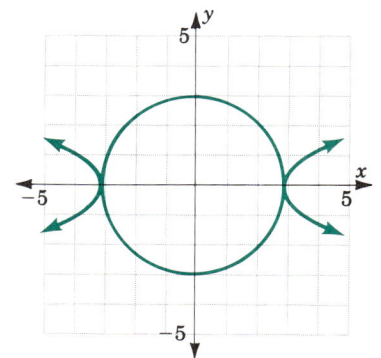

NAME

CLASS

SECTION

ANSWERS

A. In Problems 1–16 solve the system and check by graphing.

1. $x^2 + y^2 = 16$
$\quad x + y = 4$

2. $x^2 + y^2 = 9$
$\quad x + y = 3$

1. _____

2. _____

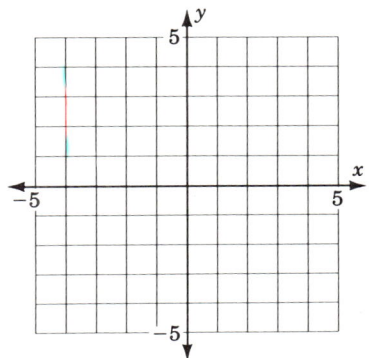

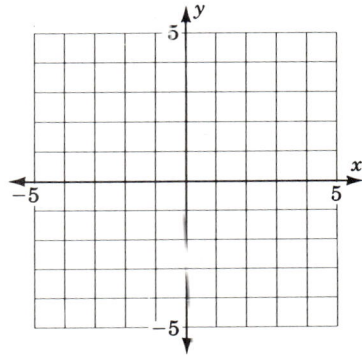

3. $x^2 + y^2 = 25$
$\quad y - x = 5$

4. $x^2 + y^2 = 9$
$\quad y - x = 3$

3. _____

4. _____

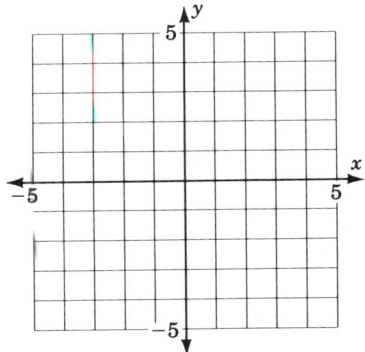

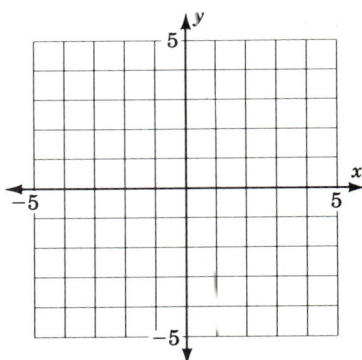

5. $x^2 + y^2 = 25$
$\quad y - x = 1$

6. $x^2 + y^2 = 5$
$\quad y - x = 1$

5. _____

6. _____

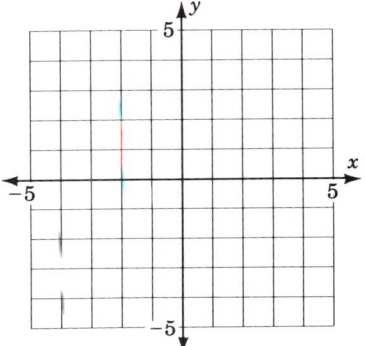

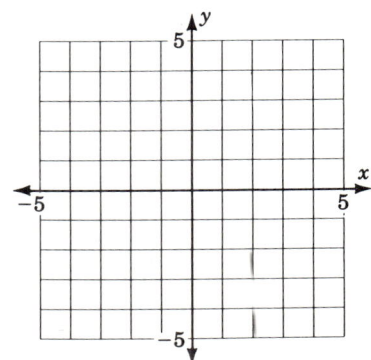

7. $y = x^2 - 5x + 4$
    $x - y = 1$

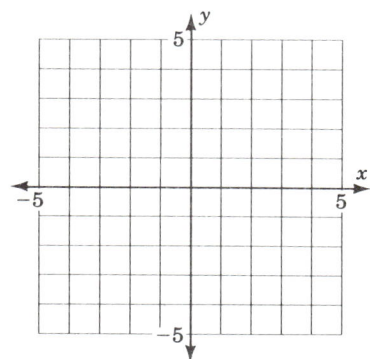

7. _____

8. $y = x^2 - 2x + 1$
    $3x - 3y = 3$

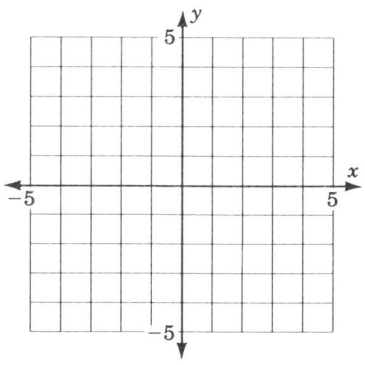

8. _____

9. $y = (x - 1)^2$
    $y - x = 1$

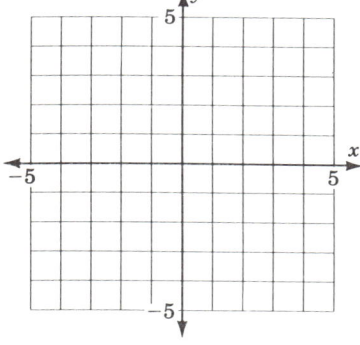

9. _____

10. $y = (x + 3)^2$
     $x + y = -1$

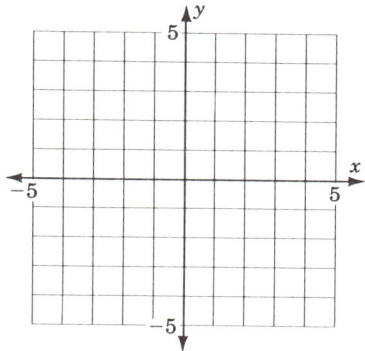

10. _____

11. $4x^2 + 9y^2 = 36$
     $3y - 2x = 6$

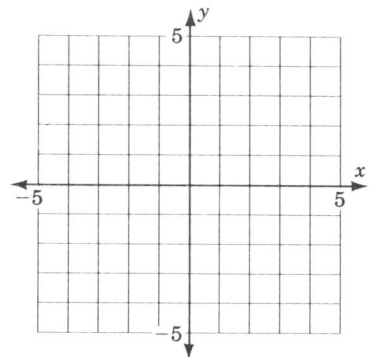

11. _____

12. $4x^2 + 9y^2 = 36$
     $3y + 2x = 6$

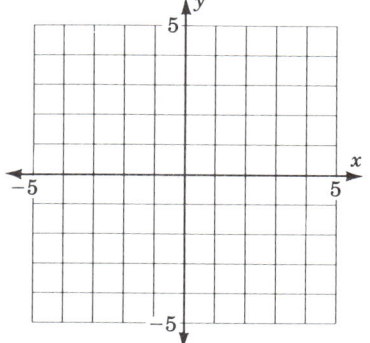

12. _____

13. $x^2 - y^2 = 16$

$x + 4y = 4$

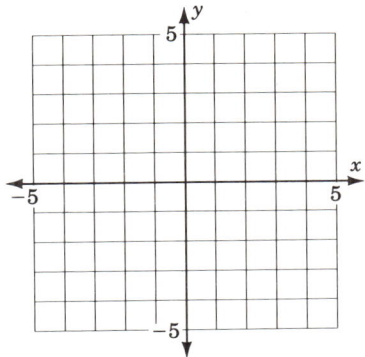

14. $x^2 - y^2 = 9$

$x + 3y = 3$

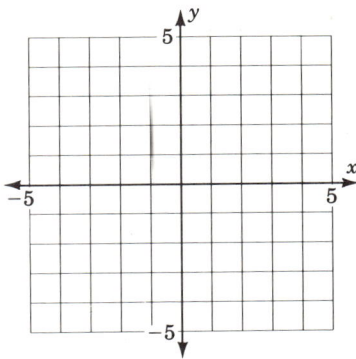

15. $x^2 + y^2 = 4$

$y - x = 5$

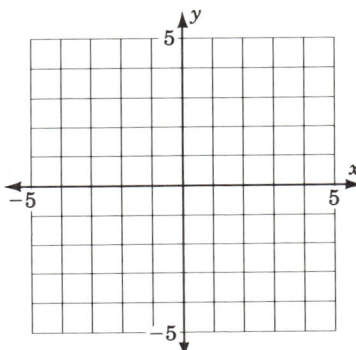

16. $x^2 + y^2 = 4$

$y - x = 3$

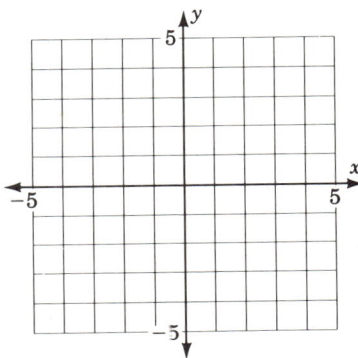

B. In Problems 17–30 solve the system and check by graphing.

17. $y = 4 - x^2$

$y = x^2 - 4$

18. $y = 2 - x^2$

$y = x^2 - 2$

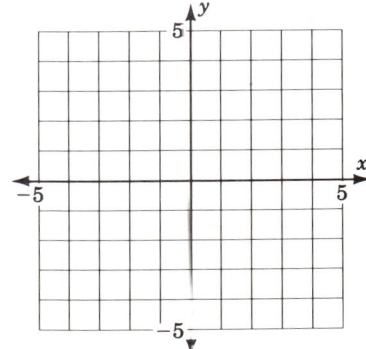

13. _____

14. _____

15. _____

16. _____

17. _____

18. _____

19. $x^2 + y^2 = 25$
$x^2 - y^2 = 7$

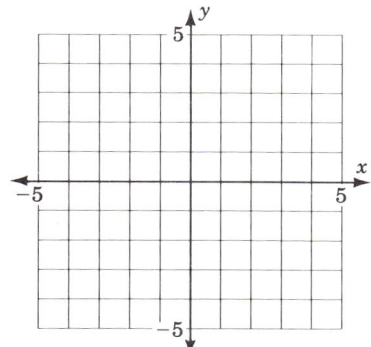

20. $x^2 + y^2 = 20$
$x^2 - y^2 = 2$

19. _____

20. _____

21. $x^2 + y^2 = 16$
$x^2 + 16y^2 = 16$

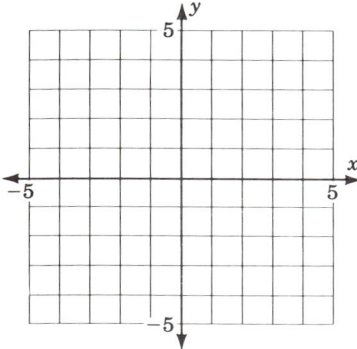

22. $x^2 + y^2 = 9$
$x^2 + 9y^2 = 9$

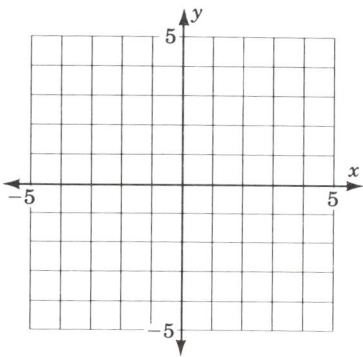

21. _____

22. _____

23. $3x^2 - y^2 = 2$
$x^2 + 2y^2 = 3$

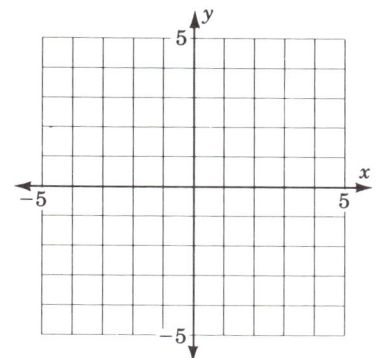

24. $x^2 - 4y^2 = 4$
$9x^2 + 4y^2 = 36$

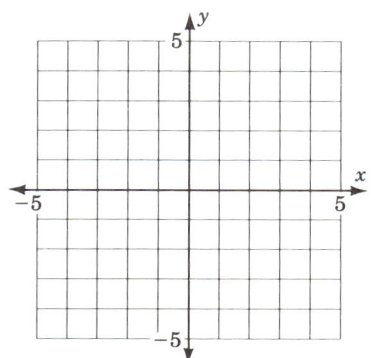

23. _____

24. _____

25. $x^2 + 2y^2 = 11$
$2x^2 + y^2 = 19$

26. $4x^2 + 9y^2 = 52$
$9x^2 + 4y^2 = 52$

25. _____

26. _____

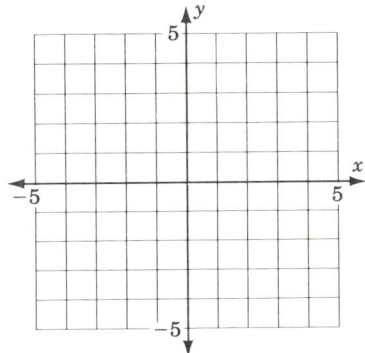

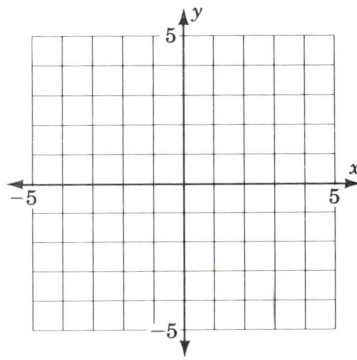

27. $x^2 + y^2 = 4$
$x^2 - y^2 = 9$

28. $x^2 + y^2 = 9$
$x^2 - y^2 = 16$

27. _____

28. _____

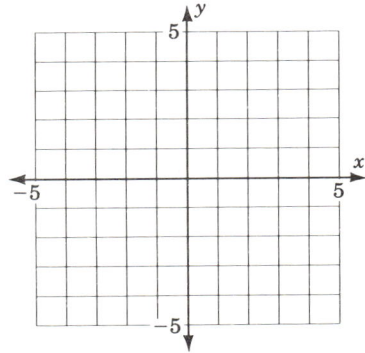

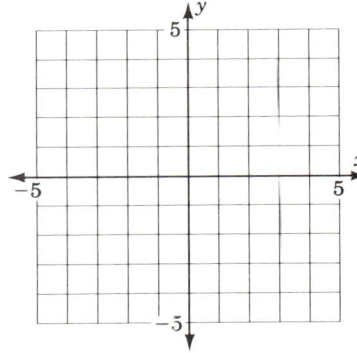

29. $x^2 + y^2 = 1$
$4x^2 + 9y^2 = 36$

30. $x^2 + y^2 = 25$
$4x^2 + 9y^2 = 36$

29. _____

30. _____

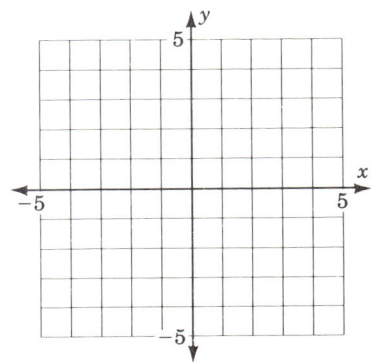

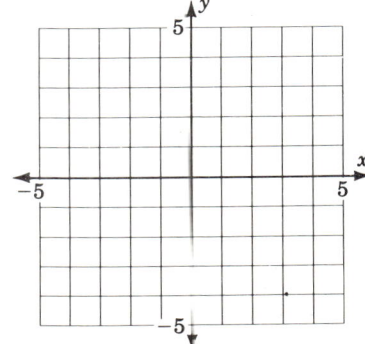

C. Applications

31. The total cost C (in thousands of dollars) for manufacturing and selling x (thousand) items of a product each month is given by $C = x + 4$, and the revenue is $R = 6x - x^2$. How many items must be manufactured and sold for the company to break even?

31. _____

32. The total cost C (in thousands of dollars) for manufacturing and selling x (thousand) items of a product each month is given by $C = 2x + 6$, and the revenue is $R = 150 + 20x - x^2$. How many items must be manufactured and sold for the company to break even?

32. _____

33. There are two very interesting numbers. Their sum is 15 and the difference of their squares is also 15. Find the numbers.

33. _____

34. Two squares differ in area by 108 ft^2 and their sides differ by 6 ft. Find the dimensions of these squares. (Make a picture!)

34. _____

✓ SKILL CHECKER

Graph.

35. $y = 2x - 4$

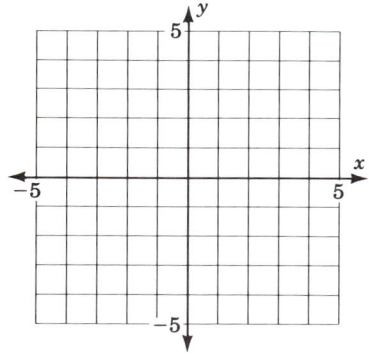

36. $x^2 + y^2 = 4$

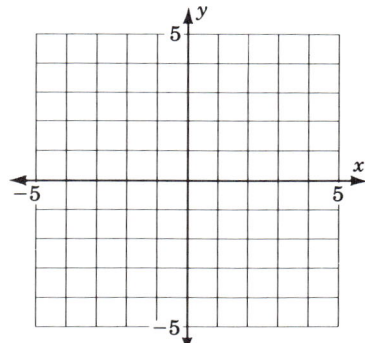

37. $y = x^2 + 1$

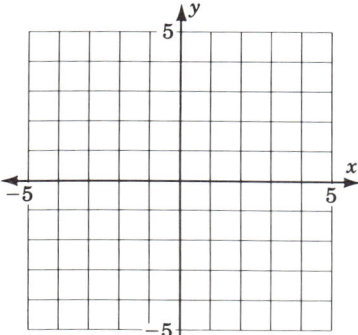

38. $y = x^2 - 1$

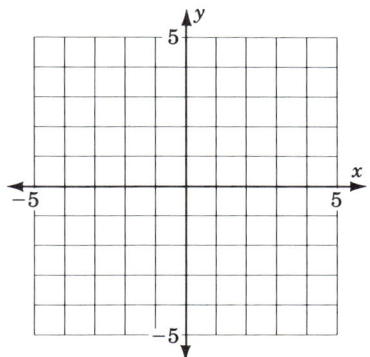

39. $4x^2 + 9y^2 = 36$

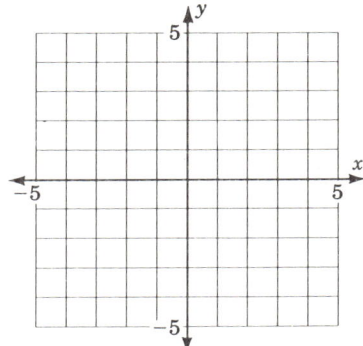

40. $9x^2 + 4y^2 = 36$

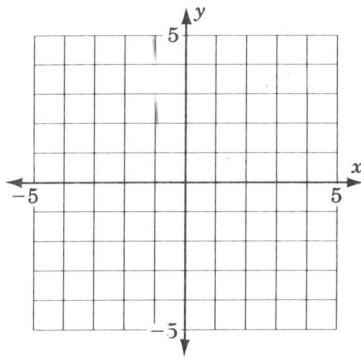

9.5 USING YOUR KNOWLEDGE

Do you think the algebra we have been studying is relatively new? What does it have to do with the price of eggs? Here are a couple of problems from *Elementary Algebra,* published by Macmillan in 1894. See if you have the knowledge to solve these problems.

1. "The number of eggs which can be bought for 25 cents is equal to twice the number of cents which 8 eggs cost. How many eggs can be bought for 25 cents?" *Hint:* If you let p be the price of eggs, then $\dfrac{25}{p}$ is the number of eggs you can buy for 25¢.

1. _____

2. "One half of the number of cents which a dozen apples cost is greater by 2 than twice the number of apples which can be bought for 30 cents. How many can be bought for \$2.50?" *Hint:* If you let d be the cost of a dozen apples, then $\dfrac{d}{12}$ is the cost of one apple and $\dfrac{30}{\frac{d}{12}}$ is the number of apples that can be bought for 30¢.

2. _____

SUMMARY

SECTION	ITEM	MEANING	EXAMPLE
9.1	Conic section	Curves obtained by slicing a cone with a plane	Parabolas, circles, ellipses, and hyperbolas
9.1A	Parabola (vertical)	A curve with equation $y = a(x - h)^2 + k$	$y = x^2$, $y = 3(x - 1)^2 + 2$, and $y = -4(x + 1)^2 - 3$
9.1A	Vertex	The highest or lowest point of a parabola opening vertically	The vertex of $y = x^2 + 2$ is at $(0, 2)$.
9.1C	Parabola (horizontal)	A curve with equation $x = a(y - k)^2 + h$	$x = y^2$, $x = 3(y - 1)^2 + 2$, and $x = -4(y + 1)^2 - 3$
9.2A	Circle	A curve with the equation $(x - h)^2 + (y - k)^2 = r^2$	$(x - 3)^2 + (y + 4)^2 = 9$ is a circle with center at $(3, -4)$ and radius 3.
9.2C	Ellipse	A curve with equation $\dfrac{x^2}{a^2} + \dfrac{y^2}{b^2} = 1$	$\dfrac{x^2}{16} + \dfrac{y^2}{9} = 1$ is an ellipse.

(Continued)

SUMMARY *(Continued)*

SECTION	ITEM	MEANING	EXAMPLE
9.3A	Hyperbola	A curve with equation $\dfrac{x^2}{a^2} - \dfrac{y^2}{b^2} = 1$ or $\dfrac{y^2}{a^2} - \dfrac{x^2}{b^2} = 1$	$\dfrac{x^2}{9} - \dfrac{y^2}{16} = 1$ is a hyperbola.
9.3A	Asymptotes	Lines through the opposite corners of the rectangle associated with a hyperbola whose sides pass through $\pm a$ and $\pm b$	
9.4	Second-degree inequality	An inequality containing at least one second-degree term	$y \le x^2 + 2$ and $x^2 + y^2 > 9$ are second-degree inequalities.
9.5	Nonlinear systems	A system of equations containing at least one second-degree equation	$x^2 + y^2 = 9$ $x - y = 3$ is a nonlinear system.
9.5A	Substitution method	A method of solving nonlinear systems in which substitution is made from one of the equations into the other	To solve $x^2 + y^2 = 9$ $x = y + 3$ by substitution, replace x by $y + 3$ in $x^2 + y^2 = 9$.

NAME

CLASS

SECTION

ANSWERS

(If you need help with these exercises, look in the section indicated in brackets.)

1. [9.1A] Graph.
   a. $y = 9x^2$

1. **a.**

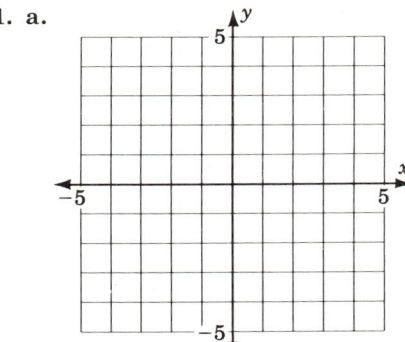

   b. $y = -9x^2$

b.

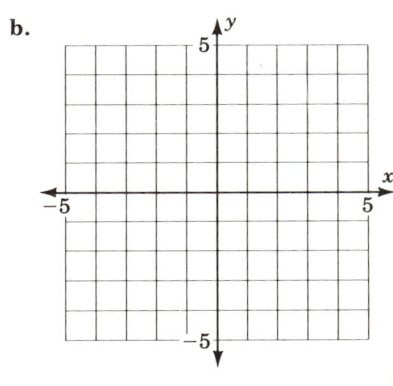

2. [9.1A] Find the vertex and graph.
   a. $y = (x - 1)^2 - 2$

2. **a.** _____

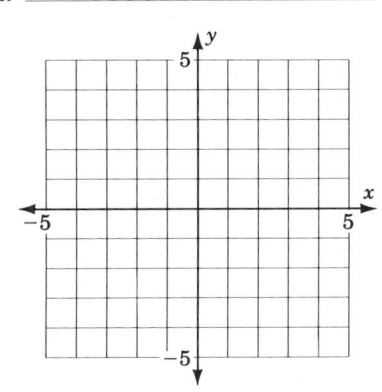

b. $y = -(x - 1)^2 + 2$

b. _____

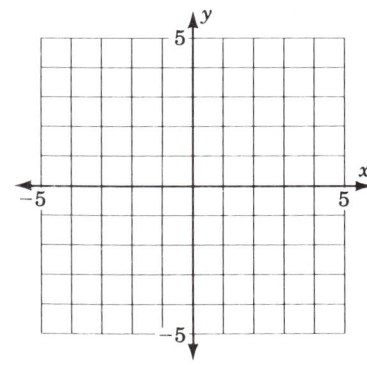

3. [9.1B] Find the vertex and graph.
   a. $y = x^2 - 4x + 2$

3. **a.** _____

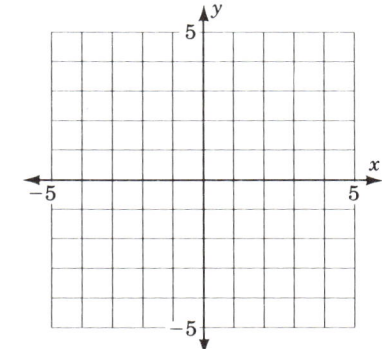

b. $y = -x^2 + 6x - 5$

b. _____

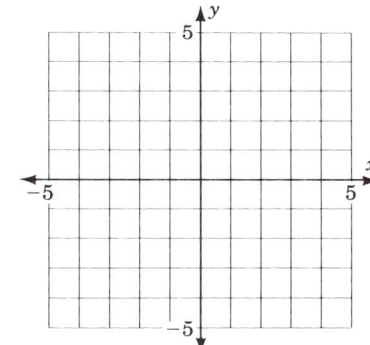

4. [9.1B] Find the vertex and graph.
   a. $y = 2x^2 - 4x + 3$

4. **a.** _____

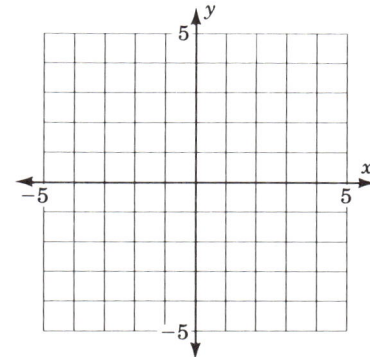

b. $y = -2x^2 + 4x - 5$

b. _____

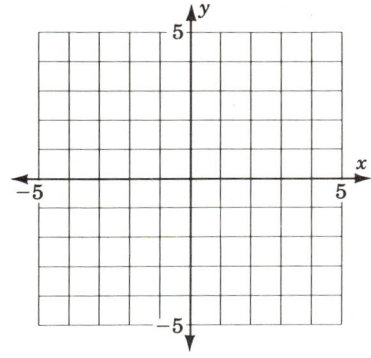

5. [9.1C] Find the vertex and graph.
a. $x = 2(y - 2)^2 - 2$

5. a. _____

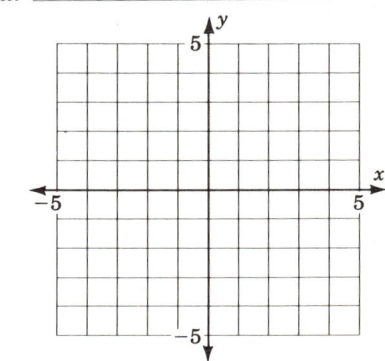

b. $x = -2(y - 3)^2 + 1$

b. _____

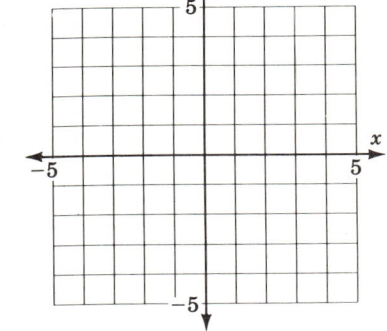

6. [9.1C] Find the vertex and graph.
a. $x = y^2 - 4y + 1$

6. a. _____

b. $x = y^2 - 2y + 3$

b. _____

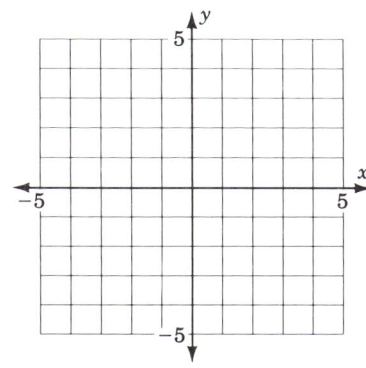

7. [9.1D] Find the value of x that gives the maximum revenue R if
   a. $R = 20x - 0.01x^2$
   b. $R = 10x - 0.02x^2$

7. a. _____
   b. _____

8. [9.2A] Find an equation of the circle of radius 3 and with center at
   a. $(-2, 2)$
   b. $(3, -2)$

8. a. _____
   b. _____

9. [9.2A] Find an equation of the circle with center at the origin and of radius
   a. 5
   b. 8

9. a. _____
   b. _____

10. [9.2B] Find the center and the radius and sketch the graph.
   a. $(x + 2)^2 + (y - 1)^2 = 4$

10. a. _____

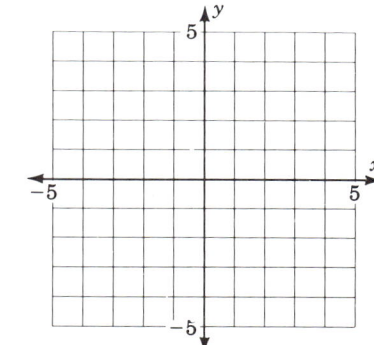

b. $(x - 1)^2 + (y + 2)^2 = 9$

b. _____

11. [9.2B] Sketch the graph.
   a. $x^2 + y^2 = 4$

11. **a.**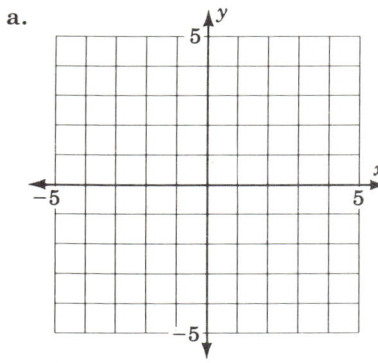

   b. $x^2 + y^2 = 25$

b.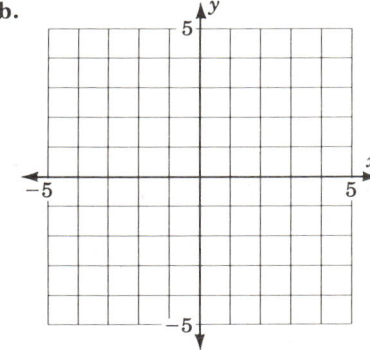

12. [9.2B] Find the center and the radius and sketch the graph.
   a. $x^2 + y^2 + 2x + 2y - 2 = 0$

12. **a.** _____

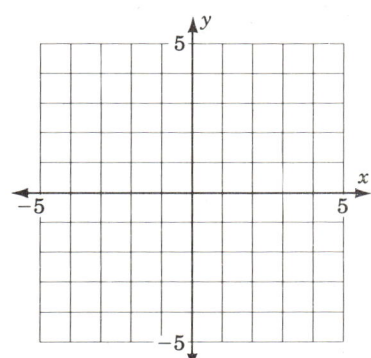

   b. $x^2 + y^2 - 4x + 6y + 9 = 0$

b. _____

13. [9.2C] Graph.

   a. $4x^2 + 9y^2 = 36$

13. a.

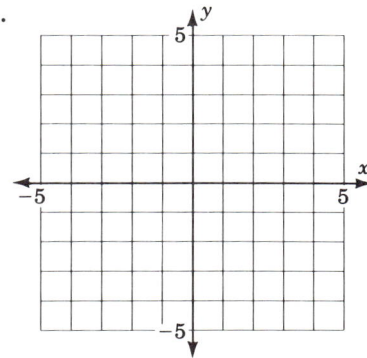

   b. $9x^2 + y^2 = 9$

b.

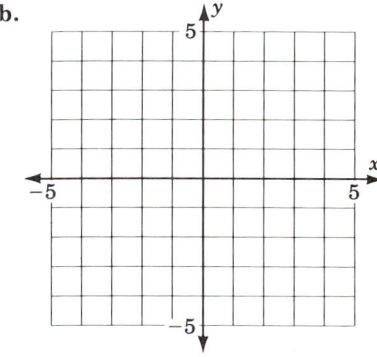

14. [9.2C] Graph.

   a. $\dfrac{(x-1)^2}{4} + \dfrac{(y-2)^2}{9} = 1$

14. a.

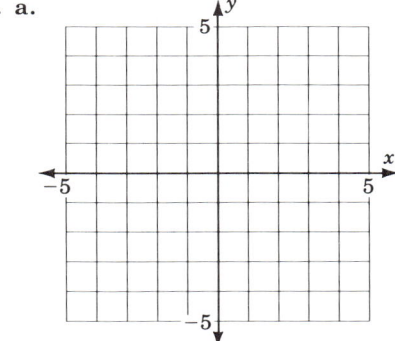

   b. $\dfrac{(x+2)^2}{9} + \dfrac{(y-1)^2}{4} = 1$

b.

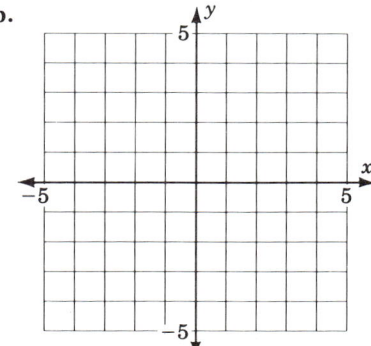

15. [9.3A] Graph.

    a. $\dfrac{x^2}{9} - \dfrac{y^2}{16} = 1$

15. a.

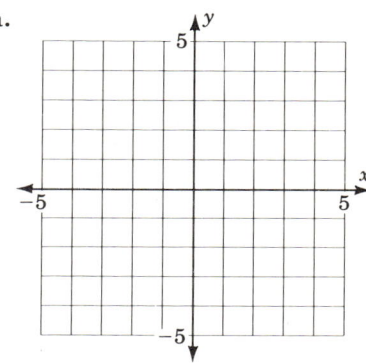

    b. $\dfrac{x^2}{16} - \dfrac{y^2}{9} = 1$

b.

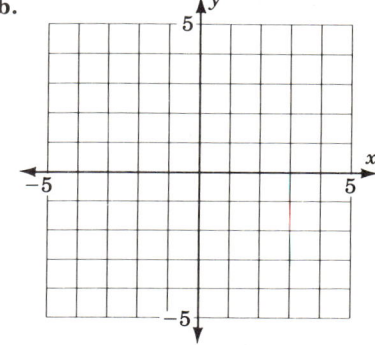

16. [9.3A] Graph.

    a. $\dfrac{y^2}{9} - \dfrac{x^2}{16} = 1$

16. a.

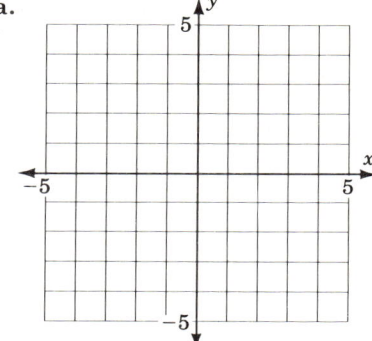

    b. $\dfrac{y^2}{16} - \dfrac{x^2}{9} = 1$

b.

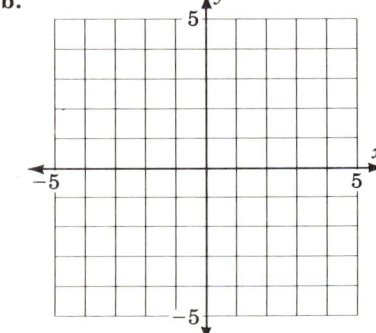

17. [9.3B] Identify each of the curves.
   a. $x = 1 - y^2$
   b. $x^2 = 4y^2 - 4$
   c. $y^2 = 9 - x^2$
   d. $4y^2 = 36 - 9x^2$

17. **a.** _____
    b. _____
    c. _____
    d. _____

18. [9.4A] Graph the inequality
   a. $y \leq 1 - x^2$

18. **a.**

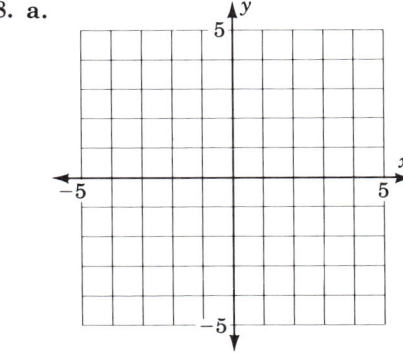

   b. $x \leq 4 - y^2$

b.

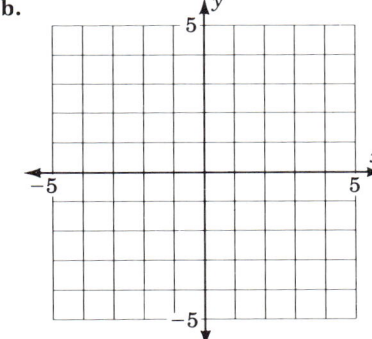

19. [9.4A] Graph the inequality.
   a. $x^2 + y^2 \leq 4$

19. **a.**

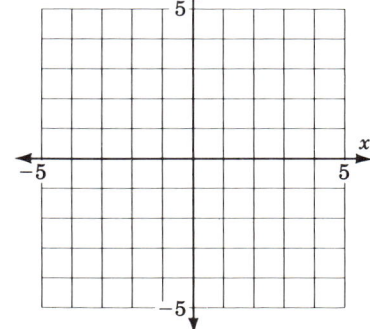

b. $x^2 + y^2 > 9$

b.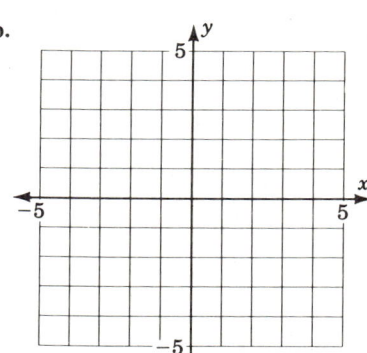

20. [9.4A] Graph the inequality.
a. $4x^2 - y^2 \leq 4$

20. a.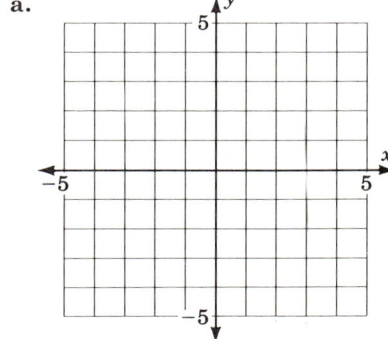

b. $x^2 - 4y^2 \leq 4$

b.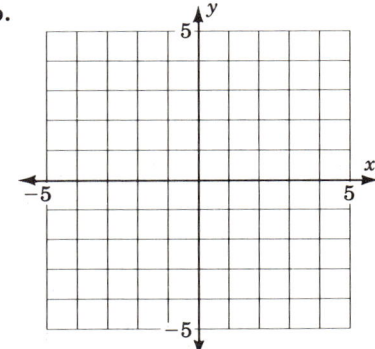

21. [9.4B] Graph the system.
a. $x^2 + y^2 \leq 4$ and $y \leq 2 - x^2$

21. a.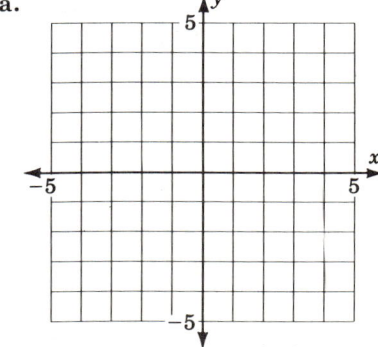

b. $x^2 + y^2 \leq 4$ and $y \geq 4x^2$

b.

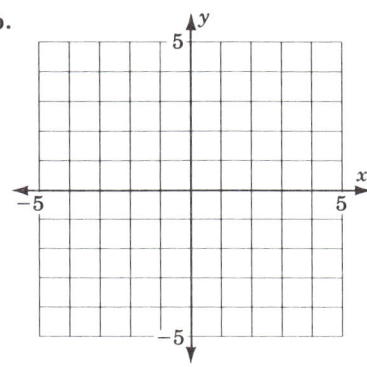

22. [9.5A] Solve the system by the substitution method.

    a. $x^2 + y^2 = 1$
           $x + y = 1$

    b. $x^2 + y^2 = 10$
           $x + y = 4$

22. a. _____
    b. _____

23. [9.5a] Solve the system by the substitution method.

    a. $x^2 - y^2 = 16$
            $2x = y$

    b. $x^2 + y^2 = 4$
           $x + y = 3$

23. a. _____
    b. _____

24. [9.5B] Solve the system.

    a. $x^2 - y^2 = 5$
          $x^2 + 2y^2 = 17$

    b. $x^2 - y^2 = 3$
         $2x^2 + y^2 = 9$

24. a. _____
    b. _____

25. [9.5C]

    a. The cost C of manufacturing and selling x units of a product is $C = 10x + 400$, and the corresponding revenue R is $R = x^2 - 200$. Find the break-even point.

    b. Repeat part a if $C = 6x + 80$ and $R = x^2 - 200$.

25. a. _____
    b. _____

NAME

CLASS

SECTION

(Answers on pages 756–761)

ANSWERS

1. Graph the parabola $y = -x^2 - 4$.

1.

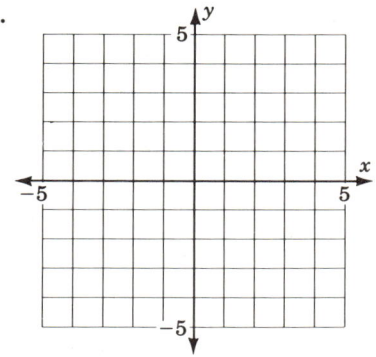

2. Find the vertex and graph the parabola $y = (x - 2)^2 + 2$.

2. _____

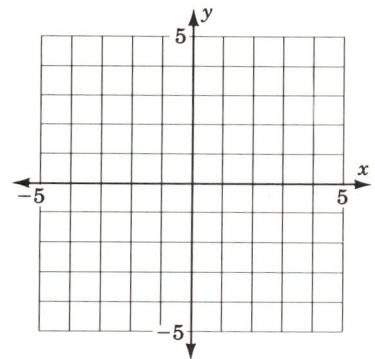

3. Find the vertex and graph the parabola $y = -x^2 - 4x - 1$.

3. _____

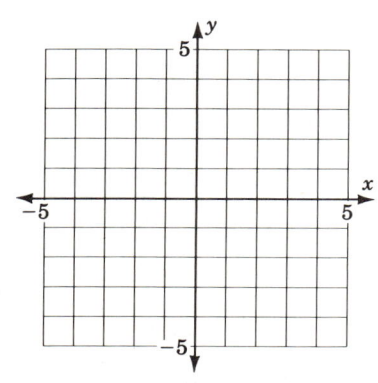

4. Find the vertex and graph the parabola $y = -2x^2 + 4x + 1$.

4. _____

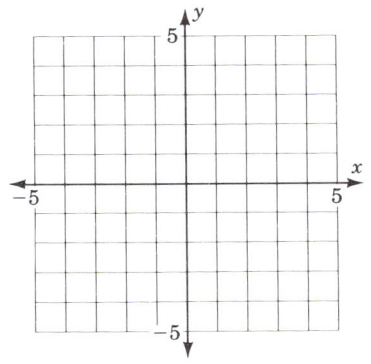

5. Find the vertex and graph the parabola $x = 2(y + 1)^2 - 1$.

5. _____

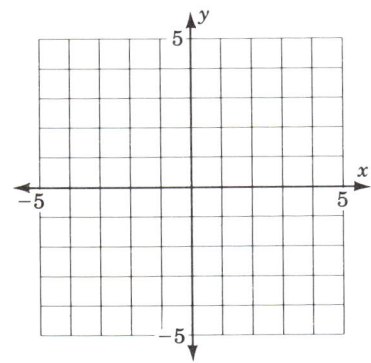

6. Find the vertex and graph the parabola $x = y^2 + 2y - 1$.

6. _____

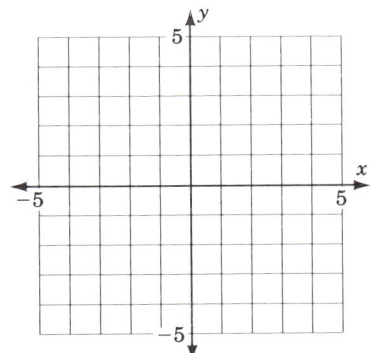

7. If the revenue is given by $R = 60x - 0.03x^2$, find the value of x that yields the maximum revenue.

7. _____

8. Find an equation of the circle of radius 2 with its center at $(1, -2)$.

8. _____

9. Find an equation of the circle of radius 4 with center at the origin.

9. _____

10. Find the center and the radius and sketch the graph of $(x + 1)^2 + (y - 2)^2 = 9$.

10.

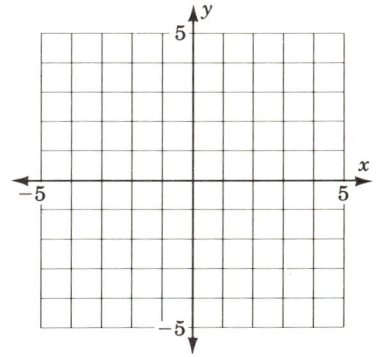

11. Sketch the graph of $x^2 + y^2 = 16$.

11.

12. Find the center and the radius and sketch the graph of $x^2 + y^2 + 4x - 2y - 4 = 0$.

12.

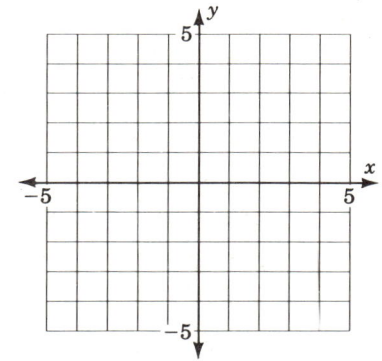

13. Graph $x^2 + 9y^2 = 9$.

13.

14. Graph $\dfrac{(x-2)^2}{4} + \dfrac{(y+1)^2}{9} = 1$.

14.

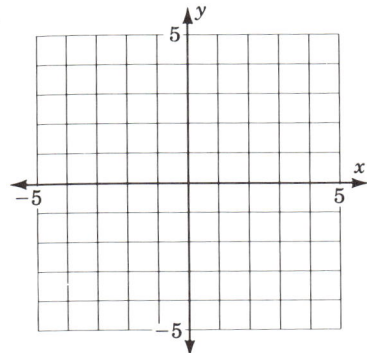

15. Graph $\dfrac{y^2}{9} - \dfrac{x^2}{25} = 1$.

15.

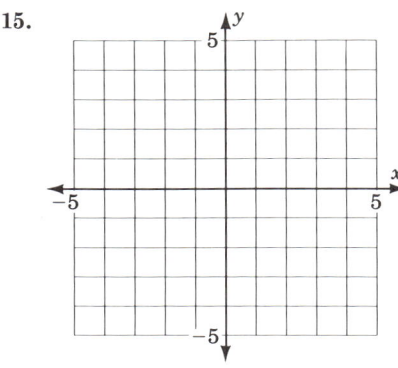

16. Graph $\dfrac{x^2}{9} - \dfrac{y^2}{25} = 1$.

16.

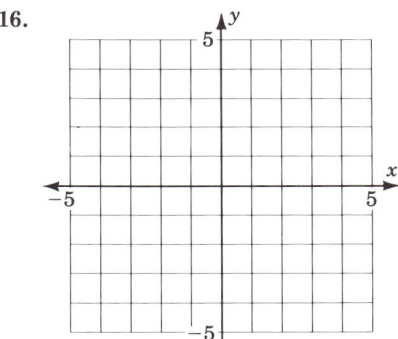

17. Identify each of the following curves.
   a. $x = y^2 - 4$
   c. $9y^2 = 144 + 16x^2$
   b. $16y^2 = 144 - 9x^2$
   d. $x^2 = 16 - y^2$

17. a. _____   c. _____
    b. _____   d. _____

18. Graph the inequality $y \le -x^2 - 1$.

18.

19. Graph the inequality $x^2 + y^2 < 9$.

19.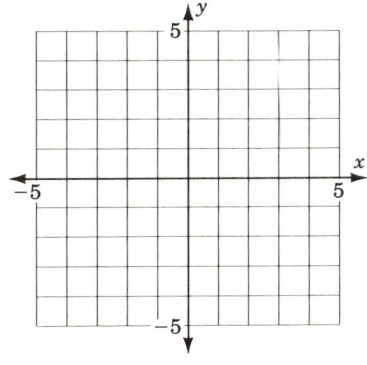

20. Graph the inequality $y^2 - 4x^2 \leq 4$.

20.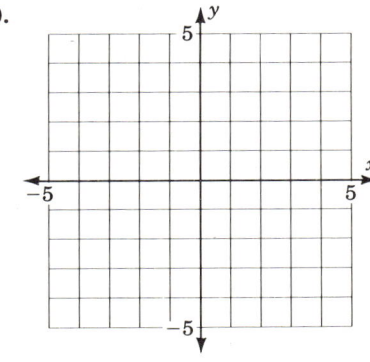

21. Graph the system $x^2 + y^2 \leq 4$ and $y \leq 2 - x^2$.

21.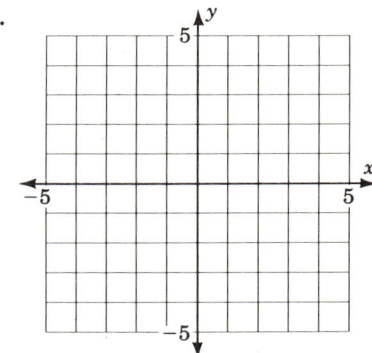

22. Use the substitution method to solve the system.

$$x^2 + y^2 = 4$$
$$x + y = 2$$

22. _____

23. Use the substitution method to solve the system.

$$x^2 + y^2 = 1$$
$$x + y = 2$$

23. _____

24. Solve the system.

$$x^2 + y^2 = 20$$
$$x^2 - 2y^2 = 8$$

24. _____

25. The cost C of manufacturing and selling x units of a product is $C = 20x + 50$, and the corresponding revenue R is $R = x^2 - 75$. Find the break-even value of x.

25. _____

IF YOU MISSED QUESTION	SECTION	EXAMPLES	PAGE	ANSWERS
1	9.1A	1, 2	672–674	1.
2	9.1A	3	674	2. Vertex (2, 2)
3	9.1B	4	676–677	3. Vertex (−2, 3)

IF YOU MISSED QUESTION	SECTION	EXAMPLES	PAGE	ANSWERS
4	9.1B	5	678	**4.** Vertex $(1, 3)$
5	9.1C	6	679	**5.** Vertex $(-1, -1)$
6	9.1C	7	679	**6.** Vertex $(-2, -1)$
7	9.1D	8	680	**7.** $x = 1000$
8	9.2A	1	689	**8.** $(x - 1)^2 + (y + 2)^2 = 4$
9	9.2A	2	690	**9.** $x^2 + y^2 = 16$

IF YOU MISSED QUESTION	SECTION	EXAMPLES	PAGE	ANSWERS
10	9.2B	3	690	**10.** Center $(-1, 2)$; $r = 3$
11	9.2B	4	691	**11.**
12	9.2B	5	691	**12.** Center $(-2, 1)$; $r = 3$

IF YOU MISSED QUESTION	SECTION	EXAMPLES	PAGE	ANSWERS
13	9.2C	6	692	**13.**
14	9.2C	7	693	**14.**
15	9.3A	1a	705	**15.**
16	9.3A	1b	705	**16.**

IF YOU MISSED QUESTION	SECTION	EXAMPLES	PAGE	ANSWERS
17a	9.3B	2	707	**17. a.** Parabola
17b	9.3B	2	707	**b.** Ellipse
17c	9.3B	2	707	**c.** Hyperbola
17d	9.3B	2	707	**d.** Circle
18	9.4A	1	717–718	**18.**
19	9.4A	2	718	**19.**
20	9.4A	3	718–719	**20.**

IF YOU MISSED QUESTION	SECTION	EXAMPLES	PAGE	ANSWERS
21	9.4B	4	719	21.
22	9.5A	1	728–729	22. $(2, 0)$ and $(0, 2)$
23	9.5A	2	729–730	23. $\left(1 + \dfrac{\sqrt{2}}{2}i,\ 1 - \dfrac{\sqrt{2}}{2}i\right),$ $\left(1 - \dfrac{\sqrt{2}}{2}i,\ 1 + \dfrac{\sqrt{2}}{2}i\right)$
24	9.5B	3	730–731	24. $(4, 2), (4, -2), (-4, 2), (-4, -2)$
25	9.5C	4	731–732	25. $x = 25$

CHAPTER 10

RELATIONS AND FUNCTIONS

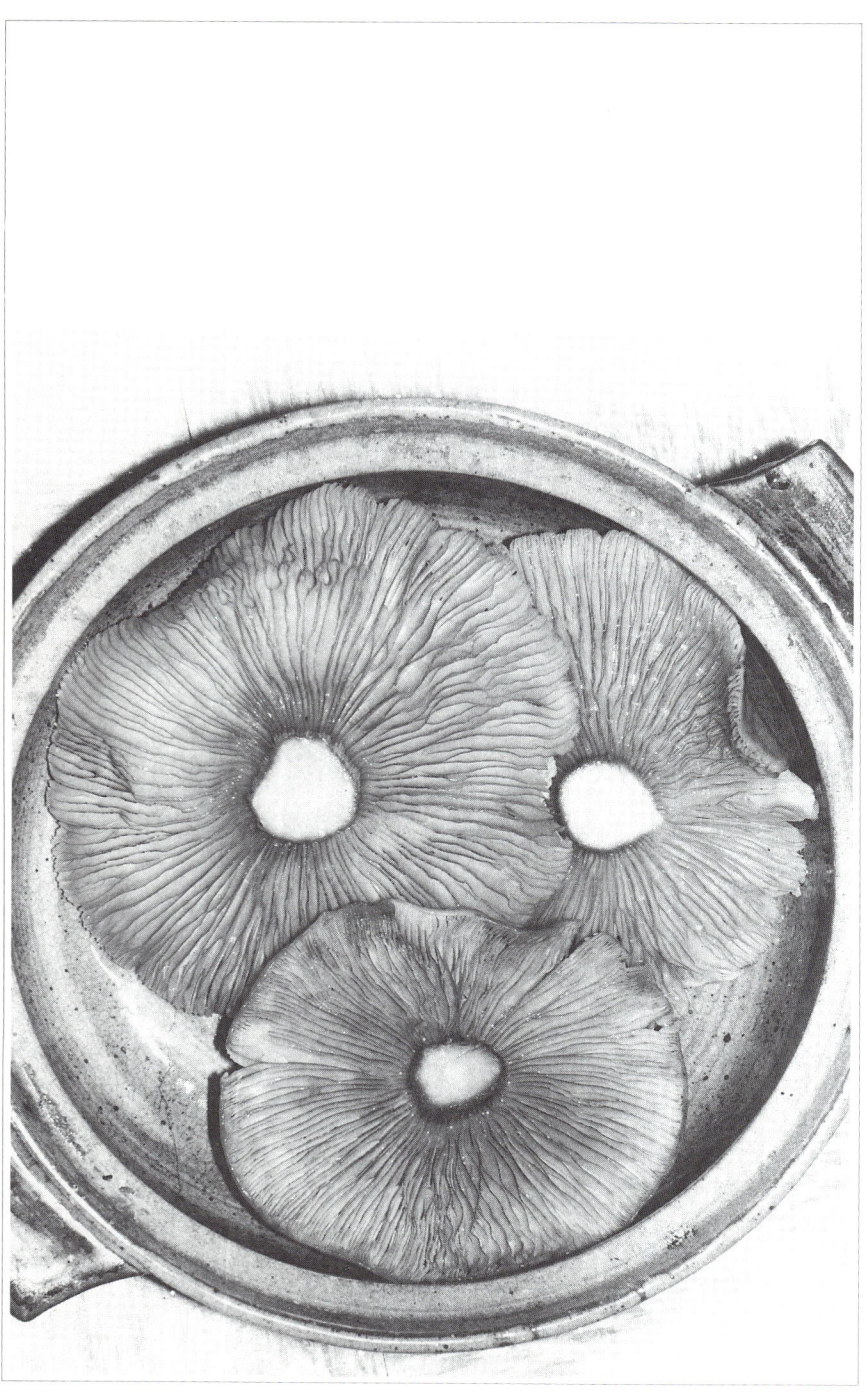

In this chapter we study one of the most important ideas in mathematics: functions. We relate our previous work with constant, linear, and quadratic equations to these types of functions. We then study polynomial and exponential functions and end the chapter with a discussion of the inverse of a function.

NAME

CLASS

SECTION

(Answers on pages 768–772)

ANSWERS

1. Find the domain and range of the relation
   $\{(0, 4), (2, 8), (3, 9), (5, 7)\}$.

1. _____

2. Find the domain and the range and graph the relation
   $\{(x, y) | y = 4 - x\}$.

2. _____

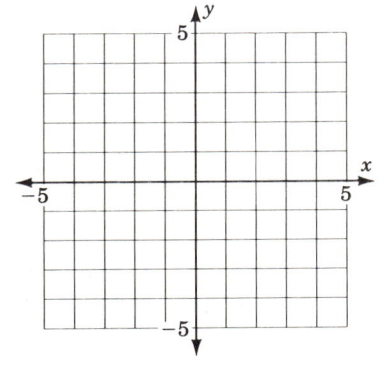

3. Find the domain and the range and graph.
   a. $A = \{(x, y) | x^2 + y^2 = 1\}$

3. **a.** _____

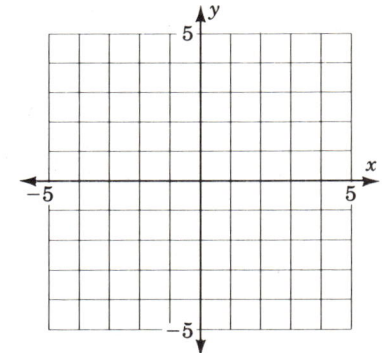

b. $B = \{(x, y) | y = \sqrt{1 - x^2}\}$

b. _____

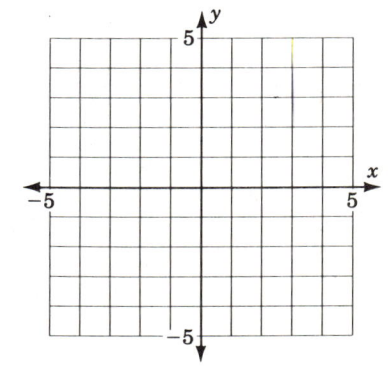

c. $C = \{(x, y) \mid y = -\sqrt{1 - x^2}\}$

c. _____

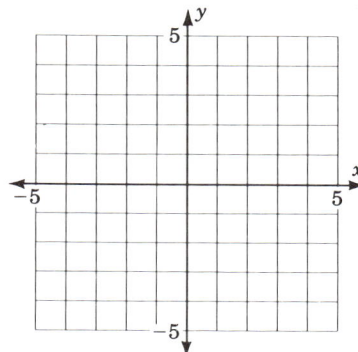

4. Use the vertical line test to determine if the relation defined by the given equation is a function.
   a. $y = 4 + x^2$
   b. $x^2 + 4y^2 = 4$

4. a. _____
   b. _____

5. Find the domain of the function defined by
   a. $y = \dfrac{2}{x - 3}$
   b. $y = \sqrt{x - 2}$

5. a. _____
   b. _____

6. Let $f(x) = 3x - 4$. Find.
   a. $f(2)$
   b. $f(1)$
   c. $f(2) - f(1)$

6. a. _____
   b. _____
   c. _____

7. Let $f = \{(1, 0), (2, 3), (3, -1)\}$. Find.
   a. $f(2)$
   b. $f(1)$
   c. $f(2) - f(1)$

7. a. _____
   b. _____
   c. _____

8. Let $f(x) = 1 - x^2$ and $g(x) = 1 + x$. Find.
   a. $(f + g)(x)$
   b. $(f - g)(x)$
   c. $(fg)(x)$
   d. $\left(\dfrac{f}{g}\right)(x)$

8. a. _____
   b. _____
   c. _____
   d. _____

9. If $f(x) = 4x + 1$ find
   $$\dfrac{f(x) - f(a)}{x - a}, \qquad x \neq a$$

9. _____

10. If $f(x) = x^3$ and $g(x) = 1 - x$, find
    a. $f \circ g$
    b. $g \circ f$
    c. $(g \circ f)(2)$

10. a. _____
    b. _____
    c. _____

11. The lower limit L (heartbeats per minute) of a person's target zone is given by $L(a) = -\frac{2}{3}a + 150$, where a is the person's age (in years). Find L for a person who is
    a. 12 yr old
    b. 24 yr old

11. a. _____
    b. _____

12. The revenue (in dollars) obtained from selling x units of a product is $R(x) = 100x - 0.02x^2$ and the cost is
$C(x) = 20,000 + 40x$.
  a. Find the profit function, $P(x) = R(x) - C(x)$.
  b. How many units must be made and sold to yield the maximum profit?

12. a. _____
    b. _____

13. Classify the function as constant, linear, or quadratic and graph.
  a. $f(x) = -2$

13. a. _____

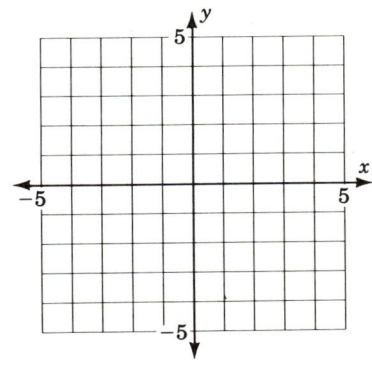

  b. $f(x) = 3 - 2x$

b. _____

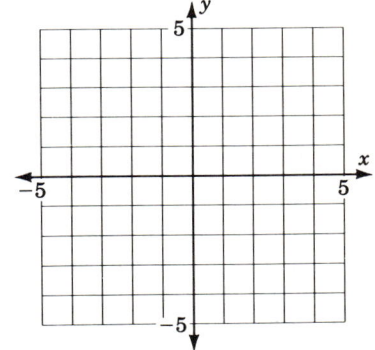

  c. $f(x) = x^2 - 4x$

c. _____

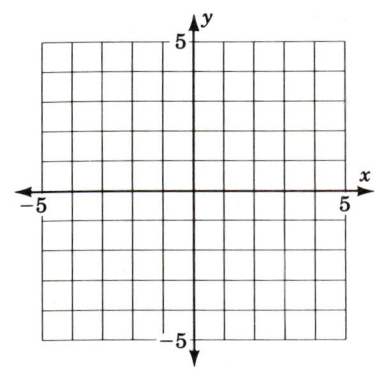

14. Graph $f(x) = x^3 - 4x$

14.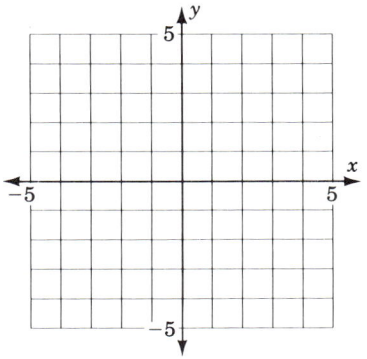

15. Graph $f(x) = \dfrac{x+2}{x-1}$.

15.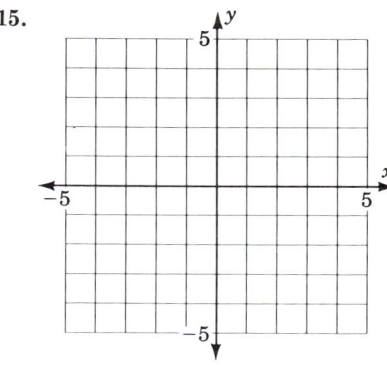

16. Graph $f(x) = \left(\dfrac{1}{2}\right)^x$.

16.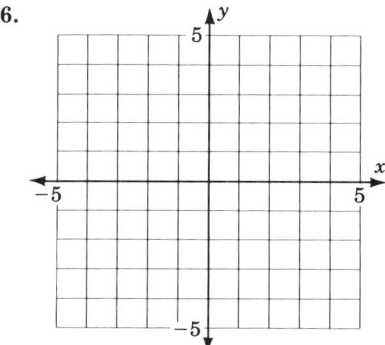

17. Let $S = \{(3, 4), (5, 6), (7, 8)\}$. Find.
   a. The domain and range of S
   b. S^{-1}
   c. The domain and range of S^{-1}
   d. The graph of S and S^{-1}

17. **a.** _____
    b. _____
    c. _____
    d.

18. Let $f(x) = y = 2x - 2$.
   a. Find $f^{-1}(x)$.
   b. Graph f and its inverse.

18. a. _____
   b.

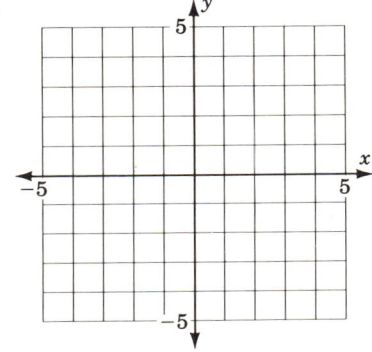

19. Find the inverse of $f(x) = y = 2x^2$. Is the inverse a function?

20. Graph $y = 2^x$. Is the inverse a function?

19. _____

20. _____

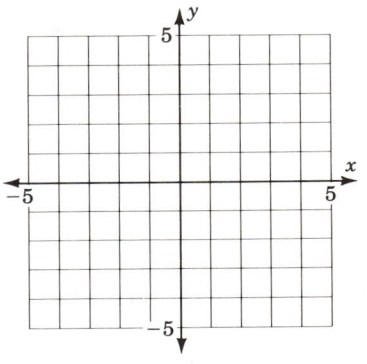

IF YOU MISSED QUESTION	SECTION	EXAMPLES	PAGE	ANSWERS
1	10.1A	1	774	**1.** $D = \{0, 2, 3, 5\}$; $R = \{4, 7, 8, 9\}$
2	10.1A	2	774	**2.** The domain and range are the set of real numbers.
3a	10.1A	3	775	**3. a.** $D = \{x \mid -1 \le x \le 1\}$; $R = \{y \mid -1 \le y \le 1\}$
3b	10.1A	3	775	**b.** $D = \{x \mid -1 \le x \le 1\}$; $R = \{y \mid 0 \le y \le 1\}$

IF YOU MISSED QUESTION	SECTION	EXAMPLES	PAGE	ANSWERS
3c	10.1A	3	775	**c.** $D = \{x \mid -1 \le x \le 1\}$; $R = \{y \mid -1 \le y \le 0\}$
				$y = -\sqrt{1-x^2}$
4a	10.1B	4	776	**4. a.** A function
4b	10.1B	4	776	**b.** Not a function
5a	10.1C	5	777	**5. a.** All real numbers except 3
5b	10.1C	5	777	**b.** All real numbers greater than or equal to 2
6a	10.2A	1	788	**6. a.** 2
6b	10.2A	1	788	**b.** -1
6c	10.2A	1	788	**c.** 3
7a	10.2A	2	788	**7. a.** 3
7b	10.2A	2	788	**b.** 0
7c	10.2A	2	788	**c.** 3
8a	10.2B	3	789	**8. a.** $-x^2 + x + 2$
8b	10.2B	3	789	**b.** $-x^2 - x$
8c	10.2B	3	789	**c.** $-x^3 - x^2 + x + 1$
8d	10.2B	3	789	**d.** $1 - x$
9	10.2B	4	789	**9.** 4
10a	10.2C	5	790	**10. a.** $(1-x)^3$
10b	10.2C	5	790	**b.** $1 - x^3$
10c	10.2C	5	790	**c.** -7
11a	10.2D	6	790–791	**11. a.** 142
11b	10.2D	6	790–791	**b.** 134
12a	10.2D	7	791	**12. a.** $R(x) = -0.02x^2 + 60x - 20{,}000$
12b	10.2D	7	791	**b.** 1500

IF YOU MISSED QUESTION	SECTION	EXAMPLES	PAGE	ANSWERS
13a	10.3A	1	800–801	**13. a.** A constant function
13b	10.3A	1	800–801	**b.** A linear function
13c	10.3A	1	800–801	**c.** A quadratic function

13. a. A constant function

$f(x) = -2$

b. A linear function

$f(x) = 3 - 2x$

c. A quadratic function

$f(x) = x^2 - 4x$

IF YOU MISSED QUESTION	SECTION	EXAMPLES	PAGE	ANSWERS
14	10.3B	2	802	**14.**
15	10.3C	3	803	**15.**
16	10.3D	4	804	**16.**
17a	10.4A	1	816	**17. a.** $D = \{3, 5, 7\}, \ R = \{4, 6, 8\}$
17b	10.4A	1	816	**b.** $\{(4, 3), (6, 5), (8, 7)\}$
17c	10.4A	1	816	**c.** $D = \{4, 6, 8\}, \ R = \{3, 5, 7\}$

14.

$f(x) = x^3 - 4x$

15.

$f(x) = \dfrac{x + 2}{x - 1}$

$y = 1$

$x = 1$

16.

$f(x) = \left(\tfrac{1}{2}\right)^x$

IF YOU MISSED QUESTION	SECTION	EXAMPLES	PAGE	ANSWERS
17d	10.4A	1	816	**d.**
18a	10.4B	2	817	**18. a.** $f^{-1}(x) = \dfrac{x+2}{2}$
18b	10.4B	2	817–818	**b.**
19	10.4C	3	819	**19.** $y = \pm\sqrt{\dfrac{x}{2}}$; no
20	10.4C	4	819	**20.** Yes

10.1 RELATIONS AND FUNCTIONS

OBJECTIVES

REVIEW

Before starting this section, you should know:

1. How to graph a conic section.
2. How to graph a line.

OBJECTIVES

You should be able to:

A. Find the domain and range of a relation.
B. Use the vertical line test to determine if a relation is a function.
C. Find the domain of a function defined by an equation.

Did you know that there is a relationship between the temperature and the rate of travel (speed) of certain ants? If y is the speed in centimeters per second (cm/sec) and x is the temperature in degrees Celsius, the relationship is

$$y = \frac{1}{6}(x - 4)$$

Thus if the temperature is $10°C$, the speed is

$$y = \frac{1}{6}(10 - 4) = \frac{1}{6} \cdot 6 = 1 \text{ cm/sec}$$

If the temperature is $16°C$, the speed is

$$y = \frac{1}{6}(16 - 4) = \frac{1}{6} \cdot 12 = 2 \text{ cm/sec}$$

We can make a table showing two related sets of numbers, one for the temperature x and the other for the speed y, as shown.

TEMPERATURE (CELSIUS)	RATE OF TRAVEL
4	0
10	1
16	2
22	3

(By the way, x has to be greater than 4 and less than $35°C$.)

The numbers in the first column are called values of the **independent** variable because they are chosen independently of the second number. The numbers in the second column are called values of the **dependent** variable because they *depend* on the values of the numbers in the first column. The numbers in our table can also be written as ordered pairs, as shown.

(TEMPERATURE, SPEED)
(4, 0)
(10, 1)
(16, 2)
(22, 3)

These ordered pairs can be written as the set

$$\{(4, 0), (10, 1), (16, 2), (22, 3)\}$$

A. DOMAIN AND RANGE

Any set of ordered pairs is a relation, which we define as follows.

> A **relation** is a set of ordered pairs. The set of all first coordinates is the **domain** of the relation and the set of all second coordinates is the **range** of the relation.

Now, suppose $S = \{(1, 5), (2, 7), (3, 9)\}$. The domain D is the set of all first coordinates—that is, $D = \{1, 2, 3\}$—and the range R is the set of all second coordinates—that is, $R = \{5, 7, 9\}$.

The relation S can also be written by giving the rule used to obtain the ordered pairs. Thus

$$S = \{(x, y)\,|\,y = 2x + 3\}$$

Here $x = 1, 2,$ or 3.

EXAMPLE 1 Find the domain and range of the relation

$$A = \{(1, 2), (2, 3), (3, 4)\}.$$

Solution The domain of A is the set of first coordinates, so $D = \{1, 2, 3\}$. The range of A is the set of second coordinates, so $R = \{2, 3, 4\}$. ▲

Since a relation is a set of ordered pairs, relations can be graphed in the Cartesian plane. We can then identify the domain and range by examining the graph of the relation. When no domain is specified, the domain is assumed to be the set of all real numbers for which the function is defined, as shown next.

EXAMPLE 2 Find the graph, the domain, and the range of the relation

$$\{(x, y)\,|\,y = 2x - 4\}$$

Solution The graph of the relation is the graph of the equation $y = 2x - 4$, shown in the figure. The domain of this relation is the set of all real numbers, since any real number x can be used as the first coordinate. Similarly, the range of y is the set of all real numbers. Some of the ordered pairs in the relation are $(0, -4)$, $(1, -2)$, and $(2, 0)$.

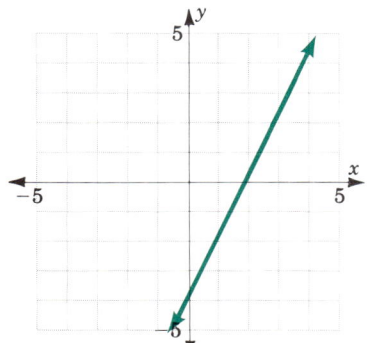

Problem 1 Find the domain and range of the relation
$$B = \{(7, 8), (8, 9), (9, 10)\}.$$

Problem 2 Find the graph, the domain, and the range of the relation
$$\{(x, y)\,|\,y = 2x + 4\}.$$

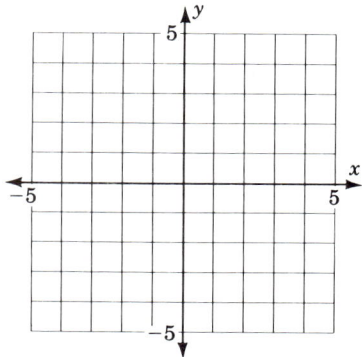

▲

EXAMPLE 3 Find the graph, the domain, and the range of the following relations.

a. $A = \{(x, y) \mid x^2 + y^2 = 4\}$ **b.** $B = \{(x, y) \mid y = \sqrt{4 - x^2}\}$
c. $C = \{(x, y) \mid y = -\sqrt{4 - x^2}\}$

Solution

a. The graph of $x^2 + y^2 = 4$ is a circle of radius 2 centered at the origin, as shown in the figure. From the figure, it is clear that x and y can be any real numbers between -2 and 2, inclusive. Thus the domain of A is $D = \{x \mid -2 \le x \le 2\}$ and the range is $R = \{y \mid -2 \le y \le 2\}$.

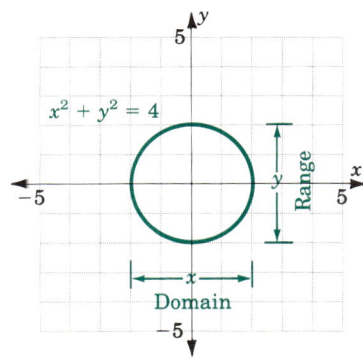

b. The graph of $y = \sqrt{4 - x^2}$ is the top half of the circle in (a), as shown in the figure. The domain of B is $D = \{x \mid -2 \le x \le 2\}$, and the range is $R = \{y \mid 0 \le y \le 2\}$.

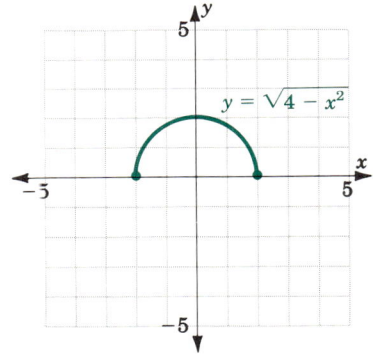

c. The graph of $y = -\sqrt{4 - x^2}$ is the bottom half of the circle in (a), as shown in the figure. The domain of C is $D = \{x \mid -2 \le x \le 2\}$, and the range is $R = \{y \mid -2 \le y \le 0\}$.

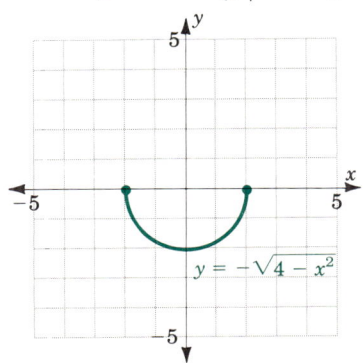

Problem 3 Find the graph, the domain, and the range of
a. $A = \{(x, y) \mid x^2 + y^2 = 9\}$

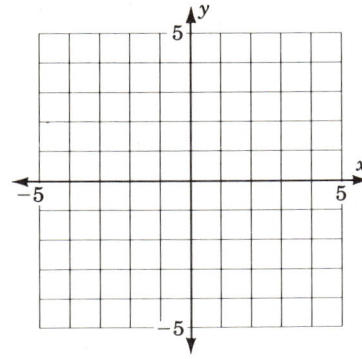

b. $B = \{(x, y) \mid y = \sqrt{9 - x^2}\}$

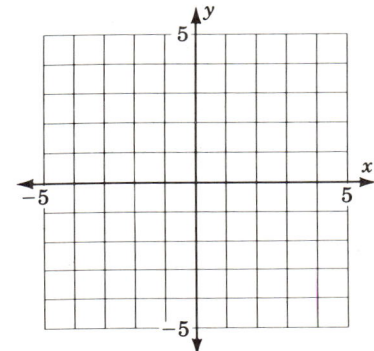

c. $C = \{(x, y) \mid y = -\sqrt{9 - x^2}\}$

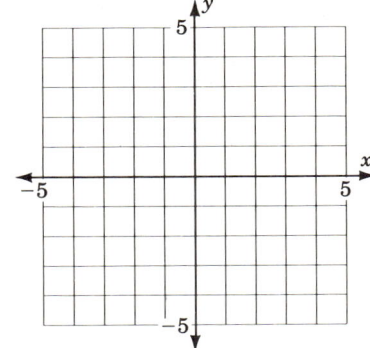

B. THE VERTICAL LINE TEST

The relation in Example 3(a) allows *two* values of y for the same value of x. For instance if $x = 0$, then $y = \pm 2$. On the other hand, the relations in 3(b) and 3(c) allow only *one* value of y for each value of x. These two relations are *functions*. Here is the definition.

> A **function** is a relation in which no two different ordered pairs have the same first coordinate.

Thus the relation $\{(1, 2), (2, 3), (3, 4)\}$ is a function, since no two ordered pairs have the same first coordinate. On the other hand, the relation $\{(1, 2), (2, 3), (1, 3)\}$ is not a function, since $(1, 2)$ and $(1, 3)$ have the same first coordinate.

The graph of a relation can be used to determine if the relation is a function. Since any two points with the same first coordinate will be on a vertical line parallel to the y-axis, if any vertical line intersects the graph more than once, the relation is *not* a function. Testing to see if a relation is a function by determining if a vertical line crosses the graph more than once is called the **vertical line test.** Using this test, we can see that the relation $\{(x, y) \mid x^2 + y^2 = 4\}$ in 3(a) is *not* a function. (A vertical line crosses the graph in more than one place). On the other hand, the graphs in 3(b) and 3(c) represent functions.

EXAMPLE 4 Use the vertical line test to determine if the relation defined by the given equation is a function.

a. $y = x^2 + 1$ **b.** $9x^2 + 4y^2 = 36$

Solution

a. The graph of $y = x^2 + 1$ is a parabola whose vertex is at $(0, 1)$. The relation is a function, since no vertical line crosses the graph more than once.

b. The graph of $9x^2 + 4y^2 = 36$ is the ellipse shown. Since we can draw a vertical line that crosses the graph in more than one point, the relation is not a function.

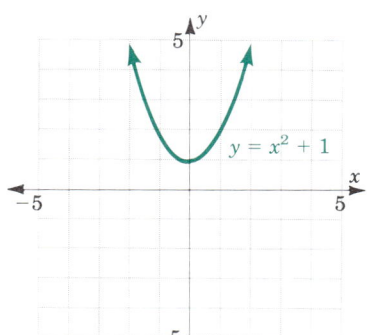

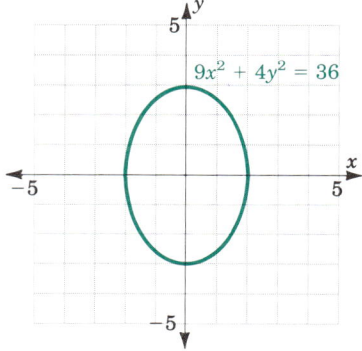

C. FINDING THE DOMAIN OF A FUNCTION

When relations are defined by means of an equation, the domain is the set of all possible replacements for the variable x that result in real numbers for y. Thus we cannot replace x by values that will produce

Problem 4   Do Example 4 for the equations
a. $y = x^2 - 1$
b. $4x^2 + 9y^2 = 36$

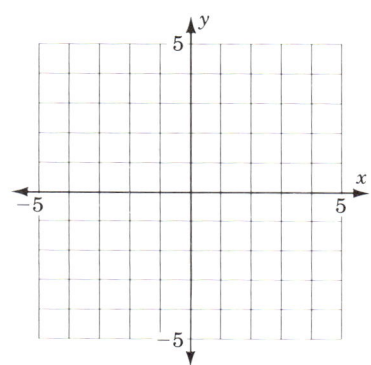

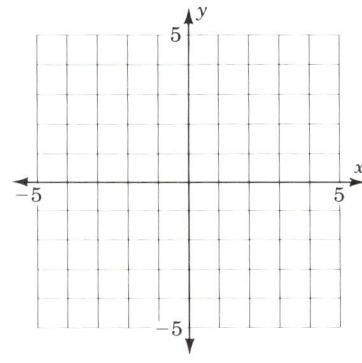

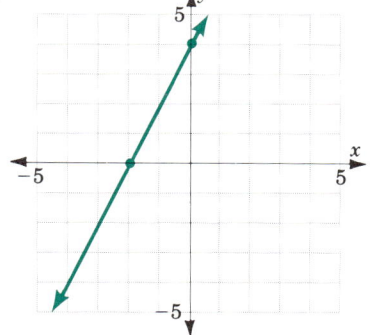

0 in the denominator or the square root of a negative number. For example, the domain of $y = \dfrac{1}{x}$ is the set of all real numbers *except* 0 and the domain of $y = \sqrt{x}$ is the set of all real numbers $x \geq 0$.

EXAMPLE 5   Find the domain of the function defined by

a. $y = \dfrac{1}{x-2}$ **b.** $y = \sqrt{x-3}$

Solution

a. Since we cannot replace x by values that will produce 0 in the denominator, we must avoid the case in which

$$x - 2 = 0$$
$$x = 2$$

Thus the domain of $y = \dfrac{1}{x-2}$ is the set of all real numbers except 2.

b. Since the square root of a negative number is not a real number, we must make the expression under the radical, $x - 3$, nonnegative.

$$x - 3 \geq 0$$
$$x \geq 3$$

The domain is $\{x \mid x \geq 3\}$.

Problem 5   Find the domain of the function defined by

a. $y = \dfrac{1}{x+2}$ **b.** $y = \sqrt{x+3}$

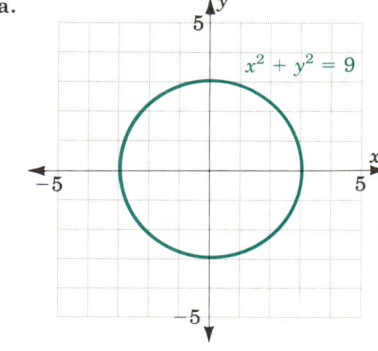
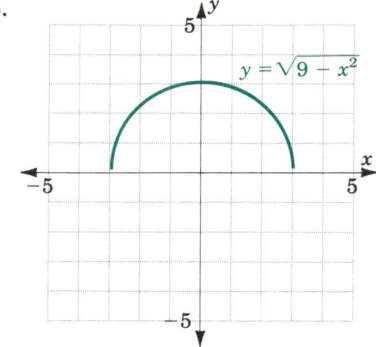
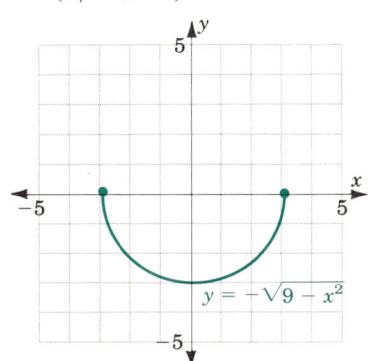

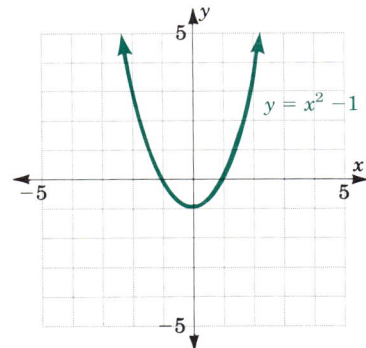
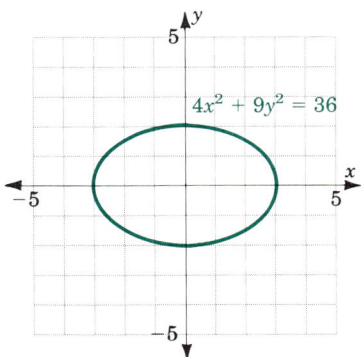

_____
_____
_____
_____
_____

A. In Problems 1–10 find (a) the domain, (b) the range, and (c) if the relation is a function.

ANSWERS

1. $\{(-3, 0), (-2, 1), (-1, 2)\}$

   1. a. _____
      b. _____
      c. _____

2. $\{(-1, -2), (0, -1), (1, 0)\}$

   2. a. _____
      b. _____
      c. _____

3. $\{(3, 0), (4, 0), (5, 0)\}$

   3. a. _____
      b. _____
      c. _____

4. $\{(0, 1), (0, 2), (0, 3)\}$

   4. a. _____
      b. _____
      c. _____

5. $\{(1, 2), (1, 3), (2, 2), (2, 3)\}$

   5. a. _____
      b. _____
      c. _____

6. $\{(2, 1), (1, 2), (3, 4), (4, 3)\}$

   6. a. _____
      b. _____
      c. _____

7. $\{(1, -1), (3, -1), (5, -1), (7, -1)\}$

   7. a. _____
      b. _____
      c. _____

8. $\{(-3, 2), (4, 3), (5, 7), (9, 8)\}$

   8. a. _____
      b. _____
      c. _____

9. $\{(2, 1), (2, 0),(2, -1), (2, -2)\}$

   9. a. _____
      b. _____
      c. _____

10. $\{(-3, -1), (-3, 0), (-3, 1), (-3, 2)\}$

  10. a. _____
      b. _____
      c. _____

A, B. In Problems 11–30 graph the relation, give the domain and the range, and use the vertical line test to determine if the relation is a function.

11. $\{(x, y) \mid x^2 + y^2 = 25\}$

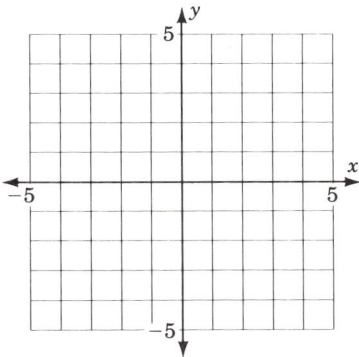

12. $\{(x, y) \mid x^2 + y^2 = 16\}$

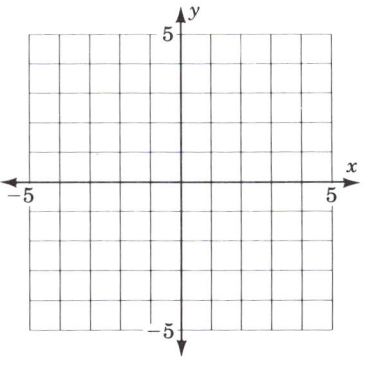

11. _____

12. _____

13. $\{(x, y) \mid y = \sqrt{25 - x^2}\}$

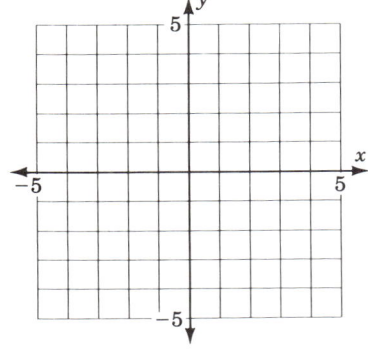

14. $\{(x, y) \mid y = -\sqrt{25 - x^2}\}$

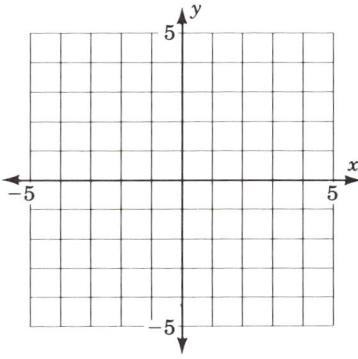

13. _____

14. _____

15. $\{(x, y) \mid x = \sqrt{25 - y^2}\}$

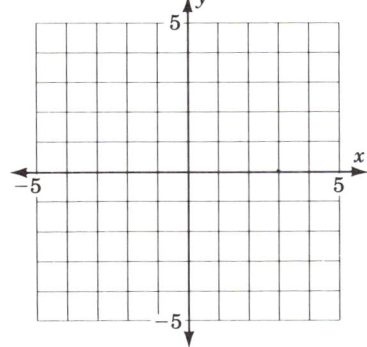

16. $\{(x, y) \mid x = -\sqrt{25 - y^2}\}$

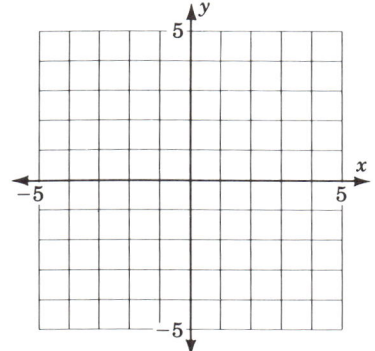

15. _____

16. _____

17. $\{(x, y) \mid y = x^2 - 1\}$

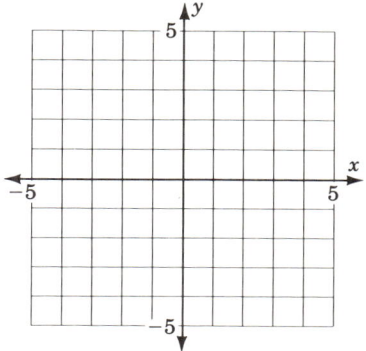

18. $\{(x, y) \mid y = x^2 + 3\}$

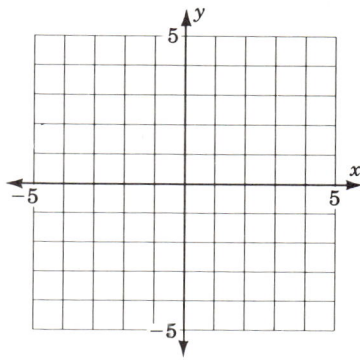

17. _____

18. _____

19. $\{(x, y) \mid y = -x^2\}$

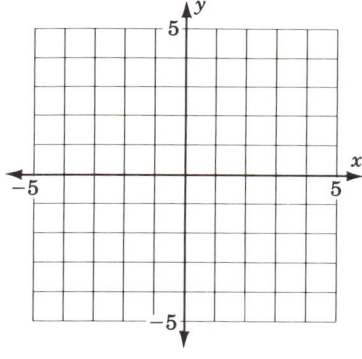

20. $\{(x, y) \mid y = -x^2 + 1\}$

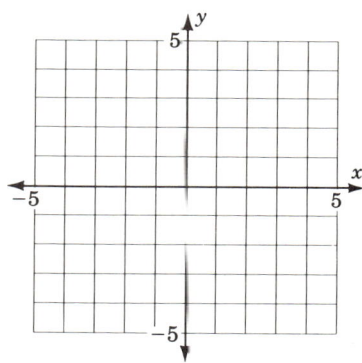

19. _____

20. _____

21. $\{(x, y) \mid x = y^2\}$

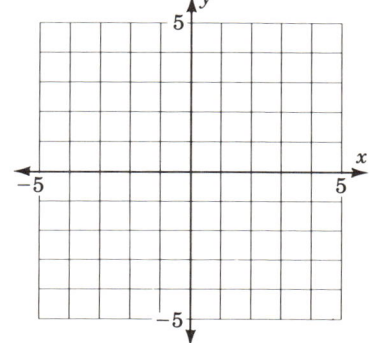

22. $\{(x, y) \mid x = y^2 + 1\}$

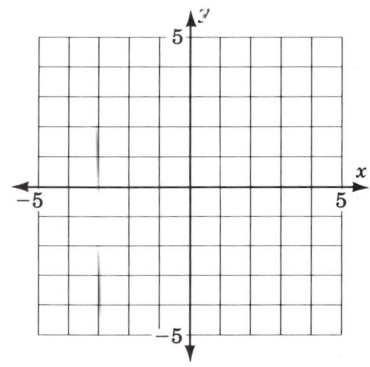

21. _____

22. _____

23. $\{(x, y) \mid y = -x + 5\}$

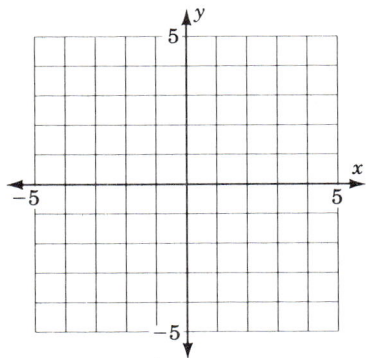

24. $\{(x, y) \mid y = -2x + 4\}$

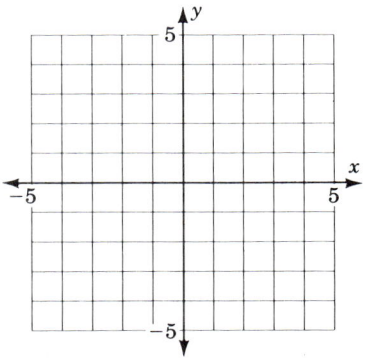

25. $\{(x, y) \mid y = x + 2\}$

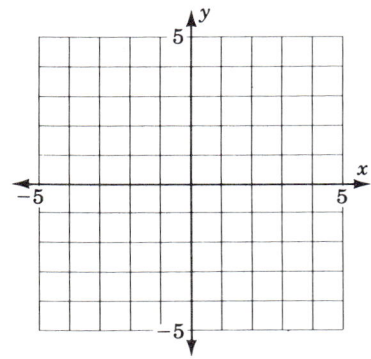

26. $\{(x, y) \mid y = 2x - 2\}$

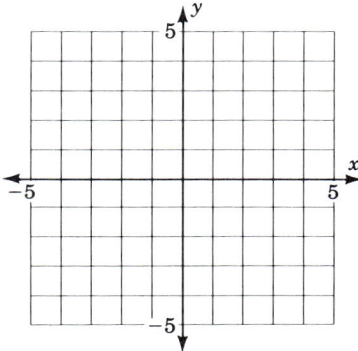

27. $\{(x, y) \mid 4x^2 + 9y^2 = 36\}$

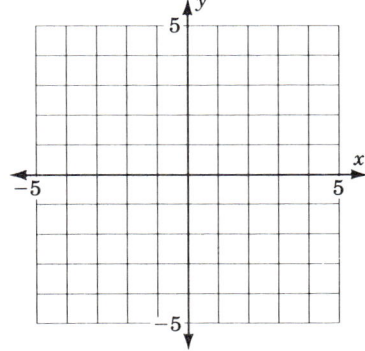

28. $\{(x, y) \mid 9x^2 + 4y^2 = 36\}$

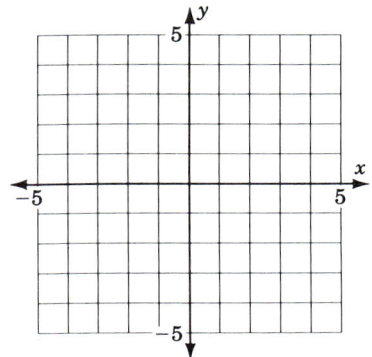

23. _____

24. _____

25. _____

26. _____

27. _____

28. _____

29. $\{(x, y) \mid 9x^2 - 4y^2 = 36\}$

30. $\{(x, y) \mid 9y^2 - 4x^2 = 36\}$

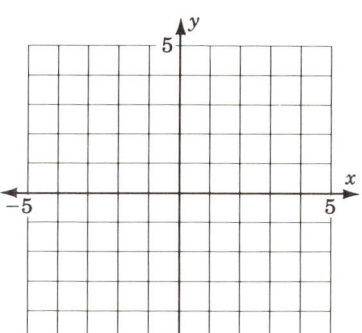

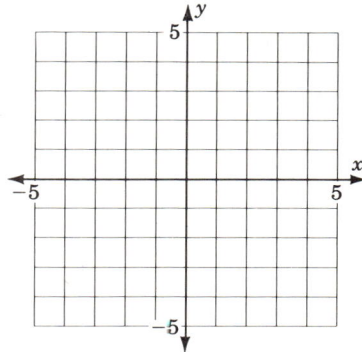

29. _____

30. _____

B. In Problems 31–36 use the vertical line test to determine if the graphs represent functions.

31.

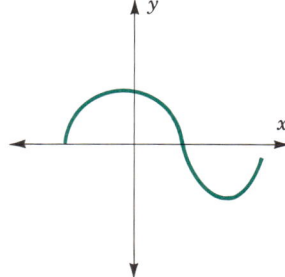

32.

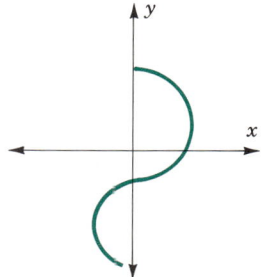

31. _____

32. _____

33.

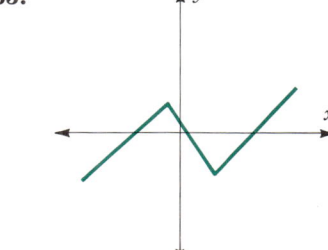

34.

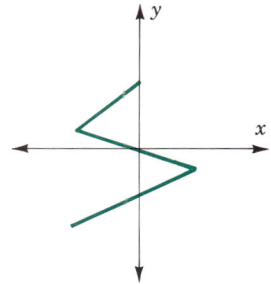

33. _____

34. _____

35.

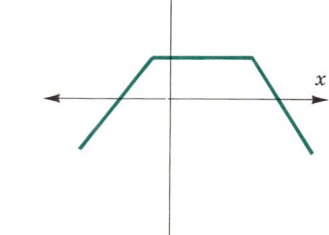

36.

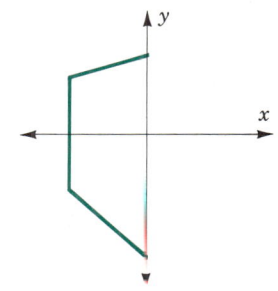

35. _____

35. _____

C. In Problems 37–50 find the domain of the function defined by the given equation.

37. $y = \sqrt{x - 5}$ **38.** $y = \sqrt{x + 5}$

39. $y = \sqrt{4 - 2x}$ **40.** $y = \sqrt{6 + 3x}$

41. $y = \sqrt{x^2 + 1}$ **42.** $y = \sqrt{x^2 + 2}$

43. $y = \dfrac{1}{x - 5}$ **44.** $y = \dfrac{1}{x + 5}$

45. $y = \dfrac{x + 2}{x + 5}$ **46.** $y = \dfrac{x + 3}{x - 6}$

47. $y = \dfrac{x}{x^2 + 3x + 2}$ **48.** $y = \dfrac{x}{x^2 + 5x + 6}$

49. $y = \dfrac{x - 1}{x^2 - 16}$ **50.** $y = \dfrac{x + 2}{x^2 - 9}$

37. _____
38. _____
39. _____
40. _____
41. _____
42. _____
43. _____
44. _____
45. _____
46. _____
47. _____
48. _____
49. _____
50. _____

D. Applications

In Problems 51–60 write an equation representing the relationship between the given quantities and state the domain for the resulting relation.

51. The circumference C of a circle and the length r of its radius

52. The area A of a square and the length r of its side

53. The perimeter P of a square and the length s of its side.

54. The distance D traveled by a car moving at a constant rate r for a period of 2 hr.

55. The distance D traveled by a car moving at 50 mi/hr and the time t (hours) traveled.

56. The daily cost y of renting a car costing \$30 per day plus 15¢ per mile for x miles traveled.

57. The daily cost y of renting a car costing \$35 per day plus 10¢ per mile for x miles traveled.

58. The distance D traveled by a car that gets 20 mi to the gallon when y gallons of gas have been used.

59. The number c of chirps made by a cricket that in one minute chirps four times the difference between the temperature t in degrees Fahrenheit and 40.

60. The temperature t in degrees Fahrenheit when a cricket chirps c times in a minute (see Problem 59).

51. _____
52. _____
53. _____
54. _____
55. _____
56. _____
57. _____
58. _____
59. _____
60. _____

EXERCISE 10.1

SKILL CHECKER

Evaluate $p = 62.5x$ when

61. $x = 2$

62. $x = 4$

Evaluate $y = x^2 - 1$ when

63. $x = -2$

64. $x = 2$

65. $x = 3$

66. $x = -3$

61. _____

62. _____

63. _____

64. _____

65. _____

66. _____

10.1 USING YOUR KNOWLEDGE

A special kind of relation that is important in mathematics is called an **equivalence relation.** A relation R is an equivalence relation if it has the following three properties:

a. Reflexive property. If a is an element of the domain of R, then (a, a) is an element of R.
b. Symmetric property. If (a, b) is an element of R, then (b, a) is an element of R.
c. Transitive property. If (a, b) and (b, c) are both elements of R, then (a, c) is an element of R.

A very simple example of an equivalence relation is

$$R = \{(x, y) \mid y = x, x \text{ an integer}\}$$

To show that R is an equivalence relation, we check the above three properties:

a. Reflexive. If a is an integer, then $a = a$, so (a, a) is an element of R.
b. Symmetric. Suppose that (a, b) belongs to R. Then, by the definition of R, a and b are integers and $b = a$. But if $b = a$, then $a = b$, so (b, a) also belongs to R.
c. Transitive. Suppose that (a, b) and (b, c) both belong to R. Then $b = a$ and $c = b$, so $c = a$. Hence, (a, c) also belongs to R.

Because R has all three properties, it is an equivalence relation.
  The pairs in a relation do not have to be numbers, and some interesting relations occur outside the field of numbers. For example,

$$R = \{(x, y) \mid y \text{ is a member of the same family as } x, x \text{ is a person}\}$$

is a relation. Is R an equivalence relation?
  We check the three properties as before:

a. Reflexive. Given a person A, is (A, A) an element of R? Yes, A is obviously a member of the same family as A.
b. Symmetric. Suppose that (A, B) is an element of P. Then B is a member of the same family as A. But then A is a member of the same family as B, so (B, A) is an element of R.

c. Transitive. Suppose that (A, B) and (B, C) both belong to R. Then A, B, C are all members of the same family. Thus, (A, C) is an element of R.

Again, we see that R has all three properties, so it is an equivalence relation.

Can you discover which of the following are equivalence relations?

1. $R = \{(x, y) \mid x$ and y are triangles and y is similar to $x\}$
   [*Note:* Here, *is similar to* means *has the same shape as.*]

2. $R = \{(x, y) \mid x$ and y are integers and $y > x\}$

3. $R = \{(x, y) \mid x$ and y are positive integers and y has the same parity as $x\}$; that is, y is odd if x is odd and y is even if x is even

4. $R = \{(x, y) \mid x$ and y are boys and y is the brother of $x\}$

5. $R = \{(x, y) \mid x$ and y are positive integers and when x and y are divided by 3, y leaves the same remainder as $x\}$

6. $R = \{(x, y) \mid x$ is a fraction a/b, where a and b are integers $(b \neq 0)$, and y is an equivalent fraction ma/mb, where m is a nonzero integer$\}$

1. _____

2. _____

3. _____

4. _____

5. _____

6. _____

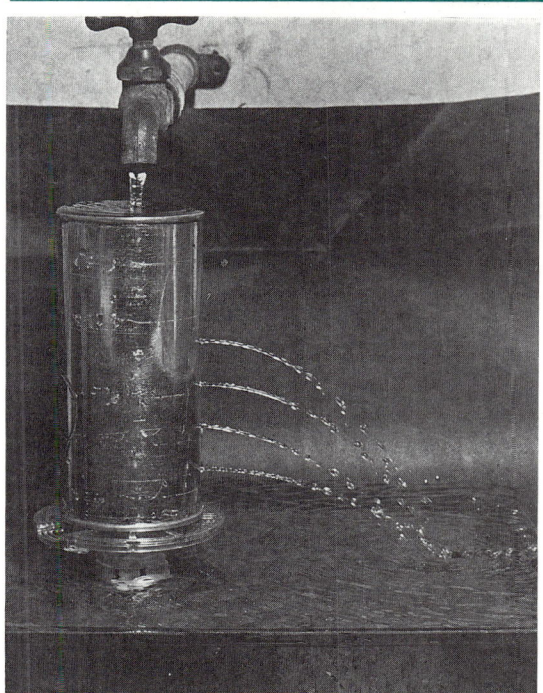

Photo by Fotoart, Tampa, Florida.

OBJECTIVES

REVIEW

Before starting this section, you should know:

1. How to evaluate an expression.
2. How to add, subtract, multiply, and divide polynomials.

OBJECTIVES

You should be able to:

A. Find the value of a function at a point.
B. Find the sum, difference, product, and quotient of two functions.
C. Find the composite of two functions.
D. Use the ideas studied in applications involving functions.

The photo illustrates the fact that the water pressure is a *function* of the depth. The higher pressure at the lower holes of the can makes water squirt out in a flat trajectory, whereas the lower pressure at the upper holes produces only a weak stream. It is known that the pressure y of the water (in pounds per square foot) at a depth of x feet (x a positive number) is

$$y = 62.5x$$

If we wish to emphasize that the pressure y is a function of the depth x we can use the **function notation** $f(x)$ (read "f of x") and write

$$f(x) = 62.5x$$

To find the pressure at 2 ft below the surface, we find the value of y when $x = 2$. Thus if $x = 2$, then $y = (62.5)(2) = 125$. Using functional notation, we would write $f(2) = (62.5)(2) = 125$. Here is a table illustrating both notations.

Y IN TERMS OF X $y = 62.5x$	FUNCTION NOTATION $f(x) = 62.5x$
If $x = 2$, then $y = (62.5)(2) = 125$	$f(2) = (62.5)(2) = 125$
If $x = 4$, then $y = (62.5)(4) = 250$	$f(4) = (62.5)(4) = 250$
If $x = 5$, then $y = (62.5)(5) = 312.5$	$f(5) = (62.5)(5) = 312.5$

A.   FINDING THE VALUE OF A FUNCTION

If $y = f(x)$, the symbols $f(x)$ and y are interchangeable; they both represent the value of the function for the given value of x. Thus, if

$$f(x) = 2x + 3$$

then

$$f(1) = 2(1) + 3 = 5$$
$$f(0) = 2(0) + 3 = 3$$
$$f(-6) = 2(-6) + 3 = -9$$
$$f(4) = 2(4) + 3 = 11$$
$$f(a) = 2(a) + 3 = 2a + 3$$
$$f(w + 2) = 2(w + 2) + 3 = 2w + 7$$

and so on. Whatever appears between the parentheses in $f(\ )$ is to be substituted for x in the rule that defines $f(x)$.

Instead of describing a function in set notation, we frequently say "the function defined by $f(x) = \ldots,$" where the three dots are to be replaced by the expression for the value of the function. For instance, "the function defined by $f(x) = 2x + 3$" has the same meaning as "the function $f = \{(x, y) \mid y = 2x + 3\}$."

EXAMPLE 1 Let $f(x) = 3x + 5$. Find.

a. $f(4)$ **b.** $f(2)$ **c.** $f(2) + f(4)$ **d.** $f(x + 1)$

Solution

a. Since $f(x) = 3x + 5$,

$$f(4) = 3 \cdot 4 + 5 = 12 + 5 = 17$$

b. $f(2) = 3 \cdot 2 + 5 = 6 + 5 = 11$

c. Since $f(2) = 11$ and $f(4) = 17$,

$$f(2) + f(4) = 11 + 17 = 28$$

d. $f(x + 1) = 3(x + 1) + 5 = 3x + 8$ ▲

EXAMPLE 2 If $f = \{(3, 2), (4, -2), (5, 0)\}$, find
a. $f(3)$ **b.** $f(4)$ **c.** $f(3) + f(4)$

Solution

a. $f(3) = 2$
b. $f(4) = -2$
c. $f(3) + f(4) = 2 + (-2) = 0$ ▲

Problem 1 Do Example 1 if $f(x) = 2x - 3$.

Problem 2 If
$f = \{(4, 3), (5, -1), (6, 0)\}$, find
a. $f(4)$ **b.** $f(5)$ **c.** $f(4) + f(5)$

B. OPERATIONS WITH FUNCTIONS

Functions can be added, subtracted, multiplied, and divided to obtain other functions. Here are the definitions we need to do it.

> If f and g are functions, then
>
> $(f + g)(x) = f(x) + g(x)$ The sum of f and g
>
> $(f - g)(x) = f(x) - g(x)$ The difference of f and g
>
> $(fg)(x) = f(x) \cdot g(x)$ The product of f and g
>
> $\left(\dfrac{f}{g}\right)(x) = \dfrac{f(x)}{g(x)}, \quad g(x) \neq 0$ The quotient of f and g

EXAMPLE 3 If $f(x) = x^2 - 4$ and $g(x) = x + 2$, find

a. $(f + g)(x)$ **b.** $(f - g)(x)$ **c.** $(fg)(x)$ **d.** $\left(\dfrac{f}{g}\right)(x)$

Problem 3 Do Example 3 if $g(x) = x - 2$.

Solution

a. By definition,

$$(f + g)(x) = \quad f(x) \quad + \quad g(x)$$

$$= (x^2 - 4) + (x + 2)$$
$$= x^2 + x - 2$$

b. $(f - g)(x) = \quad f(x) \quad - \quad g(x)$

$$= (x^2 - 4) - (x + 2)$$
$$= x^2 - 4 - x - 2$$
$$= x^2 - x - 6$$

c. $(fg)(x) = \quad f(x) \quad \cdot \quad g(x)$

$$= (x^2 - 4)(x + 2)$$
$$= x^3 + 2x^2 - 4x - 8$$

d. $\left(\dfrac{f}{g}\right)(x) = \dfrac{f(x)}{g(x)}$

$$= \dfrac{x^2 - 4}{x + 2}$$

$$= \dfrac{(x + 2)(x - 2)}{x + 2}$$

$$= x - 2, \quad x \neq -2 \qquad \blacktriangle$$

EXAMPLE 4 If $f(x) = 3x + 1$, find $\dfrac{f(x) - f(a)}{x - a}$, $x \neq a$.

Problem 4 Do Example 4 if $f(x) = 2x + 1$.

Solution   Since $f(x) = 3x + 1$, and $f(a) = 3a + 1$, we have

$$\dfrac{f(x) - f(a)}{x - a} = \dfrac{(3x + 1) - (3a + 1)}{x - a}$$

$$= \dfrac{3x - 3a}{x - a}$$

$$= \dfrac{3(x - a)}{x - a}$$

$$= 3, \qquad x \neq a \qquad \blacktriangle$$

C. COMPOSITE FUNCTIONS

Sometimes it is necessary to break a complicated function into a "composite" of simpler functions. For example, let us look at the function

$$h(x) = \sqrt{x + 1}$$

The function h is a combination of a square root function and a linear function. To see this, we let

$$f(u) = \sqrt{u}$$
$$g(x) = x + 1$$

Then

$$f[g(x)] = f(x + 1) = \sqrt{x + 1} = h(x)$$

The function h is said to be the *composite* of f with g. The expression $f[g(x)]$ is sometimes written as $f \circ g$. Here is the definition.

If f and g are functions, then

$$f \circ g = f[g(x)]$$

is the **composite of f with g.**

Thus if $f(x) = x^2$ and $g(x) = x + 2$,

$$f \circ g = f(g(x)) = f(x + 2)$$
$$= (x + 2)^2$$

EXAMPLE 5 If $f(x) = x^3$ and $g(x) = x - 1$, find

a. $f \circ g$ **b.** $g \circ f$ **c.** $(f \circ g)(3)$

Problem 5 Do Example 5 if $g(x) = x + 1$.

Solution

a. Substituting $x - 1$ for $g(x)$ in $f[g(x)]$, we have

$$f \circ g = f[g(x)] = f(x - 1)$$
$$= (x - 1)^3$$

b. Substituting x^3 for $f(x)$ in $g[f(x)]$, we have

$$g \circ f = g[(f(x)] = g(x^3)$$
$$= x^3 - 1$$

c. Since $f \circ g = f[g(x)] = (x - 1)^3$,

$$(f \circ g)(3) = f[g(3)] = (3 - 1)^3 = 8 \qquad \blacktriangle$$

D. APPLICATIONS

In recent years, aerobic exercises such as jogging, swimming, and bicycling have been taken up by millions of Americans. To see if you are exercising too hard (or not hard enough), you should stop from time to time and take your pulse to determine your heart rate. The idea is to keep your rate within a range known as the **target zone,** which is determined by your age. The next example explains how to find the *lower limit* of your target zone.

EXAMPLE 6 The lower limit L (heartbeats per minute) of a person's target zone is a function of age a (in years) and is given by

$$L(a) = -\frac{2}{3}a + 150$$

Find the value of L for a person who is

a. 30 yr old   **b.** 45 yr old

Problem 6 Find L for a person who is

a. 18 years old
b. 60 years old

Solution

a. We need to find $L(30)$:

$$L(a) = -\frac{2}{3}a + 150$$

$$L(30) = -\frac{2}{3}(30) + 150$$

$$= -20 + 150 = 130$$

This result means that a 30-yr-old person should try to attain at least 130 heartbeats per minute while exercising.

b. Here, we want to find $L(45)$. Proceeding as before, we obtain

$$L(45) = -\frac{2}{3}(45) + 150$$

$$= -30 + 150 = 120$$

(Find the value of L for your own age.)　▲

Functions are used in business. For example, suppose the cost C of making x items and the resulting revenue R are given as functions of x. When does one make a profit? Since the profit (or loss) is the difference between the revenue and the cost, we can write for the profit, P,

$$P(x) = R(x) - C(x)$$

EXAMPLE 7　The revenue (in dollars) obtained from selling x units of a product is given by $R(x) = 200x - \dfrac{x^2}{30}$ and the cost is given by $C(x) = 72{,}000 + 60x$.

a. Find the profit function, $P(x)$.

b. How many units must be made and sold to yield the maximum profit? Find this profit.

Solution

a. $P(x) = R(x) - C(x)$

$$= \left(200x - \frac{x^2}{30}\right) - (72{,}000 + 60x)$$

$$= -\frac{x^2}{30} + 140x - 72{,}000$$

b. To find the value of x that gives the maximum profit, we note that the graph of the profit function is a parabola opening downward. Thus the maximum value of P is at the vertex. The vertex is at the point where

$$x = -\frac{b}{2a} = -\frac{140}{-\frac{2}{30}} = (15)(140) = 2100$$

Thus 2100 units must be made and sold to give the maximum profit. This profit is P dollars, where P is given by

$$P(2100) = -\frac{(2100)^2}{30} + 140(2100) - 72{,}000$$

$$= 75{,}000$$

So the maximum profit is \$75,000.　▲

Problem 7　Do Example 7 if $C(x) = 50{,}000 + 50x$.

NAME

CLASS

SECTION

A.

ANSWERS

1. A function f is defined by $f(x) = 3x + 1$. Find.
   a. $f(0)$
   b. $f(2)$
   c. $f(-2)$

2. A function g is defined by $g(x) = -2x + 1$. Find.
   a. $g(0)$
   b. $g(1)$
   c. $g(-1)$

3. A function F is defined by $F(x) = \sqrt{x - 1}$. Find.
   a. $F(1)$
   b. $F(5)$
   c. $F(26)$

4. A function G is defined by $G(x) = x^2 + 2x - 1$. Find.
   a. $G(0)$
   b. $G(2)$
   c. $G(-2)$

5. A function f is defined by $f(x) = \dfrac{1}{3x + 1}$. Find.

   a. $f(1)$
   b. $f(1) - f(2)$

   c. $\dfrac{f(1) - f(2)}{3}$

6. A function f is defined by $f(x) = \dfrac{x - 2}{x + 3}$. Find.

   a. $f(2)$
   b. $f(3)$
   c. $f(3) - f(2)$

The functions defined by $f(x) = 3x - 4$ and $g(x) = x^2 + 2x + 4$ will be used in Problems 7–10.

7. Find.
   a. $f(3)$
   b. $g(3)$
   c. $f(3) + g(3)$

8. Find.
   a. $f(4)$
   b. $g(4)$
   c. $f(4) - g(4)$

9. Find.
   a. $f(-2)$
   b. $g(-3)$
   c. $f(-2) \cdot g(-3)$

1. a. _____
   b. _____
   c. _____

2. a. _____
   b. _____
   c. _____

3. a. _____
   b. _____
   c. _____

4. a. _____
   b. _____
   c. _____

5. a. _____
   b. _____
   c. _____

6. a. _____
   b. _____
   c. _____

7. a. _____
   b. _____
   c. _____

8. a. _____
   b. _____
   c. _____

9. a. _____
   b. _____
   c. _____

10. Find.
  a. $f(-1)$
  b. $g(-2)$

  c. $\dfrac{f(-1)}{g(-2)}$

The functions $f = \{(1, 3), (-1, 5), (-3, 7), (-5, 9)\}$ and
$g = \{(-2, 4), (0, 6), (2, 8), (4, 10)\}$ will be used in Problems 11–14.

11. Find.
  a. $f(1)$
  b. $g(-2)$
  c. $f(1) + g(-2)$

12. Find.
  a. $f(-1)$
  b. $g(0)$
  c. $f(-1) - g(0)$

13. Find.
  a. $f(-3)$
  b. $g(2)$
  c. $f(-3) \cdot g(2)$

14. Find.
  a. $f(-5)$
  b. $g(4)$

  c. $\dfrac{f(-5)}{g(4)}$

B. If $f(x) = x - 4$, $g(x) = x^2 - 5x + 4$, and $h(x) = x^2 - 16$, find.

15. $f + g$ **16.** $f - g$

17. hg **18.** $\dfrac{h}{f}$

19. $\dfrac{f}{h}$ **20.** $f + g - h$

15. _____
16. _____
17. _____
18. _____
19. _____
20. _____

C. In Problems 21–28, find
  a. $(f \circ g)(x)$
  b. $(g \circ f)(x)$

21. $f(x) = x^2$, $g(x) = \sqrt{x}$, $x > 0$ **22.** $f(x) = x - 1$, $g(x) = x^2$

23. $f(x) = 3x - 2$, $g(x) = x + 1$ **24.** $f(x) = x^2$, $g(x) = x - 1$

25. $f(x) = \sqrt{x + 1}, \quad g(x) = x^2 - 1$

25. a. _____
    b. _____

26. $f(x) = \sqrt{x^2 + 1}, \quad g(x) = 2x + 1$

26. a. _____
    b. _____

27. $f(x) = 3, \quad g(x) = -1$ **28.** $f(x) = ax, \quad g(x) = bx$

27. a. _____
    b. _____

28. a. _____
    b. _____

In Problems 29–34, find $\dfrac{f(x) - f(a)}{x - a}, \; x \neq a.$

29. $f(x) = 3x - 2$ **30.** $f(x) = 5x - 1$

29. _____

30. _____

31. $f(x) = x^2$ **32.** $f(x) = x^3$

31. _____

32. _____

33. $f(x) = x^2 + 3x$ **34.** $f(x) = x^2 - 2x$

33. _____

34. _____

D. Applications

35. The revenue obtained from selling x textbooks is given by $R(x) = 30x - 0.0005x^2$. The cost of producing the books is $C(x) = 100,000 + 6x$.
  a. Find the profit function $P(x)$.
  b. Find the maximum profit.

35. a. _____
    b. _____

36. The Fahrenheit temperature reading F is a function of the Celsius temperature reading C. This function is given by

$$F(C) = \tfrac{9}{5}C + 32$$

  a. If the temperature is 15°C, what is the Fahrenheit temperature?
  b. Water boils at 100°C. What is the corresponding Fahrenheit temperature?
  c. The freezing point of water is 0°C or 32°F. How many Fahrenheit degrees below freezing is a temperature of -10°C?
  d. The lowest temperature attainable is -273°C; this is the zero point on the absolute temperature scale. What is the corresponding Fahrenheit temperature?

36. a. _____
    b. _____
    c. _____
    d. _____

37. Refer to Example 6. The *upper limit U* of a person's target zone when exercising is also a function of age a (in years) and is given by

$$U(a) = -a + 190$$

Find the highest safe heart rate for a person who is
  a. 50 yr old
  b. 60 yr old

37. a. _____
    b. _____

38. Refer to Example 6 and Problem 37. The target zone for a person a years old consists of all the heart rates between $L(a)$ and $U(a)$, inclusive. Thus, if a person's heart rate is R, that person's target zone is described by $L(a) \leq R \leq U(a)$. Find the target zone for a person who is
  a. 30 yr old
  b. 45 yr old

38. a. _____
    b. _____

39. The ideal weight w (in pounds) of a man is a function of his height h (in inches). This function is defined by

$$w(h) = 5h - 190$$

  a. If a man is 70 in. tall, what should his weight be?
  b. If a man weighs 200 lb, what should his height be?

39. a. _____
    b. _____

40. The cost C in dollars of renting a car for 1 day is a function of the number m of miles traveled. For a car renting for $20 per day and 20¢ per mile, this function is given by

$$C(m) = 0.20m + 20$$

  a. Find the cost of renting a car for 1 day and driving 290 mi.
  b. If an executive paid $60.60 after renting a car for 1 day, how many miles did she drive?

40. a. _____
    b. _____

41. The pressure P (in pounds per square foot) at a depth of d feet below the surface of the ocean is a function of the depth. This function is given by

$$P(d) = 63.9d$$

What is the pressure on a submarine at a depth of
  a. 10 ft
  b. 100 ft

41. a. _____
    b. _____

42. If a ball is dropped from a point above the surface of the earth, the distance s (in meters) that the ball falls in t seconds is a function of t. This function is given by

$$s(t) = 4.9t^2$$

Find the distance that the ball falls in
  a. 2 sec
  b. 5 sec

42. a. _____
    b. _____

✓ **SKILL CHECKER**

Evaluate x^3 if x is

43. **a.** 1
    b. 2

43. a. _____
    b. _____

44. **a.** -1
    b. -2

44. a. _____
    b. _____

10.2   USING YOUR KNOWLEDGE

1. There are many interesting functions that can be defined using the ideas of this section. For example, did you know that the frequency with which a cricket chirps is a function of the temperature? The table shows the number of chirps per minute and the temperature in degrees Fahrenheit. If f is the function that relates the number (c) of chirps per minute and the temperature x, find

   a. $f(40)$
   b. $f(42)$
   c. $f(44)$

TEMPERATURE (°F)	40	41	42	43	44
CHIRPS PER MINUTE	0	4	8	12	16

2. The function relating the number of chirps per minute of the cricket and the temperature is given by $f(x) = 4(x - 40)$. If the temperature is 80°F, how many chirps per minute will you hear from your friendly house cricket?

3. An interesting function in physics was discovered by Galileo Galilei. This function relates the distance an object (dropped from a given height) travels and the time elapsed. The table in the margin shows the time (in seconds) and the distance (in feet) traveled by a rock dropped from a tall building. Can you find the relationship between the number of seconds elapsed, t, and the distance traveled, $f(t)$?

TIME ELAPSED (sec)	DISTANCE (ft)
1	$16 = 16 \times 1$
2	$64 = 16 \times 4$
3	$144 = 16 \times 9$
4	$256 = 16 \times 16$
5	$400 = 16 \times 25$
6	$576 = 16 \times 36$

4. Assume that a rock took 10 sec to reach the ground when dropped from the top of a building. Using the results of Problem 3, can you find the height of the building?

1. a. _____
   b. _____
   c. _____

2. _____

3. _____

4. _____

10.3 CLASSIFICATION OF FUNCTIONS

REVIEW

Before starting this section, you should know:

1. How to graph linear and quadratic equations.
2. How to recognize the standard form of a linear and a quadratic equation.

OBJECTIVES

You should be able to:

A. Classify a function as constant, linear, quadratic, or polynomial.
B. Graph a polynomial function of degree three or less.
C. Graph rational functions.
D. Graph exponential functions.

Have you heard the saying "A watched pot never boils"? The temperature at which water boils at sea level is always the same, a constant 212°F. If we express this as a function,

$$f(t) = 212$$

the function $f(t) = 212$ is a **constant function.** Its graph is in the margin.

On the other hand, the number of chirps a cricket makes in a minute is a function of the temperature t in degrees Fahrenheit and given by

$$f(t) = 4(t - 40) = 4t - 160$$

The function $f(t) = 4t - 160$ is a **linear function.** It tells us that if the temperature is 70°F, the number of chirps a cricket makes is 120. The graph is in the margin.

If an object is dropped from a certain height, the distance traveled by the object is a function of time and is given by

$$f(t) = 16t^2$$

This function is a **quadratic function.** Its graph is a parabola.

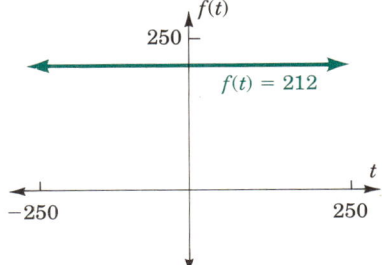

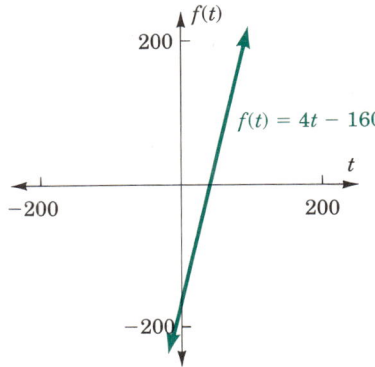

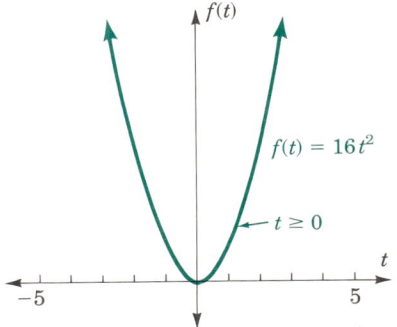

Here is a table summarizing this information.

FUNCTION	NAME	SHAPE OF GRAPH	GRAPH
$f(x) = k$	Constant function	Horizontal line	
$f(x) = mx + b$	Linear function	A line with slope m, and y-intercept b.	
$f(x) = ax^2 + bx + c$	Quadratic function, opening up if $a > 0$, down if $a < 0$. Vertex at $x = -\dfrac{b}{2a}$	Parabola	

EXAMPLE 1 Classify and graph.

a. $f(x) = -3$ **b.** $f(x) = -x^2 - 2x + 3$ **c.** $f(x) = 2x - 4$

Solution

a. $f(x) = -3$ is a constant function. Its graph is the horizontal line shown.

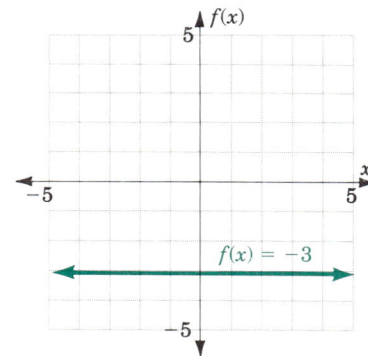

b. $f(x) = -x^2 - 2x + 3$ is of the form $f(x) = ax^2 + bx + c$, where $a = -1$, $b = -2$, and $c = -3$. Thus it is a quadratic function. Its graph is a parabola with vertex at $x = -1$,

$$y = f(-1) = -(-1)^2 - 2(-1) + 3$$
$$= -1 + 2 + 3$$
$$= 4$$

Thus the vertex is at $(-1, 4)$. For $x = 0$, $f(x) = y = 3$, the y-intercept. The x-intercepts occur when $y = f(x) = 0$, that is, when

$$-x^2 - 2x + 3 = 0$$

Problem 1 Classify and graph.
a. $f(x) = -x^2 + 2x + 3$
b. $f(x) = 2$
c. $f(x) = -2x + 4$

or

$$x^2 + 2x - 3 = 0 \quad \text{Multiply by } -1.$$
$$(x + 3)(x - 1) = 0 \quad \text{Factor.}$$
$$x + 3 = 0 \quad \text{or} \quad x - 1 = 0$$
$$x = -3 \quad \text{or} \quad x = 1$$

Its graph is shown.

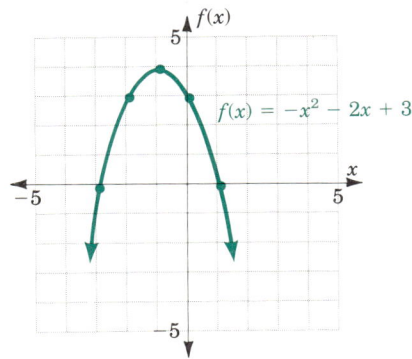

c. $f(x) = 2x - 4$ is of the form $f(x) = mx + b$, where $m = 2$ and $b = -4$. It is a linear function with slope 2 and y-intercept -4, as shown.

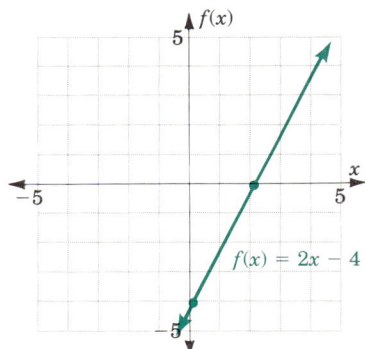

B. POLYNOMIAL FUNCTIONS

The linear and quadratic equations we have studied are special cases of the polynomials we studied in Chapter 3. We now discuss *polynomial functions*. Here is the definition.

> A **polynomial function** of degree n is a function that can be written in the form
>
> $$f(x) = c_n x^n + c_{n-1} x^{n-1} + c_{n-2} x^{n-2} + \cdots + c_1 x + c_0$$
>
> where the coefficients $c_n, c_{n-1}, \ldots, c_1, c_0, c_n \neq 0$, are real numbers.

A polynomial of the *third* degree is called a **cubic.** To graph $f(x) = x^3$, we choose values for x and find the corresponding $f(x) = y$ values as shown in the following table.

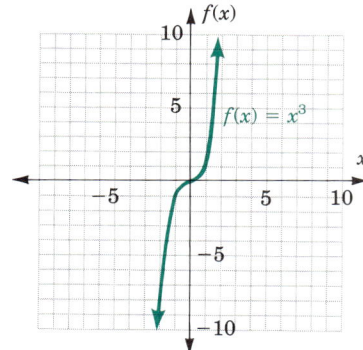

x	$f(x) = y$
-2	-8
-1	-1
0	0
1	1
2	8

We then draw a smooth curve through these points, obtaining the graph of $f(x) = x^3$. The shape of the graph is typical of cubic functions.

EXAMPLE 2   Graph $f(x) = x^3 - 2x$.

Solution   We choose different values of x and find the corresponding values $f(x) = y$, as shown in the table. We then drawn a smooth curve through these points, obtaining the desired graph.

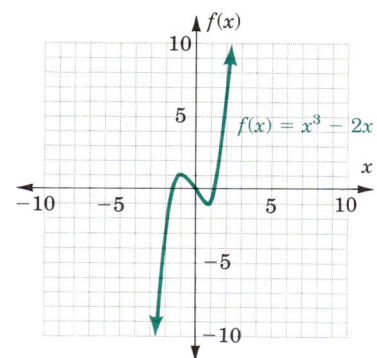

x	$f(x) = y$
-2	-4
-1	1
0	0
1	-1
2	4

C. RATIONAL FUNCTIONS

You may recall that after studying polynomials in Chapter 3, we studied rational expressions, expressions that were quotients of polynomials, in Chapter 4. In the same manner, we generalize the idea of polynomial functions to rational functions. Here is the definition.

RATIONAL FUNCTION

A **rational function** is a function that can be written in the form

$$f(x) = \frac{P(x)}{Q(x)}$$

where $P(x)$ and $Q(x)$ are polynomials and $Q(x) \neq 0$.

Problem 2   Graph $f(x) = x^3 - 3x$.

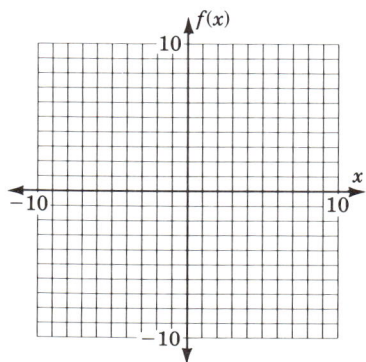

ANSWER   (to problem on page 800)
1. **a.** A quadratic function
   b. A constant function
   c. A linear function
   The graphs are shown.

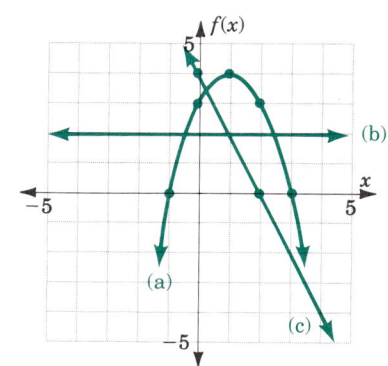

EXAMPLE 3 Graph $f(x) = \dfrac{x-3}{x-1}$.

Solution We proceed by steps:

1. To find the *y*-intercept, let $x = 0$. Then $f(x) = y = 3$.

2. To find the *x*-intercept, let $y = 0$. Then $0 = \dfrac{x-3}{x-1}$. This happens only if the numerator is 0—that is, when $x = 3$.

3. The denominator of the function is 0 when $x = 1$. This means that $f(x) = \dfrac{x-3}{x-1}$ is not defined when $x = 1$. The graph has a vertical asymptote at $x = 1$.

4. For very large values of x, $x - 3$ and $x - 1$ will both be very large. Their quotient will be very near 1. (Use a calculator and try $x = 10{,}000$.) This means that for very large values of x, the function will be very near $y = 1$. This means that $y = 1$ is a horizontal asymptote for the function.

5. We then try the convenient values $x = -1$ and $x = 2$ and obtain the corresponding *y* values shown in the table. The graph is obtained by drawing a smooth curve through the points we have obtained, as shown.

x	$f(x) = y$
0	3
3	0
1	undefined
−1	2
2	−1

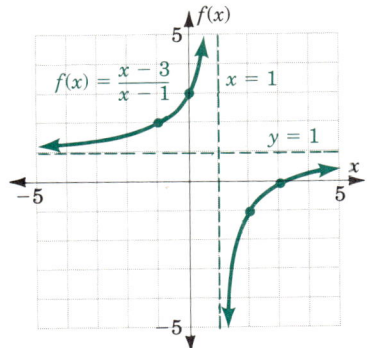

Problem 3 Graph $f(x) = \dfrac{x-2}{x-1}$.

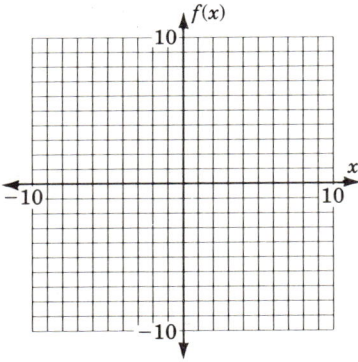

D. EXPONENTIAL FUNCTIONS

So far, we have never used a variable as an exponent. There is a very important type of function, called an *exponential function*, which has a variable as an exponent. We discuss exponential functions more extensively in the next chapter. For now, here is the definition we need.

EXPONENTIAL FUNCTIONS

An **exponential function** is a function that can be written in the form

$$f(x) = b^x$$

where b is a positive real number other than 1.

Thus, $f(x) = 2^x$, $g(x) = (\frac{1}{2})^x$, $h(x) = 3^x$, and $i(x) = (\frac{1}{3})^x$ are exponential functions. To obtain the graph of these functions, we proceed as usual: Give some values to x and find the corresponding values $f(x) = y$. Thus to graph $f(x) = 2^x$, we let $x = 0$, $x = 1$, and $x = 2$, obtaining

$$f(0) = 2^0 = 1$$
$$f(1) = 2^1 = 2$$
$$f(2) = 2^2 = 4$$

If we use negative values of x, we must recall the definition of a negative exponent. In general,

$$a^{-n} = \frac{1}{a^n}$$

Thus for $x = -2$ and -1, we have

$$f(-1) = 2^{-1} = \frac{1}{2^1} = \frac{1}{2}$$

$$f(-2) = 2^{-2} = \frac{1}{2^2} = \frac{1}{4}$$

Here are all our results entered in a table:

x	$f(x) = 2^x$
-2	$\frac{1}{4}$
-1	$\frac{1}{2}$
0	1
1	2
2	4

Note that 2^x is *positive* for all values of x.

EXAMPLE 4   Graph $f(x) = 2^x$.

Solution   We use the ordered pairs $(-2, \frac{1}{4})$, $(-1, \frac{1}{2})$, $(0, 1)$, $(1, 2)$, and so on from the preceding table and then join them with a smooth curve as shown in the graph.

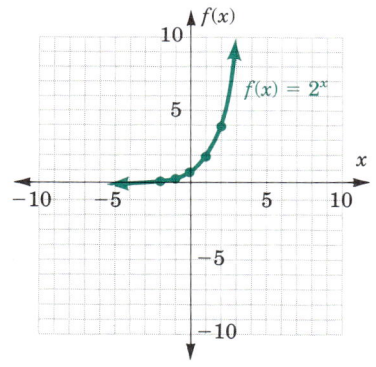

Problem 4   Graph $g(x) = 3^x$.

ANSWERS   (to problems on pages 802–803)
2.

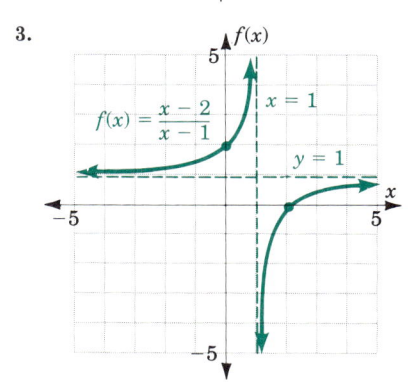

3.

NAME

CLASS

SECTION

ANSWERS

1. _____

2. _____

A. In Problems 1–12 classify and graph the function.

1. $f(x) = -2$

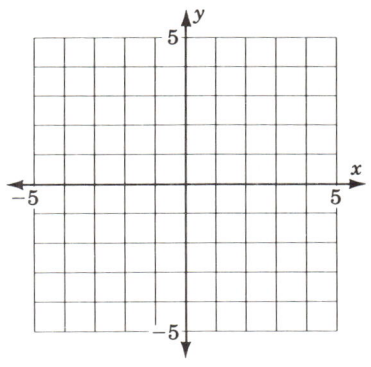

2. $f(x) = 3$

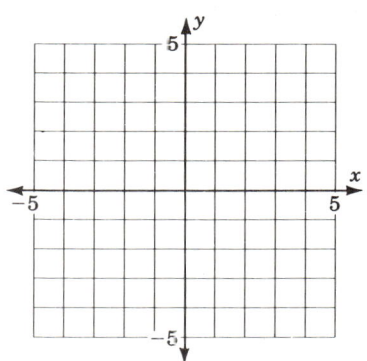

3. $f(x) = -2x - 4$

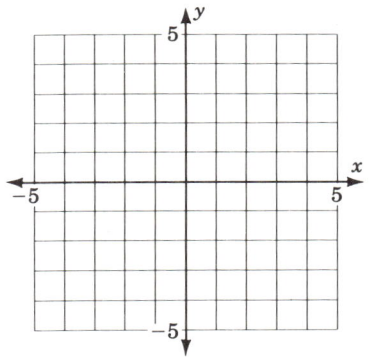

4. $f(x) = 2x + 5$

3. _____

4. _____

5. $f(x) = \dfrac{5}{2}$

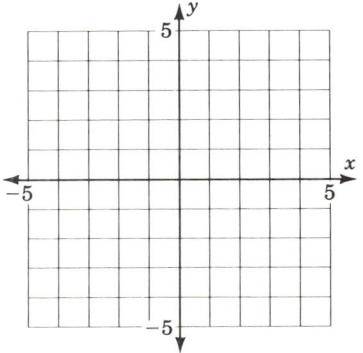

6. $f(x) = -\dfrac{7}{2}$

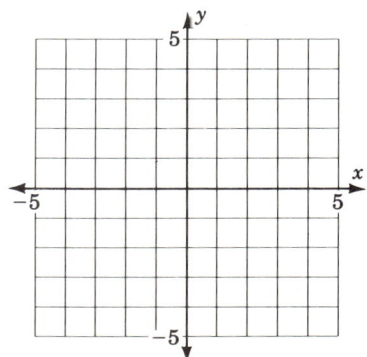

5. _____

6. _____

7. $f(x) = x^2 + x - 2$

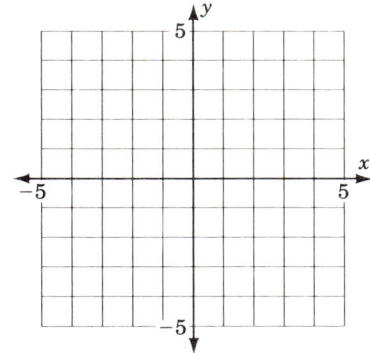

8. $f(x) = -x^2 - 2x - 3$

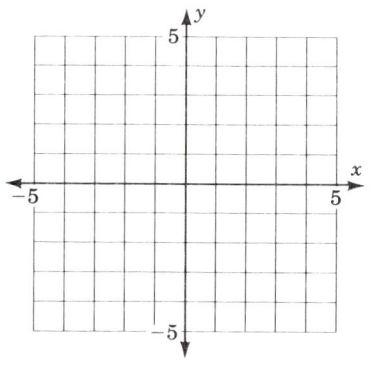

7. _____

8. _____

9. $f(x) = -x^2 + 2x + 3$

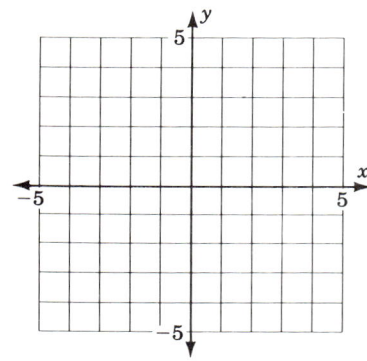

10. $f(x) = x^2 - 2x - 3$

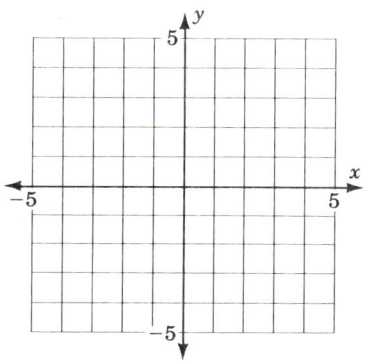

9. _____

10. _____

11. $f(x) = x^2 - 2x - 1$

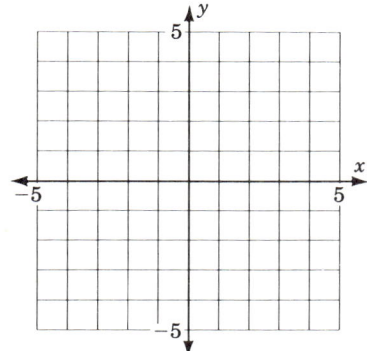

12. $f(x) = x^2 - 4x - 1$

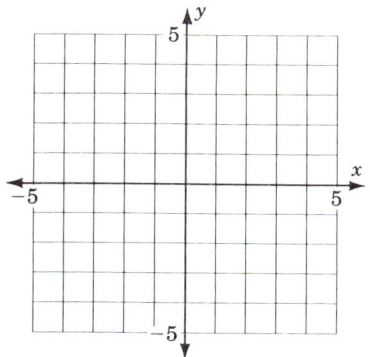

11. _____

12. _____

B. In Problems 13–22 graph the function.

13. $f(x) = \dfrac{1}{2}x^3$

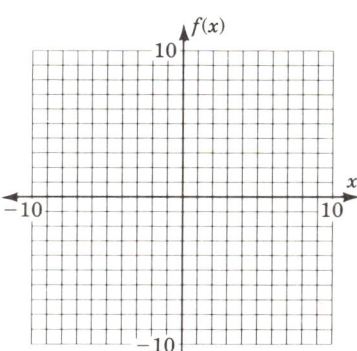

14. $f(x) = \dfrac{1}{4}x^3$

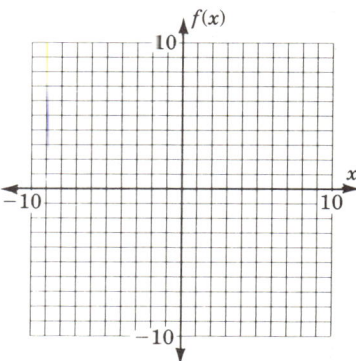

15. $f(x) = -x^3 + 2x$

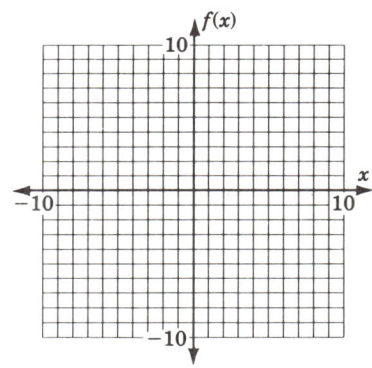

16. $f(x) = -x^3 + 3x$

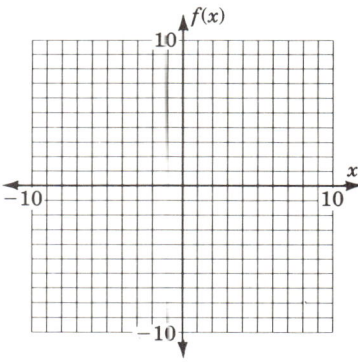

17. $f(x) = x^3 - 2x^2$

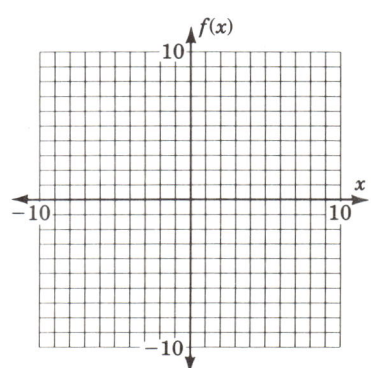

18. $f(x) = x^3 - 3x^2$

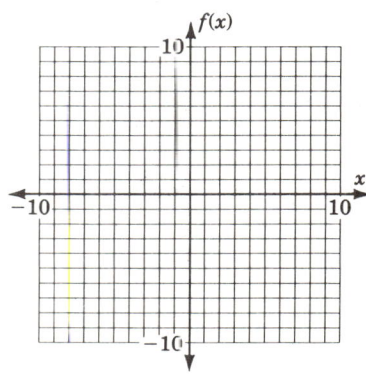

ANSWER (to problem on page 804)

4.

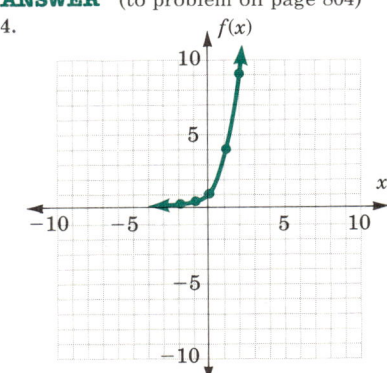

19. $f(x) = -x^3 + 3x^2$

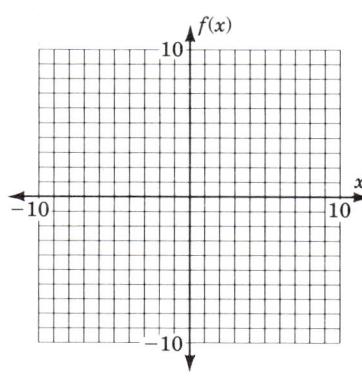

20. $f(x) = -x^3 + 2x^2$

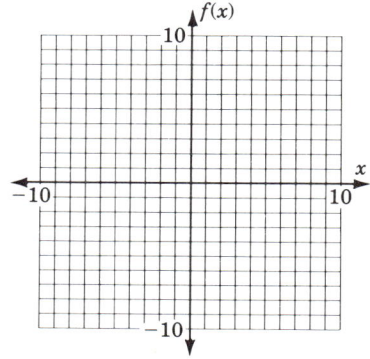

21. $f(x) = x^3 - x^2 + x - 1$

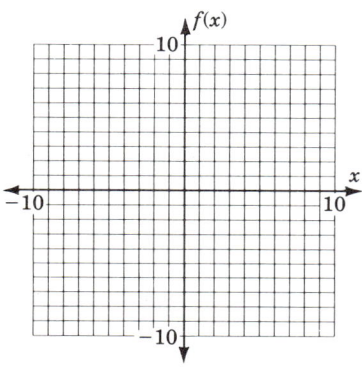

22. $f(x) = x^3 - x^2 + x - 2$

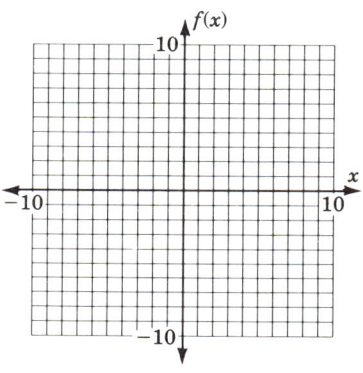

C. In Problems 23–32 graph the function.

23. $f(x) = \dfrac{1}{x}$

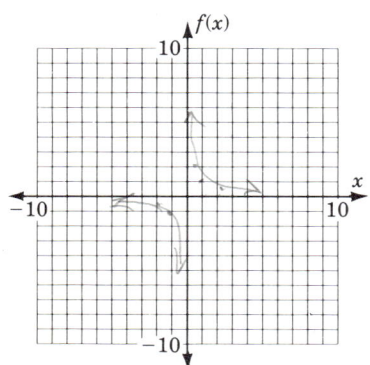

24. $f(x) = \dfrac{1}{x - 1}$

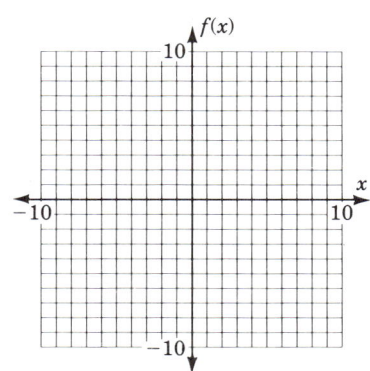

25. $f(x) = \dfrac{2}{x + 1}$

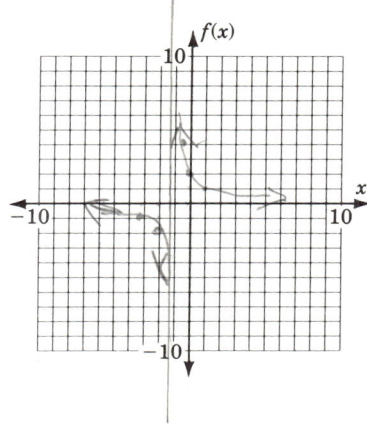

26. $f(x) = \dfrac{3}{x + 2}$

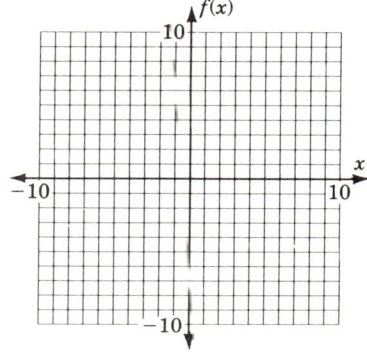

27. $f(x) = \dfrac{x + 2}{x + 1}$

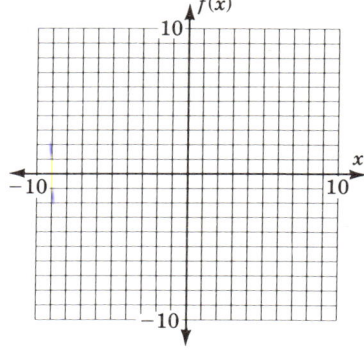

28. $f(x) = \dfrac{x + 3}{x + 1}$

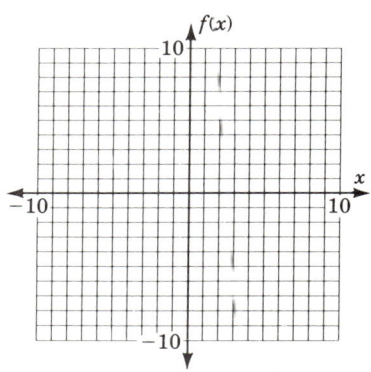

29. $f(x) = \dfrac{x - 2}{x + 1}$

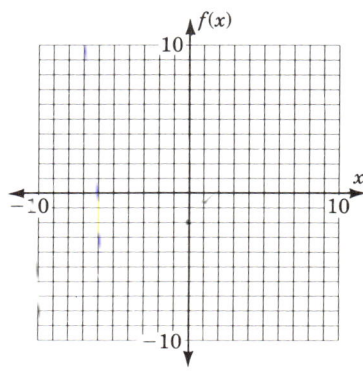

30. $f(x) = \dfrac{x - 3}{x + 1}$

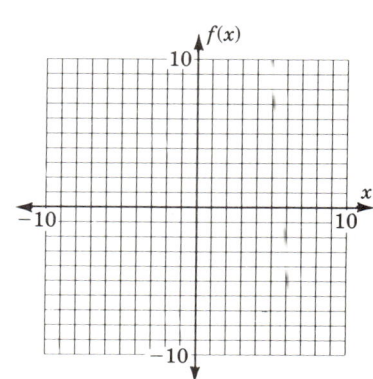

31. $f(x) = \dfrac{x + 2}{x - 1}$

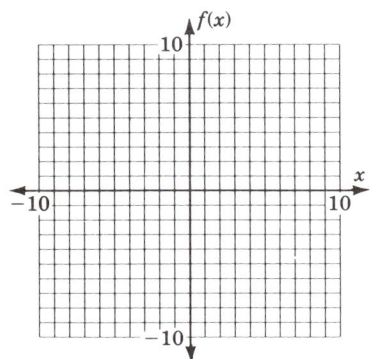

32. $f(x) = \dfrac{x + 3}{x - 1}$

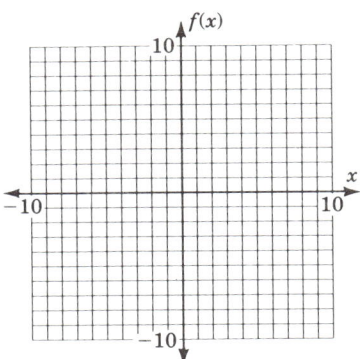

D. In Problems 33–40 graph the function.

33. $f(x) = 2^{x + 1}$

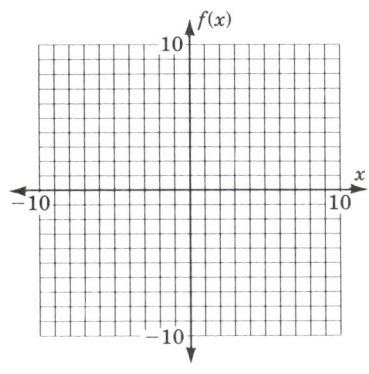

34. $f(x) = 3^{x + 1}$

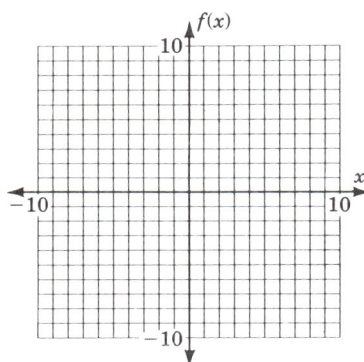

35. $f(x) = \left(\dfrac{1}{2}\right)^{x}$

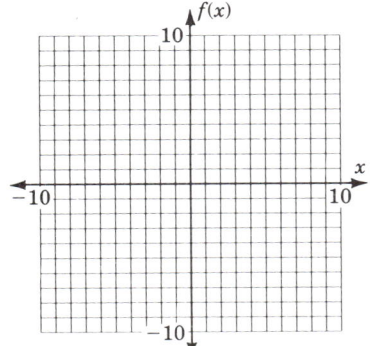

36. $f(x) = \left(\dfrac{1}{3}\right)^{x}$

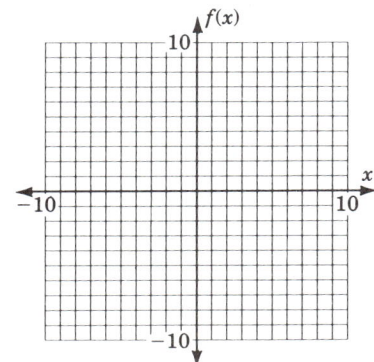

37. $f(x) = 2^{-x}$

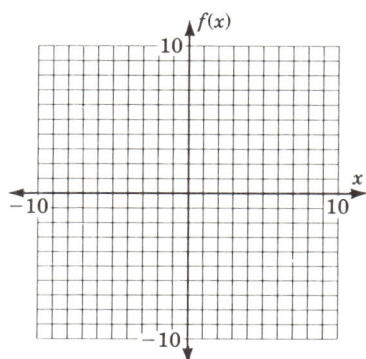

38. $f(x) = 3^{-x}$

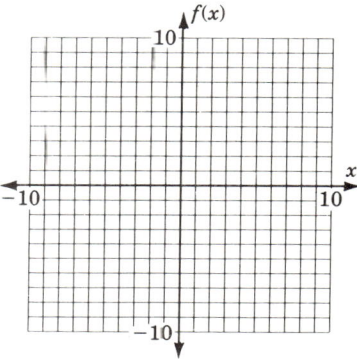

39. $f(x) = 4^{-x}$

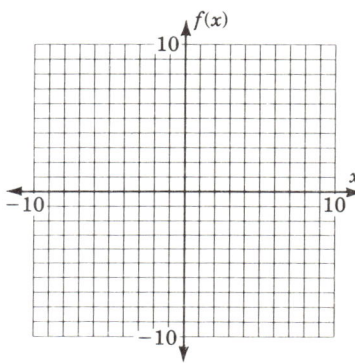

40. $f(x) = 5^{-x}$

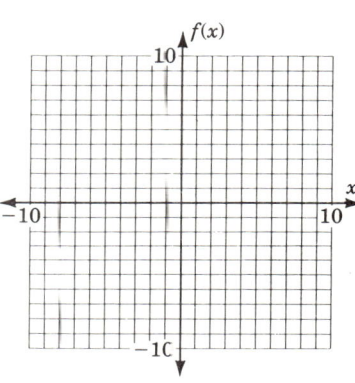

E. Applications

41. The gross profit P (in thousands of dollars) obtained by producing and selling x (thousand) units of a certain product is given by $P(x) = -x^2 + 4x - 1$.
   a. What kind of function is $P(x)$?
   b. What is the number of units that have to be produced to maximize the profits?

41. a. _____
    b. _____

42. The winning time in the Olympic women's 400-meter track relay is a function of the year in which the event is run and is given by $f(x) = -0.12x + 280$ (seconds).
   a. What kind of function is $f(x)$?
   b. Predict the winning time for 1988 (the actual time was 41.98 sec).
   c. Predict the winning time for 1992.

42. a. _____
    b. _____
    c. _____

43. The winning time in the Olympic women's 200-meter dash is a function of the year in which the event is run, starting with 1948, and is given by $f(x) = -0.27x + 24.32$ (seconds), where $x = 1$ corresponds to the 1948 Olympics, $x = 2$ corresponds to the 1952 Olympics, and so on.
   a. What kind of function is $f(x)$?
   b. Predict the winning time for the 1984 Olympics ($x = 10$) (The actual time was 21.62 sec).
   c. Predict the winning time for the 1992 Olympics.

43. a. _____
    b. _____
    c. _____

44. If we start with a culture of 10 bacteria, the number of bacteria present is a function of the number h of hours elapsed and is given by $f(x) = 10 \cdot 2^h$.
   a. What kind of function is $f(x)$?
   b. How many bacteria would there be after 3 hr?

44. a. _____

    b. _____

✓ **SKILL CHECKER**

Solve for y.

45. $x = 4y - 2$ **46.** $x = 2y - 5$

47. $x = 5y + 3$ **48.** $x = 3y + 4$

49. $x = 3y + 1$ **50.** $x = 4y + 8$

45. _____

46. _____

47. _____

48. _____

49. _____

50. _____

10.3 USING YOUR KNOWLEDGE

In this section we have discussed constant, linear, quadratic, cubic, exponential, and rational functions. All these functions can be classified further as *odd* or *even*. Here is the definition.

> The function f is **even** if $f(-x) = f(x)$ for every value of x.
> The function f is **odd** if $f(-x) = -f(x)$ for every value of x.

For example, the function $f(x) = x^2$ is even. Let us see why.

$$f(-x) = (-x)^2 = x^2 = f(x)$$

Thus $f(-x) = f(x)$. On the other hand, $f(x) = x^3$ is odd because $f(-x) = (-x)^3 = -x^3 = -f(x)$.

1. Is $f(x) = x^4$ odd or even?

2. Is $f(x) = x^5$ odd or even?

3. Is $f(x) = x^2 + x^3$ odd, even, or neither?

1. _____

2. _____

3. _____

In Problems 4–6 we show that $y = \dfrac{a}{c}$ is a horizontal asymptote for the graph of $y = \dfrac{ax + b}{cx + d}$.

4. Divide each term in the numerator and denominator of $\dfrac{ax + b}{cx + d}$ by x. What do you get?

4. _____

5. If

$$y = \frac{a + \dfrac{b}{x}}{c + \dfrac{d}{x}}$$

what will happen to $\dfrac{b}{x}$ and $\dfrac{d}{x}$ as x becomes very large in absolute value?

6. Your answer to Problem 5 indicates that $y = \dfrac{c}{c}$ is a horizontal asymptote for the graph of $y = \dfrac{ax + b}{cx + d}$. Explain why.

5. _____

6. _____

FOREIGN EXCHANGE

Country	Foreign Currency In U.S. $ Fri.	U.S. $ In Foreign Currency Fri.	Foreign Currency In U.S. $ Last Fri.	U.S. $ In Foreign Currency Last Fri.
Argentina (Austral)	.0005587	1790.03	.0006803	1470.00
Australia (Dollar)	.7600	1.3158	.7937	1.2599
Austria (Schilling) . · · · ·	.08427	11.87	.08316	12.03
Bahrain (Dinar)	2.5971	.3851	2.6525	.3770
Belgium (Franc)				
Commercial rate	.02837	35.25	.02798	35.74
Financial rate	.02836	35.26	.2791	14.70
Brazil (New Cruzado) . .	.06090	16.42	.06805	.6073
Britain (Pound)	1.6640	.6010	1.6465	.6108
30-Day Forward	1.6552	.6042	1.6372	.6172
90-Day Forward	1.6374	.6107	1.6202	.6233
180-Day Forward	1.6133	.6198	1.6044	1.1832
Canada (Dollar)	.8370	1.1947	.8452	1.1872
30-Day Forward	.8341	1.1989	.8423	1.1938
90-Day Forward	.8289	1.2064	.8377	1.2008
180-Day Forward	.8226	1.2156	.8328	

OBJECTIVES

REVIEW

Before starting this section, you should know:

1. How to find the range and domain of a relation.
2. How to solve an equation for a specified variable (Section 2.5).
3. How to graph linear and quadratic equations.

OBJECTIVES

You should be able to:

A. Find the inverse of a function when the function is given as a set of ordered pairs.

B. Find the equation of the inverse of a function.

C. Graph a function and its inverse and determine if the inverse is a function.

Are you planning on traveling somewhere? It is a good idea to know how much your dollars are worth in the country you plan to visit. The second column in the table shows the value of $1 (U.S.) in different currencies. If you look at the first column, the value of a Canadian dollar in U.S. dollars is $0.8370. This means that the number of dollars D you get for C Canadian dollars is given by

$$D = 0.8370C$$

Here is a table giving the dollar value of 1, 10, 100, and 1000 Canadian dollars.

CANADIAN DOLLARS	U.S. DOLLARS
1	0.8370
10	8.37
100	83.70
1000	837.00

A. THE INVERSE OF A FUNCTION

If we think of D as the function $f(C)$, we can write the function f as

$$f = \{(1, 0.8370), (10, 8.370), (100, 83.70), (1000, 837)\}$$

On the other hand, if you exchange U.S. currency for Canadian, the number of Canadian dollars you get for $0.8370, $8.370, $83.70, and $837 (U.S.), respectively, corresponds to the set of ordered pairs (D, C) and is given by

$$g = \{(0.8370, 1), (8.370, 10), (83.70, 100), (837, 1000)\}$$

The relation g obtained by reversing the order of the coordinates in each ordered pair in f is called the *inverse* of f. As you can see, the domain of f is the range of g and the range of f is the domain of g. Here is the definition we need.

If f is a relation, then the **inverse** of f, denoted by f^{-1} (read "f inverse," or "the inverse of f") is the relation obtained by reversing the order of the coordinates in each ordered pair in f.

EXAMPLE 1   Let $S = \{(1, 2), (3, 4), (5, 4)\}$. Find.

a. The domain and range of S
b. S^{-1}
c. The domain and range of S^{-1}
d. The graphs of S and S^{-1} on the same coordinate axes

Solution

a. The domain of S is $\{1, 3, 5\}$. The range is $\{2, 4\}$.

b. $S^{-1} = \{(2, 1), (4, 3), (4, 5)\}$

c. The domain of S^{-1} is $\{2, 4\}$; the range is $\{1, 3, 5\}$.

d. The graphs of S (in color) and S^{-1} (in black) are shown in the figure. As you can see, the two graphs are symmetric with respect to the line $y = x$ (shown dashed).

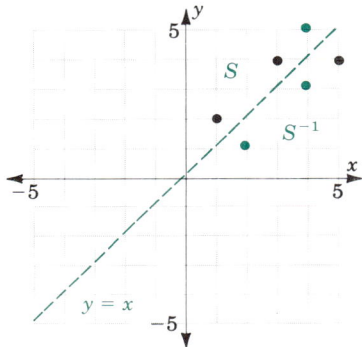

Problem 1   Let $S = \{(4, 3), (3, 2), (2, 1)\}$. Find.

a. The domain and range of S
b. S^{-1}
c. The domain and range of S^{-1}
d. The graphs of S and S^{-1}

B. FINDING THE EQUATION OF THE INVERSE FUNCTION

The equation $D = 0.8370C$ gives the U.S. dollar value of C Canadian dollars. To find the inverse of

$$f = \{(1, 0.8370), (10, 8.370), (100, 83.70), (1000, 837)\}$$

we reverse the *order* of the coordinates in each ordered pair in f. To find the inverse of $D = 0.8370C$, we interchange the *variables,* obtaining

$C = 0.8370D$

$D = \dfrac{1}{0.8370}C$ Solve for D.

$D = 1.1947C$ Divide 1 by 0.8370.

This means that \$1 (U.S.) is worth \$1.1947 (Canadian). You can verify this by looking at the second column in the table on page 815.

Thus to find the equation for the inverse of f, we interchange the roles of x and y in the equation for f and then solve for y.

For example, consider the relation

$$y = 4x - 4 \qquad (1)$$

The inverse of this relation is obtained by interchanging the x- and y-coordinates—that is, by writing $x = 4y - 4$. Solving for y, we get

$$y = \frac{1}{4}(x + 4) \qquad (2)$$

The graphs of (1) (in black) and its inverse (2) (in color) are shown in the figure. Clearly, the graphs are symmetric to each other with respect to the line $y = x$, shown dotted. This is to be expected, since one relation was obtained from the other by interchanging x and y.

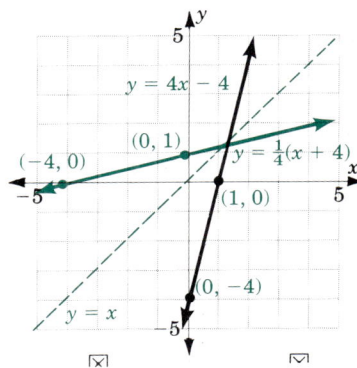

EXAMPLE 2 Let $f(x) = y = 4x - 2$.

a. Find $f^{-1}(x)$.   **b.** Graph f and its inverse.

Solution

a. Since $y = 4x - 2$, we interchange the variables x and y, obtaining

$$x = 4y - 2$$
$$x + 2 = 4y$$
$$y = \frac{x + 2}{4}$$

Thus, the inverse of f is $f^{-1}(x) = \dfrac{x + 2}{4}$.

b. We can graph $y = 4x - 2$ in the usual way. Letting $x = 0$, $y = -2$. For $y = 0$, $x = \frac{1}{2}$. The graph is shown in black. We then graph

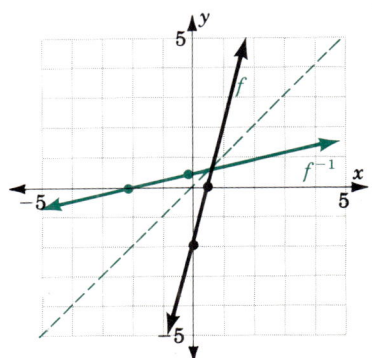

Problem 2   Let $f(x) = y = 2x - 4$.
a. Find $f^{-1}(x)$.
b. Graph f and its inverse.

$y = \dfrac{x + 2}{4}$ in a similar manner. The graph is in color and is symmetric to the graph of $y = 4x - 2$ with respect to the line $y = x$, shown dashed. Note that we could have obtained the graph of the inverse function by reflecting the graph of $y = 4x - 2$ about the line $y = x$. ▲

C. FUNCTIONS WHOSE INVERSES ARE ALSO FUNCTIONS

Since every function is a set of ordered pairs (a relation), every function has an inverse. Is this inverse always a function? We can very quickly see that it is not. For example, if S is the function defined by $S = \{(1, 2,), (3, 2)\}$, the inverse of S is $S^{-1} = \{(2, 1), (2, 3)\}$, which is *not* a function, since two distinct ordered pairs have the same first component, 2. On the other hand, if G is the function defined by $G = \{(3, 4), (5, 6)\}$, the inverse is $G^{-1} = \{(4, 3), (6, 5)\}$, which *is* a function. The reason that the inverse of S is not a function is that S has two ordered pairs with the same *second* component. A function in which no two distinct ordered pairs have the same second component is called a **one-to-one** function. The inverse of such a function is always a function. We summarize this discussion as follows.

If the function $y = f(x)$ is one-to-one, then the inverse of f is also a function and is denoted by $y = f^{-1}(x)$.

Thus to determine if the inverse of a function is a function, we must ascertain if the original function is one-to-one. In order to do this, we must return to the definition of a one-to-one function—a one-to-one function *cannot* have two ordered pairs with the same second component. Since any two points with the same second coordinate will be on a *horizontal* line parallel to the x-axis, if any horizontal line intersects the graph of a function more than once, the function will not be one-to-one. Its inverse will not be a function. Thus, we have the following **horizontal line test.**

THE HORIZONTAL LINE TEST

If a horizontal line intersects the graph of a function more than once, the inverse of the function is not a function.

Thus linear functions have inverses that are functions (horizontal lines will intersect the graph only once), but quadratic functions do not have inverses that are functions because they have graphs that can be intersected by a horizontal line more than once.

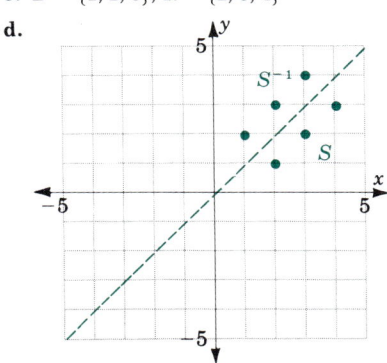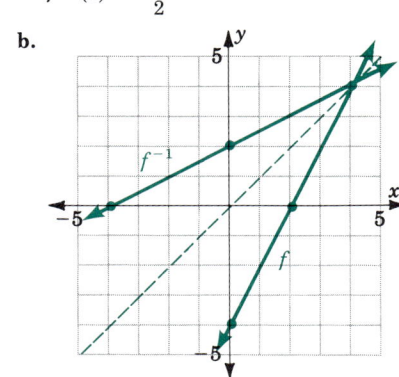

EXAMPLE 3 Find the inverse of $f(x) = y = x^2$. Is the inverse a function?

Solution Since $y = x^2$, we interchange the variables x and y to obtain

$$x = y^2$$
$$y = \pm\sqrt{x}$$

The inverse of $f(x)$ is *not* a function, since we can draw a horizontal line that intersects the graph of $y = x^2$ at more than one point.

The function $f(x) = x^2$ (in black) and its inverse (in color) are shown in the graph.

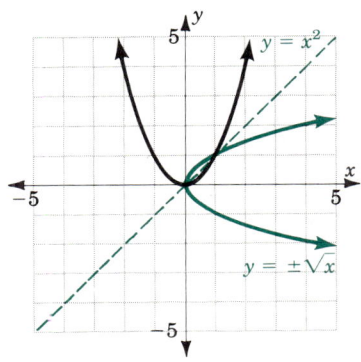

▲

EXAMPLE 4 Graph the function $f(x) = y = 3^x$. Is the inverse a function?

Solution We graphed $y = 3^x$ in the preceding section. If we examine the graph of $y = 3^x$ (in color), we can see that any horizontal line will intersect the graph only once. Thus the inverse is a function. To try to find the inverse, we interchange the x and y in $y = 3^x$, obtaining $x = 3^y$.

Unfortunately, we are not able to solve for y at this time (we will have to wait until the next chapter to do so). However, we can still graph the inverse $x = 3^y$ by giving values to y and finding the corresponding x-values, as shown in the table. The graph of the inverse, $x = 3^y$, is in black.

x	y
$\frac{1}{3}$	-1
1	0
3	1

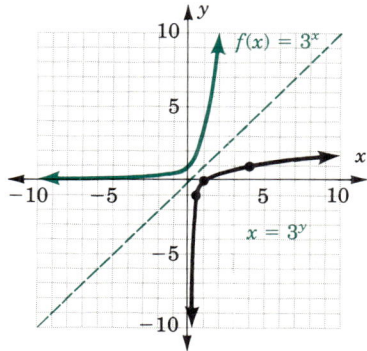

▲

Problem 3 Find the inverse of $y = f(x) = x^3$. Is the inverse a function? Graph the function.

Problem 4 Graph $f(x) = y = 2^x$. Is the inverse a function?

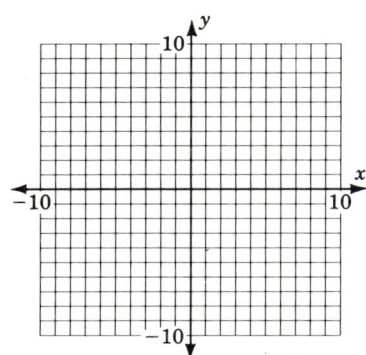

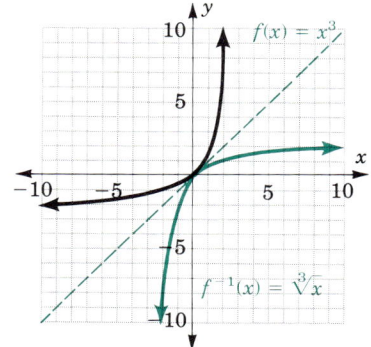
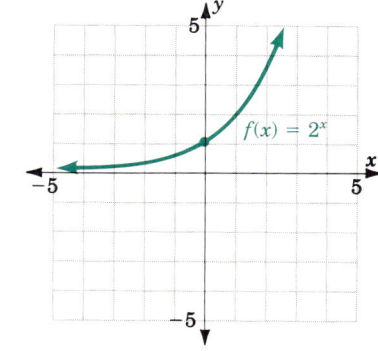

A, B. In Problems 1–4, find f^{-1}, draw the graph of f and f^{-1} on the same axes and determine if f^{-1} is a function.

1. $f = \{(1, 3), (2, 4), (3, 5)\}$

2. $f = \{(2, 3), (3, 4), (4, 5)\}$

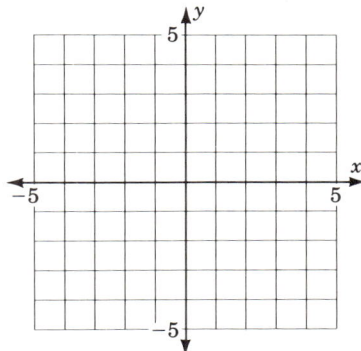

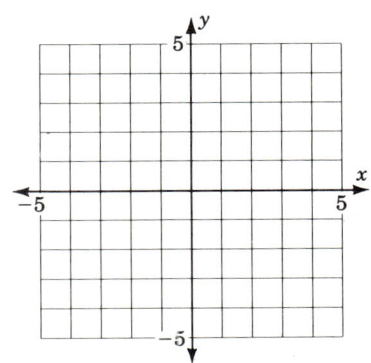

3. $f = \{(-1, 5), (-3, 4), (-4, 4)\}$

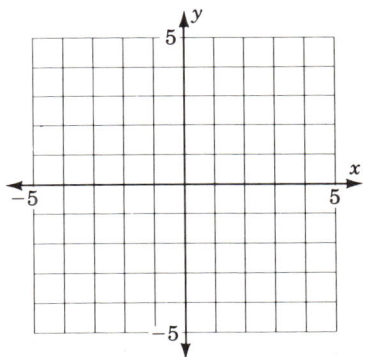

4. $f = \{(-2, 4), (-3, 3), (-5, 3)\}$

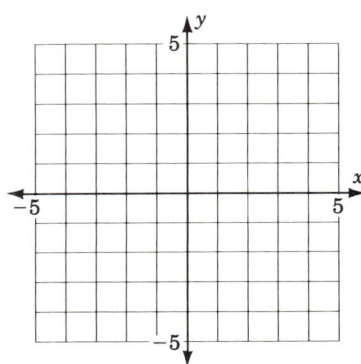

B, C. In Problems 5–14 find the equation of the inverse, graph it, and state if the inverse is a function.

5. $\{(x, y)\,|\,y = 3x + 3\}$

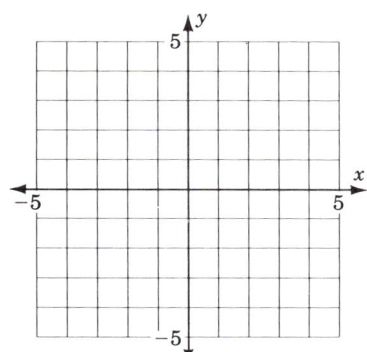

6. $\{(x, y)\,|\,y = 2x + 4\}$

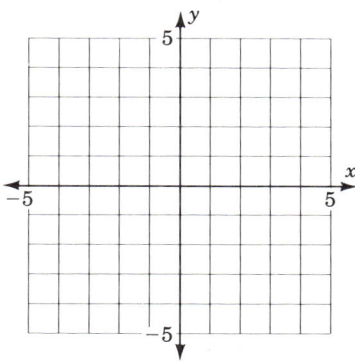

7. $\{(x, y)\,|\,y = 2x - 4\}$

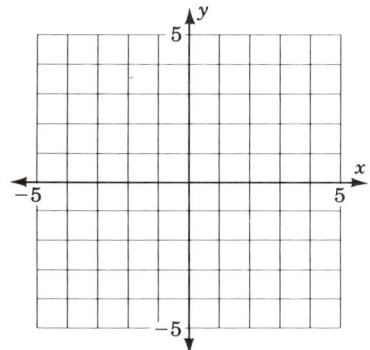

8. $\{(x, y)\,|\,y = 3x - 3\}$

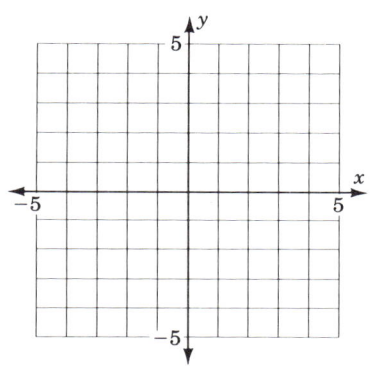

9. $\{(x, y)\,|\,y = 2x^2\}$

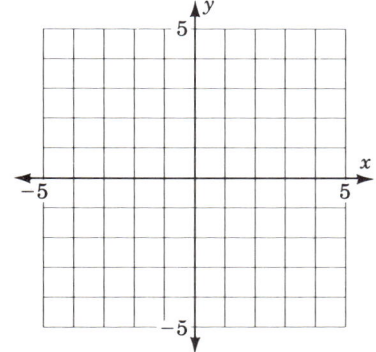

10. $\{(x, y)\,|\,y = x^2 + 1\}$

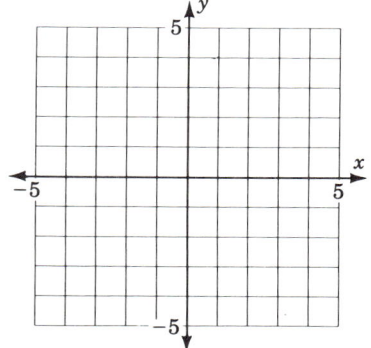

5. _____

6. _____

7. _____

8. _____

9. _____

10. _____

11. $\{(x, y) | y = x^2 - 1\}$

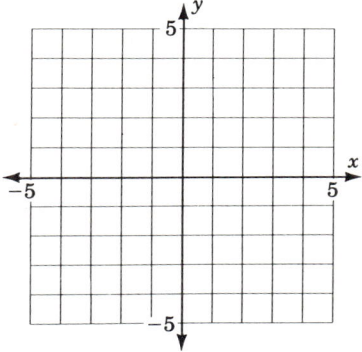

12. $\{(x, y) | y = x^3 - 1\}$

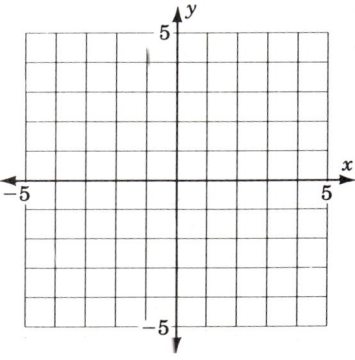

13. $\{(x, y) | y = -x^3\}$

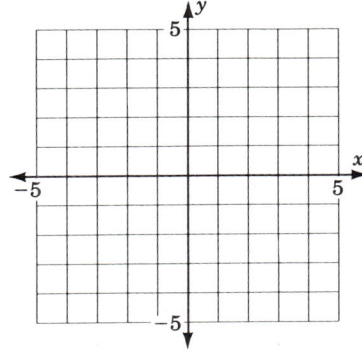

14. $\{(x, y) | y = -2x^3\}$

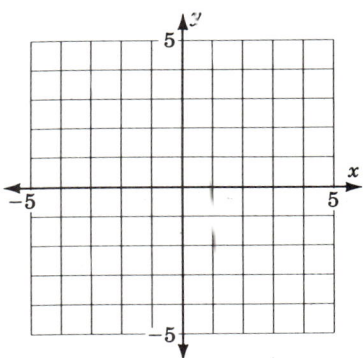

C. In Problems 15–20, graph the function and its inverse and determine if the inverse is a function.

15. $y = f(x) = 2^{x+1}$

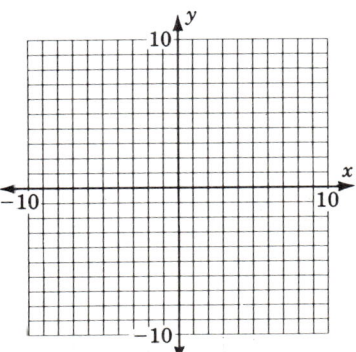

16. $y = f(x) = 3^{x+1}$

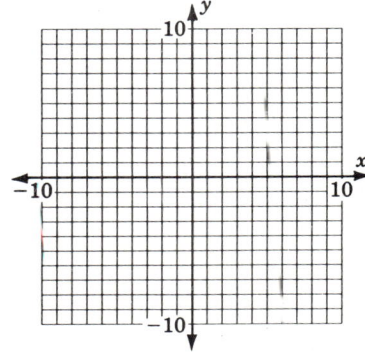

17. $y = f(x) = \left(\dfrac{1}{3}\right)^x$

18. $y = f(x) = \left(\dfrac{1}{2}\right)^x$

17. _____

18. _____

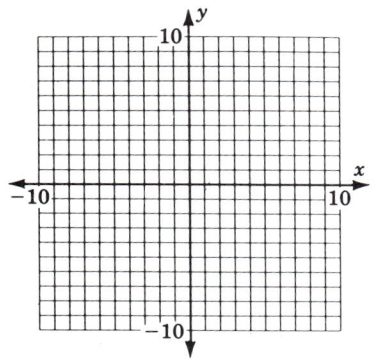

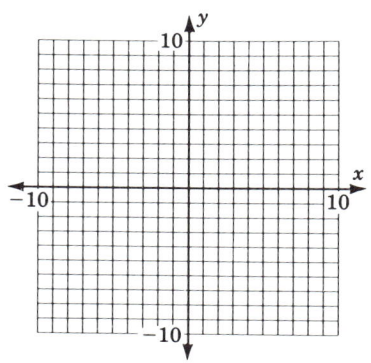

19. $y = f(x) = 2^{-x}$

20. $y = f(x) = 3^{-x}$

19. _____

20. _____

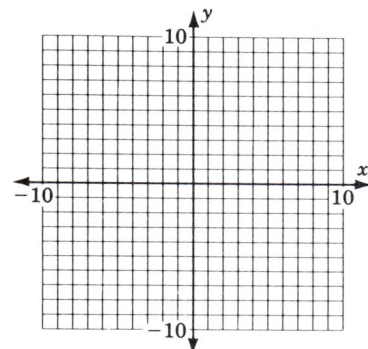

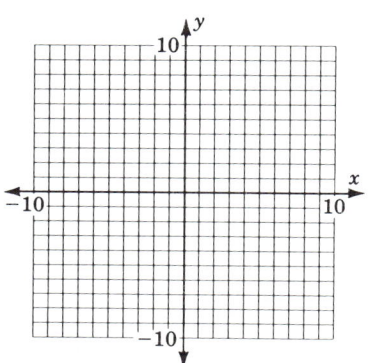

21. If $f(x) = 4x + 4$, $f^{-1}(x) = \dfrac{x - 4}{4}$. Find.

    a. $f((f^{-1}(3))$
    b. $f^{-1}(f(-1))$

21. **a.** _____
    b. _____

22. If $f(x) = 2x - 2$, $f^{-1}(x) = \dfrac{x + 2}{2}$. Find.

    a. $f(f^{-1}(-1))$
    b. $f^{-1}(f(x))$

22. **a.** _____
    b. _____

23. If $y = f(x) = \dfrac{1}{x}$, find $f^{-1}(x)$.

23. _____

24. If $y = f(x) = \dfrac{2}{x}$, find $f^{-1}(x)$.

24. _____

✓ SKILL CHECKER

Perform the indicated operations.

25. $x^2 \cdot x^8$

26. $x^3 \cdot x^9$

25. _____

26. _____

EXERCISE 10.4

27. $x^3 \cdot x^{-6}$

28. $x^4 \cdot x^{-5}$

29. $\dfrac{x^{10}}{x^6}$

30. $\dfrac{x^{12}}{x^9}$

27. _____

28. _____

29. _____

30. _____

10.4 USING YOUR KNOWLEDGE

The number of chirps c a cricket makes in 1 min is
$f(F) = c = 4(F - 40)$, F the temperature in degrees Fahrenheit.

1. Find $f^{-1}(c)$.
2. Can you find the temperature when a cricket is chirping 160 times a minute?

1. _____

2. _____

SUMMARY

SECTION	ITEM	MEANING	EXAMPLE
10.1A	Relation	A set of ordered pairs	$\{(2, -1), (4, 6), (5, 2)\}$
10.1A	Domain	The set of first coordinates of a relation	$\{2, 4, 5\}$ is the domain of the preceding relation.
10.1A	Range	The set of second coordinates of a relation	$\{-1, 2, 6\}$ is the range of the preceding relation.
10.1B	Function	A relation in which no two different ordered pairs have the same first coordinate	$\{(1, 2), (2, 4), (3, 6)\}$ is a function.
10.1B	Vertical line test	If any vertical line intersects the graph of a relation more than once, the relation is *not* a function.	
10.2A	Function notation	Use of a letter such as f to denote a function and $f(x)$ to mean the value of the function for the given value of x.	$f = \{(x, y) \mid y = x^2\}$. For this function, $f(x) = x^2$, $f(2) = 4$, and $f(-3) = 9$.
10.2B	$(f + g)(x)$	$f(x) + g(x)$	If $f(x) = x^2$ and $g(x) = x + 1$, then $(f + g)(x) = x^2 + x + 1$.
10.2B	$(f - g)(x)$	$f(x) - g(x)$	If $f(x) = x^2$ and $g(x) = x + 1$, then $(f - g)(x) = x^2 - x - 1$.
10.2B	$(fg)(x)$	$f(x) \cdot g(x)$	If $f(x) = x^2$ and $g(x) = x + 1$, then $(fg)(x) = x^3 + x^2$.
10.2B	$\left(\dfrac{f}{g}\right)(x)$	$\dfrac{f(x)}{g(x)}, \quad g(x) \neq 0$	If $f(x) = x^2$ and $g(x) = x + 1$, then $\left(\dfrac{f}{g}\right)(x) = \dfrac{x^2}{x + 1}$.
10.2C	$f \circ g$	$f(g(x))$, the composite of f with g	If $f(x) = x^2$ and $g(x) = x + 1$, then $f \circ g = f[g(x)] = (x + 1)^2$
10.3A	Constant function	$f(x) = k$, k a constant	$f(x) = 10$
10.3A	Linear function	$f(x) = mx + b$, m and b constants	$f(x) = 10x - 27$
10.3A	Quadratic function	$f(x) = ax^2 + bx + c$, a, b, and c constants, $a \neq 0$	$f(x) = 10x^2 + 9x - 1$

(Continued)

SECTION	ITEM	MEANING	EXAMPLE
10.3B	Polynomial function	$f(x) = c_n x^n + c_{n-1} x^{n-1} + \cdots$ $+ c_1 x + c_0$	$f(x) = 2x^3 + 4x^2 - 7x + 3$
10.3B	Cubic	A polynomial function of degree three	$f(x) = 5x^3 - 10x + 2$
10.3C	Rational function	A function of the form $f(x) = \dfrac{P(x)}{Q(x)}$, where $P(x)$ and $Q(x)$ are polynomials and $Q(x) \neq 0$	$f(x) = \dfrac{x^2}{x+1}$
10.3D	Exponential function	A function of the form $f(x) = b^x$, where b is a positive real number $\neq 1$	$f(x) = 4^x$, $g(x) = \left(\dfrac{1}{3}\right)^x$
10.4A	f^{-1}	The inverse of a relation, obtained by reversing the order of the coordinates in each ordered pair in f.	If $f = \{(1, 2), (4, 6), (6, 9)$, then $f^{-1} = \{(2, 1), (6, 4), (9, 6)\}$
10.4B	Horizontal line test	If any horizontal line intersects the graph of $f(x)$ more than once, then f^{-1} is not a function.	$f(x) = x^2$. The graph is a parabola that opens upward. Thus, any horizontal line $y = b > 0$ cuts the graph in two points. The inverse, $f^{-1} = \pm\sqrt{x}$, is not a function.

NAME

CLASS

SECTION

ANSWERS

(If you need help with these exercises, look in the section indicated in brackets.)

1. [10.1A] Find the domain and range.
   a. $\{(0, 5), (2, 9), (3, 10), (5, 8)\}$
   b. $\{(0, 6), (2, 10), (3, 11), (5, 9)\}$

2. [10.1A] Find the domain and the range and graph the relation.
   a. $\{(x, y) \mid y = 1 - x\}$

   b. $\{(x, y) \mid y = 2 - x\}$

   c. $\{(x, y) \mid y = 3 - x\}$

1. a. _____
   b. _____

2. a. _____

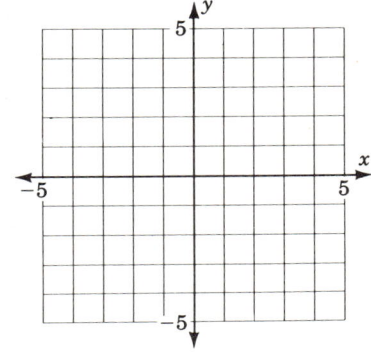

b. _____

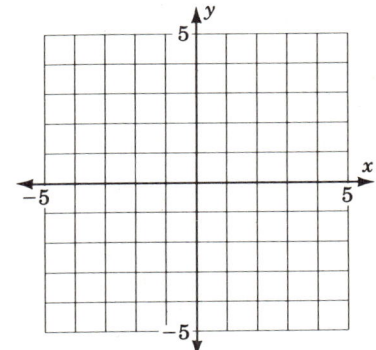

c. _____

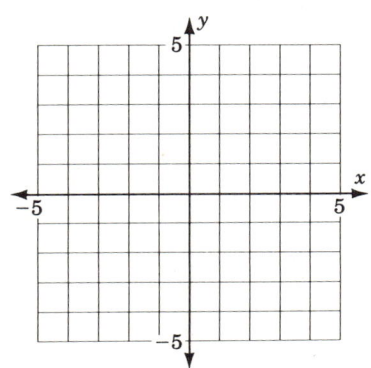

3. [10.1B] Find the domain and the range and graph.
  a. $A = \{(x, y) \mid x^2 + y^2 = 4\}$

3. a. _____

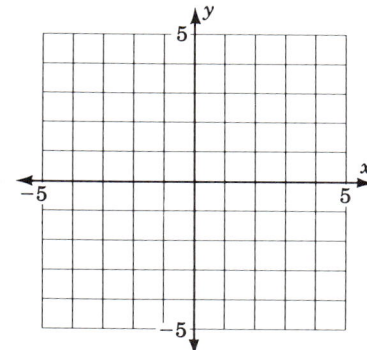

  b. $A = \{(x, y) \mid x^2 + y^2 = 9\}$

b. _____

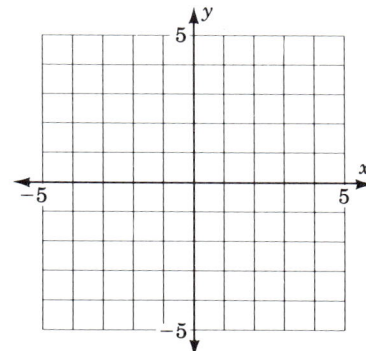

4. [10.1B] Find the domain and the range and graph.
  a. $B = \{(x, y) \mid y = \sqrt{4 - x^2}\}$

4. a. _____

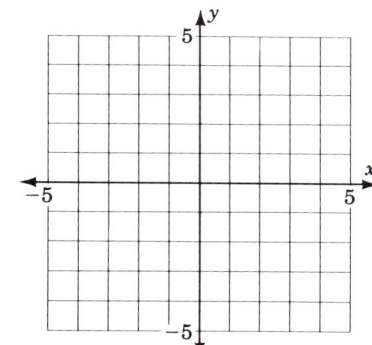

  b. $B = \{(x, y) \mid y = \sqrt{9 - x^2}\}$

b. _____

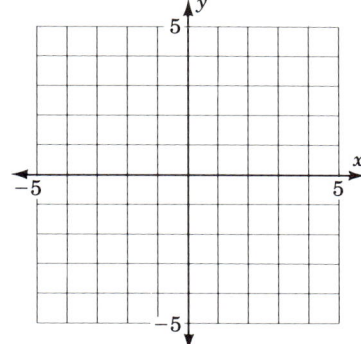

5. [10.1B] Find the domain and the range and graph.

   a. $C = \{(x, y) \,|\, y = -\sqrt{4 - x^2}\}$

5. **a.** _____

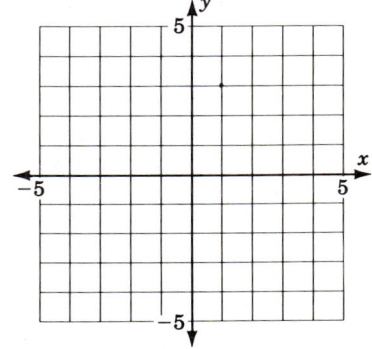

   b. $C = \{(x, y) \,|\, y = -\sqrt{9 - x^2}\}$

b. _____

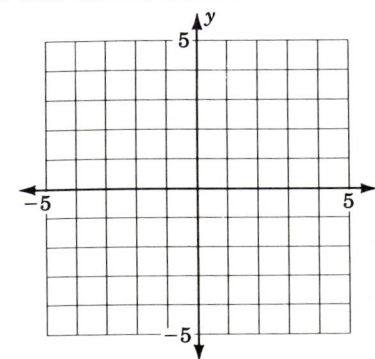

6. [10.1B] Use the vertical line test to determine if the relation defined by the equation is a function.

   a. $y = 1 + x^2$

6. **a.** _____

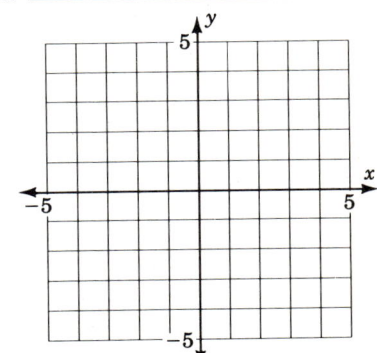

   b. $y = 2 + x^2$

b. _____

7. [10.1C] Find the domain of the function.

    a. $y = \dfrac{2}{x - 1}$

    b. $y = \dfrac{2}{x - 2}$

7. a. _____
   b. _____

8. [10.1C] Find the domain of the function.
    a. $y = \sqrt{x - 3}$
    b. $y = \sqrt{x - 4}$

8. a. _____
   b. _____

9. [10.2A] Let $f(x) = x - 4$. Find.
    a. $f(2)$
    b. $f(1)$
    c. $f(2) - f(1)$

9. a. _____
   b. _____
   c. _____

10. [10.2A] Let $f(x) = x^2 - 3$. Find.
    a. $f(2)$
    b. $f(1)$
    c. $f(2) - f(1)$

10. a. _____
    b. _____
    c. _____

11. [10.2A] Let $f = \{(2, 0), (3, 3), (1, -1)\}$. Find.
    a. $f(2)$
    b. $f(1)$
    c. $f(2) - f(1)$

11. a. _____
    b. _____
    c. _____

12. [10.2A] Let $f = \{(2, 1), (3, 4), (1, 0)\}$. Find.
    a. $f(2)$
    b. $f(1)$
    c. $f(2) - f(1)$

12. a. _____
    b. _____
    c. _____

13. [10.2B] Let $f(x) = 2 - x^2$ and $g(x) = 2 + x$. Find.
    a. $(f + g)(x)$
    b. $(f - g)(x)$
    c. $(fg)(x)$

    d. $\left(\dfrac{f}{g}\right)(x)$

13. a. _____
    b. _____
    c. _____
    d. _____

14. [10.2B] Let $f(x) = 3 - x^2$ and $g(x) = 3 + x$. Find.
    a. $(f + g)(x)$
    b. $(f - g)(x)$
    c. $(fg)(x)$

    d. $\left(\dfrac{f}{g}\right)(x)$

14. a. _____
    b. _____
    c. _____
    d. _____

15. [10.2B] If $f(x) = 6x + 1$, find

    $\dfrac{f(x) - f(a)}{x - a}, \quad x \neq a$

15. _____

16. [10.2B] If $f(x) = 7x + 1$, find

    $\dfrac{f(x) - f(a)}{x - a}, \quad x \neq a$

16. _____

17. [10.2C] If $f(x) = x^3$ and $g(x) = 2 - x$, find
    a. $f \circ g$
    b. $g \circ f$
    c. $(g \circ f)(2)$

17. a. _____
    b. _____
    c. _____

18. [10.2C] If $f(x) = x^3$ and $g(x) = 3 - x$, find
    a. $f \circ g$
    b. $g \circ f$
    c. $(g \circ f)(2)$

18. a. _____
    b. _____
    c. _____

19. [10.2D] The lower limit L (heartbeats per minute) of a person's target zone is given by $L(a) = -\frac{2}{3}a + 150$, where a is age (in years). Find L for a person who is
    a. 21 yr old
    b. 30 yr old

19. a. _____
    b. _____

20. [10.2D] The revenue (in dollars) obtained from selling x units of a product is $R(x) = 100x - 0.02x^2$ and the cost is $C(x) = 30{,}000 + 30x$.
    a. Find the profit function, $P(x) = R(x) - C(x)$.
    b. How many units must be made and sold to yield the maximum profit?

20. a. _____
    b. _____

21. [10.2D] The revenue (in dollars) obtained from selling x units of a product is $R(x) = 100x - 0.02x^2$ and the cost is $C(x) = 40{,}000 + 40x$.
    a. Find the profit function, $P(x) = R(x) - C(x)$
    b. How many units must be made and sold to yield the maximum profit?

21. a. _____
    b. _____

22. [10.3A] Classify and graph.
    a. $f(x) = -2$

22. a. _____

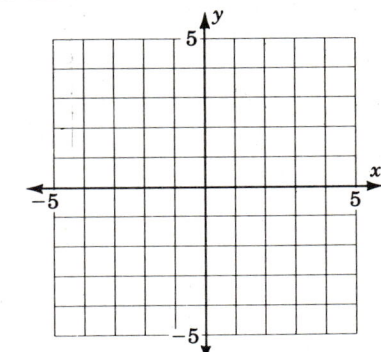

    b. $f(x) = -3$

b. _____

23. [10.3A] Classify and graph.
  a. $f(x) = 3 - 3x$

23. a. _____

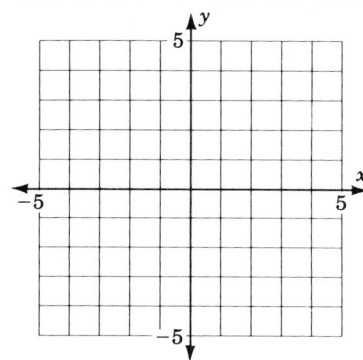

  b. $f(x) = 4 - 4x$

b. _____

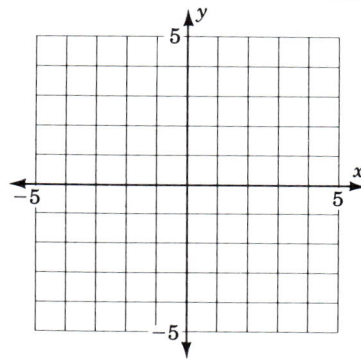

24. [10.3A] Classify and graph.
  a. $f(x) = x^2 - 2x$

24. a. _____

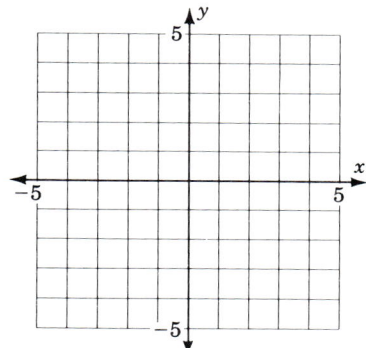

  b. $f(x) = x^2 - 3x$

b. _____

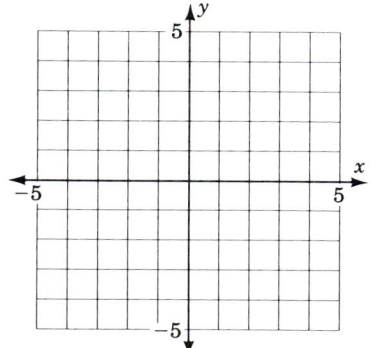

25. [10.3B] Graph.

   a. $f(x) = x^3 - x$

25. **a.**

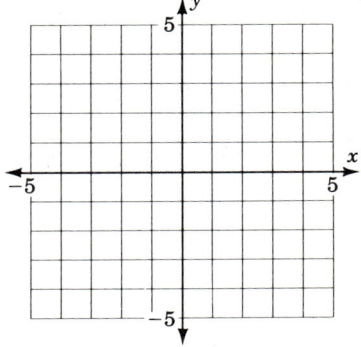

   b. $f(x) = x^3 - 2x$

b.

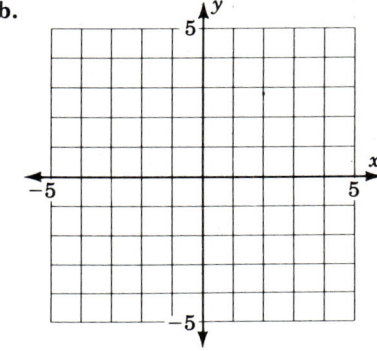

26. [10.3C] Graph.

   a. $f(x) = \dfrac{x + 3}{x - 1}$

26. **a.**

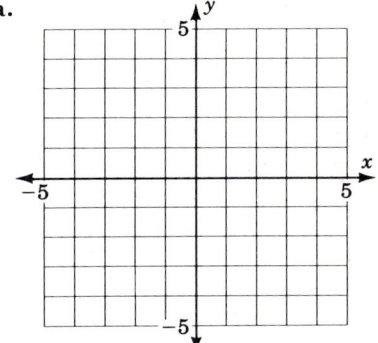

   b. $f(x) = \dfrac{x + 4}{x - 1}$

b.

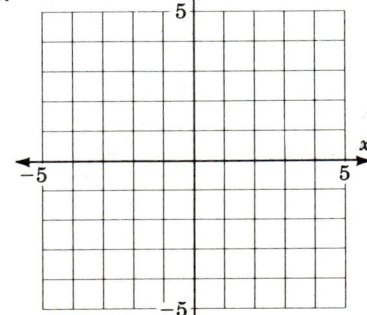

27. [10.3D] Graph.

   a. $f(x) = (\frac{1}{3})^x$

27. a.

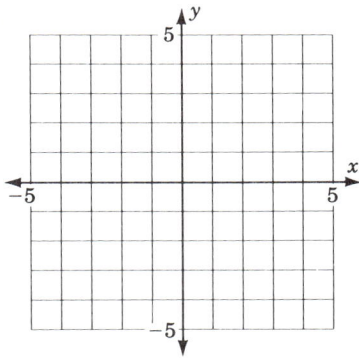

   b. $f(x) = (\frac{1}{4})^x$

b.

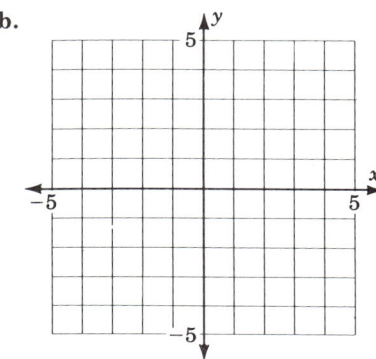

28. [10.4A] Let $S = \{(4, 4), (6, 6), (8, 8)\}$. Find.
   a. The domain and range of S
   b. S^{-1}
   c. The domain and range of S^{-1}
   d. The graph of S and S^{-1}

28. a. _____
   b. _____
   c. _____
   d.

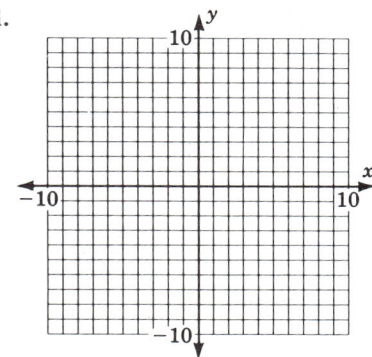

29. [10.4A] Let $S = \{(4, 5), (6, 7), (8, 9)\}$. Find.
   a. The domain and range of S
   b. S^{-1}
   c. The domain and range of S^{-1}
   d. The graph of S and S^{-1}

29. a. _____
   b. _____
   c. _____
   d.

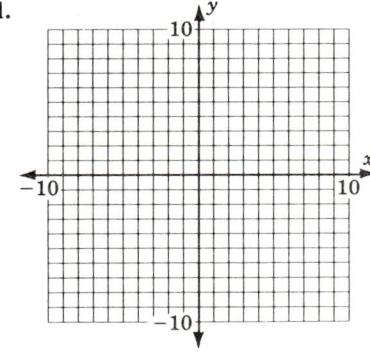

30. [10.4B] Let $f(x) = y = 3x - 3$.
   a. Find $f^{-1}(x)$.
   b. Graph f and its inverse.

30. a. _____
   b.

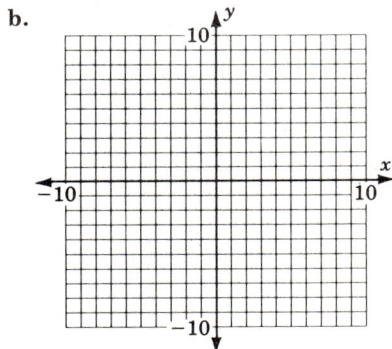

31. [10.4B] Let $f(x) = y = 4x - 4$.
   a. Find $f^{-1}(x)$.
   b. Graph f and its inverse.

31. a. _____
   b.

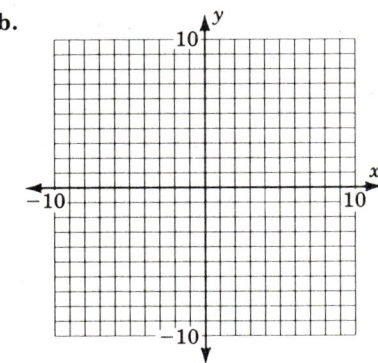

32. [10.4C] Find the inverse of $f(x) = y = 4x^2$. Is the inverse a function?

32. _____

33. [10.4C] Find the inverse of $f(x) = y = 5x^2$. Is the inverse a function?

33. _____

34. [10.4C] Graph $y = 3^x$. Is the inverse a function?

34. _____

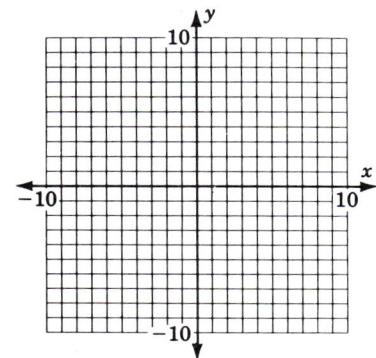

35. [10.4C] Graph $y = 4^x$. Is the inverse a function?

35. _____

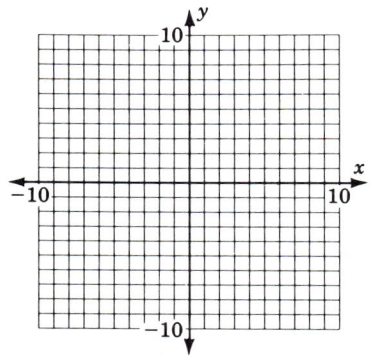

NAME

CLASS

SECTION

ANSWERS

(Answers on pages 842–846)

1. Find the domain and range of the relation
   $\{(1, 3), (2, 5), (3, 7), (4, 9)\}$.

2. Find the domain and the range and graph the relation
   $\{(x, y) \mid y = 4 + x\}$.

3. Find the domain and the range and graph.
   a. $A = \{(x, y) \mid x^2 + y^2 = 4\}$

   b. $B = \{(x, y) \mid y = \sqrt{4 - x^2}\}$

1. _____

2. _____

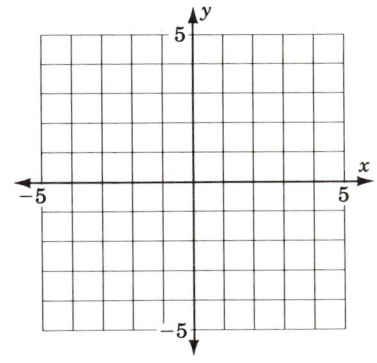

3. a. _____

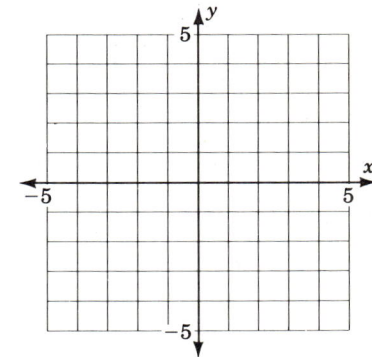

b. _____

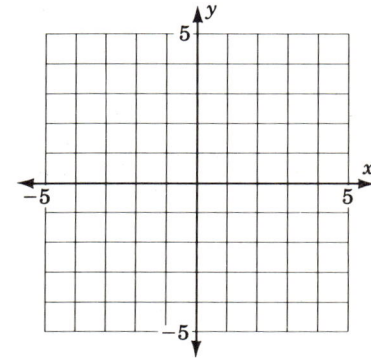

c. $C = \{(x, y) \mid y = -\sqrt{4 - x^2}\}$

c. _____

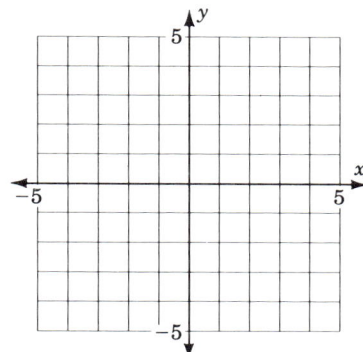

4. Use the vertical line test to determine if the relation defined by the given equation is a function.
   a. $y = 4 - x^2$
   b. $4x^2 + y^2 = 4$

4. a. _____
   b. _____

5. Find the domain of the function.

   a. $y = \dfrac{2}{x + 3}$

   b. $y = \sqrt{x + 2}$

5. a. _____
   b. _____

6. Let $f(x) = 4x - 3$. Find.
   a. $f(2)$
   b. $f(1)$
   c. $f(2) - f(1)$

6. a. _____
   b. _____
   c. _____

7. Let $f = \{(1, 4), (2, -1), (3, 2)\}$. Find.
   a. $f(2)$
   b. $f(1)$
   c. $f(2) - f(1)$

7. a. _____
   b. _____
   c. _____

8. Let $f(x) = 1 - x^2$ and $g(x) = 1 - x$. Find.
   a. $(f + g)(x)$
   b. $(f - g)(x)$
   c. $(fg)(x)$
   d. $\left(\dfrac{f}{g}\right)(x)$

8. a. _____
   b. _____
   c. _____
   d. _____

9. If $f(x) = 5x + 1$, find

   $\dfrac{f(x) - f(a)}{x - a}, \qquad x \neq a$

9. _____

10. If $f(x) = x^3$ and $g(x) = 1 + x$, find
    a. $f \circ g$
    b. $g \circ f$
    c. $(g \circ f)(2)$

10. a. _____
    b. _____
    c. _____

11. The lower limit L (heartbeats per minute) of a person's target zone is given by $L(a) = -\frac{2}{3}a + 150$, where a is age (in years). Find L for a person who is
    a. 15 yr old
    b. 21 yr old

11. a. _____
    b. _____

12. The revenue (in dollars) obtained from selling x units of a product is $R(x) = 100x - 0.02x^2$ and the cost is $C(x) = 10,000 + 20x$.
   a. Find the profit function, $P(x) = R(x) - C(x)$.
   b. How many units must be made and sold to yield the maximum profit?

12. a. _____
    b. _____

13. Classify the function as constant, linear, or quadratic and graph.
   a. $f(x) = 1$

13. a. _____

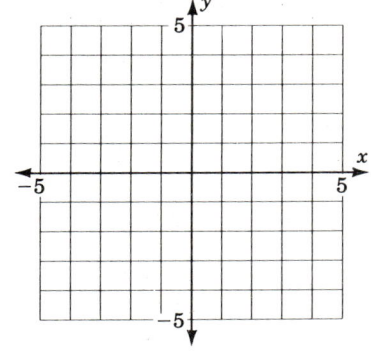

   b. $f(x) = 4x - 2x^2$

b. _____

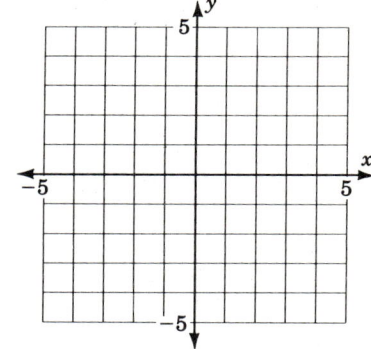

   c. $f(x) = 2x + 3$

c. _____

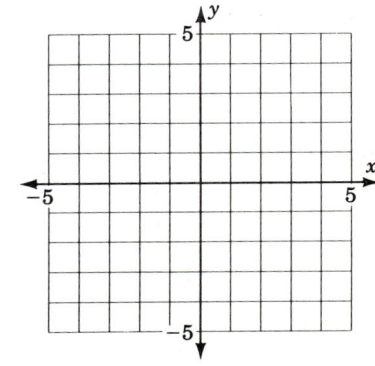

14. Graph $f(x) = x^3 - 5x$.

14.

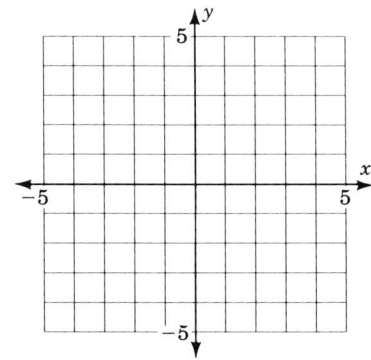

15. Graph $f(x) = \dfrac{x-1}{x-2}$.

15.

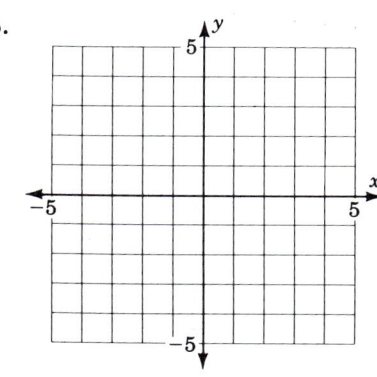

16. Graph $f(x) = \left(\dfrac{1}{3}\right)^x$.

16.

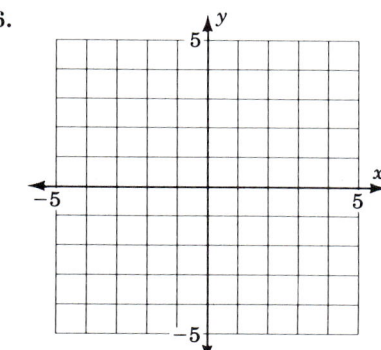

17. Let $S = \{(3, 5), (5, 7), (7, 9)\}$. Find.
   a. The domain and range of S
   b. S^{-1}
   c. The domain and range of S^{-1}
   d. The graph of S and S^{-1}.

17. **a.** _____
   b. _____
   c. _____

   d.

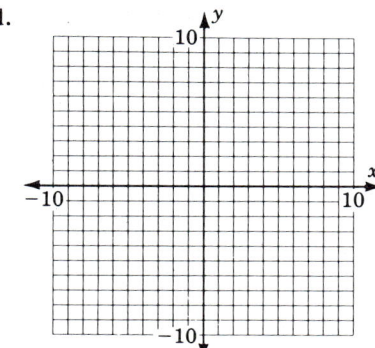

18. Let $f(x) = y = 4x - 4$.
    a. Find $f^{-1}(x)$.
    b. Graph f and its inverse.

18. a. _____
    b.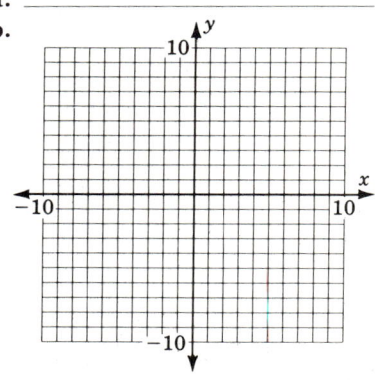

19. Find the inverse of $f(x) = y = 3x^2$. Is the inverse a function?

20. Graph $y = 3^x$. Is the inverse a function?

19. _____

20. _____

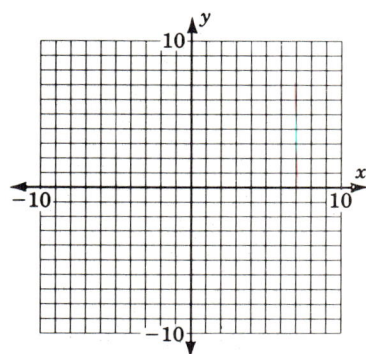

IF YOU MISSED QUESTION	SECTION	EXAMPLES	PAGE	ANSWERS
1	10.1A	1	774	**1.** $D = \{1, 2, 3, 4\}; R = \{3, 5, 7, 9\}$
2	10.1A	2	774	**2.** The domain and range are the set of real numbers.

$y = 4 + x$

3a	10.1A	3	775	**3. a.** $D = \{x \mid -2 \leq x \leq 2\};$ $R = \{y \mid -2 \leq y \leq 2\}$

$x^2 + y^2 = 4$

3b	10.1A	3	775	**b.** $D = \{x \mid -2 \leq x \leq 2\};$ $R = \{y \mid 0 \leq y \leq 2\}$

$y = \sqrt{4 - x^2}$

IF YOU MISSED QUESTION	SECTION	EXAMPLES	PAGE	ANSWERS
3c	10.1A	3	775	$D = \{x \mid -2 \le x \le 2\}$; **c.** $R = \{y \mid -2 \le y \le 0\}$

IF YOU MISSED QUESTION	SECTION	EXAMPLES	PAGE	ANSWERS
4a	10.1B	4	776	**4. a.** A function
4b	10.1B	4	776	**b.** Not a function
5a	10.1C	5	777	**5. a.** All real numbers except -3
5b	10.1C	5	777	**b.** All real numbers greater than or equal to -2
6a	10.2A	1	788	**6. a.** 5
6b	10.2A	1	788	**b.** 1
6c	10.2A	1	788	**c.** 4
7a	10.2A	2	788	**7. a.** -1
7b	10.2A	2	788	**b.** 4
7c	10.2A	2	788	**c.** -5
8a	10.2B	3	789	**8. a.** $-x^2 - x + 2$
8b	10.2B	3	789	**b.** $-x^2 + x$
8c	10.2B	3	789	**c.** $x^3 - x^2 - x + 1$
8d	10.2B	3	789	**d.** $1 + x, x \neq 1$
9	10.2B	4	789	**9.** 5
10a	10.2C	5	790	**10. a.** $(1 + x)^3$
10b	10.2C	5	790	**b.** $1 + x^3$
10c	10.2C	5	790	**c.** 9
11a	10.2D	6	790–791	**11. a.** 140
11b	10.2D	6	790–791	**b.** 136
12a	10.2D	7	791	**12. a.** $R(x) = -0.02x^2 + 80x - 10{,}000$
12b	10.2D	7	791	**b.** 2000

IF YOU MISSED QUESTION	SECTION	EXAMPLES	PAGE	ANSWERS
13a	10.3A	1	800–801	
13b	10.3A	1	800–801	
13c	10.3A	1	800–801	

13. a. A constant function

$f(x) = 1$

b. A quadratic function

$f(x) = 4x - 2x^2$

c. A linear function

$f(x) = 2x + 3$

IF YOU MISSED QUESTION	SECTION	EXAMPLES	PAGE	ANSWERS
14	10.3B	2	802	**14.**
15	10.3C	3	803	**15.**
16	10.3D	4	804	**16.**
17a	10.4A	1	816	**17. a.** $D = \{3, 5, 7\}; R = \{5, 7, 9\}$
17b	10.4A	1	816	**b.** $\{(5, 3), (7, 5), (9, 7)\}$
17c	10.4A	1	816	**c.** $D = \{5, 7, 9\}; R = \{3, 5, 7\}$

IF YOU MISSED QUESTION	SECTION	EXAMPLES	PAGE	ANSWERS
17d	10.4A	1	816	d.
18a	10.4B	2	817–818	18. a. $f^{-1}(x) = \dfrac{x+4}{4}$
18b	10.4B	2	817–818	b.
19	10.4C	3	819	19. $y = \pm\sqrt{\dfrac{x}{3}}$; no
20	10.4C	4	819	20. Yes

CHAPTER 11

EXPONENTIAL AND LOGARITHMIC FUNCTIONS

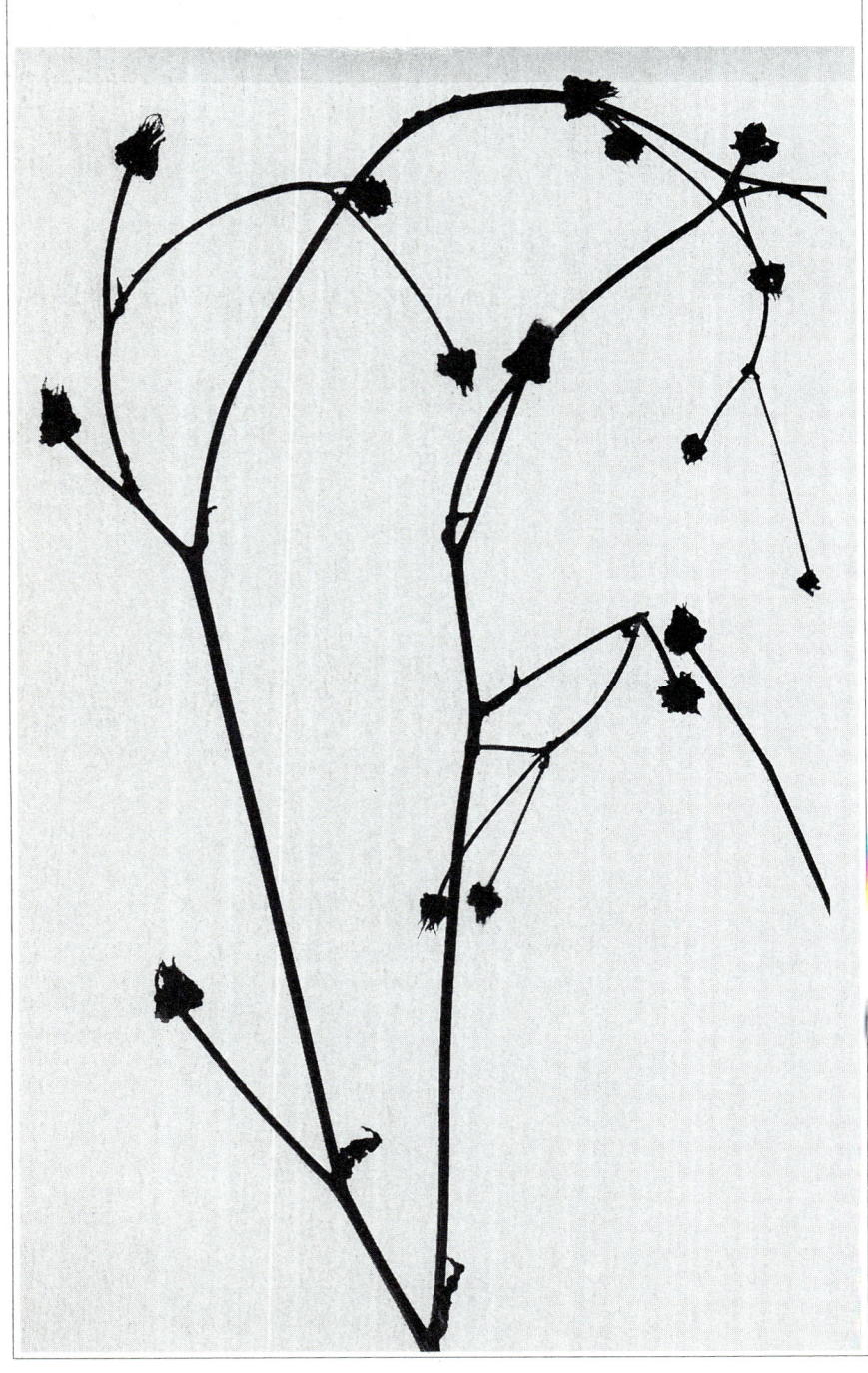

We start this chapter by defining a new type of function with the variable in the exponent, an exponential function. We graph these functions and apply the ideas studied to problems of growth and decay and compound interest. We then discuss logarithms, their properties, and their use in computation; we end the chapter by studying exponential equations and natural logarithms.

NAME

CLASS

SECTION

(Answers on pages 852–853)

ANSWERS

1. Graph the function $f(x) = 2^x$.

1.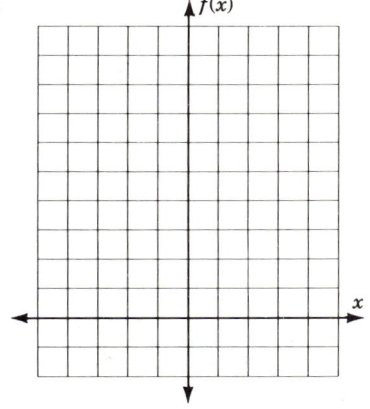

2. Graph the function $g(x) = \left(\dfrac{1}{2}\right)^x$.

2.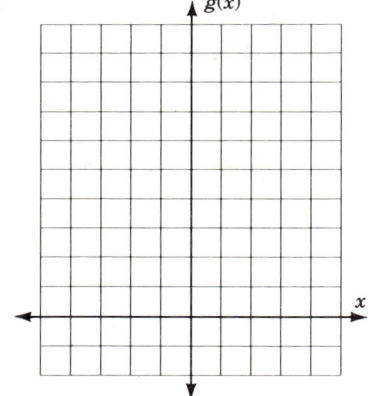

3. Find $\log_2 16$.

3. _____

4. Find $\log_{10} 0.1$.

4. _____

5. Write $3^5 = 243$ in logarithmic form.

5. _____

6. Write $\log_8 16 = \dfrac{4}{3}$ in exponential form.

6. _____

7. Show that $\log_b \sqrt{\dfrac{43}{97}} = \dfrac{1}{2}(\log_b 43 - \log_b 97)$.

7. _____

8. Show that $\log\left(1 - \dfrac{1}{x^3}\right) = \log(x - 1) + \log(x^2 + x + 1) - 3\log x$.

8. _____

9. If $1000 is invested at 12% compounded monthly for 5 yr, the compound amount is given by $A = 1000(1.01)^{60}$. Given that $\log 1.01 = 0.0043$, find $\log A$. (Note that $1000 = 10^3$.)

9. _____

The following table is to be used in Problems 10–15.

Some Common Logarithms

x	0	1	2	3	4	5	6	7	8	9
3.0	.4771	.4786	.4800	.4814	.4829	.4843	.4857	.4871	.4886	.4900
3.1	.4914	.4928	.4942	.4955	.4969	.4983	.4997	.5011	.5024	.5038
3.2	.5051	.5065	.5079	.5092	.5105	.5119	.5132	.5145	.5159	.5172
3.3	.5185	.5198	.5211	.5224	.5237	.5250	.5263	.5276	.5289	.5302
3.4	.5315	.5328	.5340	.5353	.5366	.5378	.5391	.5403	.5416	.5428
3.5	.5441	.5453	.5465	.5478	.5490	.5502	.5514	.5527	.5539	.5551
3.6	.5563	.5575	.5587	.5599	.5611	.5623	.5635	.5647	.5658	.5670
3.7	.5682	.5694	.5705	.5717	.5729	.5740	.5752	.5763	.5775	.5786
3.8	.5798	.5809	.5821	.5832	.5843	.5855	.5866	.5877	.5888	.5899
3.9	.5911	.5922	.5933	.5944	.5955	.5966	.5977	.5988	.5999	.6010

10. Find.
    a. $\log 375$
    b. $\log 0.00375$

10. a. _____
    b. _____

11. Use interpolation to find.
    a. $\log 3.755$
    b. $\log 0.03756$

11. a. _____
    b. _____

12. Find.
    a. antilog 3.5159
    b. antilog $7.5198 - 10$

12. a. _____
    b. _____

13. Use interpolation to find.
    a. antilog 2.5419
    b. antilog $7.5815 - 10$

13. a. _____
    b. _____

14. Find $(3.05)^{1.2}$ to four digits.

14. _____

15. Find $\sqrt{31.2 \times 0.00397}$ to four digits.

15. _____

16. Find x if $e^{10.5x} = 2$ and you are given $\ln 2 = 0.69315$.

16. _____

17. Given the change of base formula,

$$\ln M = \frac{\log M}{0.43429} = 2.3026 \log M$$

find $\ln 100$.

17. _____

18. The compound amount with continuous compounding is given by $A = Pe^{rt}$, where P is the principal, r is the rate, and t is the time in years. If the rate is 10%, find how long it takes for the money to double—that is, for A to equal $2P$ ($\ln 2 = 0.69315$).

18. _____

19. The number of bacteria in a culture after t minutes is given by $N = 1000e^{kt}$. If there are 2117 bacteria after 3 min, find k. (Use the following table.)

19. _____

x	e^x	e^{-x}
0.55	1.7333	0.5769
0.60	1.8221	0.5488
0.65	1.9155	0.5220
0.70	2.0138	0.4966
0.75	2.1170	0.4724

20. A radioactive substance decays so that the amount A present at time t (years) is $A = A_0 e^{-0.25t}$. Find the half-life (time for half to decay) of this substance ($\ln 2 = 0.69315$).

20. _____

IF YOU MISSED QUESTION	SECTION	EXAMPLES	PAGE	ANSWERS
1	11.1	1a	856	**1.**
2	11.1	1b	857	**2.**
3	11.2	1a	866	**3.** 4
4	11.2	1d	866	**4.** -1
5	11.2	2	866	**5.** $\log_3 243 = 5$
6	11.2	3	866	**6.** $16 = 8^{4/3}$
7	11.2	4	867	**7.** $\log \sqrt{\dfrac{43}{97}} = \dfrac{1}{2} \log \dfrac{43}{97}$ $= \dfrac{1}{2}(\log 43 - \log 97)$
8	11.2	5	867	$1 - \dfrac{1}{x^3} = \dfrac{x^3 - 1}{x^3} = \dfrac{(x-1)(x^2 + x + 1)}{x^3};$ $\log\left(1 - \dfrac{1}{x^3}\right) = \log(x - 1)$ **8.** $+ \log(x^2 + x + 1) - 3\log x$
9	11.2	7	868	**9.** 3.2580
10a	11.3	1a	875	**10. a.** 2.5740
10b	11.3	1b	875	**b.** $7.5740 - 10$

IF YOU MISSED QUESTION	SECTION	EXAMPLES	PAGE	ANSWERS
11a	11.3	2	875	**11. a.** 0.5746
11b	11.3	2	875	**b.** 8.5747 − 10
12a	11.3	3	876	**12. a.** 3280
12b	11.3	3	876	**b.** 0.00331
13a	11.3	4	876	**13. a.** 348.3
13b	11.3	4	876	**b.** 0.003815
14	11.3	5	877	**14.** 3.813
15	11.3	6	877	**15.** 0.3520
16	11.4	1	884–885	**16.** 0.06601
17	11.4	2	885–886	**17.** 4.6052
18	11.4	3	886	**18.** About 6.93 yr
19	11.4	5	887	**19.** 0.25
20	11.4	6	887–888	**20.** About 2.773 yr

11.1 EXPONENTIAL FUNCTIONS

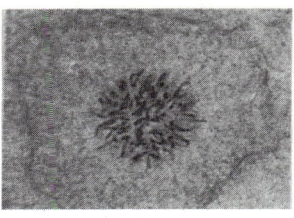
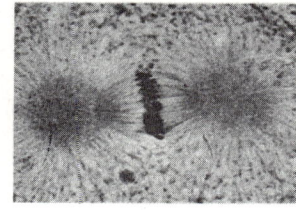
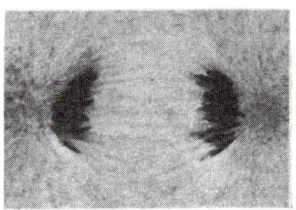
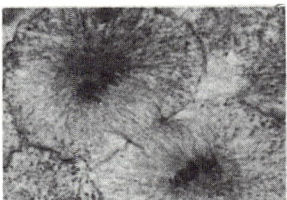

Photo credit: General Biological Supply.

The photograph at the beginning of this section shows a cell reproducing by a process called *mitosis*. In mitosis, a single cell or bacterium divides and forms two identical daughter cells. Each daughter cell then doubles in size and divides. As you can see, the number of bacteria present is a function of time. If we start with one cell and assume that each cell divides after 10 min, then the number of bacteria present at the end of the first 10-min period ($t = 10$) is

$$2 = 2^1 = 2^{10/10}$$

At the end of the second 10-min period ($t = 20$), the two cells divide, and the number of bacteria present is

$$4 = 2^2 = 2^{20/10}$$

Similarly, at the end of the third 10-min period ($t = 30$), the number is

$$8 = 2^3 = 2^{30/10}$$

Thus, we can see that the number of bacteria present at the end of t minutes is given by the function

$$f(t) = 2^{t/10}$$

Note that this also gives the correct result for $t = 0$, because $2^0 = 1$.

The function $f(t) = 2^{t/10}$ is called an *exponential function* because the variable t is in the exponent. Here are some more examples of exponential functions:

$$f(x) = 3^x, \qquad F(y) = \left(\frac{1}{2}\right)^y, \qquad H(z) = (1.02)^{z/2}$$

An **exponential function** is a function defined for all real values of x by

$$f(x) = b^x, \qquad b > 0, b \neq 1$$

In this definition, b is a constant called the **base** and the **exponent** x is the variable. It is proved in more advanced courses that b^x, for $b > 0$, has a unique real value for each real value of x. We assume this in all the following work.

An exponential function is frequently not in the form given in our definition but can be put in that form. For example, $f(t) = 2^{t/10}$ can be written as $(2^{1/10})^t$, so that the base is $2^{1/10}$ and the exponent is t.

A. GRAPHS OF EXPONENTIAL FUNCTIONS

The exponential function defined by $f(t) = 2^{t/10}$ can be graphed and used to predict the number of bacteria present after a period of time t. To make this graph, we first construct a table giving the value of the function for certain convenient times, as shown.

t	0	10	20	30
$f(t) = 2^{t/10}$	1	2^1	2^2	2^3

The corresponding points can then be graphed and joined with a smooth curve, as in Figure 1.

In general, we graph an exponential function by plotting several points calculated from the function and then drawing a smooth curve through these points.

EXAMPLE 1 Graph the given functions on the same coordinate system.

a. $f(x) = 2^x$ **b.** $g(x) = \left(\dfrac{1}{2}\right)^x$

Problem 1 Graph the given functions on the same coordinate system.

a. $f(x) = 3^x$

b. $g(x) = \left(\dfrac{1}{3}\right)^x$

Solution

a. We first make a table with convenient values for x and then find the corresponding values for $f(x)$, as shown below. We graph the points and connect them with a smooth curve, as shown in black in Figure 2.

x	-2	-1	0	1	2
$f(x) = 2^x$	$2^{-2} = \frac{1}{4}$	$2^{-1} = \frac{1}{2}$	$2^0 = 1$	$2^1 = 2$	$2^2 = 4$

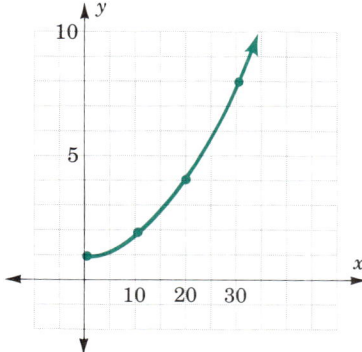

FIG. 1

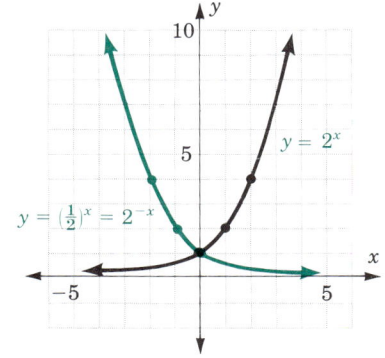

FIG. 2

b. We note that $g(x) = (\frac{1}{2})^x = (2^{-1})^x = 2^{-x}$. Thus, $g(x)$ can be obtained from $f(x) = 2^x$ by replacing x by $-x$—that is, $g(x) = f(-x)$. This means that we can use the values in the table for $f(x)$ by just interchanging the function values for the corresponding positive and negative values of x. For example,

$$g(-2) = f(2) = 4$$

and $g(1) = f(-1) = \frac{1}{2}$. The table for $g(x)$ is as follows:

x	-2	-1	0	1	2
$g(x) = (\frac{1}{2})^x$	4	2	1	$\frac{1}{2}$	$\frac{1}{4}$

The graph of $g(x) = (\frac{1}{2})^x = 2^{-x}$ is shown in color in Figure 2. ▲

Notice that the two graphs in Figure 2 are symmetric to each other with respect to the y-axis. This is a consequence of the fact that $g(x) = f(-x)$, which means that if (x_1, y_1) is a point on the graph of $f(x)$, then $(-x_1, y_1)$ is the corresponding point on the graph of $g(x)$.

B. INCREASING AND DECREASING FUNCTIONS

If the graph of a function goes *up* to the right, the function is an **increasing** function. If the graph goes *down* to the right, the function is a **decreasing** function. Thus we see that $f(x) = 2^x$ is an increasing function and $g(x) = (\frac{1}{2})^x$ is a decreasing function.

In our definition of the function b^x, it was required only that $b > 0$ and $b \neq 1$. For many practical applications, however, there is a particularly important base, the number e. The value of e is approximately 2.718282. The reasons for using this base are made clear in more advanced mathematics courses, but we note here that e is defined as the value that the quantity $\left(1 + \dfrac{1}{n}\right)^n$ approaches as n increases indefinitely.

For example,

when $n = 1{,}000$, $\qquad \left(1 + \dfrac{1}{n}\right)^n = 2.716924$

when $n = 10{,}000$, $\qquad \left(1 + \dfrac{1}{n}\right)^n = 2.718146$

when $n = 100{,}000$, $\qquad \left(1 + \dfrac{1}{n}\right)^n = 2.718255$

when $n = 1{,}000{,}000$, $\qquad \left(1 + \dfrac{1}{n}\right)^n = 2.718282$

Values of the exponential functions e^x and e^{-x} are given in the table in Appendix 3. This table was used to get the approximate values in the next example.

EXAMPLE 2 Use the values in the given tables to graph (on the same coordinate system) $f(x) = e^x$ and $g(x) = e^{-x}$.

x	-2	-1	0	1	2
e^x	0.1353	0.3679	1	2.7183	7.3891

x	-2	-1	0	1	2
e^{-x}	7.3891	2.7183	1	0.3679	0.1353

Solution Plotting the given values, we obtain the graphs of $f(x) = e^x$ and $g(x) = e^{-x}$ shown in Figure 3 (page 858).

Problem 2 Graph the functions $f(x) = e^{x/2}$ and $g(x) = e^{-x/2}$ on the same coordinate system.

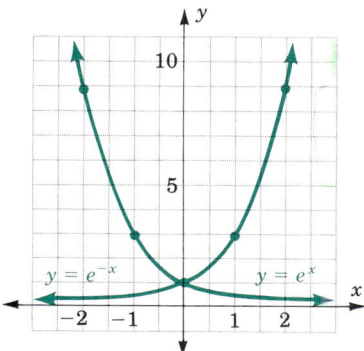

FIG. 3 ▲

C. APPLICATIONS

Do you have some money invested? Is it earning interest, compounded annually, quarterly, monthly, or daily? Does the frequency of compounding make a difference? Recently, some financial concerns have instituted what is called *continuous interest compounding*. The formula for this type of compounding is

$$A = Pe^{rt}$$

where

$A =$ the compound amount
$P =$ the principal
$r =$ the rate
$t =$ the time in years

EXAMPLE 3 Find the compound amount for $100 compounded continuously for 18 months at 6%.

Solution Here, $P = 100$, $r = 0.06$, and $t = 1.5$, so

$$A = 100e^{(0.06)(1.5)}$$
$$= 100e^{0.09}$$

The table (Appendix 3) gives the value

$$e^{0.09} = 1.0942$$

Thus

$$A = (100)(1.0942) = 109.42 \qquad \blacktriangle$$

and the compound amount is $1090.42.

(The compound amount for $100 at the same rate, compounded quarterly, is given by $A = 100(1.015)^6 = 109.34$. So the compound amount is $109.34, only 8¢ less! For other comparisons, see Using Your Knowledge 11.1.)

EXAMPLE 4 A radioactive substance decays so that G, the number of grams present, is given by

$$G = 1000e^{-1.2t}$$

Problem 3 Do Example 3 if time is 30 mo.

Problem 4 Do Example 4 if the time is 18 mo.

where t is the time in years. Find, to the nearest gram, the amount of the substance present

a. At the start   **b.** In 2 yr

Solution

a. Here, $t = 0$, so $G = 1000e^0 = 1000(1) = 1000$.

b. Since $t = 2$, $G = 1000e^{-2.4}$. To evaluate G, we use the same table as in Example 3, which gives

$$e^{-2.4} = 0.090718$$

so that

$$G = (1000)(0.090718)$$
$$= 90.0718$$

So there are 90 g present in 2 yr.   ▲

ANSWERS (to problems on pages 856–857

1. The graph of $f(x) = 3^x$ is shown in black and that of $g(x) = (\frac{1}{3})^x$ is in color in the figure.

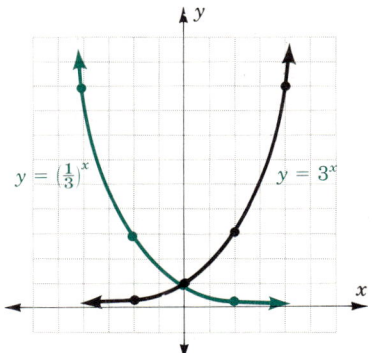

2. The graph of $f(x) = e^{x/2}$ is shown in black and that of $g(x) = e^{-x/2}$ is in color in the figure.

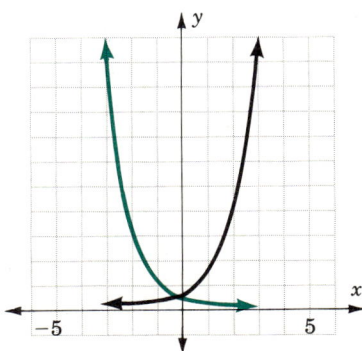

NAME

CLASS

SECTION

A. In Problems 1–6 find the value of the given exponential for the indicated values of the variable.

ANSWERS

1. 5^x
    a. $x = -1$
    b. $x = 0$
    c. $x = 1$

1. **a.** _____
  b. _____
  c. _____

2. 5^{-x}
    a. $x = -1$
    b. $x = 0$
    c. $x = 1$

2. **a.** _____
  b. _____
  c. _____

3. 3^t
    a. $t = -2$
    b. $t = 0$
    c. $t = 2$

3. **a.** _____
  b. _____
  c. _____

4. 3^{-t}
    a. $t = -2$
    b. $t = 0$
    c. $t = 2$

4. **a.** _____
  b. _____
  c. _____

5. $10^{t/2}$
    a. $t = -2$
    b. $t = 0$
    c. $t = 2$

5. **a.** _____
  b. _____
  c. _____

6. $10^{-t/2}$
    a. $t = -2$
    b. $t = 0$
    c. $t = 2$

6. **a.** _____
  b. _____
  c. _____

B. In Problems 7–12 graph the functions given in parts a and b on the same coordinate system. State whether the function is increasing or decreasing.

7. **a.** $f(x) = 5^x$
    b. $g(x) = 5^{-x}$

8. **a.** $f(t) = 3^t$
    b. $g(t) = 3^{-t}$

7. **a.** _____
  b. _____

8. **a.** _____
  b. _____

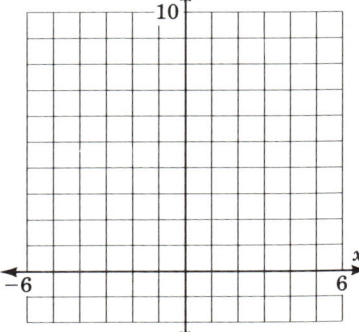

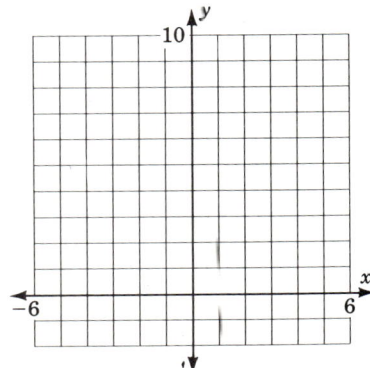

9. **a.** $f(x) = 10^x$
  b. $g(x) = 10^{-x}$

10. **a.** $f(t) = 10^{t/2}$
  b. $g(t) = 10^{-t/2}$

9. a. _____
   b. _____

10. a. _____
    b. _____

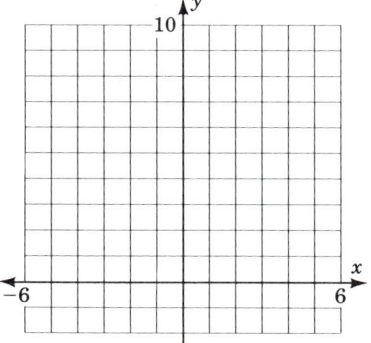

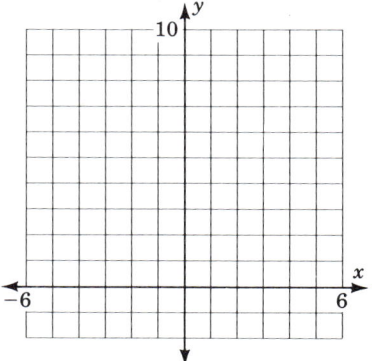

11. **a.** $f(x) = e^{2x}$
  b. $g(x) = e^{-2x}$

12. **a.** $f(t) = e^{t/4}$
  b. $g(t) = e^{-t/4}$

11. a. _____
    b. _____

12. a. _____
    b. _____

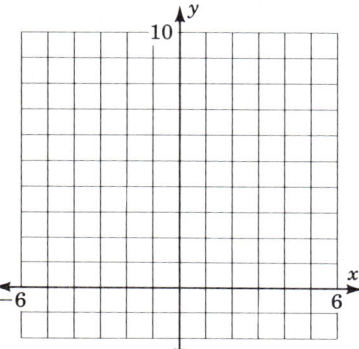

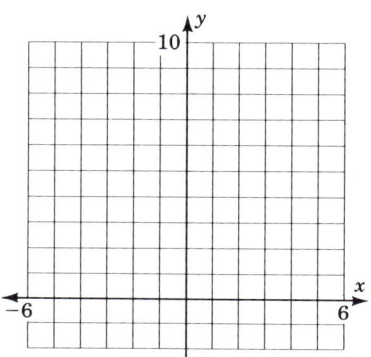

C. In Problems 13–16 find the compound amount if the compounding is continuous.

13. $1000 at 9% for 10 yr

14. $1000 at 9% for 20 yr

15. $1000 at 6% for 10 yr

16. $1000 at 6% for 20 yr

13. _____

14. _____

15. _____

16. _____

17. The population of a town is given by the equation $P = 2000(2^{0.2t})$, where t is the time in years from 1985. Find the population in
  a. 1985
  b. 1990
  c. 1995

17. a. _____
    b. _____
    c. _____

18. A colony of bacteria grows so that their number, B, is given by the equation $B = 1200(2^t)$, where t is in days. Find the number of bacteria
  a. At the start ($t = 0$)
  b. In 5 days
  c. In 10 days

18. a. _____
    b. _____
    c. _____

19. A radioactive substance decays so that G, the number of grams present, is given by $G = 2000e^{-1.05t}$, where t is the time in years. Find the amount of the substance present

    a. At the start

    b. In 1 yr

    c. In 2 yr

20. Do Problem 19 if the equation is $G = 2000e^{-1.1t}$.

19. a. _____

    b. _____

    c. _____

20. a. _____

    b. _____

    c. _____

✓ **SKILL CHECKER**

In Problems 21–24, state the law of exponents that is being applied.

21. $(10^x)(10^y) = 10^{x+y}$

22. $\dfrac{10^x}{10^y} = 10^{x-y}$

23. $(10^x)^3 = 10^{3x}$

24. $[(2)(10^x)]^y = (2^y)(10^{xy})$

21. _____

22. _____

23. _____

24. _____

11.1 USING YOUR KNOWLEDGE

In this section, you learned that for continuous compounding, the compound amount is given by $A = Pe^{rt}$. You learned earlier that for ordinary compound interest, the compound amount is given by $A = P(1 + i)^n$, where i is the rate per period and n is the number of periods. Suppose you have $1000 to put into an account where the interest rate is 6%. How much more would you have at the end of 2 yr for continuous compounding than for monthly compounding?

For continuous compounding, the amount is given by

$$A = 1000e^{(0.06)(2)} = 1000e^{0.12}$$

$$= (1000)(1.1275) \qquad \text{Use the table in Appendix 3.}$$

$$= 1127.50$$

Thus the amount is $1127.50. For monthly compounding, the amount is given by

$$A = 1000(1 + 0.005)^{24} = 1000(1.005)^{24}$$

Compound interest tables or your calculator will give the value

$$(1.005)^{24} = 1.1271598$$

so that $A = 1127.16$ (to the nearest hundredth). Thus the amount is $1127.16. Continuous compounding gives you only 34¢ more! But, do the next problems.

1. Make the same comparison if the time is 10 yr.

2. Make the same comparison if the time is 20 yr.

1. _____

2. _____

11.2 LOGARITHMS

The photo shows part of the damage done by the 1989 earthquake in San Francisco. If I_0 denotes the minimum intensity of an earthquake (used for comparison purposes), the intensity of this quake was $10^{7.1} \cdot I_0$—that is, $10^{7.1}$ times as intense as the minimum. The magnitude of an earthquake is usually measured on the Richter scale. On this scale, the magnitude is reported as 7.1. Read on for an explanation of the Richter scale.

A. DEFINITION OF THE LOGARITHM

In the preceding section, we studied exponential functions of the type given by $y = 2^t$, $y = 5^t$, $y = 10^x$, and so on. In many practical applications, it is important to study the exponent when the value of y, the function, is given. This exponent is called the **logarithm of y to the given base,** and we write

$$t = \log_2 y \quad \text{if} \quad y = 2^t$$
$$t = \log_5 y \quad \text{if} \quad y = 5^t$$
$$x = \log_{10} y \quad \text{if} \quad y = 10^x$$

and so on. For example,

$$\log_2 16 = 4 \quad \text{because} \quad 16 = 2^4 \quad \text{The logarithm is the exponent.}$$
$$\log_5 25 = 2 \quad \text{because} \quad 25 = 5^2 \quad \text{The logarithm is the exponent.}$$
$$\log_{10} 0.001 = -3 \quad \text{because} \quad 0.001 = 10^{-3}$$

The Richter scale is a logarithmic scale. In this scale the magnitude R of an earthquake of intensity $I = 10^{6.9} \cdot I_0$ is reported as

$$R = \log_{10} \frac{I}{I_0} = \log_{10} \frac{10^{6.9} \cdot I_0}{I_0} = \log_{10} 10^{6.9}$$

That is, $R = 6.9$.

REVIEW

Before starting this section, you should know:

1. The rules of exponents in multiplication, division, and raising to a power.
2. How to simplify algebraic expressions.

OBJECTIVES

You should be able to:

A. Write an exponential equation in logarithmic form and a logarithmic equation in exponential form.
B. Use the properties of logarithms to simplify logarithms of products, quotients, and powers.
C. Solve applications of these ideas.

EXAMPLE 1 Find the following logarithms.
a. $\log_2 32$ **b.** $\log_2(\frac{1}{4})$ **c.** $\log_{10} 1000$ **d.** $\log_{10} 0.01$

Solution

a. $\log_2 32 = 5$ because $32 = 2^5$.
b. $\log_2(\frac{1}{4}) = -2$ because $\frac{1}{4} = 2^{-2}$.
c. $\log_{10} 1000 = 3$ because $1000 = 10^3$.
d. $\log_{10} 0.01 = -2$ because $0.01 = 10^{-2}$. ▲

Using this idea, we have the following definition.

DEFINITION OF LOGARITHM

$$\log_b x = y \quad \text{means} \quad x = b^y$$

EXAMPLE 2 Write the equation $5^3 = 125$ in logarithmic form.

Solution By the definition of $\log_b x$,

$$x = b^y \text{ means } \log_b x = y$$

with exponent pointing to y and base pointing to b.

Thus $125 = 5^3$ means $\log_5 125 = 3$. ▲

EXAMPLE 3 Write the equation $\log_{32} 64 = \frac{6}{5}$ in exponential form and check its accuracy.

Solution By the definition, $\log_b x = y$ means $x = b^y$. Hence $\log_{32} 64 = \frac{6}{5}$ means $32^{6/5} = 64$. Because $32^{6/5} = (\sqrt[5]{32})^6 = 2^6 = 64$, the equation $\log_{32} 64 = \frac{6}{5}$ is correct. ▲

B. PROPERTIES OF LOGARITHMS

Logarithms have three important properties that are the counterparts of the corresponding properties of exponents. By Definition (1), if $x = \log_b M$ and $y = \log_b N$, then $M = b^x$ and $N = b^y$. From the properties of exponents, it follows that

$$MN = b^x b^y = b^{x+y}$$

so that

$$\log_b MN = x + y = \log_b M + \log_b N$$

This means that the *logarithm of a product is the sum of the logarithms of its factors*. Similarly,

$$\frac{M}{N} = \frac{b^x}{b^y} = b^{x-y}$$

which shows that

$$\log_b \frac{M}{N} = x - y = \log_b M - \log_b N$$

Problem 1 Find.
a. $\log_2 64$ **b.** $\log_2(\frac{1}{28})$
c. $\log_{10} 100$ **d.** $\log_{10} 0.1$

Problem 2 Write the equation $2^{10} = 1024$ in logarithmic form.

Problem 3 Write the equation $\log_4 32 = \frac{5}{2}$ in exponential form and check its accuracy.

Thus the *logarithm of M divided by N is the logarithm of M minus the logarithm of N*. If we have a power of a number such as M^r, then the fact that

$$M^r = (b^x)^r = b^{rx}$$

means that $\log_b M^r = r \log_b M$. In words, the *logarithm of M^r is r times the logarithm of M*. We summarize these results as follows.

PROPERTIES OF LOGARITHMS

$\log_b MN = \log_b M + \log_b N$ Product

$\log_b \dfrac{M}{N} = \log_b M - \log_b N$ Quotient

$\log_b M^r = r \log_b M$ Power

EXAMPLE 4   Show that $\log_b \sqrt{\frac{61}{37}} = \frac{1}{2}(\log_b 61 - \log_b 37)$.

Solution

$$\log_b \sqrt{\frac{61}{37}} = \log_b \left(\frac{61}{37}\right)^{1/2}$$

$$= \frac{1}{2} \log_b \frac{61}{37} \qquad \text{Power property}$$

$$= \frac{1}{2}(\log_b 61 - \log_b 37) \quad \text{Quotient property} \qquad \blacktriangle$$

Problem 4   Show that
$\log_b \sqrt[3]{\frac{4}{5}} = \frac{1}{3}(\log_b 4 - \log_b 5)$.

EXAMPLE 5   Use the properties of logarithms to show that

$$\log\left(x - \frac{1}{x}\right) = \log(x + 1) + \log(x - 1) - \log x$$

where base 10 is understood throughout.

Solution   Since

$$x - \frac{1}{x} = \frac{x^2 - 1}{x} = \frac{(x + 1)(x - 1)}{x}$$

we can apply the product and the quotient properties to write

$$\log\left(x - \frac{1}{x}\right) = \log \frac{(x + 1)(x - 1)}{x}$$

$$= \log(x + 1) + \log(x - 1) - \log x \qquad \blacktriangle$$

Problem 5   As in Example 5, show that

$$\log\left(x^2 + \frac{1}{x}\right) = \log(x + 1)$$
$$+ \log(x^2 - x + 1) - \log x$$

C. APPLICATIONS

EXAMPLE 6   A recent California earthquake was $10^{6.4}$ times as intense as that of a quake of minimum intensity, I_0. What was the magnitude of that quake on the Richter scale?

Solution   We are given that $I = 10^{6.4}I_0$, so the magnitude on the Richter scale is

$$R = \log_{10} \frac{I}{I_0} = \log_{10} 10^{6.4} = 6.4 \qquad \blacktriangle$$

Problem 6   Do Example 6 with $10^{6.4}$ replaced by $10^{7.2}$

EXAMPLE 7 In 1626, Peter Minuit bought the island of Manhattan from the Indians for the equivalent of about $24. If this money were invested at 5% compounded annually, then in 2001 the money would be worth $24(1.05)^{375}$. If it is known that $\log_{10} 24 = 1.3802$ and $\log_{10} 1.05 = 0.0212$, find $\log_{10} 24(1.05)^{375}$.

Solution

$$
\begin{aligned}
\log_{10} 24(1.05)^{375} &= \log_{10} 24 + \log_{10}(1.05)^{375} \\
&= \log_{10} 24 + 375(\log_{10} 1.05) \\
&= 1.3802 + 375(0.0212) \\
&= 1.3802 + 7.9500 \\
&= 9.3302 \quad \text{(log of the accumulated amount)}
\end{aligned}
$$

The accumulated amount here is approximately $2,140,000,000. ▲

Problem 7 Do Example 7 if the rate is changed to 6% if $\log_{10} 1.06 = 0.0253$.

ANSWERS (to problems on pages 866–867)

1. a. 6 **b.** -3 **c.** 2 **d.** -1

2. $10 = \log_2 1024$

3. $4^{5/2} = 32$ $[(\sqrt{4})^5 = 32]$

4.
$$
\begin{aligned}
\log_b \sqrt[3]{\frac{4}{5}} &= \log_b \left(\frac{4}{5}\right)^{1/3} \\
&= \left(\frac{1}{3}\right) \log_b \frac{4}{5} \\
&= \left(\frac{1}{3}\right)(\log_b 4 - \log_b 5)
\end{aligned}
$$

5.
$$
\begin{aligned}
x^2 + \frac{1}{x} &= \frac{x^3 + 1}{x} \\
&= \frac{(x+1)(x^2 - x + 1)}{x}
\end{aligned}
$$
$$
\log\left(x^2 + \frac{1}{x}\right) = \log(x+1) \\
+ \log(x^2 - x + 1) - \log x
$$

6. 7.2

NAME

CLASS

SECTION

ANSWERS

A. In Problems 1–8 find the value of the logarithm.

1. $\log_2 256$

2. $\log_2 128$

3. $\log_3 81$

4. $\log_3 243$

5. $\log_2 \dfrac{1}{8}$

6. $\log_3 \dfrac{1}{27}$

7. $\log_{10} 1,000,000$

8. $\log_{10} 0.001$

In Problems 9–18 write the equation in logarithmic form.

9. $2^7 = 128$

10. $3^4 = 81$

11. $10^3 = 1000$

12. $10^{-3} = 0.001$

13. $81^{1/2} = 9$

14. $16^{1/2} = 4$

15. $216^{1/3} = 6$

16. $64^{1/6} = 2$

17. $N = b^7$

18. $M = a^{-5}$

In Problems 19–24 write the equation in exponential form and check its accuracy.

19. $\log_9 729 = 3$

20. $\log_7 343 = 3$

21. $\log_2 \dfrac{1}{256} = -8$

22. $\log_5 \dfrac{1}{125} = -3$

23. $\log_{81} 27 = \dfrac{3}{4}$

24. $\log_{64} 32 = \dfrac{5}{6}$

B. In Problems 25–34 use the properties of logarithms to transform the left side into the right side of the stated equation. Assume the logarithms are all to base 10.

25. $\log \dfrac{26}{7} - \log \dfrac{15}{63} + \log \dfrac{5}{26} = \log 3$

ANSWERS

1. _____
2. _____
3. _____
4. _____
5. _____
6. _____
7. _____
8. _____
9. _____
10. _____
11. _____
12. _____
13. _____
14. _____
15. _____
16. _____
17. _____
18. _____

19. _____
20. _____
21. _____
22. _____
23. _____
24. _____

25. _____

26. $\log 9 - \log 8 - \log \sqrt{75} + \log \sqrt{\dfrac{25}{27}} = -3 \log 2$

26. _____

27. $\log b^3 + \log 2 - \log \sqrt{b} + \log \dfrac{\sqrt{b^3}}{2} = 4 \log b$

27. _____

28. $\log k^2 - \log k^{-2} - \log \sqrt{k} - \log k^{-1} = \dfrac{9}{2} \log k$

28. _____

29. $\log k^{3/2} + \log r - \log k - \log r^{3/4} = \dfrac{1}{4}(\log k^2 r)$

29. _____

30. $\log a - \dfrac{1}{6} \log b - \dfrac{1}{2} \log a + \dfrac{1}{3} \log b = \dfrac{1}{6} \log a^3 b$

30. _____

31. $\log\left(y - \dfrac{1}{y^2}\right)^3 = 3 \log(y - 1) + 3 \log(y^2 + y + 1) - 6 \log y$

31. _____

32. $\log \dfrac{x^2(x + 5)^{3/2}}{x - 5} = 2 \log x + \dfrac{3}{2} \log(x + 5) - \log(x - 5)$

32. _____

33. $\log \dfrac{(x^2 - 4)\sqrt{x^2 + 2x + 4}}{(x^3 - 8)^2} = \log(x + 2) - \log(x - 2) - \dfrac{3}{2} \log(x^2 + 2x + 4)$

33. _____

34. $\log\left[\dfrac{1}{12(z - 3)^2} - \dfrac{1}{12(z + 3)^2}\right] = \log z - 2 \log(z + 3) - 2 \log(z - 3)$

34. _____

C. Applications

35. The worst earthquake ever recorded occurred in the Pacific Ocean near Colombia. The intensity of this earthquake was $10^{8.9}$ as great as that of an earthquake of minimum intensity I_0. What was the magnitude of this earthquake on the Richter scale?

35. _____

36. The San Francisco earthquake of 1906 was $10^{8.3}$ times as intense as an earthquake of minimum intensity I_0. What was the magnitude of the San Francisco earthquake on the Richter scale?

36. _____

37. When Johnny was born, his father deposited $5000 in an account that paid 6% compounded monthly. This was to help pay Johnny's college expenses starting on his eighteenth birthday. The compound amount in this account was $A = \$5000(1.005)^{216}$. If $\log 5000 = 3.69897$ and $\log 1.005 = 0.00217$, find $\log A$ to five decimal places.

37. _____

38. Suppose that the account in Problem 37 paid 10% compounded quarterly. The compound amount would then be given by $A = \$5000(1.025)^{72}$. If $\log 1.025 = 0.01072$, find $\log A$ to five decimal places.

38. _____

✓ SKILL CHECKER

In Problems 39–46 write the number in scientific notation.

39. 32.68

40. 326.8

39. _____

40. _____

41. 461.2

42. 46,120

43. 0.002387

44. 0.0004392

45. 0.0000569

46. 0.000006731

41. _____

42. _____

43. _____

44. _____

45. _____

46. _____

11.2 USING YOUR KNOWLEDGE

One way of comparing the compound amounts in Problems 37 and 38 is to find their ratio. If we denote this ratio by R, then

$$R = \frac{5000(1.005)^{216}}{5000(1.025)^{72}} = \frac{(1.005)^{216}}{(1.025)^{72}}$$

so that

$$\log R = \log(1.005)^{216} - \log(1.025)^{72}$$
$$= 216 \log 1.005 - 72 \log 1.025$$

Using the values of the logarithms given in Problems 37 and 38, we get

$$\log R = (216)(0.00217) - (72)(0.01072) = -0.30312$$

The negative value for $\log R$ means that R is less than 1. Can you see why? This means that the compound amount in Problem 37 is less than that in Problem 38. (A calculator gives the value of R as 0.49760, so that the first amount is less than half the second!)

1. Compare an investment of $1000 at 8% compounded quarterly with the same amount invested at 8.5% compounded annually for a period of 10 yr. You will need the values log 1.02 = 0.00860 and log 1.085 = 0.03543.

2. Compare an investment of $1000 at 6% compounded quarterly with the same amount invested at 6.25% compounded annually for a period of 10 yr. You will need the values log 1.015 = 0.00647 and log 1.0625 = 0.02633.

1. _____

2. _____

ANSWER (to problem on page 868)

7. 10.8677

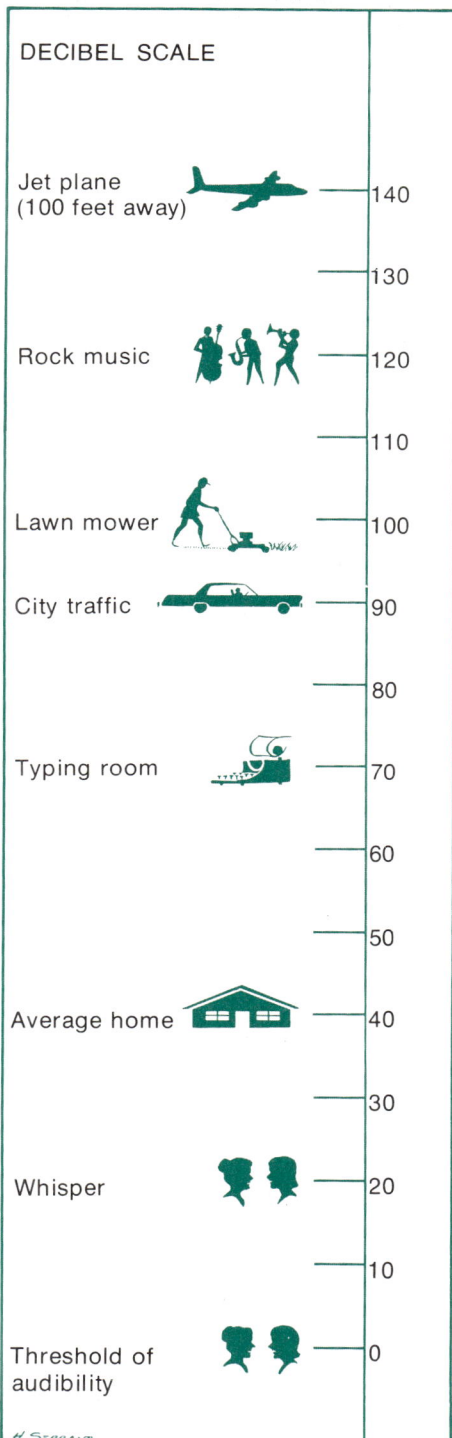

DECIBEL SCALE

Jet plane (100 feet away) —— 140

—— 130

Rock music —— 120

—— 110

Lawn mower —— 100

City traffic —— 90

—— 80

Typing room —— 70

—— 60

—— 50

Average home —— 40

—— 30

Whisper —— 20

—— 10

Threshold of audibility —— 0

H. SERRANO

A decibel is a unit based on the faintest sound a person can hear. The scale is logarithmic, so that a sound 10 times as intense as another is 1 bel louder.

REVIEW

Before starting this section, you should know how to:

1. Write a number in scientific notation.
2. Transform a radical expression into exponential form.

OBJECTIVES

You should be able to:

A. Use a table to find logarithms and antilogarithms.
B. Compute with common logarithms.
C. Solve applications using these ideas.

The chart at the beginning of this section shows a scale for measuring the loudness of sounds. This scale is called the decibel scale. The loudness L of a sound of intensity I measured in decibels (dB) is expressed as

$$L = 10 \log_{10} \frac{I}{I_0}$$

where I_0 is the minimum intensity detectable by the human ear. For example, the sound of a riveting machine 30 ft away is 10^{10} times as intense as the minimum intensity I_0, and hence its loudness in decibels is expressed as

$$L = 10 \log_{10} \frac{10^{10} I_0}{I_0} = 10 \log_{10} 10^{10} = (10 \cdot 10) \log_{10} 10$$

$$= 100 \text{ dB}$$

A. COMMON LOGARITHMS

The decibel equation is just an example of the many applications of logarithms that use the base 10. Logarithms to base 10 are called **common logarithms;** it is customary to omit the base when working with these logarithms. Be sure to keep in mind that when the base is omitted,

$$\log M = \log_{10} M$$

Recall that any positive real number M can be written in scientific notation as

$$M = N \times 10^p$$

where $1 \leq N < 10$ and p is an integer. If we use this form, then we have

$$\log M = \log(N \times 10^p) = \log N + \log 10^p$$

Since $\log 10^p = p \log 10$ and $\log 10 = 1$ (because $10 = 10^1$), we have the important result that

$$\log M = p + \log N$$

For example, if $M = 396.7 = 3.967 \times 10^2$, then $\log 396.7 = \log 3.967 + 2 = 2 + \log 3.967$. If $M = 0.00259 = 2.59 \times 10^{-3}$, then $\log 0.00259 = \log 2.59 - 3 = -3 + \log 2.59$. Thus the logarithm of any positive real number can be written as the sum of an integer and the logarithm of a number between 1 and 10. Consequently, a table of common logarithms can be restricted to the logarithms of numbers between 1 and 10. Because $\log 1 = 0$ and $\log 10 = 1$, the logarithms in the table will all be decimals between 0 and 1. The table in Appendix 1 is such a table; it lists four-place values of the logarithms of the numbers from 1.00 to 9.99 at intervals of 0.01. To read the logarithm of one of these numbers, such as 3.16, we go down the first column to the entry 3.1 and then across to the column under 6. This gives

$$\log 3.16 = 0.4997$$

Now we can use our general result to write

$$\log 3160 = \log (10^3 \times 3.16)$$
$$= 3 + \log 3.16 = 3.4997$$
$$\log 0.00316 = \log (10^{-3} \times 3.16)$$
$$= -3 + \log 3.16 = -3 + 0.4997$$

When the integer part of the logarithm is negative, it is usually not combined with the decimal part but instead is often written in the so-called -10 notation by adding and subtracting 10. In this notation,

$$\log 0.00316 = 7.4997 - 10$$

The integer part of the logarithm is called the **characteristic** and the decimal part is called the **mantissa.** Thus, log 0.00316 has characteristic -3 (or $7 - 10$) and mantissa 0.4997.

Note that the characteristic is the exponent of 10 when the number is written in scientific notation. Thus *for any number with a whole-number part, the characteristic is 1 less than the number of digits to the left of the decimal point. For a pure decimal, the characteristic is the negative of 1 more than the number of zeros before the first significant digit.* For example,

log 3592.5 has characteristic 3

    ↓

   (4 digits to left of point)

log 0.00947 has characteristic -3 or $(7 - 10)$

    ↓

   (2 zeros between point and digit 9)

EXAMPLE 1 Use the table to find

a. log 42,500   **b.** log 0.000425

Solution

a. The characteristic here is 4 (1 less than the number of digits to the left of the decimal point). The mantissa is read from the table as log 4.25 = 0.6284. Therefore, log 42,500 = 4.6284.

b. The characteristic here is -4 (the negative of 1 more than the number of zeros between the decimal point and the first significant digit). The mantissa is the same as in part a, 0.6284. Thus

log 0.000425 = $-4 + 0.6284 = 6.6284 - 10$ ▲

EXAMPLE 2 Find log 6.257.

Solution For logarithms of numbers such as 6.257, with more than two decimal places, we **interpolate** in the table as follows:

Difference   Difference

$$10\left\{7\left\{\begin{array}{l}\log 6.250 = 0.7959\\ \log 6.257 = \quad ?\\ \log 6.260 = 0.7966\end{array}\right\}0.0007\right.$$

Since 6.257 is 0.7 of the way from 6.250 to 6.260, we take 0.7 of the difference between the two logarithms, round to four places, and add the result to log 6.250.

$(0.7)(0.0007) = 0.00049 \approx 0.0005$

Thus log 6.257 = 0.7959 + 0.0005 = 0.7964. You can check this by referring to a more accurate table or by using a calculator with a log key. The calculator gives

log 6.257 = 0.7963662

which rounds to the answer we obtained. (Since the table entries are rounded to four places and there is rounding in the interpolation, you must not expect to get perfect agreement every time.)   ▲

Problem 1 Use the table to find
a. log 734,000   **b.** 0.0000734

Problem 2 Find log 3.746.

The number that corresponds to a given logarithm is usually referred to as the **antilogarithm.** Thus if

$$\log x = 2.5105$$

then

$$x = \text{antilog } 2.5105$$

To find x, we refer to the table and go down the first column of mantissas to the greatest number that is less than 0.5105. This is the number 0.5051 in the row labeled 3.2. Then we go across this row until we come to 0.5105, which occurs in the column labeled 4. This means that 0.5051 is the logarithm of 3.24. Since the characteristic of $\log x$ is 2, we see that

$$x = \text{antilog } 2.5105 = 324$$

EXAMPLE 3 Find antilog $7.8176 - 10$.

Solution Here, the mantissa is 0.8176, so we go down the first column of mantissas to 0.8129, the largest entry less than 0.8176. This is the row labeled 6.5. We go across this row to the entry 0.8176, which is in the column labeled 7. This means that $\log 6.57 = 0.8176$. Because the given characteristic is $7 - 10 = -3$, we know that

$$\text{antilog}(7.8176 - 10) = 0.00657 \qquad \blacktriangle$$

If the given mantissa is not in the table, we interpolate to find the antilogarithm, as in the next example.

EXAMPLE 4 Find antilog 3.8179.

Solution Since the mantissa is 0.8179, we go down the first column of mantissas to 0.8129 (the largest entry less than 0.8179). This is the row labeled 6.5. We read across the row to 0.8176 (the largest entry less than 0.8129). The next entry is 0.8182, so we interpolate as follows:

Difference   Difference

$$6\left\{3\begin{cases}0.8176 = \log 6.57\\0.8179 = \log x\\0.8182 = \log 6.58\end{cases}\right\}\ 0.01$$

Since $\frac{3}{6} = \frac{1}{2}$, we take $\frac{1}{2}$ of the difference, 0.01, between the antilogarithms, which gives 0.005 to be added to the smaller number. Thus we obtain antilog $0.8179 = 6.575$. The characteristic of the given logarithm is 3, so that the final result is

$$\text{antilog } 3.8179 = 6575$$

(You can check this on a calculator by keying in the number 3.8179 and then pressing the $\boxed{\text{INV}}$ and the $\boxed{\log}$ keys. The calculator shows the answer 6575.0642, which agrees to four significant digits with our answer.) $\qquad \blacktriangle$

B. CALCULATING WITH LOGARITHMS

We can use tables and the properties of logarithms to do many different kinds of calculations. This idea is illustrated in the next examples.

Problem 3 Find antilog $6.7059 - 10$.

Problem 4 Find antilog 2.7814.

EXAMPLE 5 Use logarithms to calculate.

a. $25^{2.56}$ **b.** $\sqrt[3]{16.4}$ **c.** $(0.147)^{-2.6}$

Solution

a. $\log 25^{2.56} = 2.56 \log 25$

$\qquad\qquad\quad = 2.56(1.3979)$

$\qquad\qquad\quad = 3.5786$

From the table, we find antilog $0.5786 = 3.79$. Thus

$25^{2.56} = $ antilog $3.5786 = 3790$

b. Since $\sqrt[3]{16.4} = (16.4)^{1/3}$,

$\log \sqrt[3]{16.4} = \dfrac{1}{3} \log 16.4$

$\qquad\qquad\quad = \dfrac{1}{3}(1.2148)$

$\qquad\qquad\quad = 0.4049$

Therefore,

$\sqrt[3]{16.4} = $ antilog 0.4049

$\qquad\quad = 2.541$

by interpolation in the table.

c. $\log(0.147)^{-2.6} = -2.6 \log 0.147$

$\qquad\qquad\qquad = -2.6(9.1673 - 10)$

$\qquad\qquad\qquad = -23.83498 + 26$

$\qquad\qquad\qquad = 2.1650 \qquad$ (to four places)

Thus

$(0.147)^{-2.6} = $ antilog 2.1650

$\qquad\qquad\quad = 146.3$

by interpolation in the table. ▲

EXAMPLE 6 Use logarithms to calculate $\sqrt[4]{\dfrac{(25.65)(139.7)}{16.38}}$.

Solution Since $\sqrt[4]{N} = N^{1/4}$, $\log \sqrt[4]{N} = \frac{1}{4} \log N$. So if the final answer is called x, then

$\log x = \dfrac{1}{4}(\log 25.65 + \log 139.7 - \log 16.38)$

We organize the calculation as follows:

$\qquad\quad \log 25.65 = 1.4091$

$\underline{(+) \log 139.7 = 2.1452}$

$\qquad\qquad\qquad\quad 3.5543$

$\underline{(-) \log 16.38 = 1.2143}$

$\qquad\qquad\quad 4 \overline{)\,2.3400}$

$\qquad\quad \log x = 0.5850$

$\qquad\qquad\quad x = 3.846$

(to four significant digits). ▲

Problem 5 Use logarithms to calculate.

a. $35^{1.25}$ **b.** $\sqrt[5]{16.4}$

c. $(0.024)^{-2.7}$

Problem 6 Use logarithms to calculate $\sqrt[5]{\dfrac{(47.25)(56.43)}{22.76}}$.

C. APPLICATIONS

EXAMPLE 7 Find the approximate compound amount at the end of 5 yr if $500 is invested at 5% compounded quarterly.

Solution We know that if a principal P earns interest for n periods at the rate r per period, the compound amount A is given by

$$A = P(1 + r)^n$$

Therefore,

$$\log A = \log P + n \log(1 + r)$$

In this problem, $P = 500$, $n = 20$ (four periods per year for 5 yr), and $r = \dfrac{0.05}{4} = 0.0125$, so

$$
\begin{aligned}
\log A &= \log 500 + 20 \log(1.0125) \\
&= 2.6990 + 20(0.0054) \\
&= 2.6990 + 0.1080 \\
&= 2.8070 \\
A &= \text{antilog } 2.8070 = 641.2
\end{aligned}
$$

Thus, the compound amount is approximately $641.20. (A calculator gives $641.02 to the nearest cent, so we are off by 18¢; but don't forget that we are using only four-place logarithms!) ▲

EXAMPLE 8 In chemistry, the PH (hydrogen potential) of a solution is defined by the formula PH $= -\log[H^+]$, where $[H^+]$ is the hydrogen ion concentration of the solution in moles per liter. Find the PH of a solution for which $[H^+] = 8 \times 10^{-6}$.

Solution The PH is defined to be $-\log[H^+]$, so for this solution

$$
\begin{aligned}
\text{PH} &= -\log(8 \times 10^{-6}) \\
&= -(\log 8 + \log 10^{-6}) \\
&= -(\log 8 - 6) \\
&= 6 - \log 8 \\
&= 6 - 0.9031 \\
&= 5.0969
\end{aligned}
$$

▲

Problem 7 Do Example 7 if the time is changed to 10 yr.

Problem 8 Find the PH of a solution for which $[H^+] = 4.2 \times 10^{-7}$

A. In Problems 1–12 use the table (Appendix 1) to find the logarithm of the given number.

1. 74.5

2. 952

3. 1840

4. 3.05

5. 0.0437

6. 0.0673

7. 50.18

8. 94.44

9. 0.01238

10. 0.01004

11. 0.008606

12. 0.0004632

In Problems 13–20 find the indicated antilogarithm.

13. antilog 1.2672

14. antilog 2.4409

15. antilog(7.7672 − 10)

16. antilog(6.5955 − 10)

17. antilog 1.4630

18. antilog 2.9408

19. antilog(9.8660 − 10)

20. antilog(8.7725 − 10)

B. In Problems 21–30 use logarithms to find the value to four significant digits.

21. $2.78 \times 56.9 \times 97.3$

22. $43.9 \times 1.87 \times 108$

23. $\dfrac{56.23 \times 0.002962}{3.589}$

24. $\dfrac{0.0002691 \times 376.5}{0.02948}$

25. $\sqrt{\dfrac{29.12}{5.931}}$

26. $\sqrt{\dfrac{2.759}{42.38}}$

27. $\sqrt[3]{21.83 \times 596.2}$

28. $\sqrt[4]{3212 \times 0.5926}$

29. $\sqrt{\log 4.12}$

30. $\sqrt{\log 31.5}$

27. _____

28. _____

29. _____

30. _____

C. In Problems 31–34 find the approximate compound amount for the principal, compounded annually at the given rate for the given time.

	PRINCIPAL	RATE	TIME
31.	$100	7%	10 yr
32.	$1000	6%	20 yr
33.	$500	5%	30 yr
34.	$200	10%	10 yr

In Problems 35–40 find the PH of a solution with the given $[H^+]$.

35. $[H^+] = 7 \times 10^{-7}$

36. $[H^+] = 1.5 \times 10^{-9}$

37. Eggs whose $[H^+]$ is 1.6×10^{-8}

38. Tomatoes whose $[H^+]$ is 6.3×10^{-5}

39. Milk whose $[H^+]$ is 4×10^{-7}

40. $[H^+] = 5 \times 10^{-8}$

31. _____

32. _____

33. _____

34. _____

35. _____

36. _____

37. _____

38. _____

39. _____

40. _____

✓ **SKILL CHECKER**

In Problems 41–44, solve for k.

41. $\log 3 = k \log 2$

42. $\log 25 = 0.15k$

43. $25 = 10^k$

44. $100 = 5^k$

41. _____

42. _____

43. _____

44. _____

11.3   USING YOUR KNOWLEDGE

You can use your knowledge of common logarithms to compare investments of money at compound interest as follows: Suppose $1000 can be invested at 10.25% compounded annually or at 10% compounded quarterly. Which is the better investment? To make the comparison, we use the formula $A = P(1 + r)^n$ for the compound amount of P dollars at the rate r per period for n periods. For n years, the 10.25% investment would amount to $\$1000(1.1025)^n$ and the 10% investment would amount to $\$1000(1.025)^{4n}$. (Keep in mind that 10% compounded quarterly means that $r = 0.025$ is the quarterly rate.) Thus we need to compare $(1.1025)^n$ with $(1.025)^{4n} = [(1.025)^4]^n$. Logarithms can be used to find the value of $(1.025)^4$. If you use the table in Appendix 1, you will find that $(1.025)^4 = 1.1036$. This shows that the 10% compounded quarterly is a much better deal.

ANSWERS   (to problems on page 878)
7. $822.20   **8.** 6.3768

Use logarithms to compare the following investments.

1. 6.25% compounded annually and 6% compounded monthly.

2. 9.25% compounded annually and 9% compounded monthly.

3. Find the nominal rate that compounded monthly is equivalent to 8% compounded annually. (*Hint:* You will first have to solve the equation $(1 + r)^{12} = 1.08$, which you can do with logarithms.)

4. Find the nominal rate that compounded quarterly is equivalent to 8% compounded annually. (See the hint in Problem 3.)

1. _____

2. _____

3. _____

4. _____

11.4 EXPONENTIAL EQUATIONS AND NATURAL LOGARITHMS

OBJECTIVES

REVIEW

Before starting this section you should know:

1. The basic laws of exponents.
2. The corresponding properties of logarithms.

OBJECTIVES

You should be able to:

A. Use natural logarithms.
B. Calculate natural logarithms using the change-of-base formula.
C. Solve applications of logarithms and exponential functions.

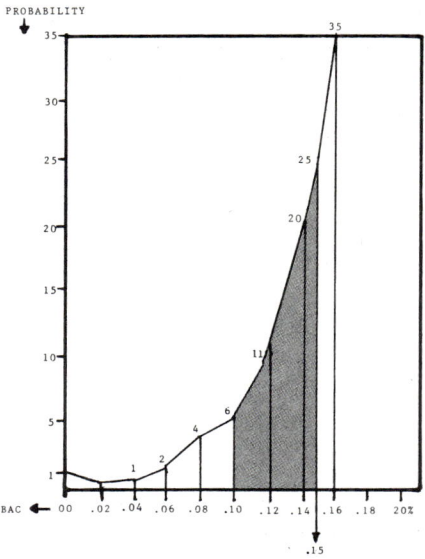

Indiana University Institute for Research in Public Safety.

The chart at the beginning of this section shows the probability of having an accident as a function of blood alcohol level (BAC). The exact formula relating the probability P (in percent) of having an accident when the percent of alcohol in the blood is b is expressed as

$$P(b) = e^{kb} \qquad (1)$$

As you can see from the chart, this probability is 25% when the alcohol level in the blood is 0.15%. Can we find k? To do this, we let $P(b) = 25$ and $b = 0.15$ in Equation 1. We then have

$$25 = e^{0.15k} \qquad (2)$$

Equation 2 is an example of an *exponential* equation.

> An **exponential equation** is an equation in which a variable occurs in an exponent.

In order to solve Equation 2, we make use of a fundamental property of logarithms.

> If M, N, and b are all positive numbers and $b \neq 1$, then
>
> $\log_b M = \log_b N$ *if and only if* $M = N$

This statement means that

if $\log_b M = \log_b N$, then $M = N$

and, conversely,

if $M = N$, then $\log_b M = \log_b N$

A. NATURAL LOGARITHMS

We discussed the number $e = 2.718282$ in Section 11.1A. This number is used as the base of an important system of logarithms called **natural logarithms.**

It is customary to use the abbreviation ln (pronounced *el-en*) for $\log_e$:

$$\ln x = \log_e x \qquad \text{for all } x > 0$$

We are now ready to solve Equation 2. Since

$$25 = e^{0.15k}$$
$$\ln 25 = \ln e^{0.15k} = 0.15k \ln e.$$

Now $\ln e = \log_e e = 1$ because $e = e^1$. Thus the equation becomes

$$\ln 25 = 0.15k$$

Solving for k gives

$$k = \frac{\ln 25}{0.15}$$

The value of ln 25 can be found in the table of natural logarithms in Appendix 2 or by using a calculator with an $\boxed{\ln}$ key. To four decimal places, $\ln 25 = 3.2189$. With this value, we find

$$k = \frac{3.2189}{0.15} = 21.459$$
$$= 21.5 \qquad \text{(to the nearest tenth)}$$

EXAMPLE 1 If we substitute $k = 21.5$ into Equation 2, the formula for $P(b)$ becomes

$$P(b) = e^{21.5b}$$

At what blood alcohol level b will the probability of having an accident be 100%?

Solution We want to find b so that $P(b) = 100$. To do this we must solve the equation

$$100 = e^{21.5b}$$

First, we take natural logarithms to get

$$\ln 100 = \ln e^{21.5b}$$
$$\qquad = 21.5b \ln e \qquad \text{Use the power property.}$$
$$\qquad = 21.5b \qquad \ln e = 1$$
$$b = \frac{\ln 100}{21.5} \qquad \text{Solve for } b.$$
$$\qquad = \frac{2 \ln 10}{21.5} \qquad 100 = 10^2.$$

Problem 1 At what blood alcohol level will the probability of having an accident be 50%?

The value of ln 10 can be found in the table (Appendix 2) or with a calculator. To five decimal places, ln 10 = 2.30259. Thus

$$b = \frac{4.60518}{21.5} = 0.214$$

Thus when the blood alcohol level is about 0.21%, the probability of an accident is 100%. Of course, if your blood alcohol level is 0.21%, you are probably not sober enough to drive!! ▲

B. CHANGE OF BASE

Although any positive number other than 1 can be used as the base of a system of logarithms, 10 and e are the only bases that are important in applications. So we limit our study to the relationship between common and natural logarithms. It is possible to use a table of common logarithms to calculate the corresponding natural logarithms.

Suppose a positive number M has a common logarithm c and a natural logarithm n. That is, $\log M = c$, so that $M = 10^c$ and $\ln M = n$ so that $M = e^n$. Then, we must have

$$e^n = 10^c$$

If we take common logarithms of both sides, the result is

$$n \log e = c \log 10 = c$$

$$n = \frac{c}{\log e}$$

Replacing c by $\log M$ and n by $\ln M$, we get the change-of-base formula.

$$\ln M = \frac{\log M}{\log e} \tag{1}$$

A five-place table of common logarithms or a calculator will give the value $\log e = 0.43429$, so that we have the following approximate formula.

$$\ln M = \frac{\log M}{0.43429} \tag{2}$$

EXAMPLE 2 Use Equation 2 to find
a. ln 100   **b.** ln 0.253

Solution

a. Equation 2 with $M = 100$ becomes

$$\ln 100 = \frac{\log 100}{0.43429}$$

$$= \frac{2}{0.43429} = 4.6052$$

Problem 2 Use Equation 2 to find
a. ln 10   **b.** ln 0.035

b. For $M = 0.253$, equation 2 becomes

$$\ln 0.253 = \frac{\log 0.253}{0.43429}$$

$$= \frac{9.4031 - 10}{0.43429} \quad \text{Use the table in Appendix 1.}$$

$$= -\frac{0.5969}{0.43429} = -1.374 \quad \text{Round to four significant digits.}$$

(You can check these results on a calculator that has the $\boxed{\ln}$ key. If you key in 0.253 and press the $\boxed{\ln}$ key, the calculator shows -1.3743658.)

▲

C. APPLICATIONS

Exponential and logarithmic equations have many applications in such areas as business, engineering, social science, psychology, and science.

EXAMPLE 3 In Section 11.1, we learned that with continuous compounding, a principal of P dollars accumulates to an amount A given by the equation

$$A = Pe^{rt}$$

where r is the rate and t is the time in years. If the interest rate is 6%, how long does it take for a sum of money to double?

Solution With $A = 2P$ and $r = 0.06$, the equation becomes

$$2P = Pe^{0.06t}$$

or

$$2 = e^{0.06t} \quad \text{Divide by } P.$$

We want to solve this equation for t, so we take natural logarithms of both sides:

$$\ln 2 = \ln e^{0.06t} = 0.06t \ln e = 0.06t$$

Thus

$$t = \frac{\ln 2}{0.06}$$

Using the table (Appendix 2) or a calculator, we find $\ln 2 = 0.69315$, so that

$$t = \frac{0.69315}{0.06} = 11.6$$

This means that it takes about 11.6 yr for the money to double.   ▲

EXAMPLE 4 In 1984 the population of the world was about 4.8 billion and the yearly growth rate was 2%. The equation giving the population P in terms of the time t is

$$P = 4.8e^{0.02t}$$

Estimate the population in 1994.

Problem 3 Do Example 3 if the rate is 8%.

Problem 4 Use the data in Example 4 to estimate the population in the year 2000.

Solution Since $P = 4.8$ for $t = 0$, the equation shows that t is measured from the year 1984. To estimate the population in 1994, we put $t = 10$ in the given equation:

$$P = 4.8e^{(0.02)(10)} = 4.8e^{0.2}$$

The value $e^{0.2} = 1.2214$ can be found in the table (Appendix 3), so

$$P = (4.8)(1.2214) = 5.9$$

Thus our estimate for the population in 1994 is about 5.9 billion. ▲

EXAMPLE 5 If B is the number of bacteria present in a laboratory culture after t minutes, then, under ideal conditions,

$$B = Ke^{0.05t}$$

If the initial number of bacteria is 1000, find how long it would take for there to be 50,000 bacteria present.

Solution Since $B = 1000$ for $t = 0$, we have

$$1000 = Ke^0 = K$$

The equation for B is then

$$B = 1000e^{0.05t}$$

Now let $B = 50,000$:

$$50,000 = 1000e^{0.05t}$$
$$50 = e^{0.05t}$$

To solve for t, take natural logarithms of both sides:

$$\ln 50 = \ln e^{0.05t}$$
$$\ln 50 = 0.05t \qquad (\ln e = 1)$$

Thus

$$t = \frac{\ln 50}{0.05} = \frac{3.9120}{0.05} = 78.2 \text{ min}$$ ▲

EXAMPLE 6 The element cesium 137 decays at the rate of 2.3% per year. Find the half-life of this element.

Solution The half-life of a substance is found by using the equation

$$A(t) = A_0 e^{-kt}$$

where $A(t)$ is the amount present at time t (years), k is the decay rate, and A_0 is the initial amount of the substance present. In this problem,

$$k = 2.3\% = 0.023, \qquad A(t) = \frac{1}{2}A_0$$

and we want to find t. With this information, the basic equation becomes

$$\frac{1}{2}A_0 = A_0 e^{-0.023t}$$

$$\frac{1}{2} = e^{-0.023t} \qquad \text{Divide by } A_0.$$

Problem 5 Use the data in Example 5 to find how long it would take for there to be 100,000 bacteria present.

Problem 6 Find the half-life of an element that decays at the rate of 4% per year.

Now take natural logarithms of both sides:

$$\ln \frac{1}{2} = \ln e^{-0.023t}$$

$$-\ln 2 = -0.023t$$

$$t = \frac{\ln 2}{0.023}$$

$$= \frac{0.69315}{0.023} = 30.14$$

Thus the half-life of cesium 137 is about 30.14 yr.

NAME

CLASS

SECTION

A. In Problems 1–6 use the procedure of Example 1 to find the blood alcohol level at which the probability of having an accident is as given.

1. 60%   **2.** 70%

3. 75%   **4.** 80%

5. 90%   **6.** 95%

B. In Problems 7–12 use common logarithms and the change-of-base formula to find each of the following to the nearest four significant digits.

7. ln 7.14   **8.** ln 8.26

9. ln 49.6   **10.** ln 98.5

11. ln 0.271   **12.** ln 0.0493

C. In Problems 13–16 assume continuous compounding and follow the procedure in Example 3 to find how long it takes a given amount to double at the given interest rate.

13. Rate = 5%   **14.** Rate = 7%

15. Rate = 6.5%   **16.** Rate = 7.5%

17. Suppose the population of the world grows at the rate of 1.5% and that the population in 1984 was about 4.8 billion. Follow the procedure of Example 4 to estimate the population in 1994.

18. Repeat Problem 17 if the growth rate is 1.75%.

In Problems 19–22 assume that the number of bacteria present in a culture after t minutes is given by $B = 1000e^{0.04t}$. Find the time it takes for the number of bacteria present to be

19. 2000   **20.** 5000

1. _____
2. _____
3. _____
4. _____
5. _____
6. _____

7. _____
8. _____
9. _____
10. _____
11. _____
12. _____

13. _____
14. _____
15. _____
16. _____
17. _____

18. _____

19. _____
20. _____

21. 25,000   **22.** 50,000

21. _____

22. _____

23. When a bacteria-killing solution is introduced into a certain culture, the number of live bacteria is given by the equation $B = 100,000e^{-0.2t}$, where t is the time in hours. Find the number of live bacteria present at the following times.
   a. $t = 0$
   b. $t = 2$
   c. $t = 10$
   d. $t = 20$

23. a. _____
   b. _____
   c. _____
   d. _____

24. The number of honey bees in a hive is growing according to the equation $N = N_0e^{0.015t}$, where t is the time in days. If the bees swarm when their number is tripled, find how many days till this hive swarms.

24. _____

In Problems 25–28 follow the procedure of Example 6 to find the half-life of the given substance.

25. Plutonium, whose decay rate is 0.003% per year

25. _____

26. Krypton, whose decay rate is 6.3% per year

26. _____

27. A radioactive substance whose decay rate is 5.2% per year

27. _____

28. A radioactive substance whose decay rate is 0.2% per year

28. _____

29. The atmospheric pressure P in pounds per square inch at an altitude of h feet above the earth is given by the equation $P = 14.7e^{-0.00005h}$. Find the pressure at an altitude of
   a. 0 ft
   b. 5000 ft
   c. 10,000 ft

29. a. _____
   b. _____
   c. _____

30. If the atmospheric pressure in Problem 29 is measured in inches of mercury, then $P = 30e^{-0.207h}$, where h is the altitude in miles. Find the pressure
   a. At sea level
   b. At 5 mi above sea level

30. a. _____
   b. _____

11.4 USING YOUR KNOWLEDGE

In Using Your Knowledge 11.2, you saw how to use common logarithms to compare regular compound interest investments. Now, with your knowledge of natural logarithms and exponential equations, you can compare continuous with periodic compounding.

In Section 11.1B, you learned that the formula for continuous compounding is

$$A = Pe^{rt}$$

where A is the compound amount, P is the principal, r is the rate, and t is the time in years. Suppose you want to compare the amount at 6% compounded monthly for 2 yr with the amount under continuous compounding for the same period. For monthly compounding, the amount—for instance, say A_1—would be

$$A_1 = P(1.005)^{24}$$

For continuous compounding the amount—say A_2—would be

$A_2 = Pe^{(0.06)(2)} = Pe^{0.12}$

The table in Appendix 3 gives $e^{0.12} = 1.1275$. You can find the value of $(1.005)^{24}$ by using common logarithms. Thus

$\log(1.005)^{24} = 24 \log 1.005 = (24)(0.00215) = 0.0516$

and

antilog $0.0516 = 1.1262$.

For a principal of \$1000, $A_1 \approx \$1126.20$ and $A_2 \approx \$1127.50$. This shows that the continuous compounding is a little better than the monthly compounding. A greater difference would occur over a longer time.

1. Compare continuous compounding at 9% with monthly compounding at the same annual rate for 5 yr for $P = \$1000$.

2. Compare continuous compounding at 12% with monthly compounding at the same annual rate for 5 yr for $P = \$1000$.

1. _____

2. _____

SUMMARY

SECTION	ITEM	MEANING	EXAMPLE
11.1	Exponential function	A function defined for all real values of x by $f(x) = b^x$, $b > 0$, $b \neq 1$	$f(x) = 2^x$
11.1A	Increasing function	A function whose graph goes up to the right	$f(x) = 2^x$
11.1A	Decreasing function	A function whose graph goes down to the right	$f(x) = 2^{-x}$
11.2A	Logarithm	$\log_b x = y$ means $x = b^y$.	$\log_2 8 = 3$ because $2^3 = 8$
11.2B	Logarithm of a product	$\log_b MN = \log_b M + \log_b N$	$\log_2(4 \times 8) = \log_2 4 + \log_2 8$
11.2B	Logarithm of a quotient	$\log_b \dfrac{M}{N} = \log_b M - \log_b N$	$\log_2 \dfrac{4}{8} = \log_2 4 - \log_2 8$
11.2B	Logarithm of a power	$\log_b M^r = r \log_b M$	$\log_2 4^3 = 3 \log_2 4$
11.3A	Common logarithm	The logarithm to base 10	$\log 10 = 1$, $\log 100 = 2$
11.3A	Characteristic	The integer part of the common log	$\log 200 = 2.3010$ has characteristic 2. $\log 0.002 = 7.3010 - 10$ has characteristic $7 - 10$, or -3.
11.3A	Mantissa	The decimal part of the common log	Both $\log 2$ and $\log 0.002$ have mantissa 0.3010.
11.3A	Antilogarithm	The number that corresponds to a given logarithm	antilog $2.3010 = 200$
11.4	Exponential equation	An equation in which the variable occurs in an exponent	$10^{2x} = 5$
11.4A	Natural logarithms	Logarithms to base e	$\ln 2 = 0.69315$
11.4B	Change-of-base formula	$\ln M = \dfrac{\log M}{\log e}$	$\ln 2 = \dfrac{\log 2}{\log e} = \dfrac{0.30103}{0.43429}$ $= 0.69315$

(If you need help with these exercises, look in the section indicated in brackets.)

1. [11.1A] For $-6 \le x \le 6$, graph the function:
   a. $f(x) = 2^{x/2}$ **b.** $f(x) = 2^{-x/2}$

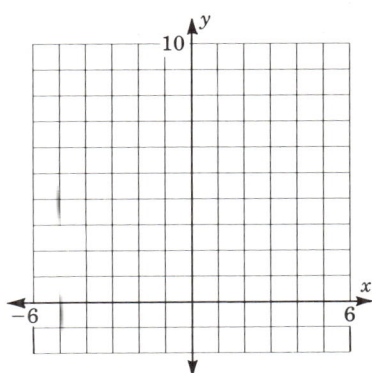 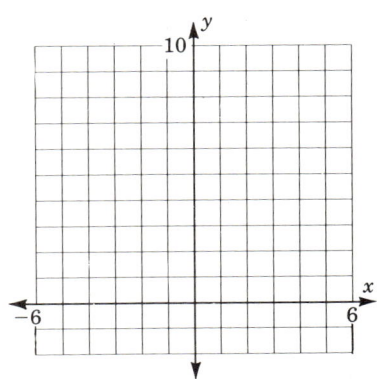

2. [11.1A] For $-6 \le x \le 6$, graph the function.

   a. $g(x) = \left(\dfrac{1}{2}\right)^{x/2}$ **b.** $g(x) = \left(\dfrac{1}{2}\right)^{-x/2}$

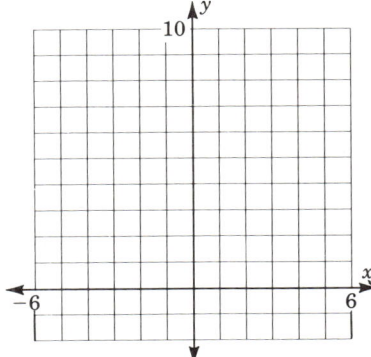 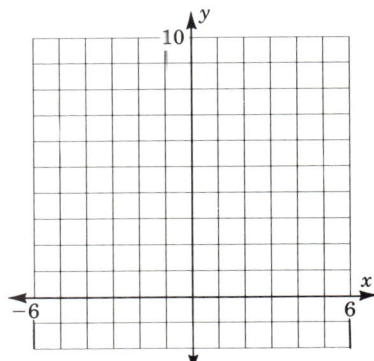

3. [11.2A] Find.
   a. $\log_2 4$
   b. $\log_2 8$

4. [11.2A] Find.

   a. $\log_2 \dfrac{1}{4}$

   b. $\log_2 \dfrac{1}{8}$

3. a. _____
   b. _____

4. a. _____
   b. _____

5. [11.2A] Write in logarithmic form.
   a. $2^6 = 64$
   b. $3^4 = 81$

5. a. _____
   b. _____

6. [11.2A] Write in exponential form.

   a. $\log_{16} 8 = \dfrac{3}{4}$

   b. $\log_{27} \dfrac{1}{9} = -\dfrac{2}{3}$

6. a. _____
   b. _____

7. [11.2B] Show that

   a. $\log_b \sqrt{\dfrac{37}{91}} = \dfrac{1}{2}(\log_b 37 - \log_b 91)$

   b. $\log_b \sqrt[3]{\dfrac{23 \times 41}{59}} = \dfrac{1}{3}(\log_b 23 + \log_b 41 - \log_b 59)$

7. a. _____
   b. _____

8. [11.2B] Show that

   a. $\log\left(1 - \dfrac{1}{x^2}\right) = \log(x - 1) + \log(x + 1) - 2 \log x$

   b. $\log\left(\dfrac{1}{x^4} - 1\right) = \log(1 - x) + \log(1 + x) + \log(1 + x^2) - 4 \log x$

8. a. _____
   b. _____

9. [11.2C] If \$1000 is invested at 12% compounded monthly for n years, the compound amount is given by $A = 1000(1.01)^{12n}$. Knowing that $\log 1.01 = 0.0043$, find $\log A$ for the given value of n.
   a. $n = 4$
   b. $n = 10$

9. a. _____
   b. _____

Use the table of common logarithms (Appendix 1) in Problems 10–20.

10. [11.3A] Find.
   a. $\log 975$
   b. $\log 837$

10. a. _____
    b. _____

11. [11.3A] Find.
   a. $\log 0.00759$
   b. $\log 0.000648$

11. a. _____
    b. _____

12. [11.3A] Use interpolation to find.
   a. $\log 975.3$
   b. $\log 83.72$

12. a. _____
    b. _____

13. [11.3A] Use interpolation to find.
   a. $\log 0.07955$
   b. $\log 0.6489$

13. a. _____
    b. _____

14. [11.3A] Find.
   a. antilog 2.8215
   b. antilog $6.6096 - 10$

14. a. _____
    b. _____

15. [11.3A] Use interpolation to find.
   a. antilog 3.8484
   b. antilog $7.3191 - 10$

15. a. _____
    b. _____

16. [11.3B] Find to four significant digits.
   a. $(5.02)^{2.3}$
   b. $(5.02)^{-3.2}$

16. a. _____
    b. _____

17. [11.3B] Find to four significant digits.
   a. $\sqrt{86.1 \times 0.203}$
   b. $\sqrt[3]{983 \times 0.502}$

17. a. _____
    b. _____

18. [11.3B] Find to four significant digits.

   a. $\sqrt{\dfrac{793}{0.662}}$

   b. $\sqrt[3]{\dfrac{41.3}{0.00247}}$

18. a. _____
    b. _____

19. [11.4] Find x if
   a. $10^{2.5x} = 3$
   b. $100^{0.2x} = 4$

19. a. _____
    b. _____

20. [11.4] Find x if
   a. $2^x = 3$
   b. $5^{2x} = 2.5$

20. a. _____
    b. _____

21. [11.4A] Use $\ln 2 = 0.69315$ and find x if
   a. $e^{5.6x} = 2$
   b. $e^{-0.33x} = 2$

21. a. _____
    b. _____

22. [11.4B] Use the change-of-base formula,

   $$\ln M = \frac{\log M}{0.43429} = 2.3026 \log M$$

   to find
   a. $\ln 1000$
   b. $\ln 0.001$

22. a. _____
    b. _____

23. [11.4C] The compound amount with continuous compounding is given by $A = Pe^{rt}$, where P is the principal, r the rate, and t the time in years. Find how long it takes for the money to double—that is, for A to equal $2P$—if $\ln 2 = 0.69315$ and the rate is
   a. 5%
   b. 8.5%

23. a. _____
    b. _____

24. [11.4C] Use the table of exponential functions (Appendix 3) in this problem. The number of bacteria in a culture after t minutes is given by $N = 1000e^{kt}$. If there are 1804 bacteria after the given time, find k.
   a. 2 min
   b. 4 min

24. a. _____
    b. _____

25. [11.4C] A radioactive substance decays so that the amount present in t years is given by $A = A_0 e^{-kt}$. Use $\ln 2 = 0.69315$ and find the half-life if
   a. $k = 0.5$
   b. $k = 0.02$

25. a. _____
    b. _____

NAME

CLASS

SECTION

(Answers on pages 900–901)

ANSWERS

1. Graph the function $f(x) = 3^x$.

1.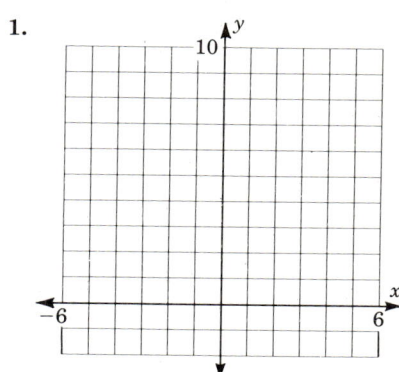

2. Graph the function $g(x) = \left(\dfrac{1}{3}\right)^x$.

2.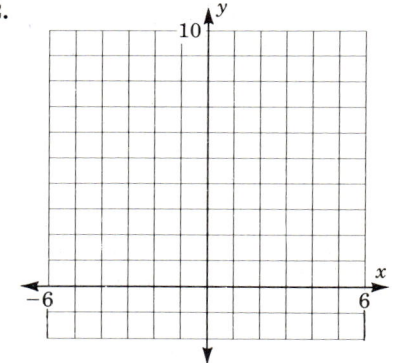

3. Find $\log_2 32$.

3. _____

4. Find $\log_{10} 0.01$.

4. _____

5. Write $3^4 = 81$ in logarithmic form.

5. _____

6. Write $\log_8 4 = \dfrac{2}{3}$ in exponential form.

6. _____

7. Show that
$$\log_b \sqrt{\dfrac{53}{79}} = \dfrac{1}{2}(\log_b 53 - \log_b 79)$$

7. _____

8. Show that $\log\left(1 + \dfrac{1}{x^3}\right) = \log(x + 1) + \log(x^2 - x + 1) - 3 \log x$.

8. _____

9. If $1000 is invested at 9% compounded monthly for 5 yr, the compound amount is given by $A = 1000(1.0075)^{60}$. Given that log 1.0075 = 0.00325, find log A. (Note that $1000 = 10^3$.)

9. _____

The following table is to be used in Problems 10–15.

Some Common Logarithms

x	0	1	2	3	4	5	6	7	8	9
3.0	.4771	.4786	.4800	.4814	.4829	.4843	.4857	.4871	.4886	.4900
3.1	.4914	.4928	.4942	.4955	.4969	.4983	.4997	.5011	.5024	.5038
3.2	.5051	.5065	.5079	.5092	.5105	.5119	.5132	.5145	.5159	.5172
3.3	.5185	.5198	.5211	.5224	.5237	.5250	.5263	.5276	.5289	.5302
3.4	.5315	.5328	.5340	.5353	.5366	.5378	.5391	.5403	.5416	.5428
3.5	.5441	.5453	.5465	.5478	.5490	.5502	.5514	.5527	.5539	.5551
3.6	.5563	.5575	.5587	.5599	.5611	.5623	.5635	.5647	.5658	.5670
3.7	.5682	.5694	.5705	.5717	.5729	.5740	.5752	.5763	.5775	.5786
3.8	.5798	.5809	.5821	.5832	.5843	.5855	.5866	.5877	.5888	.5899
3.9	.5911	.5922	.5933	.5944	.5955	.5966	.5977	.5988	.5999	.6010

10. Find.
    a. log 325
    b. log 0.00325

10. a. _____
    b. _____

11. Use interpolation to find.
    a. log 3.255
    b. log 0.03256

11. a. _____
    b. _____

12. Find.
    a. antilog 3.5502
    b. antilog 7.5809 − 10

12. a. _____
    b. _____

13. Use interpolation to find.
    a. antilog 2.5505
    b. antilog 7.5804 − 10

13. a. _____
    b. _____

14. Find $(3.05)^{1.1}$ to four digits.

14. _____

15. Find $\sqrt{32.2 \times 0.00387}$ to four digits.

15. _____

16. Find x if $e^{9.6x} = 2$ and you are given ln 2 = 0.69315.

16. _____

17. Given the change-of-base formula,

$$\ln M = \frac{\log M}{0.43429} = 2.3026 \log M$$

find ln 1000.

17. _____

18. The compound amount with continuous compounding is given by $A = Pe^{rt}$, where P is the principal, r is the rate, and t is the time in years. If the rate is 8%, find how long it takes for the money to double—that is, for A to equal $2P$ (ln 2 = 0.69315).

18. _____

19. The number of bacteria in a culture after t minutes is given by $N = 1000e^{kt}$. If there are 2117 bacteria after 4 min, find k. (Use the following table.)

x	e^x	e^{-x}
0.55	1.7333	0.5769
0.60	1.8221	0.5488
0.65	1.9155	0.5220
0.70	2.0138	0.4966
0.75	2.1170	0.4724

19. _____

20. A radioactive substance decays so that the amount A present at time t (years) is $A = A_0 e^{-0.5t}$. Find the half-life (time for half to decay) of this substance ($\ln 2 = 0.69315$).

20. _____

IF YOU MISSED QUESTION	SECTION	EXAMPLES	PAGE	ANSWERS
1	11.1	1a	856	1.
2	11.1	1b	857	2.
3	11.2	1a	866	3. 5
4	11.2	1d	866	4. -2
5	11.2	2	866	5. $\log_3 81 = 4$
6	11.2	3	866	6. $4 = 8^{2/3}$
7	11.2	4	867	7. $\log\sqrt{\dfrac{53}{79}} = \dfrac{1}{2}\log\dfrac{53}{79}$ $= \dfrac{1}{2}(\log 53 - \log 79)$
8	11.2	5	867	8. $1 + \dfrac{1}{x^3} = \dfrac{x^3 + 1}{x^3}$ $= \dfrac{(x+1)(x^2 - x + 1)}{x^3}$ $\log\left(1 + \dfrac{1}{x^3}\right) = \log(x+1)$ $+ \log(x^2 + x + 1) - 3\log x$
9	11.2	7	868	9. 3.1950

IF YOU MISSED QUESTION	SECTION	EXAMPLES	PAGE	ANSWERS
10a	11.3	1a	875	**10. a.** 2.5119
10b	11.3	1b	875	**b.** $7.5119 - 10$
11a	11.3	2	875	**11. a.** 0.5126
11b	11.3	2	875	**b.** $8.5128 - 10$
12a	11.3	3	876	**12. a.** 3550
12b	11.3	3	876	**b.** 0.00381
13a	11.3	4	876	**13. a.** 355.2
13b	11.3	4	876	**b.** 0.003805
14	11.3	5	877	**14.** 3.409
15	11.3	6	877	**15.** 0.3530
16	11.4	1	884–885	**16.** 0.072200
17	11.4	2	885–886	**17.** 6.9078
18	11.4	3	886	**18.** About 8.66 yr
19	11.4	5	887	**19.** 0.1875
20	11.4	6	887–888	**20.** About 1.386 yr

CHAPTER 12

SEQUENCES, SERIES, AND THE BINOMIAL THEOREM

In this chapter we study sequences of numbers and their applications. Two special types of sequences are arithmetic progressions, in which the difference between any term and the preceding term is a constant, and geometric progressions, in which each term is a constant times the preceding term. Infinite geometric series, formed by adding terms in geometric progression, are considered. The chapter closes with a discussion of the expansion of a binomial of the form $(a + b)^n$.

NAME

CLASS

SECTION

(Answers on pages 906–907)

ANSWERS

1. For the sequence of odd natural numbers 3, 5, 7, 9, . . . , find.
   a. a_{10}, the tenth term
   b. a_n, the general term

   1. a. _____
      b. _____

2. Each card in a pack of four index cards is cut in half, and the halves are put in a single pile. This step is repeated again and again.
   a. Write the sequence that gives the number of cards in the deck after each step.
   b. How many cards are in the pack after the fourth step?
   c. How many cards are in the pack after the nth step?

   2. a. _____
      b. _____
      c. _____

3. Find the first three terms and the eighth term of the sequence whose general term is

   $$a_n = \frac{n(n + 3)}{2}$$

   3. _____

4. Find the sequence that corresponds to the function $a(n) = 3n + 1$, $n = 1, 2, 3, \ldots$.

   4. _____

5. A sculpture by a famous artist doubles in value every 100 yr. Find the value of the sculpture in the year 2000 if it was worth $500 in the year 1500.

   5. _____

6. For the arithmetic progression 7, 10, 13, 16, . . . , find.
   a. d
   b. a_{10}

   6. a. _____
      b. _____

7. For the arithmetic progression 7, 10, 13, 16, . . . , find a_n.

   7. _____

8. Sally draws a sequence of circles. She starts with a row of 4 circles, then adds 2 circles to get the second row, adds 2 more circles to get the third row, and so on. Find how many circles she would have in the
   a. fifth row
   b. tenth row
   c. nth row

   8. a. _____
      b. _____
      c. _____

9. The distance (in feet) that a free-falling body falls in each second, starting with the first second, is given by the arithmetic progression 16, 48, 80, 112, ... Find the distance that the body falls in 6 sec.

   9. _____

10. The sum of the first eight terms of an arithmetic progression is 196 and the eighth term is 35. Find.
    a. a_1, the first term
    b. d, the common difference

10. a. _____
    b. _____

11. A piece of machinery valued at $50,000 depreciates $8000 the first year, $7800 the second year, $7600 the third year, and so on. Find the value of this piece of machinery at the end of 6 yr.

11. _____

12. José was trying to save his pennies. He saved 5¢ the first week, 7¢ the second week, 9¢ the third week, and so on, in arithmetic progression. After a few weeks, José had saved a total of $1.17. For how many weeks had he been saving?

12. _____

13. For the geometric progression $9, 3, 1, \frac{1}{3}, \ldots$, find.
    a. r
    b. a_6

13. a. _____
    b. _____

14. For the progression in Problem 13, find a_n.

14. _____

15. For the geometric progression $1, -3, 9, -27, \ldots$, find.
    a. r
    b. a_8
    c. S_8

15. a. _____
    b. _____
    c. _____

16. A rubber ball, dropped on a hard surface, takes a sequence of bounces, each one $\frac{4}{5}$ as high as the preceding one. If this ball is dropped from a height of 10 ft, how far will it have traveled when it hits the surface the fifth time?

16. _____

17. A certain sum of money is invested in a business. In each year this investment earns $1\frac{1}{2}$ times as much as in the preceding year. If the investment earned a total of $16,250 in 4 yr, how much did it earn in the fourth year?

17. _____

18. Find the sum of the geometric series $16 - 8 + 4 - 2 + \cdots$.

18. _____

19. Find the sum of the geometric series $1.002 + (1.002)^2 + (1.002)^3 + \cdots$ if the sum exists.

19. _____

20. Find the fraction that is equivalent to the repeating decimal $0.213213213\ldots$.

20. _____

21. If the ball of Problem 16 is assumed to continue bouncing indefinitely, find the total distance it would travel.

22. Expand $(a - 2b)^4$.

23. Find the fourth term in the expansion of $(2a - b)^7$.

24. Seven coins are tossed. In how many ways can exactly three heads turn up?

25. A fair coin is tossed 12 times. Find the probability of getting exactly 5 heads.

21. _____

22. _____

23. _____

24. _____

25. _____

IF YOU MISSED QUESTION	SECTION	EXAMPLES	PAGE	ANSWERS
1a	12.1	1	909–910	**1. a.** $a_{10} = 21$
1b	12.1	1	909–910	**b.** $a_n = 2n + 1$
2a	12.1	2, 3	910	**2. a.** $4, 8, 16, 32, \ldots$
2b	12.1	2, 3	910	**b.** 64
2c	12.1	2, 3	910	**c.** $4 \cdot 2^n$, or 2^{n+2}
3	12.1	4	910–911	**3.** $a_1 = 2, a_2 = 5, a_3 = 9, a_8 = 44$
4	12.1	5	911	**4.** $4, 7, 10, 13, \ldots$
5	12.1	6	911	**5.** $16,000
6a	12.2	1	918	**6. a.** $d = 3$
6b	12.2	1	918	**b.** $a_{10} = 34$
7	12.2	1	918	**7.** $a_n = 3n + 4$
8a	12.2	1	918	**8. a.** 12
8b	12.2	1	918	**b.** 22
8c	12.2	1	918	**c.** $2n + 2$
9	12.2	2	919	**9.** 576 ft
10a	12.2	3	920	**10. a.** $a_1 = 14$
10b	12.2	3	920	**b.** $d = 3$
11	12.2	4	920	**11.** $5000
12	12.2	5	921	**12.** 9
13a	12.3	1	927–928	**13. a.** $r = \dfrac{1}{3}$
13b	12.3	1	927–928	**b.** $a_6 = \dfrac{1}{27}$
14	12.3	1	927–928	**14.** $a_n = \left(\dfrac{1}{3}\right)^{n-3}$
15a	12.3	2	929	**15. a.** $r = -3$
15b	12.3	2	929	**b.** $a_8 = -2187$
15c	12.3	2	929	**c.** $S_8 = -1640$
16	12.3	4	930	**16.** $57\dfrac{29}{125}$ ft, or 57.232 ft
17	12.3	5	930–931	**17.** $6750
18	12.4	1	938	**18.** $\dfrac{32}{3}$ or $10\dfrac{2}{3}$
19	12.4	2	938	**19.** Sum does not exist.

IF YOU MISSED QUESTION	SECTION	EXAMPLES	PAGE	ANSWERS
20	12.4	3	939	**20.** $\dfrac{71}{333}$
21	12.4	4	939	**21.** 90 ft
22	12.5	1	945	**22.** $a^4 - 8a^3b + 24a^2b^2 - 32ab^3 + 16b^4$
23	12.5	2	945–946	**23.** $-560a^4b^3$
24	12.5	3	946	**24.** 35
25	12.5	4	946	**25.** $\dfrac{99}{512}$

12.1 SEQUENCES

OBJECTIVES

REVIEW

Before starting this section, you should know how to:

1. Simplify an algebraic expression.
2. Evaluate a formula.

OBJECTIVES

You should be able to:

A. Identify terms in a given sequence.
B. Find specified terms in a sequence when the general term is given.
C. Solve applications using the ideas discussed.

Leonardo Fibonacci was one of the greatest mathematicians of the Middle Ages. The following is a problem that interested Fibonacci.

> Let us suppose you have a 1 mo-old pair of rabbits. Assume that in the second month, and every month thereafter, they produce a new pair. If the new pair does the same and none of the rabbits die, how many pairs of rabbits will there be at the beginning of each month?

Here is the solution that you can check just by counting:

Beginning month number	1	2	3	4	5	6	7	...
Number of pairs	1	1	2	3	5	8	13	...

A. TERMS OF A SEQUENCE

The numbers 1, 1, 2, 3, 5, 8, 13, . . . form a *sequence,* called the **Fibonacci sequence.** In mathematics, a **sequence** is a set of numbers arranged according to some given law. For example, the natural numbers 1, 2, 3, 4, 5, 6, 7, 8, 9, form a sequence. The numbers in a sequence are the **terms** of the sequence and are called the *first term, second term, third term,* and so on. The terms of a sequence are usually denoted by using subscripts. Thus a_1, a_2, and a_3 denote the first, second, and third terms of a sequence. For the sequence of counting numbers 1, 2, 3, 4, . . .

$$a_1 = 1$$
$$a_2 = 1 + 1 = 2$$
$$a_3 = 2 + 1 = 3$$
$$a_4 = 3 + 1 = 4$$
$$\vdots$$

This pattern shows that the nth term, called the **general term,** is n. Thus

$$a_{31} = 31, \qquad a_{52} = 52, \quad \text{and} \quad a_n = n$$

EXAMPLE 1 For the sequence of even counting numbers 2, 4, 6, 8, . . . , find.

a. a_2 and a_4 **b.** a_{10}, the tenth term   **c.** a_n, the nth (general) term

Solution

a. By inspection, $a_2 = 4$ and $a_4 = 8$.

Problem 1 For the sequence of multiples of 3, that is, 3, 6, 9, 12, . . . , find.
a. a_2 and a_4 **b.** a_{10} **c.** a_n

b. The given terms show that $a_1 = 2 \cdot 1$, $a_2 = 2 \cdot 2$, $a_3 = 2 \cdot 3$, $a_4 = 2 \cdot 4$, and so on. Thus, $a_{10} = 2 \cdot 10 = 20$.

c. From the pattern in part b, $a_n = 2n$. ▲

EXAMPLE 2 For the sequence of even integers with alternating signs,

2, −4, 6, −8, . . .

find a_{10} and a_n.

Solution To get the correct signs, we can use the factor $(-1)^{n-1}$, which gives, alternately, $+1$ and -1. Thus for a_{10} and a_n, we can use the answers of example 1(b) and 1(c). We have $a_{10} = (-1)^{10-1}20 = (-1)^9 20 = -20$ and $a_n = (-1)^{n-1}(2n)$. ▲

EXAMPLE 3 In the publishing industry, large printed pages are folded to make the pages of a book. If sheets are folded once, this makes 2 pages, or a *folio*. If the sheets are folded twice, this makes 4 pages, or a *quarto*. If sheets are folded three times, this makes 8 pages, or an *octavo*. Further folding produces units called 16 mo, 32 mo, and so on.

a. Write the sequence that gives the number of pages after each fold.
b. If a sheet is folded six times, how many pages are there?
c. How many pages result after n folds?

Solution

a. $a_1 = 2$, $a_2 = 4$, $a_3 = 8$, $a_4 = 16$, ... Thus the sequence is 2, 4, 8, 16, ...
b. The terms of the sequence can be written as $a_1 = 2^1$, $a_2 = 2^2$, $a_3 = 2^3$, $a_4 = 2^4$, and so on. If the sheet is folded 6 times, the number of pages is $a_6 = 2^6 = 64$.
c. If the sheet is folded n times, the number of pages is $a_n = 2^n$. ▲

B. THE GENERAL TERM

In Examples 1 and 2, we found the general term of a given sequence. However, you must understand that if only a finite number of successive terms are given without a rule that defines the general term, a *unique* general term cannot be obtained. For example,

$$a_n = 2n \quad \text{and} \quad a_n = 2n + \frac{1}{2}(n-1)(n-2)(n-3)(n-4)$$

both give $a_1 = 2$, $a_2 = 4$, $a_3 = 6$, $a_4 = 8$, the first four terms of the sequence in Example 1. However, the first formula gives $a_5 = 10$ and the second formula gives $a_5 = 10 + \frac{1}{2}(4)(3)(2)(1) = 22$. Thus you can see that the first four terms do not determine the fifth term. This procedure can be extended to show that no finite number of terms can define the next term uniquely.

The next example shows how to find specified terms of a sequence when the general term is given.

EXAMPLE 4 Find the first three terms and the ninth term of the sequence whose general term is

$$a_n = \frac{1}{2}n(n-1)$$

Problem 2 For the sequence of multiples of 3 with alternating signs, $-3, 6, -9, 12, \ldots$, find a_{10} and a_n.

Problem 3 Suppose each card in a pack of 5 index cards is cut in half and all the halves are put in a single pile. Then this is repeated again and again.
a. Write the sequence that gives the number of cards in the pack after each cut.
b. How many cards are in the pack after the fourth cut?
c. How many cards are in the pack after the nth cut?

Problem 4 Find the first three terms and the tenth term of the sequence whose general term is

$$a_n = \frac{1}{2}(n^2 - 1)$$

Solution   We find the required terms by substituting the corresponding values of n into the given formula. Thus

$$a_1 = \frac{1}{2}(1)(1-1) = 0$$

$$a_2 = \frac{1}{2}(2)(2-1) = 1$$

$$a_3 = \frac{1}{2}(3)(3-1) = 3$$

$$a_9 = \frac{1}{2}(9)(9-1) = 36$$

▲

Sometimes function notation rather than subscript notation is used to define the terms of a sequence. For instance, in Example 4, we could write

$$a(n) = \frac{1}{2}n(n-1)$$

so that

$$a(1) = \frac{1}{2}(1)(1-1) = 0$$

$$a(2) = \frac{1}{2}(2)(2-1) = 1$$

and so on.

EXAMPLE 5   Consider the function

$$a(n) = 2n + 1, \qquad n = 1, 2, 3, \ldots$$

Find the sequence corresponding to this function.

Solution

for $n = 1$, $a(1) = 2(1) + 1 = 3$
for $n = 2$, $a(2) = 2(2) + 1 = 5$
for $n = 3$, $a(3) = 2(3) + 1 = 7$
for $n = 4$, $a(4) = 2(4) + 1 = 9$

Thus the sequence is $3, 5, 7, 9, \ldots$.

▲

C.   APPLICATIONS

The preceding ideas are illustrated in the next example.

EXAMPLE 6   Suppose a painting by a famous artist doubles in value every 50 yr. Find the value of the painting in the year 2000 if it was worth $1000 in the year 1500.

Solution   If we write a_n for the value of the painting at the end of the nth 50-yr period, then $a_1 = 2 \times \$1000$, $a_2 = 4 \times \$1000$, $a_3 = 8 \times \$1000$, and so on. Thus, at the end of the nth period, $a_n = 2^n \times \$1000$. Since there are ten 50-yr periods between 1500 and 2000, the value of the painting would be

$$a_{10} = 2^{10} \times \$1000 = 1024 \times \$1000$$
$$= \$1,024,000$$

Problem 5   Consider the function

$$a(n) = n(n-1), \qquad n = 1, 2, 3, \ldots$$

Find the sequence that corresponds to this function.

Problem 6   Do Eample 6 if the painting doubles in value every 100 yr.

A. In Problems 1–22 find a tenth term and an nth term to fit the given sequence.

1. $1, 4, 7, 10, \ldots$

2. $5, 7, 9, 11, \ldots$

3. $5, 8, 11, 14, \ldots$

4. $3, 8, 13, 18, \ldots$

5. $20, 25, 30, 35, \ldots$

6. $15, 18, 21, 24, \ldots$

7. $50, 45, 40, 35, \ldots$

8. $30, 28, 26, 24 \ldots$

9. $\dfrac{1}{2}, \dfrac{1}{3}, \dfrac{1}{4}, \dfrac{1}{5}, \ldots$

10. $\dfrac{1}{2}, \dfrac{2}{3}, \dfrac{3}{4}, \dfrac{4}{5}, \ldots$

11. $1, -1, 1, -1, \ldots$

12. $-1, 2, -4, 8, \ldots$

13. $x, x^2, x^3, x^4, \ldots$

14. $x^2, x^4, x^6, x^8, \ldots$

15. $x, -x^3, x^5, -x^7, \ldots$

16. $-x, x^2, -x^4, x^8, \ldots$

17. $x, -x, x, -x, \ldots$

18. $-x, x, -x, x, \ldots$

19. $x, \dfrac{x^2}{2}, \dfrac{x^3}{3}, \dfrac{x^4}{4}, \ldots$

20. $\dfrac{x}{5}, \dfrac{x^2}{10}, \dfrac{x^3}{15}, \dfrac{x^4}{20}, \ldots$

21. $\dfrac{x}{2}, -\dfrac{x^2}{4}, \dfrac{x^3}{8}, -\dfrac{x^4}{16}, \ldots$

22. $\dfrac{x}{2}, -\dfrac{x^3}{4}, \dfrac{x^5}{8}, -\dfrac{x^7}{16}, \ldots$

B. In Problems 23–36 find the first three terms of the sequence with the given general term.

23. $a_n = 2n - 3$

24. $a_n = 2n + 3$

25. $a_n = \dfrac{n(n-2)}{2}$

26. $a_n = \dfrac{n(n+2)}{2}$

27. $a(n) = 1 - \dfrac{1}{n}$

28. $a(n) = 1 + \dfrac{2}{n}$

ANSWERS

1. _____
2. _____
3. _____
4. _____
5. _____
6. _____
7. _____
8. _____
9. _____
10. _____
11. _____
12. _____
13. _____
14. _____
15. _____
16. _____
17. _____
18. _____
19. _____
20. _____
21. _____
22. _____

23. _____
24. _____
25. _____
26. _____
27. _____
28. _____

29. $a_n = n^2$

30. $a_n = -n^3$

31. $a(n) = \dfrac{n}{2n+1}$

32. $a(n) = \dfrac{n}{3n-1}$

33. $a_n = (-1)^n$

34. $a_n = (-2)^{n-1}$

35. $a(n) = (-1)^n 2^{-n}$

36. $a(n) = (-1)^{n-1} 3^n$

29. _____

30. _____

31. _____

32. _____

33. _____

34. _____

35. _____

36. _____

C. Applications

37. A property valued at \$30,000 will depreciate \$1380 the first year, \$1340 the second year, \$1300 the third year, and so on. What will be the depreciation during the given years?
  a. The eighth year
  b. The tenth year

37. **a.** _____
    b. _____

38. Strikers at a plant were ordered to return to work and were told they would be fined \$50 the first day they failed to do so, \$75 the second day, \$100 the third day, and so on. If the strikers stayed out for 6 days, what was the fine for the sixth day?

38. _____

39. When dropped on a hard surface, a Super Ball takes a sequence of bounces, each one about $\frac{9}{10}$ as high as the preceding one. If a Super Ball is dropped from a height of 10 ft, find how high it will bounce on the
  a. First bounce
  b. Third bounce
  c. nth bounce

39. **a.** _____
    b. _____
    c. _____

40. An ancient legend says that the King of Persia offered the inventor of chess anything he wished as a reward for his invention. The man asked for 1 grain of wheat to be placed on the first square of the chessboard, 2 grains on the second, 4 grains on the third, and so on. How many grains would there be on each of the following squares?
  a. The fifth square
  b. The ninth square
  c. The nth square

40. **a.** _____
    b. _____
    c. _____

41. A colony of bacteria starts with 100 members and doubles every hour. How many bacteria are there after each given time?
  a. 2 hr
  b. 4 hr
  c. n hr

41. **a.** _____
    b. _____
    c. _____

42. A free-falling body falls about 16 ft the first second, 48 ft the next second, 80 ft the third second, and so on. How far does it fall during each of the following?
  a. The eighth second
  b. The nth second

42. **a.** _____
    b. _____

43. A salesman sold $100 worth of goods on Monday and doubled his sales each day thereafter for a week. What was the amount of sales on Saturday?

43. _____

44. A racer runs 6 m in the first second of a certain race and increases her speed by 25 cm/sec in each succeeding second. (This means that she goes 6 m 25 cm the second second, 6 m 50 cm the third second, and so on.) How far does she go during each time?
  a. The eighth second
  b. The nth second

44. **a.** _____
    b. _____

✔ **SKILL CHECKER**

In Problems 45–48, simplify the given expression.

45. $7 + (n-1)(3)$

46. $16 + (n-1)(2)$

45. _____

46. _____

47. $\dfrac{1}{2}n(16 + 32n - 16)$

48. $\dfrac{n}{2}(3 + 5n - 2)$

47. _____

48. _____

In Problems 49–50, factor the given quadratic.

49. $5n^2 + n - 328$

50. $7n^2 - n - 336$

49. _____

50. _____

12.1 USING YOUR KNOWLEDGE

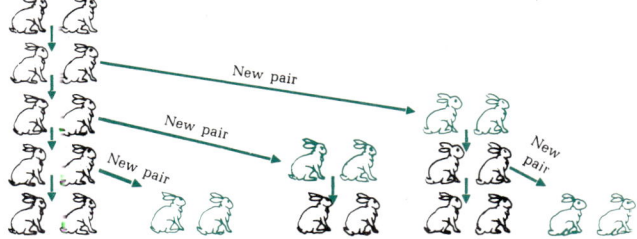

New pair
New pair
New pair
New pair
New pair

We return to the rabbit problem that was introduced at the beginning of this section. The figure illustrates what happens in the first 5 mo. As we saw, the numbers of rabbits at the beginning of each month are the terms of the Fibonacci sequence 1, 1, 2, 3, 5, 8, 13, ...

Notice that, starting with the third term, each term is the sum of the two preceding terms: $2 = 1 + 1$, $3 = 1 + 2$, $5 = 2 + 3$, and so on. This leads us to the general formula

$a_n = a_{n-2} + a_{n-1}$

Use these ideas to write the following terms of the Fibonacci sequence.

1. The eighth term

2. The ninth term

1. _____

2. _____

3. The tenth term

4. The eleventh term

3. _____

4. _____

5. The twelfth term

5. _____

12.2 ARITHMETIC PROGRESSIONS

Courtesy of Bernard Liebman

OBJECTIVES

REVIEW

Before starting this section, you should know:

1. What is meant by a sequence.
2. How to use the formula for the general term of a sequence.

OBJECTIVES

You should be able to:

A. Recognize an arithmetic progression and find the common difference and the general term.
B. Find the sum of an arithmetic progression.
C. Solve applications using these ideas.

Do you recognize the illustration? It is "A partridge in a pear tree," a line from the song "The Twelve Days of Christmas." According to this song, you will receive

1 gift the first day

1 + 2 gifts the second day

1 + 2 + 3 gifts the third day and so on

Can you find how many gifts you will receive on the tenth day?

A. TERMS OF AN ARITHMETIC PROGRESSION

This problem asks you to add the terms of the sequence 1, 2, 3, . . . , 10. This sequence is an example of an *arithmetic progression*. An **arithmetic progression** is a sequence in which each term after the first is obtained by adding a quantity d, called the **common difference,** to the preceding term. Thus, the sequence 1, 2, 3, . . . is an arithmetic progression with the common difference

$$d = 1$$

Since each term after the first is formed by adding 1 to the preceding term, it follows that the general term of this sequence is

$$a_n = 1 + (n - 1)(1) = n$$

For example, $a_{10} = 10$.

In general, if we let the first term be a_1 and the following terms be formed by the rule

$$a_{n+1} = a_n + d$$

then the sequence is an arithmetic progression with the general term

$$a_n = a_1 + (n - 1)d$$

For example, the sequence 10, 15, 20, . . . , for which $a_1 = 10$ and $a_{n+1} = a_n + 5$ is an arithmetic progression, where each term after the first is formed by adding 5 to the preceding term. The general term of this progression is thus seen to be

$$a_n = 10 + (n-1)(5)$$

or

$$a_n = 5 + 5n$$

You can check this formula by direct substitution. Thus

$$a_1 = 5 + 5(1) = 10$$
$$a_2 = 5 + 5(2) = 15$$
$$a_3 = 5 + 5(3) = 20$$
$$\vdots$$

Note that the common difference is just the difference between any two consecutive terms.

EXAMPLE 1   For the arithmetic progression 7, 10, 13, 16, . . . , find

a. a_1 **b.** d **c.** a_n

Solution

a. By inspection, you see that the first term is $a_1 = 7$.

b. The common difference is the difference between any term and the preceding term. Thus $d = 10 - 7 = 3$.

c. The formula for the general term is

$$a_n = a_1 + (n-1)d$$
$$= 7 + (n-1)(3)$$
$$= 7 + 3n - 3$$
$$a_n = 4 + 3n \qquad \blacktriangle$$

Problem 1   For the arithmetic progression 5, 9, 13, 17, . . . , find
a. a_1 **b.** d **c.** a_n

B.   SUM OF AN ARITHMETIC PROGRESSION

A free-falling body near the surface of the earth falls about 16 ft in the first second, 48 ft in the next second, 80 ft in the third second, and so on. Thus the number of feet fallen in each successive second is given by the arithmetic progression

16, 48, 80, 112, 144, . . .

for which the common difference is $48 - 16$, or $80 - 48$, and so on; that is,

$$d = 32$$

The most interesting question to ask about the falling body is: How far does it fall in n seconds? To answer this question, we must find the sum of the first n terms of the sequence. The easiest way to do this is first to find a general formula for the sum of an arithmetic progression. Let us write S_n for this sum. Then

$$S_n = a_1 + (a_1 + d) + (a_2 + 2d) + \cdots + (a_n - 2d) + (a_n - d) + a_n$$

We now write the same equation with the terms of the progression in reverse order:

$$S_n = a_n + (a_n - d) + (a_n - 2d) + \cdots + (a_1 + 2d) + (a_1 + d) + a_1$$

If we add the corresponding terms of these two equations, we see that all the d's drop out; the result is

$$2S_n = (a_1 + a_n) + (a_1 + a_n) + \cdots + (a_1 + a_n) + (a_1 + a_n)$$

or

$$2S_n = n(a_1 + a_n)$$

Thus we obtain the general formula for the sum of an arithmetic progression:

$$S_n = \frac{1}{2}n(a_1 + a_n)$$

We return to the falling-body problem in the next example.

EXAMPLE 2 Find the distance the body falls in

a. 5 sec   **b.** 10 sec   **c.** n sec

Solution To do this problem, we first find the distance fallen in the nth second. Since $a_1 = 16$ and $d = 32$, we get

$$a_n = a_1 + (n - 1)d$$
$$= 16 + (n - 1)(32)$$

or

$$a_n = 32n - 16$$

a. For $n = 5$, $a_5 = (32)(5) - 16 = 160 - 16 = 144$, so that

$$S_5 = \frac{1}{2}n(a_1 + a_n)$$

$$= \frac{5}{2}(16 + 144) = \frac{5}{2}(160) = 400$$

Thus the body falls 400 ft in 5 sec.

b. For $n = 10$, $a_{10} = (32)(10) - 16 = 320 - 16 = 304$, so that

$$S_{10} = \frac{10}{2}(16 + 304) = 5(320) = 1600$$

Thus the body falls 1600 ft in 10 sec.

c. Here, we use the general formulas for S_n and a_n.

$$S_n = \frac{1}{2}n(a_1 + a_n) \quad \text{and} \quad a_n = 32n - 16$$

Now, we substitute $a_1 = 16$ and $a_n = 32n - 16$ into the formula for S_n to get

$$S_n = \frac{1}{2}n(16 + 32n - 16) = \frac{1}{2}n(32n)$$

or

$$S_n = 16n^2$$

You can check that this surprisingly simple formula gives the same answers that we found in parts a and b. ▲

Problem 2 Return to the gift problem at the start of this section and find
a. The number of gifts you receive on the tenth day
b. The number of gifts you receive on the nth day

EXAMPLE 3 The sum of the first 10 terms of an arithmetic progression is 205 and the tenth term is 34. Find.

a. a_1, the first term
b. d, the common difference

Solution

a. We use the formula for the sum with $n = 10$:

$$S_{10} = \frac{10}{2}(a_1 + a_{10}) = 5(a_1 + a_{10})$$

We then substitute $S_{10} = 205$ and $a_{10} = 34$ to get

$205 = 5(a_1 + 34)$
$41 = a_1 + 34$ Divide by 5.
$7 = a_1$ Subtract 34.

b. Now we use the formula for the nth term,

$$a_n = a_1 + (n - 1)d$$

With $n = 10$, $a_1 = 7$, and $a_{10} = 34$, to get

$34 = 7 + 9d$
$27 = 9d$ Subtract 7.
$3 = d$ Divide by 9.   ▲

C. APPLICATIONS

EXAMPLE 4 A heavy-duty truck valued at $50,000 depreciates $5000 the first year, $4800 the second year, $4600 the third year, and so on. What will be the value of the truck at the end of 8 y?

Solution The yearly depreciations form an arithmetic progression:

5000, 4800, 4600, . . .

so that $a_1 = 5000$ and $d = -200$. Thus

$a_8 = a_1 + (n - 1)d$
    $= 5000 + (7)(-200)$
    $= 3600$

The total depreciation will be the sum of the first eight terms of the progression:

$$S_8 = \frac{8}{2}(a_1 + a_8)$$

$= 4(5000 + 3600)$
$= 4(8600)$
$= 34,400$

So the total depreciation is $34,400 and the remaining value is

$50,000 - \$34,400 = \$15,600$ ▲

Problem 3 The sum of the first eight terms of an arithmetic progression is 136 and the eighth term is 24. Find.
a. The first term
b. The common difference

Problem 4 Do Example 4 if the truck depreciates $6000 the first year, $5800 the second year, $5600 the third year, and so on.

EXAMPLE 5 Alice started a savings campaign. She put aside 3¢ the first day, 8¢ the second day, 13¢ the third day, and so on in arithmetic progression. After a few days, Alice found that she had saved $1.64. For how many days had she been saving?

Solution Here, we know that $a_1 = 3$, $d = 5$, and $S_n = 164$, and we want to find n. So we use the formula for the sum

$$S_n = \frac{n}{2}(a_1 + a_n)$$

and the formula for the nth term

$$a_n = a_1 + (n - 1)d$$

Substituting the known values, we get the two equations

$$164 = \frac{n}{2}(3 + a_n)$$

$$a_n = 3 + (n - 1)(5)$$
$$= 5n - 2$$

We next substitute for a_n in the first equation to obtain

$$164 = \frac{n}{2}(3 + 5n - 2)$$

$$328 = n(5n + 1) \qquad \text{Multiply by 2 and simplify.}$$

We rewrite this equation in standard quadratic form:

$$5n^2 + n - 328 = 0$$
$$(5n + 41)(n - 8) = 0 \quad \text{Factor.}$$

Since n must be positive, the solution is $n = 8$. Thus it took 8 days for Alice to get up to $1.64. ▲

Problem 5 In Example 5, suppose Alice increased her daily savings amount by 10¢ instead of 5¢, so that the progression is 3, 13, 23, ... How many days would it take her to get a total of $3.04?

ANSWERS (to problems on pages 920–921)
3. a. 10 **b.** 2   **4.** $7600   **5.** 8 days

EXERCISE 12.2

A. In Problems 1–10 an arithmetic progression is given. Find.

    a. a_1, the first term,
    b. d, the common difference
    c. a_n, the nth term

1. 5, 8, 11, 14, . . .   **2.** 5, 10, 15, 20, . . .

1. a.	_____
b.	_____
c.	_____
2. a.	_____
b.	_____
c.	_____

3. 11, 6, 1, -4, . . .   **4.** 43, 32, 2¯, 10, . . .

3. a.	_____
b.	_____
c.	_____
4. a.	_____
b.	_____
c.	_____

5. 3, -1, -5, -9, . . .   **6.** 0.6, 0.2, -0.2, -0.6, . . .

5. a.	_____
b.	_____
c.	_____
6. a.	_____
b.	_____
c.	_____

7. $\dfrac{1}{2}, \dfrac{1}{4}, 0, -\dfrac{1}{4}, \ldots$ **8.** $\dfrac{2}{3}, \dfrac{5}{6}, 1, \dfrac{7}{6}, \ldots$

7. a.	_____
b.	_____
c.	_____
8. a.	_____
b.	_____
c.	_____

9. $-\dfrac{5}{6}, -\dfrac{1}{3}, \dfrac{1}{6}, \dfrac{2}{3}, \ldots$ **10.** $-\dfrac{1}{4}, \dfrac{1}{4}, \dfrac{3}{4}, \dfrac{5}{4}, \ldots$

9. a.	_____
b.	_____
c.	_____
10. a.	_____
b.	_____
c.	_____

B. In Problems 11–20 some values for an arithmetic progression are given. Find the other indicated values.

11. $a_1 = 7$, $n = 15$, $d = 6$; a_{15}, S_{15}

12. $a_1 = -2$, $d = -5$, $a_n = -72$; n, S_n

13. a_1, d, S_8 for the sequence 4, 10, 16, 22, . . .

11.	_____
12.	_____
13.	_____

14. a_1, d, S_n for the sequence 3, -1, -5, -9, . . .

15. $a_1 = 3$, $a_6 = 8$; d, S_6

16. $a_1 = -1$, $a_{10} = -4$; d, S_{10}

17. $a_1 = 6$, $S_{14} = -280$; d, a_{14}

18. $a_1 = 15$, $a_n = -25$, $S_n = -85$; d, n

19. $d = 40$, $S_{40} = 40$; a_1, a_{40}

20. $a_1 = 4$, $d = 2$, $a_n = 30$; n, S_n

14. _____

15. _____

16. _____

17. _____

18. _____

19. _____

20. _____

C. Applications

21. A certain property valued at $30,000 will depreciate $1380 the first year, $1340 the second year, $1300 the third year, and so on, the annual depreciation decreasing $40 per year. What will the property be worth at the end of 20 yr?

21. _____

22. Strikers at a certain plant were ordered to return to work and told that their union would be fined $50 the first day they refused to do so, $60 the second day, $70 the third day, and so on. If their union paid a $680 fine, after how many days did they go back to work?

22. _____

23. The diagram shows a sequence of crosses. Starting with a single square, a cross is constructed by adding a square to each side. Then the crosses are extended by adding a square to both ends of the vertical and the horizontal pieces. Find the total number of squares in the first 10 elements of this sequence.

23. _____

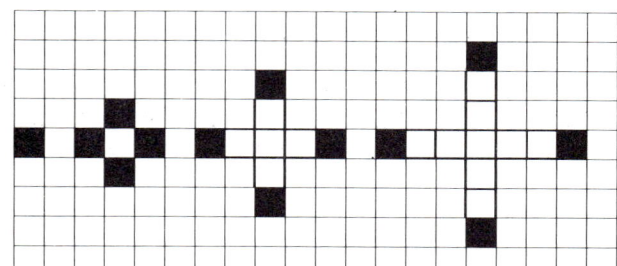

24. The diagram shows a sequence of staircases. Starting with a single square, more squares are added at each stage, as shown by the black squares in the diagram. Find the number of squares in the tenth staircase.

24. _____

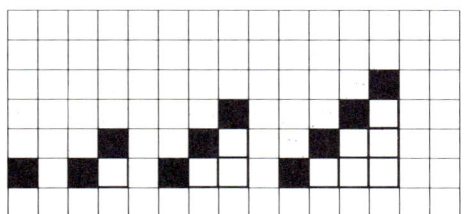

25. Show that the sum of the first n natural numbers is $\dfrac{n(n+1)}{2}$.

26. Show that the sum of the first n odd natural numbers is n^2.

27. Show that the sum of the first n even natural numbers is $n^2 + n$.

28. Find the sum of the natural numbers between 50 and 100 that are divisible by 3.

✓ **SKILL CHECKER**

Evaluate each of the following.

29. r^{n-1} for $r = 2$, $n = 6$

30. r^{n-1} for $r = 3$, $n = 4$

31. $\dfrac{1 - r^n}{1 - r}$ for $r = 2$, $n = 6$

32. $\dfrac{1 - r^n}{1 - r}$ for $r = 3$, $n = 4$

33. $\dfrac{10}{1 - r}$ for $r = \dfrac{1}{2}$

34. $\dfrac{10}{1 - r}$ for $r = \dfrac{1}{4}$

12.2 USING YOUR KNOWLEDGE

1. You have a piece of property valued at \$35,000, which is to be depreciated to a value of \$5000 in 5 yr. Suppose that the first year's depreciation is \$10,000, and the depreciation for the successive years decreases by a fixed amount each year. Find the depreciation for each of the remaining 4 yr.

2. In Problem 1, suppose the first year's depreciation is \$5000 and the depreciation for the successive years increases by a fixed amount. Find the depreciation for each of the remaining 4 yr.

25. _____

26. _____

27. _____

28. _____

29. _____

30. _____

31. _____

32. _____

33. _____

34. _____

1. _____

2. _____

12.3 GEOMETRIC PROGRESSIONS

REVIEW

Before starting this section, you should know:

1. What a sequence is.
2. How to recognize and work with an arithmetic progression.

OBJECTIVES

You should be able to:

A. Recognize a geometric progression and find the common ratio and the general term.
B. Find the sum of a geometric progression.
C. Solve applications using these ideas.

The diagram shows the power of multiplication by 2. Notice how rapidly the strips of squares are extended. The next strip (not shown) would have 32 squares and the one after that would have 64 squares. Keep this in mind as you consider the following two offers. One will pay you $500 per day for 25 days and the other will pay you 1¢ the first day, 2¢ the second day, 4¢ the third day, and so on. The number of cents on consecutive days would form the squence

1, 2, 4, 8, 16, . . .

like the sequence of strips of squares in the diagram. Which of the two offers would pay you the larger total amount? Think about it! We will answer this question later in this section.

A. TERMS OF A GEOMETRIC PROGRESSION

A sequence of terms in which each term after the first is formed by multiplying the preceding term by a constant r is called a **geometric progression.** The sequence 1, 2, 4, 8, 16, . . . is an example of such a progression in which $a_1 = 1$ and $r = 2$. The number r is the ratio of a term to the preceding term and so is called the **common ratio.** Another example of a geometric progression is the sequence

$$6, -1, \frac{1}{6}, -\frac{1}{36}, \ldots$$

Here, $a_1 = 6$ and $r = -\frac{1}{6}$. In general, a geometric progression is of the form

$$a_1, a_1 r, a_1 r^2, a_1 r^3, \ldots$$

Since multiplication by r starts with the second term, you can see that the general term is

$$a_n = a_1 r^{n-1}$$

EXAMPLE 1 For the geometric progression

1, 3, 9, 27, . . .

find

a. a_1 **b.** r **c.** a_6 **d.** a_n

Solution

a. By inspection, we see that the first term is $a_1 = 1$.
b. The common ratio r can be found by taking the ratio of any term to the preceding term. Thus using the ratio of the third term to the

Problem 1 For the geometric progression 2, 1, $\frac{1}{2}$, $\frac{1}{4}$, . . . , find
a. a_1 **b.** r **c.** a_6 **d.** a_n

second, we find

$$r = \frac{9}{3} = 3$$

c. The formula $a_n = a_1 r^{n-1}$ gives, for $n = 6$,

$$a_6 = (1)(3^{6-1}) = 3^5 = 243$$

d. The general term is obtained with $r = 3$.

$$a_n = (1)(3^{n-1}) = 3^{n-1} \qquad \blacktriangle$$

B. SUM OF A GEOMETRIC PROGRESSION

We return to the two offers presented at the beginning of this section. The first offer was for $500 per day for 25 days. So there is no difficulty in finding the total amount. It is

$$(25)(\$500) = \$12,500$$

To get the total number of cents in the second offer, we must add the terms of the sequence

$$1, 2, 4, 8, \ldots, 2^{22}, 2^{23}, 2^{24}$$

Thus

$$S_{25} = 1 + 2 + 4 + \cdots + 2^{22} + 2^{23} + 2^{24}$$

Instead of evaluating the individual terms and adding, it is simpler to use the following clever trick: We multiply S_{25} by the common ratio 2 to get

$$2S_{25} = 2 + 4 + 8 + \cdots + 2^{23} + 2^{24} + 2^{25}$$

Then we subtract the first equation from the second, which gives

$$2S_{25} - S_{25} = -1 + (2 - 2) + (4 - 4) + \cdots$$
$$+ (2^{23} - 2^{23}) + (2^{24} - 2^{24}) + 2^{25}$$

or

$$S_{25} = 2^{25} - 1$$

You can find the value of 2^{25} on your calculator if you have a $\boxed{y^x}$ key (or an equivalent). Otherwise, you can write

$$2^{25} = (2^{10})(2^{10})(2^5)$$
$$= (1024)(1024)(32)$$
$$= 33,554,432$$

Using this number, we get

$$S_{25} = 33,554,431$$

In 25 days, you would get this number of cents, which is equivalent to $335,544.31, a much larger amount than $12,500, as in the first offer!

We can use a similar procedure to find the sum of a geometric progression in general. Recall that the first n terms of such a progression are

$$a_1, a_1 r, a_1 r^2, \ldots, a_1 r^{n-2}, a_1 r^{n-1}$$

Thus

$$S_n = a_1 + a_1 r + a_1 r^2 + \cdots + a_1 r^{n-2} + a_1 r^{n-1} \tag{1}$$

We multiply by r to get

$$rS_n = a_1 r + a_1 r^2 + a_1 r^3 + \cdots + a_1 r^{n-1} + a_1 r^n \tag{2}$$

Subtracting Equation 2 from Equation 1 gives

$$S_n - rS_n = a_1 - a_1 r^n$$

or

$$S_n(1 - r) = a_1(1 - r^n)$$

If $r \neq 1$, we can divide by $1 - r$ to obtain

$$S_n = \frac{a_1(1 - r^n)}{1 - r}$$

EXAMPLE 2 For the geometric progression

$$4, -8, 16, -32, \ldots$$

find

a. r **b.** a_{10} **c.** S_{10}

Solution

a. Since r is the common ratio, $r = \dfrac{-8}{4} = -2$.

b. We have $a_1 = 4$ and $r = -2$, so

$$\begin{aligned}
a_{10} &= a_1 r^{n-1} \\
&= (4)(-2)^9 = (4)(-512) \\
&= -2048
\end{aligned}$$

c. Using the formula for S_n with $n = 10$, we get

$$\begin{aligned}
S_{10} &= \frac{4[1 - (-2)^{10}]}{1 - (-2)} = \frac{4(1 - 2^{10})}{1 + 2} \\
&= \frac{4(1 - 1024)}{3} = -1364
\end{aligned}$$

▲

EXAMPLE 3 Given a geometric progression with $a_1 = 2$ and $S_3 = 26$, find a_3 and r.

Solution Using the formulas for the nth term and the sum of n terms, we have

$$a_3 = a_1 r^2 = 2r^2 \quad \text{and} \quad S_3 = \frac{a_1(1 - r^3)}{1 - r} = \frac{2(1 - r^3)}{1 - r}$$

Since $1 - r^3 = (1 - r)(1 + r + r^2)$, the formula for S_3 can be simplified to

$$S_3 = 2(1 + r + r^2)$$

Thus, we have

$$26 = 2(1 + r + r^2)$$
$$13 = 1 + r + r^2$$

Problem 2 For the progression in Example 2, find
a. a_n **b.** S_n

Problem 3 Do Example 3 if $S_3 = 42$.

which gives a quadratic equation for r:

$r^2 + r - 12 = 0$

By factoring, we get

$(r + 4)(r - 3) = 0$

so that $r = -4$ or $r = 3$. If $r = -4$, then $a_3 = 2(-4)^2 = 32$, and if $r = 3$, then $a_3 = 2(3^2) = 18$.

We can check these results by writing the three terms of the progression:

For $r = -4$: $2, -8, 32$, which add to 26.

For $r = 3$: $2, 6, 18$, which add to 26.

Hence the correct answers are $a_3 = 32$, $r = -4$ or $a_3 = 18$, $r = 3$.   ▲

C. APPLICATIONS

EXAMPLE 4   When dropped on a hard surface, a Super Ball takes a sequence of bounces, each one about $\frac{9}{10}$ as high as the preceding one. If a Super Ball is dropped from a height of 10 ft, how far will the ball have traveled when it hits the surface the sixth time?

Solution   The ball traveled 10 ft when it was dropped, then bounced up 9 ft, dropped 9 ft, bounced up $(9)(\frac{9}{10}) = \frac{81}{10}$ ft, and so on. Thus, we must find the sum

$$10 + 2\left[(10)\left(\frac{9}{10}\right) + (10)\left(\frac{9}{10}\right)^2 + 10\left(\frac{9}{10}\right)^3 + (10)\left(\frac{9}{10}\right)^4 + (10)\left(\frac{9}{10}\right)^5\right]$$

We can factor $(10)(\frac{9}{10}) = 9$ out of the quantity in brackets to get

$$10 + 18\left[1 + \frac{9}{10} + \left(\frac{9}{10}\right)^2 + \left(\frac{9}{10}\right)^3 + \left(\frac{9}{10}\right)^4\right]$$

or $10 + 18[1 + 0.9 + (0.9)^2 + (0.9)^3 + (0.9)^4]$. The terms in the brackets form a geometric progression with

$a_1 = 1,$ $r = 0.9,$ $n = 5$

so the sum of these terms is

$$S_n = (1)\frac{1 - (0.9)^5}{1 - 0.9}$$

$$= \frac{1 - 0.59049}{1 - 0.9}$$

$$= \frac{0.40951}{0.1} = 4.0951$$

Thus the distance the ball traveled is

$10 + 18(4.0951) = 83.7118$ ft   ▲

EXAMPLE 5   A sum of money is invested in a small business. In the second year, the investment earned $1\frac{1}{4}$ times as much as in the first year, in the third year it earned $1\frac{1}{4}$ times as much as in the second year, and so on. If the investment earned a total of \$92,250 in 4 yr, how much did it earn in the third and fourth years?

Problem 4   Replace the Super Ball by a rubber ball, where each bounce is one-half as high as the preceding one, and answer the same question.

Problem 5   Do Example 4 with the total sum replaced by \$130,000 and the factor $1\frac{1}{4}$ replaced by $1\frac{1}{2}$.

Solution If we let a dollars be the amount earned in the first year, then the total amount earned in 4 yr is the sum of the geometric progression $a, \frac{5}{4}a, (\frac{5}{4})^2a, (\frac{5}{4})^3a$. Using the formula for the sum of such a progression, we get

$$S_4 = a \cdot \frac{1 - (\frac{5}{4})^4}{1 - \frac{5}{4}} = a \cdot \frac{1 - \frac{625}{256}}{1 - \frac{5}{4}}$$

$$= a \cdot \frac{256 - 625}{256 - 320} \qquad \text{Multiply numerator and denominator by 256.}$$

$$= \frac{369a}{64}$$

Since this sum is 92,250, we have the equation

$$\frac{369a}{64} = 92{,}250$$

Thus

$$a = \frac{(64)(92{,}250)}{369} = 16{,}000$$

This means that the amount earned the first year was \$16,000. Then we multiply by $\frac{5}{4}$ to get the second year's earnings, and multiply again by the same fraction to get the third year's earnings, and so on. This gives

$$\left(\frac{5}{4}\right)(\$16{,}000) = \$20{,}000, \qquad \text{second year}$$

$$\left(\frac{5}{4}\right)(\$20{,}000) = \$25{,}000, \qquad \text{third year}$$

$$\left(\frac{5}{4}\right)(\$25{,}000) = \$31{,}250, \qquad \text{fourth year}$$

As a check, you can add \$16,000 + \$20,000 + \$25,000 + \$31,250 to get the total, \$92,250. ▲

EXERCISE 12.3

NAME

CLASS

SECTION

ANSWERS

A., B. In Problems 1–10 a geometric progression is given. Find.
  a. a_1
  b. r
  c. a_n
  d. S_n

1. 3, 6, 12, 24, . . .

2. $\dfrac{1}{3}$, 1, 3, 9, . . .

3. 8, 24, 72, 216, . . .

4. $\dfrac{1}{5}$, $\dfrac{1}{10}$, $\dfrac{1}{20}$, $\dfrac{1}{40}$, . . .

5. 16, -4, 1, $-\dfrac{1}{4}$, . . .

6. 3, -1, $\dfrac{1}{3}$, $-\dfrac{1}{9}$, . . .

7. $-\dfrac{3}{5}$, $\dfrac{3}{2}$, $-\dfrac{15}{4}$, $\dfrac{75}{8}$, . . .

8. 60, -6, $\dfrac{6}{10}$, $-\dfrac{6}{100}$, . . .

9. $-\dfrac{3}{4}$, $-\dfrac{1}{4}$, $-\dfrac{1}{12}$, $-\dfrac{1}{36}$, . . .

10. $-\dfrac{5}{6}$, $-\dfrac{1}{3}$, $-\dfrac{2}{15}$, $-\dfrac{4}{75}$, . . .

1. a. _____	b. _____		
c. _____	d. _____		
2. a. _____	b. _____		
c. _____	d. _____		
3. a. _____	b. _____		
c. _____	d. _____		
4. a. _____	b. _____		
c. _____	d. _____		
5. a. _____	b. _____		
c. _____	d. _____		
6. a. _____	b. _____		
c. _____	d. _____		
7. a. _____	b. _____		
c. _____	d. _____		
8. a. _____	b. _____		
c. _____	d. _____		
9. a. _____	b. _____		
c. _____	d. _____		
10. a. _____	b. _____		
c. _____	d. _____		

In Problems 11–20 some values for a geometric progression are given. Find the other indicated values.

11. $a_1 = 1$, $S_3 = \dfrac{7}{4}$; a_3, r

12. $a_1 = 4$, $S_3 = 7$; a_3, r

13. $a_1 = 3$, $S_3 = 21$; a_3, r

14. $a_1 = \dfrac{1}{2}$, $S_3 = \dfrac{39}{50}$; a_3, r

15. $r = 2$, $S_8 = 1785$; a_1, a_8

16. $a_6 = -\dfrac{16}{27}$, $r = -\dfrac{1}{3}$; a_1, S_6

11. _____

12. _____

13. _____

14. _____

15. _____

16. _____

17. $a_1 = -4$, $a_n = 108$, $S_n = 80$; r, n

17. _____

18. $a_1 = \dfrac{3}{4}$, $a_n = 192$, $S_n = 225\dfrac{3}{4}$; r, n

18. _____

19. $a_1 = \dfrac{16}{125}$, $r = \dfrac{5}{2}$, $a_n = \dfrac{25}{2}$; n, S_n

19. _____

20. $a_1 = 7$, $r = 2$, $a_n = 896$; n, S_n

20. _____

C. Applications

21. The population of a certain town increases at the rate of 4% per year. If the present population is 20,000, what will the population be at the end of 5 yr?

21. _____

22. The number of bacteria in a culture increased from 320,000 at the beginning of the first day to 2,430,000 at the end of the fifth day. Find the daily rate of increase if this rate is assumed to be constant—that is, if the starting number and the numbers at the ends of the successive days form a geometric progression.

22. _____

23. The distance traveled in any swing by a point on a compound pendulum is 20% less than in the preceding swing. If the length of the first swing is 62.5 cm, find the total distance the point has traveled at the end of the fourth swing.

23. _____

24. A small business makes a net profit of $10,000 in its first year. If the net profit increases by 25% each year for the next 4 yr, what is the total net profit for these 5 yr?

24. _____

25. If the rate of increase in Problem 24 is 50%, what is the total net profit for the 5 yr?

25. _____

26. A polluted tank holds 100 gal of a poisonous chemical that mixes readily with water. After 25 gal of the chemical are drawn off, the tank is refilled with water. Then 25 gal of the mixture are drawn off, and the tank is again refilled with water. If this operation is performed until five batches have been drawn from the tank, how much of the original chemical remains?

26. _____

27. If the first three terms in an arithmetic progression are increased by 1, 3, and 13, respectively, the resulting numbers are in geometric progression. Find the original terms if their sum is 9.

27. _____

28. If the first three terms in an arithmetic progression are increased by 9, 7, and 9, respectively, the resulting numbers are in geometric progression. Find the original terms if their sum is 3.

28. _____

29. Roberto and Jimmy are golf pals. Yesterday, Roberto persuaded Jimmy to bet 1¢ on the first hole, 2¢ on the second hole, 4¢ on the third hole, and so on (doubling the bet on each successive hole). Roberto did not have very good luck! He won the first hole, lost the second hole, and continued winning a hole and losing a hole in that order for the remainder of the game. How much did Roberto lose in all? (*Hint:* You will need to use the number $2^{18} = 262,144$.)

29. _____

30. Refer to Problem 29. Suppose Roberto and Jimmy play again, making the same bets as before. This time Roberto wins the first two holes, loses the next two holes, and continues winning two holes and losing two holes in that order for the remainder of the game. How did Roberto come out this time?

30. _____

✓ **SKILL CHECKER**

Find the value of the fraction $\dfrac{a_1}{1-r}$ for the given values of a_1 and r.

31. $a_1 = 4, r = -\dfrac{1}{2}$ **32.** $a_1 = 9, r = -\dfrac{1}{3}$

33. $a_1 = -10, r = \dfrac{1}{5}$ **34.** $a_1 = -20, r = -\dfrac{1}{4}$

31. _____

32. _____

33. _____

34. _____

12.3 USING YOUR KNOWLEDGE

Geometric progressions occur in many important financial problems. For example, suppose a person deposits \$100 at the beginning of each month in a savings account that pays interest at the rate of 6%, compounded monthly. How much will be in this account at the end of 5 yr? In order to answer this question, you must recall that if P dollars earns interest for n periods at the rate i per period, the accumulated amount (principal plus interest) is

$$A = P(1 + i)^n$$

Interest at the nominal rate of 6% compounded monthly is equivalent to $\frac{6}{12}\% = 0.5\%$ per month. With these ideas in mind, we can set up the answer to our question.

The first \$100 deposited will earn interest for 60 mo (5 yr) and will accumulate to $\$100(1 + 0.005)^{60}$. The second deposit will earn interest for 59 mo and so will accumulate to $\$100(1 + 0.005)^{59}$. Similarly, the third deposit will accumulate to $\$100(1 + 0.005)^{58}$, and so on, until the last deposit, which accumulates for 1 mo to $\$100(1 + 0.005)$. It is convenient to start with the last deposit and to add the terms as follows to get the final number of dollars:

$$100(1.005) + 100(1.005)^2 + 100(1.005)^3 + \cdots$$
$$+ 100(1.005)^{59} + 100(1.005)^{60}$$

By factoring out $100(1.005)$, we can rewrite this sum as

$$100(1.005)[1 + 1.005 + (1.005)^2 + \cdots + (1.005)^{58} + (1.005)^{59}] \qquad (1)$$

Now you can see that the quantity in brackets is a geometric progression with $a_1 = 1$, $r = 1.005$, and $n = 60$, so you can use your knowledge to find the sum.

$$S_{60} = \frac{1 - (1.005)^{60}}{1 - 1.005} = \frac{(1.005)^{60} - 1}{0.005}$$

A calculator or an interest table will give you the value of $(1.005)^{60}$ to seven places as 1.3488501. Thus

$$S_{60} = \frac{0.3488501}{0.005} = 69.77003$$

The required total will then be

$$T = 100(1.005)(69.77003) = 7{,}011.89 \quad \text{See (1).}$$

and the final answer is \$7,011.89

1. Do the preceding problem if the interest rate is 4.8% compounded monthly.

2. Do the preceding problem if the interest rate is 7.2% compounded monthly.

1. _____

2. _____

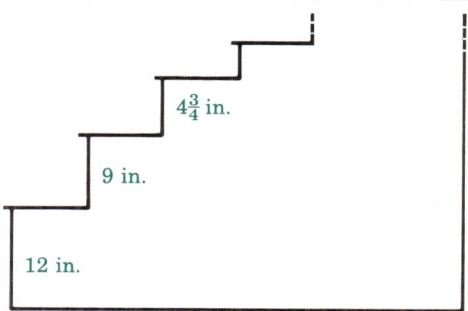

REVIEW

Before starting this section, you should know:

1. What a geometric progression is.
2. How to find the common ratio and the general term of a geometric progression.

OBJECTIVES

You should be able to:

A. Decide whether the sum of infinite geometric series exists or not and find it when it does.
B. Solve applications using the ideas discussed.

Look at the diagram. It shows a side view of a plan for a peculiar set of stairs. The first step is 1 ft high, and each succeeding step is three-fourths the height of the preceding step. Five of these steps would reach a height of a little over 3 ft. You can check this by taking the sum of the progression

$$1 + \frac{3}{4} + \left(\frac{3}{4}\right)^2 + \left(\frac{3}{4}\right)^3 + \left(\frac{3}{4}\right)^4$$

Now, consider this: How many steps do you think it would take to get to a height of 4 ft? Read on for the answer to this question.

In the preceding section, we considered the sum of a geometric progression with a finite number n of terms. Now, we ask what happens if n is allowed to increase without bound. To indicate that the number of terms is **infinite** (not finite), we write

$$a_1 + a_1 r + a_1 r^2 + \cdots + a_1 r^{n-1} + \cdots$$

(The $+ \cdots$ at the end means that the terms go on without stopping.) This indicated sum is called an **infinite geometric series.**

A. THE SUM OF A GEOMETRIC SERIES

The sum of the first n terms of a geometric series can be found by using the formula

$$S_n = \frac{a_1(1 - r^n)}{1 - r}$$

Now, if the absolute value of r is less than 1—that is, if $-1 < r < 1$—then r^n becomes smaller and smaller in absolute value as n increases. For example, if $r = 0.6$, then a calculation with logarithms or with a computer gives the approximate results:

$$r^{10} = (0.6)^{10} = 6.05 \times 10^{-3}$$
$$r^{100} = (0.6)^{100} = 6.53 \times 10^{-23}$$
$$r^{1000} = (0.6)^{1000} = 1.4 \times 10^{-222}$$

(The first of these has 2 zeros before the first significant digit, and the second and third have 22 and 221 zeros, respectively, before the first significant digit.) So we can make r^n as small as we like by taking n large enough.

Thus you can see that if $|r| < 1$, the factor $(1 - r^n)$ in the sum formula can be made as close to 1 as we wish by taking n large enough. Hence, as n becomes greater and greater, the sum S_n is more and more closely approximated by the expression

$$\frac{a_1}{1 - r}$$

We summarize this discussion by making the following definition.

> If $|r| < 1$, then the sum S of the geometric series with first term a_1 and common ratio r is defined to be
>
> $$S = a_1 + a_1 r + a_1 r^2 + \cdots + a_1 r^{n-1} + \cdots$$
>
> $$= \frac{a_1}{1 - r}$$

If $|r| \geq 1$, the sum of n terms does not get closer and closer to any number as n becomes larger and larger without bound. In this case, we say that S *does not exist*.

Let us return to the staircase question that was raised at the beginning of this section. We consider the infinite geometric series that would result if the staircase were extended indefinitely:

$$1 + \frac{3}{4} + \left(\frac{3}{4}\right)^2 + \left(\frac{3}{4}\right)^3 + \left(\frac{3}{4}\right)^4 + \cdots$$

In this series, $a_1 = 1$ and $r = \frac{3}{4}$, so that the sum is

$$S = \frac{a_1}{1 - r} = \frac{1}{1 - \frac{3}{4}} = \frac{1}{\frac{1}{4}} = 4$$

What does this mean? Just that we can never reach 4 ft with a finite number of steps! However, we can get as close to 4 ft as we wish by using enough steps.

EXAMPLE 1 Find the sum of the geometric series $4 - 2 + 1 - \frac{1}{2} + \cdots$.

Solution In this series, $a_1 = 4$ and $r = -\frac{1}{2}$, so

$$S = \frac{a_1}{1 - r} = \frac{4}{1 - \dfrac{-1}{2}}$$

$$= \frac{8}{2 + 1} = \frac{8}{3} \qquad \blacktriangle$$

EXAMPLE 2 If the following geometric series has a sum, find it.

$$(1.01) + (1.01)^2 + (1.01)^3 + \cdots$$

Solution For this series the ratio r is

$$\frac{(1.01)^2}{(1.01)} = 1.01$$

which is greater than 1. So the sum of this series does not exist. $\blacktriangle$

Problem 1 Find the sum of the geometric series $9 - 3 + 1 - \frac{1}{3} + \cdots$.

Problem 2 Do as in Example 2 for the geometric series $(1.01) - (1.01)^2 + (1.01)^3 - \cdots$.

B. APPLICATIONS

Geometric series can be used to express nonterminating repeating decimals as fractions. For example, the decimal

$$0.333\ldots = 0.\overline{3}$$

can be written as

$$\frac{3}{10} + \frac{3}{100} + \frac{3}{1000} + \cdots$$

which is a geometric series with

$$a_1 = \frac{3}{10} \quad \text{and} \quad r = \frac{1}{10}$$

Thus

$$S = \frac{\frac{3}{10}}{1 - \frac{1}{10}} = \frac{3}{9} = \frac{1}{3}$$

EXAMPLE 3 Find the fraction equivalent to the repeating decimal $0.414141\ldots$.

Solution We can write this decimal as

$$\frac{41}{100} + \frac{41}{(100)^2} + \frac{41}{(100)^3} + \cdots$$

which is a geometric series with

$$a_1 = \frac{41}{100} \quad \text{and} \quad r = \frac{1}{100}$$

Thus the sum of this series is

$$S = \frac{a_1}{1 - r} = \frac{\frac{41}{100}}{1 - \frac{1}{100}}$$

$$= \frac{\frac{41}{100}}{\frac{99}{100}} = \frac{41}{99} \qquad \blacktriangle$$

Problem 3 Do Example 3 for the decimal $0.412412412\ldots$.

EXAMPLE 4 Suppose a Super Ball is dropped from a height of 10 ft and takes a sequence of bounces, each one 0.9 as high as the preceding one. If the ball is assumed to continue bouncing indefinitely, find the total distance it travels.

Solution The total distance the ball travels can be written as the series

$$10 + 2[(10)(0.9) + (10)(0.9)^2 + 10(0.9)^3 + \cdots]$$

Since the series in the brackets is an infinite geometric progression with first term $a = (10)(0.9)$ and ratio $r = 0.9$, the sum is

$$S = \frac{(10)(0.9)}{1 - 0.9} = \frac{9}{0.1} = 90$$

Thus the distance traveled is

$$10 + (2)(90) = 190 \text{ ft} \qquad \blacktriangle$$

Problem 4 A rubber ball is dropped from a height of 12 ft and takes a sequence of bounces, each one 0.6 as high as the preceding one. If the ball continues bouncing indefinitely, find the total distance it travels.

A. In Problems 1–10 an infinite geometric series is given. Find the sum if it exists.

1. $6 + 3 + 1\frac{1}{2} + \cdots$

2. $12 + 4 + 1\frac{1}{3} + \cdots$

3. $(-6) + (-3) + \left(-\frac{3}{2}\right) + \cdots$

4. $(-8) + (-4) + (-2) + \cdots$

5. $2 - 1 + \frac{1}{2} - \frac{1}{4} + \cdots$

6. $9 - 3 + 1 - \frac{1}{3} + \cdots$

7. $4 - 8 + 16 - 32 + \cdots$

8. $(-5) + (-10) + (-20) + \cdots$

9. $\frac{1}{10} + \frac{1}{5} + \frac{2}{5} + \cdots$

10. $0.0001 - 0.001 + 0.01 - \cdots$

B. In Problems 11–20 find a fraction equivalent to the given repeating decimal.

11. $0.555\ldots$

12. $0.666\ldots$

13. $0.181818\ldots$

14. $0.242424\ldots$

15. $4.050505\ldots$

16. $2.313131\ldots$

17. $2.3161616\ldots$

18. $4.1272727\ldots$

19. $0.140140140\ldots$

20. $1.123123123\ldots$

21. A rubber ball is dropped from a height of 8 ft. It makes a sequence of bounces, each three-fourths the height of the preceding bounce. About how far does the ball travel before coming to rest?

22. A pendulum on each separate swing describes an arc whose length is 98% of the length of the preceding swing. If the first arc is 12 in. long, about how far does the pendulum travel before it comes to rest?

✓ SKILL CHECKER

Expand each of the following.

23. $(a + b)^2$

24. $(x - y)^2$

25. $(x + y)^3$

26. $(a - b)^3$

27. $(y - 2z)^3$

28. $(2a + b)^3$

29. $\left(\dfrac{1}{x} - \dfrac{1}{y}\right)^3$

30. $\left(\dfrac{2}{y} - \dfrac{1}{2}\right)^3$

23. _____

24. _____

25. _____

26. _____

27. _____

28. _____

29. _____

30. _____

12.4 USING YOUR KNOWLEDGE

In the preceding sections, you have learned what we mean by a sequence, an arithmetic progression, and a geometric progression. You can use this knowledge in the following problems.

In Problems 1–12, the first four terms of a series are given. Determine if these terms form an *arithmetic progression,* a *geometric progression,* or *neither.* If the terms form a geometric progression and the series is extended to form an infinite geometric series, find the sum if it has one.

1. $1 + \dfrac{2}{5} + \dfrac{4}{25} + \dfrac{8}{125} + \cdots$

2. $1 - \dfrac{1}{2} + \dfrac{1}{3} - \dfrac{1}{4} + \cdots$

3. $1 + \dfrac{2}{5} - \dfrac{1}{5} - \dfrac{4}{5} - \cdots$

4. $1 + 3 + 4 + 7 + \cdots$

5. $2 + \dfrac{7}{4} + \dfrac{49}{32} + \dfrac{343}{256} + \cdots$

6. $2 + 1\dfrac{3}{4} + 1\dfrac{1}{2} + 1\dfrac{1}{4} + \cdots$

7. $5 + 4 + 2 - 1 - \cdots$

8. $4 + 2 + 0 - 2 - \cdots$

9. $\dfrac{1}{2} + \dfrac{2}{3} + \dfrac{3}{4} + \dfrac{4}{5} + \cdots$

10. $-12 + 4 - \dfrac{4}{3} + \dfrac{4}{9} - \cdots$

11. $4 - 2 + 1 - \dfrac{1}{2} + \cdots$

12. $-6 - 3 + 0 + 3 + \cdots$

1. _____

2. _____

3. _____

4. _____

5. _____

6. _____

7. _____

8. _____

9. _____

10. _____

11. _____

12. _____

12.5 THE BINOMIAL EXPANSION

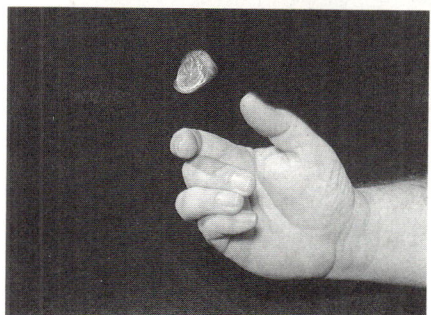

REVIEW

Before starting this section, you should know how to:

1. Simplify an algebraic expression.
2. Expand the binomials $(a + b)^2$ and $(a + b)^3$.

OBJECTIVES

You should be able to:

A. Use the binomial theorem to expand and simplify a power of a binomial.

B. Solve applications using the ideas discussed.

If a fair coin is tossed, it can turn up in one of two ways: heads (H) or tails (T). If two coins are tossed, the outcome can be

HH, HT, TH, TT

Similarly, for three coins, the possible outcomes are

HHH, HHT, HTH, THH, HTT, THT, TTH, TTT

Suppose we make a table of the number of ways in which heads (H) can turn up.

For one coin: H T
   Number of heads: 1   0
   Number of ways: 1   1

For two coins: HH HT, TH TT
   Number of heads: 2   1   0
   Number of ways: 1   2   1

For three coins: HHH HHT, HTH, THH HTT, THT, TTH TTT
   Number of heads: 3   2   1   0
   Number of ways: 1   3   3   1

Now, let us arrange the numbers of ways in which the coins can fall as follows:

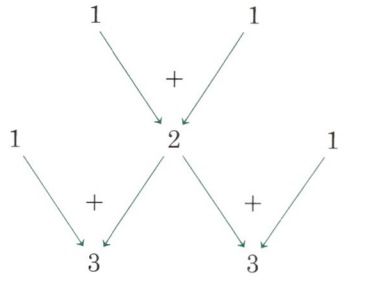

Notice that the 2 in the second row is the sum of the two elements indicated by the arrows in the first row. The first 3 in the third row is the sum of the two elements indicated by the arrows in the second row. It is the same for the second 3. If we continue to construct rows

in the same way, we get the following triangular array:

$$
\begin{array}{ccccccc}
 & & & 1 & 1 & & & \\
 & & 1 & 2 & 1 & & \\
 & 1 & 3 & 3 & 1 & \\
 1 & 4 & 6 & 4 & 1 \\
1 & 5 & 10 & 10 & 5 & 1 \\
 & & \vdots & & & &
\end{array}
$$

This array of numbers, known as **Pascal's triangle,** is named after Blaise Pascal, a seventeenth-century French mathematician. Pascal wrote a book about this triangle and its properties. One of these interesting properties is that the nth row gives the number of ways in which n coins can fall. For example, for 5 coins we have:

Number of heads:	5	4	3	2	1	0
Number of ways:	1	5	10	10	5	1

A. THE BINOMIAL EXPANSION

We will see next a somewhat different way of arriving at the numbers in Pascal's triangle. From our previous work, we know that

$$(a + b)^1 = a + b$$
$$(a + b)^2 = a^2 + 2ab + b^2$$
$$(a + b)^3 = a^3 + 3a^2b + 3ab^2 + b^3$$

If we multiply the last equation by $(a + b)$ to form $(a + b)^4$, we get

$$
\begin{array}{r}
a^3 + 3a^2b + 3ab^2 + b^3 \\
a + b \\
\hline
a^4 + 3a^3b + 3a^2b^2 + \ ab^3 \\
a^3b + 3a^2b^2 + 3ab^3 + b^4 \\
\hline
a^4 + 4a^3b + 6a^2b^2 + 4ab^3 + b^4
\end{array}
$$

Notice that the coefficients, 1, 4, 6, 4, 1, are exactly the numbers in the fourth row of Pascal's triangle. Also, an examination of the multiplication shows that these coefficients are obtained by doing the additions indicated in our construction of this triangle. By proceeding in the same way, you can see that

$$(a + b)^5 = a^5 + 5a^4b + 10a^3b^2 + 10a^2b^3 + 5ab^4 + b^5$$

We have now obtained five special cases of the expansion of $(a + b)^n$. Before trying to write the general formula, let us see what seems to be common to these five special cases. If $n = 1, 2, 3, 4,$ or 5

1. The first term is a^n and the last term is b^n.
2. The exponents of a decrease by 1 from term to term; the exponents of b increase by 1 from term to term; and the sum of the two exponents in each term is n.
3. If the coefficient of any term is multiplied by the exponent of a in that term and is divided by the number of the term, the coefficient of the next term is obtained.

To verify (3), note that the coefficient of the second term in each case is $\frac{n}{1}$—that is, the exponent of a in the first term divided by 1. The coefficient of the third term is $\dfrac{n(n-1)}{1 \cdot 2}$, which is

$$\dfrac{2(2-1)}{1 \cdot 2} = 1, \qquad \text{for } n = 2$$

$$\dfrac{3(3-1)}{1 \cdot 2} = 3, \qquad \text{for } n = 3$$

$$\dfrac{4(4-1)}{1 \cdot 2} = 6, \qquad \text{for } n = 4$$

$$\dfrac{5(5-1)}{1 \cdot 2} = 10, \qquad \text{for } n = 5$$

Assuming that these characteristics are true for all positive integer values of n, we have

$$(a + b)^n = a^n + \frac{n}{1}a^{n-1}b + \frac{n(n-1)}{1 \cdot 2}a^{n-2}b^2$$
$$+ \frac{n(n-1)(n-2)}{1 \cdot 2 \cdot 3}a^{n-3}b^3 + \cdots + b^n$$

This identity is called the **binomial expansion.** It has been proved that this identity holds for all positive integer exponents.

You should note that the expansion has $n + 1$ terms and that the coefficients are the same counting from the left or the right. Note also that the exponent of b in each term is equal to the last factor in the denominator of the coefficient and that the numerator and the denominator of each coefficient (starting with the second term) have the same number of factors. Let us check the formula for $n = 5$.

$$(a + b)^5 = a^5 + \frac{5}{1}a^4b^1 + \frac{5 \cdot 4}{1 \cdot 2}a^3b^2 + \frac{5 \cdot 4 \cdot 3}{1 \cdot 2 \cdot 3}a^2b^3$$
$$+ \frac{5 \cdot 4 \cdot 3 \cdot 2}{1 \cdot 2 \cdot 3 \cdot 4}ab^4 + \frac{5 \cdot 4 \cdot 3 \cdot 2 \cdot 1}{1 \cdot 2 \cdot 3 \cdot 4 \cdot 5}b^5$$

which simplifies to

$$(a + b)^5 = a^5 + 5a^4b + 10a^3b^2 + 10a^2b^3 + 5ab^4 + b^5$$

This agrees with our previous result.

EXAMPLE 1 Expand $(x - 2y)^4$.

Solution Here $n = 4$ and we have x in place of a and $-2y$ in place of b. If we make these replacements in the binomial expansion, we find

$$(x - 2y)^4 = x^4 + \frac{4}{1}x^3(-2y) + \frac{4 \cdot 3}{1 \cdot 2}x^2(-2y)^2$$
$$+ \frac{4 \cdot 3 \cdot 2}{1 \cdot 2 \cdot 3}x(-2y)^3 + \frac{4 \cdot 3 \cdot 2 \cdot 1}{1 \cdot 2 \cdot 3 \cdot 4}(-2y)^4$$
$$= x^4 + 4x^3(-2y) + 6x^2(-2y)^2 + 4x(-2y)^3 + (-2y)^4$$
$$= x^4 - 8x^3y + 24x^2y^2 - 32xy^3 + 16y^4 \qquad \blacktriangle$$

EXAMPLE 2 Find the fourth term in the expansion of $(2x - 1)^7$.

Solution Here, $n = 7$, $a = 2x$, $b = -1$, and we want the fourth term. This will be

$$\frac{n(n-1)(n-2)}{1 \cdot 2 \cdot 3}a^4b^3$$

Problem 1 Expand $(2x - y)^4$.

Problem 2 Find the fifth term in the expansion of $(x - 2)^7$.

which becomes

$$\frac{7(7-1)(7-2)}{1 \cdot 2 \cdot 3}(2x)^4(-1)^3 = \frac{7 \cdot 6 \cdot 5}{1 \cdot 2 \cdot 3}(16x^4)(-1)$$

$$= -560x^4 \qquad \blacktriangle$$

B. APPLICATIONS

The binomial coefficients for $(a + b)^n$ furnish us with a row of Pascal's triangle. We have seen that the numbers in such a row tell us in how many ways a stated number of heads can come up if n coins are tossed. For example, the first coefficient, 1, is the number of ways that n heads can occur; the second coefficient, n, is the number of ways that $n - 1$ heads can occur; the third coefficient, $\frac{n(n-1)}{1 \cdot 2}$, is the number of ways that $n - 2$ heads can occur, and so on.

EXAMPLE 3 Six coins are tossed. In how many ways can exactly two heads come up?

Solution Since $2 = 6 - 4$, the answer is given by

$$\frac{n(n-1)(n-2)(n-3)}{1 \cdot 2 \cdot 3 \cdot 4} \qquad \text{for } n = 6, \text{ that is,}$$

$$\frac{6 \cdot 5 \cdot 4 \cdot 3}{1 \cdot 2 \cdot 3 \cdot 4} = 15 \qquad \blacktriangle$$

It is shown in probability theory that if p is the probability of an event occurring favorably in one trial, then the probability of its occurring favorably r times in n trials is given by the formula

$$\frac{n(n-1) \cdots (n-r+1)}{1 \cdot 2 \cdots r} p^r(1-p)^{n-r}$$

For example, if a fair coin is tossed, the probability of its coming up heads is $\frac{1}{2}$. So if the coin is tossed 6 times, the probability of getting exactly 3 heads is

$$\frac{6 \cdot 5 \cdot 4}{1 \cdot 2 \cdot 3}\left(\frac{1}{2}\right)^3\left(1 - \frac{1}{2}\right)^3 = 20\left(\frac{1}{2}\right)^6 = \frac{20}{64} = \frac{5}{16}$$

EXAMPLE 4 A fair coin is tossed 8 times. Find the probability of getting exactly 4 heads.

Solution Here, $n = 8$, $r = 4$, and $p = \frac{1}{2}$. Substituting in the probability formula, we get

$$\frac{8 \cdot 7 \cdot 6 \cdot 5}{1 \cdot 2 \cdot 3 \cdot 4}\left(\frac{1}{2}\right)^4\left(1 - \frac{1}{2}\right)^4 = 70\left(\frac{1}{2}\right)^8$$

$$= \frac{70}{256} = \frac{35}{128} \qquad \blacktriangle$$

Problem 3 Seven coins are tossed. In how many ways can exactly three heads come up?

Problem 4 A fair coin is tossed 10 times. Find the probability of getting exactly 5 heads.

ANSWERS (to problems on page 945)
1. $16x^4 - 32x^3y + 24x^2y^2 - 8xy^3 + y^4$
2. $560x^3$

NAME

CLASS

SECTION

A. In Problems 1–12 use the binomial expansion to expand the given binomial.

ANSWERS

1. $(a + 3b)^4$

2. $(a - 3b)^4$

3. $(x + 4)^4$

4. $(4x - 1)^4$

5. $(2x - y)^5$

6. $(x + 2y)^5$

7. $(2x + 3y)^5$

8. $(3x - 2y)^5$

9. $\left(\dfrac{1}{x} - \dfrac{y}{2}\right)^4$

10. $\left(\dfrac{x}{2} + \dfrac{3}{y}\right)^4$

11. $(x + 1)^6$

12. $(y - 1)^6$

1. _____

2. _____

3. _____

4. _____

5. _____

6. _____

7. _____

8. _____

9. _____

10. _____

11. _____

12. _____

In Problems 13–20 find the indicated term in the expansion of the given binomial.

13. $(x - 3)^6$; 4th term

14. $(x + 2)^6$; 4th term

15. $(x + 2y)^7$; 3rd term

16. $(y - 2z)^7$; 3rd term

17. $(2x - 1)^8$; 5th term

18. $(y + 2z)^8$; 5th term

19. $\left(\dfrac{a}{2} - 1\right)^5$; 3rd term

20. $\left(y + \dfrac{z}{2}\right)^5$; 4th term

13. _____

14. _____

15. _____

16. _____

17. _____

18. _____

19. _____

20. _____

B. Applications

21. Six coins are tossed. In how many ways can exactly three heads come up?

22. Six coins are tossed. In how many ways can exactly two heads come up?

23. Nine coins are tossed. In how many ways can exactly three heads come up?

24. Nine coins are tossed. In how many ways can exactly four heads come up?

21. _____

22. _____

23. _____

24. _____

25. A fair coin is tossed six times. Find the probability of getting exactly two heads.

25. _____

26. A fair coin is tossed six times. Find the probability of getting exactly three heads.

26. _____

27. A fair coin is tossed nine times. Find the probability of getting exactly three heads.

27. _____

28. A fair coin is tossed nine times. Find the probability of getting exactly four heads.

28. _____

29. A single die (plural, dice) is rolled four times. Find the probability that a 4 comes up exactly twice. (*Note:* The probability that a specified number, 1, 2, 3, 4, 5, or 6, comes up in a single throw is $\frac{1}{6}$.)

29. _____

30. Do Problem 29 if the die is rolled five times.

30. _____

12.5 USING YOUR KNOWLEDGE

In problems involving compound interest, we often need the value of the quantity $(1 + i)^n$, where i is the periodic interest rate and n is the number of periods. For example, if a sum of money earns interest for one year at 12% compounded monthly, the final accumulated amount is the original sum multiplied by $(1.01)^{12}$.

We can use our knowledge of the binomial expansion to calculate the decimal value of $(1 + i)^n$ if i is a small rate and n is not too large. Thus we can write $(1.01)^{12} = (1 + 0.01)^{12}$ and let $a = 1$, $b = 0.01$, $n = 12$ to get

$$(1.01)^{12} = 1 + \frac{12}{1}(0.01) + \frac{12 \cdot 11}{1 \cdot 2}(0.01)^2 + \frac{12 \cdot 11 \cdot 10}{1 \cdot 2 \cdot 3}(0.01)^3$$

$$+ \frac{12 \cdot 11 \cdot 10 \cdot 9}{1 \cdot 2 \cdot 3 \cdot 4}(0.01)^4 + \cdots + (0.01)^{12}$$

$$= 1 + 0.12 + 0.0066 + 0.00022 + 0.00000495 + \cdots$$

$$= 1.126825 \quad \text{(rounded to six decimal places)}$$

Use the binomial expansion to find the value of each of the following correct to five decimal places.

1. $(1.01)^8$ **2.** $(1.01)^{10}$

1. _____

2. _____

3. $(1.02)^6$ **4.** $(1.02)^9$

3. _____

4. _____

5. $(1.04)^8$ **6.** $(1.03)^8$

5. _____

6. _____

7. $(0.98)^6$
   (*Hint:* $0.98 = 1 - 0.02$.)   **8.** $(0.99)^6$

7. _____

8. _____

SUMMARY

SECTION	ITEM	MEANING	EXAMPLE
12.1A	Sequence	A set of numbers arranged according to some given law	2, 4, 6, . . .
12.1A	Terms a_1, a_2, a_3, and so on	The numbers of a sequence	The terms of 2, 4, 6, . . . are $c_1 = 2$, $a_2 = 4$, $a_3 = 6$, and so on.
12.1B	General term	The formula that generates the terms of the sequence	In the sequence 2, 4, 6, . . . , the general term is $a_n = 2n$.
12.2A	Arithmetic progression	A sequence in which each term after the first is obtained by adding a quantity d to the preceding term	2, 4, 6, . . . is an arithmetic progression.
12.2A	Common difference d	The difference between two successive terms of an arithmetic progression	In the arithmetic progression 3, 6, 9, the common difference is $d = 3$.
12.2A	General term of an arithmetic progression	$a_n = a_1 + (n - 1)d$	In the progression 8, 12, 16, . . . , the general term is $a_n = 8 + (n - 1)(4)$.
12.2B	Sum S_n of an arithmetic progression	$S_n = \dfrac{1}{2}n(a_1 + a_n)$	In the progression 2, 4, 6, . . . , $S_6 = \dfrac{1}{2} \cdot 6(2 + 12) = 42.$
12.3A	Geometric progression	A sequence in which each term after the first is formed by multiplying the preceding term by a constant r	3, 6, 12, . . . is a geometric progression.
12.3A	Common ratio r	The ratio of two successive terms of a geometric progression	In the geometric progression 3, 6, 12, . . . , $r = 2$.
12.3A	General term of a geometric progression	$a_n = a_1 r^{n-1}$	In the geometric progression 3, 6, 12, . . . , $a_n = 3 \cdot 2^{n-1}$
12.3B	Sum S_n of a geometric progression.	$S_n = \dfrac{a_1(1 - r^n)}{1 - r}$	For the geometric progression 3, 6, 12, . . . , $S_6 = \dfrac{3(1 - 2^6)}{1 - 2}$ $= \dfrac{3 \cdot (-63)}{-1}$ $= 189.$
12.4	Infinite geometric series	A series of the form $a_1 + a_1 r + a_1 r^2 + \cdots + a_1 r^{n-1} + \cdots$	$2 + 1 + \dfrac{1}{2} + \cdots$, where $a_1 = 2$ and $r = \dfrac{1}{2}$

(Continued)

SECTION	ITEM	MEANING	EXAMPLE
12.4A	Sum S of a geometric series with $\lvert r \rvert < 1$	$S = \dfrac{a_1}{1-r}$	For $2 + 1 + \dfrac{1}{2} + \cdots$, $$S = \dfrac{2}{1 - \frac{1}{2}} = 4$$
12.5	Pascal's triangle	An arrangement of numbers of the form $$\begin{array}{ccccccc} & & 1 & & 1 & & \\ & 1 & & 2 & & 1 & \\ 1 & & 3 & & 3 & & 1 \\ & & & \vdots & & & \end{array}$$	
12.5A	Binomial expansion $(a+b)^n$	An expansion in which the general term is $$\dfrac{n(n-1)\cdots(n-r+1)}{1 \cdot 2 \cdot 3 \cdots r} \cdot a^{n-r+1}b^{r-1}$$	$(a+b)^3$ $$= a^3 + \dfrac{3}{1}a^2b + \dfrac{3 \cdot 2}{1 \cdot 2}ab^2$$ $$+ \dfrac{3 \cdot 2 \cdot 1}{1 \cdot 2 \cdot 3}b^3$$ $$= a^3 + 3a^2b + 3ab^2 + b^3$$

ANSWERS

(If you need help with these exercises, look in the section indicated in brackets.)

1. [12.1A] For the given sequence, find a_6 and a_n.
   a. The sequence of odd counting numbers: 1, 3, 5, 7, . . .
   b. The sequence of multiples of 3: 3, 6, 9, 12, . . .

   1. a. _____
      b. _____

2. [12.1A] Each card in a pack of four index cards is cut in half, and the halves are put in a single pile. This step is repeated again and again.
   a. How many cards are in the pack after the fourth cut?
   b. How many cards are in the pack after the nth cut?

   2. a. _____
      b. _____

3. [12.1B] Find the first three terms and the tenth term of the sequence whose general term is

   a. $\frac{1}{3}(n^2 + 1)$

   b. $\frac{1}{4}(n^2 + n)$

   3. a. _____
      b. _____

4. [12.1B] Find the sequence that corresponds to the function.
   a. $a(n) = n^2 + 2$, $n = 1, 2, 3, \ldots$
   b. $a(n) = 2n^2 - 1$, $n = 1, 2, 3, \ldots$

   4. a. _____
      b. _____

5. [12.1C] A painting doubles in value every 100 yr. Find the value of this painting in the year 2000 if it was worth $1000 in the year
   a. 1400
   b. 1600

   5. a. _____
      b. _____

6. [12.2A] For the given arithmetic progression, find d and a_{10}.
   a. 3, 6, 9, 12, . . .
   b. 4, 8, 12, 16, . . .

   6. a. _____
      b. _____

7. [12.2A] For the given arithmetic progression, find a_n.
   a. 3, 6, 9, 12, . . .
   b. 4, 8, 12, 16, . . .

   7. a. _____
      b. _____

8. [12.2B] The distance (in feet) that a free-falling body falls in each second, starting with the first second, is given by the arithmetic progression 16, 48, 80, 112, ... Find what distance the body falls in the
   a. Sixth second
   b. Eighth second

   8. a. _____
      b. _____

9. [12.2B] The sixth term of an arithmetic progression is 24. Find the first term and the common difference if the sum of the first six terms is
   a. 114
   b. 99

   9. a. _____
      b. _____

10. [12.2B] The sum of the first six terms of an arithmetic progression is 54. Find the first term and the common difference if the sixth term is
    a. 14
    b. 16.5

10. a. _____
    b. _____

11. [12.2C] A machine valued at $80,000 depreciates $8000 the first year, $7800 the second year, $7600 the third year, and so on. Find the value of the machine at the end of
    a. 8 yr
    b. 10 yr

11. a. _____
    b. _____

12. [12.2C] Juan is saving his pennies. He saved 5¢ the first week, 9¢ the second week, 13¢ the third week, and so on, in arithmetic progression. How many weeks would it take Juan to get a total savings of
    a. $2.30
    b. $3.24

12. a. _____
    b. _____

13. [12.3A] For the given geometric progression, find r and a_6.
    a. 3, 6, 12, 24, . . .

    b. $4, 2, 1, \frac{1}{2}, \ldots$

13. a. _____
    b. _____

14. [12.3A] For the given geometric progression, find r and a_n.
    a. 1, 4, 16, 64, . . .

    b. $4, -2, 1, -\frac{1}{2}, \ldots$

14. a. _____
    b. _____

15. [12.3B] For the given geometric progression, find a_n and S_n.
    a. $2, -4, 8, -16, \ldots$

    b. $1, -\frac{1}{2}, \frac{1}{4}, -\frac{1}{8}, \ldots$

15. a. _____
    b. _____

16. [12.3C] A rubber ball, dropped on a hard surface, takes a sequence of bounces, each one-half as high as the preceding one. If the ball is dropped from a height of 12 ft, find how far the ball will have traveled when it hits the surface the
    a. Sixth time
    b. Eighth time

16. a. _____
    b. _____

17. [12.3C] A sum of money, invested in a business, earned $1\frac{1}{2}$ times as much each year as in the preceding year. How much did the investment earn in the fourth year if the total earned in 4 yr was
    a. $16,250
    b. $24,375

17. a. _____
    b. _____

18. [12.4A] Find the sum (if it exists) of the geometric series.
    a. $32 - 16 + 8 - 4 + \cdots$

    b. $18 - 6 + 2 - \frac{2}{3} + \cdots$

18. a. _____
    b. _____

19. [12.4A] Find the sum (if it exists) of the geometric series.
    a. $1 + (1.005) + (1.005)^2 + (1.005)^3 + \cdots$
    b. $1 - (1.001) + (1.001)^2 - (1.001)^3 + \cdots$

19. a. _____
    b. _____

20. [12.4B] Find the fraction equivalent to the repeating decimal.
  a. 0.313131 . . .
  b. 0.324324324 . . .

20. a. _____
    b. _____

21. [12.4B] A rubber ball is dropped and takes a sequence of bounces, each one 0.5 as high as the preceding one. If the ball continues bouncing indefinitely, find the total distance it travels, given that it was dropped from a height of
  a. 8 ft
  b. 6 ft

21. a. _____
    b. _____

22. [12.5A] Expand.
  a. $(a - 3b)^4$
  b. $(3a + 2b)^4$

22. a. _____
    b. _____

23. [12.5A] Find the fifth term in the expansion of
  a. $(x + 2)^8$
  b. $(3x - 1)^7$

23. a. _____
    b. _____

24. [12.5B] Seven coins are tossed. In how many ways can exactly
  a. 4 heads come up?
  b. 3 heads come up?

24. a. _____
    b. _____

25. [12.5B] A fair coin is tossed 7 times. Find the probability of getting exactly
  a. 4 heads
  b. 3 heads

25. a. _____
    b. _____

NAME

CLASS

SECTION

(Answers on pages 958–959)

ANSWERS

1. For the sequence of odd natural numbers $1, 3, 5, 7, \ldots$, find
   a. a_{10}, the tenth term
   b. a_n, the general term

   1. a. _____
      b. _____

2. Each card in a pack of three index cards is cut in half, and the halves are put in a single pile. This step is repeated again and again.
   a. Write the sequence that gives the number of cards in the deck after each step.
   b. How many cards are in the pack after the fourth step?
   c. How many cards are in the pack after the nth step?

   2. a. _____
      b. _____
      c. _____

3. Find the first three terms and the eighth term of the sequence whose general term is

   $$a_n = \frac{n(n + 1)}{2}$$

   3. _____

4. Find the sequence that corresponds to the function $a(n) = 3n - 1$, $n = 1, 2, 3, \ldots$.

   4. _____

5. A sculpture by a famous artist doubles in value every 50 yr. Find the value of the sculpture in the year 2000 if it was worth $500 in the year 1500.

   5. _____

6. For the arithmetic progression $5, 8, 11, 14, \ldots$, find
   a. d
   b. a_{10}

   6. a. _____
      b. _____

7. For the arithmetic progression $5, 8, 11, 14, \ldots$, find a_n.

   7. _____

8. Sally draws a sequence of circles. She starts with a row of three circles, then adds two circles to get the second row, adds two more circles to get the third row, and so on. Find how many circles she would have in the
   a. Fifth row
   b. Tenth row
   c. nth row

   8. a. _____
      b. _____
      c. _____

9. The distance (in feet) that a free-falling body falls in each second, starting with the first second, is given by the arithmetic progression $16, 48, 80, 112, \ldots$. Find the distance that the body falls in 7 sec.

   9. _____

10. The sum of the first eight terms of an arithmetic progression is 172 and the eighth term is 32. Find.
  a. a_1, the first term
  b. d, the common difference

10. a. _____
 b. _____

11. A piece of machinery valued at $50,000 depreciates $8000 the first year, $7500 the second year, $7000 the third year, and so on. Find the value of this piece of machinery at the end of 6 yr.

11. _____

12. Natasha was trying to save her pennies. She saved 5¢ the first week, 8¢ the second week, 11¢ the third week, and so on, in arithmetic progression. After a few weeks, Natasha had saved a total of $1.85. For how many weeks had she been saving?

12. _____

13. For the geometric progression $3, 1, \frac{1}{3}, \frac{1}{9}, \ldots$, find.
  a. r
  b. a_6

13. a. _____
 b. _____

14. For the progression in Problem 13, find a_n.

14. _____

15. For the geometric progression $2, -4, 8, -16, \ldots$, find.
  a. r
  b. a_8
  c. S_8

15. a. _____
 b. _____
 c. _____

16. A rubber ball, dropped on a hard surface, takes a sequence of bounces, each one $\frac{2}{3}$ as high as the preceding one. If this ball is dropped from a height of 9 ft, how far will it have traveled when it hits the surface the fifth time?

16. _____

17. A certain sum of money is invested in a business. In each year this investment earns $1\frac{1}{2}$ times as much as in the preceding year. If the investment earned a total of $26,000 in four years, how much did it earn in the fourth year?

17. _____

18. Find the sum of the geometric series $8 - 4 + 2 - 1 + \cdots$.

18. _____

19. Find the sum of the geometric series $1.001 + (1.001)^2 + (1.001)^3 + \cdots$ if the sum exists.

19. _____

20. Find the fraction that is equivalent to the repeating decimal $0.312312312\ldots$.

20. _____

21. If the ball of Problem 16 is assumed to continue bouncing indefinitely, find the total distance it would travel.

22. Expand $(2a - b)^4$.

23. Find the fourth term in the expansion of $(a - 2b)^7$.

24. Seven coins are tossed. In how many ways can exactly four heads turn up?

25. A fair coin is tossed 12 times. Find the probability of getting exactly 6 heads.

21. _____

22. _____

23. _____

24. _____

25. _____

IF YOU MISSED QUESTION	SECTION	EXAMPLES	PAGE	ANSWERS
1a	12.1	1	909–910	**1. a.** $a_{10} = 19$
1b	12.1	1	909–910	**b.** $a_n = 2n - 1$
2a	12.1	2, 3	910	**2. a.** $6, 12, 24 \ldots$
2b	12.1	2, 3	910	**b.** 48
2c	12.1	2, 3	910	**c.** $3 \cdot 2^n$
3	12.1	4	910–911	**3.** $a_1 = 1, a_2 = 3, a_3 = 6, a_8 = 36$
4	12.1	5	911	**4.** $2, 5, 8, 11 \ldots$
5	12.1	6	911	**5.** $\$512{,}000$
6a	12.2	1	918	**6. a.** $d = 3$
6b	12.2	1	918	**b.** $a_{10} = 32$
7	12.2	1	918	**7.** $a_n = 3n + 2$
8a	12.2	1	918	**8. a.** 11
8b	12.2	1	918	**b.** 21
8c	12.2	1	918	**c.** $2n + 1$
9	12.2	2	919	**9.** 784 ft
10a	12.2	3	920	**10. a.** $a_1 = 11$
10b	12.2	3	920	**b.** $d = 3$
11	12.2	4	920	**11.** $\$9500$
12	12.2	5	921	**12.** 10
13a	12.3	1	927–928	**13. a.** $r = \dfrac{1}{3}$
13b	12.3	1	927–928	**b.** $a_6 = \dfrac{1}{81}$
14	12.3	1	927–928	**14.** $a_n = \left(\dfrac{1}{3}\right)^{n-2}$
15a	12.3	2	929	**15. a.** $r = -2$
15b	12.3	2	929	**b.** $a_8 = -256$
15c	12.3	2	929	**c.** $S_8 = -170$
16	12.3	4	929	**16.** $37\dfrac{8}{9} \text{ ft}$, or about 37.9 ft
17	12.3	5	930–931	**17.** $\$10{,}800$
18	12.4	1	938	**18.** $\dfrac{16}{3}$, or $5\dfrac{1}{3}$
19	12.4	2	938	**19.** Sum does not exist.

IF YOU MISSED QUESTION	SECTION	EXAMPLES	PAGE	ANSWERS
20	12.4	3	939	20. $\dfrac{104}{333}$
21	12.4	4	939	21. 45 ft
22	12.5	1	945	22. $16a^4 - 32a^3b + 24a^2b^2 - 8ab^3 + b^4$
23	12.5	2	945–946	23. $-280a^4b^3$
24	12.5	3	946	24. 35
25	12.5	4	946	25. $\dfrac{231}{1024}$

ANSWERS

EXERCISE 1.1

A 1. {1, 2}
   3. {5, 6, 7}
   5. {−1, −2, −3}
   7. {0, 1, 2, 3}
   9. {1, 2, 3, . . .}
  11. Rational, Real
  13. Irrational, Real
  15. Rational, Real
B 17. $0.\overline{6}$
  19. 0.875
  21. $2.\overline{5}$
  23. $1.1\overline{6}$
  25. $\frac{31}{100}$
  27. $\frac{34}{100}$ or $\frac{17}{50}$
  29. $\frac{418}{1000}$ or $\frac{209}{500}$
C 31. $0.\overline{3}$
  33. 1.5
  35. 0.98, 0.91, 0.90
  37. 0.046875
  39. 0.252 in.

1.1 USING YOUR KNOWLEDGE

1. The denominators are products of powers of 2 and 5.
3. $\frac{5}{9}$
5. $\frac{55}{99}$

EXERCISE 1.2

A 1.

   3.
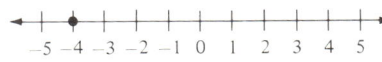
   5.
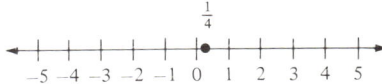
   7.
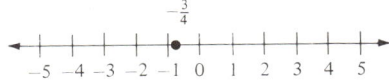
   9.
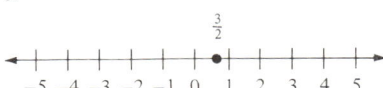

B 11. −8
  13. 7
  15. $-\frac{3}{4}$
  17. $\frac{1}{5}$
  19. −0.5
  21. $-0.\overline{2}$
  23. $1.\overline{36}$
  25. −π
C 27. 10
  29. 17
  31. $\frac{3}{5}$
  33. $0.\overline{5}$
  35. $3.\overline{61}$
  37. $\sqrt{2}$
  39. π
D 41. <
  43. >
  45. >
  47. <
  49. <

✓ SKILL CHECKER

51. {6, 7, 8, . . .}
53. {0, 1, 2, 3, 4}
55. {1, 2, 3}

1.2 USING YOUR KNOWLEDGE

1. 0.11; The Burger King
3. 0.625; The Yankees
5. Monjane

EXERCISE 1.3

A 1. $\frac{2}{5}$
   3. −0.1
   5. 2
   7. −0.8
   9. $\frac{1}{5}$
B 11. −7
  13. −0.3
  15. $-\frac{2}{7}$
  17. −14
  19. −0.6
  21. $-\frac{5}{7}$
  23. −1
  25. −4
  27. −0.1
  29. −0.1
  31. $\frac{2}{7}$
C 33. Commutative Law
  35. Commutative Law

37. Associative Law
39. Associative Law
D 41. 3500°C
  43. $46
  45. 14°C

✓ SKILL CHECKER

47. −7
49. 2.3

1.3 USING YOUR KNOWLEDGE

1. 799
3. 284
5. 49

EXERCISE 1.4

A 1. −40
   3. −12
   5. 50
   7. 60
   9. 40
  11. 30
  13. −40
  15. −7.26
  17. 2.86
  19. $-\frac{25}{42}$
  21. $\frac{1}{4}$
  23. $-\frac{15}{4}$
B 25. −2
  27. −4
  29. 2
  31. 0
  33. Not defined
  35. −2
  37. 9
  39. 3
  41. 1
  43. −4
  45. −7
  47. $-\frac{21}{20}$
  49. $\frac{4}{7}$
  51. $-\frac{5}{7}$
  53. $-\frac{1}{2}$
  55. $\frac{1}{6}$
C 57. The commutative law
  59. The associative law
  61. The commutative law
  63. The multiplicative identity
  65. The associative law
  67. 38,325

69. a. 54 ft
    b. 72 ft
    c. 99 ft
71. 160

73. Additive inverse -7
    Reciprocal $\frac{1}{7}$
75. Additive inverse 0
    Reciprocal not defined

1.4 USING YOUR KNOWLEDGE

1. -1.35
3. -3.045
5. 3.875

EXERCISE 1.5

A 1. $4x - 4y$
   3. $-9a + 9b$
   5. $1.2x - 0.6$
   7. $-\dfrac{3a}{2} + \dfrac{6}{7}$
   9. $-2x + 6y$
   11. $-2.1 - 3y$
   13. $-4a - 20$
   15. $-6x - xy$
   17. $-8x + 8y$
   19. $-6a + 21b$
   21. $0.5x + 0.5y - 1.0$
   23. $-\frac{6}{5}a + \frac{6}{5}b - 6$
   25. $-2x + 2y - 6z - 10$
   27. $-0.3x - 0.3y + 0.6z + 1.8$
   29. $-\frac{5}{2}a + 5b - \frac{5}{2}c - 5d + 5$
B 31. $9x - 6$
   33. $-9x - 8$
   35. $11L - 4W$
   37. $-2x$
   39. $\dfrac{x}{9} + 2$
   41. $6a + 2b$
   43. $3x - 4y$
   45. $x - 5y + 36$
C 47. $-2x^3 + 9x^2 - 3x + 12$
   49. $\frac{2}{7}x^2 + \frac{4}{5}x - \frac{3}{4}$
   51. $4a - 11$
   53. $-7a + 10b - 3$
   55. $-4.8x + 3.4y + 5$

57. 16
59. -2

1.5 USING YOUR KNOWLEDGE

1. $v_a = \frac{1}{2}v_1 + \frac{1}{2}v_2$
3. K.E. $= \frac{1}{2}mv_1^2 + \frac{1}{2}mv_2^2$

1.5 CALCULATOR CORNER

1. 270
3. 150

EXERCISE 1.6

A 1. -16
   3. 25
   5. -125
   7. 1296
   9. -32
B 11. $\frac{1}{16}$
   13. $\frac{1}{125}$
   15. $\frac{1}{81}$
   17. $\dfrac{1}{x^6}$
   19. $\dfrac{1}{a^8}$
C 21. $\frac{1}{64}$
   23. $12x^2$
   25. $-15y^2$
   27. $\dfrac{20}{a^5}$
   29. $-\dfrac{30y^3}{x^2}$
   31. $-\dfrac{24y^6}{x^4}$
   33. $-\dfrac{40}{a^2b^3}$
   35. -30
   37. $2x^4$
   39. $\dfrac{a^2}{2}$
   41. $-2x^3y^2$
   43. $-\dfrac{x}{2}$
   45. $\dfrac{2}{3a^3}$
   47. $\frac{3}{4}$
   49. $\dfrac{3b^3}{2a^6}$
D 51. $\dfrac{8x^9}{y^6}$
   53. $\dfrac{4y^6}{x^4}$
   55. $-\dfrac{1}{27x^9y^6}$
   57. $\dfrac{1}{x^{12}y^6}$
   59. $x^{12}y^{12}$
E 61. $\dfrac{a^2}{b^6}$
   63. $-\dfrac{8b^6}{27a^3}$
   65. a^8b^4
   67. $\dfrac{1}{x^{15}y^6}$
   69. $x^{27}y^6$
   71. 2.68×10^8
   73. 2.4×10^{-4}

75. $8,000,000$
77. 0.23
79. 2×10^3
81. 3×10^8
83. 31 yr

85. a. -9.856
    b. -2.772

1.6 USING YOUR KNOWLEDGE

1. 3.34×10^5
3. a. 7.3   10
   b. 1.23 -07

EXERCISE 1.7

A 1. a. -26
      b. -70
   3. a. 27
      b. 3
   5. -47
   7. 20
   9. -15
   11. -13
   13. -36
   15. -10
   17. -4
   19. 0
   21. 1
   23. 57
   25. 3
   27. -6
   29. 1
B 31. -20
   33. 8
   35. -24
   37. -33
   39. 11,800

41. $-2x - 12$
43. $-9x + 15$

1.7 USING YOUR KNOWLEDGE

1. $(3 + 4)/(5 + 9)$
3. $3^2 + 4^2$
5. $2*3^3 - 5*4^2$
7. $5*8/(6*9)$
9. 4
11. 2
13. 20
15. 16
17. 500

REVIEW EXERCISES CHAPTER 1

1. a. $\{4, 5, 6, 7, 8\}$
   b. $\{5, 6, 7\}$
2. 0.3 Rational, Real
   0 Whole, Integer, Rational, Real

$-\frac{3}{4}$ Rational, Real
-5 Integer, Rational, Real
$\sqrt{3}$ Irrational, Real

3. a. 0.2
   b. 0.4
4. a. 0.$\overline{1}$
   b. 0.$\overline{2}$
5. a. $\frac{31}{100}$
   b. $\frac{41}{100}$
6. a.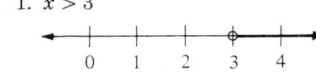
   b.
7. a. -5
   b. 8
8. a. 3.5
   b. $-\frac{3}{4}$
9. a. 9
   b. 4.2
10. a. $\frac{1}{8}$
    b. 0.$\overline{4}$
11. a. -11
    b. -3
12. a. $-\frac{2}{7}$
    b. -0.6
13. a 12
    b. 4
14. a. Commutative law of addition
    b. Associative law of addition
15. a. -36
    b. -14.4
16. a. $-\frac{21}{32}$
    b. $\frac{5}{21}$
17. a. 0
    b. Not defined
18. a. $-\frac{5}{3}$
    b. $\frac{1}{0.3}$ or $\frac{10}{3}$
19. a. $-\frac{9}{4}$
    b. -3
20. a. $-3x + 21$
    b. $2x + 17$
21. a. $-x + 10y - 16$
    b. $3x^2 + 3x + 8$
22. a. 81
    b. -81
23. a. $-\frac{1}{512}$
    b. $\frac{1}{x^{10}}$
24. a. $-\frac{15y^{10}}{x^4}$
    b. $-\frac{24}{x^{11}y^8}$
25. a. $\frac{3}{x^2}$
    b. $-4x^{11}$
26. a. $-\frac{x}{3}$
    b. $-\frac{2}{x^{11}}$

27. a. $-\frac{8x^{21}}{y^{18}}$
    b. $\frac{16}{x^{24}y^{24}}$
28. a. $\frac{1}{x^{24}y^{12}}$
    b. $x^{25}y^{15}$
29. a. 3.4×10^5
    b. 4.7×10^{-5}
30. a. 37,000
    b. 0.0078
31. a. -75
    b. 50
32. a. 9
    b. 20
33. a. -7
    b. -73

EXERCISE 2.1

A 1. Yes
  3. Yes
  5. Yes
  7. No
  9. No
B 11. 4
  13. 1
  15. 2
  17. 2
  19. 7
  21. -1
  23. -8
  25. -4
  27. 0
  29. 0
  31. $\frac{1}{13}$
  33. 8
  35. $-\frac{11}{80}$
C 37. 24
  39. 1
  41. 15
  43. 10
  45. 4
  47. 10
  49. -5
  51. 0
  53. -1
  55. $-\frac{10}{11}$
  57. $\frac{5}{2}$
  59. An identity
  61. No solution

✓ SKILL CHECKER

63. 5
65. 0.3
67. -0.9
69. $-\sqrt{5}$

2.1 USING YOUR KNOWLEDGE

1. 360
3. 130

EXERCISE 2.2

A 1. 4
  3. -6
  5. 57
  7. 0
  9. 16
  11. $\frac{43}{2}$ or 21.5
  13. 1500
  15. 6
B 17. $-13, 13$
  19. $-2.3, 2.3$
  21. 0
  23. No solution
  25. $-9, -5$
  27. $-2, 6$
  29. $-\frac{6}{5}, 2$
  31. $-20, 4$
  33. $-9, 18$
  35. -3
  37. $-5, -\frac{1}{3}$
  39. $-7, 5$
  41. No solution
  43. All real numbers
  45. All real numbers
  47. a. True for all real numbers
      b. $|9 - 5| = |4| = 4$ and $|5 - 9| = |-4| = 4$
      c. $|0 + 4| = |4| = 4$ and $|-4 - 0| = |-4| = 4$

✓ SKILL CHECKER

49. a. $<$
    b. $>$
    c. $<$

2.2 USING YOUR KNOWLEDGE

1. 250
3. 625

EXERCISE 2.3

A 1. $x > 3$
  3. $x \leq -3$
  5. $x \geq 3$
  7. $x \geq -1$
  9. $x \leq 2$
B 11. $x \leq 1$

13. $y \le 3$

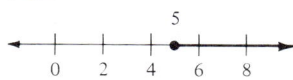

15. $x \ge 5$

17. $a \le 2$

18. $z \le -4$

19. $z \le -4$

21. $x \le -1$

23. $x \le 0$

25. $x \le -\frac{11}{80}$

27. $x \ge -20$

29. $x \ge -2$

C 31. $h \le 29{,}028$
33. $e \ge 2$
35. $n \ge 4 \times 10^{25}$

✓ **SKILL CHECKER**

37. $>$
39. $<$
41. 9
43. 0.34

2.3 USING YOUR KNOWLEDGE

1. $[-4, \infty)$
3. $(-\infty, -6]$

EXERCISE 2.4

A 1. $-2 \le x \le 4$

3. $2 < x \le 6$

5. $1 < x < 3$

7. $x < -5$ or $x > 5$

9. $1 \le x \le 4$

11. $2 < x < 3$

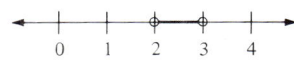

13. $1 < x < 3$

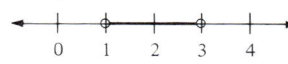

15. $-2 < x < 5$

17. $-5 < x < 1$

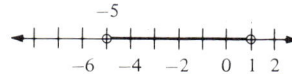

19. $3 < x < 4$

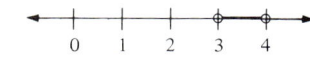

21. $-2 < x < 4$

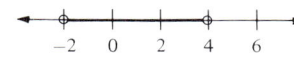

23. $-6 < y < 1$

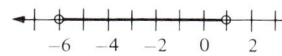

25. $4 \le y \le 6$

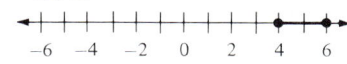

27. $-2 < x < 4$

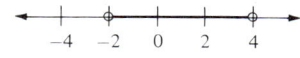

29. $-3 < a < 3$

B 31. $-4 < x < 4$

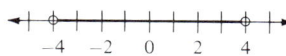

33. $-2.4 \le z \le 2.4$

35. $-4 \le a \le 4$

37. $-1 < x < 3$

39. No solution
41. $-2 \le x \le -1$

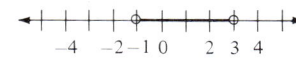

43. $-2 < x < 1$

45. $x < -2$ or $x > 2$

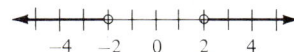

47. $z \le -1.4$ or $z \ge 1.4$

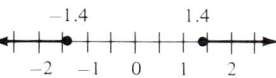

49. $a \le -3$ or $a \ge 3$

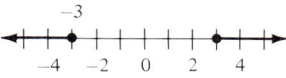

51. $x < 0$ or $x > 2$

53. All real numbers

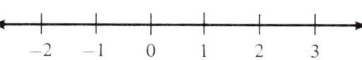

55. $x \le -2$ or $x \ge -1$

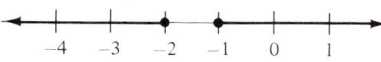

57. $x < -1$ or $x > \frac{5}{2}$

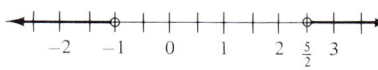

59. $x \le -2$ or $x \ge 1$

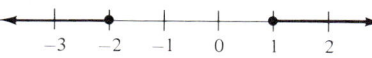

61. $a < 2$ or $a > 6$

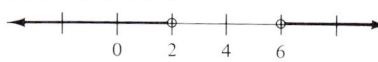

✓ **SKILL CHECKER**

63. 2
65. $\frac{5}{2}$

2.4 USING YOUR KNOWLEDGE

1. $|x| < 5,\ -5 < x < 5$

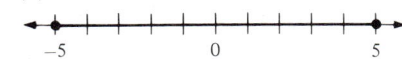

3. $|z| \ge 3,\ z \le -3$ or $z \ge 3$

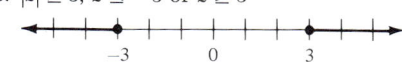

EXERCISE 2.5

1. $h = \dfrac{V}{\pi r^2}$

3. $W = \dfrac{V}{LH}$

5. $b = P - s_1 - s_2$

7. $s = \dfrac{A}{\pi r} - r = \dfrac{A - \pi r^2}{\pi r}$

9. $V_2 = \dfrac{P_1 V_1}{P_2}$

11. $P_2 = \dfrac{P_1 V_1}{V_2}$

13. a. $T = \dfrac{D}{R}$

   b. 4 hr

15. a. $A = 34 - 2H$

   b. 18 yr

17. a. $CGS = (OPM)(NS) - OE$

   b. $29,500

19. a. $U = \dfrac{F - PF}{P}$

   b. 1000

✓ **SKILL CHECKER**

21. 3500
23. 2000

2.5 USING YOUR KNOWLEDGE

1. a. $L = \dfrac{H - 32}{1.88}$

   b. No

   c. Yes

3. a. 18.09 in.

   b. 19.15 in.

   c. 18.97 in.

   d. 21.03 in.

EXERCISE 2.6

A 1. 44, 46, 48
    3. $-11, -9, -7$
    5. 87 and 92
    7. 47
    9. 937
    11. 168 lb

B 13. 12 ft by 5000 ft
    15. 518 ft by 716 ft

C 17. 1920
    19. $95.04 (billion)
    21. $6000 at 7%, $9000 at 5%
    23. $8000 at 7.5%, $17,000 at 6%

D 25. 360 km
    27. 60 mi
    29. 300 mi

E 31. 6
    33. 30 lb copper, 40 lb zinc
    35. 20
    37. 10
    39. $33\frac{1}{3}$

✓ **SKILL CHECKER**

41. a^3
43. a^3
45. $\dfrac{a^{12}}{b^{20}}$
47. $\dfrac{b^8}{a^{12}}$
49. $\dfrac{1}{a^{12}b^{12}}$

REVIEW EXERCISES CHAPTER 2

1. a. no
   b. no
   c. yes

2. a. 12
   b. 15
   c. 18

3. a. 2
   b. 3
   c. 4

4. a. 14
   b. 21
   c. 28

5. a. 3
   b. 3
   c. 3

6. a. $P = 500$
   b. $P = 2000$
   c. $P = 4000$

7. a. 7 or -21
   b. 14 or -28
   c. 21 or -35

8. a. 2
   b. 4
   c. 6

9. a. $x \geq -2$

9. b. $x \geq -3$

9. c. $x \geq -4$

10. a. $x \geq -3$

10. b. $x \geq -2$

10. c. $x \geq -2$

11. a. $x > 3$

11. b. $x > 3$

11. c. $x > 3$

12. a. $-1 < x < 2$

12. b. $-2 < x < 3$

13. c. $-3 < x < 4$
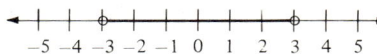

13. a. $x < -2$ or $x \geq 3$
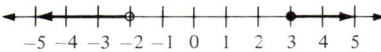

13. b. $x < -3$ or $x \geq 2$

13. c. $x < -4$ or $x \geq 1$
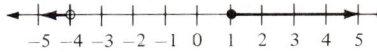

14. a. $-2 < x \leq 2$
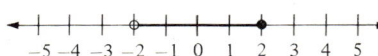

14. b. $-3 < x \leq 3$

14. c. $-2 < x \leq 4$
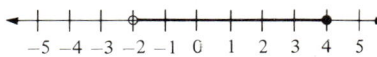

15. a. $-5 < x \leq -1$

15. b. $-4 < x \leq -1$
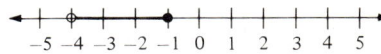

15. c. $-3 < x \leq 1$

16. a. $-\frac{1}{3} \leq x \leq 1$

16. b. $-\frac{1}{2} \leq x \leq 1$

16. c. $-\frac{3}{5} \leq x \leq 1$

17. a. $x \leq -\frac{1}{3}$ or $x \geq 1$

17. b. $x \leq -\frac{1}{2}$ or $x \geq 1$

17. c. $x \leq -\frac{3}{5}$ or $x \geq 1$

18. a. $h = \dfrac{H - 72.48}{2.5};\ 4$

   b. $h = \dfrac{H - 77.48}{2.5};\ 2$

c. $h = \dfrac{H - 84.98}{2.5}$; -1

19. a. $A = \dfrac{7B + 14}{2}$

    b. $A = \dfrac{7B + 21}{3}$

    c. $A = \dfrac{7B + 28}{4}$

20. a. 40 ft. by 50 ft
    b. 50 ft. by 60 ft
    c. 60 ft. by 70 ft
21. a. 49, 51, 53
    b. 51, 53, 55
    c. 67, 69, 71
22. a. \$20,000
    b. \$30,000
    c. \$15,000
23. a. \$5000 in bonds,
       \$15,000 in CD's
    b. \$17,000 in bonds,
       \$3000 in CD's
    c. \$10,000 in bonds,
       \$10,000 in CD's
24. a. 200 miles from town
    b. 300 miles from town
    c. 120 miles from town
25. a. 100 liters
    b. 25 liters
    c. 0 liters

EXERCISE 3.1

A,B 1. Monomial, degree 4
    3. Binomial, degree 3
    5. Trinomial, degree 3
    7. Trinomial, degree 5
    9. Zero polynomial. No degree
B,C 11. $x^3 + y^4$; 4
    13. $x^4 + y^4 + z^4$; 4
    15. $3x^2 - 4y^3 + 5x^2y^2$; 4
D 17. -8
    19. 4
    21. -1
    23. 16
    25. -136
    27. a. 0
       b. -9
       c. -9
    29. a. 8
       b. 2
       c. 6

✓ SKILL CHECKER

31. $7x^2 + 5x - 3$
32. $3x^2 - 5x - 3$

3.1 USING YOUR KNOWLEDGE

1. a. 48 ft
   b. 64 ft
3. a. \$42,500
   b. \$70,000
5. a. \$50,000
   b. \$0 (No value)

3.1 CALCULATOR CORNER

1. -9.317184
3. 3.423744
5. 0.907099

EXERCISE 3.2

A,B 1. $6x^2 - 5$
    3. $x^2 - 2x - 4$
    5. $3x^2 + 4x - 9$
    7. $-11y^2 - y - 3$
    9. $4x^3 - 12x^2 + 9x - 6$
    11. $7y^2 + 12y - 11$
    13. $3v^3 - 2v^2 + v - 7$
    15. $-5u^3 - 5u^2 - 3u + 10$
    17. $4x^3 + 2xy - 1$
    19. $x^3 - x^2 - 4$
    21. $-2a^2 + 5a$
    23. $4y$
    25. $3x^2 + 7y$
    27. $P(0) = 3$
       $Q(0) = -1$
       $P(0) + Q(0) = 3 + (-1) = 2$
    29. $P(x) - P(x) = 0$
    31. Commutative Law
    33. Distributive Law
    35. Associative Law
    37. Commutative Law
    39. Distributive Law

APPLICATIONS

C 41. \$20
    43. \$1000
    45. \$30,000

✓ SKILL CHECKER

47. $18x^3y^2$
49. $14x^4y^2$

3.2 USING YOUR KNOWLEDGE

1. $2x^2 + 6x$
3. $3x^2 + 4x$
5. $27x^2 + 10x$

EXERCISE 3.3

A 1. $12x^2 - 6x$
    3. $-3x^3 + 9x^2$
    5. $-24x^3 + 16x^2 - 8x$
    7. $-18x^3y^2 - 9xy^4 + 21xy^2$
    9. $6x^3y^6 - 10x^2y^5 + 2x^2y^4$
B 11. $x^3 + 4x^2 + 8x + 15$
    13. $x^3 + 3x^2 - x + 12$
    15. $x^3 + 2x^2 - 5x - 6$
    17. $x^3 - 8$
    19. $x^4 - x^3 + x^2 + x - 2$
C 21. $9x^2 + 9x + 2$
    23. $5x^2 + 11x - 12$
    25. $3a^2 + 14a - 5$
    27. $2y^2 + 7y - 15$
    29. $x^2 - 8x + 15$
    31. $6x^2 - 7x + 2$
    33. $4x^2 + 4ax - 15a^2$

35. $x^2 + 15x + 56$
37. $4a^2 + 10ab + 4b^2$
D,E 39. $16u^2 + 8uv + v^2$
    41. $4y^2 + 4yz + z^2$
    43. $9a^2 - 6ab + b^2$
    45. $a^2 - b^2$
    47. $25x^2 - 4y^2$
    49. $b^2 - 9a^2$
    51. $3x^3 + 9x^2 + 6x$
    53. $-3x^3 + 12x^2 - 9x$
    55. $x^3 + 6x^2 + 9x$
    57. $-2x^3 + 4x^2 - 2x$
    59. $4x^2y^2 - y^4$
F 61. a. $R = 1000p - 30p^2$
       b. \$8000
    63. $T_1^4 - T_2^4$
    65. $Kt_n^2 - 2Kt_nt_a + Kt_a^2$

✓ SKILL CHECKER

67. $5(x + y)$
69. $3a(b + c)$

3.3 USING YOUR KNOWLEDGE

1. a. 9
   b. 5
   c. No
3. a. x^2
   b. xy
   c. y^2
   d. xy
5. $(x + y)^2 = x^2 + 2xy + y^2$

EXERCISE 3.4

A 1. $8(x + 2)$
    3. $9(y - 2)$
    5. $-5(y - 5)$
    7. $-8(x + 3)$
    9. $4x(x + 9)$
    11. $6x(1 - 7x^2)$
    13. $-5x(1 + 7x^2)$
    15. $3x(x^2 + 2x + 13)$
    17. $9y(7y^2 - 2y + 3)$
    19. $6x^2(6x^4 + 2x^3 - 3x^2 + 5)$
    21. $8y^3(6y^5 + 2y^2 - 3y + 1)$
    23. $\frac{1}{7}(4x^3 + 3x^2 - 9x + 3)$
    25. $\frac{1}{8}y^2(7y^7 + 3y^4 - 5y^2 + 5)$
B 27. $(x + 2)(x^2 + 1)$
    29. $(y - 3)(y^2 + 1)$
    31. $(2x + 3)(2x^2 + 1)$
    33. $(3x - 1)(2x^2 + 1)$
    35. $(y + 2)(4y^2 + 1)$
    37. $(a^4 + 1)(2a^2 + 3)$
    39. $(x^2 + 4)(3x^3 + 1)$
    41. $(2y^2 + 3)(3y^3 + 1)$
    43. $y^2(y^2 + 3)(4y^3 + 1)$
    45. $a^2(a^2 - 2)(3a^3 - 2)$
    47. $\alpha L(t_2 - t_1)$
    49. $(R - 1)(R - 1)$

✓ SKILL CHECKER

51. $x^2 + 7x + 12$
53. $x^2 + 3x - 10$

55. $25x^2 + 20xy + 4y^2$
57. $25x^2 - 20xy + 4y^2$
59. $u^2 - 36$

3.4 USING YOUR KNOWLEDGE

1. $-w(l - z)$
3. $a(a + 2s)$
5. $-16(t^2 - 5t - 15)$

EXERCISE 3.5

A   1. $(x + 2)(x + 3)$
   3. $(a + 2)(a + 5)$
   5. $(x + 4)(x - 3)$
   7. $(x + 2)(x - 1)$
   9. $(x + 1)(x - 2)$
  11. $(x + 2)(x - 5)$
  13. $(a - 7)(a - 9)$
  15. $(y - 2)(y - 11)$
B,C 17. $(9x + 1)(x + 4)$
  19. $(3a + 1)(a - 2)$
  21. $(2y + 5)(y - 4)$
  23. $(4x - 3)(x - 2)$
  25. $(2x + 3)(3x - 4)$
  27. $(3a + 2)(7a - 1)$
  29. $(2x + 3y)(3x - y)$
  31. $x^2(7x - 3y)(x - y)$
  33. $y^3(3x + y)(5x - 2y)$
  35. $xy^2(15x^2 - 2xy - 2y^2)$
  37. $(2g + 7)(g - 3)$
  39. $(2R - 1)(R - 1)$

✓ SKILL CHECKER

41. $4a^2 + 4ab + b^2$
43. $a^2 - 4ab + 4b^2$
45. $a^2 - b^2$
47. $4x^2 - 9y^2$

3.5 USING YOUR KNOWLEDGE

1. $(L - 3)(2L - 3)$
3. $(5t - 7)(t - 1)$

EXERCISE 3.6

A   1. $(x + 1)^2$
   3. $(y + 11)^2$
   5. $(1 + 2x)^2$
   7. $(3x + 5y)^2$
   9. $(6a + 4)^2 = 4(3a + 2)^2$
  11. $(y - 1)^2$
  13. $(7 - x)^2$
  15. $(7a - 2x)^2$
  17. $(4x - 3y)^2$
  19. $(3x^2 + 2)^2$
  21. $(4x^2 - 3)^2$
  23. $(1 + x^2)^2$
B 25. $(y + 8)(y - 8)$
  27. $(a + \frac{1}{3})(a - \frac{1}{3})$
  29. $(8 + b)(8 - b)$
  31. $(6a + 7b)(6a - 7b)$
  33. $\left(\frac{x}{3} + \frac{y}{4}\right)\left(\frac{x}{3} - \frac{y}{4}\right)$
  35. $(a + 2b + c)(a + 2b - c)$

37. $(2x - y + 1)(2x - y - 1)$
39. $(3y - 2x + 5)(3y - 2x - 5)$
41. $(4a + x + 3y)(4a - x - 3y)$
43. $(y + a - b)(y - a + b)$
C 45. $(x + 5)(x^2 - 5x + 25)$
47. $(1 + a)(1 - a + a^2)$
49. $(2x + y)(4x^2 - 2xy + y^2)$
51. $(x - 1)(x^2 + x + 1)$
53. $(5a - 2b)(25a^2 + 10ab + 4b^2)$
55. $(x+2)(x-2)(x^2+2x+4)(x^2-2x+4)$
57. $\left(x + \frac{1}{2}\right)\left(x - \frac{1}{2}\right)\left(x^2 + \frac{1}{2}x + \frac{1}{4}\right)$
   $\times \left(x^2 - \frac{1}{2}x + \frac{1}{4}\right)$
59. $\left(\frac{x}{2} + 1\right)\left(\frac{x}{2} - 1\right)\left(\frac{x^2}{4} + \frac{1}{2}x + 1\right)$
   $\times \left(\frac{x^2}{4} - \frac{1}{2}x + 1\right)$

✓ SKILL CHECKER

61. $x^2 - 2x - 15$
63. $x^2 - 6x - 16$
65. $4x^2 - 12xy + 9y^2$

3.6 USING YOUR KNOWLEDGE

1. $(10 + x)(10 - x)$
3. $(2x + 1)(4x^2 - 2x + 1)$

EXERCISE 3.7

   1. $3x^2(x + 2)(x - 3)$
   3. $5x^2(x + 4y)(x - 2y)$
   5. $-3x^4(x^2 + 2x + 7)$
   7. $2x^4y(x^2 - 2xy - 5y^2)$
   9. $-2x^4(2x^2 + 6xy + 9y^2)$
  11. $2y^2(3x^2 + 1)(x + 2)$
  13. $-3xy(3x^2 + 2)(x + 1)$
  15. $-2x(2x^2 - y)(x + y)$
  17. $3y^2(x + 4y)^2$
  19. $-2k(3x + 2y)^2$
  21. $4xy^2(2x - 3y)^2$
  23. Not factorable
  25. $3x^3(x + 2y)^2$
  27. $2x^4(3x + y)^2$
  29. $3x^2y^2(2x - 3y)^2$
  31. $6(x + 2)(x + 1)(x - 1)$
  33. $7(x^2 + y^2)(x + y)(x - y)$
  35. $2x^2(x^2 + 4y^2)(x + 2y)(x - 2y)$
  37. $-2(x + 3)^2$
  39. $-3(x + 2)^2$
  41. $-x^2(2x + y)^2$
  43. $-y^2(3x + 2y)^2$
  45. $-2y^2(2x - 3y)^2$
  47. $-2x(3x + 2y)^2$
  49. $-2x(3x + 5y)^2$
  51. $-x(x + y)(x - y)$
  53. $-x^2(x + 2y)(x - 2y)$
  55. $-x^2(2x + 3y)(2x - 3y)$
  57. $-2x(2x + 3y)(2x - 3y)$
  59. $-2x^2(3x + 2y)(3x - 2y)$
  61. $x^2(3 - x)(9 + 3x + x^2)$

63. $x^4(x - 2)(x^2 + 2x + 4)$
65. $x^4(3 + 2x)(9 - 6x + 4x^2)$
67. $x^4(3x + 4y)(9x^2 - 12xy + 16y^2)$

✓ SKILL CHECKER

69. $(2x + 1)(3x - 2)$
71. $(3x - 1)(4x + 1)$

3.7 USING YOUR KNOWLEDGE

1. $\dfrac{2\pi A}{360}(IR + Kt)$

3. $\dfrac{3S}{2bd^3}(d + 2z)(d - 2z)$

EXERCISE 3.8

A 1. $-1, -2$
   3. $1, -4, -3$
   5. $\frac{1}{2}, \frac{1}{3}$
   7. $0, 3$
   9. $8, -8$
  11. $9, -9$
  13. $0, -6$
  15. $0, 3$
  17. $3, 9$
  19. $-1, -5$
  21. $5, -3$
  23. $-\frac{2}{3}, -1$
  25. $1, \frac{1}{2}$
  27. $1, -\frac{1}{2}$
  29. $2, -6$
  31. $1, \frac{1}{2}$
  33. $-2, -4$
  35. $4, 1$
  37. $-2, -\frac{5}{2}$
  39. 1
  41. $2, -2, -4$
  43. $5, 3, -3$
  45. $2, -2, -1$
B 47. $6, 8, 10$
  49. 10 in., 24 in., 26 in.
  51. 1 sec
  53. 1 sec

✓ SKILL CHECKER

55. $\dfrac{1}{5x^2}$

57. $2x^{10}$

59. $\dfrac{2}{x^{12}}$

3.8 USING YOUR KNOWLEDGE

1. 119, 120, 169

REVIEW CHAPTER 3

1. a. Binomial, degree 6
   b. Monomial, degree 8
   c. Trinomial, degree 5

2. a. $8x^3y^3 - 3x^3y^2 - 7z^3$, degree 6
   b. $3x^2y^2z^3 + 4x^2y^2 - 4z^5$, degree 7
   c. $6x^2y^2z^2 + 3xyz^3 - 4x^2y^3z^3$, degree 8
3. a. 9
   b. 3
   c. 23
4. a. $x^3 + 8x^2 - 10x + 7$
   b. $5x^3 - 3x^2 - 3x + 8$
   c. $3x^3 + 5x^2 - 7x + 6$
5. a. $-3x^3 + 7x^2 - 3x + 3$
   b. $10x^2 + x - 6$
   c. $-7x^3 + 13x^2 - 12x + 4$
6. a. $-2x^4y - 6x^3y^2 + 4x^2y^4$
   b. $-3x^4y^2 - 9x^3y^3 + 6x^2y^5$
   c. $-4x^3y^2 - 12x^2y^3 + 8xy^5$
7. a. $x^3 - 4x^2 + x + 2$
   b. $x^3 - 5x^2 + 4x + 4$
   c. $x^3 - 6x^2 + 7x + 6$
8. a. $8x^2 + 2xy - 15y^2$
   b. $6x^2 + 5xy - 6y^2$
   c. $10x^2 + 9xy - 9y^2$
9. a. $4x^2 + 28xy + 49y^2$
   b. $9x^2 + 42xy + 49y^2$
   c. $16x^2 + 56xy + 49y^2$
10. a. $9x^2 - 42xy + 49y^2$
    b. $16x^2 - 56xy + 49y^2$
    c. $25x^2 - 70xy + 49y^2$
11. a. $9x^2 - 4y^2$
    b. $16x^2 - 9y^2$
    c. $25x^2 - 9y^2$
12. a. $5x^2(3x^3 - 4x^2 + 2x + 5)$
    b. $3x^2(3x^3 - 4x^2 + 2x + 5)$
    c. $2x^2(3x^3 - 4x^2 + 2x + 5)$
13. a. $x(2x^3 + 5)(3x^2 - 1)$
    b. $x(2x^3 + 5)(3x^2 - 4)$
    c. $x(2x^2 + 3)(3x^2 - 2)$
14. a. $(x - 6y)(x + 3y)$
    b. $(x - 6y)(x + 2y)$
    c. $(x - 6y)(x + y)$
15. a. $(2x + 5y)(x - 6y)$
    b. $(2x + 5y)(x - 4y)$
    c. $(2x + 5y)(x - 5y)$
16. a. $3x^2y(3x + 2y)(2x - y)$
    b. $5x^2y(3x + 2y)(2x - y)$
    c. $6x^2y(3x + 2y)(2x - y)$
17. a. $(2x - 7y)^2$
    b. $(3x - 7y)^2$
    c. $(4x - 7y)^2$
18. a. $(3x + 4y)^2$
    b. $(3x + 5y)^2$
    c. $(3x + 6y)^2 = 9(x + 2y)^2$
19. a. $(9x^2 + y^2)(3x + y)(3x - y)$
    b. $(x^2 + 4y^2)(x + 2y)(x - 2y)$
    c. $(9x^2 + 4y^2)(3x + 2y)(3x - 2y)$
20. a. $(x - 2 + y)(x - 2 - y)$
    b. $(x - 3 + y)(x - 3 - y)$
    c. $(x - 4 + y)(x - 4 - y)$
21. a. $(3x + 2y)(9x^2 - 6xy + 4y^2)$
    b. $(3x + 4y)(9x^2 - 12xy + 16y^2)$
    c. $(4x + 3y)(16x^2 - 12xy + 9y^2)$
22. a. $(3x - 2y)(9x^2 + 6xy + 4y^2)$
    b. $(3x - 4y)(9x^2 + 12xy + 16y^2)$
    c. $(4x - 3y)(16x^2 + 12xy + 9y^2)$
23. a. $x^3(3x - 2y)(9x^2 + 6xy + 4y^2)$
    b. $x^4(3x - 4y)(9x^2 + 12xy + 16y^2)$
    c. $x^5(4x - 3y)(16x^2 + 12xy + 9y^2)$

24. a. $3x^4(9x^2 + 1)$
    b. $4x^4(x^2 + 16)$
    c. $2x^4(x^2 + 9)$
25. a. $3x^2(3x + 2y)^2$
    b. $4x^2(3x + 2y)^2$
    c. $5x^2(3x + 2y)^2$
26. a. $3x^2(3x - 2y)^2$
    b. $4x^2(3x - 2y)^2$
    c. $5x^2(3x - 2y)^2$
27. a. $4xy(3x + y)(x - 4y)$
    b. $5xy(3x + y)(x - 4y)$
    c. $6xy(3x + y)(x - 4y)$
28. a. $(x + 1)(x - 1)(2x - 1)$
    b. $(3x + 1)(3x - 1)(2x - 1)$
    c. $(4x + 1)(4x - 1)(2x - 1)$
29. a. 3 or -4
    b. 4 or -5
    c. 4 or -6
30. a. $\frac{1}{3}$ or $-\frac{1}{2}$
    b. $\frac{1}{4}$ or $-\frac{1}{2}$
    c. $\frac{1}{5}$ or $-\frac{1}{2}$
31. a. 1 or -1 or -2
    b. 1 or -1 or -4
    c. 3 or -2 or -3
32. a. 6 units, 8 units, 10 units
    b. 9 units, 12 units, 15 units
    c. 12 units, 16 units, 20 units

EXERCISE 4.1

A 1. $\dfrac{4xy^2}{6y^3}$

3. $\dfrac{a(x - y)}{(x - y)^2}$ or $\dfrac{ax - ay}{(x - y)^2}$

5. $\dfrac{x(x - y)}{x^2 - y^2}$ or $\dfrac{x^2 - xy}{x^2 - y^2}$

7. $\dfrac{-x(y + x)}{y^2 - x^2}$ or $\dfrac{-(xy + x^2)}{y^2 - x^2}$

   or $\dfrac{-xy - x^2}{y^2 - x^2}$

9. $\dfrac{-x(2x + 3y)}{4x^2 - 9y^2}$ or $\dfrac{-(2x^2 + 3xy)}{4x^2 - 9y^2}$

11. $\dfrac{4x(x - 2)}{x^2 - x - 2}$ or $\dfrac{4x^2 - 8x}{x^2 - x - 2}$

13. $\dfrac{-5x(x - 2)}{x^2 + x - 6}$ or $\dfrac{-5x^2 + 10x}{x^2 + x - 6}$

15. $\dfrac{3(x^2 - xy + y^2)}{x^3 + y^3}$ or $\dfrac{3x^2 - 3xy + 3y^2}{x^3 + y^3}$

17. $\dfrac{x(x^2 + xy + y^2)}{x^3 - y^3}$ or $\dfrac{x^3 + x^2y + xy^2}{x^3 - y^3}$

19. $\dfrac{x(x + y)}{x^2 + y^3}$ or $\dfrac{x^2 + xy}{x^3 + y^3}$

B 21. $\dfrac{y}{2}$

23. $\dfrac{x}{5 - x}$

25. $\dfrac{-2x}{5y}$

27. $\dfrac{x + y}{x - y}$

29. $\dfrac{1}{x - 2}$

C 31. $\dfrac{x^3}{y^3}$

33. 3

35. $\dfrac{1}{3x + 2y}$

37. $\dfrac{(x - y)^2}{x + y}$

39. $y - 1$

41. $\dfrac{x + y}{x - y}$

43. $\dfrac{y - 5}{y + 6}$

45. -1
47. $-(x + 3)$
49. $-(y^2 + 2y + 4)$
51. -1
53. $-(x + 5)$
55. $2 - x$

57. $\dfrac{-1}{x + 6}$

59. $\dfrac{1}{x - 2}$

✓ **SKILL CHECKER**

61. $2x^2$
63. $-2x^2$

4.1 USING YOUR KNOWLEDGE

1. x is not a factor of the numerator
3. $\frac{10}{3}$
5. $\frac{24}{11}$

EXERCISE 4.2

A 1. $x^2 + 3x - 2$
   3. $-2x^2 + x - 3$
   5. $-2y^2 + 8y - 3$

   7. $5x^2 + 4x - 8 + \dfrac{3}{x}$

   9. $3xy - 2 + \dfrac{3}{xy}$

B 11. $x + 3$
   13. $y + 5$
   15. $x^2 - x - 1$
   17. $x^2 + 5x + 6$
   19. $x^2 - 2x - 8$
   21. $x^2 + 4x + 3$ R 1 or

      $x^2 + 4x + 3 + \dfrac{1}{2x - 1}$

   23. $y^2 - y - 1$
   25. $4x^2 + 3x + 7$ R 12 or

      $4x^2 + 3x + 7 + \dfrac{12}{2x - 3}$

   27. $x^2 - 2x + 4$

29. $4y^2 + 8y + 16$
31. $a^2 - a - 1$
33. $x^3 + 2x^2 - x$
35. $2x^3 - 3x^2 + x - 4$ R -1
C 37. $(x + 1)(x - 3)(x - 2)$
39. $(x - 2)(x - 2)(x + 1)(x - 1)$
41. $(x + 5)(x + 4)(x^2 - 3x + 7)$
D 43. $\dfrac{500}{x} + 4$

✓ **SKILL CHECKER**

45. $(x + 3)(x - 3)$
47. $(x + 3)^2$
49. $(x + 2)(x + 1)$

4.2 USING YOUR KNOWLEDGE

1. $x + 1,\ x + 2,\ x + 3,\ x + 6$

EXERCISE 4.3

A 1. $\dfrac{3}{10}$

3. $\dfrac{2x}{3}$

5. $\dfrac{2y}{21x^2z^4}$

7. $\dfrac{4}{x + 1}$

9. 1

11. $\dfrac{-5}{x(x + y)}$ or $\dfrac{-5}{x^2 + xy}$

13. $\dfrac{3y - 2}{y - 1}$

15. 1

17. $1 - x$

19. $\dfrac{a + b}{a - b}$

B 21. $\dfrac{27}{50}$

23. $\dfrac{5x}{3}$

25. $\dfrac{9ad}{c}$

27. $\dfrac{3x}{x + 1}$

29. $\dfrac{4(y + 5)}{3(y + 2)}$ or $\dfrac{4y + 20}{3y + 6}$

31. $\dfrac{a + b}{a - b}$

33. $\dfrac{-(4a^2 + 2a + 1)}{2u^2w^2}$ or $\dfrac{-4a^2 - 2a - 1}{2u^2w^2}$

35. $\dfrac{-(1 + x)}{5x}$ or $\dfrac{-1 - x}{5x}$

37. $\dfrac{2y + 3}{3y - 5}$

39. $\dfrac{2x - 3}{5 - 2x}$

C 41. $\dfrac{1}{x + 1}$

43. $\dfrac{x - 1}{x - 4}$

45. -1

47. $\dfrac{x}{x + y}$

49. $\dfrac{xy + 2y^2 + 2x + 4y}{xy - 3y^2 - 2x + 6y}$

51. $\dfrac{x^2 + 4x + 3}{x^2 - x + 1}$

53. $\dfrac{1}{x^3 + 2x^2 + 4x}$

D 55. $\dfrac{450}{x}$

57. $\dfrac{4t(t + 3)}{t^2 + 9}$ or $\dfrac{4t^2 + 12t}{t^2 + 9}$

✓ **SKILL CHECKER**

59. $\dfrac{x^2 + x - 2}{x^2 - x - 6}$

61. $\dfrac{x - 2}{x^3 - 8}$

4.3 USING YOUR KNOWLEDGE

1. $\dfrac{RR_T}{R - R_T}$

3. $RC = \dfrac{60{,}000 + 9000x}{x}$

EXERCISE 4.4

A 1. $\dfrac{3x}{5}$

3. $\dfrac{5x}{3}$

5. $\dfrac{3 + 2x}{5(x + 2)}$

7. $\dfrac{x + 2}{2(x + 1)}$

9. $\dfrac{2x + 5}{3(x - 1)}$

B 11. $\dfrac{2x^2 - 5x}{(x + 4)(x - 4)(x - 1)}$

13. $\dfrac{5x^2 + 9x}{(x + 5)(x - 2)(x + 3)}$

15. $\dfrac{6x - 4y}{(x + y)^2(x - y)}$

17. $\dfrac{10 - x}{x^2 - 25}$

19. $\dfrac{-6x - 10}{(x + 1)(x + 2)(x + 3)}$

21. $\dfrac{2x - 5}{(x - 2)(x - 3)}$

23. $\dfrac{2}{x + 3}$

25. $\dfrac{2a + 6}{(a + 2)(a + 4)}$

27. $\dfrac{2}{a + 4}$

29. $\dfrac{15 - a^2}{(a + 5)(a + 3)}$

31. $\dfrac{y^2 + 2y}{(y + 1)(y - 1)}$

33. $\dfrac{-2y^2 - 4y + 5}{(y + 4)(y - 4)}$

35. $\dfrac{50x^2 + 8y^2}{(5x - 2y)(5x + 2y)}$

37. $\dfrac{5x^2}{(2x - y)(3x + y)}$

39. $\dfrac{2x^2 + x + 5}{(x - 2)(x + 1)^2}$

41. $\dfrac{4 - 3x}{(x - 2)(x - 4)(x - 1)}$

43. $\dfrac{-45}{(x + 5)(x - 5)}$

45. $\dfrac{-x}{(x - y)(x - 2)}$ or $\dfrac{x}{(y - x)(x - 2)}$ or $\dfrac{-x}{(y - x)(2 - x)}$

47. $\dfrac{18b^2 + 12ab + 12a - 18b}{(2a + 3b)(2a - 3b)}$

49. $\dfrac{1}{x^2 - 5x + 25}$

C 51. $\dfrac{-w_0x^3 + 3w_0L^2x - 2w_0L^3}{6L}$

53. $\dfrac{p^2 - 2gm^2Mr}{2mr^2}$

✓ **SKILL CHECKER**

55. 20
57. $24x + 18y$
59. $x^2 - 1$

4.4 USING YOUR KNOWLEDGE

1. $P(x + h) = x^2 + 2xh + h^2$
3. $\dfrac{P(x + h) - P(x)}{h} = 2x + h$

EXERCISE 4.5

A 1. $\frac{3}{4}$

3. $\dfrac{a}{c}$

5. $\dfrac{z}{xy}$

7. $\dfrac{2z}{5y}$

9. $\frac{1}{3}$

11. $\dfrac{ab - a}{b + a}$

13. $\dfrac{x}{xy - 2}$

15. $\dfrac{x - y}{x^2 y^2}$

17. $\dfrac{9}{5}$

19. $\dfrac{2a^2 - a}{2a + 1}$

21. $\dfrac{x^2}{2x - 1}$

23. $\dfrac{30}{43}$

25. $\dfrac{-(x^2 + 1)}{2x}$

27. $R = \dfrac{R_1 R_2}{R_1 + R_2}$

29. $\dfrac{c + v}{c - v}$

31. 4
33. 3

4.5 USING YOUR KNOWLEDGE

1. $\frac{6}{25}$ years
3. $11\frac{43}{50}$ years

EXERCISE 4.6

1. 6
3. -5
5. $\frac{1}{3}$
7. -4
9. $\frac{26}{9}$
11. $-\frac{1}{12}$
13. No solution
15. Any real number ($x \neq -\frac{1}{3}$)
17. -11
19. 7
21. 2
23. -3
25. $-\frac{3}{2}$
27. No solution
29. 0
31. $\frac{4}{5}$
33. 2
35. -4

37. 21, 23, 25
39. 60 gallons

4.6 USING YOUR KNOWLEDGE

1. $h = \dfrac{2A}{b_1 + b_2}$

3. $Q_1 = \dfrac{PQ_2}{1 + P}$

5. $f = \dfrac{ab}{a + b}$

EXERCISE 4.7

A 1. 8
3. 5 and 10
5. 6 and 8
7. $\frac{2}{7}$
9. $20,000
B 11. $1\frac{7}{8}$ hours
13. $4\frac{14}{19}$ hours
15. 6 hours and 3 hours
17. $15\frac{3}{4}$ hours
19. $5\frac{1}{4}$ hours
21. 36 seconds
23. 18 hours
C 25. 150 miles per hour
27. 30 miles per hour
29. auto: 25 mph,
plane: 125 mph

31. $\dfrac{1}{x^8}$
33. x^{20}
35. $-8x^3 y^6$
37. $\dfrac{1}{x^2}$
39. $\dfrac{1}{a^8 b^6}$

4.7 USING YOUR KNOWLEDGE

1. $F = \dfrac{f_1 f_2}{f_1 + f_2}$

3. $R = \dfrac{2E - 2ri}{i}$

REVIEW EXERCISES CHAPTER 4

1. a. $\dfrac{8x^2 y^3}{36y^7}$

b. $\dfrac{10x^2 y^4}{45y^8}$

c. $\dfrac{12x^2 y^5}{54y^9}$

2. a. $\dfrac{2x^2 + 9x + 4}{x^2 + 5x + 4}$

b. $\dfrac{2x^2 + 11x + 5}{x^2 + 6x + 5}$

c. $\dfrac{2x^2 + 13x + 6}{x^2 + 7x + 6}$

3. a. $\dfrac{6}{y}$

b. $\dfrac{7}{y}$

c. $\dfrac{8}{y}$

4. a. $\dfrac{y - x}{6}$

b. $\dfrac{y - x}{7}$

c. $\dfrac{y - x}{8}$

5. a. $x^3 y^5$
b. $x^3 y^6$
c. $x^3 y^7$

6. a. $\dfrac{y^2}{x - y}$

b. $\dfrac{y^3}{x - y}$

c. $\dfrac{y^4}{x - y}$

7. a. $\dfrac{-(x - 2y)}{x^2 - 2xy + 4y^2}$

b. $\dfrac{-(x + 2y)}{x^2 + 2xy + 4y^2}$

c. $\dfrac{-(x + 3y)}{x^2 + 3xy + 9y^2}$

8. a. $3x^3 - 2x + 1$

b. $3x^2 - 2 + \dfrac{1}{x}$

c. $3x - \dfrac{2}{x} + \dfrac{1}{x^2}$

9. a. $x^2 - x - 1$ R(-6)
b. $x^2 - x - 1$ R(-7)
c. $x^2 - x - 1$ R(-8)
10. a. $(x - 1)(x - 2)(x - 3)$
b. $(x - 1)(x - 2)(x - 3)$
c. $(x - 1)(x - 2)(x - 3)$

11. a. $\dfrac{3x + 2y}{3x + 1}$

b. $\dfrac{3x + 2y}{4x + 1}$

c. $\dfrac{3x + 2y}{5x + 1}$

12. a. $\dfrac{1}{(x - 2)(x + 4)}$

b. $\dfrac{1}{(x - 2)(x + 5)}$

c. $\dfrac{1}{(x - 2)(x + 6)}$

13. a. $\dfrac{-(x^2 - 3x + 9)}{x^2 + 2x + 4}$

b. $\dfrac{-(x^2 - 3x + 9)}{x^2 + 2x + 4}$

c. $\dfrac{-(x^2 - 3x + 9)}{x^2 + 2x + 4}$

14. a. $\dfrac{1}{x - 2}$

b. $\dfrac{1}{x-3}$

c. $\dfrac{1}{x-4}$

15. a. $\dfrac{1}{x+3}$

   b. $\dfrac{1}{x+4}$

   c. $\dfrac{1}{x+5}$

16. a. $\dfrac{2x^2+9x+11}{(x+1)(x-1)(x+2)}$

   b. $\dfrac{2x^2+10x+13}{(x+1)(x-1)(x+2)}$

   c. $\dfrac{2x^2+11x+15}{(x+1)(x-1)(x+2)}$

17. a. $\dfrac{-4x-14}{(x+3)(x-3)(x+2)}$

   b. $\dfrac{-3x-11}{(x+3)(x-3)(x+2)}$

   c. $\dfrac{-x-5}{(x+3)(x-3)(x+2)}$

18. a. $\dfrac{x(x^2-x+1)}{(x^2+1)(x-1)}$

   b. $\dfrac{x(x^2-x+1)}{(x^2+1)(x-1)}$

   c. $\dfrac{x(x^2-x+1)}{(x^2+1)(x-1)}$

19. a. $\dfrac{80+20a+a^2}{20+4a}$

   b. $\dfrac{150+30a+a^2}{30+5a}$

   c. $\dfrac{252+42a+a^2}{42+6a}$

20. a. No solution
   b. No solution
   c. No solution

21. a. -9
   b. -11
   c. -13

22. a. 10 and 12
   b. 12 and 14
   c. 14 and 16

23. a. $2\frac{2}{9}$ hours
   b. $2\frac{2}{5}$ hours
   c. $2\frac{6}{11}$ hours

24. a. 225 mph
   b. 275 mph
   c. 350 mph

EXERCISE 5.1

A 1. 2
   3. 2
   5. -2
   7. $-\frac{1}{4}$
   9. 2
   11. 2

B 13. 3
   15. Not a real number
   17. 3
   19. 3
   21. $-\frac{1}{2}$
   23. Not a real number
   25. 9
   27. 25
   29. $\frac{1}{4}$
   31. 16
   33. 16
   35. -16
   37. $\frac{1}{16}$
   39. 7

C 41. $x^{3/7}$
   43. $\dfrac{1}{x^{5/9}}$
   45. $x^{2/5}$
   47. z
   49. x^2
   51. $\dfrac{1}{z^2}$
   53. $\dfrac{1}{b^{4/5}}$
   55. $\dfrac{1}{a^8 b^9}$
   57. $\dfrac{b^9}{a^{10}}$
   59. $x^{16}y^{30}$
   61. $x + x^{1/3}y^{1/2}$
   63. $x^{1/2}y^{3/4} - y^{5/4}$
   65. $\dfrac{1}{x}$
   67. $\dfrac{x^2}{y}$
   69. $x^3 y^{11}$

D 71. $v = 15$ m/sec
   73. $v = 28$ ft/sec
   75. 3
   77. 13
   79. $\frac{2}{3}$

✓ **SKILL CHECKER**

81. $a^2 - b^2$
83. $x^3 - y^3$
85. $\dfrac{48x^3y^2}{16x^4y^4}$

5.1 USING YOUR KNOWLEDGE

1. $x^{1/4}$
3. a. $x^{1/3}$
   b. $x^{1/6}$
   c. 3

CALCULATOR CORNER

1. a. $-\frac{1}{2}$
   b. -5
   c. $\frac{1}{2}$

EXERCISE 5.2

A 1. 5
   3. -4
   5. $|-x|$
   7. $|x+6|$
   9. $|3x-2|$
   11. $4|xy|\sqrt{xy}$
   13. $2x\sqrt[3]{5xy}$
   15. $|xy|\sqrt[4]{xy^3}$
   17. $-3a^2b^3\sqrt[5]{b^2}$
   19. $\dfrac{\sqrt{13}}{7}$
   21. $\dfrac{\sqrt{17}}{2|x|}$
   23. $\dfrac{\sqrt[3]{3}}{4x}$

B 25. $\dfrac{\sqrt{6}}{3}$
   27. $\dfrac{-\sqrt{14}}{7}$
   29. $\dfrac{\sqrt{10a}}{2a}$
   31. $\dfrac{\sqrt{10ab}}{8ab}$
   33. $-\dfrac{\sqrt{6ab}}{2a^2b^2}$
   35. xy
   37. $-\dfrac{\sqrt[3]{21}}{3}$
   39. $\dfrac{\sqrt[3]{12x}}{4x}$

C 41. $\sqrt[3]{3}$
   43. $\sqrt{2a}$
   45. $\sqrt{5x^3y}$
   47. $\sqrt{7x^5y^3}$ or $x^2y\sqrt{7xy}$
   49. $\sqrt{2ab}$
   51. $\dfrac{\sqrt[3]{a^2b^2}}{b^2}$
   53. $\dfrac{2\sqrt{6ab}}{3b^2}$
   55. $\dfrac{\sqrt{2abx}}{2x}$

D 57. a. $\dfrac{\sqrt[3]{6V\pi^2}}{2\pi}$
   b. 3 ft
   59. $m = \dfrac{m_0 c\sqrt{c^2-v^2}}{c^2-v^2}$

✓ **SKILL CHECKER**

61. $7x+1$
63. $4x^2 - y^2$
65. $1 + 2x$
67. 7
69. 8

5.2 USING YOUR KNOWLEDGE

1. $m = m_0 c(c^2 - v^2)^{-1/2}$

3. $T = \dfrac{(2\pi Lg)^{1/2}}{g}$

5. $v = \dfrac{\sqrt{3kTm}}{m}$

EXERCISE 5.3

A 1. $15\sqrt{2}$
   3. $9\sqrt{5a}$
   5. $\sqrt{2}$
   7. $-5a\sqrt{2}$
   9. $-26\sqrt{3}$
   11. $7\sqrt[3]{5}$
   13. $-12\sqrt[3]{3}$
   15. $3\sqrt[3]{3}$
   17. $4\sqrt[3]{3a}$
   19. $\dfrac{7\sqrt[3]{3}}{6}$
   21. $\dfrac{3\sqrt{2} + 2\sqrt{3} + \sqrt{6}}{6}$
   23. $\dfrac{15\sqrt{3} + 28\sqrt{15}}{30}$
   25. $\dfrac{\sqrt{xy}(x + y - 1)}{xy}$
   27. $3\sqrt[3]{75}$
   29. $\dfrac{4\sqrt[3]{18x^2} + \sqrt[3]{6x^2}}{12x}$

B 31. $15 - 3\sqrt{2}$
   33. $2 + 3\sqrt[3]{2}$
   35. $14\sqrt{15} + 30$
   37. $6\sqrt[3]{15} - 15$
   39. $-8\sqrt{21} + 20\sqrt{14}$
   41. $55 + 13\sqrt{15}$
   43. $42 + 21\sqrt{2}$
   45. $-441 + \sqrt{35}$
   47. -1
   49. -23
   51. $5 + 2\sqrt{6}$
   53. $a^2 + 2a\sqrt{b} + b$
   55. $5 - 2\sqrt{6}$
   57. $a^2 - 2a\sqrt{b} + b$
   59. $a - 2\sqrt{ab} + b$
   61. $1 + \sqrt{2}$
   63. $\dfrac{2 - \sqrt{3}}{4}$

C 65. $\dfrac{3\sqrt{2} + \sqrt{6}}{2}$
   67. $\dfrac{6 + 2\sqrt{2}}{7}$
   69. $3a + a\sqrt{5}$
   71. $\dfrac{9a - 3a\sqrt{2} + 6b - 2b\sqrt{2}}{7}$
   73. $\dfrac{a + 2b\sqrt{a} + b^2}{a - b^2}$

75. $\dfrac{a + 2\sqrt{2ab} + 2b}{a - 2b}$

77. 11
79. 2 or 1

5.3 USING YOUR KNOWLEDGE

1. $\dfrac{1}{\sqrt{5} - \sqrt{2}}$

3. $\dfrac{x - 2}{5(\sqrt{x} + \sqrt{2})}$

5. $\dfrac{x - y}{x(\sqrt{x} - \sqrt{y})}$ or $\dfrac{x - y}{x\sqrt{x} - x\sqrt{y}}$

7. $\dfrac{x - y}{\sqrt{x}(\sqrt{x} - \sqrt{y})}$ or $\dfrac{x - y}{x - \sqrt{xy}}$

9. $\dfrac{x - y}{\sqrt{x}(\sqrt{x} + \sqrt{y})}$ or $\dfrac{x - y}{x + \sqrt{xy}}$

EXERCISE 5.4

1. 16
3. 43
5. 18
7. 7
9. 3
11. 0
13. 6
15. -16
17. 11
19. 9
21. 0
23. 1
25. 0
27. 1
29. 4
31. $a + b^2$
33. $\dfrac{a - c^3}{b}$
35. ab^2
37. $\dfrac{b^2 - a}{b}$
39. $\dfrac{b}{3}$
41. $5\sqrt{3}$ ft
43. a. $d = \dfrac{gt^2}{2}$
    b. 144.9 ft
45. a. $L = \dfrac{gt^2}{4\pi^2}$
    b. $L = \dfrac{392}{121}$ ft ≈ 3.2 ft
47. $11 + 6x$
49. $-1 + 8x$
51. $\dfrac{1 + 5\sqrt{2}}{7}$

53. $\dfrac{4 + \sqrt{2}}{7}$

55. $\dfrac{x + y - 2\sqrt{xy}}{x - y}$

5.4 USING YOUR KNOWLEDGE

1. $r = \dfrac{625}{9}$ ft or $69.\overline{4}$ ft
3. $r = \dfrac{1225}{9}$ ft or $136.\overline{1}$ ft

EXERCISE 5.5

A 1. $5i$
   3. $5i\sqrt{2}$
   5. $24i\sqrt{2}$
   7. $-12i\sqrt{2}$
   9. $8i\sqrt{7} + 3$

B 11. $6 + 4i$
   13. $-2 - 6i$
   15. $-5 - 6i$
   17. $-2 + 5i$
   19. $-7 + 4i$
   21. $8 - i$
   23. $10 + 5i$
   25. $-3 - 2i\sqrt{2}$
   27. $-1 + 2i\sqrt{2}$
   29. $-7 + 3i\sqrt{5}$

C 31. $12 + 6i$
   33. $-12 + 20i$
   35. $-4 + 6i$
   37. $-3 + 3i\sqrt{3}$
   39. $-6 + 9i$
   41. $28 + 12i$
   43. $-20 + 20i$
   45. $3 + 11i$
   47. 13
   49. $24 + 7i$
   51. 31
   53. $-3i$
   55. $6i$
   57. $\dfrac{2}{5} + \dfrac{1}{5}i$
   59. $-\dfrac{6}{5} + \dfrac{3}{5}i$
   61. $-1 + 2i$
   63. $\dfrac{17}{13} - \dfrac{6}{13}i$
   65. $-\dfrac{3}{2}i$
   67. $\dfrac{12 + \sqrt{10}}{18} + \dfrac{4\sqrt{5} - 3\sqrt{2}}{18}i$
   69. $\dfrac{3 + \sqrt{6}}{12} + \dfrac{3\sqrt{2} - \sqrt{3}}{12}i$

D 71. 1
   73. $-i$
   75. i
   77. 1
   79. i

E 81. $Z_1 + Z_2 = 8 + i$ ohms
   83. $Z_t = \dfrac{167 - 29i}{65}$

85. 2 or -2
87. 5 or -5

5.5 USING YOUR KNOWLEDGE

1. 5
3. $\sqrt{13}$

REVIEW EXERCISES CHAPTER 5

1. a. Not a real number
   b. -4
2. a. Not a real number
   b. -5
3. a. -3
   b. -4
4. a. $\frac{1}{2}$
   b. $\frac{1}{4}$
5. a. 25
   b. 16
6. a. Not a real number
   b. Not a real number
7. a. $\frac{1}{4}$
   b. $\frac{1}{16}$
8. a. $\frac{1}{4}$
   b. $\frac{1}{16}$
9. a. $x^{8/15}$
   b. $x^{9/20}$
10. a. $\dfrac{1}{x^{9/20}}$
    b. $\dfrac{1}{x^{8/15}}$
11. a. $\dfrac{1}{x^5 y^6}$
    b. $\dfrac{1}{x^5 y^{12}}$
12. a. $x^{2/5} + x^{3/5}y^{3/5}$
    b. $x^{3/5} + x^{4/5}y^{3/5}$
13. a. 7
    b. 6
14. a. $|-x|$
    b. $|-x|$
15. a. $2\sqrt[3]{6}$
    b. $2\sqrt[3]{7}$
16. a. $2xy^2\sqrt[3]{2x}$
    b. $2x^2y^5\sqrt[3]{2x^2}$
17. a. $\dfrac{\sqrt{15}}{27}$
    b. $\dfrac{\sqrt{5}}{32}$
18. a. $\dfrac{1}{x}$
    b. $\dfrac{\sqrt[3]{5}}{x}$
19. a. $\dfrac{\sqrt{55}}{11}$
    b. $\dfrac{\sqrt{65}}{13}$
20. a. $\dfrac{\sqrt{10x}}{5x}$
    b. $\dfrac{\sqrt{15x}}{5x}$

21. a. $\dfrac{\sqrt[3]{25x^2}}{5x}$
    b. $\dfrac{\sqrt[3]{49x^2}}{7x}$
22. a. $\dfrac{\sqrt[3]{2x^2}}{2x}$
    b. $\dfrac{\sqrt[3]{10x^2}}{2x}$
23. a. $\frac{4}{3}$
    b. $\frac{5}{3}$
24. a. $\sqrt[3]{9c^2d^2}$
    b. $\sqrt[3]{25c^2d^2}$
25. a. $\dfrac{\sqrt[3]{3ac}}{3c^3}$
    b. $\dfrac{\sqrt[3]{9ac}}{3c^3}$
26. a. $6\sqrt{2}$
    b. $7\sqrt{2}$
27. a. $\sqrt{7}$
    b. $\sqrt{7}$
28. a. $\dfrac{3\sqrt{2}}{4}$
    b. $\dfrac{5\sqrt{2}}{4}$
29. a. $\dfrac{9\sqrt[3]{6x^2}}{4x}$
    b. $\dfrac{11\sqrt[3]{6x^2}}{4x}$
30. a. $6 + \sqrt{6}$
    b. $8 + \sqrt{6}$
31. a. $2x\sqrt[3]{6} - 3\sqrt[3]{6x^2}$
    b. $2x\sqrt[3]{6} - 3\sqrt[3]{3x^2}$
32. a. $30 + 12\sqrt{6}$
    b. $24 + 10\sqrt{6}$
33. a. $19 + 8\sqrt{3}$
    b. $28 + 10\sqrt{3}$
34. a. $12 - 6\sqrt{3}$
    b. $19 - 8\sqrt{3}$
35. a. 3
    b. 4
36. a. $4 - \sqrt{2}$
    b. $6 - \sqrt{2}$
37. a. $\dfrac{\sqrt{xy} + y}{x - y}$
    b. $\dfrac{x + \sqrt{xy}}{x - y}$
38. a. No real number solution
    b. No real number solution
39. a. 4
    b. 10
40. a. 4 or 3
    b. 5 or 4
41. a. 4
    b. 9
42. a. 30
    b. 67
43. a. $10i$

b. $11i$
44. a. $6i\sqrt{2}$
    b. $5i\sqrt{2}$
45. a. $10 + 3i$
    b. $6 + 3i$
46. a. $-4 + 7i$
    b. $2 + 11i$
47. a. $21 + i$
    b. $23 - 2i$
48. a. $24\sqrt{2} + 16i$
    b. $36\sqrt{2} + 24i$
49. a. $\dfrac{17 + 6i}{25}$ or $\dfrac{17}{25} + \dfrac{6}{25}i$
    b. $\dfrac{27 - 11i}{25}$ or $\dfrac{27}{25} - \dfrac{11}{25}i$
50. a. -1
    b. $-i$
51. a. -1
    b. i

EXERCISE 6.1

A 1. ± 8
3. $\pm 11i$
5. ± 13
7. $\pm 2i$
9. $\pm\frac{7}{6}$
11. $\pm\frac{9}{2}i$
13. $\pm\dfrac{5\sqrt{3}}{3}$
15. $\pm\dfrac{6\sqrt{5}}{5}i$
17. $\pm\dfrac{10\sqrt{3}}{3}$
19. $\pm\dfrac{9\sqrt{13}}{13}i$

B 21. -3 or -7
23. $-2 \pm 5i$
25. $6 \pm 3\sqrt{2}$
27. $1 \pm 2\sqrt{7}i$
29. $1 \pm 5\sqrt{2}$
31. $5 \pm 4\sqrt{2}$
33. $9 \pm 8i$
35. $-1 + 4\sqrt{2}$
37. $2 \pm 5\sqrt{2}$
39. $5 \pm 3\sqrt{3}i$

C 41. -1 or -5
43. -5 or -3
45. $-3 \pm i$
47. 4 or 6
49. 3 or 7
51. $4 \pm i$
53. $-1 \pm \dfrac{\sqrt{2}}{2}i$ or $\dfrac{-2 \pm \sqrt{2}i}{2}$
55. $-1 \pm 5i$
57. $\frac{3}{5}$ or $\frac{2}{5}$
59. $\frac{1}{2} \pm i$
61. $\frac{1}{2} \pm \sqrt{2}$ or $\dfrac{1 \pm 2\sqrt{2}}{2}$
63. $1 \pm \dfrac{\sqrt{2}}{2}$ or $\dfrac{2 \pm \sqrt{2}}{2}$

65. 1 or -2

67. $\dfrac{-5 \pm 3\sqrt{3}}{2}$

69. $\dfrac{-3 \pm \sqrt{29}}{2}$

D 71. 2 seconds

73. 10%

✓ SKILL CHECKER

75. 0 or 9

77. $1 \pm \sqrt{3}$

79. $\frac{1}{8}$ or -1

81. $\dfrac{4 \pm \sqrt{5}i}{3}$

83. $-1 \pm \sqrt{5}i$

6.1 USING YOUR KNOWLEDGE

1. a. 2 thousand
   b. $2
3. 3

EXERCISE 6.2

A 1. 1 or -2

3. $-2 \pm \sqrt{3}$

5. $\dfrac{3 \pm \sqrt{17}}{2}$

7. $\frac{5}{7}$ or 1

9. $\dfrac{-4 \pm 3i}{5}$

11. $-\frac{3}{2}$ or -2

13. $\frac{3}{2}$ or 1

15. $-\frac{1}{2}$ or -3

17. $\dfrac{3 \pm \sqrt{5}}{4}$

19. -1

21. $\dfrac{3 \pm \sqrt{5}}{4}$

23. $\dfrac{-3 \pm \sqrt{21}}{6}$

25. ± 2

27. $\pm 3\sqrt{5}$

29. $\pm \dfrac{2\sqrt{3}}{3}$

31. 2 or $-1 \pm i\sqrt{3}$

33. $\frac{1}{2}$ or $\dfrac{-1 \pm \sqrt{3}i}{4}$

B 35. $5

37. $\dfrac{5 + \sqrt{37}}{6} \approx \$1.85$

39. $10 \pm 2\sqrt{15}$

41. 1

✓ SKILL CHECKER

43. 4

45. $\sqrt{23}i$

47. $6x^2 - 5x - 4$

49. $12x^2 - 19x - 21$

6.2 USING YOUR KNOWLEDGE

1. Multiply each side by $4a$.
3. To complete the square on the left side, add b^2 to each side.
5. Take the square root of each side.
7. Divide each side by $2a$.

EXERCISE 6.3

A 1. $D = 49$
   The two roots are rational numbers.
3. $D = 0$
   There is one rational root.
5. $D = 0$
   There is one rational root.
7. $D = -23$
   The two roots are imaginary numbers.
9. $D = 57$
   The two roots are irrational numbers.
11. ± 4
13. -5
15. ± 8
17. ± 20
19. -16

B 21. Not factorable
23. $(4x - 3)(3x - 2)$
25. $(9x - 7)(3x + 8)$
27. $(5x - 6)(3x + 14)$
29. $(4x - 15)(3x - 4)$

C 31. a. Sum $= \frac{6}{4} = \frac{3}{2}$
     b. Product $= \frac{5}{4}$
     c. No
33. a. Sum $= -\frac{13}{5}$
    b. Product $= -\frac{6}{5}$
    c. Yes
35. a. Sum $= -\frac{5}{2}$
    b. Product $= 1$
    c. No
37. $\frac{8}{3}$
39. 15

✓ SKILL CHECKER

41. 1

43. -1 or -5

EXERCISE 6.4

A 1. 0 or $-\frac{5}{2}$
3. 7 or -3
5. 0
7. 2 or 1
9. $-\frac{16}{5}$ or -1

B 11. ± 3 or ± 2
13. $\pm \frac{1}{2}$ or $\pm 3i$
15. $\pm \sqrt{2}$ or $\pm \dfrac{\sqrt{3}}{3}i$
17. $1, -2, 1 \pm \sqrt{3}i, \dfrac{-1 \pm \sqrt{3}i}{2}$

19. 7 or -6

21. $\dfrac{1 \pm \sqrt{37}}{2}, \dfrac{1 \pm \sqrt{3}i}{2}$

23. 16

25. 8 or 27

27. 4

29. $2 \pm \sqrt{29}, 2 \pm \sqrt{13}$

31. 6

33. 1 or $-\frac{1}{3}$

35. $\frac{16}{81}$ or 1

C 37. 10 hours, 15 hours

✓ SKILL CHECKER

39. $x \geq -2$

6.4 USING YOUR KNOWLEDGE

1. $-\frac{2}{3}, \frac{1}{2}$ and -4 (given)
3. $-1, 2,$ and 3 (given)

EXERCISE 6.5

A 1. $x < -1$ or $x > 3$
3. $-4 \leq x \leq 0$
5. $-1 \leq x \leq 2$
7. $x \leq 0$ or $x \geq 3$
9. $1 < x < 2$
11. $-3 < x < 1$
13. $x = -5$
15. All real values
17. $x = 1.6$ and -0.6
19. $x = 2.55$ and -1.55

B 21. $x \geq 2$ or $-3 \leq x \leq -1$

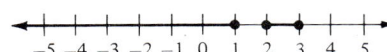

23. $x \leq 1$ or $2 \leq x \leq 3$

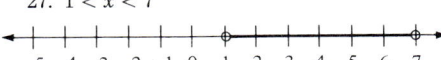

C 25. $x > 2$

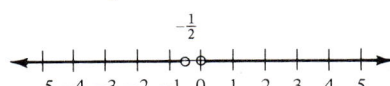

27. $1 < x < 7$

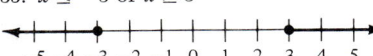

29. $\frac{1}{2} < x < 3$

31. $-2 < x < 1$

33. $x < -\frac{1}{2}$ or $x > 0$

35. $x \leq -3$ or $x \geq 3$

37. $x \le 1$ or $x \ge 5$

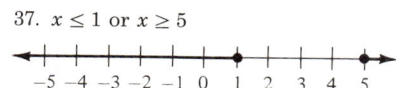

D 39. $R > 4$

41. $10 < T < 100$

43. $1 < t < 2$

45. 12

47. 2

49. −2

6.5 USING YOUR KNOWLEDGE

1. $23.2 \le v \le 26.1$

3. $v = 128.2$ mph

REVIEW CHAPTER 6

1. a. $\pm\frac{7}{4}$
   b. $\pm\frac{4}{5}$
2. a. $\pm\sqrt{6}i$
   b. $\pm\sqrt{7}i$
3. a. $3 \pm 4\sqrt{2}$
   b. $5 \pm 5\sqrt{2}$
4. a. $2 \pm \frac{5\sqrt{2}i}{2}$ or $\frac{4 \pm 5\sqrt{2}i}{2}$
   b. $3 \pm \frac{8\sqrt{3}i}{3}$ or $\frac{9 \pm 8\sqrt{3}i}{3}$
5. a. $1 \pm \frac{2\sqrt{15}}{5}$ or $\frac{5 \pm 2\sqrt{15}}{5}$
   b. $-\frac{1}{2} \pm \frac{2\sqrt{3}}{3}$ or $\frac{-3 \pm 4\sqrt{3}}{6}$
6. a. 9 or −1
   b. −4 or −8
7. a. $\frac{1}{2}, -\frac{3}{2}$
   b. $\frac{3 \pm \sqrt{2}}{4}$
8. a. $\frac{1}{3}$ or −2
   b. 2 or $-\frac{1}{5}$
9. a. $\frac{1 \pm \sqrt{13}}{3}$
   b. $\frac{3 \pm \sqrt{21}}{4}$
10. a. 0 or 16
    b. 0 or 12
11. a. $\frac{-1 \pm \sqrt{301}}{30}$
    b. $\frac{1}{5}$ or −2
12. a. $\frac{1 \pm \sqrt{2}i}{3}$
    b. $\frac{1 \pm \sqrt{19}i}{5}$
13. a. $\frac{5}{2}, \frac{-5 \pm 5\sqrt{3}i}{4}$
    b. $\frac{2}{5}, \frac{-1 \pm \sqrt{3}i}{5}$

14. a. 3
    b. 1
15. a. $D = k^2 - 64$; $k = \pm 8$
    b. $D = k^2 - 64$; $k = \pm 8$
16. a. Not factorable
    b. $(2x + 1)(9x + 2)$
17. a. $(6x - 5)(3x + 1)$
    b. Not factorable
18. a. Sum $= -\frac{4}{15}$, Prod $= -\frac{1}{5}$
    b. Sum $= \frac{4}{3}$, Prod $= -\frac{5}{9}$
19. a. Yes
    b. No
20. a. −5
    b. 0 or 2
21. a. $1, -2, \dfrac{-1 \pm \sqrt{15}i}{2}$
    b. $1, 2, \dfrac{3 \pm \sqrt{33}}{2}$
22. a. 1 or 81
    b. 16
23. a. 27 or −64
    b. 216 or −1
24. a. 9
    b. 25
25. a. $-3 < x < 2$

25. b. $-2 < x < 3$

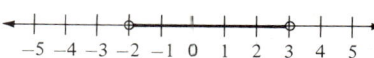

26. a. $x \ge 0$ or $x \le -4$

26. b. $x \ge 3$ or $x \le 0$

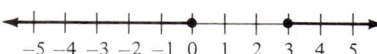

27. a. $x \le -2 - 2\sqrt{3}$ or $x \ge -2 + 2\sqrt{3}$

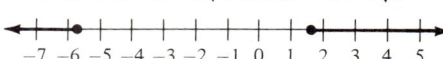

27. b. $x \le -3 - 3\sqrt{3}$ or $x \ge -3 + 3\sqrt{3}$

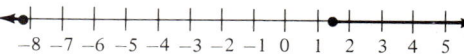

28. a. $x \le 1$ or $2 \le x \le 3$

28. b. $x \le -2$ or $-1 \le x \le 3$

29. a. $x \ge 3$ or $-2 \le x \le -1$

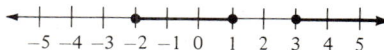

29. b. $x \ge 1$ or $-3 \le x \le -2$

30. a. $x < 2$ or $x \ge 6$

30. b. $-4 \le x < -2$

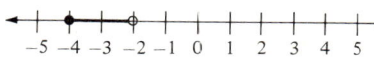

EXERCISE 7.1

A

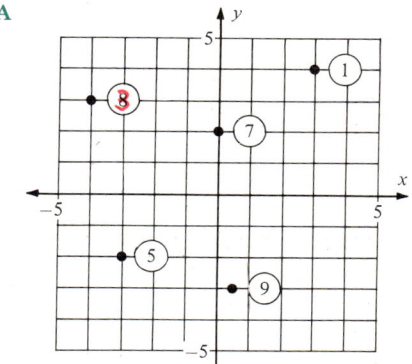

B 11. $y = x + 3$,
   $(-2, 1), (-1, 2), (0, 3), (1, 4),$
   $(2, 5)$

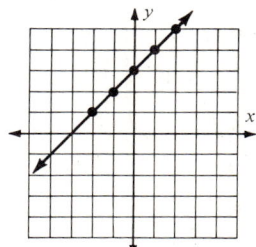

13. $x - y = 4$
   $(-1, -5), (0, -4), (1, -3)$

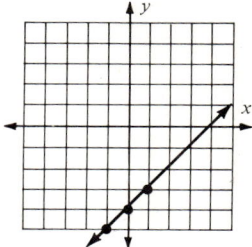

15. $2x - y - 3 = 0$
   $(-1, -5), (0, -3), (1, -1)$

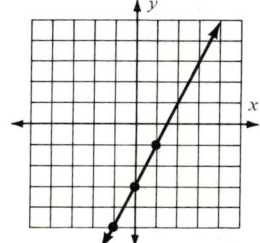

C 17. $y = x - 5$

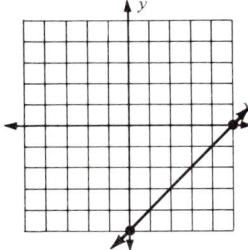

19. $2x + 3y = 6$

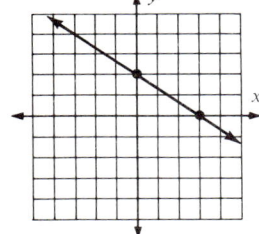

21. $2x - y = 4$

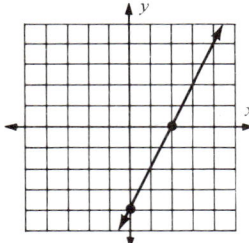

23. $2x + y - 4 = 0$

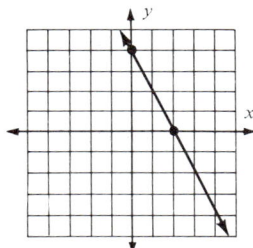

25. $y + 2x = 4$

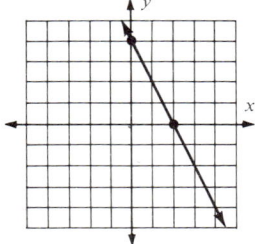

27. $2x - 5y = -10$

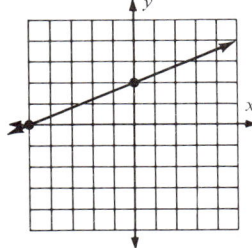

29. $x - y - 5 = 0$

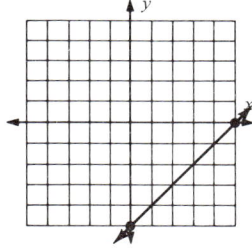

D 31. $-\frac{7}{2}x = 14$

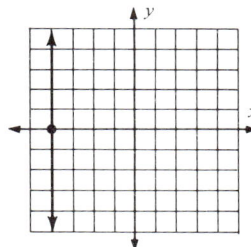

33. $\frac{3}{2}x = 6$

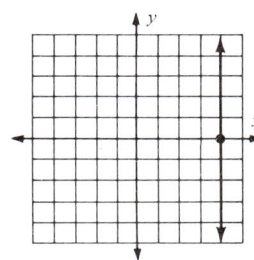

35. $-\frac{3}{4}x = 3$

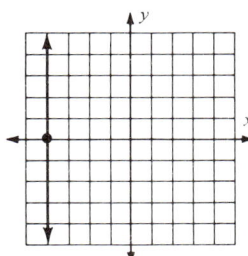

37. $-\frac{1}{3} + y = \frac{2}{3}$

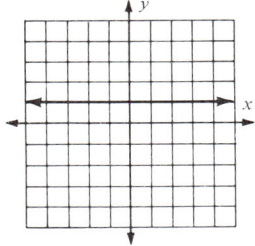

39. $\frac{2}{3} = x - \frac{4}{3}$

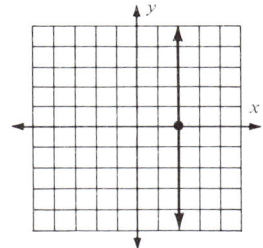

E 41. a. Yes
    b. 80
    c. 160
43. (20, 140)
45. (45, 148)

✓ **SKILL CHECKER**

47. $-\frac{3}{2}$
49. -3
51. $\frac{2}{3}$

7.1 USING YOUR KNOWLEDGE

1. $86°F$
3. Less than 10%
5. $12°F$

EXERCISE 7.2

A,B 1. Distance 5; slope $\frac{4}{3}$

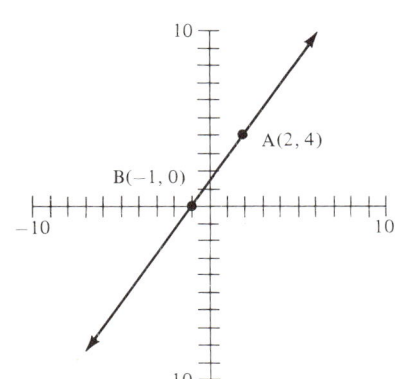

3. Distance $\sqrt{73}$; slope $\frac{8}{3}$

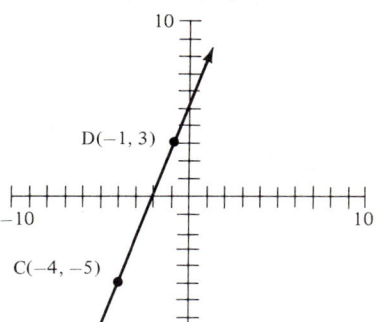

D(−1, 3)

C(−4, −5)

5. Distance $3\sqrt{10}$; slope 3

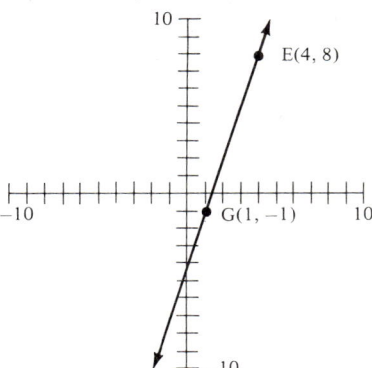

E(4, 8)

G(1, −1)

7. Distance 5; slope 0

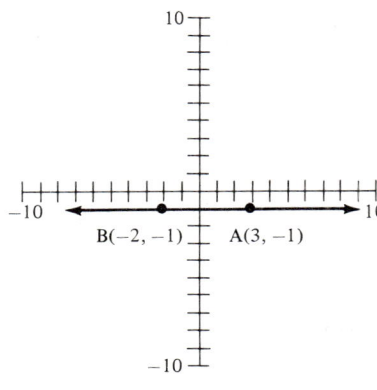

B(−2, −1)   A(3, −1)

9. Distance 6; slope undefined

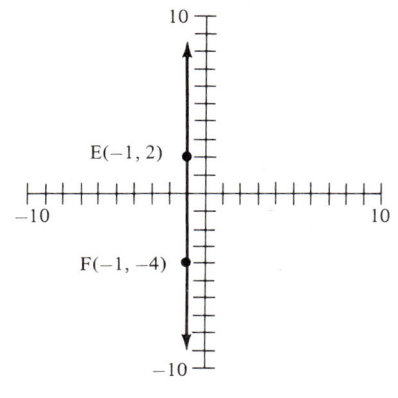

E(−1, 2)

F(−1, −4)

C
11. Parallel
13. Perpendicular
15. Neither
17. Perpendicular
19. Parallel
21. 2
23. 4
25. $-\frac{16}{3}$
27. −8
29. 3

D 31.

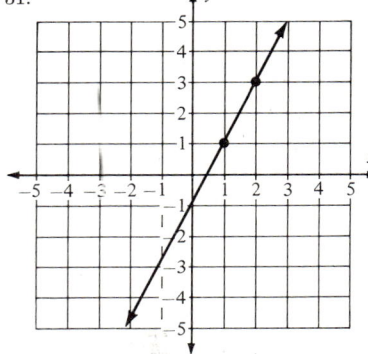

33.

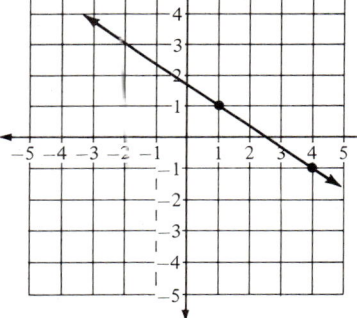

35.

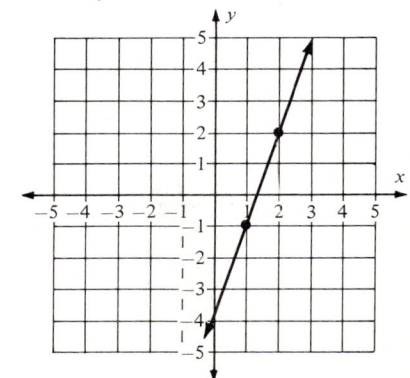

37.

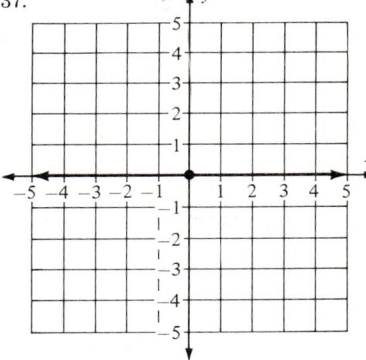

39.

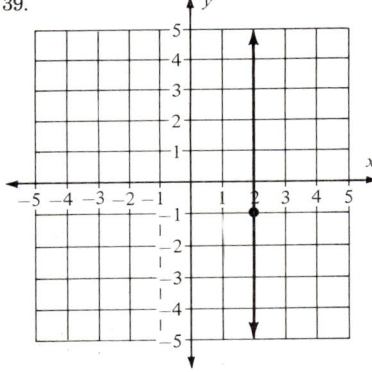

E 41. No (584 ≠ 499 + 13)
43. Yes (650 = 637 + 13)
45. Yes (584 = 292 + 292)

✓ **SKILL CHECKER**

47. $y = -2x + 4$
49. $y = \frac{2}{3}x + 4$

7.2 USING YOUR KNOWLEDGE

1. $\frac{3}{2}$
3. $-2\sqrt{3}$
5. $\frac{\sqrt{2}}{4}$
7. $\frac{1 + \sqrt{13}}{2}$ or $\frac{1 - \sqrt{13}}{2}$
9. 2 or −1

EXERCISE 7.3

A 1. $3x - y = 4$

3. $x + y = 5$

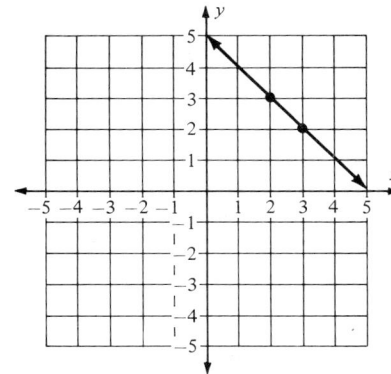

B 5. $y - 5 = 2(x + 3)$

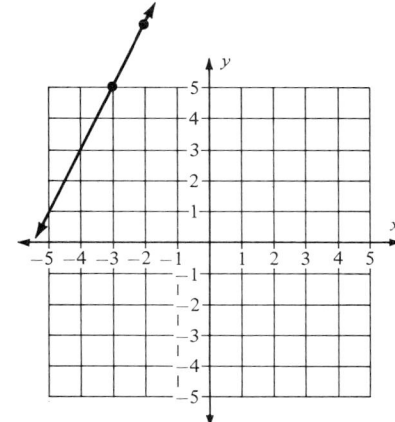

7. $y + 2 = -3(x + 1)$

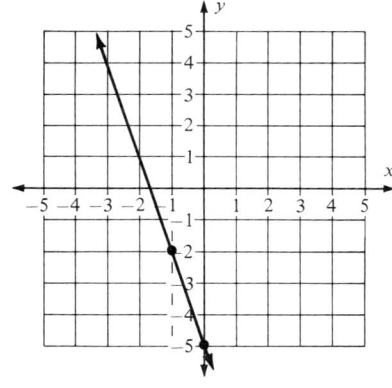

C 9. $y = 5x + 2$
11. $y = -\frac{1}{5}x - \frac{1}{3}$
D 13. $y = 2x - 4$
15. $y = -3x - 12$
17. $y = -2x + 3$
19. $y = -x - 6$
21. $y = \frac{1}{3}x - \frac{5}{3}$

23. $y = -\frac{5}{2}x + 3$
25. $y = -\frac{1}{8}x + \frac{1}{2}$
27. $y = 4$
29. $x = -2$
31. $x = -2$
33. $y = 2$
35. $y = -2x + 4$
37. $y = \frac{2}{3}x + 2$
E 39. a. $d + 2p = 14$
  b. 10
41. a. $2p - s = 20$
  b. $10
  c. 60
43. $4

45. x-int. 4; y-int. -2
47. x-int. -3; y-int. 6
49. $x \geq -2$

7.3 USING YOUR KNOWLEDGE

1. $y = 2x + 50$
3. a. $2
  b. $75

EXERCISE 7.4

A Graph
1. $x + 2y > 4$

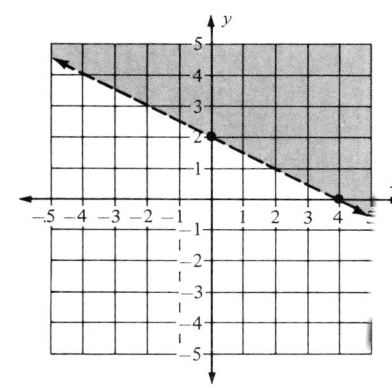

3. $-2x - 5y \leq -10$

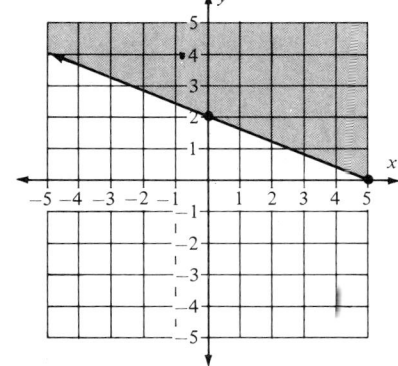

5. $y \geq 2x - 2$

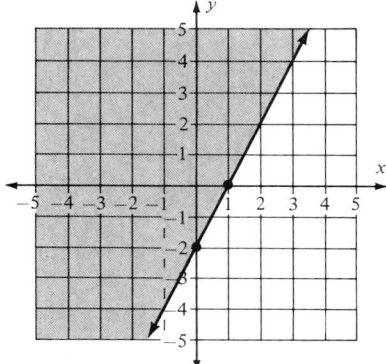

7. $6 < 3x - 2y$

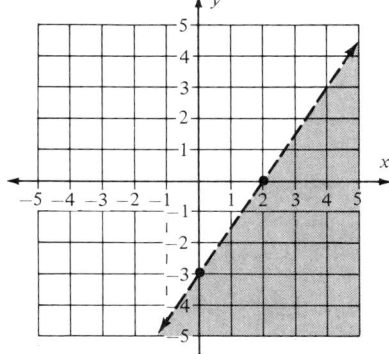

9. $4x + 3y \geq 12$

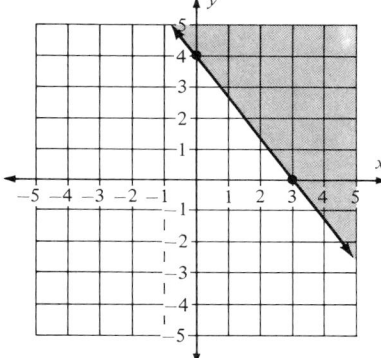

11. $10 < -5x + 2y$

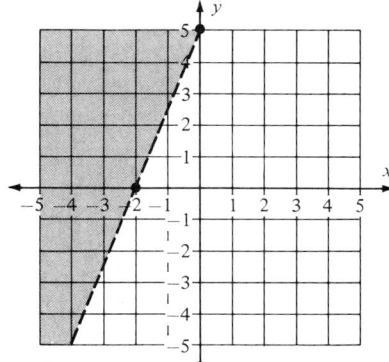

13. $2x \geq 2y - 4$

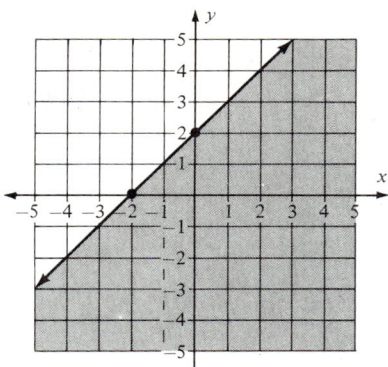

21. $|x| < 1$

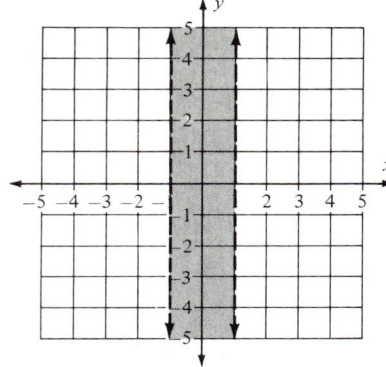

29. $|x + 2| < 1$

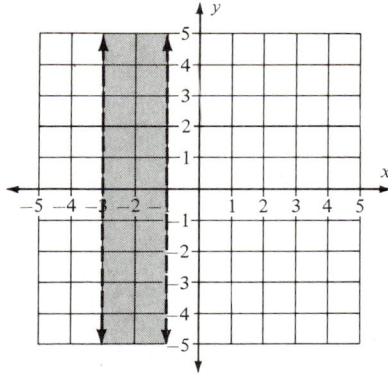

15. $2y < -4x + 8$

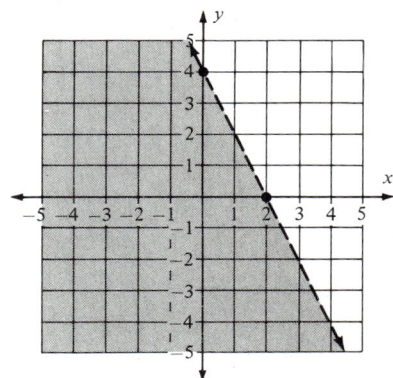

23. $|y| < 2$

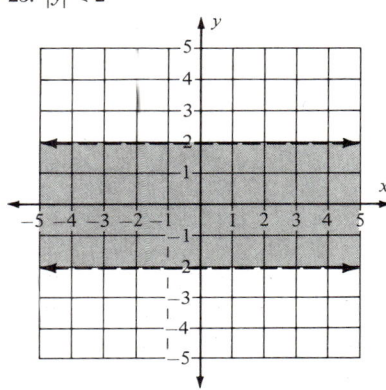

31. $|y + 2| < 1$

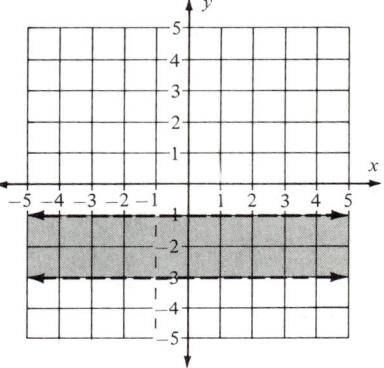

17. $x \geq -3$

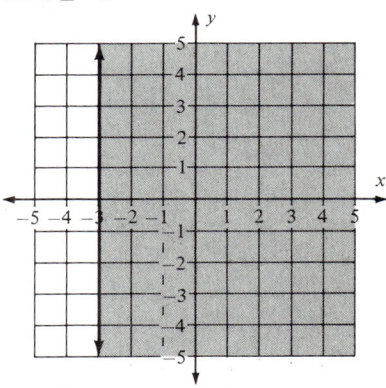

25. $|x| \geq 1$

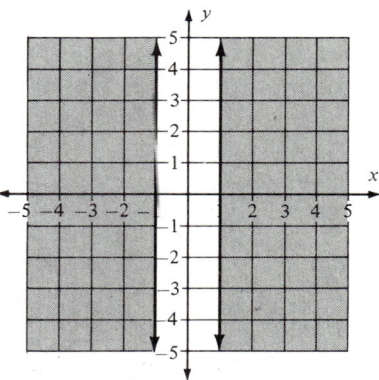

33. $|x + 1| \geq 3$

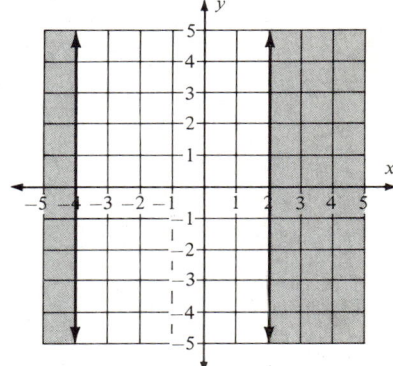

19. $y < 3$

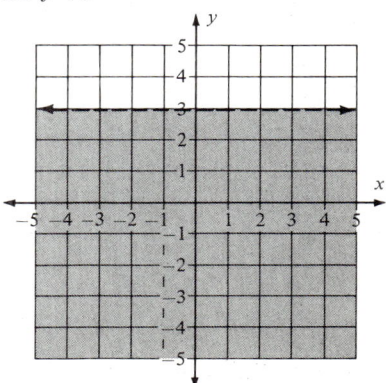

27. $|y| \geq 2$

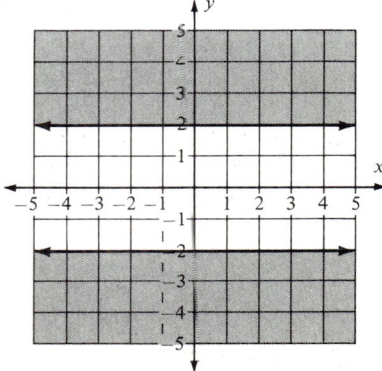

35. $|x - 1| \leq 2$

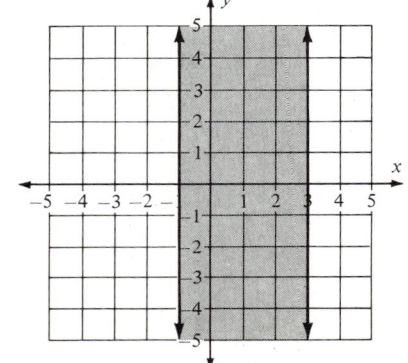

37. $|y - 2| < 1$

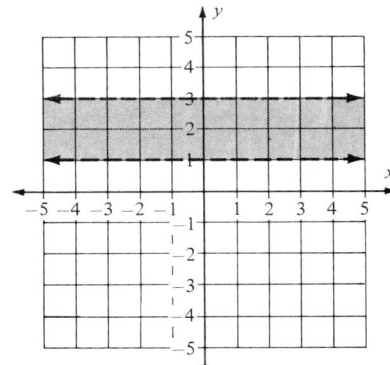

EXERCISE 7.4

✓ **SKILL CHECKER**

39. $k = 4$
41. $k = 180$

7.4 USING YOUR KNOWLEDGE

1. a. 33 miles
   b. Rental A (It is $10 cheaper)
3. If you plan to drive more than
   33 miles, Rental A is the cheapest.

EXERCISE 7.5

A 1. $T = ks$
   3. $W = kh^3$
   5. $W = kB$
B 7. $R = \dfrac{k}{D^2}$
C 9. $I = kPr$
   11. $A = ksv^2$
   13. $V = kdw$
   15. $I = \dfrac{ki}{d^2}$
   17. $R = \dfrac{kL}{A}$
   19. $W = \dfrac{k}{d^2}$
   21. a. $I = km$
       b. $k = 0.055$ or 5.5%
       c. $41.25
   23. a. $d = ks^2$
       b. $k = 0.06$
       c. 216 feet
   25. a. $S = \dfrac{k}{y}$
       b. 30 songs
   27. a. $W = \dfrac{k}{d^2}$
       b. $k = 121(3960)^2$
       c. 81 pounds

29. $2x - y = 2$

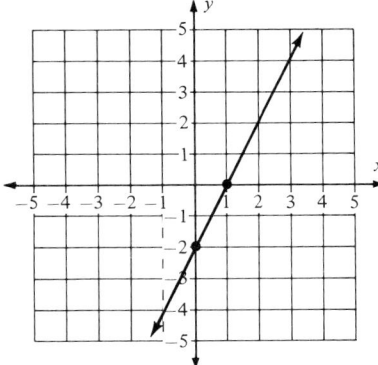

31. $y = -x - 3$

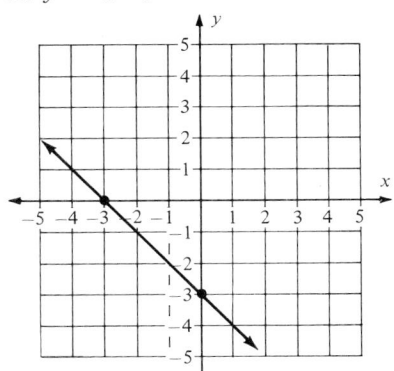

7.5 USING YOUR KNOWLEDGE

1.

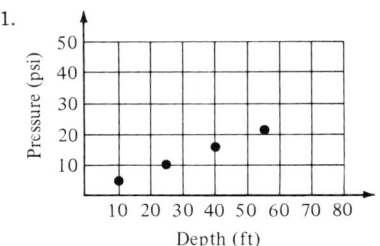

3. $p = kd$
5. They are equal.

REVIEW EXERCISES CHAPTER 7

1. a.

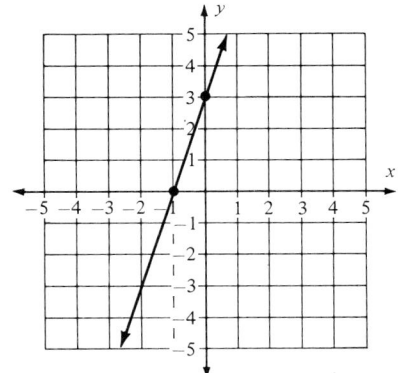

b.

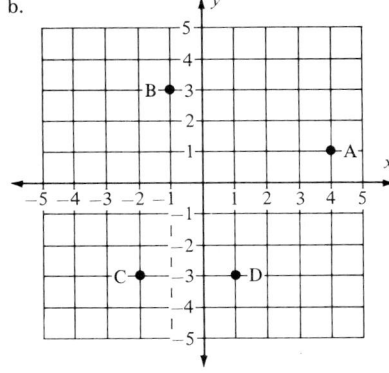

2. a.

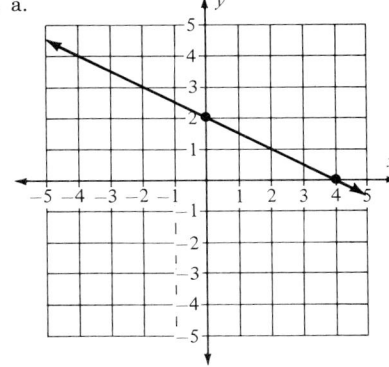

b.

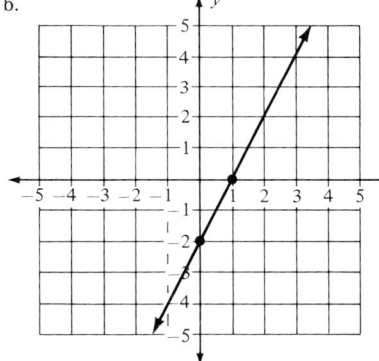

3. a. x-int. -1; y-int. 3

b. *x*-int. 2; *y*-int. -4

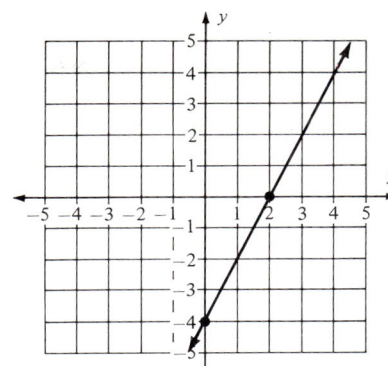

12. a.

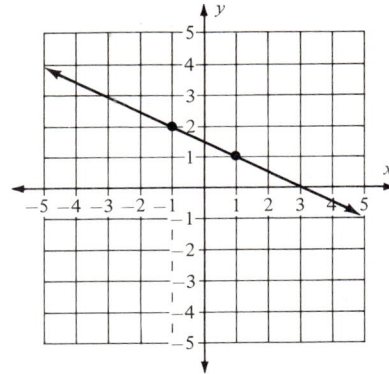

b.

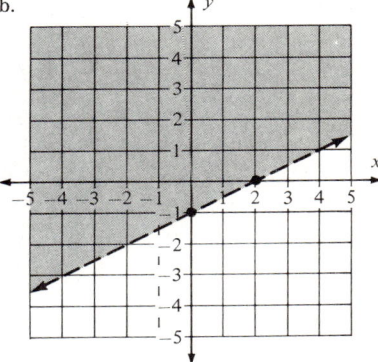

4.

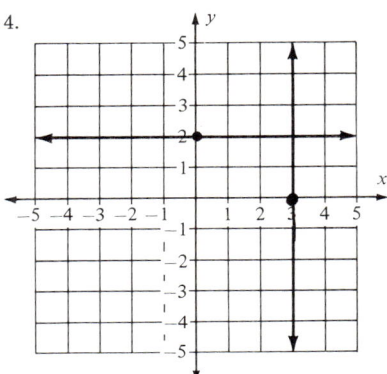

b.

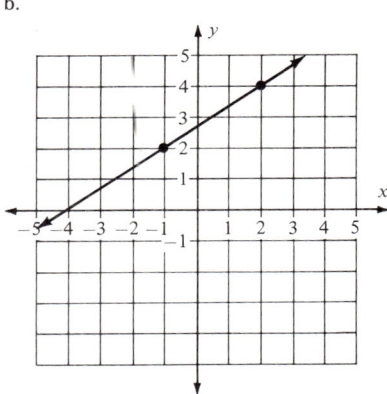

21. a.

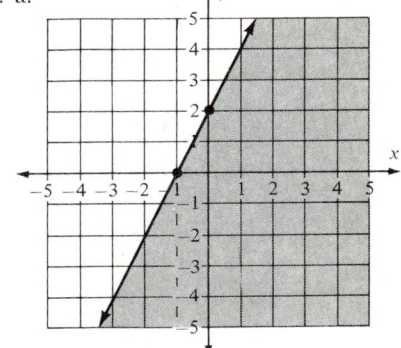

5.

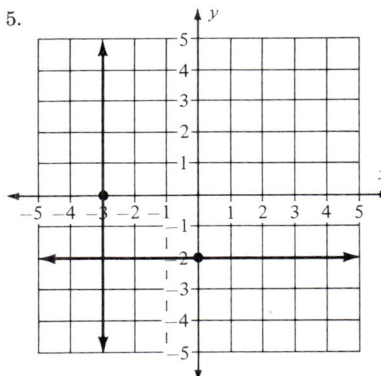

b.

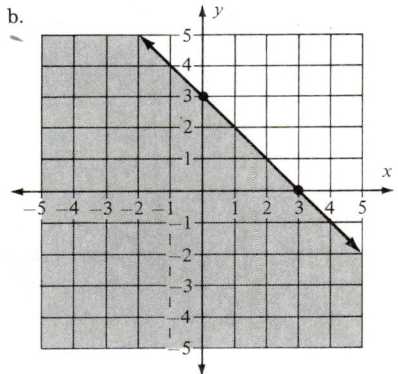

13. a. $x - y = -3$
    b. $5x + 2y = -14$
14. a. $2x - y = 1$
    b. $2x - y = 10$
15. a. $y = 3x + 2$
    b. $y = -3x + 4$
16. a. Slope 2; *y*-int -4
    b. Slope $-\frac{1}{2}$; *y*-int. 2
17. a. Slope -2; *y*-int. 4
    b. Slope $\frac{4}{3}$; *y*-int. 4
18. a. $2x + y = 5$
    b. $3x - y = 5$
19. a. $3x - 2y = 4$
    b. $2x + 3y = 7$

20. a.

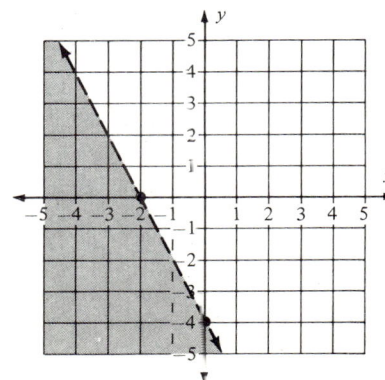

22. a.

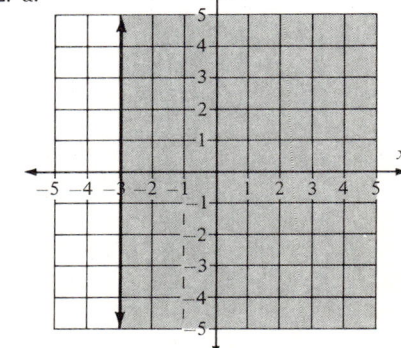

6. a. 13
   b. $2\sqrt{5}$
   c. 4
7. a. $-\frac{1}{2}$
   b. Undefined
8. a. -1
   b. $-\frac{3}{2}$
9. a. Perpendicular
   b. Parallel
10. a. Parallel
    b. Neither
11. a. $y = 2$
    b. $\frac{11}{2}$

b.

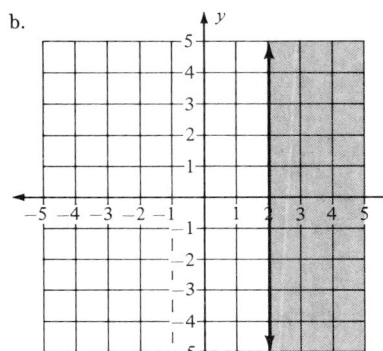

b.

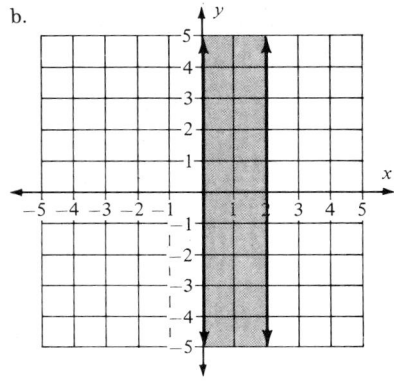

5. Consistent

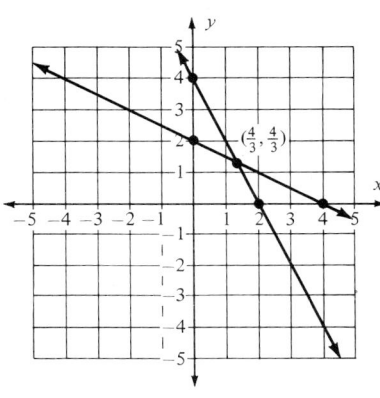

23. a.

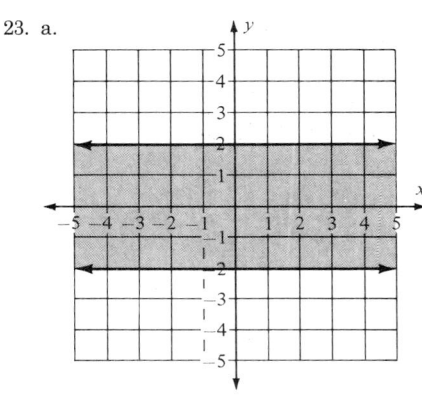

25. a. $P = kT$

   b. $k = \dfrac{1}{120}$

26. a. $P = \dfrac{k}{V}$

   b. $k = 3200$

27. a. $F = \dfrac{k}{d^2}$

   b. $k = 1.92 \times 10^9$

28. 76.8 lb.

29. a. $h = kd^3r$

   b. $k = \frac{1}{20}$

30. 600 hp

b.

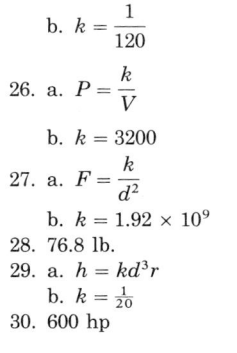

EXERCISE 8.1

A 1. Consistent

7. Consistent

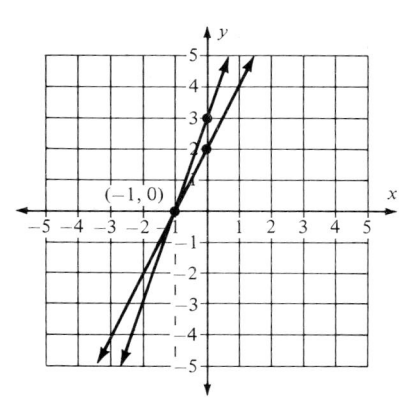

24. a.

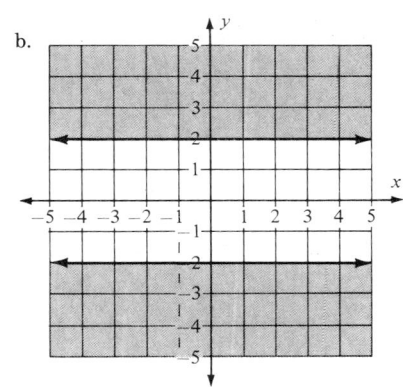

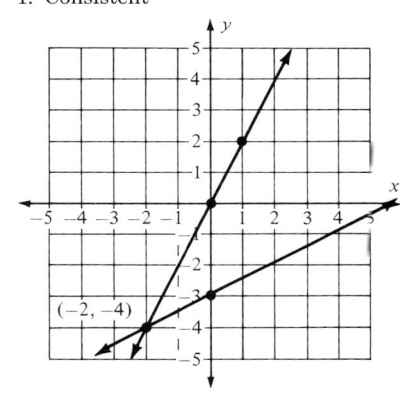

3. Consistent

9. Inconsistent

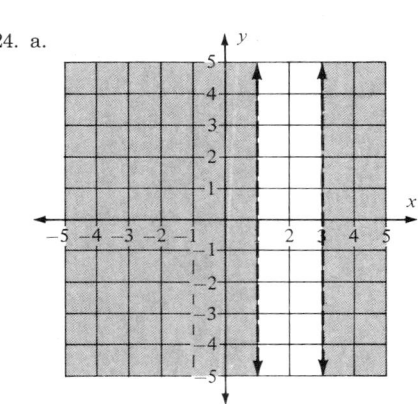

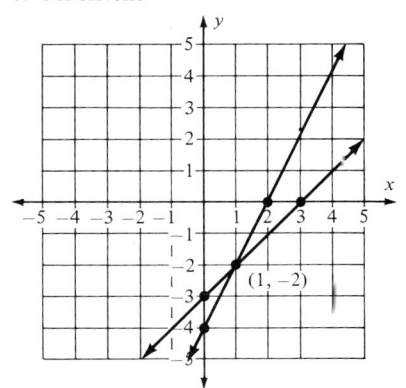

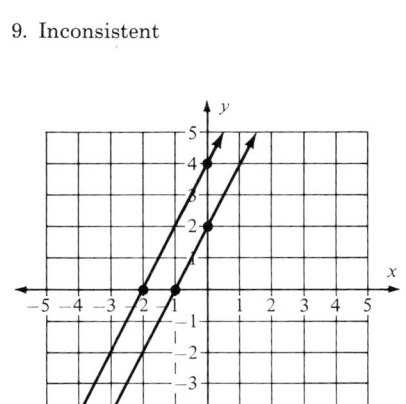

11. Consistent

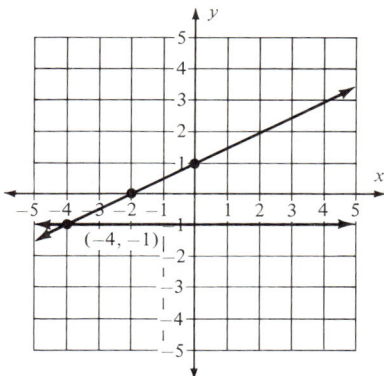
$(-4, -1)$

13. Consistent

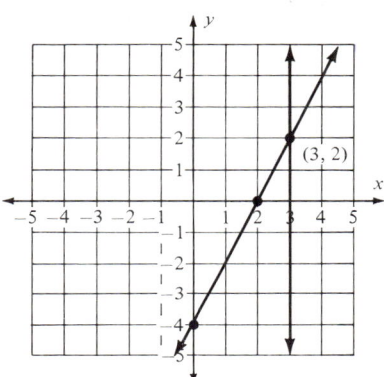
$(3, 2)$

15. Consistent

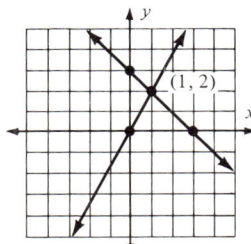
$(1, 2)$

17. Consistent

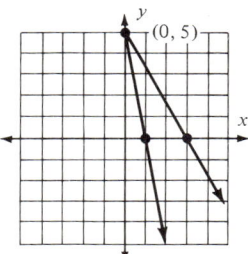
$(0, 5)$

19. Dependent

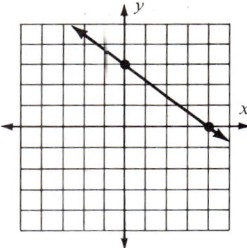

EXERCISE 8.1

B,C 21. $(2, 0)$; consistent
23. $(2, 3)$; consistent
25. No solution; inconsistent
27. No solution; inconsistent
29. Infinitely many solutions; dependent
31. $(-1, -1)$; consistent
33. No solution; inconsistent
35. $(4, 1)$; consistent
D 37. 4
39. 5

✓ SKILL CHECKER

41. 50
43. 2
45. 1

8.1 USING YOUR KNOWLEDGE

1. $L_1 = 80$, $L_2 = 820$
3. a. $5x = 4x + 500$
   b. 500

EXERCISE 8.2

A,B 1. $(5, 3)$ Consistent
3. $(0, \frac{1}{2})$ Consistent
5. $(0, 1)$ Consistent
7. $(2, 3)$ Consistent
9. $(\frac{5}{2}, -\frac{1}{2})$ Consistent
11. No solution; Inconsistent
13. $(3, 2)$ Consistent
15. $(2, 1)$ Consistent
17. $(\frac{1}{3}, 2)$ Consistent
19. $(5, -2)$ Consistent
21. $(-\frac{1}{2}, -\frac{2}{3})$ Consistent
23. $(8, -12)$ Consistent
25. $(6, 8)$ Consistent
27. $(4, -3)$ Consistent
29. $(4, 2)$ Consistent

C 31. 36 and 34
33. Antenna = 222 ft
    Building = 1250 ft
35. 660 lb, 640 lb

✓ SKILL CHECKER

37. x-int. = 4, y-int. = 6
39. x-int. = $-\frac{4}{3}$, y-int. = 2

8.2 USING YOUR KNOWLEDGE

1. Dee = $120\frac{2}{3}$ lb, Dum = $119\frac{2}{3}$ lb

EXERCISE 8.3

1. $(5, 3, 4)$ Consistent
3. $(-1, 1, 4)$ Consistent
5. $(3, 4, 1)$ Consistent
7. No solution; inconsistent
9. $(\frac{1}{2}, \frac{1}{4}, \frac{1}{3})$ Consistent
11. No solution; Inconsistent
13. No solution; Inconsistent
15. $(\frac{9}{2}, \frac{1}{2}, \frac{5}{2})$ Consistent
17. $(6, 3, -1)$ Consistent
19. $(-2, -3, -4)$ Consistent
21. a must be $\dfrac{c - b}{2}$
23. $(-1, 0, 2)$
25. $k = -2$

✓ SKILL CHECKER

27. -473
29. $\dfrac{6}{47}$

8.3 USING YOUR KNOWLEDGE

1. $F = \dfrac{54d + 196}{5}$
3. $2\frac{2}{3}$ cm/sec

EXERCISE 8.4

A 1. 2
3. 7
5. 6
7. $\frac{1}{2}$
9. $-\frac{7}{40}$
B 11. $(2, 3)$
13. $(4, 5)$
15. $(3, -1)$

17. (4, 5)

19. Dependent equations; infinitely many solutions

21. $(-2, -3)$

23. No solution; inconsistent equations

25. (10, 1)

27. (5, 2)

29. $(-1, -1)$

C 31. -7

33. 0

35. -1

37. -4

39. -9

D,E 41. (1, 2, 3)

43. $(3, -1, -2)$

45. (3, 0, 4)

47. $(-5, 1, 5)$

49. $(-6, 2, 5)$

51. $\begin{vmatrix} a & b & 0 \\ c & d & 0 \\ e & f & 0 \end{vmatrix}$

$= a\begin{vmatrix} d & 0 \\ f & 0 \end{vmatrix} - b\begin{vmatrix} c & 0 \\ e & 0 \end{vmatrix} + 0\begin{vmatrix} d & 0 \\ f & 0 \end{vmatrix}$

$= a \cdot 0 - b \cdot 0 + 0 = 0$

53. $\begin{vmatrix} a & b & c \\ 1 & 2 & 3 \\ a & b & c \end{vmatrix} = a\begin{vmatrix} 2 & 3 \\ b & c \end{vmatrix} - b\begin{vmatrix} 1 & 3 \\ a & c \end{vmatrix}$

$\qquad + c\begin{vmatrix} 1 & 2 \\ a & b \end{vmatrix}$

$= a(2c - 3b) - b(c - 3a)$
$\quad + c(b - 2a)$
$= 2ac - 3ab - bc + 3ab$
$\quad + bc - 2ac$
$= 0$

55. $\begin{vmatrix} 1 & 2 & 3 \\ 3 & 1 & 2 \\ k & 2k & 3k \end{vmatrix} \overset{?}{=} k\begin{vmatrix} 1 & 2 & 3 \\ 3 & 1 & 2 \\ 1 & 2 & 3 \end{vmatrix}$

$1\begin{vmatrix} 1 & 2 \\ 2k & 3k \end{vmatrix} - 2\begin{vmatrix} 3 & 2 \\ k & 3k \end{vmatrix} + 3\begin{vmatrix} 3 & 1 \\ k & 2k \end{vmatrix}$

$\overset{?}{=} k\left[1\begin{vmatrix} 1 & 2 \\ 2 & 3 \end{vmatrix} - 2\begin{vmatrix} 3 & 2 \\ 1 & 3 \end{vmatrix} + 3\begin{vmatrix} 3 & 1 \\ 1 & 2 \end{vmatrix}\right]$

$1(3k - 4k) - 2(9k - 2k) + 3(6k - k)$
$\overset{?}{=} k[1(3 - 4) - 2(9 - 2) + 3(6 - 1)]$
$-k - 14k + 15k$
$\qquad \overset{?}{=} k[-1 - 14 + 15]$
$0 \cdot k \overset{?}{=} k \cdot 0$
$\qquad 0 = 0$

57. $kb_1\begin{vmatrix} b_2 & 2 \\ b_3 & 3 \end{vmatrix} - b_1\begin{vmatrix} kb_2 & 2 \\ kb_3 & 3 \end{vmatrix} + 1\begin{vmatrix} kb_2 & b_2 \\ kb_3 & b_3 \end{vmatrix}$

$= kb_1(3b_2 - 2b_3) - b_1(3kb_2 - 2kb_3)$
$\quad + 1(kb_2b_3 - kb_2b_3)$
$= 3kb_1b_2 - 2kb_1b_3 - 3kb_1b_2$
$\quad + 2kb_1b_3 + 0$
$= 0$

59. $1\begin{vmatrix} a & a \\ b & b \end{vmatrix} - 1\begin{vmatrix} 2 & a \\ 3 & b \end{vmatrix} + 1\begin{vmatrix} 2 & a \\ 3 & b \end{vmatrix} =$
$1(ab - ab) + 0 = 0$

✓ **SKILL CHECKER**

61. 12,000 at 5%,
8,000 at 7%

63. 32 gallons

65. 250 mph

8.4 USING YOUR KNOWLEDGE

1. $2x - y + 3 = 0$

3. $2x + 9y - 34 = 0$

5. $bx + ay - ab = 0$

EXERCISE 8.5

A 1. 25 nickels, 50 dimes

3. 10 nickels, 20 quarters

5. 5 pennies, 5 nickels

7. 20 tens, 5 twenties

9. 13 nickels, 9 dimes, 22 quarters

B 11. 43 and 59

13. 21 and 105

15. 4 and 20

17. Longs is 14,255 ft high, Pikes is 14,110 ft high

19. butterscotch: $2033\frac{1}{3}$ lb
caramel: $2033\frac{1}{3}$ lb
chocolate: $2633\frac{1}{3}$ lb

C 21. plane: 210 mph, wind: 30 mph

23. boat: 12 mph, current: 3 mph

25. plane: 450 mph, wind: 50 mph

D 27. $6000 at 8%, $4000 at 6%

29. $10,000 at 6%,
$5000 at 8%
$10,000 at 10%

✓ **SKILL CHECKER**

31. $(x + 3)(x + 1)$

33. $(x + 3)(x - 1)$

35. $-1(x - 3)(x - 1)$

8.5 USING YOUR KNOWLEDGE

1. $15a + 5b + 20c = 170$

3. $10a + 15b + 10c = 110$

REVIEW EXERCISES CHAPTER 8

1. a.

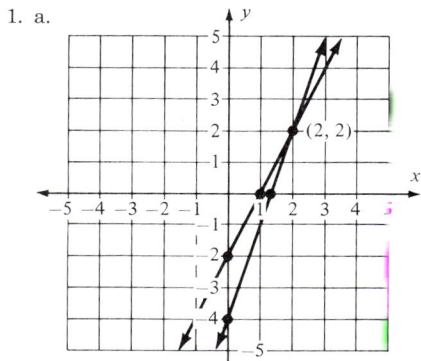

3. a.

---

b.

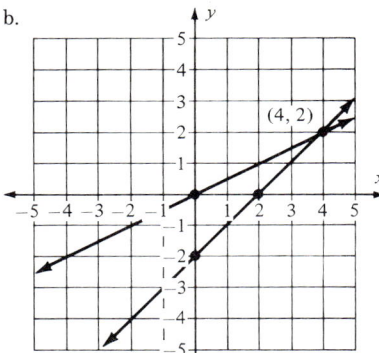

2. a.

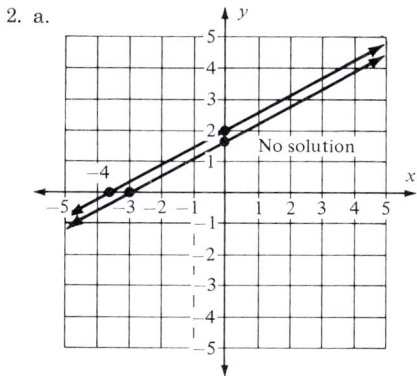

b.

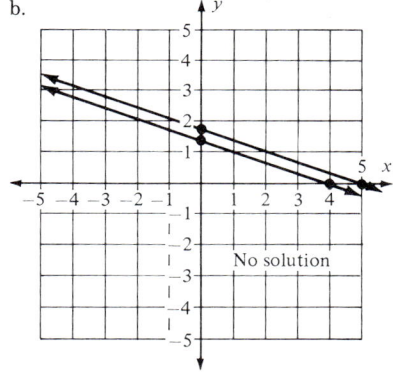

3. a.

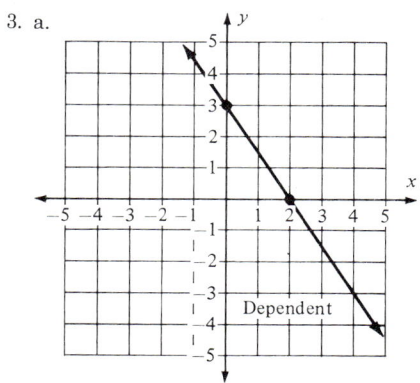

---

b.

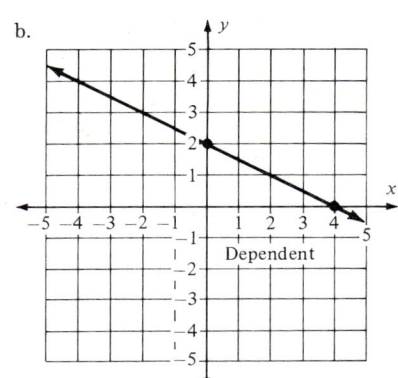
Dependent

b. 10 nickels
   15 dimes
23. a. 180 ft
    b. 160 ft
24. a. 6 mph
    b. 3 mph
25. a. $10,000 at 4%
       $10,000 at 6%
       $20,000 at 8%
    b. $10,000 at 4%
       $15,000 at 6%
       $20,000 at 8%

7.

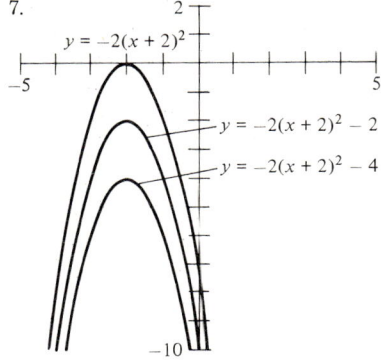

EXERCISE 9.1

4. a. $(3, 2)$
   b. $\left(\frac{7}{5}, \frac{12}{5}\right)$
5. a. Inconsistent. No solution
   b. Inconsistent. No solution
6. a. Dependent. Infinitely many
      solutions
   b. Dependent. Infinitely many
      solutions
7. a. $(4, -1)$
   b. $(-1, 2)$
8. a. Inconsistent. No solution
   b. Inconsistent. No solution
9. a. Dependent. Infinitely many
      solutions
   b. Dependent. Infinitely many
      solutions
10. a. $(2, -3)$
    b. $(1, 1)$
11. a. Dependent. Infinitely many
       solutions
    b. Dependent. Infinitely many
       solutions
12. a. Inconsistent. No solution
    b. Inconsistent. No solution
13. a. Dependent. Infinitely many
       solutions
    b. Dependent. Infinitely many
       solutions
14. a. -22
    b. 14
15. a. $D = -23, D_x = -23, D_y = 46$
       $x = 1, y = -2$
    b. $D = -28, D_x = -14, D_y = 14$
       $x = \frac{1}{2}, y = -\frac{1}{2}$
16. a. 15
    b. -18
17. a. $D = -10, D_x = -20, D_y = -30, D_z = 20$
       $x = 2, y = 3, z = -2$
    b. $D = 6, D_x = 6, D_y = -12, D_z = -6$
       $x = 1, y = -2, z = -1$
18. a. Inconsistent. No solution
    b. Inconsistent. No solution
19. a. 9
    b. 45
20. a. -10
    b. -20
21. a. -1
    b. -4
22. a. 30 nickels
       25 dimes

A 1.

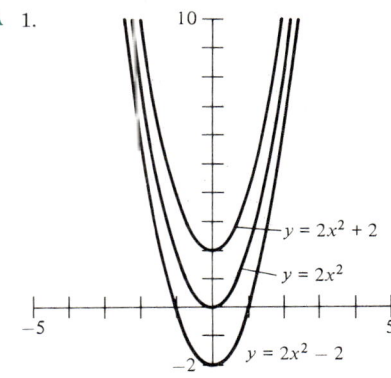

9.

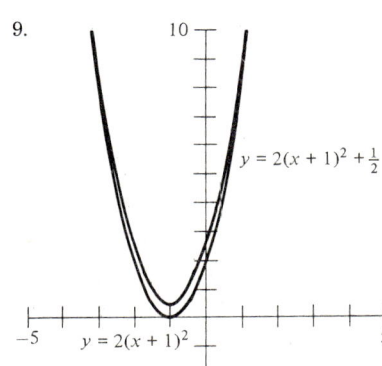

3.

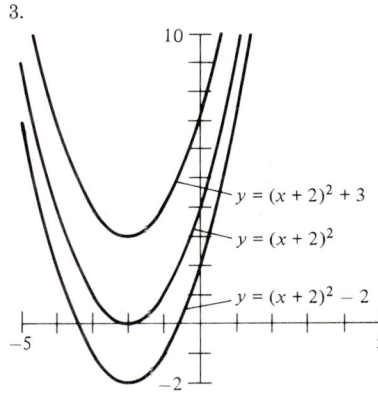

B 11.

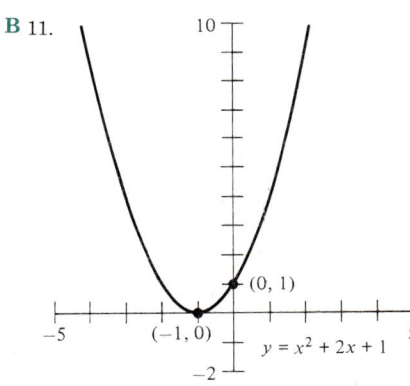

5.

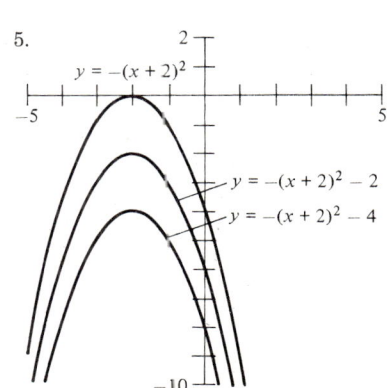

13.

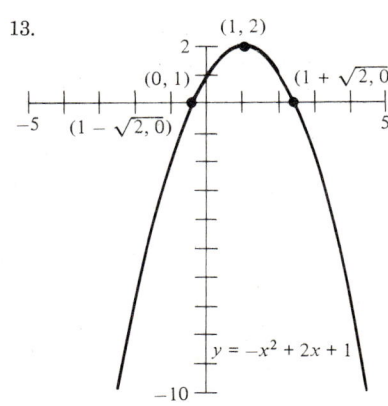

15.
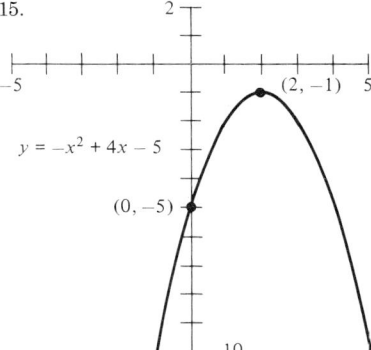
$y = -x^2 + 4x - 5$
$(2, -1)$
$(0, -5)$

17.
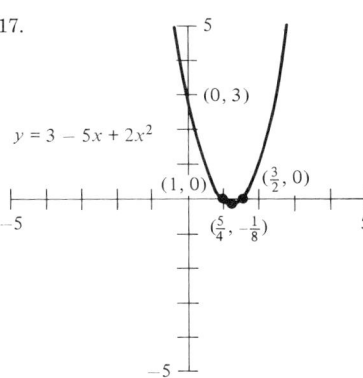
$y = 3 - 5x + 2x^2$
$(0, 3)$
$(1, 0)$
$(\frac{3}{2}, 0)$
$(\frac{5}{4}, -\frac{1}{8})$

19.
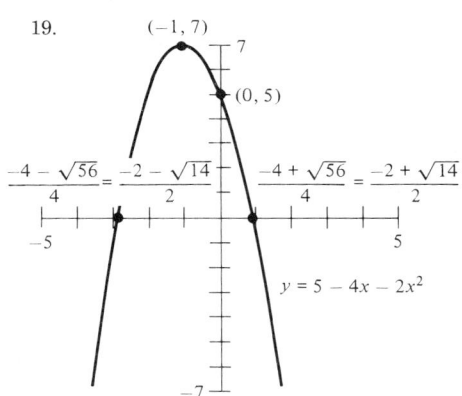
$(-1, 7)$
$(0, 5)$
$\dfrac{-4 - \sqrt{56}}{4} = \dfrac{-2 - \sqrt{14}}{2}$
$\dfrac{-4 + \sqrt{56}}{4} = \dfrac{-2 + \sqrt{14}}{2}$
$y = 5 - 4x - 2x^2$

21.
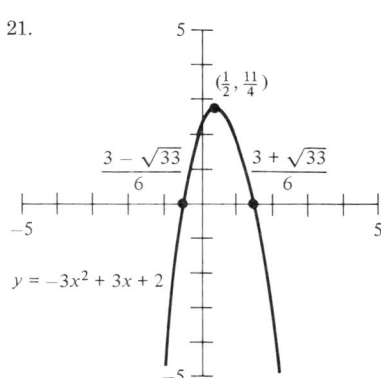
$(\frac{1}{2}, \frac{11}{4})$
$\dfrac{3 - \sqrt{33}}{6}$
$\dfrac{3 + \sqrt{33}}{6}$
$y = -3x^2 + 3x + 2$

C 23.
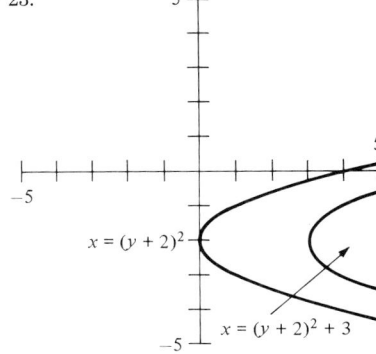
$x = (y + 2)^2$
$x = (y + 2)^2 + 3$

25.
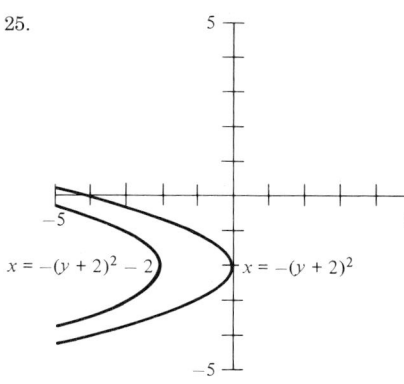
$x = -(y + 2)^2 - 2$
$x = -(y + 2)^2$

27.
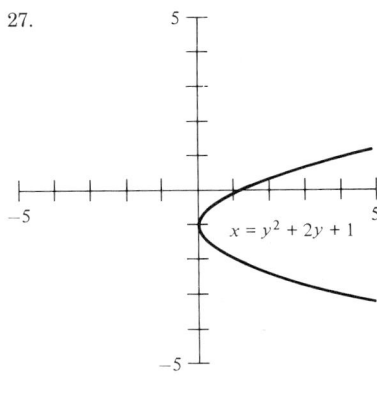
$x = y^2 + 2y + 1$

29.
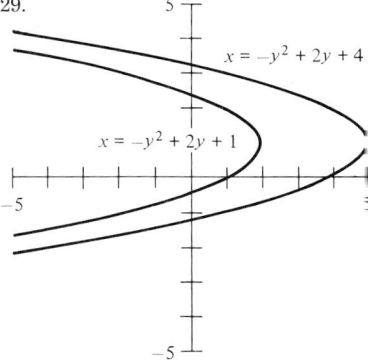
$x = -y^2 + 2y + 4$
$x = -y^2 + 2y + 1$

D 31. $x = 4000$, $P = \$11,000$
   33. $25(thousand) or $25,000
   35. 400 ft

37. The equation for the price p is:
$p = (600 + 100w)(1 - 0.10w)$, where
w is the number of weeks elapsed.
The maximum for p occurs when
$w = 2$.

✓ SKILL CHECKER

39. 5 units

9.1 USING YOUR KNOWLEDGE

1. $\sqrt{x^2 + (y - p)^2}$
3. $x^2 = 4py$
5. $y^2 = 12.5x$. Focus (3.125, 0)

EXERCISE 9.2

A 1. $(x - 3)^2 + (y - 8)^2 = 4$
   3. $(x + 3)^2 + (y - 4)^2 = 25$
   5. $(x + 3)^2 + (y + 2)^2 = 16$
   7. $(x - 2)^2 + (y + 4)^2 = 5$
   9. $x^2 + y^2 = 9$
B 11. Center at (1, 2), radius 3

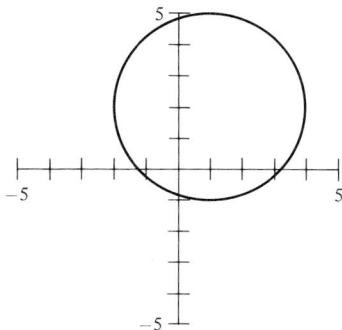

13. Center at $(-1, 2)$, radius 2

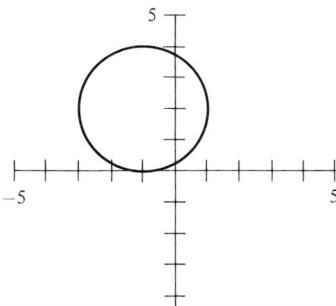

15. Center at (1, -2), radius 1

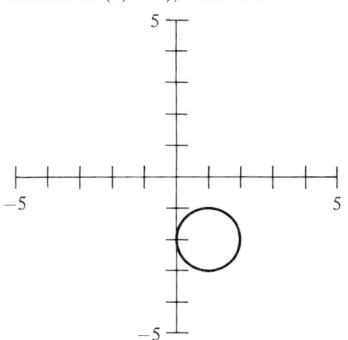

17. Center at $(-2, -1)$, radius 3

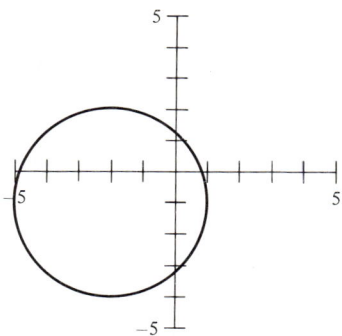

19. Center at $(1, 1)$, radius $\sqrt{7}$

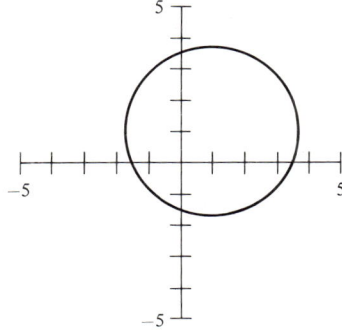

21. Center at $(3, 2)$, radius 2

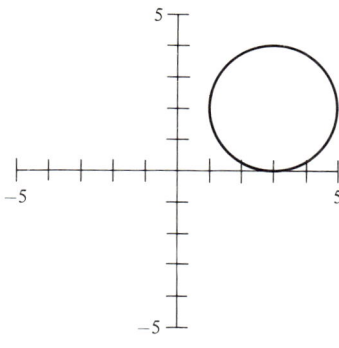

23. Center at $(2, -1)$, radius 3

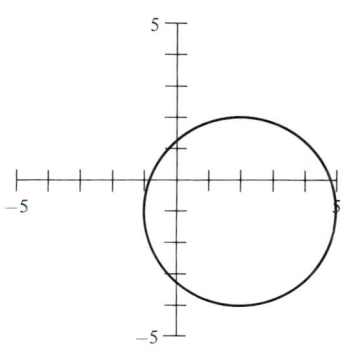

25. Center at $(0, 0)$, radius 5

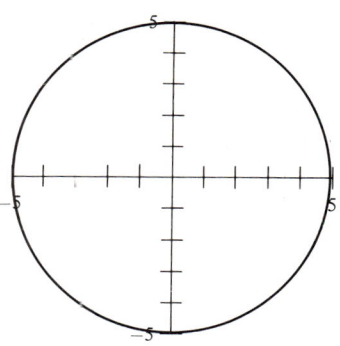

27. Center at $(0, 0)$, radius $\sqrt{7}$

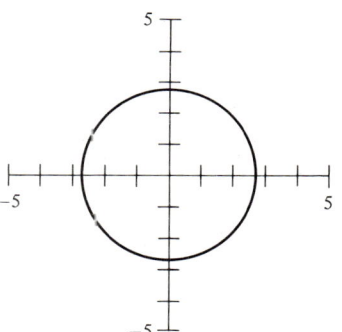

29. Center at $(-3, 1)$, radius 2

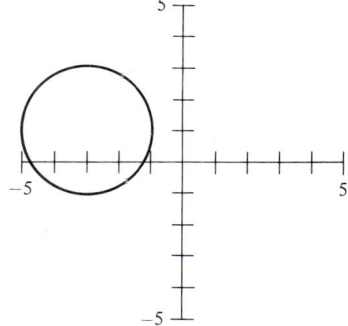

31. Center at $(3, 1)$, radius 2

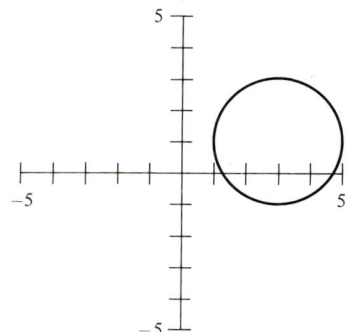

C 33.

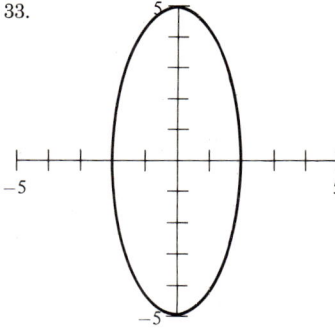

35.

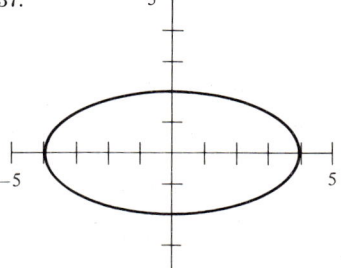

37.

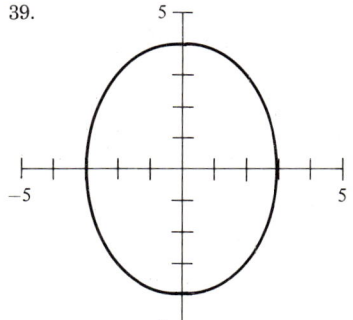

39.

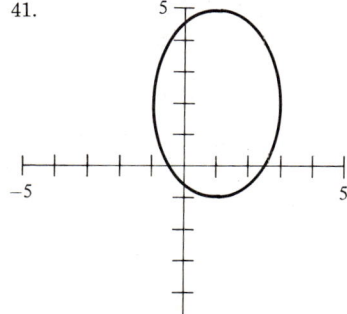

41.

43.

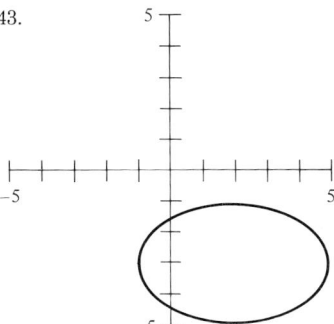

45.

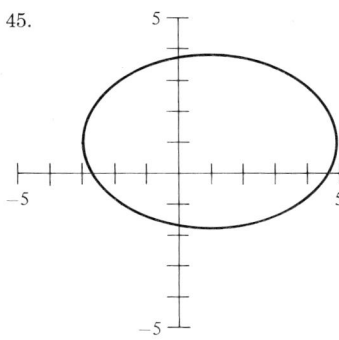

47. $x^2 + y^2 = 25$
49. $x^2 + y^2 = 169$
51. $x^2 + y^2 = 9$

✓ **SKILL CHECKER**

53. Right half of the circle

9.2 USING YOUR KNOWLEDGE

1. $\sqrt{(x-c)^2 + y^2}$
3. $-4xc = 4a^2 - 4a\sqrt{(x+c)^2 + y^2}$
   or $cx + a^2 = a\sqrt{(x+c)^2 + y^2}$
5. $b^2x^2 + a^2y^2 = a^2b^2$

EXERCISE 9.3

A 1. $\dfrac{x^2}{25} - \dfrac{y^2}{9} = 1$

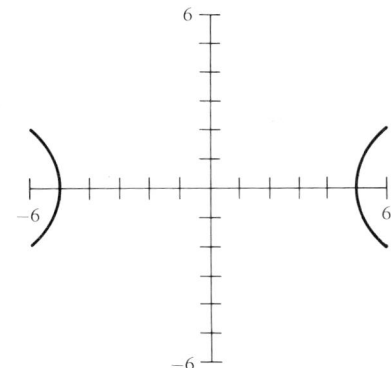

3. $\dfrac{y^2}{9} - \dfrac{x^2}{9} = 1$

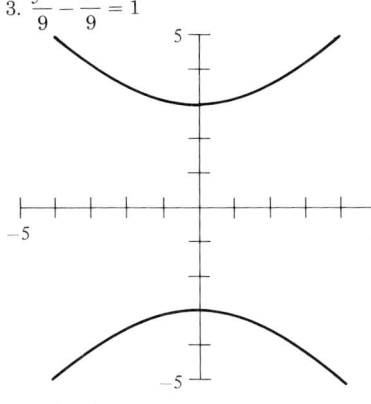

5. $\dfrac{x^2}{9} - \dfrac{y^2}{1} = 1$

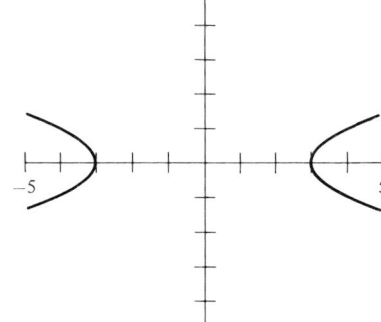

7. $\dfrac{x^2}{64} - \dfrac{y^2}{49} = 1$

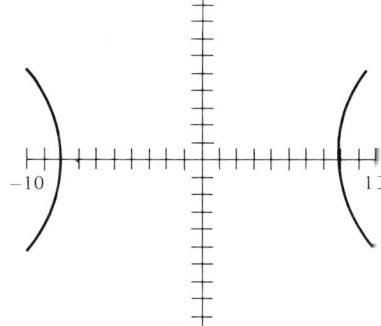

9. $\dfrac{y^2}{\frac{16}{9}} - \dfrac{x^2}{\frac{9}{16}} = 1$

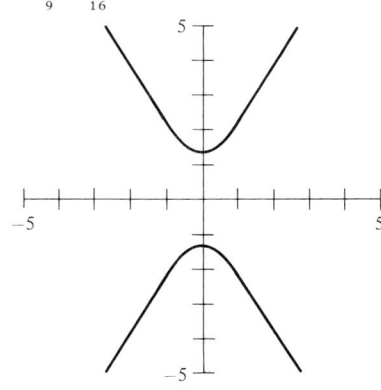

11. $x^2 - 9y^2 = 9$

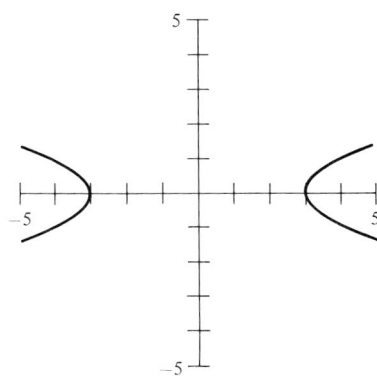

B 13. Circle $(5, 0)$, $(0, 5)$, $(-5, 0)$, $(0, -5)$
   15. Hyperbola $(6, 0)$, $(-6, 0)$
   17. Parabola $(0, -9)$
   19. Parabola $(-4, 0)$
   21. Circle $(2, 0)$, $(0, 2)$, $(-2, 0)$, $(0, -2)$
   23. Hyperbola $(2, 0)$, $(-2, 0)$
   25. Ellipse $(3, 0)$, $(0, 1)$, $(-3, 0)$, $(0, -1)$
   27. $\dfrac{(x-1)^2}{4} - \dfrac{(y+1)^2}{9} = 1$

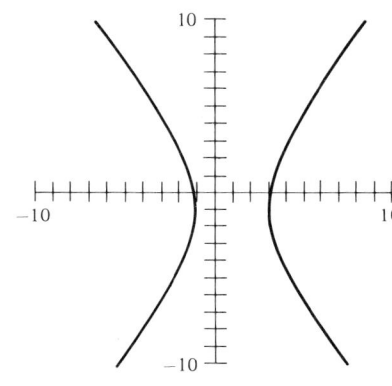

29. $\dfrac{(x-1)^2}{9} - \dfrac{(y-2)^2}{4} = 1$

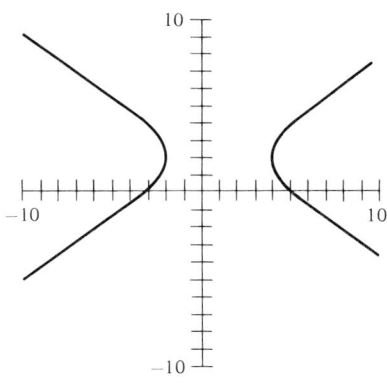

31. a. A hyperbola

b. $\dfrac{D^2}{8} - \dfrac{d^2}{4} = 1$

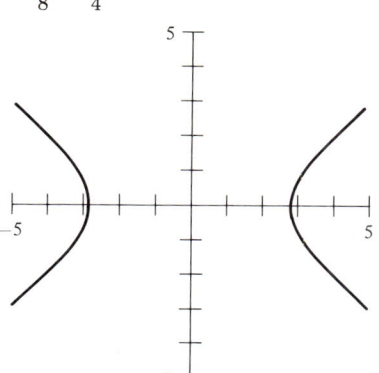

39. $y \geq 2x + 4$

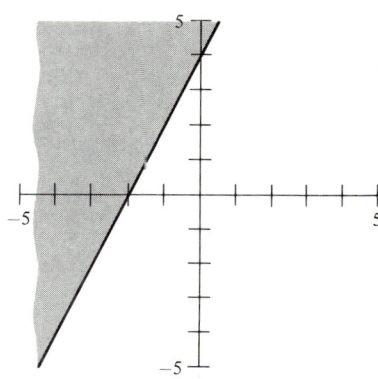

5.

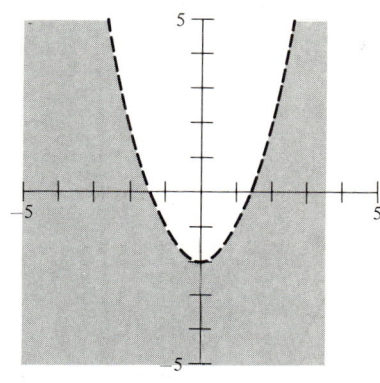

33. $4v^2 + 9\omega^2 = 144$

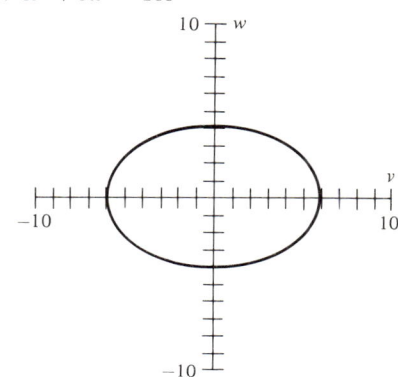

9.3 USING YOUR KNOWLEDGE

1. $\sqrt{(x-c)^2 + y^2}$
3. $cx + a^2 = a\sqrt{(x+c)^2 + y^2}$
5. $b^2x^2 - a^2y^2 = a^2b^2$
7. $\left(\dfrac{x}{a} + \dfrac{y}{b}\right)\left(\dfrac{x}{a} - \dfrac{y}{b}\right) = 1$
9. The denominator becomes very large and the complex fraction becomes very small.
11. $y = -\dfrac{b}{a}x$

7.

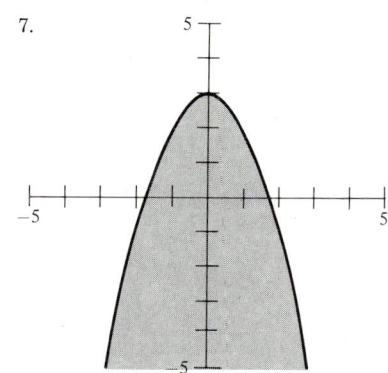

35. $x - y < 4$

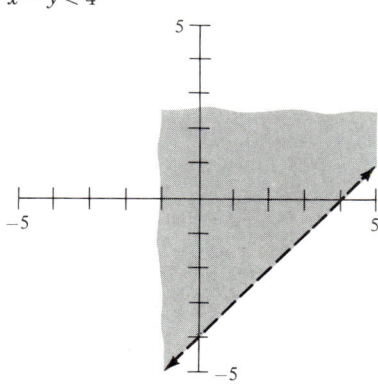

EXERCISE 9.4

A 1.

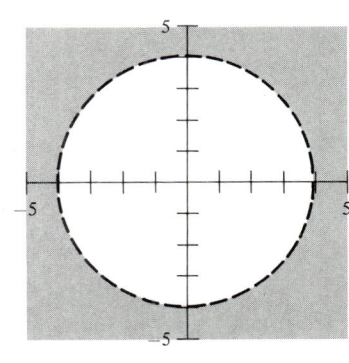

9.

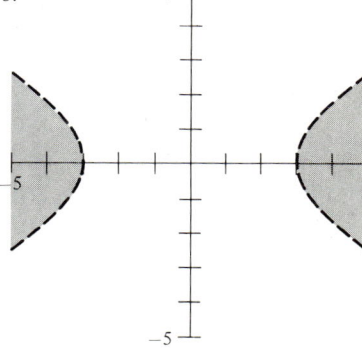

37. $2x - 3y \geq 6$

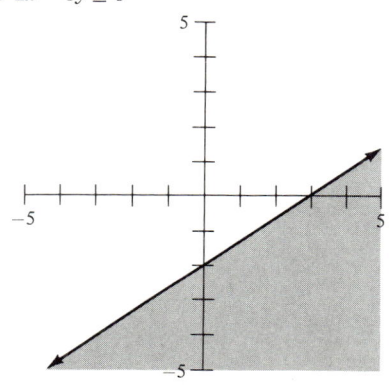

3.

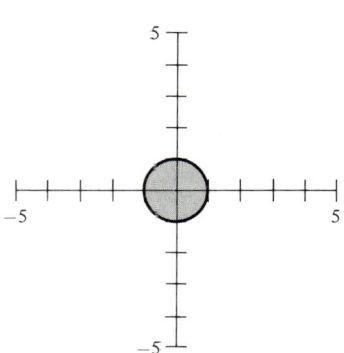

11.

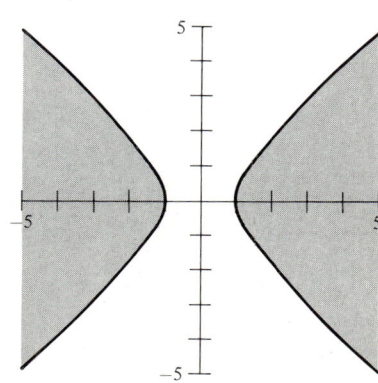

B 13.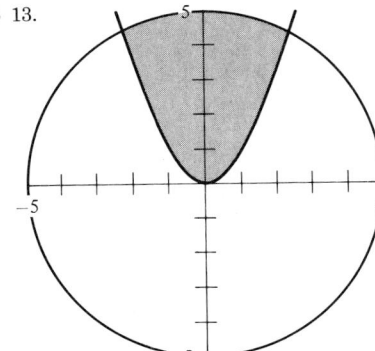

15.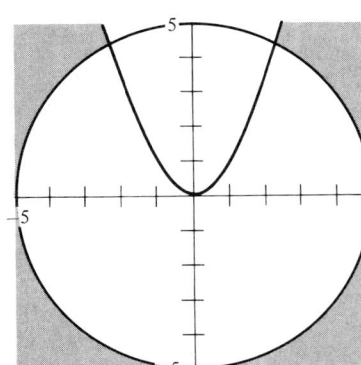

17.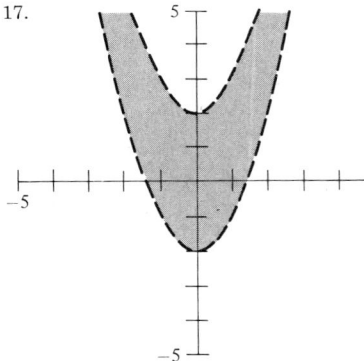

19.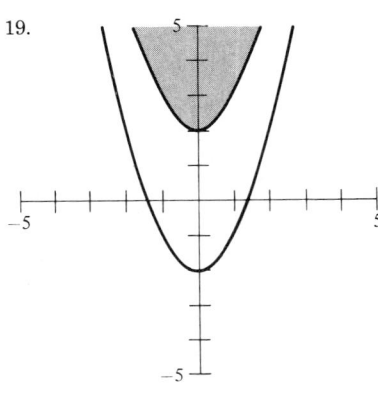

21.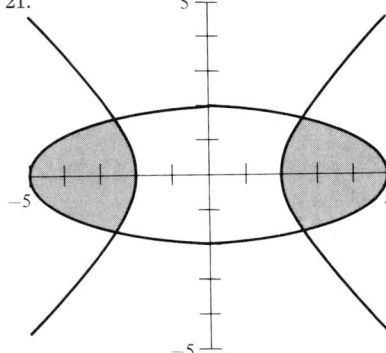

23.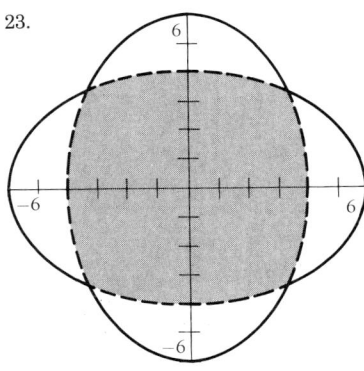

25. a. $\dfrac{x^2}{36} + \dfrac{y^2}{16} > 1$

$\dfrac{x^2}{16} + \dfrac{y^2}{16} < 1$

b. $\dfrac{x^2}{36} + \dfrac{y^2}{16} \geq 1$

$\dfrac{x^2}{16} + \dfrac{y^2}{16} \leq 1$

✓ **SKILL CHECKER**

27. $(3, 1)$
29. $(-9, 4)$

9.4 USING YOUR KNOWLEDGE

1. $C = 432{,}000 - 1800p$
3. $p = 180$ or $p = 80$

EXERCISE 9.5

A 1. The solutions are $(0, 4)$ and $(4, 0)$
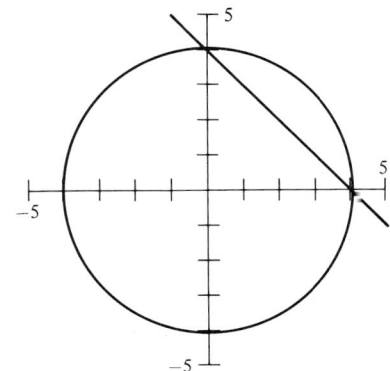

3. The solutions are $(-5, 0)$ and $(0, 5)$
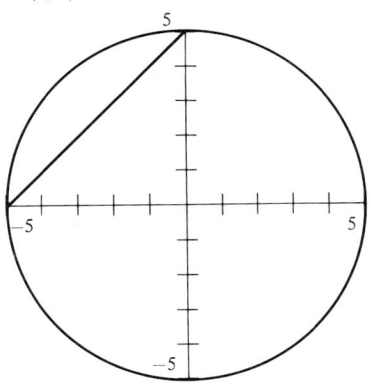

5. The solutions are $(-4, -3)$ and $(3, 4)$
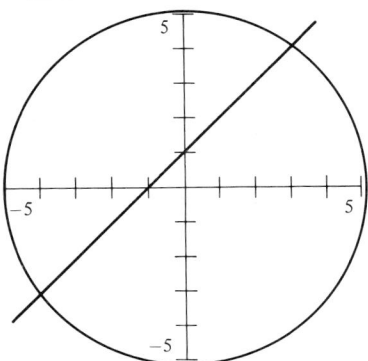

7. The solutions are $(1, 0)$ and $(5, 4)$
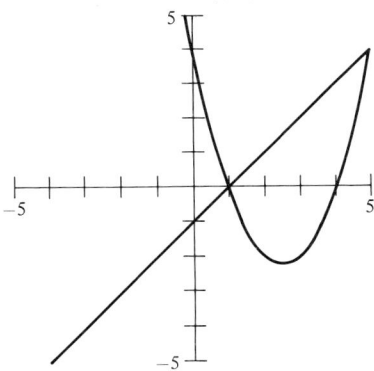

9. The solutions are $(0, 1)$ and $(3, 4)$
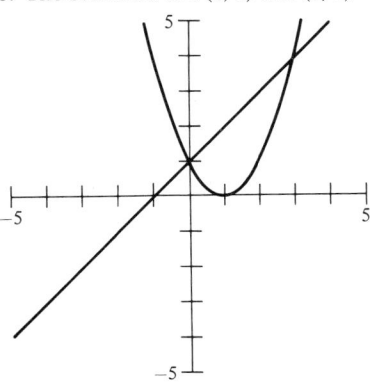

11. The solutions are $(-3, 0)$ and $(0, 2)$

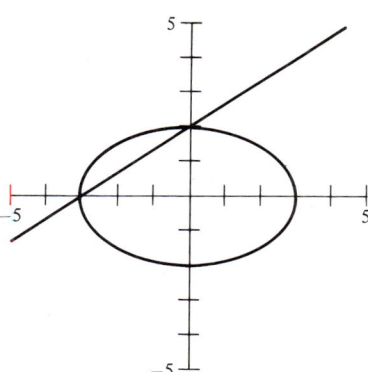

13. The solutions are $(4, 0)$ and $\left(-\frac{68}{15}, \frac{32}{15}\right)$

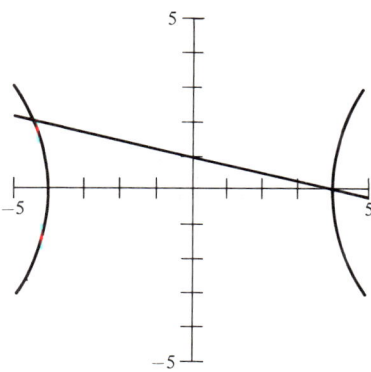

15. There is no real solution $\left(\frac{-5 + \sqrt{15}i}{2}, \frac{5 + \sqrt{15}i}{2}\right)$, $\left(\frac{-5 - \sqrt{15}i}{2}, \frac{5 - \sqrt{15}i}{2}\right)$

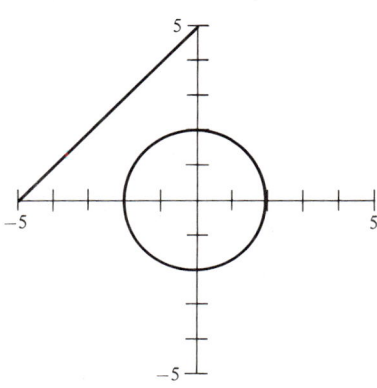

B 17. The solutions are $(-2, 0)$ and $(2, 0)$

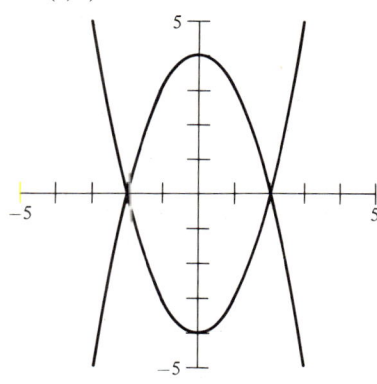

19. The solutions are $(-4, \pm 3)$ and $(4, \pm 3)$

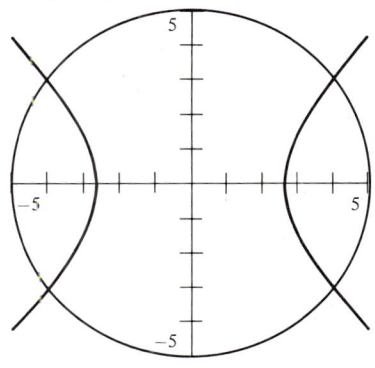

21. The solutions are $(-4, 0)$ and $(4, 0)$

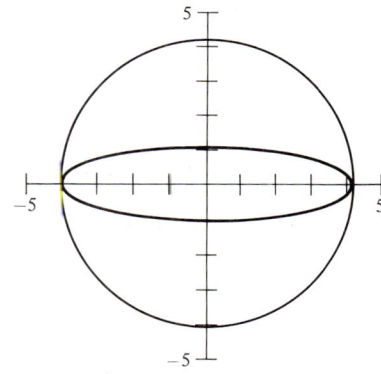

23. The solutions are $(\pm 1, \pm 1)$

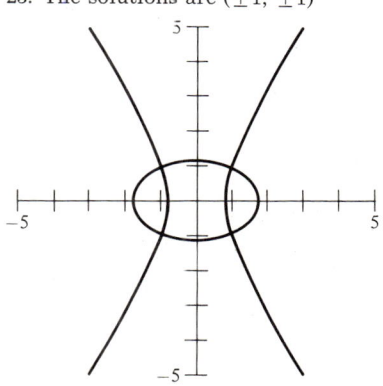

25. The solutions are $(\pm 3, \pm 1)$

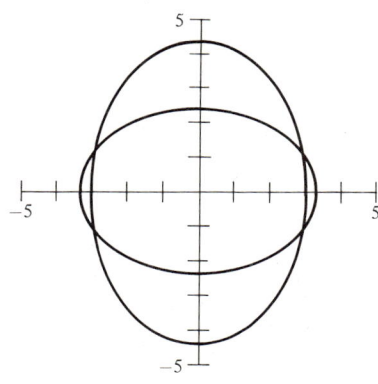

27. There is no real solution. $\left(\pm\frac{\sqrt{26}}{2}, \pm\frac{\sqrt{10}i}{2}\right)$

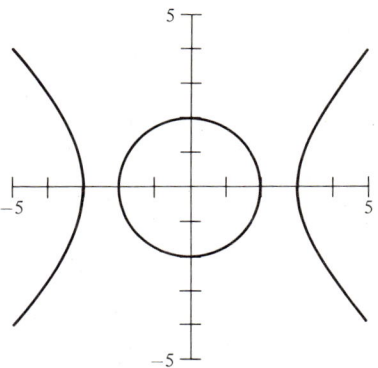

29. There is no real solution. $\left(\pm\frac{3\sqrt{15}i}{5}, \pm\frac{4\sqrt{10}}{5}\right)$

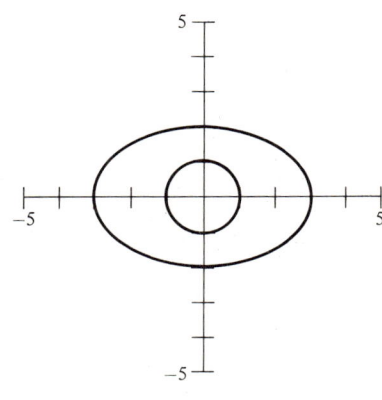

C 31. 1000 or 4000
33. 8 and 7

35.

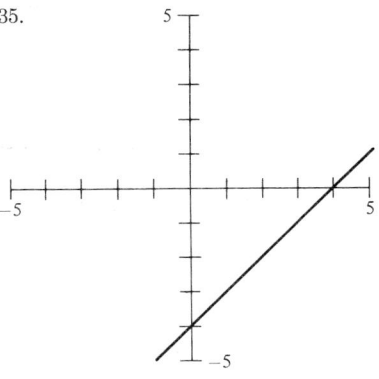

37.

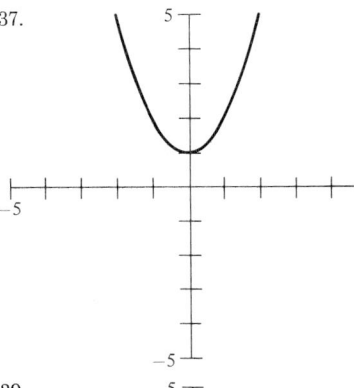

39.

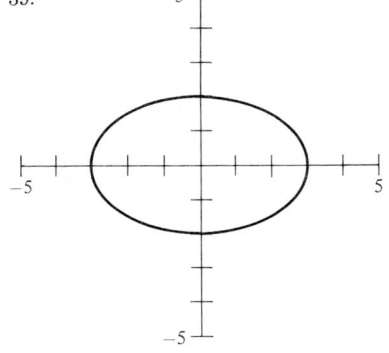

9.5 USING YOUR KNOWLEDGE

1. 20 eggs

REVIEW CHAPTER 9

1. a.

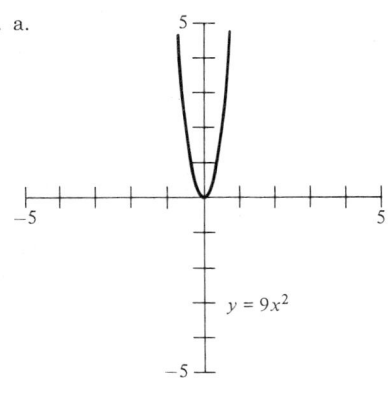

$y = 9x^2$

b.

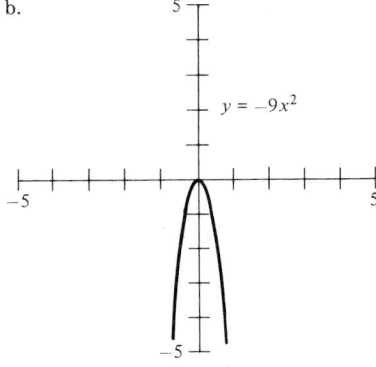

$y = -9x^2$

2. a. Vertex at $(1, -2)$

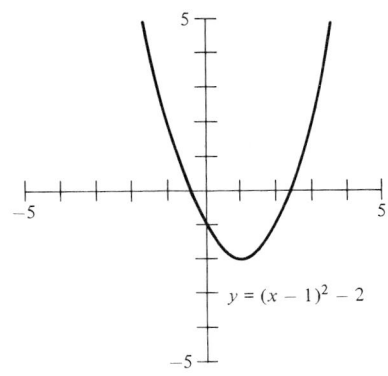

$y = (x - 1)^2 - 2$

b. Vertex at $(1, 2)$

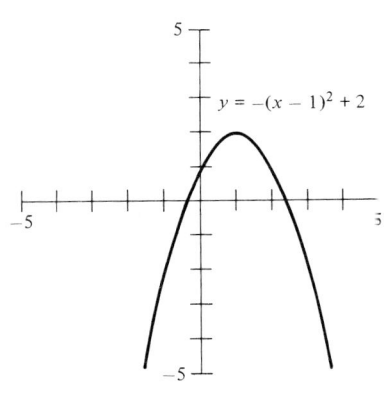

$y = -(x - 1)^2 + 2$

3. a. Vertex at $(2, -2)$

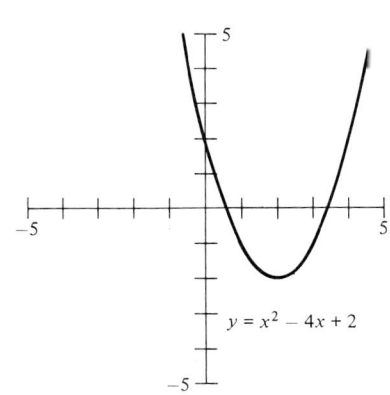

$y = x^2 - 4x + 2$

b. Vertex at $(3, 4)$

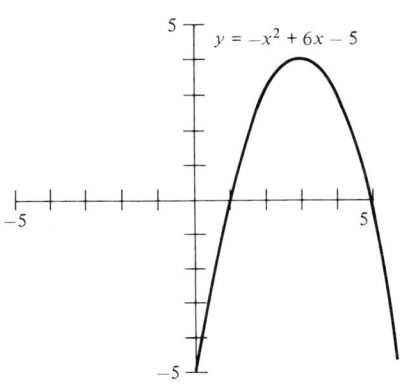

$y = -x^2 + 6x - 5$

4. a. Vertex at $(1, 1)$

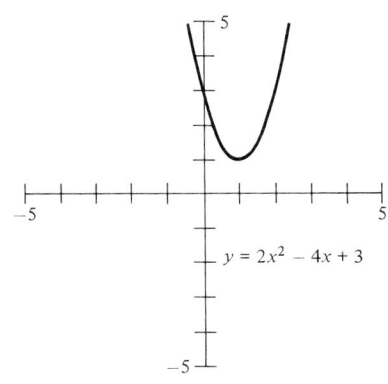

$y = 2x^2 - 4x + 3$

b. Vertex at $(1, -3)$

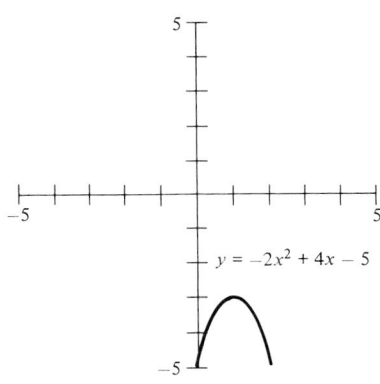

$y = -2x^2 + 4x - 5$

5. a. Vertex at $(-2, 2)$

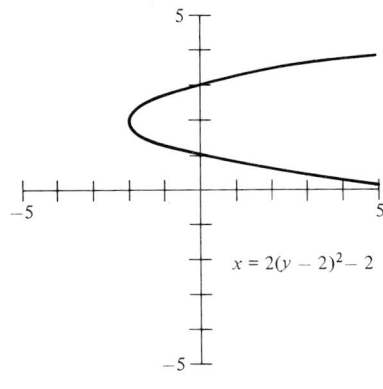

$x = 2(y - 2)^2 - 2$

b. Vertex at (1, 3)

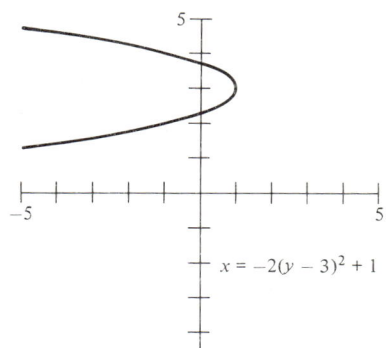

$$x = -2(y - 3)^2 + 1$$

6. a. Vertex at (−3, 2)

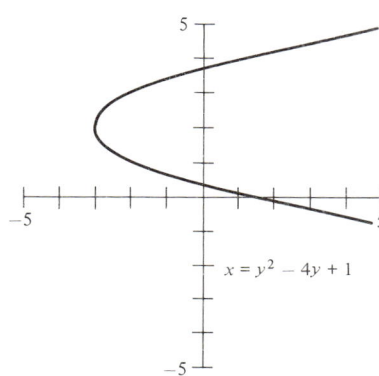

$$x = y^2 - 4y + 1$$

b. Vertex at (2, 1)

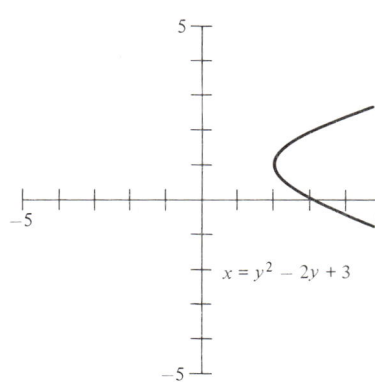

$$x = y^2 - 2y + 3$$

7. a. $x = 1000$
   b. $x = 250$
8. a. $(x + 2)^2 + (y - 2)^2 = 9$
   b. $(x - 3)^2 + (y + 2)^2 = 9$
9. a. $x^2 + y^2 = 25$
   b. $x^2 + y^2 = 64$

10. a. Center at (−2, 1), radius 2

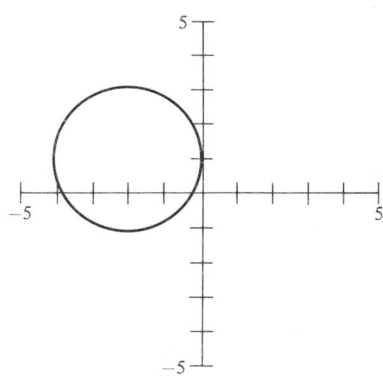

b. Center at (1, −2), radius 3

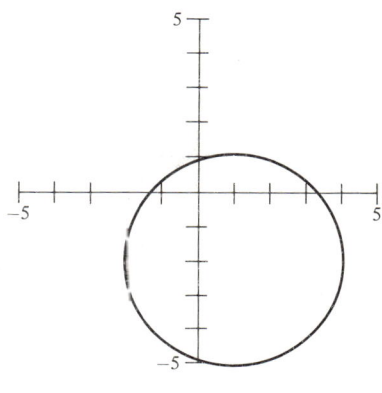

11. a.

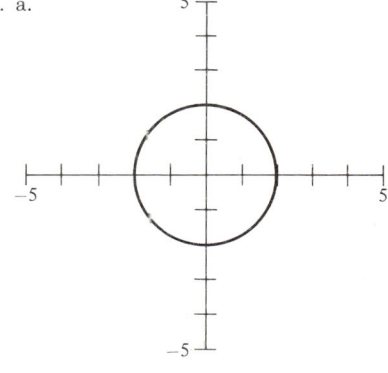

b.

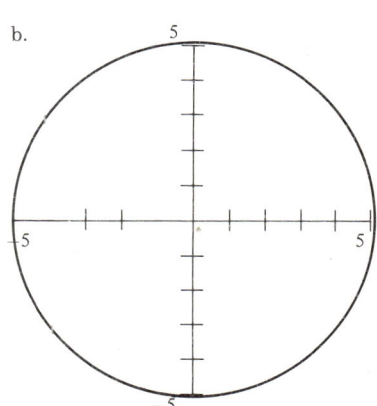

12. a. Center at (−1, −1), radius 2

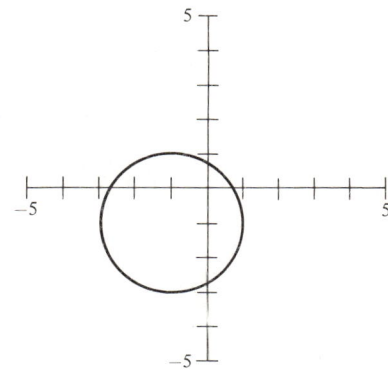

b. Center at (2, −3), radius 2

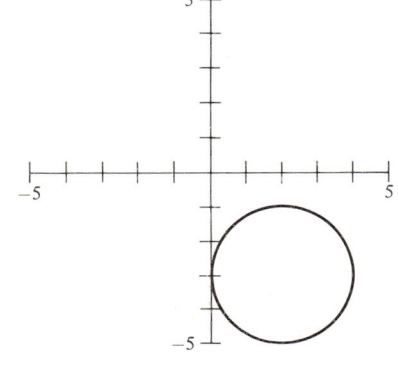

13. a.

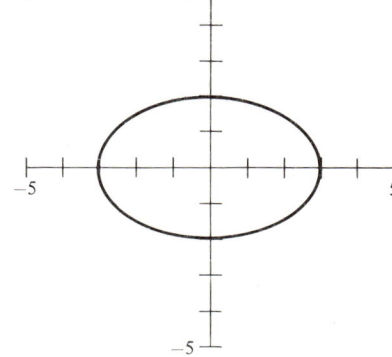

b.

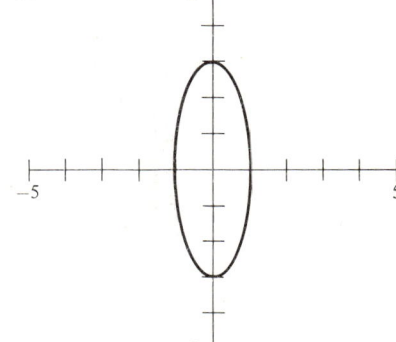

14. a.

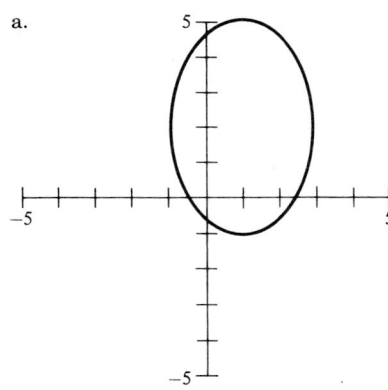

b.

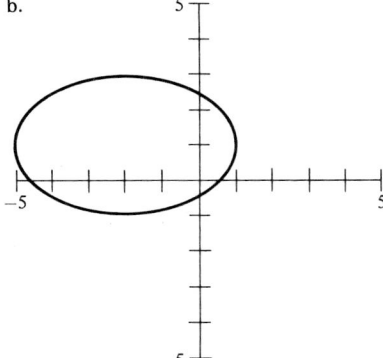

15. a. $\dfrac{x^2}{9} - \dfrac{y^2}{16} = 1$

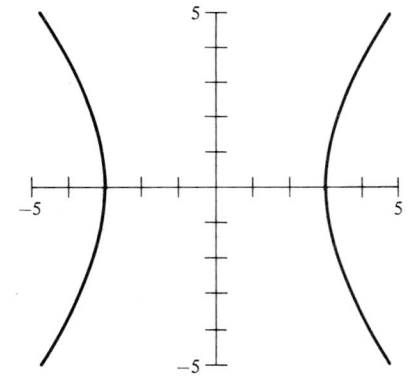

b. $\dfrac{x^2}{16} - \dfrac{y^2}{9} = 1$

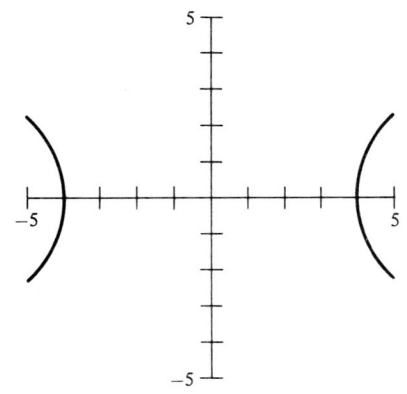

16. a. $\dfrac{y^2}{9} - \dfrac{x^2}{16} = 1$

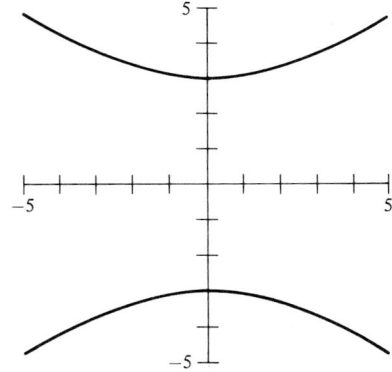

b. $\dfrac{y^2}{16} - \dfrac{x^2}{9} = 1$

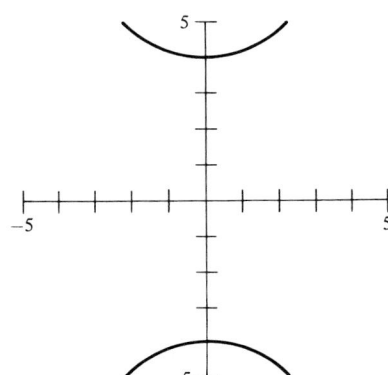

17. a. A parabola
b. A hyperbola
c. A circle
d. An ellipse

18. a.

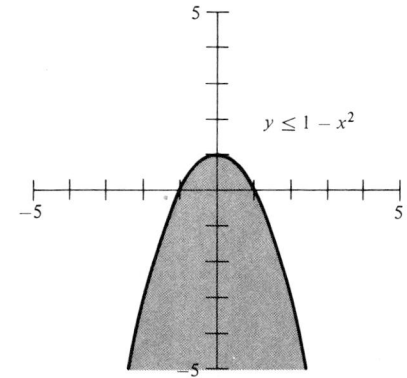

$y \le 1 - x^2$

b.

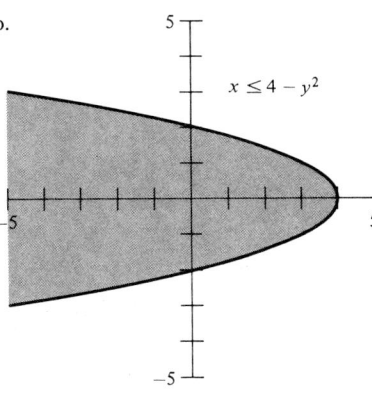

$x \le 4 - y^2$

19. a.

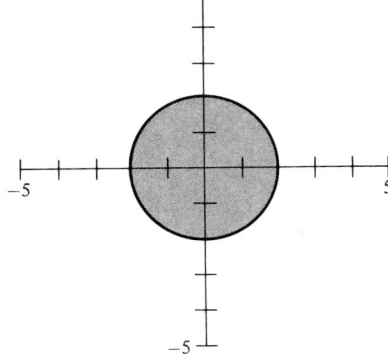

b.

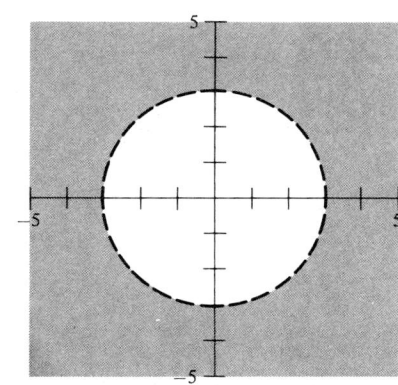

20. a. $4x^2 - y^2 \le 4$

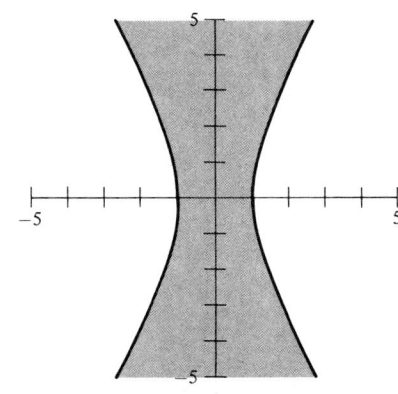

b. $x^2 - 4y^2 \leq 4$

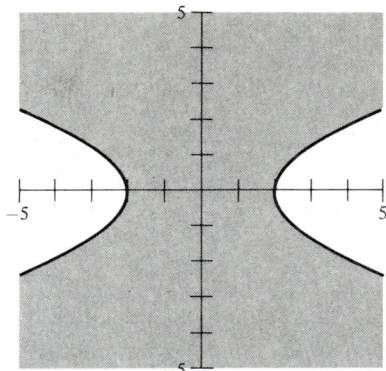

21. a.

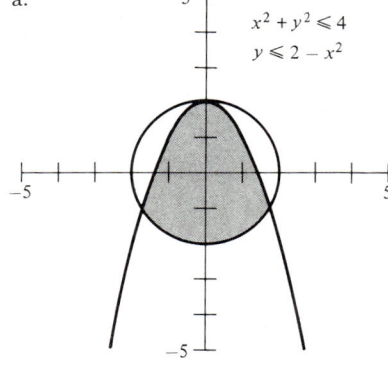

$x^2 + y^2 \leq 4$
$y \leq 2 - x^2$

b.

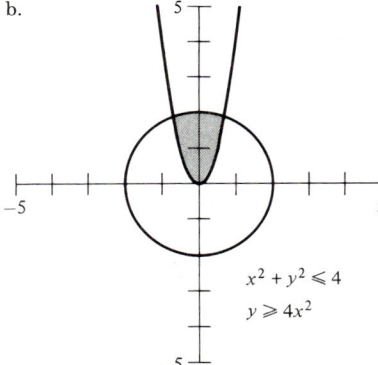

$x^2 + y^2 \leq 4$
$y \geq 4x^2$

22. a. (0, 1) and (1, 0)
    b. (1, 3) and (3, 1)

23. a. $\left(\dfrac{4\sqrt{3}}{3} i, \dfrac{8\sqrt{3}}{3} i\right), \left(-\dfrac{4\sqrt{3}}{3} i, -\dfrac{8\sqrt{3}}{3} i\right)$

    b. $\left(\dfrac{3+i}{2}, \dfrac{3-i}{2}\right), \left(\dfrac{3-i}{2}, \dfrac{3+i}{2}\right)$

24. a. $(\pm 3, \pm 2)$
    b. $(\pm 2, \pm 1)$

25. a. $x = 30$
    b. $x = 20$

EXERCISE 10.1

A 1. a. $D = \{-3, 2, -1\}$
      b. $R = \{0, 1, 2\}$
      c. Yes, it is a function.
   3. a. $D = \{3, 4, 5\}$

b. $R = \{0\}$
c. Yes

5. a. $D = \{1, 2\}$
   b. $R = \{2, 3\}$
   c. No

7. a. $D = \{1, 3, 5, 7\}$
   b. $R = \{-1\}$
   c. Yes

9. a. $D = \{2\}$
   b. $R = \{1, 0, -1, -2\}$
   c. No

11.

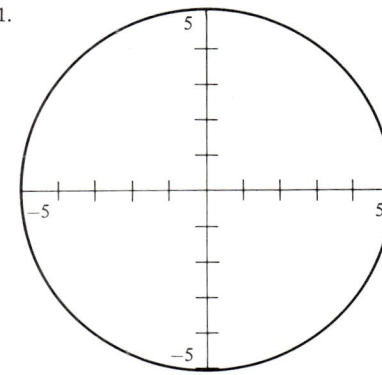

$D = \{x \mid -5 \leq x \leq 5\}$
$R = \{y \mid -5 \leq y \leq 5\}$
Not a function

13.

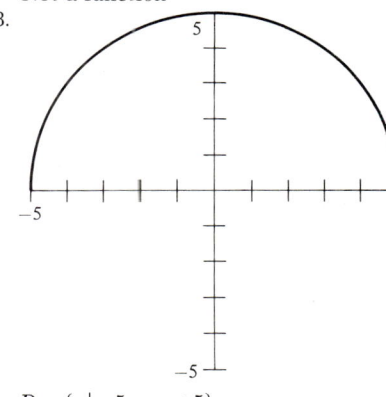

$D = \{x \mid -5 \leq x \leq 5\}$
$R = \{y \mid 0 \leq y \leq 5\}$
Function

15.

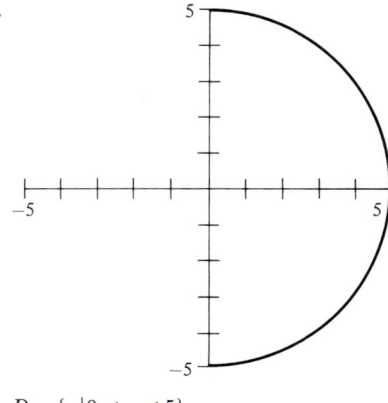

$D = \{x \mid 0 \leq x \leq 5\}$
$R = \{y \mid -5 \leq y \leq 5\}$
Not a function

17.

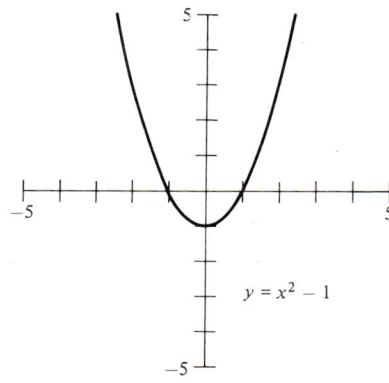

$y = x^2 - 1$

$D = \{x \mid x \text{ is a real number}\}$
$R = \{y \mid y \geq -1\}$
Function

19.

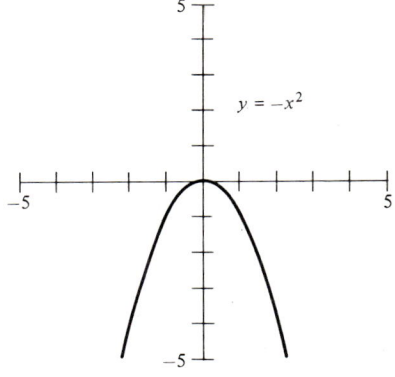

$y = -x^2$

$D = \{x \mid x \text{ is a real number}\}$
$R = \{y \mid y \leq 0\}$
Function

21.

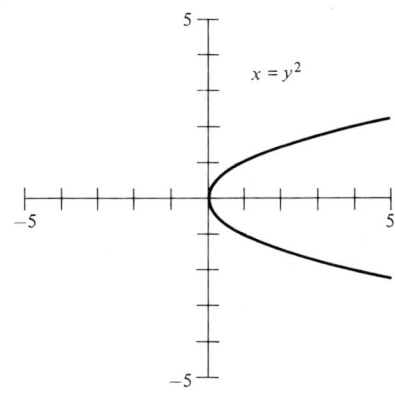

$x = y^2$

$D = \{x \mid x \geq 0\}$
$R = \{y \mid y \text{ is a real number}\}$
Not a function

23.

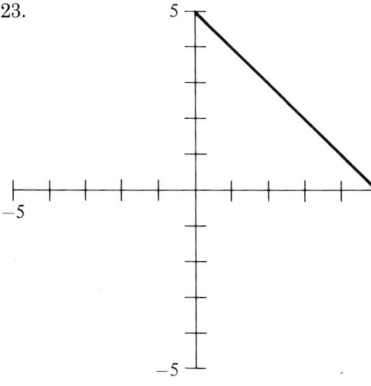

$D = \{x | x \text{ is a real number}\}$
$R = \{y | y \text{ is a real number}\}$
Function

25.

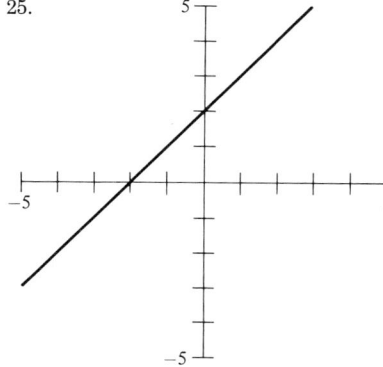

$D = \{x | x \text{ is a real number}\}$
$R = \{y | y \text{ is a real number}\}$
Function

27.

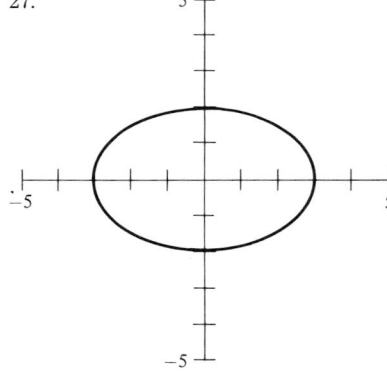

$D = \{x | -3 \leq x \leq 3\}$
$R = \{y | -2 \leq y \leq 2\}$
Not a function

29.

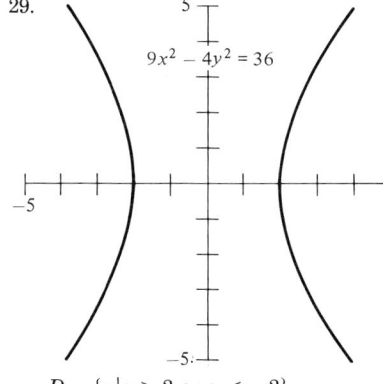

$9x^2 - 4y^2 = 36$

$D = \{x | x \geq 2 \text{ or } x \leq -2\}$
$R = \{y | y \text{ is a real number}\}$
Not a function

B 31. Yes
33. Yes
35. Yes
C 37. $D = \{x | x \geq 5\}$
39. $D = \{x | x \leq 2\}$
41. $D = \{x | x \text{ is a real number}\}$
43. $D = \{x | x \text{ is a real number}, x \neq 5\}$
45. $D = \{x | \text{ is a real number}, x \neq -5\}$
47. $D = \{x | x \text{ is a real number}, x \neq -1 \text{ or } -2\}$
49. $D = \{x | x \text{ is a real number}, x \neq \pm 4\}$
D 51. $C = \pi r$
   $D = \{r | r \geq 0\}$
53. $P = 4s$
   $D = \{s | s \geq 0\}$
55. $D = 50t$
   $D = \{t | t \geq 0\}$
57. $y = 35 + 0.10x$
   $D = \{x | x \geq 0\}$
59. $C = 4(t - 40)$
   $D = \{t | t \geq 40\}$

✓ **SKILL CHECKER**

61. $p = 125$
63. $y = 3$
65. $y = 8$

10.1 USING YOUR KNOWLEDGE

1. Yes
3. Yes
5. Yes

EXERCISE 10.2

A 1. a. $f(0) = 1$
      b. $f(2) = 7$
      c. $f(-2) = -5$
   3. a. $F(1) = 0$
      b. $F(5) = 2$
      c. $F(26) = 5$
   5. a. $f(1) = \frac{1}{4}$
      b. $f(1) - f(2) = \frac{3}{28}$
      c. $\dfrac{f(1) - f(2)}{3} = \dfrac{1}{28}$

7. a. $f(3) = 5$
   b. $g(3) = 19$
   c. $f(3) + g(3) = 24$
9. a. $f(-2) = -10$
   b. $g(-3) = 7$
   c. $f(-2) \cdot g(-3) = -70$
11. a. $f(1) = 3$
    b. $g(-2) = 4$
    c. $f(1) + g(-2) = 7$
13. a. $f(-3) = 7$
    b. $g(2) = 8$
    c. $f(-3) \cdot g(2) = 56$
B 15. $f + g = x^2 - 4x$
   17. $hg = x^4 - 5x^3 - 12x^2 + 80x - 64$
   19. $\dfrac{f}{h} = \dfrac{1}{x + 4}; \ x \neq \pm 4$
C 21. a. $f \circ g(x) = x$
      b. $g \circ f(x) = x$
   23. a. $f \circ g = 3x + 1$
       b. $g \circ f = 3x - 1$
   25. a. $f \circ g = x$
       b. $g \circ f = x$
   27. a. $f \circ g = 3$
       b. $g \circ f = -1$
   29. 3
   31. $x + a$
   33. $x + a + 3$
D 35. a. $P(x) = -0.0005x^2 + 24x - 100,000$
      b. Maximum profits equals
         \$188,000
   37. a. $U(50) = 140$
       b. $U(60) = 130$
   39. a. 160 lb
       b. 78 inches
   41. a. 639 lb/ft^2
       b. 6390 lb/ft^2

✓ **SKILL CHECKER**

43. a. 1
    b. 8

10.2 USING YOUR KNOWLEDGE

1. a. 0
   b. 8
   c. 16
3. $f(t) = 16t^2$

EXERCISE 10.3

A 1. A constant function

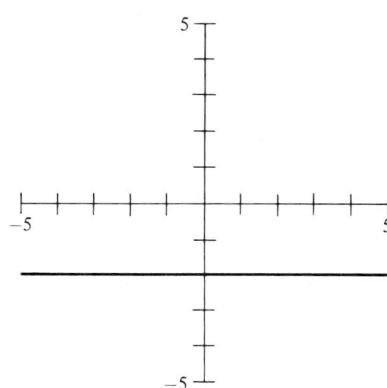

3. A linear function

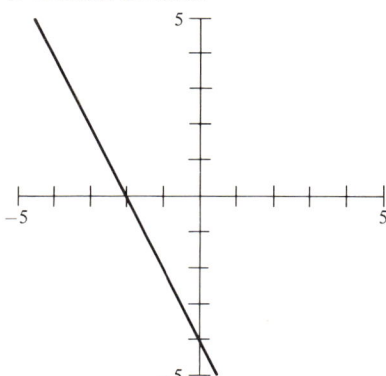

11. A quadratic function

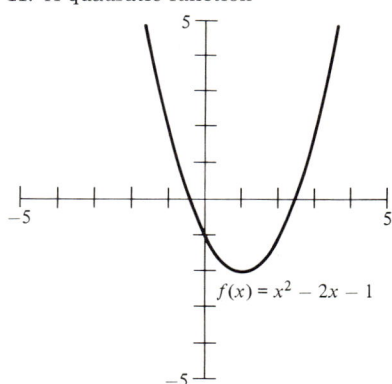

$f(x) = x^2 - 2x - 1$

19.

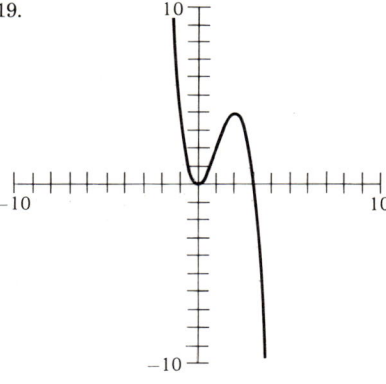

5. A constant function

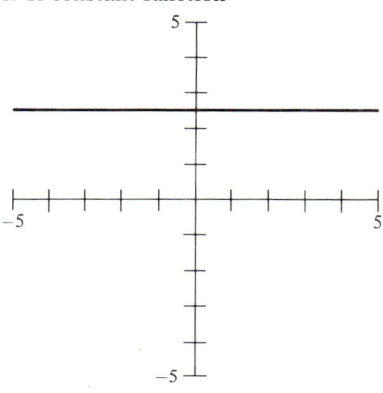

B 13.

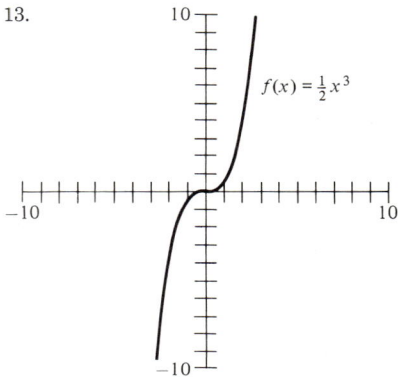

$f(x) = \frac{1}{2}x^3$

21.

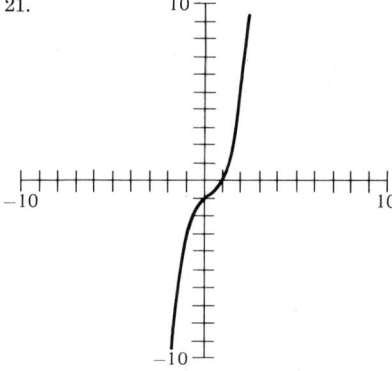

7. A quadratic function

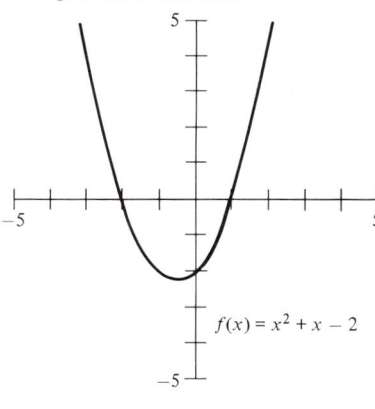

$f(x) = x^2 + x - 2$

15.

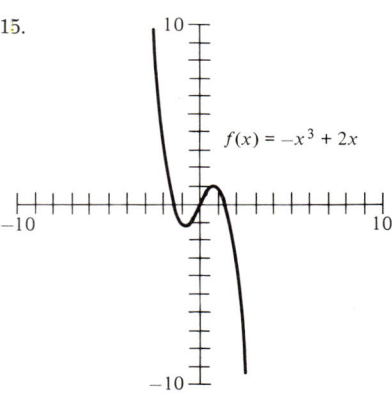

$f(x) = -x^3 + 2x$

C 23.

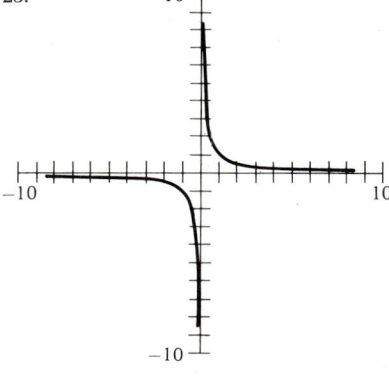

9. A quadratic function

$f(x) = -x^2 + 2x + 3$

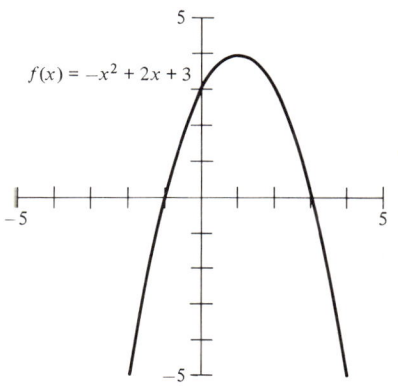

17.

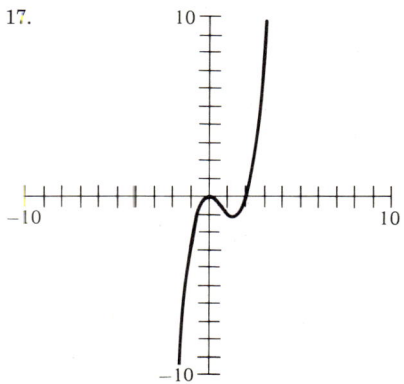

25.

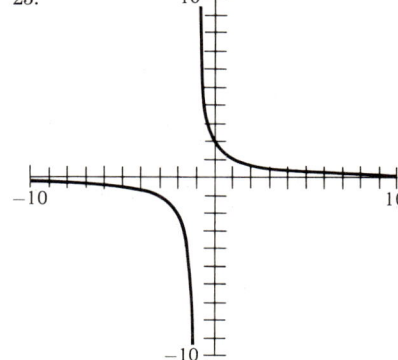

27.

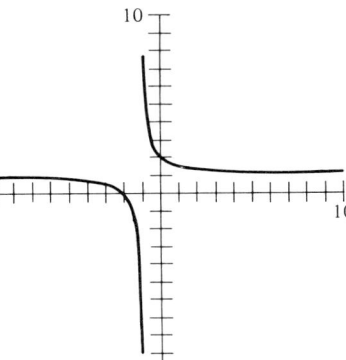

29.

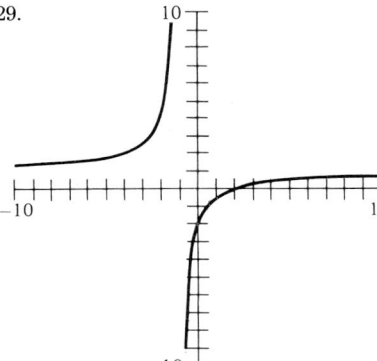

31.

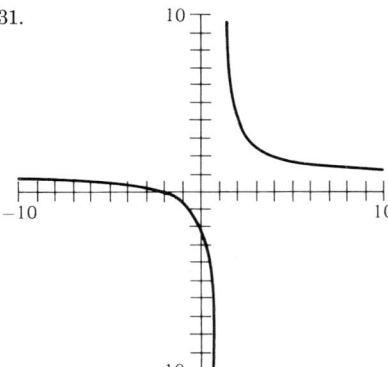

D 33.

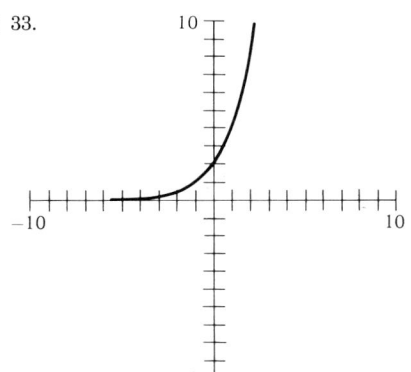

35.

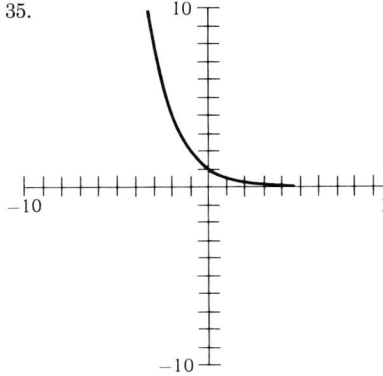

37.

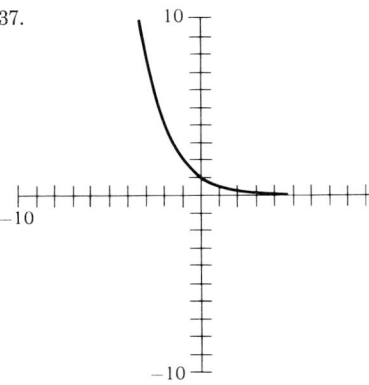

39.

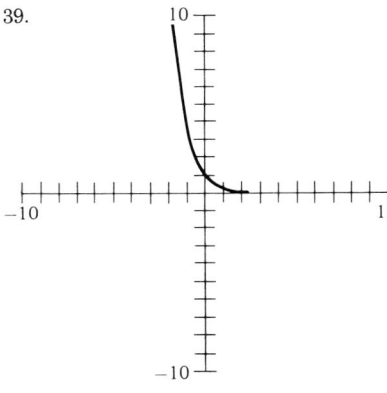

E 41. a. Quadratic
     b. 2 thousand
   43. a. Linear
     b. 21.62 seconds
     c. 21.08 seconds

✓ **SKILL CHECKER**

45. $y = \dfrac{x+2}{4}$ or $y = \dfrac{1}{4}x + \dfrac{1}{2}$

47. $y = \dfrac{x-3}{5}$ or $y = \dfrac{1}{5}x - \dfrac{3}{5}$

49. $y = \dfrac{x-1}{3}$ or $y = \dfrac{1}{3}x - \dfrac{1}{3}$

10.3 USING YOUR KNOWLEDGE

1. Even
3. Neither
5. They will be near zero

EXERCISE 10.4

A,B 1. $f^{-1} = \{(3, 1), (4, 2), (5, 3)\}$

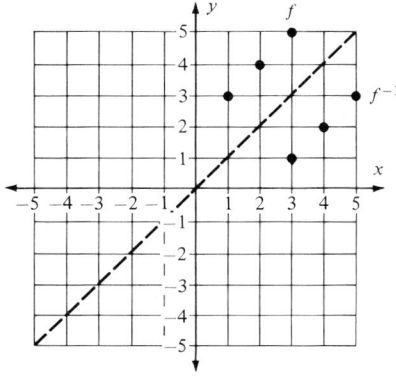

Yes, f^{-1} is a function.

3. $f^{-1} = \{(5, -1), (4, -3), (4, -4)\}$

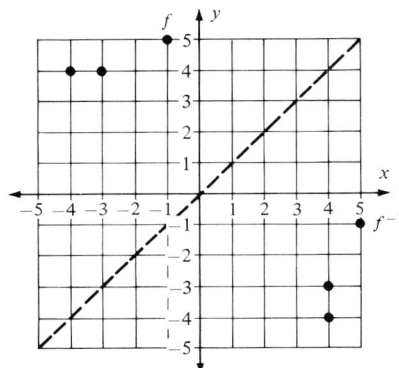

No, f^{-1} is not a function.

B,C 5. $y = \dfrac{x-3}{3}$

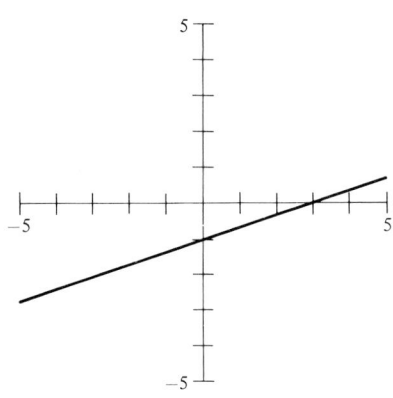

Yes

7. $y = \dfrac{x + 4}{2}$

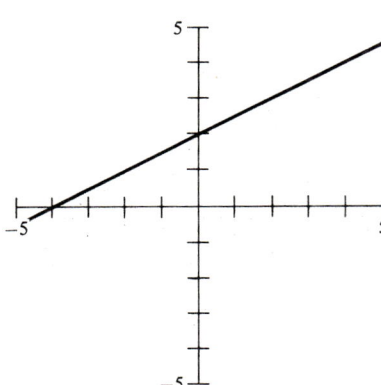

Yes

9. $y = \pm\dfrac{\sqrt{2x}}{2}$

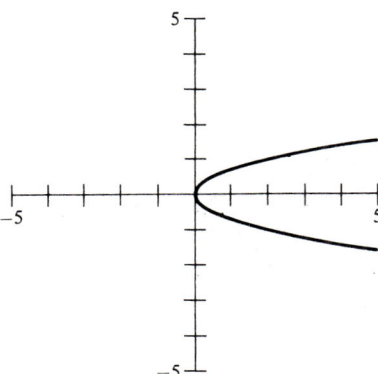

No

11. $y = \pm\sqrt{x + 1}$

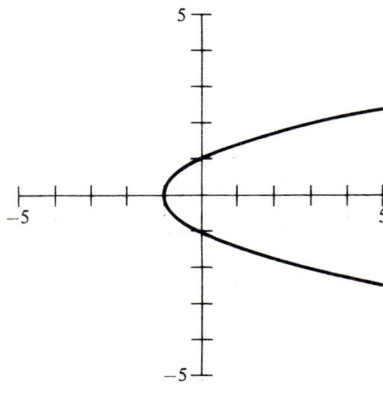

No

13. $y = -\sqrt[3]{x}$. A function

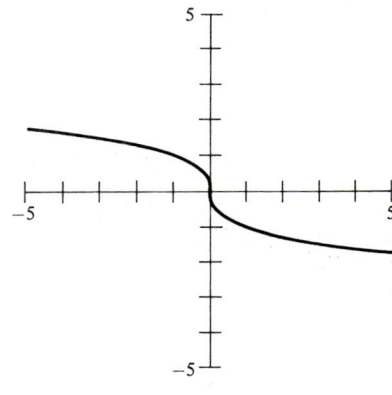

C 15. The inverse $x = 2^{y+1}$ is a function

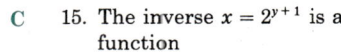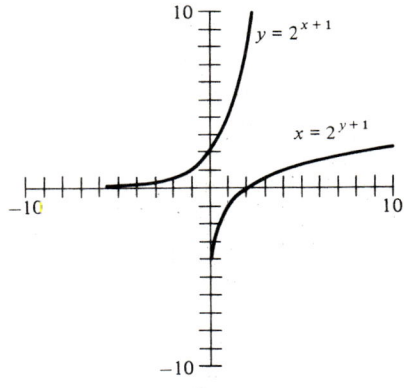

17. The inverse $x = \left(\tfrac{1}{3}\right)^y$ is a function

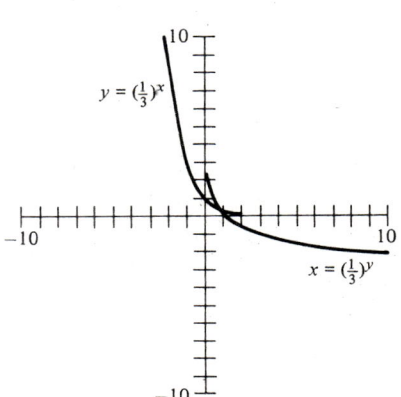

19. The inverse $x = 2^{-y}$ is a function

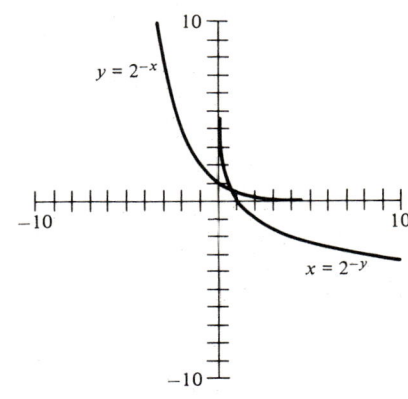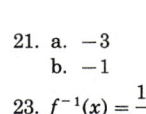

21. a. -3
    b. -1

23. $f^{-1}(x) = \dfrac{1}{x}$

✓ **SKILL CHECKER**

25. x^{10}

27. x^{-3} or $\dfrac{1}{x^3}$

29. x^4

10.4 USING YOUR KNOWLEDGE

1. $f^{-1}(c) = \tfrac{1}{4}F + 40$

REVIEW EXERCISES CHAPTER 10

1. a. $D = \{0, 2, 3, 5\}$
    $R = \{5, 8, 9, 10\}$
  b. $D = \{0, 2, 3, 5\}$
    $R = \{6, 9, 10, 11\}$

2. a. $D = \{x \mid x \text{ is a real number}\}$
    $R = \{x \mid x \text{ is a real number}\}$

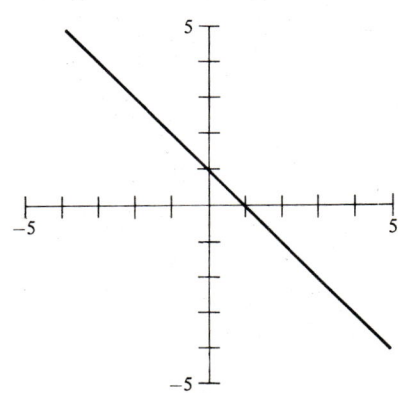

b. $D = \{x \mid x$ is a real number$\}$
  $R = \{x \mid x$ is a real number$\}$

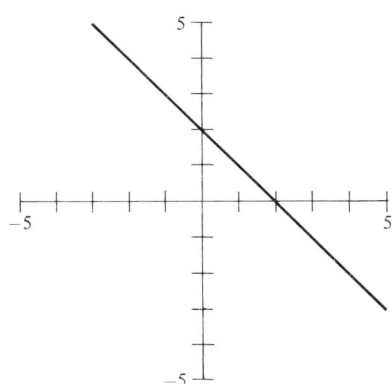

c. $D = \{x \mid x$ is a real number$\}$
  $R = \{x \mid x$ is a real number$\}$

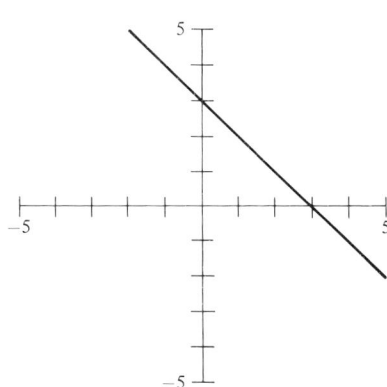

3. a. $D = \{x \mid -2 \le x \le 2\}$
   $R = \{y \mid -2 \le y \le 2\}$

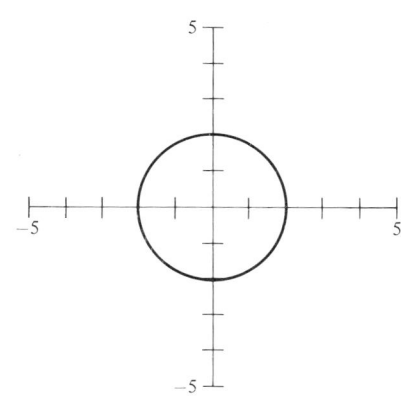

b. $D = \{x \mid -3 \le x \le 3\}$
  $R = \{y \mid -3 \le y \le 3\}$

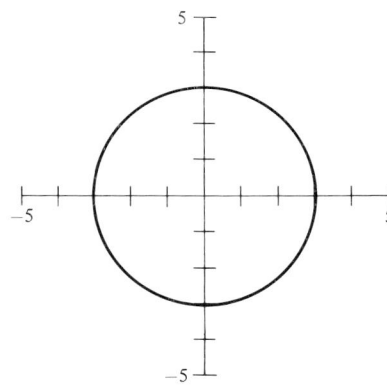

4. a. $D = \{x \mid -2 \le x \le 2\}$
   $R = \{y \mid 0 \le y \le 2\}$

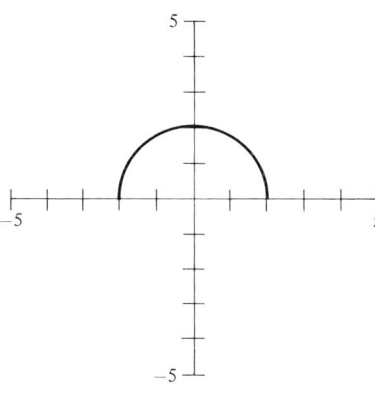

b. $D = \{x \mid -3 \le x \le 3\}$
  $R = \{y \mid 0 \le y \le 3\}$

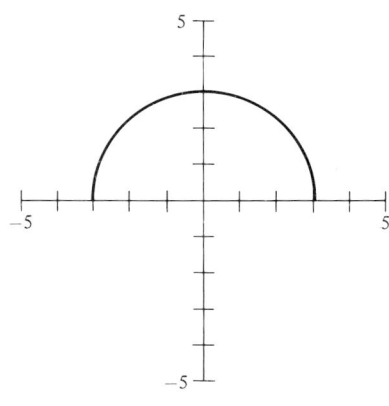

5. a. $D = \{x \mid -2 \le x \le 2\}$
   $R = \{y \mid -2 \le y \le 0\}$

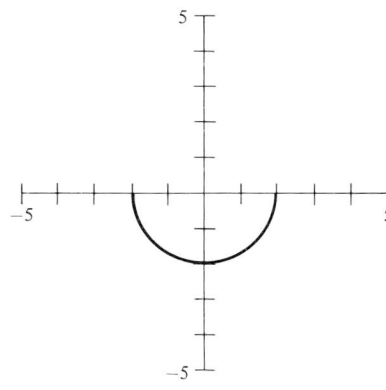

b. $D = \{x \mid -3 \le x \le 3\}$
  $R = \{y \mid -3 \le y \le 0\}$

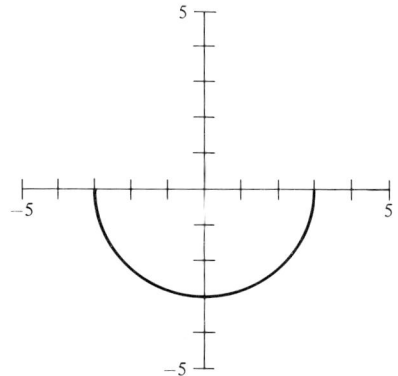

6. a. Yes

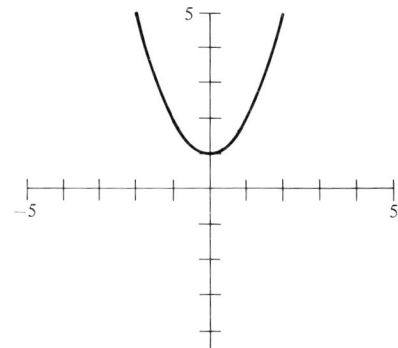

b. Yes

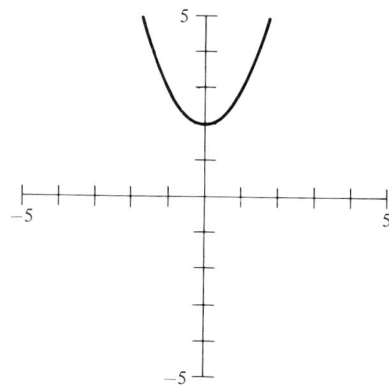

7. a. $D = \{x \mid x \text{ is a real number, } x \neq 1\}$
   $R = \{x \mid x \text{ is a real number, } x \neq 0\}$
   b. $D = \{x \mid x \text{ is a real number, } x \neq 2\}$
   $R = \{x \mid x \text{ is a real number, } x \neq 0\}$
8. a. $D = \{x \mid x \geq 3\}$
   $R = \{y \mid y \geq 0\}$
   b. $D = \{x \mid x \geq 4\}$
   $R = \{y \mid y \geq 0\}$
9. a. $f(2) = -2$
   b. $f(1) = -3$
   c. $f(2) - f(1) = 1$
10. a. $f(2) = 1$
    b. $f(1) = -2$
    c. $f(2) - f(1) = 3$
11. a. $f(2) = 0$
    b. $f(1) = -1$
    c. $f(2) - f(1) = 1$
12. a. $f(2) = 1$
    b. $f(1) = 0$
    c. $f(2) - f(1) = 1$
13. a. $f + g = 4 + x - x^2$
    b. $f - g = -x - x^2$
    c. $fg = 4 + 2x - 2x^2 - x^3$
    d. $\dfrac{f}{g} = \dfrac{2 - x^2}{2 + x}$
14. a. $f + g = 6 + x - x^2$
    b. $f - g = -x - x^2$
    c. $fg = 9 + 3x - 3x^2 - x^3$
    d. $\dfrac{f}{g} = \dfrac{3 - x^2}{3 + x}$
15. 6
16. 7
17. a. $f \circ g = (2 - x)^3$
    b. $g \circ f = 2 - x^3$
    c. $(g \circ f)(2) = -6$
18. a. $f \circ g = (3 - x)^3$
    b. $g \circ f = 3 - x^3$
    c. $(g \circ f) = -5$
19. a. 136 beats per minute
    b. 130 beats per minute
20. a. $P(x) = -0.02x^2 + 70x - 30,000$
    b. 1750
21. a. $P(x) = -0.02x^2 + 60x - 40,000$
    b. 1500
22. a. A constant function

b. A constant function

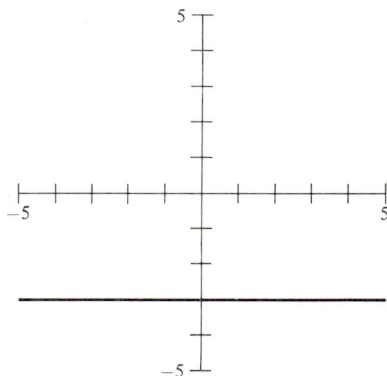

23. a. A linear function

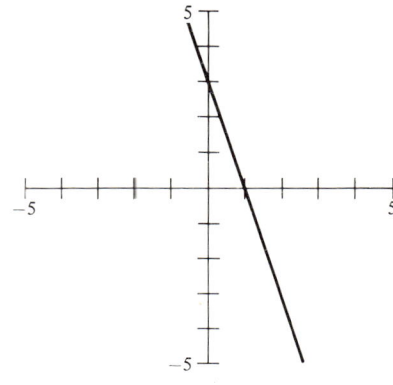

b. A linear function

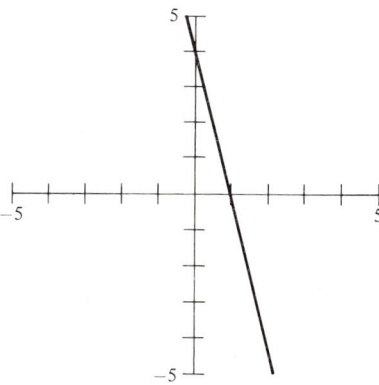

24. a. A quadratic function

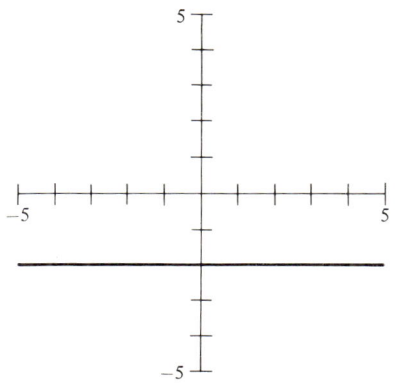

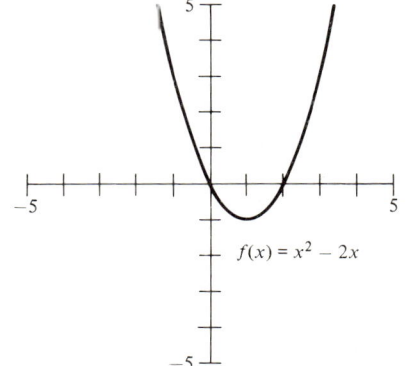

$f(x) = x^2 - 2x$

b. A quadratic function

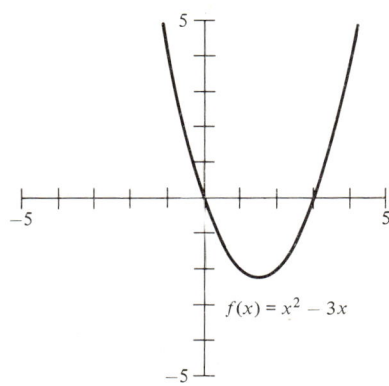

$f(x) = x^2 - 3x$

25. a.

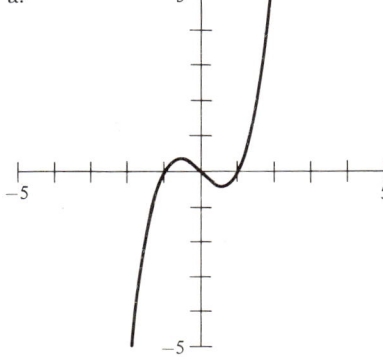

b.

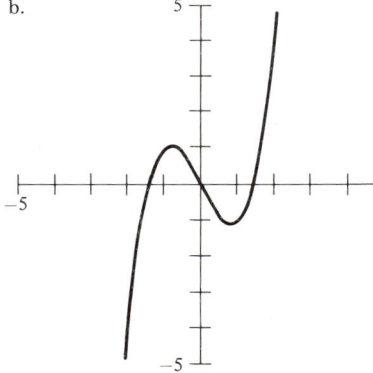

26. a.

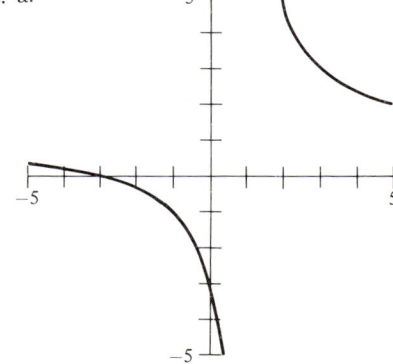

b.

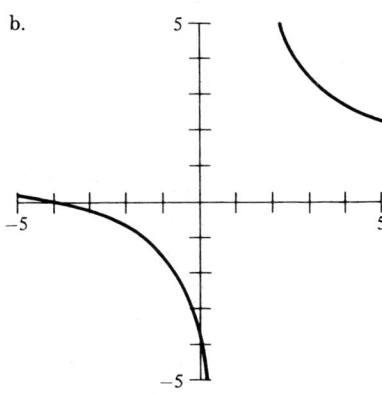

27. a.

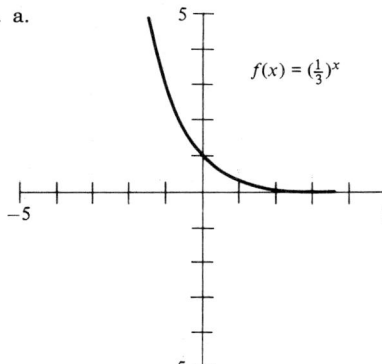

$$f(x) = (\tfrac{1}{3})^x$$

b.

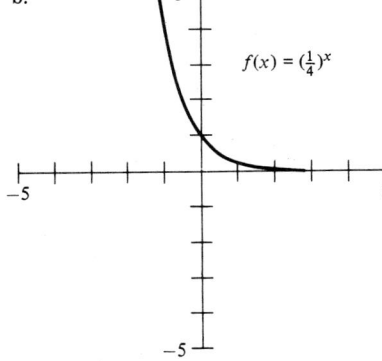

$$f(x) = (\tfrac{1}{4})^x$$

28. a. $D = \{4, 6, 8\}$
$\quad R = \{4, 6, 8\}$
b. $S^{-1} = \{(4, 4), (6, 6), (8, 8)\}$
c. $D = \{4, 6, 8\}$
$\quad R = \{4, 6, 8\}$
d.

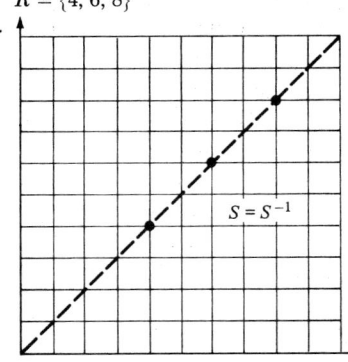

$S = S^{-1}$

29. a. $D = \{4, 6, 8\}$
$\quad R = \{5, 7, 9\}$
b. $S^{-1} = \{(5, 4), (7, 6), (9, 8)\}$
c. $D = \{5, 7, 9\}$
$\quad R = \{4, 6, 8\}$
d.

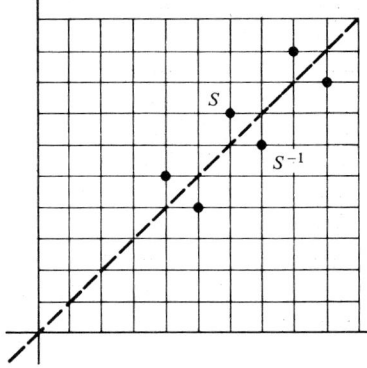

S

S^{-1}

30. a. $f^{-1}(x) = \dfrac{x + 3}{3}$

b.

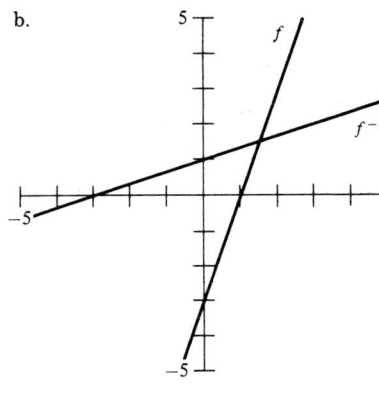

f

f^{-1}

31. a. $f^{-1}(x) = \dfrac{x + 4}{4}$

b.

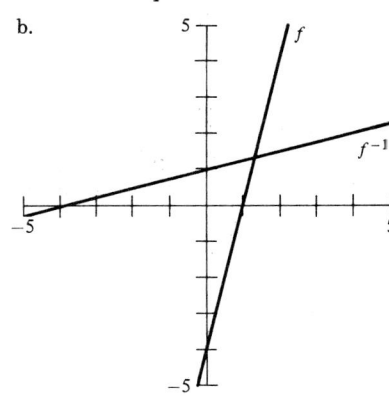

f

f^{-1}

32. $y = \pm\dfrac{\sqrt{x}}{2}$. No

33. $y = \pm\dfrac{\sqrt{5x}}{5}$. No

34.

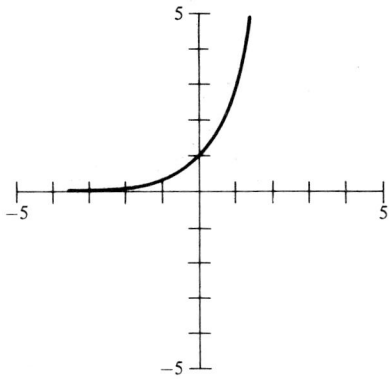

The inverse $x = 3^y$ is a function

35.

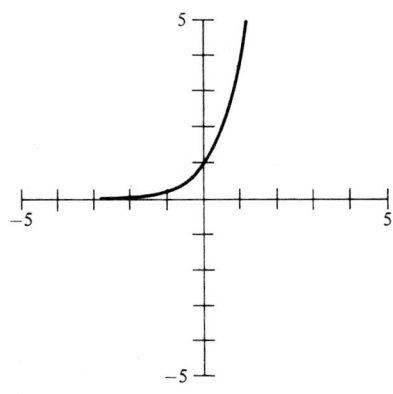

The inverse $x = 4^y$ is a function

EXERCISE 11.1

1. a. $\tfrac{1}{5}$
   b. 1
   c. 5
3. a. $\tfrac{1}{9}$
   b. 1
   c. 9
5. a. $\tfrac{1}{10}$
   b. 1
   c. 10
7. a. $f(x)$ increasing
   b. $g(x)$ decreasing

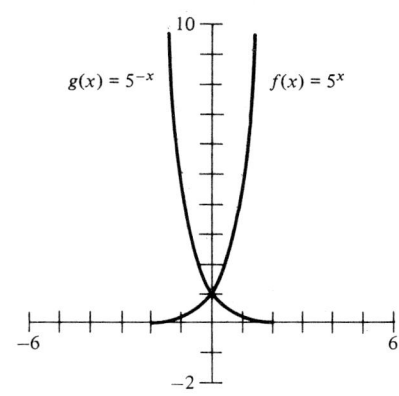

$g(x) = 5^{-x}$

$f(x) = 5^x$

9.

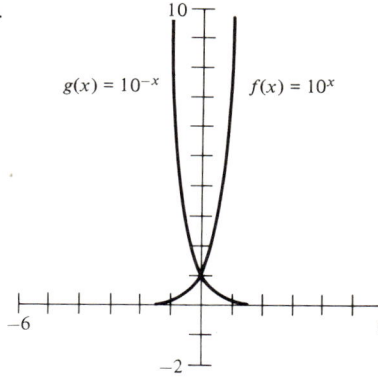

$g(x) = 10^{-x}$ $f(x) = 10^x$

a. $f(x)$ increasing
b. $g(x)$ decreasing

11.

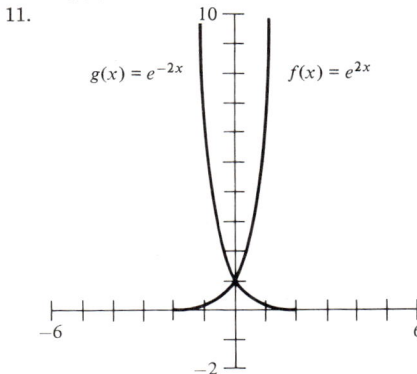

$g(x) = e^{-2x}$ $f(x) = e^{2x}$

a. $f(x)$ increasing
b. $g(x)$ decreasing
13. $2459.60
15. $1822.10
17. a. 2000
    b. 400
    c. 8000
19. a. 2000 g
    b. 699.9 g
    c. 244.9 g
21. Product Law
23. Power Law

11.1 USING YOUR KNOWLEDGE

1. Continuous: $1822.10
   Monthly: $1819.40
   Continuous compounding gives about $2.70 more.

EXERCISE 11.2

1. 8
3. 4
5. −3
7. 6
9. $\log_2 128 = 7$
11. $\log_{10} 1000 = 3$
13. $\log_{81} 9 = \frac{1}{2}$
15. $\log_{216} 6 = \frac{1}{3}$

17. $\log_b N = 7$
19. $9^3 = 729$, OK
21. $2^{-8} = \frac{1}{256}$, OK
23. $81^{3/4} = 27$, OK
25. $\log(\frac{26}{7} \times \frac{63}{15} \times \frac{5}{26}) = \log 3$
27. $\log\left(b^3 \times 2 \times \dfrac{1}{\sqrt{b}} \times \dfrac{\sqrt{b^3}}{2} \right)$

   $= \log b^4 = 4 \log b$

29. $\log\left(k^{3/2} \times r \times \dfrac{1}{k} \times \dfrac{1}{r^{3/4}} \right)$

   $= \log k^{1/2} r^{1/4} = \log k^{2/4} r^{1/4}$
   $= \log(k^2 r)^{1/4} = \frac{1}{4} \log k^2 r$

31. $3 \log \dfrac{y^3 - 1}{y^2}$

   $= 3 \log \dfrac{(y-1)(y^2 + y + 1)}{y^2}$

   $= 3 \log(y - 1) + 3 \log(y^2 + y + 1)$
   $\quad - 6 \log y$

33. $\log \dfrac{(x^2 - 4)\sqrt{x^2 + 2x + 4}}{(x^3 - 8)^2}$

   $= \log \dfrac{(x - 2)(x + 2)(x^2 + 2x + 4)^{1/2}}{(x - 2)^2 (x^2 + 2x + 4)^2}$

   $= \log(x + 2) - \log(x - 2)$
   $\quad - \frac{3}{2} \log(x^2 + 2x + 4)$
35. $R = 8.9$
37. $\log A = 4.16769$
39. 3.268×10^1
41. 4.612×10^2
43. 2.387×10^{-3}
45. 5.69×10^{-5}

11.2 USING YOUR KNOWLEDGE

1. $R = \dfrac{(1.02)^{40}}{(1.085)^{10}}$, $\log R = -0.01030$
   The 8.5% annually is better.

EXERCISE 11.3

1. 1.8722
3. 3.2648
5. 8.6405 − 10
7. 1.7005
9. 8.0927 − 10
11. 7.9348 − 10
13. 18.5
15. 0.00585
17. 29.04
19. 0.7345
21. 15,390
23. 0.04641
25. 2.216
27. 23.52
29. 0.7842
31. $197, rounded from $196.72
33. $2161, rounded from $2160.97
35. 6.2
37. 7.8
39. 6.4

41. $k = \dfrac{\log 3}{\log 2} \approx 1.585$
43. $k = \log 25 \approx 1.3979$

11.3 USING YOUR KNOWLEDGE

1. Compare 1.0625 with $(1.005)^{12}$. Since $(1.005)^{12} \approx 1.0617$, the 6.25% compounded annually is better.
3. 7.72%

EXERCISE 11.4

1. 0.19%
3. 0.20%
5. 0.209%
7. 1.966
9. 3.904
11. −1.306
13. 13.86 yr
15. 10.66 yr
17. About 5.6 billion
19. About 17.3 min
21. About 80.5 min
23. a. 100,000
    b. 67,032
    c. 13,533
    d. 1832
25. About 23,100 yr
27. About 13.3 yr
29. a. 14.7 lb/in.²
    b. 11.4 lb/in.²
    c. 8.91 lb/in.²

11.4 USING YOUR KNOWLEDGE

1. $A_1 = \$1000(1.0075)^{60} \approx \$1566$
   $A_2 = \$1000 e^{0.45} \approx \$1568$
   Continuous compounding is better by about $2.

REVIEW EXERCISES CHAPTER 11

1. a.

$f(x) = 2^{x/2}$

b.

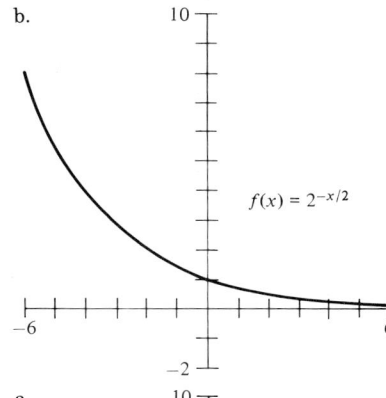

$f(x) = 2^{-x/2}$

2. a.

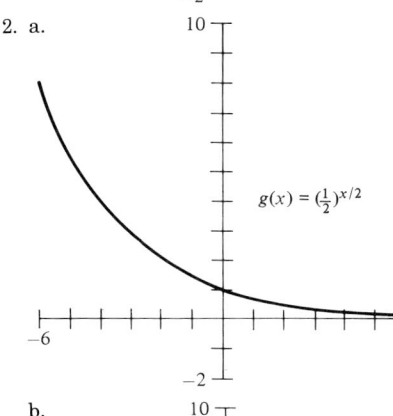

$g(x) = (\frac{1}{2})^{x/2}$

b.

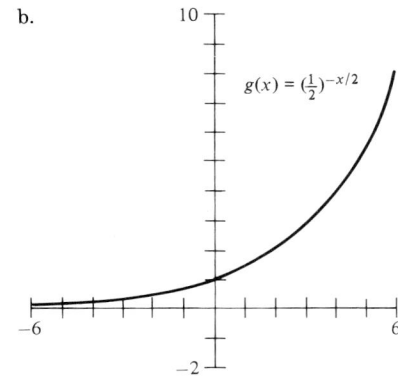

$g(x) = (\frac{1}{2})^{-x/2}$

3. a. 2
   b. 3
4. a. -2
   b. -3
5. a. $\log_2 64 = 6$
   b. $\log_3 81 = 4$
6. a. $16^{3/4} = 8$
   b. $27^{-2/3} = \frac{1}{9}$
7. a. $\log_b \sqrt{\frac{37}{91}} = \frac{1}{2} \log_b \frac{37}{91}$

   $= \frac{1}{2}(\log_b 37 - \log_b 91)$

   b. $\log_b \sqrt[3]{\frac{23 \times 41}{59}}$

   $= \frac{1}{3} \log_b \frac{23 \times 41}{59}$

   $= \frac{1}{3}(\log_b 23 + \log_b 41 - \log_b 59)$

8. a. $1 - \frac{1}{x^2} = \frac{x^2 - 1}{x^2} = \frac{(x-1)(x+1)}{x^2}$

   Thus, $\log\left(1 - \frac{1}{x^2}\right)$

   $= \log(x - 1) + \log(x + 1) - 2 \log x$

   b. $\frac{1}{x^4} - 1 = \frac{1 - x^4}{x^4}$

   $= \frac{(1-x)(1+x)(1+x^2)}{x^4}$

   Thus, $\log\left(\frac{1}{x^4} - 1\right)$

   $= \log(1 - x) + \log(1 + x)$
   $\quad + \log(1 + x^2) - 4 \log x$

9. a. 3.2064
   b. 3.5160
10. a. 2.9890
    b. 2.9227
11. a. $7.8802 - 10$
    b. $6.8116 - 10$
12. a. 2.9891
    b. 1.9228
13. a. $8.9007 - 10$
    b. $9.8121 - 10$
14. a. 663
    b. 0.000407
15. a. 7053
    b. 0.002085
16. a. 40.89
    b. 0.005725
17. a. 4.181
    b. 7.903
18. a. 34.61
    b. 25.58
19. a. 0.1908
    b. 1.505
20. a. 1.585
    b. 0.2846
21. a. 0.1238
    b. -2.100
22. a. 6.9078
    b. -6.9078
23. a. About 13.9 yr
    b. About 8.2 yr
24. a. 0.2950
    b. 0.1475
25. a. About 1.39 yr
    b. About 34.7 yr

EXERCISE 12.1

A 1. $a_{10} = 28$; $a_n = 3n - 2$
   3. $a_{10} = 32$; $a_n = 3n + 2$
   5. $a_{10} = 65$; $a_n = 5n + 15$
   7. $a_{10} = 5$; $a_n = 55 - 5n$
   9. $a_{10} = \frac{1}{11}$; $a_n = \frac{1}{n+1}$
   11. $a_{10} = -1$; $a_n = (-1)^{n-1}$
   13. $a_{10} = x^{10}$; $a_n = x^n$
   15. $a_{10} = -x^{19}$; $a_n = (-1)^{n+1}(x)^{2n-1}$
   17. $a_{10} = -x$; $a_n = (-1)^{n-1}(x)$
   19. $a_{10} = \frac{x^{10}}{10}$; $a_n = \frac{x^n}{n}$

21. $a_{10} = -\frac{x^{10}}{1024}$; $a_n = (-1)^{n-1}\left(\frac{x}{2}\right)^n$
B 23. $-1, 1, 3$
   25. $-\frac{1}{2}, 0, \frac{3}{2}$
   27. $0, \frac{1}{2}, \frac{2}{3}$
   29. 1, 4, 9
   31. $\frac{1}{3}, \frac{2}{5}, \frac{3}{7}$
   33. $-1, 1, -1$
   35. $-\frac{1}{2}, \frac{1}{4}, -\frac{1}{8}$
C 37. a. $a_8 = \$1100$
       b. $a_{10} = \$1020$
   39. a. 9 ft
       b. $\frac{729}{100}$ ft or 7.29 ft
       c. $\frac{9^n}{10^{n-1}}$
   41. a. 400
       b. 1600
       c. $100(2)^n$
   43. $\$3200$

✓ **SKILL CHECKER**

45. $4 + 3n$
47. $16n^2$
49. $(5n + 41)(n - 8)$

12.1 USING YOUR KNOWLEDGE

1. $a_8 = 21$
3. $a_{10} = 55$
5. $a_{12} = 144$

EXERCISE 12.2

A 1. a. $a_1 = 5$
       b. $d = 3$
       c. $a_n = 3n + 2$
   3. a. $a_1 = 11$
       b. $d = -5$
       c. $a_n = 16 - 5n$
   5. a. $a_1 = 3$
       b. $d = -4$
       c. $a_n = 7 - 4n$
   7. a. $a_1 = \frac{1}{2}$
       b. $d = -\frac{1}{4}$
       c. $a_n = \frac{3}{4} - \frac{1}{4}n$ or $\frac{3-n}{4}$
   9. a. $a_1 = -\frac{5}{6}$
       b. $d = \frac{1}{2}$
       c. $a_n = \frac{1}{2}n - \frac{4}{3}$ or $\frac{3n-8}{6}$
B 11. $a_{15} = 91$; $S_{15} = 735$
   13. $a_1 = 4$; $d = 6$; $S_8 = 200$
   15. $d = 1$; $S_6 = 33$
   17. $d = -4$; $a_{14} = -46$
   19. $a_1 = -779$; $a_{40} = 781$
C 21. $\$10,000$
   23. 190
   25. Natural numbers are 1, 2, 3 . . .
       $a_1 = 1$; $a_n = n$
       $S_n = \frac{n}{2}(1 + n) = \frac{n(n+1)}{2}$

27. Even natural numbers are
2, 4, 6 . . .
$a_1 = 2$; $a_n = 2n$
$$S_n = \frac{n}{2}(2 + 2n) = n^2 + n$$

✓ **SKILL CHECKER**

29. 32
31. 63
33. 20

12.2 USING YOUR KNOWLEDGE

1. $8000; $6000; $4000; $2000

EXERCISE 12.3

A,B 1. a. $a_1 = 3$
b. $r = 2$
c. $a_n = 3(2^{n-1})$
d. $S_n = 3(2^n - 1)$
3. a. $a_1 = 8$
b. $r = 3$
c. $a_n = 8(3^{n-1})$
d. $S_n = 4(3^n - 1)$
5. a. $a_1 = 16$
b. $r = -\frac{1}{4}$
c. $a_n = (-4)^{3-n}$
d. $S_n = \frac{64}{5}\left[1 - \left(-\frac{1}{4}\right)^n\right]$
7. a. $a_1 = -\frac{3}{5}$
b. $r = -\frac{5}{2}$
c. $a_n = \left(-\frac{3}{5}\right)\left(-\frac{5}{2}\right)^{n-1}$
d. $S_n = -\frac{6}{35}\left[1 - \left(-\frac{5}{2}\right)^n\right]$
$= \frac{6}{35}\left[\left(-\frac{5}{2}\right)^n - 1\right]$
9. a. $a_1 = -\frac{3}{4}$
b. $r = \frac{1}{3}$
c. $a_n = -\frac{1}{4}(3^{2-n}) = -\frac{1}{4(3^{n-2})}$
d. $S_n = \left(-\frac{9}{8}\right)\left(1 - \left(\frac{1}{3}\right)^n\right)$
or $S_n = \left(-\frac{9}{8}\right)\left(\frac{3^n - 1}{3^n}\right)$
11. $a_3 = \frac{1}{4}$, $r = \frac{1}{2}$
or $a_3 = \frac{9}{4}$, $r = -\frac{3}{2}$
13. $a_3 = 12$, $r = 2$
or $a_3 = 27$, $r = -3$
15. $a_1 = 7$, $a_8 = 896$
17. $r = -3$; $n = 4$
19. $n = 6$, $S_n = S_6 = \frac{5187}{250}$
C 21. 24,333
23. $S_5 = 184.5$ cm
25. $131,875
27. 1, 3, 5 or 17, 3, −11
29. 87,381 cents or $873.81

✓ **SKILL CHECKER**

31. $\frac{8}{3}$
33. $-\frac{25}{2}$

12.3 USING YOUR KNOWLEDGE

1. $6793.08

EXERCISE 12.4

A 1. 12
3. −12
5. $\frac{4}{3}$ or $1\frac{1}{3}$
7. Sum does not exist, $|r| > 1$
9. Sum does not exist, $|r| > 1$
B 11. $\frac{5}{9}$
13. $\frac{2}{11}$
15. $\frac{401}{99}$
17. $\frac{2293}{990}$
19. $\frac{140}{999}$
21. 56 ft

✓ **SKILL CHECKER**

23. $a^2 + 2ab + b^2$
25. $x^3 + 3x^2y + 3xy^2 + y^3$
27. $y^3 - 6y^2z + 12yz^2 - 8z^3$
29. $\frac{1}{x^3} - \frac{3}{x^2y} + \frac{3}{xy^2} - \frac{1}{y^3}$

12.4 USING YOUR KNOWLEDGE

1. Geometric Progression,
$|r| < 1$, $\therefore S = \frac{5}{3}$
3. Arithmetic Progression
5. Geometric Progression
$|r| < 1$, $\therefore S = 16$
7. Neither AP nor GP
9. Neither AP nor GP
11. GP, $|r| < 1$, $\therefore S = \frac{8}{3}$

EXERCISE 12.5

A 1. $a^4 + 12a^3b + 54a^2b^2$
$+ 108ab^3 + 81b^4$
3. $x^4 + 16x^3 + 96x^2 + 256x + 256$
5. $32x^5 - 80x^4y + 80x^3y^2$
$-40x^2y^3 - 10xy^4 - y^5$
7. $32x^5 + 240x^4y + 720x^3y^2$
$+ 1080x^2y^3 + 810xy^4 + 243y^5$
9. $\frac{1}{x^4} - \frac{2y}{x^3} + \frac{3y^2}{2x^2} - \frac{y^3}{2x} + \frac{y^4}{16}$
11. $x^6 + 6x^5 + 15x^4 + 20x^3$
$+ 15x^2 + 6x + 1$
13. $-540x^3$
15. $84x^5y^2$
17. $1120x^4$
19. $\frac{5a^3}{4}$
B 21. 20
23. 84
25. $\frac{15}{64}$
27. $\frac{21}{128}$
29. $\frac{25}{216}$

12.5 USING YOUR KNOWLEDGE

1. 1.08286
3. 1.12616
5. 1.36857
7. 0.88584

REVIEW EXERCISES CHAPTER 12

1. a. $a_6 = 11$, $a_n = 2n - 1$
b. $a_6 = 18$, $a_n = 3n$
2. a. 64
b. 2^{n+2}
3. a. $\frac{2}{3}, \frac{5}{3}, \frac{10}{3}$; $a_{10} = \frac{101}{3}$
b. $\frac{1}{2}, \frac{3}{2}, 3$; $a_{10} = \frac{55}{2}$
4. a. 3, 6, 11, . . .
b. 1, 7, 17, . . .
5. a. $64,000
b. $16,000
6. a. $d = 3$, $a_{10} = 30$
b. $d = 4$, $a_{10} = 40$
7. a. $a_n = 3n$
b. $a_n = 4n$
8. a. 176 ft
b. 240 ft
9. a. $a_1 = 14$, $d = 2$
b. $a_1 = 9$, $d = 3$
10. a. $a_1 = 4$, $d = 2$
b. $a_1 = 1.5$, $d = 3$
11. a. $21,600
b. $9000
12. a. 10
b. 12
13. a. $r = 2$, $a_6 = 96$
b. $r = \frac{1}{2}$, $a_6 = \frac{1}{8}$
14. a. $r = 4$, $a_n = 4^{n-1}$
b. $r = -\frac{1}{2}$, $a_n = \frac{(1)^{n-1}}{2^{n-3}}$
15. a. $a_n = (-1)^{n-1}2^n$, $S_n = \frac{2}{3}[1 - (-2)^n]$
b. $a_n = (-\frac{1}{2})^{n-1}$, $S_n = \frac{2}{3}[1 - (-\frac{1}{2})^n]$
16. a. $35\frac{1}{4}$ ft
b. $35\frac{13}{16}$ ft
17. a. $6750
b. $10,125
18. a. $\frac{64}{3}$
b. $\frac{27}{2}$
19. a. Sum does not exist
b. Sum does not exist
20. a. $\frac{31}{99}$
b. $\frac{12}{37}$
21. a. 24 ft
b. 18 ft
22. a. $a^4 - 12a^3b + 54a^2b^2$
$- 108ab^3 + 81b^4$
b. $81a^4 + 216a^3b + 216a^2b^2$
$+ 96ab^3 + 16b^4$
23. a. $1120x^4$
b. $945x^3$
24. a. 35
b. 35
25. a. $\frac{35}{128}$
b. $\frac{35}{128}$

Table 1. Four-Place Logarithms

	0	1	2	3	4	5	6	7	8	9
1.0	.0000	.0043	.0086	.0128	.0170	.0212	.0253	.0294	.0334	.0374
1.1	.0414	.0453	.0492	.0531	.0569	.0607	.0645	.0682	.0719	.0755
1.2	.0792	.0828	.0864	.0899	.0934	.0969	.1004	.1038	.1072	.1106
1.3	.1139	.1173	.1206	.1239	.1271	.1303	.1335	.1367	.1399	.1430
1.4	.1461	.1492	.1523	.1553	.1584	.1614	.1644	.1673	.1703	.1732
1.5	.1761	.1790	.1818	.1847	.1875	.1903	.1931	.1959	.1987	.2014
1.6	.2041	.2068	.2095	.2122	.2148	.2175	.2201	.2227	.2253	.2279
1.7	.2304	.2330	.2355	.2380	.2405	.2430	.2455	.2480	.2504	.2529
1.8	.2553	.2577	.2601	.2625	.2648	.2672	.2695	.2718	.2742	.2765
1.9	.2788	.2810	.2833	.2856	.2878	.2900	.2923	.2945	.2967	.2989
2.0	.3010	.3032	.3054	.3075	.3096	.3118	.3139	.3160	.3181	.3201
2.1	.3222	.3243	.3263	.3284	.3304	.3324	.3345	.3365	.3385	.3404
2.2	.3424	.3444	.3464	.3483	.3502	.3522	.3541	.3560	.3579	.3598
2.3	.3617	.3636	.3655	.3674	.3692	.3711	.3729	.3747	.3766	.3784
2.4	.3802	.3820	.3838	.3856	.3874	.3892	.3909	.3927	.3945	.3962
2.5	.3979	.3997	.4014	.4031	.4048	.4065	.4082	.4099	.4116	.4133
2.6	.4150	.4166	.4183	.4200	.4216	.4232	.4249	.4265	.4281	.4298
2.7	.4314	.4330	.4346	.4362	.4378	.4393	4409	.4425	.4440	.4456
2.8	.4472	.4487	.4502	.4518	.4533	.4548	.4564	.4579	.4594	.4609
2.9	.4624	.4639	.4654	.4669	.4683	.4698	.4713	.4728	.4742	.4757
3.0	.4771	.4786	.4800	.4814	.4829	.4843	.4857	.4871	.4886	.4900
3.1	.4914	.4928	.4942	.4955	.4969	.4983	.4997	.5011	.5024	.5038
3.2	.5051	.5065	.5079	.5092	.5105	.5119	.5132	.5145	.5159	.5172
3.3	.5185	.5198	.5211	.5224	.5237	.5250	.5263	.5276	.5289	.5302
3.4	.5315	.5328	.5340	.5353	.5366	.5378	.5391	.5403	.5416	.5428
3.5	.5441	.5453	.5465	.5478	.5490	.5502	.5514	.5527	.5539	.5551
3.6	.5563	.5575	.5587	.5599	.5611	.5623	.5635	.5647	.5658	.5670
3.7	.5682	.5694	.5705	.5717	.5729	.5740	.5752	.5763	.5775	.5786
3.8	.5798	.5809	.5821	.5832	.5843	.5855	.5866	.5877	.5888	.5899
3.9	.5911	.5922	.5933	.5944	.5955	.5966	.5977	.5988	.5999	.6010
4.0	.6021	.6031	.6042	.6053	.6064	.6075	.6085	.6096	.6107	.6117
4.1	.6128	.6138	.6149	.6160	.6170	.6180	.6191	.6201	.6212	.6222
4.2	.6232	.6243	.6253	.6263	.6274	.6284	.6294	.6304	.6314	.6325
4.3	.6335	.6345	.6355	.6365	.6375	.6385	.6395	.6405	.6415	.6425
4.4	.6435	.6444	.6454	.6464	.6474	.6484	.6493	.6503	.6513	.6522
4.5	.6532	.6542	.6551	.6561	.6571	.6580	.6590	.6599	.6609	.6618
4.6	.6628	.6637	.6646	.6656	.6665	.6675	.6684	.6693	.6702	.6712
4.7	.6721	.6730	.6739	.6749	.6758	.6767	.6776	.6785	.6794	.6803
4.8	.6812	.6821	.6830	.6839	.6848	.6857	.6866	.6875	.6884	.6893
4.9	.6902	.6911	.6920	.6928	.6937	.6946	.6955	.6964	.6972	.6981

Table 1. Four-Place Logarithms (*continued*)

	0	1	2	3	4	5	6	7	8	9
5.0	.6990	.6998	.7007	.7016	.7024	.7033	.7042	.7050	.7059	.7067
5.1	.7076	.7084	.7093	.7101	.7110	.7118	.7126	.7135	.7143	.7152
5.2	.7160	.7168	.7177	.7185	.7193	.7202	.7210	.7218	.7226	.7235
5.3	.7243	.7251	.7259	.7267	.7275	.7284	.7292	.7300	.7308	.7316
5.4	.7324	.7332	.7340	.7348	.7356	.7364	.7372	.7380	.7388	.7396
5.5	.7404	.7412	.7419	.7427	.7435	.7443	.7451	.7459	.7466	.7474
5.6	.7482	.7490	.7497	.7505	.7513	.7520	.7528	.7536	.7543	.7551
5.7	.7559	.7566	.7574	.7582	.7589	.7597	.7604	.7612	.7619	.7627
5.8	.7634	.7642	.7649	.7657	.7664	.7672	.7679	.7686	.7694	.7701
5.9	.7709	.7716	.7723	.7731	.7738	.7745	.7752	.7760	.7767	.7774
6.0	.7782	.7789	.7796	.7803	.7810	.7818	.7825	.7832	.7839	.7846
6.1	.7853	.7860	.7868	.7875	.7882	.7889	.7896	.7903	.7910	.7917
6.2	.7924	.7931	.7938	.7945	.7952	.7959	.7966	.7973	.7980	.7987
6.3	.7993	.8000	.8007	.8014	.8021	.8028	.8035	.8041	.8048	.8055
6.4	.8062	.8069	.8075	.8082	.8089	.8096	.8102	.8109	.8116	.8122
6.5	.8129	.8136	.8142	.8149	.8156	.8162	.8169	.8176	.8182	.8189
6.6	.8195	.8202	.8209	.8215	.8222	.8228	.8235	.8241	.8248	.8254
6.7	.8261	.8267	.8274	.8280	.8287	.8293	.8299	.8306	.8312	.8319
6.8	.8325	.8331	.8338	.8344	.8351	.8357	.8363	.8370	.8376	.8382
6.9	.8388	.8395	.8401	.8407	.8414	.8420	.8426	.8432	.8439	.8445
7.0	.8451	.8457	.8463	.8470	.8476	.8482	.8488	.8494	.8500	.8506
7.1	.8513	.8519	.8525	.8531	.8537	.8543	.8549	.8555	.8561	.8567
7.2	.8573	.8579	.8585	.8591	.8597	.8603	.8609	.8615	.8621	.8627
7.3	.8633	.8639	.8645	.8651	.8657	.8663	.8669	.8675	.8681	.8686
7.4	.8692	.8698	.8704	.8710	.8716	.8722	.8727	.8733	.8739	.8745
7.5	.8751	.8756	.8762	.8768	.8774	.8779	.8785	.8791	.8797	.8802
7.6	.8808	.8814	.8820	.8825	.8831	.8837	.8842	.8848	.8854	.8859
7.7	.8865	.8871	.8876	.8882	.8887	.8893	.8899	.8904	.8910	.8915
7.8	.8921	.8927	.8932	.8938	.8943	.8949	.8954	.8960	.8965	.8971
7.9	.8976	.8982	.8987	.8993	.8998	.9004	.9009	.9015	.9020	.9026
8.0	.9031	.9036	.9042	.9047	.9053	.9058	.9063	.9069	.9074	.9079
8.1	.9085	.9090	.9096	.9101	.9106	.9112	.9117	.9122	.9128	.9133
8.2	.9138	.9143	.9149	.9154	.9159	.9165	.9170	.9175	.9180	.9186
8.3	.9191	.9196	.9201	.9206	.9212	.9217	.9222	.9227	.9232	.9238
8.4	.9243	.9248	.9253	.9258	.9263	.9269	.9274	.9279	.9284	.9289
8.5	.9294	.9299	.9304	.9309	.9315	.9320	.9325	.9330	.9335	.9340
8.6	.9345	.9350	.9355	.9360	.9365	.9370	.9375	.9380	.9385	.9390
8.7	.9395	.9400	.9405	.9410	.9415	.9420	.9425	.9430	.9435	.9440
8.8	.9445	.9450	.9455	.9460	.9465	.9469	.9474	.9479	.9484	.9489
8.9	.9494	.9499	.9504	.9509	.9513	.9518	.9523	.9528	.9533	.9538
9.0	.9542	.9547	.9552	.9557	.9562	.9566	.9571	.9576	.9581	.9586
9.1	.9590	.9595	.9600	.9605	.9609	.9614	.9619	.9624	.9628	.9633
9.2	.9638	.9643	.9647	.9652	.9657	.9661	.9666	.9671	.9675	.9680
9.3	.9685	.9689	.9694	.9699	.9703	.9708	.9713	.9717	.9722	.9727
9.4	.9731	.9736	.9741	.9745	.9750	.9754	.9759	.9763	.9768	.9773
9.5	.9777	.9782	.9786	.9791	.9795	.9800	.9805	.9809	.9814	.9818
9.6	.9823	.9827	.9832	.9836	.9841	.9845	.9850	.9854	.9859	.9863
9.7	.9868	.9872	.9877	.9881	.9886	.9890	.9894	.9899	.9903	.9908
9.8	.9912	.9917	.9921	.9926	.9930	.9934	.9939	.9943	.9948	.9952
9.9	.9956	.9961	.9965	.9969	.9974	.9978	.9983	.9987	.9991	.9996

Table 2. Four-Place Natural Logarithms

N	.00	.01	.02	.03	.04	.05	.06	.07	.08	.09
1.0	0.0000	0.0100	0.0198	0.0296	0.0392	0.0488	0.0583	0.0677	0.0770	0.0862
1.1	0.0953	0.1044	0.1133	0.1222	0.1310	0.1398	0.1484	0.1570	0.1655	0.1740
1.2	0.1823	0.1906	0.1989	0.2070	0.2151	0.2231	0.2311	0.2390	0.2469	0.2546
1.3	0.2624	0.2700	0.2776	0.2852	0.2927	0.3001	0.3075	0.3148	0.3221	0.3293
1.4	0.3365	0.3436	0.3507	0.3577	0.3646	0.3716	0.3784	0.3853	0.3920	0.3988
1.5	0.4055	0.4121	0.4187	0.4253	0.4318	0.4383	0.4447	0.4511	0.4574	0.4637
1.6	0.4700	0.4762	0.4824	0.4886	0.4947	0.5008	0.5068	0.5128	0.5188	0.5247
1.7	0.5306	0.5365	0.5423	0.5481	0.5539	0.5596	0.5653	0.5710	0.5766	0.5822
1.8	0.5878	0.5933	0.5988	0.6043	0.6098	0.6152	0.6206	0.6259	0.6313	0.6366
1.9	0.6419	0.6471	0.6523	0.6575	0.6627	0.6678	0.6729	0.6780	0.6831	0.6881
2.0	0.6931	0.6981	0.7031	0.7080	0.7129	0.7178	0.7227	0.7275	0.7324	0.7372
2.1	0.7419	0.7467	0.7514	0.7561	0.7608	0.7655	0.7701	0.7747	0.7793	0.7839
2.2	0.7885	0.7930	0.7975	0.8020	0.8065	0.8109	0.8154	0.8198	0.8242	0.8286
2.3	0.8329	0.8372	0.8416	0.8459	0.8502	0.8544	0.8587	0.8629	0.8671	0.8713
2.4	0.8755	0.8796	0.8838	0.8879	0.8920	0.8961	0.9002	0.9042	0.9083	0.9123
2.5	0.9163	0.9203	0.9243	0.9282	0.9322	0.9361	0.9400	0.9439	0.9478	0.9517
2.6	0.9555	0.9594	0.9632	0.9670	0.9708	0.9746	0.9783	0.9821	0.9858	0.9895
2.7	0.9933	0.9969	1.0006	1.0043	1.0080	1.0116	1.0152	1.0188	1.0225	1.0260
2.8	1.0296	1.0332	1.0367	1.0403	1.0438	1.0473	1.0508	1.0543	1.0578	1.0613
2.9	1.0647	1.0682	1.0716	1.0750	1.0784	1.0818	1.0852	1.0886	1.0919	1.0953
3.0	1.0986	1.1019	1.1053	1.1086	1.1119	1.1151	1.1184	1.1217	1.1249	1.1282
3.1	1.1314	1.1346	1.1378	1.1410	1.1442	1.1474	1.1506	1.1537	1.1569	1.1600
3.2	1.1632	1.1663	1.1694	1.1725	1.1756	1.1787	1.1817	1.1848	1.1878	1.1909
3.3	1.1939	1.1969	1.2000	1.2030	1.2060	1.2090	1.2119	1.2149	1.2179	1.2208
3.4	1.2238	1.2267	1.2296	1.2326	1.2355	1.2384	1.2413	1.2442	1.2470	1.2499
3.5	1.2528	1.2556	1.2585	1.2613	1.2641	1.2669	1.2698	1.2726	1.2754	1.2782
3.6	1.2809	1.2837	1.2865	1.2892	1.2920	1.2947	1.2975	1.3002	1.3029	1.3056
3.7	1.3083	1.3110	1.3137	1.3164	1.3191	1.3218	1.3244	1.3271	1.3297	1.3324
3.8	1.3350	1.3376	1.3403	1.3429	1.3455	1.3481	1.3507	1.3533	1.3558	1.3584
3.9	1.3610	1.3635	1.3661	1.3686	1.3712	1.3737	1.3762	1.3788	1.3813	1.3838
4.0	1.3863	1.3888	1.3913	1.3938	1.3962	1.3987	1.4012	1.4036	1.4061	1.4085
4.1	1.4110	1.4134	1.4159	1.4183	1.4207	1.4231	1.4255	1.4279	1.4303	1.4327
4.2	1.4351	1.4375	1.4398	1.4422	1.4446	1.4469	1.4493	1.4516	1.4540	1.4563
4.3	1.4586	1.4609	1.4633	1.4656	1.4679	1.4702	1.4725	1.4748	1.4770	1.4793
4.4	1.4816	1.4839	1.4861	1.4884	1.4907	1.4929	1.4951	1.4974	1.4996	1.5019
4.5	1.5041	1.5063	1.5085	1.5107	1.5129	1.5151	1.5173	1.5195	1.5217	1.5239
4.6	1.5261	1.5282	1.5304	1.5326	1.5347	1.5369	1.5390	1.5412	1.5433	1.5454
4.7	1.5473	1.5497	1.5518	1.5539	1.5560	1.5581	1.5602	1.5623	1.5644	1.5665
4.8	1.5686	1.5707	1.5728	1.5748	1.5769	1.5790	1.5810	1.5831	1.5851	1.5872
4.9	1.5892	1.5913	1.5933	1.5953	1.5974	1.5994	1.6014	1.6034	1.6054	1.6074
5.0	1.6094	1.6114	1.6134	1.6154	1.6174	1.6194	1.6214	1.6233	1.6253	1.6273
5.1	1.6292	1.6312	1.6332	1.6351	1.6371	1.6390	1.6409	1.6429	1.6448	1.6467
5.2	1.6487	1.6506	1.6525	1.6544	1.6563	1.6582	1.6601	1.6620	1.6639	1.6658
5.3	1.6677	1.6696	1.6715	1.6734	1.6752	1.6771	1.6790	1.6808	1.6827	1.6845
5.4	1.6864	1.6882	1.6901	1.6919	1.6938	1.6956	1.6974	1.6993	1.7011	1.7029
5.5	1.7047	1.7066	1.7084	1.7102	1.7120	1.7138	1.7156	1.7174	1.7192	1.7210
N	.00	.01	.02	.03	.04	.05	.06	.07	.08	.09

$$\log_e .1 = .6974 - 3 \qquad \log_e .01 = .3918 - 5 \qquad \log_e .001 = 0922 - 7$$

Table 2. Four-Place Natural Logarithms (*continued*)

N	.00	.01	.02	.03	.04	.05	.06	.07	.08	.09
5.5	1.7047	1.7066	1.7084	1.7102	1.7120	1.7138	1.7156	1.7174	1.7192	1.7210
5.6	1.7228	1.7246	1.7263	1.7281	1.7299	1.7317	1.7334	1.7352	1.7370	1.7387
5.7	1.7405	1.7422	1.7440	1.7457	1.7475	1.7492	1.7509	1.7527	1.7544	1.7561
5.8	1.7579	1.7596	1.7613	1.7630	1.7647	1.7664	1.7681	1.7699	1.7716	1.7733
5.9	1.7750	1.7766	1.7783	1.7800	1.7817	1.7834	1.7851	1.7867	1.7884	1.7901
6.0	1.7918	1.7934	1.7951	1.7967	1.7984	1.8001	1.8017	1.8034	1.8050	1.8066
6.1	1.8083	1.8099	1.8116	1.8132	1.8148	1.8165	1.8181	1.8197	1.8213	1.8229
6.2	1.8245	1.8262	1.8278	1.8294	1.8310	1.8326	1.8342	1.8358	1.8374	1.8390
6.3	1.8405	1.8421	1.8437	1.8453	1.8469	1.8485	1.8500	1.8516	1.8532	1.8547
6.4	1.8563	1.8579	1.8594	1.8610	1.8625	1.8641	1.8656	1.8672	1.8687	1.8703
6.5	1.8718	1.8733	1.8749	1.8764	1.8779	1.8795	1.8810	1.8825	1.8840	1.8856
6.6	1.8871	1.8886	1.8901	1.8916	1.8931	1.8946	1.8961	1.8976	1.8991	1.9006
6.7	1.9021	1.9036	1.9051	1.9066	1.9081	1.9095	1.9110	1.9125	1.9140	1.9155
6.8	1.9169	1.9184	1.9199	1.9213	1.9228	1.9242	1.9257	1.9272	1.9286	1.9301
6.9	1.9315	1.9330	1.9344	1.9359	1.9373	1.9387	1.9402	1.9416	1.9430	1.9445
7.0	1.9459	1.9473	1.9488	1.9502	1.9516	1.9530	1.9544	1.9559	1.9573	1.9587
7.1	1.9601	1.9615	1.9629	1.9643	1.9657	1.9671	1.9685	1.9699	1.9713	1.9727
7.2	1.9741	1.9755	1.9769	1.9782	1.9796	1.9810	1.9824	1.9838	1.9851	1.9865
7.3	1.9879	1.9892	1.9906	1.9920	1.9933	1.9947	1.9961	1.9974	1.9988	2.0001
7.4	2.0015	2.0028	2.0042	2.0055	2.0069	2.0082	2.0096	2.0109	2.0122	2.0136
7.5	2.0149	2.0162	2.0176	2.0189	2.0202	2.0215	2.0229	2.0242	2.0255	2.0268
7.6	2.0281	2.0295	2.0308	2.0321	2.0334	2.0347	2.0360	2.0373	2.0386	2.0399
7.7	2.0412	2.0425	2.0438	2.0451	2.0464	2.0477	2.0490	2.0503	2.0516	2.0528
7.8	2.0541	2.0554	2.0567	2.0580	2.0592	2.0605	2.0618	2.0631	2.0643	2.0656
7.9	2.0669	2.0681	2.0694	2.0707	2.0719	2.0732	2.0744	2.0757	2.0769	2.0782
8.0	2.0794	2.0807	2.0819	2.0832	2.0844	2.0857	2.0869	2.0882	2.0894	2.0906
8.1	2.0919	2.0931	2.0943	2.0956	2.0968	2.0980	2.0992	2.1005	2.1017	2.1029
8.2	2.1041	2.1054	2.1066	2.1078	2.1090	2.1102	2.1114	2.1126	2.1138	2.1150
8.3	2.1163	2.1175	2.1187	2.1199	2.1211	2.1223	2.1235	2.1247	2.1258	2.1270
8.4	2.1282	2.1294	2.1306	2.1318	2.1330	2.1342	2.1353	2.1365	2.1377	2.1389
8.5	2.1401	2.1412	2.1424	2.1436	2.1448	2.1459	2.1471	2.1483	2.1494	2.1506
8.6	2.1518	2.1529	2.1541	2.1552	2.1564	2.1576	2.1587	2.1599	2.1610	2.1622
8.7	2.1633	2.1645	2.1656	2.1668	2.1679	2.1691	2.1702	2.1713	2.1725	2.1736
8.8	2.1748	2.1759	2.1770	2.1782	2.1793	2.1804	2.1815	2.1827	2.1838	2.1849
8.9	2.1861	2.1872	2.1883	2.1894	2.1905	2.1917	2.1928	2.1939	2.1950	2.1961
9.0	2.1972	2.1983	2.1994	2.2006	2.2017	2.2028	2.2039	2.2050	2.2061	2.2072
9.1	2.2083	2.2094	2.2105	2.2116	2.2127	2.2138	2.2148	2.2159	2.2170	2.2181
9.2	2.2192	2.2203	2.2214	2.2225	2.2235	2.2246	2.2257	2.2268	2.2279	2.2289
9.3	2.2300	2.2311	2.2322	2.2332	2.2343	2.2354	2.2364	2.2375	2.2386	2.2396
9.4	2.2407	2.2418	2.2428	2.2439	2.2450	2.2460	2.2471	2.2481	2.2492	2.2502
9.5	2.2513	2.2523	2.2534	2.2544	2.2555	2.2565	2.2576	2.2586	2.2597	2.2607
9.6	2.2618	2.2628	2.2638	2.2649	2.2659	2.2670	2.2680	2.2690	2.2701	2.2711
9.7	2.2721	2.2732	2.2742	2.2752	2.2762	2.2773	2.2783	2.2793	2.2803	2.2814
9.8	2.2824	2.2834	2.2844	2.2854	2.2865	2.2875	2.2885	2.2895	2.2905	2.2915
9.9	2.2925	2.2935	2.2946	2.2956	2.2966	2.2976	2.2986	2.2996	2.3006	2.3016
10.0	2.3026	2.3036	2.3046	2.3056	2.3066	2.3076	2.3086	2.3096	2.3106	2.3115
N	.00	.01	.02	.03	.04	.05	.06	.07	.08	.09

$$\log_e .0001 = .7897 - 10 \qquad \log_e .00001 = .4871 - 12 \qquad \log_e .000\,001 = .1845 - 14$$

Table 2. Four-Place Natural Logarithms (*continued*)

N	.0	.1	.2	.3	.4	.5	.6	.7	.8	.9
10	2.3026	2.3125	2.3224	2.3321	2.3418	2.3514	2.3609	2.3702	2.3795	2.3888
11	2.3979	2.4069	2.4159	2.4248	2.4336	2.4423	2.4510	2.4596	2.4681	2.4765
12	2.4849	2.4932	2.5014	2.5096	2.5177	2.5257	2.5337	2.5416	2.5494	2.5572
13	2.5649	2.5726	2.5802	2.5878	2.5953	2.6027	2.6101	2.6174	2.6247	2.6319
14	2.6391	2.6462	2.6532	2.6603	2.6672	2.6741	2.6810	2.6878	2.6946	2.7014
15	2.7081	2.7147	2.7213	2.7279	2.7344	2.7408	2.7473	2.7537	2.7600	2.7663
16	2.7726	2.7788	2.7850	2.7912	2.7973	2.8034	2.8094	2.8154	2.8214	2.8273
17	2.8332	2.8391	2.8449	2.8507	2.8565	2.8622	2.8679	2.8736	2.8792	2.8848
18	2.8904	2.8959	2.9014	2.9069	2.9124	2.9178	2.9232	2.9235	2.9339	2.9392
19	2.9444	2.9497	2.9549	2.9601	2.9653	2.9704	2.9755	2.9806	2.9857	2.9907
20	2.9957	3.0007	3.0057	3.0106	3.0155	3.0204	3.0253	3.0301	3.0350	3.0397
21	3.0445	3.0493	3.0540	3.0587	3.0634	3.0681	3.0727	3.0773	3.0819	3.0865
22	3.0910	3.0956	3.1001	3.1046	3.1091	3.1135	3.1179	3.1224	3.1268	3.1311
23	3.1355	3.1398	3.1442	3.1485	3.1527	3.1570	3.1612	3.1655	3.1697	3.1739
24	3.1781	3.1822	3.1864	3.1905	3.1946	3.1987	3.2027	3.2068	3.2108	3.2149
25	3.2189	3.2229	3.2268	3.2308	3.2347	3.2387	3.2426	3.2465	3.2504	3.2542
26	3.2581	3.2619	3.2658	3.2696	3.2734	3.2771	3.2809	3.2847	3.2884	3.2921
27	3.2958	3.2995	3.3032	3.3069	3.3105	3.3142	3.3178	3.3214	3.3250	3.3286
28	3.3322	3.3358	3.3393	3.3429	3.3464	3.3499	3.3534	3.3569	3.3604	3.3638
29	3.3673	3.3707	3.3742	3.3776	3.3810	3.3844	3.3878	3.3911	3.3945	3.3979
30	3.4012	3.4045	3.4078	3.4111	3.4144	3.4177	3.4210	3.4243	3.4275	3.4308
31	3.4340	3.4372	3.4404	3.4436	3.4468	3.4500	3.4532	3.4563	3.4595	3.4626
32	3.4657	3.4689	3.4720	3.4751	3.4782	3.4812	3.4843	3.4874	3.4904	3.4935
33	3.4965	3.4995	3.5025	3.5056	3.5086	3.5115	3.5145	3.5175	3.5205	3.5234
34	3.5264	3.5293	3.5322	3.5351	3.5381	3.5410	3.5439	3.5467	3.5496	3.5525
35	3.5553	3.5582	3.5610	3.5639	3.5667	3.5695	3.5723	3.5752	3.5779	3.5807
36	3.5835	3.5863	3.5891	3.5918	3.5946	3.5973	3.6000	3.6028	3.6055	3.6082
37	3.6109	3.6136	3.6163	3.6190	3.6217	3.6243	3.6270	3.6297	3.6323	3.6350
38	3.6376	3.6402	3.6428	3.6454	3.6481	3.6507	3.6533	3.6558	3.6584	3.6610
39	3.6636	3.6661	3.6687	3.6712	3.6738	3.6763	3.6788	3.6814	3.6839	3.6864
40	3.6889	3.6914	3.6939	3.6964	3.6988	3.7013	3.7038	3.7062	3.7087	3.7111
41	3.7136	3.7160	3.7184	3.7209	3.7233	3.7257	3.7281	3.7305	3.7329	3.7353
42	3.7377	3.7400	3.7424	3.7448	3.7471	3.7495	3.7519	3.7542	3.7565	3.7589
43	3.7612	3.7635	3.7658	3.7682	3.7705	3.7728	3.7751	3.7773	3.7796	3.7819
44	3.7842	3.7865	3.7887	3.7910	3.7932	3.7955	3.7977	3.8000	3.8022	3.8044
45	3.8067	3.8089	3.8111	3.8133	3.8155	3.8177	3.8199	3.8221	3.8243	3.8265
46	3.8286	3.8308	3.8330	3.8351	3.8373	3.8395	3.8416	3.8437	3.8459	3.8480
47	3.8501	3.8523	3.8544	3.8565	3.8586	3.8607	3.8628	3.8649	3.8670	3.8691
48	3.8712	3.8733	3.8754	3.8774	3.8795	3.8816	3.8836	3.8857	3.8877	3.8898
49	3.8918	3.8939	3.8959	3.8979	3.9000	3.9020	3.9040	3.9060	3.9080	3.9100
50	3.9120	3.9140	3.9160	3.9180	3.9200	3.9220	3.9240	3.9259	3.9279	3.9299
51	3.9318	3.9338	3.9357	3.9377	3.9396	3.9416	3.9435	3.9455	3.9474	3.9493
52	3.9512	3.9532	3.9551	3.9570	3.9589	3.9608	3.9627	3.9646	3.9665	3.9684
53	3.9703	3.9722	3.9741	3.9759	3.9778	3.9797	3.9815	3.9834	3.9853	3.9871
54	3.9890	3.9908	3.9927	3.9945	3.9964	3.9982	4.0000	4.0019	4.0037	4.0055
55	4.0073	4.0091	4.0110	4.0128	4.0146	4.0164	4.0182	4.0200	4.0218	4.0236
N	.0	.1	.2	.3	.4	.5	.6	.7	.8	.9

$$\log_e 100 = 4.6052 \quad \log_e 1000 = 6.9078 \quad \log_e 10{,}000 = 9.2103$$

Table 2. Four-Place Natural Logarithms (*continued*)

N	.0	.1	.2	.3	.4	.5	.6	.7	.8	.9
55	4.0073	4.0091	4.0110	4.0128	4.0146	4.0164	4.0182	4.0200	4.0218	4.0236
56	4.0254	4.0271	4.0289	4.0307	4.0325	4.0342	4.0360	4.0378	4.0395	4.0413
57	4.0431	4.0448	4.0466	4.0483	4.0500	4.0518	4.0535	4.0553	4.0570	4.0587
58	4.0604	4.0622	4.0639	4.0656	4.0673	4.0690	4.0707	4.0724	4.0741	4.0758
59	4.0775	4.0792	4.0809	4.0826	4.0843	4.0860	4.0877	4.0893	4.0910	4.0927
60	4.0943	4.0960	4.0977	4.0993	4.1010	4.1026	4.1043	4.1059	4.1076	4.1092
61	4.1109	4.1125	4.1141	4.1158	4.1174	4.1190	4.1207	4.1223	4.1239	4.1255
62	4.1271	4.1287	4.1304	4.1320	4.1336	4.1352	4.1368	4.1384	4.1400	4.1415
63	4.1431	4.1447	4.1463	4.1479	4.1495	4.1510	4.1526	4.1542	4.1558	4.1573
64	4.1589	4.1604	4.1620	4.1636	4.1651	4.1667	4.1682	4.1698	4.1713	4.1728
65	4.1744	4.1759	4.1775	4.1790	4.1805	4.1821	4.1836	4.1851	4.1866	4.1881
66	4.1897	4.1912	4.1927	4.1942	4.1957	4.1972	4.1987	4.2002	4.2017	4.2032
67	4.2047	4.2062	4.2077	4.2092	4.2106	4.2121	4.2136	4.2151	4.2166	4.2180
68	4.2195	4.2210	4.2224	4.2239	4.2254	4.2268	4.2283	4.2297	4.2312	4.2327
69	4.2341	4.2356	4.2370	4.2384	4.2399	4.2413	4.2428	4.2442	4.2456	4.2471
70	4.2485	4.2499	4.2513	4.2528	4.2542	4.2556	4.2570	4.2584	4.2599	4.2613
71	4.2627	4.2641	4.2655	4.2669	4.2683	4.2697	4.2711	4.2725	4.2739	4.2753
72	4.2767	4.2781	4.2794	4.2808	4.2822	4.2836	4.2850	4.2863	4.2877	4.2891
73	4.2905	4.2918	4.2932	4.2946	4.2959	4.2973	4.2986	4.3000	4.3014	4.3027
74	4.3041	4.3054	4.3068	4.3081	4.3095	4.3108	4.3121	4.3135	4.3148	4.3162
75	4.3175	4.3188	4.3202	4.3215	4.3228	4.3241	4.3255	4.3268	4.3281	4.3294
76	4.3307	4.3320	4.3334	4.3347	4.3360	4.3373	4.3386	4.3399	4.3412	4.3425
77	4.3438	4.3451	4.3464	4.3477	4.3490	4.3503	4.3516	4.3529	4.3541	4.3554
78	4.3567	4.3580	4.3593	4.3605	4.3618	4.3631	4.3644	4.3656	4.3669	4.3682
79	4.3694	4.3707	4.3720	4.3732	4.3745	4.3758	4.3770	4.3783	4.3795	4.3808
80	4.3820	4.3833	4.3845	4.3858	4.3870	4.3883	4.3895	4.3907	4.3920	4.3932
81	4.3944	4.3957	4.3969	4.3981	4.3994	4.4006	4.4018	4.4031	4.4043	4.4055
82	4.4067	4.4079	4.4092	4.4104	4.4116	4.4128	4.4140	4.4152	4.4164	4.4176
83	4.4188	4.4200	4.4212	4.4224	4.4236	4.4248	4.4260	4.4272	4.4284	4.4296
84	4.4308	4.4320	4.4332	4.4344	4.4356	4.4368	4.4379	4.4391	4.4403	4.4415
85	4.4427	4.4438	4.4450	4.4462	4.4473	4.4485	4.4497	4.4509	4.4520	4.4532
86	4.4543	4.4555	4.4567	4.4578	4.4590	4.4601	4.4613	4.4625	4.4636	4.4648
87	4.4659	4.4671	4.4682	4.4694	4.4705	4.4716	4.4728	4.4739	4.4751	4.4762
88	4.4773	4.4785	4.4796	4.4807	4.4819	4.4830	4.4841	4.4853	4.4864	4.4875
89	4.4886	4.4898	4.4909	4.4920	4.4931	4.4942	4.4954	4.4965	4.4976	4.4987
90	4.4998	4.5009	4.5020	4.5031	4.5042	4.5053	4.5065	4.5076	4.5087	4.5098
91	4.5109	4.5120	4.5131	4.5142	4.5152	4.5163	4.5174	4.5185	4.5196	4.5207
92	4.5218	4.5229	4.5240	4.5250	4.5261	4.5272	4.5283	4.5294	4.5304	4.5315
93	4.5326	4.5337	4.5347	4.5358	4.5369	4.5380	4.5390	4.5401	4.5412	4.5422
94	4.5433	4.5444	4.5454	4.5465	4.5475	4.5486	4.5497	4.5507	4.5518	4.5528
95	4.5539	4.5549	4.5560	4.5570	4.5581	4.5591	4.5602	4.5612	4.5623	4.5633
96	4.5643	4.5654	4.5664	4.5675	4.5685	4.5695	4.5706	4.5716	4.5726	4.5737
97	4.5747	4.5757	4.5768	4.5778	4.5788	4.5799	4.5809	4.5819	4.5829	4.5839
98	4.5850	4.5860	4.5870	4.5880	4.5890	4.5901	4.5911	4.5921	4.5931	4.5941
99	4.5951	4.5961	4.5971	4.5981	4.5992	4.6002	4.6012	4.6022	4.6032	4.6042
100	4.6052	4.6062	4.6072	4.6082	4.6092	4.6102	4.6112	4.6121	4.6131	4.6141
N	.0	.1	.2	.3	.4	.5	.6	.7	.8	.9

$$\log_e 100{,}000 = 11.5129 \qquad \log_e 1{,}000{,}000 = 13.8155 \qquad \log_e 10{,}000{,}000 = 16.1181$$

Table 3. Exponential Functions

x	e^x	e^{-x}	x	e^x	e^{-x}	x	e^x	e^{-x}
0.00	1.0000	1.000000	**0.50**	1.6487	0.606531	**1.00**	2.7183	0.367879
0.01	1.0101	0.990050	0.51	1.6653	.600496	1.01	2.7456	.364219
0.02	1.0202	.980199	0.52	1.6820	.594521	1.02	2.7732	.360595
0.03	1.0305	.970446	0.53	1.6989	.588605	1.03	2.8011	.357007
0.04	1.0408	.960789	0.54	1.7160	.582748	1.04	2.8292	.353455
0.05	1.0513	0.951229	**0.55**	1.7333	0.576950	**1.05**	2.8577	0.349938
0.06	1.0618	.941765	0.56	1.7507	.571209	1.06	2.8864	.346456
0.07	1.0725	.932394	0.57	1.7683	.565525	1.07	2.9154	.343009
0.08	1.0833	.923116	0.58	1.7860	.559898	1.08	2.9447	.339596
0.09	1.0942	.913931	0.59	1.8040	.554327	1.09	2.9743	.336216
0.10	1.1052	0.904837	**0.60**	1.8221	0.548812	**1.10**	3.0042	0.332871
0.11	1.1163	.895834	0.61	1.8404	.543351	1.11	3.0344	.329559
0.12	1.1275	.886920	0.62	1.8589	.537944	1.12	3.0649	.326280
0.13	1.1388	.878095	0.63	1.8776	.532592	1.13	3.0957	.323033
0.14	1.1503	.869358	0.64	1.8965	.527292	1.14	3.1268	.319819
0.15	1.1618	0.860708	**0.65**	1.9155	0.522046	**1.15**	3.1582	0.316637
0.16	1.1735	.852144	0.66	1.9348	.516851	1.16	3.1899	.313486
0.17	1.1853	.843665	0.67	1.9542	.511709	1.17	3.2220	.310367
0.18	1.1972	.835270	0.68	1.9739	.506617	1.18	3.2544	.307279
0.19	1.2092	.826959	0.69	1.9937	.501576	1.19	3.2871	.304221
0.20	1.2214	0.818731	**0.70**	2.0138	0.496585	**1.20**	3.3201	0.301194
0.21	1.2337	.810584	0.71	2.0340	.491644	1.21	3.3535	.298197
0.22	1.2461	.802519	0.72	2.0544	.486752	1.22	3.3872	.295230
0.23	1.2586	.794534	0.73	2.0751	.481909	1.23	3.4212	.292293
0.24	1.2712	.786628	0.74	2.0959	.477114	1.24	3.4556	.289384
0.25	1.2840	0.778801	**0.75**	2.1170	0.472367	**1.25**	3.4903	0.286505
0.26	1.2969	.771052	0.76	2.1383	.467666	1.26	3.5254	.283654
0.27	1.3100	.763379	0.77	2.1598	.463013	1.27	3.5609	.280832
0.28	1.3231	.755784	0.78	2.1815	.458406	1.28	3.5966	.278037
0.29	1.3364	.748264	0.79	2.2034	.453845	1.29	3.6328	.275271
0.30	1.3499	0.740818	**0.80**	2.2255	0.449329	**1.30**	3.6693	0.272532
0.31	1.3634	.733447	0.81	2.2479	.444858	1.31	3.7062	.269820
0.32	1.3771	.726149	0.82	2.2705	.440432	1.32	3.7434	.267135
0.33	1.3910	.718924	0.83	2.2933	.436049	1.33	3.7810	.264477
0.34	1.4049	.711770	0.84	2.3164	.431711	1.34	3.8190	.261846
0.35	1.4191	0.704688	**0.85**	2.3396	0.427415	**1.35**	3.8574	0.259240
0.36	1.4333	.697676	0.86	2.3632	.423162	1.36	3.8962	.256661
0.37	1.4477	.690734	0.87	2.3869	.418952	1.37	3.9354	.254107
0.38	1.4623	.683861	0.88	2.4109	.414783	1.38	3.9749	.251579
0.39	1.4770	.677057	0.89	2.4351	.410656	1.39	4.0149	.249075
0.40	1.4918	0.670320	**0.90**	2.4596	0.406570	**1.40**	4.0552	0.246597
0.41	1.5068	.663650	0.91	2.4843	.402524	1.41	4.0960	.244143
0.42	1.5220	.657047	0.92	2.5093	.398519	1.42	4.1371	.241714
0.43	1.5373	.650509	0.93	2.5345	.394554	1.43	4.1787	.239309
0.44	1.5527	.644036	0.94	2.5600	.390628	1.44	4.2207	.236928
0.45	1.5683	0.637628	**0.95**	2.5857	0.386741	**1.45**	4.2631	0.234570
0.46	1.5841	.631284	0.96	2.6117	.382893	1.46	4.3060	.232236
0.47	1.6000	.625002	0.97	2.6379	.379083	1.47	4.3492	.229925
0.48	1.6161	.618783	0.98	2.6645	.375311	1.48	4.3929	.227638
0.49	1.6323	.612626	0.99	2.6912	.371577	1.49	4.4371	.225373
0.50	1.6487	0.606531	**1.00**	2.7183	0.367879	**1.50**	4.4817	0.223130

Table 3. Exponential Functions (continued)

x	e^x	e^{-x}	x	e^x	e^{-x}	x	e^x	e^{-x}
1.50	4.4817	0.223130	**2.00**	7.3891	0.135335	**2.50**	12.182	0.082085
1.51	4.5267	.220910	2.01	7.4633	.133989	2.51	12.305	.081268
1.52	4.5722	.218712	2.02	7.5383	.132655	2.52	12.429	.080460
1.53	4.6182	.216536	2.03	7.6141	.131336	2.53	12.554	.079659
1.54	4.6646	.214381	2.04	7.6906	.130029	2.54	12.680	.078866
1.55	4.7115	0.212248	**2.05**	7.7679	0.128735	**2.55**	12.807	0.078082
1.56	4.7588	.210136	2.06	7.8460	.127454	2.56	12.936	.077305
1.57	4.8066	.208045	2.07	7.9248	.126186	2.57	13.066	.076536
1.58	4.8550	.205975	2.08	8.0045	.124930	2.58	13.197	.075774
1.59	4.9037	.203926	2.09	8.0849	.123687	2.59	13.330	.075020
1.60	4.9530	0.201897	**2.10**	8.1662	0.122456	**2.60**	13.464	0.074274
1.61	5.0028	.199888	2.11	8.2482	.121238	2.61	13.599	.073535
1.62	5.0531	.197899	2.12	8.3311	.120032	2.62	13.736	.072803
1.63	5.1039	.195930	2.13	8.4149	.118837	2.63	13.874	.072078
1.64	5.1552	.193980	2.14	8.4994	.117655	2.64	14.013	.071361
1.65	5.2070	0.192050	**2.15**	8.5849	0.116484	**2.65**	14.154	0.070651
1.66	5.2593	.190139	2.16	8.6711	.115325	2.66	14.296	.069948
1.67	5.3122	.188247	2.17	8.7583	.114178	2.67	14.440	.069252
1.68	5.3656	.186374	2.18	8.8463	.113042	2.68	14.585	.068563
1.69	5.4195	.184520	2.19	8.9352	.111917	2.69	14.732	.067881
1.70	5.4739	0.182684	**2.20**	9.0250	0.110803	**2.70**	14.880	0.067206
1.71	5.5290	.180866	2.21	9.1157	.109701	2.71	15.029	.066537
1.72	5.5845	.179066	2.22	9.2073	.108609	2.72	15.180	.065875
1.73	5.6407	.177284	2.23	9.2999	.107528	2.73	15.333	.065219
1.74	5.6973	.175520	2.24	9.3933	.106459	2.74	15.487	.064570
1.75	5.7546	0.173774	**2.25**	9.4877	0.105399	**2.75**	15.643	0.063928
1.76	5.8124	.172045	2.26	9.5831	.104350	2.76	15.800	.063292
1.77	5.8709	.170333	2.27	9.6794	.103312	2.77	15.959	.062662
1.78	5.9299	.168638	2.28	9.7767	.102284	2.78	16.119	.062039
1.79	5.9895	.166960	2.29	9.8749	.101266	2.79	16.281	.061421
1.80	6.0496	0.165299	**2.30**	9.9742	0.100259	**2.80**	16.445	0.060810
1.81	6.1104	.163654	2.31	10.074	.099261	2.81	16.610	.060205
1.82	6.1719	.162026	2.32	10.176	.098274	2.82	16.777	.059606
1.83	6.2339	.160414	2.33	10.278	.097296	2.83	16.945	.059013
1.84	6.2965	.158817	2.34	10.381	.096328	2.84	17.116	.058426
1.85	6.3598	0.157237	**2.35**	10.486	0.095369	**2.85**	17.288	0.057844
1.86	6.4237	.155673	2.36	10.591	.094420	2.86	17.462	.057269
1.87	6.4883	.154124	2.37	10.697	.093481	2.87	17.637	.056699
1.88	6.5535	.152590	2.38	10.805	.092551	2.88	17.814	.056135
1.89	6.6194	.151072	2.39	10.913	.091630	2.89	17.993	.055576
1.90	6.6859	0.149569	**2.40**	11.023	0.090718	**2.90**	18.174	0.055023
1.91	6.7531	.148080	2.41	11.134	.089815	2.91	18.357	.054476
1.92	6.8210	.146607	2.42	11.246	.088922	2.92	18.541	.053934
1.93	6.8895	.145148	2.43	11.359	.088037	2.93	18.728	.053397
1.94	6.9588	.143704	2.44	11.473	.087161	2.94	18.916	.052866
1.95	7.0287	0.142274	**2.45**	11.588	0.086294	**2.95**	19.106	0.052340
1.96	7.0993	.140858	2.46	11.705	.085435	2.96	19.298	.051819
1.97	7.1707	.139457	2.47	11.822	.084585	2.97	19.492	.051303
1.98	7.2427	.138069	2.48	11.941	.083743	2.98	19.688	.050793
1.99	7.3155	.136695	2.49	12.061	.082910	2.99	19.886	.050287
2.00	7.3891	0.135335	**2.50**	12.182	0.082085	**3.00**	20.086	0.049787